Edited by
Santi Kulprathipanja

Zeolites in Industrial Separation and Catalysis

Further Reading

R. Xu, W. Pang, J. Yu, Q. Huo, J. Chen

Chemistry of Zeolites and Related Porous Materials

Synthesis and Structure

2007
ISBN: 978-0-470-82233-3

I. Chorkendorff, J.W. Niemantsverdriet

Concepts of Modern Catalysis and Kinetics

2007
ISBN: 978-3-527-31672-4

B. Cornils, W.A. Herrmann, M. Muhler, C.-H. Wong (Eds.)

Catalysis from A to Z

A Concise Encyclopedia

2007
ISBN: 978-3-527-31438-6

S.M. Roberts

Catalysts for Fine Chemical Synthesis

Volume 4. Microporous and Mesoporous Solid Catalysts

2006
ISBN: 978-0-471-49054-8

R.A. van Santen, M. Neurock

Molecular Heterogeneous Catalysis

A Conceptual and Computational Approach

2006
ISBN: 978-3-527-29662-0

J. Hagen

Industrial Catalysis

A Practical Approach

2006
ISBN: 978-3-527-31144-6

F. Schüth, K.S.W. Sing, J. Weitkamp (Eds.)

Handbook of Porous Solids

5 Volumes

2002
Hardcover
ISBN: 978-3-527-30246-8

Edited by
Santi Kulprathipanja

Zeolites in Industrial Separation and Catalysis

WILEY-VCH Verlag GmbH & Co. KGaA

The Editor

Dr. Santi Kulprathipanja
UOP, A Honeywell Company
25 E. Algonquin Road
Des Plaines, IL 60017
USA

■ All books published by Wiley-VCH are carefully produced. Nevertheless, authors, editors, and publisher do not warrant the information contained in these books, including this book, to be free of errors. Readers are advised to keep in mind that statements, data, illustrations, procedural details or other items may inadvertently be inaccurate.

Library of Congress Card No.: applied for

British Library Cataloguing-in-Publication Data
A catalogue record for this book is available from the British Library.

Bibliographic information published by the Deutsche Nationalbibliothek
The Deutsche Nationalbibliothek lists this publication in the Deutsche Nationalbibliografie; detailed bibliographic data are available on the Internet at http://dnb.d-nb.de

© 2010 WILEY-VCH Verlag GmbH & Co. KGaA, Weinheim

All rights reserved (including those of translation into other languages). No part of this book may be reproduced in any form – by photoprinting, microfilm, or any other means – nor transmitted or translated into a machine language without written permission from the publishers. Registered names, trademarks, etc. used in this book, even when not specifically marked as such, are not to be considered unprotected by law.

Cover Design   Formgeber, Eppelheim
Typesetting   Toppan Best-set Premedia Limited
Printing and Binding   Bell & Bain Ltd., Glasgow

Printed in Great Britain
Printed on acid-free paper

ISBN: 978-3-527-32505-4

Contents

Preface *XIX*
List of Contributors *XXIII*

1 **Introduction** *1*
Edith M. Flanigen, Robert W. Broach, and Stephen T. Wilson
1.1 Introduction *1*
1.1.1 Molecular Sieves and Zeolites *1*
1.1.2 Nomenclature *2*
1.1.3 Early History *3*
1.1.4 Natural Zeolites *4*
1.2 History of Molecular Sieve Materials *5*
1.2.1 Aluminosilicate Zeolites and Silica Molecular Sieves *6*
1.2.2 The Materials Explosion Since the 1980s *7*
1.2.2.1 The 1980s *7*
1.2.2.2 The 1990s *11*
1.2.2.3 The New Millennium *14*
1.3 Synthesis *15*
1.4 Applications *16*
1.5 Markets *17*
1.6 The Future *17*
1.6.1 Materials *17*
1.6.2 Applications *18*
1.7 History of International Conferences and Organizations *18*
1.8 Historical Epilog *20*
References *20*
Further Reading *26*

2 **Zeolite Types and Structures** *27*
Robert W. Broach
2.1 Introduction *27*
2.2 Building Units for Zeolite Frameworks *28*

Zeolites in Industrial Separation and Catalysis. Edited by Santi Kulprathipanja
Copyright © 2010 WILEY-VCH Verlag GmbH & Co. KGaA, Weinheim
ISBN: 978-3-527-32505-4

2.3	Zeolite Framework Types 31
2.4	Pores, Channels, Cages and Cavities 32
2.5	Materials Versus Framework Types 34
2.6	Structures of Commercially Significant Zeolites 35
2.6.1	Linde Type A (LTA) 36
2.6.2	Faujasite (FAU) 38
2.6.3	Mordenite (MOR) 40
2.6.4	Chabazite (CHA) 42
2.6.5	ZSM-5 (MFI) 45
2.6.6	Linde Type L (LTL) 47
2.6.7	Beta Polymorphs *BEA and BEC 49
2.6.8	MCM-22 (MWW) 51
2.7	Hypothetical Zeolite Frameworks 54
	Acknowledgments 55
	References 55

3 Synthesis of Zeolites and Manufacture of Zeolitic Catalysts and Adsorbents 61
Robert L. Bedard

3.1	Introduction 61
3.2	Synthesis of Zeolites and Aluminophosphate Molecular Sieves 62
3.2.1	Hydrothermal Synthesis – The Key to Metastable Phases 62
3.2.2	Typical Zeolite Syntheses 63
3.2.3	Important Synthesis Parameters – Zeolites 65
3.2.4	Typical Aluminophosphate Syntheses 66
3.2.5	Important Synthesis Parameters – Aluminophosphates 67
3.2.6	Dewatering, Filtration and Washing of Molecular Sieve Products 67
3.3	Forming Zeolite Powders into Usable Shapes 68
3.3.1	Chemical Engineering Considerations in Zeolite Forming 68
3.3.2	Ceramic Engineering Considerations in Zeolite Forming 69
3.3.3	Bound Zeolite Forms 70
3.3.4	Other Zeolite Forms – Colloids, Sheets, Films and Fibers 70
3.4	Finishing: Post-Forming Manufacturing of Zeolite Catalysts and Adsorbents 71
3.4.1	Post-Forming Crystallization 71
3.4.2	Stabilization and Chemical Modification of Zeolites 72
3.4.3	Ion Exchange and Impregnation 74
3.4.4	Drying and Firing 75
3.5	Selected New Developments in Catalyst and Adsorbent Manufacture 75
	References 77

4 Zeolite Characterization 85
Steven A. Bradley, Robert W. Broach, Thomas M. Mezza, Sesh Prabhakar, and Wharton Sinkler

4.1	Introduction 85
4.1.1	Importance of Characterization 85

4.2	Multi-Technique Methodology 86	
4.2.1	Identification of the Structure of a Newly Invented Zeolite 86	
4.3	X-Ray Powder Diffraction Characterization of Zeolitic Systems 91	
4.3.1	Interpretation of Powder Diffraction Data for Zeolites 91	
4.3.2	Phase Identification and Quantification 92	
4.3.3	Unit Cell Size Determination 94	
4.3.4	Crystallite Size 95	
4.3.5	Rietveld Refinement 96	
4.4	Electron Microscopy Characterization of Zeolitic Systems 97	
4.4.1	Importance of Electron Microscopy for Characterizing Zeolites 97	
4.4.2	Scanning Electron Microscopy 98	
4.4.2.1	Morphological Characterization 98	
4.4.2.2	Compositional Characterization 100	
4.4.3	Transmission Electron Microscopy 104	
4.4.3.1	Sample Preparation 104	
4.4.3.2	Structural Characterization 105	
4.4.3.3	Morphological Characterization 106	
4.4.3.4	Compositional Characterization 108	
4.4.3.5	STEM Application to Metals in Zeolites and Coke Analysis 109	
4.5	Infrared Spectroscopy Characterization of Zeolitic Systems 111	
4.5.1	Introduction to Infrared Spectroscopy 111	
4.5.2	Modes of Measurement 112	
4.5.3	Framework IR 114	
4.5.3.1	Zeolite Structure 115	
4.5.3.2	Tracking Zeolite Framework Si/Al 116	
4.5.3.3	Zeolite Synthesis 118	
4.5.4	Methods Requiring Sample Pretreatment 119	
4.5.5	Hydroxyl IR 120	
4.5.6	Acidity 123	
4.5.6.1	General Theory 124	
4.5.6.2	Selection of Probe Molecules 125	
4.5.6.3	Quantitation of Sites 125	
4.5.6.4	Pyridine Adsorption 127	
4.5.6.5	Ammonia 130	
4.5.6.6	Low-Temperature Acidity Probes 131	
4.5.6.7	Carbon Monoxide 132	
4.5.6.8	Nitrogen (N_2) 134	
4.5.6.9	Measurement of External Acidity 134	
4.5.6.10	Other Probes 135	
4.5.7	In Situ/In Operando Studies 136	
4.5.8	Characterization of Metal-Loaded Zeolites 136	
4.5.8.1	Cation Exchange for Adsorption/Separation 137	
4.5.8.2	Metal-Loading for Catalysis 138	
4.5.8.3	Noble Metal-Loading for Catalysis 138	
4.5.8.4	Non-Noble Metal-Loading for Catalysis 139	

4.6	NMR Characterization of Zeolitic Systems	140
4.6.1	Introduction to NMR	140
4.6.1.1	Spin-Half Nuclei	142
4.6.1.2	Cross Polarization	142
4.6.1.3	Quadrupolar Nuclei	143
4.6.1.4	Dipolar Recoupling	143
4.6.1.5	Pulsed Field Gradient NMR–diffusion	144
4.6.2	Applications	145
4.6.2.1	^{29}Si NMR	145
4.6.2.2	^{27}Al NMR	147
4.6.2.3	^{31}P NMR	149
4.6.2.4	^{1}H NMR	150
4.6.2.5	^{17}O and Other Nuclei	151
4.6.2.6	Diffusion of Hydrocarbons in Zeolites	151
4.7	Physical/Chemical Characterization	152
4.7.1	Nitrogen Physisorption	152
4.7.2	Thermal and Mechanical Analyses	154
4.7.3	Adsorption Capacity	156
4.7.4	Acid Sites	157
4.8	Conclusions	158
4.8.1	Future Characterization Directions	158
	References	160
5	**Overview in Zeolites Adsorptive Separation**	**173**
	Santi Kulprathipanja and Robert B. James	
5.1	Introduction	173
5.2	Industrial Adsorptive Separation	173
5.2.1	Gas Separation	173
5.2.2	Liquid Separation	174
5.2.2.1	Aromatics	174
5.2.2.2	Non-Aromatic Hydrocarbons	174
5.2.2.3	Non-Petrochemicals	175
5.2.2.4	Trace Component Removal	175
5.3	R&D Adsorptive Separation	176
5.3.1	Aromatic Hydrocarbon Separation	176
5.3.2	Non-Aromatic Hydrocarbon Separation	176
5.3.3	Carbohydrate Separation	176
5.3.4	Pharmaceutical Separation	176
5.3.5	Trace Impurities Removal	176
5.3.5.1	Sulfur Removal	176
5.3.5.2	Oxygenate Removal	188
5.3.5.3	Nitrogenate Removal	190
5.3.5.4	Iodide Removal	190
5.3.5.5	Aromatic Removal	190
5.3.5.6	Metal Removal	190

5.4	Summary Review of Zeolites in Adsorptive Separation	191
	Acknowledgments 192	
	References 192	
6	**Aspects of Mechanisms, Processes, and Requirements for Zeolite Separation** 203	
	Santi Kulprathipanja	
6.1	Introduction 203	
6.2	Impacts of Adsorptive Separation Versus Other Separation Processes 203	
6.3	Liquid Phase Adsorption 206	
6.3.1	Sanderson's Model of Intermediate Electronegativity 207	
6.4	Modes of Operation 208	
6.4.1	Adsorption Isotherms 209	
6.4.2	Pulse Test Procedure 209	
6.4.3	Breakthrough Procedure 210	
6.5	Zeolite Separation Processes 211	
6.5.1	Equilibrium-Selective Adsorption 211	
6.5.1.1	Zeolite Framework Structure 212	
6.5.1.2	Metal Cation Exchanged in Zeolite 212	
6.5.1.3	Zeolite SiO_2/Al_2O_3 Molar Ratio 216	
6.5.1.4	Moisture Content in Zeolite 218	
6.5.1.5	Characteristics of the Desorbent 219	
6.5.1.6	Operating Temperature 220	
6.5.1.7	Operating Pressure 221	
6.5.2	Rate-Selective Adsorption 221	
6.5.3	Shape-Selective Adsorption 222	
6.5.4	Ion Exchange 223	
6.5.4.1	Ion Exchange Capacity 224	
6.5.4.2	Ion Exchange Selectivity 224	
6.5.4.3	Kinetics 224	
6.5.5	Reactive Adsorption 224	
6.6	Summary 225	
	Acknowledgments 225	
	References 226	
7	**Liquid Industrial Aromatics Adsorbent Separation** 229	
	Stanley J. Frey	
7.1	Introduction 229	
7.2	Major Industrial Processes 231	
7.2.1	*p*-Xylene 231	
7.2.1.1	Industrial Uses and Demand 231	
7.2.1.2	Method of Production 232	
7.2.1.3	Characteristics of Zeolitic Adsorptive Process 235	
7.2.2	*m*-Xylene 241	

7.2.2.1	Industrial Uses and Demand	241
7.2.2.2	Method of Production	241
7.2.2.3	Characteristics of Zeolitic Adsorptive Process	242
7.3	Other Significant Processes	243
7.3.1	2,6-Dimethylnaphthalene	244
7.3.1.1	Industrial Uses and Demand	244
7.3.1.2	Method of Production	244
7.3.2	Ethylbenzene	244
7.3.2.1	Industrial Uses and Demand	244
7.3.2.2	Method of Production	244
7.3.3	*p*-Cresol	245
7.3.3.1	Industrial Uses and Demand	245
7.3.3.2	Method of Production: Adsorbent–Desorbent	245
7.4	Summary	245
	References	246
8	**Liquid Industrial Non-Aromatics Adsorptive Separations**	**249**
	Stephen W. Sohn	
8.1	Introduction	249
8.2	Normal Paraffin Separations	249
8.2.1	Characteristics of Adsorbent for Normal Paraffin Extraction	250
8.2.1.1	Selectivity	250
8.2.1.2	Capacity	251
8.2.1.3	Mass Transfer Rate	252
8.2.1.4	Stability	252
8.2.1.5	Zeolite Types	252
8.2.2	Desorbent Critical Characteristics	253
8.2.2.1	Selectivity	254
8.2.2.2	Compatibility	254
8.2.2.3	Ease of Separation	254
8.2.2.4	Availability	255
8.2.2.5	Reactivity	255
8.2.3	Simulated Moving Bed Operation: Sorbex Process	256
8.2.3.1	Adsorbent Allocation within the Molex Process	256
8.2.3.2	Critical Sorbex Zone Parameters	257
8.2.4	Light Normal Paraffin Separation (Gasoline Range nC_{5-6})	258
8.2.4.1	Industrial Use and Demand	258
8.2.4.2	Unique Operating Parameters	258
8.2.5	Intermediate Normal Paraffin Separation (C_{6-10})	260
8.2.5.1	Industrial Use and Demand	260
8.2.5.2	Unique Operating Parameters	260
8.2.6	Heavy Normal Paraffin Separation (C_{10-18})	261
8.2.6.1	Industrial Use and Demand	262
8.2.6.2	Unique Operating Parameters	262

8.3	Mono-Methyl Paraffins Separation (C_{10-16})	263
8.3.1	Industrial Use and Demand	263
8.3.2	Unique Operating Parameters	264
8.4	Olefin Separations	265
8.4.1	C_4 Separations	266
8.4.1.1	Industrial Use and Demand	266
8.4.1.2	Unique Operating Parameters	266
8.4.2	Detergent Range Olex C_{10-16}	267
8.4.2.1	Industrial Use and Demand	267
8.4.2.2	Unique Operating Parameters	268
8.5	Carbohydrate Separation	269
8.5.1	Industrial Use and Demand	269
8.5.2	Unique Operating Parameters	269
8.6	Liquid Adsorption Acid Separations	269
8.6.1	Citric Acid Separation	270
8.6.1.1	Industrial Use and Demand	270
8.6.1.2	Unique Operating Parameters	270
8.6.2	Free Fatty Acid Separation	270
8.6.2.1	Industrial Use and Demand	270
8.6.2.2	Unique Operating Parameters	270
8.7	Summary	271
	References	271
9	**Industrial Gas Phase Adsorptive Separations**	**273**
	Stephen R. Dunne	
9.1	Introduction	273
9.2	Regeneration	275
9.3	Adsorption Equilibrium	276
9.3.1	Henry's Law: A Linear Isotherm	277
9.3.2	Langmuir	277
9.3.3	Potential Theory	278
9.3.4	Universal Isotherm	278
9.3.5	Freundlich	279
9.3.6	Langmuir–Freundlich	279
9.3.7	Kelvin Equation and Capillary Condensation	279
9.4	Mass Transfer in Formed Zeolite Particles	280
9.4.1	Adsorption Wave Speed	282
9.4.2	Adsorption Wave Shape and Length	283
9.4.3	Linear Driving Force Approximation and Resistance Modeling	284
9.4.4	Diffusion Mechanisms in Formed Zeolites	286
9.4.4.1	Fluid Film Diffusion	286
9.4.4.2	Macro-Pore Diffusion	286
9.4.4.3	Intra-Crystalline Diffusion	287
9.5	Industrial TSA Separations (Purification)	288
9.5.1	Dehydration	289

9.5.2	De-Sulfurization	294
9.5.3	CO_2 Removal	295
9.5.4	VOC Removal	296
9.5.5	Mercury Removal	296
9.6	Industrial PSA	296
9.6.1	PSA Air Separation	297
9.6.2	PSA H2 Purification	299
9.6.3	PSA Dehydration	300
9.7	Industrial Dehydration (Bulk Removal)	301
9.7.1	Desiccant Wheels	301
9.7.2	Enthalpy Control Wheels	302
9.8	Non-Regenerable Adsorption	303
9.9	Summary	303
	Nomenclature	303
	Greek	304
	References	304

10 Zeolite Membrane Separations 307
Jessica O'Brien-Abraham and Jerry Y.S. Lin

10.1	Introduction	307
10.2	Synthesis and Properties of Zeolite Membranes	309
10.2.1	*In Situ* Crystallization	309
10.2.2	Secondary (Seeded) Growth	311
10.2.3	Characterization of Zeolite Membranes	313
10.3	Transport Theory and Separation Capability of Zeolite Membranes	314
10.3.1	Permeation Through Zeolite Membranes	314
10.3.2	Zeolite Membrane Separation Mechanisms	316
10.3.3	Influence of Zeolite Framework Flexibility	319
10.4	Zeolite Membranes in Separation and Reactive Processes	320
10.4.1	Liquid–Liquid Separation	320
10.4.2	Gas/Vapor Separation	322
10.4.3	Reactive Separation Processes	323
10.5	Summary	324
	Acknowledgment	325
	References	325

11 Mixed-Matrix Membranes 329
Chunqing Liu and Santi Kulprathipanja

11.1	Introduction	329
11.1.1	Scope of This Chapter	329
11.1.2	Polymer Membranes	329
11.1.3	Zeolite Membranes	331

11.2	Compositions of Mixed-Matrix Membranes	*332*
11.2.1	Non-zeolite/Polymer Mixed-Matrix Membranes	*333*
11.2.2	Zeolite/Polymer Mixed-Matrix Membranes	*333*
11.3	Concept of Zeolite/Polymer Mixed-Matrix Membranes	*334*
11.4	Material Selection for Zeolite/Polymer Mixed-Matrix Membranes	*336*
11.4.1	Selection of Polymer and Zeolite Materials	*336*
11.4.1.1	Selection of Polymer Materials	*336*
11.4.1.2	Selection of Zeolite Materials	*337*
11.4.1.3	Compatibility between Polymer and Zeolite Materials	*339*
11.4.2	Modification of Zeolite and Polymer Materials	*339*
11.5	Geometries of Zeolite/Polymer Mixed-Matrix Membranes	*341*
11.5.1	Mixed-Matrix Dense Films	*341*
11.5.2	Asymmetric Mixed-Matrix Membranes	*342*
11.5.2.1	Flat Sheet Asymmetric Mixed-Matrix Membranes	*343*
11.5.2.2	Hollow Fiber Asymmetric Mixed-Matrix Membranes	*345*
11.5.2.3	Thin-Film Composite Mixed-Matrix Membranes	*346*
11.6	Applications of Zeolite/Polymer Mixed-Matrix Membranes	*346*
11.6.1	Gas Separation Applications	*347*
11.6.2	Liquid Separation Applications	*347*
11.7	Summary	*348*
	References	*349*
12	**Overview and Recent Developments in Catalytic Applications of Zeolites** *355*	
	Christopher P. Nicholas	
12.1	History of Catalytic Uses of Zeolites	*355*
12.1.1	R&D Uses Versus Industrial Application of Zeolite Catalysis	*355*
12.2	Literature Review of Recent Developments in Catalytic Uses of Zeolites	*356*
12.2.1	Isomerization Reactions	*356*
12.2.1.1	Butane Isomerization	*356*
12.2.1.2	Pentane/Hexane Isomerization	*356*
12.2.1.3	C10+ Paraffin Isomerization	*358*
12.2.1.4	Light Olefin Isomerization	*358*
12.2.2	Oligomerization Reactions	*358*
12.2.2.1	Light Olefin Oligomerization	*358*
12.2.2.2	Heavier Olefin Oligomerization	*364*
12.2.3	Alkylation Reactions	*364*
12.2.3.1	Alkylation of Isobutane	*364*
12.2.3.2	Benzene Alkylation	*364*
12.2.4	Aromatics Reactions	*369*
12.2.4.1	Transalkylation of Toluene to Xylene and Benzene	*369*
12.2.4.2	Xylene and Ethylbenzene Isomerization	*369*

12.2.5	Chain-Breaking Reactions	369
12.2.5.1	Use of Zeolites in FCC Type Feeds	369
12.2.5.2	Olefin Cracking	377
12.2.6	Dehydroaromatization	377
12.2.6.1	Light Alkanes to Aromatics	377
12.2.6.2	Methane to Aromatics	377
12.2.7	Methanol to Olefins	383
12.2.8	Hydrotreating and Hydrocracking	383
12.2.9	Reactions Using Heteroatom Substituted Zeolites	387
12.2.9.1	Epoxidation	387
12.2.9.2	Other Oxidations	387
12.3	Future Trends in Catalysis by Zeolites	393
	References	393

13 Unique Aspects of Mechanisms and Requirements for Zeolite Catalysis in Refining and Petrochemicals *403*
Hayim Abrevaya

13.1	Introduction	403
13.2	Adsorption	404
13.2.1	Langmuir Isotherm and Reaction Kinetics	404
13.2.2	Nitrogen Adsorption	406
13.2.3	Hydrocarbon Adsorption	408
13.2.3.1	Pure Component Adsorption and Specificity with Respect to Zeolite Topology	408
13.2.3.2	Energetics of Adsorption	411
13.2.3.3	Adsorption of Mixtures	413
13.2.3.4	Compensation between Enthalpy and Entropy	415
13.3	Diffusion	416
13.4	Acidity	420
13.4.1	Bronsted Acidity	420
13.4.2	Significance of Acid Strength	421
13.4.3	Significance of Acid Site Density	423
13.4.4	Lewis Acidity	423
13.4.5	External acidity	424
13.5	Carbocations	425
13.5.1	Carbenium Ions and Alkoxides	426
13.5.2	Carbonium Ions	429
13.6	Elementary Steps of Hydrocarbon Conversion over Zeolites	429
13.7	Shape Selectivity	430
13.7.1	Tools	430
13.7.1.1	Footprints and Kinetic Diameters	431
13.7.1.2	Constraint Index	432
13.7.1.3	Modified Constraint Index	434
13.7.1.4	Spaciousness Index	434

13.7.2	Reactant Shape Selectivity	435
13.7.3	Transition State Shape Selectivity	435
13.7.3.1	*Meta*-Xylene Disproportionation	435
13.7.3.2	Methylcyclohexane Ring Contraction	436
13.7.3.3	Alkane Hydrosiomerization	436
13.7.3.4	Dimethylether Carbonylation	438
13.7.4	Product Shape Selectivity	438
13.7.4.1	Alkane Hydroisomerization	438
13.7.4.2	Oligomerization of Propylene	441
13.7.4.3	Alkylation of Naphthalene	443
13.7.4.4	Hydrocracking of *n*-Hexadecane	443
13.7.4.5	*Meta*-Xylene Isomerization	445
13.7.4.6	Methanol to Olefins	446
13.7.4.7	Other Examples of Product Shape Selectivity	446
13.7.5	Crystal Size Effects	446
13.8	Reaction Mechanisms	447
13.8.1	Alkene Skeletal Isomerization	447
13.8.1.1	Unimolecular Mechanism	447
13.8.1.2	Bimolecular Mechanism	448
13.8.2	Alkene Oligomerization	448
13.8.3	Alkylation	450
13.8.3.1	Isobutane Alkylation by 2-Butylene	450
13.8.3.2	Aromatic–Alkene and Aromatic–Alcohol	453
13.8.4	Alkane Cracking	455
13.8.4.1	Classic Cracking Mechanism, Bimolecular	455
13.8.4.2	Monomolecular Cracking Mechanism	456
13.8.4.3	Kinetics of Cracking	458
13.8.4.4	Effect of Pore Size and Acid Site Density on Cracking	461
13.8.5	Aromatic Transformation	462
13.8.5.1	Transalkylation and Disproportionation	462
13.8.5.2	Ethylbenzene Conversion to Xylenes	463
13.8.6	Methanol to Olefins	464
13.9	Key Remaining Questions	470
	References	470
14	**Industrial Isomerization**	**479**
	John E. Bauer, Feng Xu, Paula L. Bogdan, and Gregory J. Gajda	
14.1	Introduction	479
14.2	Metal–Zeolite Catalyzed Light Paraffin Isomerization	479
14.2.1	General Considerations	480
14.2.2	Bifunctional Paraffin Isomerization Mechanism	480
14.2.3	Zeolitic Paraffin Isomerization Catalysis	482
14.2.4	Industrial Zeolitic Isomerization Catalysts and Processes	483
14.2.5	Summary	484

14.3	Olefin Isomerization	484
14.3.1	General Considerations	484
14.3.2	Cis–Trans and Double Bond Isomerization	485
14.3.3	Skeletal Isomerization (Butenes, Pentenes, Hexenes)	486
14.3.4	Skeletal Isomerization (Longer-Chain Olefins)	488
14.3.5	Olefin Isomerization Summary	488
14.4	C8 Aromatics Isomerization	488
14.4.1	The Chemistry of C8 Aromatics Isomerization	489
14.4.1.1	Feed Composition and Characteristics	489
14.4.1.2	Reaction Product Composition and Characteristics	490
14.4.1.3	Isomerization Reactions	491
14.4.2	C8 Aromatics Isomerization Catalysts	494
14.4.2.1	General Aspects	494
14.4.2.2	Regeneration	495
14.4.3	C8 Aromatics Isomerization Processes	495
14.4.3.1	Process Variables	495
14.4.3.2	Commercial Catalysts	496
14.4.3.3	Modeling/Optimization of Commercial Units	497
14.4.3.4	Process Flow Schemes	498
14.4.4	Future Developments	499
14.4.5	C8 Aromatics Isomerization Summary	500
	References	500
15	**Processes on Industrial C–C Bond Formation**	**505**
	Deng-Yang Y. Jan and Paul T. Barger	
15.1	Introduction	505
15.2	Olefin Oligomerization	505
15.2.1	$C_2/C_3/C_4$ Olefin Oligomerization	505
15.2.1.1	Process Chemistry: Feeds, Products and Reactions	505
15.2.1.2	Catalysts	506
15.2.1.3	Physicochemical Characterization of Active Sites	507
15.3	Paraffin/Olefin Alkylation	507
15.3.1	Motor Fuel Alkylation	507
15.3.1.1	Process Chemistry: Feeds, Products and Reactions	508
15.3.1.2	Catalysts	509
15.3.1.3	Physicochemical Characterization of Active Sites	511
15.4	Benzene Alkylation	512
15.4.1	Ethylbenzene (Ethylene Alkylation), Cumene and Detergent Linear Alkylbenzene	512
15.4.1.1	Process Chemistry: Feeds, Products and Reactions	512
15.4.1.2	Catalysts	512
15.4.1.3	Physicochemical Characterization of Active Sites	513
15.4.2	Para-Xylene (Methylation of Toluene)	514
15.4.2.1	Process Chemistry: Feeds, Products and Reactions	514
15.4.2.2	Catalysts	515
15.4.2.3	Physicochemical Characterization of Active Sites	515

15.4.3	Styrene and Ethylbenzene from Methylation of Toluene *515*	
15.4.3.1	Process Chemistry: Feeds, Products and Reactions *515*	
15.4.3.2	Catalysts *516*	
15.4.3.3	Physicochemical Characterization of Active Sites *516*	
15.5	Alkylbenzene Disproportionation and *Trans*-Alkylation *517*	
15.5.1	Process Chemistry: Feeds, Products and Reactions *517*	
15.5.2	Catalysts *517*	
15.5.3	Physicochemical Characterization of Active Sites *518*	
15.6	Paraffin/Olefin to Aromatics *518*	
15.6.1	C_3/C_4 Paraffin to Aromatics and C_3/C_4 Paraffin/Olefin to Aromatics *518*	
15.6.1.1	Process Chemistry: Feeds, Products and Reactions *518*	
15.6.1.2	Catalysts *519*	
15.6.1.3	Physicochemical Characterization of Active Sites *519*	
15.6.2	C_6/C_7 Paraffin to Aromatics (Zeolitic Reforming) *520*	
15.6.2.1	Process Chemistry: Feeds, Products and Reactions *520*	
15.6.2.2	Catalysts *520*	
15.6.2.3	Physicochemical Characterization of Active Sites *520*	
15.7	Methanol to Olefins and Aromatics *521*	
15.7.1	Methanol to C_2–C_4 Olefins *521*	
15.7.2	Methanol to Aromatics *522*	
15.7.3	Catalysts *523*	
15.7.3.1	Physicochemical Characterization of Active Sites *524*	
15.7.4	Reaction Mechanism of Methanol to Hydrocarbons *527*	
15.8	Summary *528*	
	References *528*	
16	**Bond Breaking and Rearrangement** *535*	
	Suheil F. Abdo	
16.1	Introduction *535*	
16.2	Critical Zeolite Properties *536*	
16.2.1	Framework Types and Compositions *536*	
16.2.2	Stabilization Methods *539*	
16.2.3	Property–Function Relationship *542*	
16.2.3.1	Acid Strength Requirements for Product Control and Influence of Spatial Distribution on Selectivity *544*	
16.2.3.2	Pore Geometry and Framework Composition *545*	
16.3	Chemistry of Bond Scission Processes *546*	
16.3.1	Heteroatom Removal: Desulfurization Denitrogenation and Deoxygenation *546*	
16.3.2	Boiling Point Reduction *551*	
16.3.2.1	Paraffin Cracking *551*	
16.3.2.2	Aromatic and Naphthene Ring Opening *554*	
16.4	Fluidized Catalytic Cracking *556*	
16.4.1	Process Configuration and Catalysts *557*	

	16.4.2	The Changing Role of the FCC: Transportation Fuel Production or Petrochemical Feed Production *560*
	16.5	Hydrocracking and Hydroisomerization *560*
	16.5.1	Process Configurations and Catalysts *561*
	16.5.2	Catalyst Requirements for the Hydrocracker *561*
	16.5.3	The Changing Role of the Hydrocracker in a Reformulated Fuels Environment *565*
	16.6	Conclusions *566*
		References *566*

Index *571*

Preface

Zeolites have an enormous impact on our daily lives, both directly and indirectly. For example, upstream hydrocarbons such as aromatics and olefins are produced using zeolite catalysts. The aromatics or olefins are then separated from the reaction mixtures using zeolite adsorbents. The purified components produced by these zeolite-based methods are then used in downstream processes to produce products that we use daily, such as clothes, furniture, foods, construction materials and materials to build roads, automobile parts, fuels, gasoline, etc. In addition to the indirect impacts mentioned above, zeolites also have a direct impact on our daily lives. For example, zeolites are used as builders in detergent formulations.

With the important features mentioned above, I am pleased to accept Wiley-VCH's invitation to edit this book entitled "*Zeolites in Industrial Separation and Catalysis*", which explores the broader scope of zeolite technology and further examines zeolite applications. My hope is that the knowledge gained from this book will generate more innovation in the field of zeolites.

This is the first book to offer a practical overview of zeolites and their commercial applications. Each chapter is written by a globally recognized and acclaimed leader in the field. The book is organized into three parts. The first part discusses the history and chemistry of zeolites, the second part focuses on separation processes and the third part explores zeolites in the field of catalysis. All three parts are tied together by their focus on the unique properties of zeolites that allow them to function in different capabilities as an adsorbent, a membrane and a catalyst. Each of the chapters also discusses the impact of zeolites within the industry.

The first part of the book documents the history, structure, chemistry, formulation and characterizations of zeolites in Chapters 1–4. The past 60 years have seen a progression in molecular sieve materials from aluminosilicate zeolites to microporous silica polymorphs, microporous aluminophosphate-based polymorphs, metallosilicate and metallophosphate compositions, octahedral–tetrahedral frameworks, mesoporous molecular sieves and, most recently, hybrid metal organic frameworks (MOFs).

Introductory Chapter 1 provides a historical overview of molecular sieve materials. Chapter 2 covers the definition of a zeolite and describes their basic and composite building units and how they are linked in zeolite frameworks. It defines pores, channels, cages and cavities; and it gives references for finding detailed information about all framework types. Chapter 2 also describes the framework structures and cation locations for some of the commercially significant zeolites in more detail.

Chapter 3 outlines zeolite synthesis, modification and the manufacturing of zeolite-based catalysts and adsorbents. Extensive patent references are given to provide the reader with a historical perspective. Some of the pitfalls associated with the operation of synthesis and manufacturing units are also described.

Characterization is the foundation for the development and commercialization of new zeolites and zeolite-containing catalysts and adsorbents. Chapter 4 provides an overview of the most commonly employed characterization techniques and emphasizes the utility and limitations of each of these methods. An example is provided as to how a multi-technique characterization approach is necessary in order to determine the structure of a newly invented zeolite. Techniques covered in this chapter include X-ray powder diffraction, electron microscopy, infrared (IR) spectroscopy, nuclear magnetic resonance (NMR) spectroscopy and physical/chemical methods.

The second part of the book covers zeolite adsorptive separation, adsorption mechanisms, zeolite membranes and mixed matrix membranes in Chapters 5–11. Chapter 5 summarizes the literature and reports adsorptive separation work on specific separation applications organized around the types of molecular species being separated. A series of tables provide groupings for: (i) aromatics and derivatives, (ii) non-aromatic hydrocarbons, (iii) carbohydrates and organic acids, (iv) fine chemical and pharmaceuticals, (v) trace impurities removed from bulk materials. Zeolite adsorptive separation mechanisms are theorized in Chapter 6.

Chapter 7 gives a review of the technology and applications of zeolites in liquid adsorptive separation of petrochemical aromatic hydrocarbons. The application of zeolites to petrochemical aromatic production may be the area where zeolites have had their largest positive economic impact, accounting for the production of tens of millions of tonnes of high-value aromatic petrochemicals annually. The non-aromatic hydrocarbon liquid phase adsorption review in Chapter 8 contains both general process concepts as well as sufficient individual process details for one to understand both commercially practiced and academic non-aromatic separations.

In Chapter 9, the major industrial gas separations enabled by zeolite-based products are examined and classified by the method of regeneration employed. Thermal swing regeneration and pressure swing regeneration are the two most prevalent processes for gas separations. After delineating a number of useful equilibrium models that are used in the field, mass transfer in formed zeolite products is covered in some detail, providing the reader with some estimation methods that enable the determination of mass transfer from first principles and allowing for prediction of the performance of packed-bed contactors. With that background, the basics of thermal swing dehydration are used to teach the basic methodology behind gas phase adsorption process design. After dealing with dehydration, other purification processes are discussed, including CO_2 removal, sulfur removal, VOC abatement and removal of heavy metals, including mercury. Pressure swing adsorption (PSA) separations are then dealt with on a high level, highlighting differences that result in PSA from thermal swing behavior. Finally some novel rotating contactors that are making inroads into the commercial adsorption field are discussed. This book introduces novel separators such as zeolite membranes and mixed-matrix membranes. Zeolite membranes and mixed-matrix membranes are described in Chapters 10 and 11, respectively. Zeolite membranes are composed of inorganic zeolites. Chapter 10 describes how one can fabricate submicron particle-size zeolites into a membrane configuration. In contrast, mixed-matrix membranes are composed of inorganic zeolites and polymers. Chapter 11 provides a brief introduction to combining polymers and inorganic zeolites into mixed-matrix membranes. Both chapters have a comprehensive review of zeolite and zeolite/polymer mixed-matrix membranes. They cover the materials, separation mechanisms, methods, structures, properties and anticipated potential applications of the zeolite and zeolite/polymer mixed-matrix membranes.

The third and last part of the book (Chapters 12–16) deals with zeolite catalysis. Chapter 12 gives an overview of the various reactions which have been catalyzed by zeolites, serving to set the reader up for in-depth discussions on individual topics in Chapters 13–16. The main focus is on reactions of hydrocarbons catalyzed by zeolites, with some sections on oxidation catalysis. The literature review is drawn from both the patent and open literature and is presented primarily in table format. Brief notes about commonly used zeolites are provided prior to each table for each reaction type. Zeolite catalysis mechanisms are postulated in Chapter 13. The discussion includes the governing principles of performance parameters like adsorption, diffusion, acidity and how these parameters fundamentally influence zeolite catalysis. Brief descriptions of the elementary steps of hydrocarbon conversion over zeolites are also given. The intent is not to have an extensive review of the field of zeolite catalysis, but to select a sufficiently large subset of published literature through which key points can be made about reaction mechanisms and zeolitic requirements.

The chemistry of isomerization technology is discussed in Chapter 14. Isomerization technology provides the means to convert less valuable hydrocarbon isomers into more valuable ones. Zeolites, with their precise morphologies, can be made into exceptional catalysts with high selectivity. The ability to adjust zeolite chemistry through innovative synthesis or post-synthesis treatments further enhances their versatility in isomerization applications. Chapter 15 describes some of the key technologies involving carbon–carbon bond formation. In this class of reactions, zeolite catalysts are well established in industrial olefin oligomerization and aromatic alkylation and transalkylation processes. More recently, zeolites or other molecular sieves have found commercial application for paraffin cyclization to aromatics and the conversion of methanol to hydrocarbons ranging from light olefins to aromatics. In addition, research into the use of zeolites as a replacement for liquid acid catalysts for the alkylation of paraffins with olefins, which has yet to be applied on a full commercial scale, is discussed.

Chapter 16 discusses carbon–carbon bond breaking and rearrangement. The chapter attempts to identify the key properties of zeolites which must be tailored for optimum performance in the various application fields. Specifically, this chapter emphasizes the need to control a multiplicity of properties simultaneously in order to achieve the desired performance. Thus, channel geometry, acid site strength and spatial distribution as well as particle morphology all must be adjusted as an ensemble to deliver the desired catalysis. Beyond the fundamentals, this chapter provides an overview of key technologies employing the breaking and rearranging of hydrocarbon bonds.

Enjoyment and love of technical contributions to the scientific community are my motivations in organizing my thoughts, inviting experts to write the chapters and editing this book. The book's success also comes from the authors of the book chapters: Edith M. Flanigen, Robert W. Broach, Stephen T Wilson, Robert L. Bedard, Steve A.Bradley, Wharton Sinkler, Thomas M. Mezza, Sesh Prabhakar, Robert B. James, Stanley J. Frey, Stephen W. Sohn, Steve Dunne, Jerry Lin, Jessica O'Brien, Chunqing Liu, Christopher P. Nicholas, Hayim Abrevaya, John E. Bauer, Paula L. Bogdan, Feng Xu, Gregory J. Gajda, Deng-Yang Y. Jan, Paul T. Barger and Suheil F. Abdo. I am very thankful for their excellent technical contributions. Thanks are also due to the UOP's publication committee for reviewing the chapters and allowing the book to be published.

I would also like to thank my associates at UOP and my graduate students from the Petroleum and Petrochemical College at Chulalongkorn University, with whom I have had the pleasure of teaching and learning from. Finally, I would like to thank my family, especially my wife Apinya for supporting my love and enjoyment of the work.

September 20, 2009

Santi Kulprathipanja
UOP, A Honeywell Company
Des Plaines, Illinois, USA

List of Contributors

Suheil F. Abdo
Hayim Abrevaya
Paul T. Barger
John E. Bauer
Robert L. Bedard
Paula L. Bogdan
Steven A. Bradley
Robert W. Broach
Steve Dunne
Edith M. Flanigen
Stanley J. Frey
Gregory J. Gajda
Robert B. James
Deng-Yang Y. Jan
Santi Kulprathipanja
Chunqing Liu
Thomas M. Mezza
Christopher P. Nicholas
Sesh Prabhakar
Wharton Sinkler
Stephen W. Sohn
Stephen T. Wilson
Feng Xu

UOP, A Honeywell Company
50 East Alonquin Road
Des Plaines, IL 60017
USA

Jessica O'Brien-Abraham
Jerry Y.S. Lin

Arizona State University
Department of Chemical Engineering
ECG202, PO Box 876006
Tempe, AZ 85287-6006
USA

Zeolites in Industrial Separation and Catalysis. Edited by Kulprathipanja
Copyright © 2010 WILEY-VCH Verlag GmbH & Co. KGaA, Weinheim
ISBN: 978-3-527-32505-4

1
Introduction
Edith M. Flanigen, Robert W. Broach, and Stephen T. Wilson

1.1
Introduction

The past nearly six decades have seen a chronological progression in molecular sieve materials from the aluminosilicate zeolites to microporous silica polymorphs, microporous aluminophosphate-based polymorphs, metallosilicate and metallophosphate compositions, octahedral-tetrahedral frameworks, mesoporous molecular sieves and most recently hybrid metal organic frameworks (MOFs). A brief discussion of the historical progression is reviewed here. For a more detailed description prior to 2001 the reader is referred to [1]. The robustness of the field is evident from the fact that publications and patents are steadily increasing each year.

1.1.1
Molecular Sieves and Zeolites

Molecular sieves are porous solids with pores of the size of molecular dimensions, 0.3–2.0 nm in diameter. Examples include zeolites, carbons, glasses and oxides. Some are crystalline with a uniform pore size delineated by their crystal structure, for example, zeolites. Most molecular sieves in practice today are zeolites.

Zeolites are crystalline aluminosilicates of group IA and group IIA elements, such as sodium, potassium, magnesium and calcium [2]. Chemically, they are represented by the empirical formula:

$$M_{2/n}O \cdot Al_2O_3 \cdot ySiO_2 \cdot wH_2O$$

where y is 2–200, n is the cation valence and w represents the water contained in the voids of the zeolite. Structurally, zeolites are complex, crystalline inorganic polymers based on an infinitely extending three-dimensional, four-connected framework of AlO_4 and SiO_4 tetrahedra linked to each other by the sharing of oxygen ions. Each AlO_4 tetrahedron in the framework bears a net negative charge which is balanced by an extra-framework cation. The framework structure contains intracrystalline channels or interconnected voids that are occupied by the cations and water molecules. The cations are mobile and ordinarily undergo ion exchange. The water may be removed reversibly, generally by the application of heat, which leaves intact a crystalline host structure permeated by the micropores and voids which may amount to 50% of the crystals by volume. The intracrystalline channels or voids can be one-, two- or three-dimensional. The preferred type has two or three dimensions to facilitate intracrystalline diffusion in adsorption and catalytic applications.

In most zeolite structures the primary structural units, the AlO_4 or SiO_4 tetrahedra, are assembled into secondary building units which may be simple polyhedra, such as cubes, hexagonal prisms or cubo-octahedra. The final framework structure consists of assemblages of the secondary units (see Chapter 2). More than 70 novel, distinct framework structures of zeolites are known. They exhibit pore sizes from 0.3 to 1.0 nm and pore volumes from about 0.10 to 0.35 cm^3/g. Typical zeolite pore sizes include: (i) small pore zeolites with eight-ring pores, free diameters of 0.30–0.45 nm (e.g., zeolite A), (ii) medium pore zeolites with 10-ring pores, 0.45–0.60 nm in free diameter (ZSM-5), (iii) large pore zeolites with 12-ring pores of 0.6–0.8 nm (e.g., zeolites X, Y) and (iv) extra-large pore zeolites with 14-ring pores (e.g., UTD-1).

The zeolite framework should be viewed as somewhat flexible, with the size and shape of the framework and pore responding to changes in temperature and guest species. For example, ZSM-5 with sorbed neopentane has a near-circular pore of 0.62 nm, but with substituted aromatics as the guest species the pore assumes an elliptical shape, 0.45 to 0.70 nm in diameter.

Some of the more important zeolite types, most of which have been used in commercial applications, include the zeolite minerals mordenite, chabazite, erionite and clinoptilolite, the synthetic zeolite types A, X, Y, L, "Zeolon" mordenite, ZSM-5, beta and MCM-22 and the zeolites F and W.

1.1.2
Nomenclature

There is no systematic nomenclature developed for molecular sieve materials. The discoverer of a synthetic species based on a characteristic X-ray powder diffraction pattern and chemical composition typically assigns trivial symbols. The early synthetic materials discovered by Milton, Breck and coworkers at Union Carbide used the modern Latin alphabet, for example, zeolites A, B, X, Y, L. The use of the Greek alphabet was initiated by Mobil and Union Carbide with the zeolites alpha, beta, omega. Many of the synthetic zeolites which have the structural topology of mineral zeolite species were assigned the name of the mineral, for example, synthetic mordenite, chabazite, erionite and offretite. The molecular sieve literature is replete with acronyms: ZSM-5, -11, ZK-4 (Mobil), EU-1, FU-1, NU-1 (ICI), LZ-210, AlPO, SAPO, MeAPO, etc. (Union Carbide, UOP) and ECR-1 (Exxon). The one publication on nomenclature by IUPAC in 1979 is limited to the then-known zeolite-type materials [3].

The *Atlas of Zeolite Structure Types* [4], published and frequently updated by the IZA Structure Commission, assigns a three-letter code to be used for a known framework topology irrespective of composition. Illustrative codes are LTA for Linde zeolite A, FAU for molecular sieves with a faujasite topology (e.g., zeolites X, Y), MOR for the mordenite topology, MFI for the ZSM-5 and silicalite topologies and AFI for the aluminophosphate $AlPO_4$-5 topology. The acceptance of a newly determined structure of a zeolite or molecular sieve for inclusion in the official *Atlas* is reviewed and must be approved by the IZA Structure Commission. The IZA Structure Commission was given authority at the Seventh International Zeolite Conference (Tokyo, 1986) to approve and/or assign the three-letter structure code for new framework topologies.

The definition and usage of the term "zeolite" has evolved and changed, especially over the past decade, to include non-aluminosilicate compositions and structures. Beginning with the second revised edition of the *Atlas* [5], the term "zeolite and zeolite-like materials" is introduced to try and capture the range of materials of interest. The inclusion of a structure in the *Atlas* is limited to three-dimensional, tetrahedral oxide networks with a framework density less than about 21 T-atoms per 1000 Å^3 irrespective of framework composition. Similarly the term zeolite has been broadened in the mineralogy literature to include tetrahedral framework compositions with T-elements other than Al and Si but where classic zeolite properties are exhibited (e.g., structures containing open cavities in the form of channels and cages, reversible hydration–dehydration characteristics [6]). Very recently as a sign of the times the term "nanoporous" materials has been applied to zeolites and related molecular sieves [7].

1.1.3
Early History

The history of zeolites began in 1756 when the Swedish mineralogist Cronstedt discovered the first zeolite mineral, stilbite [8]. He recognized zeolites as a new class of minerals consisting of hydrated aluminosilicates of the alkali and alkaline earths. Because the crystals exhibited intumescence when heated in a blowpipe flame, Cronstedt called the mineral a "zeolite" (derived from two Greek words, *zeo* and *lithos*, meaning "to boil" and "a stone"). From 1777 through about the 1800s various authors described the properties of zeolite minerals, including adsorption properties and reversible cation exchange and dehydration. St. Claire Deville reported the first hydrothermal synthesis of a zeolite, levynite, in 1862 [9]. In 1896 Friedel developed the idea that the structure of dehydrated zeolites consists of open spongy frameworks after observing that various liquids such as alcohol, benzene and chloroform were occluded by dehydrated zeolites [10]. Grandjean in 1909 observed that dehydrated chabazite adsorbs ammonia, air, hydrogen and other molecules [11], and in 1925 Weigel and Steinhoff reported the first molecular sieve effect [12]. They noted that dehydrated chabazite crystals rapidly adsorbed water, methyl alcohol, ethyl alcohol and formic acid but essentially excluded acetone, ether or benzene. In 1927 Leonard described the first use of X-ray diffraction for identification in mineral synthesis [13]. Taylor and Pauling described the first single crystal structures of zeolite minerals in 1930 [14, 15]. In 1932 McBain established the term "molecular sieve" to define porous solid materials that act as sieves on a molecular scale [16].

Thus, by the mid-1930s the literature described the ion exchange, adsorption, molecular sieving and structural properties of zeolite minerals as well as a number of reported syntheses of zeolites. The early synthetic work remains unsubstantiated because of incomplete characterization and the difficulty of experimental reproducibility.

Richard M. Barrer began his pioneering work in zeolite adsorption and synthesis in the mid-1930s to 1940s. He presented the first classification of the then-known zeolites based on molecular size considerations in 1945 [17] and in 1948 reported the first definitive synthesis of zeolites, including the synthetic analog of the zeolite mineral mordenite [18] and a novel synthetic zeolite [19] much later identified as the KFI framework. Barrer's work in the mid- to late 1940s inspired Robert M. Milton of the Linde Division of Union Carbide Corporation to initiate studies in zeolite synthesis in search of new approaches for separation and purification of air. Between 1949 and 1954 Milton and coworker Donald W. Breck discovered a number of commercially significant zeolites, types A, X and Y. In 1954 Union Carbide commercialized synthetic zeolites as a new class of industrial materials for separation and purification. The earliest applications were the drying of refrigerant gas and natural gas. In 1955 T.B. Reed and D.W. Breck reported the structure of the synthetic zeolite A [20]. In 1959 Union Carbide marketed the "ISOSIV" process for normal–isoparaffin separation, representing the first major bulk separation process using true molecular sieving selectivity. Also in 1959 a zeolite Y-based catalyst was marketed by Carbide as an isomerization catalyst [21].

In 1962 Mobil Oil introduced the use of synthetic zeolite X as a hydrocarbon cracking catalyst. In 1969 Grace described the first modification chemistry based on steaming zeolite Y to form an "ultrastable" Y. In 1967–1969 Mobil Oil reported the synthesis of the high silica zeolites beta and ZSM-5. In 1974 Henkel introduced zeolite A in detergents as a replacement for the environmentally suspect phosphates. By 2008 industry-wide approximately 367 000 t of zeolite Y were in use in catalytic cracking [22]. In 1977 Union Carbide introduced zeolites for ion-exchange separations.

1.1.4
Natural Zeolites

For 200 years following their discovery by Cronstedt, zeolite minerals (or natural zeolites) were known to occur typically as minor constituents in vugs or cavities in basaltic and volcanic rock. Such occurrences precluded their being obtained in mineable quantities for commercial use. From the late 1950s to 1962 major geologic discoveries revealed the widespread occurrence of a number of natural zeolites in sedimentary deposits throughout the western United States. The discoveries resulted from the use of X-ray diffraction to examine very fine-grained (1–5 μm) sedimentary rock. Some zeolites occur in large, nearly mono-mineralic deposits suitable for mining. Those that have been commercialized for adsorbent applications include chabazite, erionite, mordenite and clinoptilolite [23].

Mordenite and clinoptilolite are used in small volume in adsorbent applications including air separation and in drying and purification [24]. Natural zeolites have also found use in bulk applications as fillers in paper, in pozzolanic cements and concrete, in fertilizer and soil conditioners and as dietary supplements in animal husbandry.

1.2
History of Molecular Sieve Materials

The theme of research on molecular sieve materials over the past nearly 60 years has been a quest for new structures and compositions. The major discoveries and advances in molecular sieve materials during that period are summarized in Table 1.1.

The history of commercially significant molecular sieve materials from 1954 to 2001 was reviewed in detail by one of us (E.M.F., ref [1]) Highlights from that review and the subsequent history are presented here. The reader is referred to Chapter 2 for the structures of the materials and to Chapter 3 and ref [25] for a detailed discussion on zeolite synthesis.

Table 1.1 Evolution of molecular sieve materials.

Time of initial discovery	Composition
Late 1940s to early 1950s	Low Si/Al ratio zeolites
Mid-1950s to late 1960s	High Si/Al ratio zeolites
Early 1970s	SiO_2 molecular sieves
Late 1970s	$AlPO_4$ molecular sieves
Late 1970s to early 1980s	SAPO and MeAPO molecular sieves
Late 1970s	Metallo-silicates, aluminosilicates
Early to mid-1980s	$AlPO_4$-based molecular sieves
Early to mid-1990s	Metallophosphates
	Mesoporous molecular sieves
	Octahedral–tetrahedral frameworks
Late 1990s	Metal organic frameworks
2000s	UZM aluminosilicate zeolites, Si/Al = 2–30
	Germanosilicate zeolites
	SiO_2 molecular sieves in fluoride media

1.2.1
Aluminosilicate Zeolites and Silica Molecular Sieves

The early evolution of aluminosilicate zeolites, in the 1950s to 1970s, is summarized in Table 1.2, based on increasing framework Si/Al composition. The four somewhat arbitrary categories are: (i) "low", (ii) "intermediate", (iii) "high" silica zeolites and (iv) "silica" molecular sieves.

The transition in properties accompanying the increase in the framework Si/Al are generalized here but should only be viewed as trends. The thermal stability increases from about 700 °C in the low silica zeolites to 1300 °C in the silica molecular sieves. The surface selectivity, which is highly hydrophilic in the low silica zeolites, is hydrophobic in the high silica zeolites and the silica molecular sieves. The acidity tends to increase in strength with increasing Si/Al ratio. As the Si/Al ratio increases, the cation concentration and ion exchange capacity (proportional to the aluminum content) decreases. The structures of the low silica zeolites are predominantly formed with four, six and eight rings of tetrahedra. In the intermediate silica zeolites we see the onset of five rings in mordenite and omega zeolite. In the high silica zeolite structures and the silica molecular sieves we find a predominance of five rings of tetrahedra, for example, silicalite.

The low silica zeolites represented by zeolites A and X are aluminum-saturated, have the highest cation concentration and give optimum adsorption properties in terms of capacity, pore size and three-dimensional channel systems. They represent highly heterogeneous surfaces with a strongly hydrophilic surface selectivity. The intermediate Si/Al zeolites (Si/Al of 2–5) consist of the natural zeolites erionite, chabazite, clinoptilolite and mordenite, and the synthetic zeolites Y, mordenite, omega and L. These materials are still hydrophilic in this Si/Al range.

The high silica zeolites with Si/Al of 10–100 can be generated by either thermochemical framework modification of hydrophilic zeolites or by direct synthesis. In

Table 1.2 The early evolution of aluminosilicate molecular sieve materials.

Composition and examples
"Low" Si/Al zeolites (1 to 1.5): A, X

"Intermediate" Si/Al zeolites (~2 to 5):
   A. Natural zeolites: erionite, clinoptilolite, mordenite
   B. Synthetic zeolites: Y, L, large pore mordenite, omega

"High" Si/Al zeolites (~10 to 100):
   A. By thermochemical framework modification:
       Highly siliceous variants of Y, mordenite, erionite
   B. By direct synthesis: ZSM-5, beta

Silica molecular sieves (Si/Al > 100):   Silicalite

the modification route stabilized, siliceous variants of Y, mordenite, erionite and over a half-dozen other zeolites have been prepared by steaming and acid extraction. These materials are reported to be hydrophobic and organophilic and represent a range of pore sizes from 0.4 to 0.8 nm. A very large number of high-silica zeolites prepared by direct synthesis have now been reported, including beta, ZSM-5, -11, -12, -21, -34, NU-1, FU-1 and ferrisilicate and borosilicate analogs of the aluminosilicate structures. Typical of the reported silica molecular sieves are silicalite, fluoride silicalite, silicalite-2 and TEA-silicate. ZSM-5 and silicalite have achieved commercial significance.

In summary, when we compare the properties of the low and intermediate zeolites with those of the high silica zeolites and silica molecular sieves, we find that their resulting properties allow the low and intermediate zeolites to remove water from organics and to carry out separations and catalysis on dry streams. In contrast, the hydrophobic high silica zeolites and silica molecular sieves can remove and recover organics from water streams and carry out separations and catalysis in the presence of water.

1.2.2
The Materials Explosion Since the 1980s

Overall the period since the 1980s can be described as a period of explosion in the discovery of new compositions and structures of molecular sieves. This can perhaps be seen most vividly by comparing the numbers of structure types contained in the various editions of the *Atlas of Zeolite Structure Types* [4]. The first edition (1978) contained 38 structure types, the second edition (1987) 64, the third edition (1992) 85 and the most recent edition (2007) 176. Thus 112 new structure types have been discovered since 1978. However, the reader should be cautioned that a significant number of the structure types included in the *Atlas* are not truly microporous or molecular sieve materials (i.e., they are not stable for the removal of as-synthesized guest species, typically water or organic templates) and therefore cannot reversibly adsorb molecules or carry out catalytic reactions. Unfortunately, the *Atlas* gives only limited information on the stability of the structures described.

1.2.2.1 The 1980s

In the 1980s there was extensive work carried out on the synthesis and applications of ZSM-5 and a proliferating number of other members of the high silica zeolite family. In 1982 microporous crystalline aluminophosphate molecular sieves were described by Wilson *et al.* [26] at Union Carbide, and additional members of the aluminophosphate-based molecular sieve family, for example, SAPO, MeAPO, MeAPSO, ElAPO and ElAPSO, were subsequently disclosed by 1986 [27]. Considerable effort in synthesizing metallosilicate molecular sieves was reported when the metals iron, gallium, titanium, germanium and others were incorporated during synthesis into silica or aluminosilicate frameworks, typically with the ZSM-5 (MFI) topology [28]. Additional crystalline microporous silica molecular sieves and related clathrasil structures were reported.

The 1980s saw major developments in secondary synthesis and modification chemistry of zeolites. Silicon-enriched frameworks of over a dozen zeolites were described using methods of: (i) thermochemical modification (prolonged steaming) with or without subsequent acid extraction, (ii) mild aqueous ammonium fluorosilicate chemistry, (iii) high-temperature treatment with silicon tetrachloride and (iv) low-temperature treatment with fluorine gas. Similarly, framework metal substitution using mild aqueous ammonium fluorometallate chemistry was reported to incorporate iron, titanium, chromium and tin into zeolite frameworks by secondary synthesis techniques.

Aluminophosphate-Based Molecular Sieves In 1982 a major discovery of a new class of aluminophosphate molecular sieves was reported by Wilson et al. [26]. By 1986 some 13 elements were reported to be incorporated into the aluminophosphate frameworks: Li, Be, B, Mg, Si, Ti, Mn, Fe, Co, Zn, Ga, Ge and As [27]. These new generations of molecular sieve materials, designated $AlPO_4$-based molecular sieves, comprise more than 24 structures and 200 compositions.

The >24 structures of $AlPO_4$-based molecular sieves reported to date include zeolite topological analogs and a large number of novel structures. The major structures are shown in Table 1.3. They include 15 novel structures as well as seven structures with framework topologies related to those found in the zeolites

Table 1.3 Typical structures in $AlPO_4$-based molecular sieves.

Species	Structure type	Pore size, nm	Saturation H_2O pore volume, cm^3/g	Species	Structure type	Pore size, nm	Saturation H_2O pore volume, cm^3/g
Very large pore				Small pore			
VPI-5	Novel	1.25	0.35	14	Novel	0.4	0.19
8	Novel	0.9	0.24	17	ERI	0.43	0.28
				18	Novel	0.43	0.35
				26	Novel	0.43	0.23
Large pore				33	Novel	0.4	0.23
5	Novel	0.8	0.31	34, 44, 47	CHA	0.43	0.3
36	Novel	0.8	0.31	35	LEV	0.43	0.3
37	FAU	0.8	0.35	39	Novel	0.4	0.23
40	Novel	0.7	0.33	42	LTA	0.43	0.3
46	Novel	0.7	0.28	43	GIS	0.43	0.3
				52	Novel	0.43	0.3
				56	Novel	0.43	0.3
				Very small pore			
Intermediate				16	Novel	0.3	0.3
11	Novel	0.6	0.16	20	SOD	0.3	0.24
31	Novel	0.65	0.17	25	Novel	0.3	0.17
41	Novel	0.6	0.22	28	Novel	0.3	0.21

CHA (-34, -44, -47), ERI (-17), GIS (-43), LEV (-35), LTA (-42), FAU (-37) and SOD (-20). Also shown is the pore size and saturation water pore volume for each structure type. The structures include the first very large pore molecular sieve, VPI-5, with an 18-ring one-dimensional channel with a free pore opening of 1.25 nm [29], large pore (0.7–0.8 nm), intermediate pore (0.6 nm), small pore (0.4 nm) and very small pore (0.3 nm) materials. Saturation water pore volumes vary from 0.16 to 0.35 cm^3/g, comparable to the pore volume range observed in zeolites (see Chapter 2 for detailed structures).

The addition of another element to the aluminophosphate reactants, for example, Si, metal ions Mg, Co, Mn, Fe, as well as other elements, led to the silicoaluminophosphate family "SAPO", the metalloaluminophosphate family "MeAPO" and other elements, the "ElAPO" family, where the added element is incorporated into the hypothetical AlPO$_4$ framework.

The aluminophospahate molecular sieve product composition expressed in terms of oxide ratios is:

$$xR \cdot Al_2O_3 \cdot 1.0 \pm 0.2\, P_2O_5 \cdot yH_2O$$

where R is an amine or quaternary ammonium ion. The AlPO$_4$ molecular sieve as synthesized must be calcined at 400–600 °C to remove the R and water, yielding a microporous aluminophosphate molecular sieve.

The characteristics of aluminophosphate molecular sieves include a univariant framework composition with Al/P = 1, a high degree of structural diversity and a wide range of pore sizes and volumes, exceeding the pore sizes known previously in zeolite molecular sieves with the VPI-5 18-membered ring material. They are neutral frameworks and therefore have nil ion-exchange capacity or acidic catalytic properties. Their surface selectivity is mildly hydrophilic. They exhibit excellent thermal and hydrothermal stability, up to 1000 °C (thermal) and 600 °C (steam).

The silicoaluminophosphate (SAPO) family [30] includes over 16 microporous structures, eight of which were never before observed in zeolites. The SAPO family includes a silicon analog of the 18-ring VPI-5, Si-VPI-5 [31], a number of large-pore 12-ring structures including the important SAPO-37 (FAU), medium-pore structures with pore sizes of 0.6–0.65 nm and small-pore structures with pore sizes of 0.4–0.43 nm, including SAPO-34 (CHA). The SAPOs exhibit both structural and compositional diversity.

The SAPO anhydrous composition can be expressed as 0–0.3R(Si$_x$Al$_y$P$_z$)O$_2$, where x, y and z are the mole fraction of the respective framework elements. The mole fraction of silicon, x, typically varies from 0.02 to 0.20 depending on synthesis conditions and structure type. Martens *et al.* have reported compositions with the SAPO-5 structure with x up to 0.8 [32]. Van Nordstrand *et al.* have reported the synthesis of a pure silica analog of the SAPO-5 structure, SSZ-24 [33].

The introduction of silicon into hypothetical phosphorus sites produces negatively charged frameworks with cation-exchange properties and weak to mild acidic catalytic properties. Again, as in the case of the aluminophosphate molecular sieves, they exhibit excellent thermal and hydrothermal stability.

In the metal aluminophosphate (MeAPO) family the framework composition contains metal, aluminum and phosphorus [27]. The metal (Me) species include the divalent forms of Co, Fe, Mg, Mn and Zn and trivalent Fe. As in the case of SAPO, the MeAPOs exhibit both structural diversity and even more extensive compositional variation. Seventeen microporous structures have been reported, 11 of these never before observed in zeolites. Structure types crystallized in the MeAPO family include framework topologies related to the zeolites, for example, -34 (CHA) and -35 (LEV), and to the AlPO$_4$s, e.g., -5 and -11, as well as novel structures, e.g., -36 (0.8 nm pore) and -39 (0.4 nm pore). The MeAPOs represent the first demonstrated incorporation of divalent elements into microporous frameworks.

The spectrum of adsorption pore sizes and pore volumes and the hydrophilic surface selectivity of the MeAPOs are similar to those described for the SAPOs. The observed catalytic properties vary from weakly to strongly acidic and are both metal- and structure-dependent. The thermal and hydrothermal stability of the MeAPO materials is somewhat less than that of the AlPO$_4$ and SAPO molecular sieves.

The MeAPO molecular sieves exhibit a wide range of compositions within the general formula $0-0.3R(Me_xAl_yP_z)O_2$. The value of x, the mole fraction of Me, typically varies from 0.01 to 0.25. Using the same mechanistic concepts described for SAPO, the MeAPOs can be considered as hypothetical AlPO$_4$ frameworks that have undergone substitution. In the MeAPOs the metal appears to substitute exclusively for aluminum resulting in a negative (Me^{2+}) or neutral (Me^{3+}) framework charge. Like SAPO, the negatively charged MeAPO frameworks possess ion-exchange properties and Bronsted acid sites.

The MeAPSO family further extends the structural diversity and compositional variation found in the SAPO and MeAPO molecular sieves. These quaternary frameworks have Me, Al, P and Si as framework species [27]. The MeAPSO structure types include framework topologies observed in the binary AlPO$_4$ and ternary (SAPO, MeAPO) compositional systems and the novel structure -46 with a 0.7 nm pore. The structure of -46 has been determined [34].

Quinary and senary framework compositions have been synthesized containing aluminum, phosphorus and silicon, with additional combinations of divalent (Me) metals. In the ElAPO and ElAPSO compositions the additional elements Li, Be, B, Ga, Ge, As and Ti have been incorporated into the AlPO$_4$ framework [27].

Most of the catalytic interest in the AlPO$_4$-based molecular sieves have centered on the SAPOs which have weak to moderate Bronsted acidity, and two have been commercialized: SAPO-11 in lube oil dewaxing by Chevron and SAPO-34 in methanol-to-olefins conversion by UOP/Norsk Hydro. Spurred on by the success of TS-1 in oxidation catalysis, there is renewed interest in Ti, Co, V, Mn and Cr substituted AlPO$_4$-based materials. For a review of recent developments in the AlPO$_4$-based molecular sieves see [35].

Metallosilicate Molecular Sieves A large number of metallosilicate molecular sieves have been reported, particularly in the patent literature. Those claimed include silicates containing incorporated tetrahedral iron, boron, chromium, arsenic, gallium, germanium and titanium. Most of the earlier work has been reported with structures of the MFI type. Others include metallosilicate analogs of ZSM-11, -12, THETA-1, ZSM-34 and beta. The early metallosilicate molecular sieves are reviewed in detail by Szostak [28]. More recently crystalline microporous frameworks have been reported with compositions of beryllosilicate, such as nabesite [36], lovdarite [37] and OSB-1 [38]. Zinc silicates with significant framework Zn include VPI-7 [39], VPI-8 [40], VPI-9 [41] and CIT-6 [42]. There has also been a dramatic increase in new frameworks prepared by incorporating Ge into silicate and aluminosilicate frameworks (see below). Fe and Ga incorporation has so far produced the same structure types as Al incorporation. In contrast, framework incorporation of B, Be, Ge and Zn in metallosilicate compositions can yield novel structures difficult or impossible to obtain with Al. To date only B, Be, Ga, Ge, Fe, Ti and Zn have been sufficiently characterized to confirm structural incorporation. The titanium-silicalite composition, TS-1, has achieved commercialization in selective oxidation processes and iron-silicalite in ethylbenzene synthesis.

Other Framework Compositions Crystalline microporous frameworks have been reported with compositions of: beryllophosphate [43], aluminoborate [44], aluminoarsenate [45], galloarsenate [46], gallophosphate [47], antimonosilicate [48] and germanosilicate [49].

Harvey et al. [43] reported the synthesis of alkali beryllophosphate molecular sieves with the RHO, GIS, EDI and ANA structure topologies and a novel structure, BPH. Simultaneously, the first beryllophosphate mineral species were reported: tiptopite [with the cancrinite (CAN) topology] by Peacor et al. [50] and pahasapaite (with the RHO topology) by Rouse et al. [51].

In the late 1980s Bedard et al. reported the discovery of microporous metal sulfides, based on germanium (IV) and Sn (IV) sulfide frameworks [52]. The microporous sulfides are synthesized hydrothermally in the presence of alkylammonium templating agents. The GeS_4-based compositions include one or more framework-incorporated metals: Mn, Fe, Co, Ni, Cu, Zn, Cd and Ga. Over a dozen novel structures were reported which have no analogs in the microporous oxides. Ozin et al. have extended this work to a large number of microporous sulfides and selenides [53]. It should be noted that the microporous sulfides and selenides are prone to structure collapse upon calcination to remove the template species.

1.2.2.2 The 1990s

The explosion in the discovery of new compositions and structures observed in the 1980s continued through the 1990s. Some three dozen or more novel tetrahedral structures were synthesized in the 1990s, based on aluminosilicate, silica, metallosilicate and metallophosphate frameworks. Three are especially noteworthy. The gallophosphate cloverite (-CLO) has the first 20-ring pore (0.4×1.32 nm in diameter) and the lowest observed framework density (number of T-atoms per 1000 Å^3): 11.1 [54]. The cloverite structure contains the D4R and alpha cages reminiscent of the aluminum-rich zeolite Type A (LTA), combined with the rpa cage found in the aluminophosphate structures. It is an interrupted framework structure and thus has somewhat limited thermal stability. The siliceous zeolite UTD-1(DON) contains a 14-ring pore (0.75 × 1.0 nm in diameter) and is the first aluminosilicate with a pore size larger than a 12-ring [55]. CIT-5 (CFI), a second 14-R structure with a pure silica composition and a 0.8 nm pore, was reported by Wagner et al. [56].

Stucky et al. discovered a generalized method for preparing a large number of metallo-aluminophosphate and metallo-gallophosphate frameworks containing transition metals. The method utilizes amine SDAs and high concentrations of transition metal and phosphate in mixed solvents, typically alcohol and water. Two of the novel structures (UCSB-6, UCSB-10) have multi-dimensional 12-ring channels connecting large cages. In addition numerous zeolite structure analogs were also observed [57]. Unfortunately, the high framework charge reduces structural stability when template removal is attempted.

Gier et al. reported zinc and beryllium phosphates and arsenates with the X (FAU), ABW and SOD structures reminiscent of the early aluminum-rich synthetic zeolite chemistry. The synthesis of $ZnPO_4$-X (FAU) is especially spectacular. Crystallization occurs almost instantaneously at 0 °C [58]. Concurrent with ease of synthesis, the structure is thermally unstable.

Table 1.4 lists some of the major new structures reported in the 1990s. Interestingly, as organic SDAs tended to dominate discovery of new frameworks, there were no new aluminum-rich synthetic zeolites reported in either the 1980s or the 1990s. The new aluminosilicate structures were all high silica or pure silica in composition. It awaited the 2000s for new aluminosilicate zeolite materials with low to medium Si/Al to be reported (see below).

Not to be outdone by humankind there were a number of new zeolite minerals discovered in nature during the 1990s. The zeolite mineral boggsite (BOG) has a novel framework topology with three-dimensional pores combining 10Rs and 12Rs, and it has not yet been reproduced synthetically [61]. Tschernichite is an aluminum-rich mineral analog of the synthetic zeolite beta [62]. Gottardiite is a new mineral analog of synthetic zeolite Nu-87 [63]. The zeolite mineral terranovaite (TER) has a novel structure with pentasil chains and a two-dimensional 10R channel [64]. Mutinaite is a high-silica zeolite mineral analog of ZSM-5 with the

Table 1.4 Major new synthetic structures of the 190s.

Species	Structure type	Pore size, nm	Ring size	Reference
MCM-22, 49	MWW	0.6	10	[59]
UTD-1	DON	1.0	14	[55]
CIT-5	CFI	0.8	14	[56]
EMC-2	EMT	0.7	12	[60]
Cloverite	−CLO	1.3	20	[54]

highest silica content of all known zeolite minerals (Si/Al = 7.7) [65]. The structure of the zeolite mineral perlialite was reported [66] to have the same topology as that of the synthetic zeolite L (LTL), some 35 years after the synthesis of zeolite L.

Tschortnerite (TSC) surely is the most remarkable novel zeolite mineral discovered [67]. Its unique framework topology contains five different cages: D-6Rs, D-8Rs, sodalite cages, truncated cubo-octahedra and a unique 96-membered cage. Cu-containing clusters are encapsulated within the truncated cubo-octahedra. The pore structure is three-dimensional with 8R channels, and the framework density of 12.2 is among the lowest known for zeolites. The framework is alumina-rich with Si/Al = 1, unusual for zeolite minerals.

Two major new classes of molecular sieve type materials were reported in the 1990s: (i) microporous frameworks based on mixed octahedral-tetrahedral frameworks in contrast to the previously described tetrahedral frameworks and (ii) mesoporous molecular sieves with pore sizes ranging from about 2 nm to greater than 10 nm.

Octahedral–Tetrahedral Frameworks The microporous materials described heretofore were all based on tetrahedral frameworks. Microporous titanosilicate materials with mixed octahedral–tetrahedral frameworks were reported in the 1990s. The framework linkage is through TiO_6 octahedra and SiO_4 tetrahedra. Chapman and Roe described the titanosilicate GTS-1, a structural analog of the mineral pharmacosiderite, with a three-dimensional channel system and 8-R pores [68]. Kuznicki and coworkers reported the synthesis of the titanosilicates ETS-4 and ETS-10 [69]. Their respective pore sizes are 0.4 and 0.8 nm. ETS-4 is the synthetic analog of the rare titanosilicate mineral zorite. The novel structure ETS-10 contains a three-dimensional 12-R pore system and shows a high degree of disorder (ref. [61]). ETS-10 has achieved commercial status in adsorption applications.

Mesoporous Molecular Sieves A major advance in molecular sieve materials was reported in 1992 by researchers at Mobil. Kresge *et al.* and Beck *et al.* described a new family of mesoporous silicate and aluminosilicate materials, designated M41S [70]. The members of the family include: MCM-41 (with a one-dimensional hexagonal arrangement of uniform open channels 0.2–10 nm in diameter), a cubic structure MCM-48 (with a three-dimensional channel system with pore sizes ~0.3–10 nm) and a number of lamellar structures. The order in the structure is derived from the channel arrangement. The silica or aluminosilicate wall outlining the channel is disordered and exhibits properties much like amorphous silica or silica–alumina.

Within the same time-frame and independently, Inagaki and coworkers reported a mesoporous material designated FSM-16, prepared by hydrothermal treatment of the layered sodium silicate kanemite, $NaHSi_2O_5 \cdot 3H_2O$ [71]. Chen *et al.* substantiated that FSM-16 and MCM-41 bear a strong resemblance to each other, both with narrow mesopore distributions and similar physicochemical properties, but with FSM-16 having higher thermal and hydrothermal stability due to the higher degree of condensation in the silicate walls [72].

Both mesoporous materials are synthesized hydrothermally with a surfactant liquid crystal as the template (see synthesis section below). They exhibit very high surface areas and pore volumes, of the order of $1000\,m^2/g$ and $1.5\,cm^3/g$, respectively.

Since this initial work there has been a plethora of literature on mesoporous molecular sieves. In addition to the silica and aluminosilicate frameworks similar mesoporous structures of metal oxides now include the oxides of Fe, Ti, V, Sb, Zr, Mn, W and others. Templates have been expanded to include nonionic, neutral surfactants and block copolymers. Pore sizes have broadened to the macroscopic size, in excess of 40 nm in diameter. A recent detailed review of the mesoporous molecular sieves is given in ref [73]. Vartuli and Degnan have reported a Mobil M41S mesoporous-based catalyst in commercial use, but to date the application has not been publicly identified.[74].

In a *tour de force* of detective work, Di Renzo et al. uncovered an obscure United States patent filed in 1969 and issued in 1971 to Chiola et al., describing a low-density silica. Reproduction of that patent resulted in a product having all of the properties of MCM-41 [75].

1.2.2.3 The New Millennium

Recent developments in zeolite synthesis and new materials include: (i) the use of combinatorial methodologies, microwave heating, multiple templates or SDAs and concentrated fluoride media in synthesis, (ii) synthesis using the charge density mismatch (CDM) concept, (iii) synthesis in ionothermal media, (iv) synthesis with complex "designer" templates or SDAs, (v) synthesis of nanozeolites, (vi) zeolite membranes and thin films (vii) and germanosilicate zeolites [76]. Several of these developments are discussed here.

A major return to the early lower Si/Al aluminosilicate zeolites of Milton and Breck was undertaken in the early part of this century. A novel synthetic strategy denominated the charge density mismatch (CDM) technique was pioneered by Lewis and coworkers at UOP. The method features the initial formation of a CDM aluminosilicate reaction mixture characterized by the mismatch between the charge density on the organoammonium structure-directing agent (SDA) and the charge density on the aluminosilicate network that is expected to form. For example, with a large SDA (low charge density) in an aluminosilicate reaction mixture with a low Si/Al ratio (high charge density), the crystallization of a zeolite is difficult or impossible, even with variation of hydroxide levels and crystallization temperature. Crystallization can be induced by the controlled addition of supplemental SDAs that have charge densities that are more suitably matched to that of the desired low ratio aluminosilicate network (e.g., alkali metal SDAs). Advantages of this approach are greater control over the crystallization process and reliable cooperation of multiple templates. The approach is demonstrated in the TEA-TMA template system, in which the new zeolites UZM-4, UZM-5 and UZM-9 are synthesized. with Si/Al of 2–5 [77]. The UZM family contains structures related to previously known topologies, such as UZM-4 (BPH) [78], UZM-9 (LTA) [78] and UZM-12 (ERI) [79], as well as new framework types such as UZM-5 (UFI) [80].

A broad range of high silica and pure silica molecular sieves have been synthesized by employing hydrothermal synthesis in fluoride media at low H_2O concentration, near neutral pH and alkali-free [81]. The significant new pure silica zeolite, ITQ-29, a structural analog of zeolite A (LTA) was reported by Corma at al [82]. Unlike the highly hydrophilic zeolite A, the ITQ-29 is hydrophobic.

The structure directing tendency of framework elements (e.g., Ge or Be) has also been successfully harnessed to prepare novel structures. In germanosilicates the smaller Ge–O–Ge and Ge–O–Si bond angles stabilize D4R building blocks improving the chances of preparing structures with three-dimensional channel systems. The novel zeolites IM-12 [83] and ITQ-15 [84] were all synthesized using organic SDAs and in some cases F in the media. They can be singled out for the special contribution of the Ge to the formation of secondary building units critical to structure formation. These silicogermanates have the UTL framework type which contains a unique two-dimensional channel system with channels bounded by 14-membered rings intersecting 12-membered ring channels. Structure analysis of both the IM-12 and ITQ-15 indicate that the Ge atoms are localized in the D4R units. Two recent beryllosilicate structures, OSB-1 and OSB-2, feature a high concentration of three rings and very low framework density [38]. These are made without the benefit of organic SDAs or fluoride. Thermal stability is very low.

The number of newly described metal organic frameworks (MOFs) exploded in the past decade. Since the MOFs are not strictly zeolites that family of materials is not discussed here. The reader is referred to several current reviews of the field [85].

1.3
Synthesis

The method developed by Milton in the late 1940s, involves the hydrothermal crystallization of reactive alkali metal aluminosilicate gels at high pH and low temperatures and pressures, typically 100 °C and ambient pressure. Milton, Breck and coworkers synthesis work led to over 20 zeolitic materials with low to intermediate Si/Al ratios (1–5) [86]. Chapter 3 and references [1] and [25] provide more detailed discussion of synthesis.

The advent of the addition of a quaternary ammonium cation as template or structure directing agent (SDA) to the alkaline gel by Barrer and coworkers, and Mobil Oil coworkers, led to the SiO_2 enriched zeolite A in the case of Barrer and to the high silica zeolites, Beta and ZSM-5, by the Mobil group. The latter synthesis temperature typically is 100–200 °C, higher than Milton's original work.

The addition of fluoride to the reactive gel led to more perfect and larger crystals of known molecular sieve structures as well as new structures and compositions. The fluoride ion also is reported to serve as a template or SDA in some cases. Fluoride addition extends the synthesis regime into the acidic pH region. Synthesis in fluoride media is discussed in greater detail in references [25] and [87].

The synthesis of the AlPO$_4$-based molecular sieves is similar to that of the high silica zeolites and silica molecular sieves, utilizing hydrothermal crystallization of reactive aluminophosphate gels, an amine or quaternary ammonium cation as a "template" or structure directing agent, and typical crystallization temperatures of 100–200 °C. In general they do not contain alkali or alkaline earth cations, unlike most zeolites, and the pH is typically slightly acidic to slightly basic.

In a further modification, the mesoporous materials are synthesized hydrothermally with a surfactant liquid crystal as the template.

Overall Milton's concept of hydrothermal crystallization of reactive gels has been followed with various additions and modifications for most of the molecular sieve, zeolite, and zeotype materials synthesis since the late 1940s.

1.4
Applications

Applications of zeolites and molecular sieves in the past several decades showed a growth in petroleum refining applications with emphasis on resid cracking and octane enhancement. ZSM-5 was commercialized as an octane enhancement additive in fluid catalytic cracking (FCC) where Si-enriched Y zeolites serve as the major catalytic component in high-octane FCC catalysts. The use of zeolite catalysts in the production of organic (fine) chemicals appeared as a major new direction. Zeolites in detergents as a replacement for phosphates became the single largest volume use for synthetic zeolites worldwide [22]. Zeolite ion exchange products, both synthetic and natural, were used extensively in nuclear waste cleanup after the Three Mile Island and Chernobyl nuclear accidents. New applications emerged for zeolite powders in two potentially major areas, odor removal and as plastic additives.

In adsorption and separation applications there has been a major growth in the use of pressure swing adsorption for the production of oxygen, nitrogen and hydrogen. Processes for the purification of gasoline oxygenate additives were introduced. Recent environmentally driven applications have arisen using the hydrophobic molecular sieves, highly siliceous Y zeolite and silicalite, for the removal and recovery of volatile organic compounds (VOC) that offer promise for significant market growth.

An exciting new scientific direction emerged in the 1980s and 1990s for exploring molecular sieves as advanced solid state materials. In a 1989 review, Ozin et al. [88] speculated "that zeolites (molecular sieves) as microporous molecular electronic materials with nanometer dimension window, channel and cavity architecture represent a 'new frontier' of solid state chemistry with great opportunities for innovative research and development". The applications described or envisioned included: molecular electronics, "quantum" dots/chains, zeolite electrodes, batteries, non-linear optical materials and chemical sensors. More recently there have been significant research reports on the use of zeolites as low k dielectric materials for microprocessors [89].

Zeolites have also been used as raw materials for ceramic compositions relevant to the electronic industry. Bedard *et al.* reported the high-temperature processing of zeolite B (P) to form cordierite ceramic compositions [90].

1.5
Markets

Since their introduction as a new class of industrial materials in 1954, the annual market for synthetic zeolites and molecular sieves has grown immensely, to 1.8×10^6 t worldwide in 2008 [22]. The major application areas are as adsorbents, catalysts and ion-exchange materials. The largest single market by volume (72%) is the detergent application, where zeolite A (and recently Type P) functions as an ion exchanger. In 2008, 1.27×10^6 t were estimated to be consumed in that application. Although the second largest volume use is as catalysts (17%), this is the largest value market for zeolites, about 55% of the total. Fluid catalytic cracking (FCC) catalysts, containing primarily silica-enriched forms of zeolite Y, represent more than 95% of total zeolite catalyst consumption, with smaller volumes used in hydrocracking and chemical and petrochemical synthesis. Catalyst consumption in 2008 is estimated at 303 000 t [22]. Adsorption applications are varied and include: drying and purification of natural gas, petrochemical streams (e.g., ethylene, propylene, refrigerants, insulated windows), bulk separations (e.g., xylenes, normal paraffins) and in air separation to produce oxygen by pressure swing adsorption (PSA) or vacuum pressure swing adsorption (VPSA) processes. Adsorbent consumption in 2008 is estimated at 180 000 t [22] (10% of the total by volume).

World production of natural zeolites was estimated at about 3.0×10^6 t in 2008 [22]. China and Cuba consume the largest quantity of natural zeolites, largely to enhance the strength of cement [22].

The price of zeolites varies considerably depending on the application. The typical price of catalysts in the United States varies from about US$ 3–4/kg for FCC to about US$ 20/kg for specialty catalysts, adsorbents from about US$ 5–9/kg, up to tens of dollars per kilogram for specialty adsorbents and about US$ 2/kg for detergents. Natural zeolites in bulk applications sell for US$ 0.04–0.25/kg and in industrial adsorbent applications for US$ 1.50–3.50/kg [22].

1.6
The Future

1.6.1
Materials

As noted previously there has been an explosive and accelerating increase in the discovery of new compositions and structural topologies. Based on the very high activity in this area in the past quarter of a century, we can expect a continuation of the proliferation of new molecular sieve compositions and structures. Further advances can also be expected in novel compositions derived from modification and secondary synthesis chemistry. When we consider the very large number of structures and compositions now reported in the molecular sieve area (176 in the sixth edition of the *Atlas*) and compare that with the number of commercial molecular sieves, approximately 12–15, what is the probability of future commercialization of a new material? There are many factors affecting the achievement of commercial status: (i) unique and advantageous properties of the material compared to present commercial products, (ii) market need, (iii) market size, (iv) cost of development and marketing and (v) cost and degree of difficulty in manufacturing. As a result it is likely, based on historical experience, that no more than a few of the prolific number of new molecular sieve materials of the past 25 years will achieve commercial status in the new millennium.

1.6.2
Applications

Molecular sieve adsorbents will continue to be used in the now-practiced separation and purification applications throughout the chemical process industry. New directions in the past 20 years include environmental and biopharmaceutical applications which have only recently received attention. Future trends in catalysis include: (i) a continuing accelerated discovery of new catalytic materials, (ii) an expanded use in petroleum refining particularly in the area of high octane gasoline, in the development of reformulated gasoline, the processing of heavy crudes and in the production of diesel, (iii) commercial development in conversion of alternate resources to motor fuels and base chemicals such as bio- and green fuels and (iv) as routes to organic chemical intermediates or end-products.

The large application of zeolites as ion exchangers in detergents leveled off in demand in North America, Western Europe and Japan in the late 1990s, but should continue to grow during the 2000s, particularly in Asia, Australia and Latin America [22]. The other applications of zeolites as ion exchangers in the nuclear industry, in radioactive waste storage and cleanup and in metals removal and recovery will probably remain a relatively small fraction of the worldwide market for molecular sieve materials.

Among the new application areas that could become large volume applications are the use of molecular sieves as functional powders, in odor removal, as plastic additives and in composites. The use of zeolites in solid-state applications is highly speculative. If ever practically realized that application would most probably represent a relatively small volume of the total zeolite consumption.

1.7
History of International Conferences and Organizations

In 1957 the first informal molecular sieve conference was held at Pennsylvania State University in the United States. In 1967 the first of a series of international molecular sieve conferences chaired by Professor R.M. Barrer was held in London. Subsequently, international meetings have been held every few years – 1970 in Worcester, 1973 in Zurich, 1977 in Chicago, 1980 in Naples, 1983 in Reno, 1986 in Tokyo, 1989 in Amsterdam, 1992 in Montreal, 1994 in Garmisch-Partenkirchen, 1996 in Seoul, 1998 in Baltimore, 2001 in Montpellier, 2004 in Capetown – and the most recent International Zeolite Conference was 2008 in Beijing.

An international molecular sieve organization was first formed in 1970 in conjunction with the Worcester Conference, called the International Molecular Sieve Conference (IMSC), and formalized with a constitution at the Zurich Conference in 1973. Its responsibility was to continue the organizational implementation of future international molecular sieve conferences on a regular basis. In 1977 at the Chicago Conference the name of the organization was changed to the International Zeolite Association (IZA) and its scope and purpose expanded to "promote and encourage all aspects of the science and technology of zeolitic materials", as well as organizing "International Zeolite Conferences" on a regular basis. The term zeolite in the new organization "is to be understood in its broadest sense to include both natural and synthetic zeolites as well as molecular sieves and other materials having related properties and/or structures" [91]. International Zeolite Association regional affiliates have been established and include: the British Zeolite Association (BZA, 1978), the Japan Association of Zeolites, (JAZ, 1986) and regional zeolite associations in France, Italy, Hungary, The Netherlands and Germany (late 1980s, early 1990s). In addition there are nearly a dozen other unaffiliated zeolite associations throughout the world (http://iza-online.org).

The IZA has several established Commissions. The Structure Commission formed in 1977 has published four editions of the *Atlas of Zeolite Structure Types* (1978, 1987, 1992, 1996) and two subsequent editions in the *Atlas of Zeolite Framework Types* (2001, 2007). An up-to-date version is maintained on the World Wide Web at the IZA Structure Commission web site (http://www.iza-structure.org). Subsequently commissions were established in catalysis (1998), synthesis (1992), ordered mesoporous materials (2001) and natural zeolites (2001). The synthesis commission published a volume, "*Verified Syntheses of Zeolitic Materials*", in 1998, with a second revised edition in 2001 (http://www.iza-structure.org).

In 1995, a Federation of European Zeolite Associations (FEZA) was formed and presently includes Zeolite Association members from the United Kingdom, Bulgaria, The Netherlands, France, Germany, Hungary, Italy, Romania, Spain, Poland, Georgia, the Czech Republic, Portugal and Slovakia. FEZA sponsors several workshops in Europe each year covering various aspects of zeolite science and holds International FEZA Conferences every three years (Hungary in 1999, Italy in 2002, Czech Republic in 2005, France in 2008; http://feza-online.org).

Other related organizations include the International Committee on Natural Zeolites (ICNC) and the International Mesostructured Materials Association (IMMA; (http://www.iza-structure.org).

1.8
Historical Epilog

Key factors in the growth of molecular sieve science and technology include: (i) the pioneering work of Barrer, (ii)the key discoveries of Milton and Breck and associates at Union Carbide, (iii) the rapid commercialization of the new synthetic zeolites and their applications by Union Carbide (1949–1954), (iv) the major development at Union Carbide in adsorption process design and engineering technology [92] major discoveries in hydrocarbon conversion catalysts at Union Carbide, Exxon, Mobil Oil, Shell and other industrial laboratories, (v) the discovery and commercialization of sedimentary zeolite mineral deposits in the United States in the 1960s and, last but not least, (vi) the dedication and contribution of so many high quality scientists and engineers. It is estimated that by the beginning of the twenty-first century there were over 20 000 such scientists and engineers in industry and academia dedicating a significant portion of their work to zeolite and molecular sieve science and technology.

References

1 Flanigen, E.M. (2001). Zeolites and molecular sieves. An historical perspective, in *Introduction to Zeolite Science and Practice*, 2nd edn (eds H. Van Bekkum, E.M. Flanigen, P.A. Jacobs, and J.C. Jensen), Stud. Surf. Sci. Catal., vol. 137, Elsevier, Amsterdam, pp. 11–35. Some content was reprinted from this review with permission of Elsevier.

2 Breck, D.W. (1974) *Zeolite Molecular Sieves, Structure, Chemistry and Use*, John Wiley & Sons, Inc., New York; reprinted by Krieger, Malabar, Florida, 1984.

3 Barrer, R.M. (1979) Chemical nomenclature and formulation of compositions of synthetic and natural zeolites. *Pure Appl. Chem.*, **51** (5), 1091–1100.

4 Baerlocher, C., McCusker, L., and Olson, D.H. (2007) *Atlas of Zeolite Framework Types*, 6th Revised edn, Elsevier, Amsterdam, and previous editions.

5 Meier, W.M., and Olson, D.H. (1988) *Atlas of Zeolite Structure Types*, 2nd Revised edn, Butterworth and Co Ltd, University Press, Cambridge, UK.

6 Coombs, D.S., Alberti, A., Armbruster, T., Artioli, G., Colella, C., Galli, E., Grice, J.D., Liebau, F., Mandarino, J.A., Minato, H., Nickel, E.H., Passaglia, E., Peacor, D.R., Quartieri, S., Rinaldi, R., Ross, M., Sheppard, R.A., Tillmanns, E., and Vezzalini, G. (1997) Recommended nomenclature for zeolite minerals; report of the Subcommittee on Zeolites of the International Mineralogical Association, Commission on New Minerals and Mineral Names. *Can. Mineral.*, **35**, 1571–1606.

7 Nanoporous Materials Gordon Research Conference, June 15–20, 2008, Colby College, Waterville, ME., now includes "zeolites, mesoporous systems, metal organic materials" as subject matter. It was formerly known as the "Zeolites and Layered Materials Gordon Research Conference".

8 Cronstedt, A.F. (1756) Ron och beskriting om en obekant bärg ant, som kallas zeolites. *Akad. Handl. Stockholm*, **18**, 120–130.

9 de St Claire Deville, H. (1862) *Comptes Rendus Acad. Sci.*, **54**, 324.

10 Friedel, G. (1896) New experiments on zeolites. *Bull. Soc. Fr. Mineral. Cristallogr.*, **19**, 363–390; Friedel, G. (1896) Zeolites and the substitution of various substances for the water which they contain. *C. R. Acad. Sci., Paris*, **122**, 948–951.

11 Grandjean, F. (1910) Optical study of the absorption of the heavy vapors by certain zeolites. *C. R. Acad. Sci., Paris*, **149**, 866–868.
12 Weigel, O. and Steinhoff, E. (1925) Adsorption of organic liquid vapors by chabazite. *Z. Kristallogr.*, **61**, 125–154.
13 Leonard, R.J. (1927) The hydrothermal alteration of certain silicate minerals. *Econ. Geol.*, **22**, 18–43.
14 Taylor, W.H. (1930) The crystal structure of analcite (NaAlSi2O6•H2O). *Z. Kristallogr.*, **74**, 1.
15 Pauling, L. (1930) The structure of some sodium and calcium aluminosilicates. *Proc. Natl. Acad. Sci. U. S. A.*, **16**, 453; Pauling, L. (1930) The structure of sodalite and helvite. *Z. Kristallogr.*, **74**, 213.
16 McBain, J.W. (1932) *The Sorption of Gases and Vapors by Solids*, Rutledge and Sons, London, Chapter 5.
17 Barrer, R.M. (1945) Separation of mixtures using zeolites as molecular sieves. I. Three classes of molecular-sieve zeolite. *J. Soc. Chem. Ind.*, **64**, 130.
18 Barrer, R.M. (1948) Synthesis and reactions of mordenite. *J. Chem. Soc.*, 2158.
19 Barrer, R.M. (1948) Synthesis of a zeolitic mineral with chabazite-like sorptive properties. *J. Chem. Soc.*, 127; Barrer, R.M. and Riley, D.W. (1948) Sorptive and molecular sieve properties of a new zeolitic mineral. *J. Chem. Soc.*, 133.
20 Reed, T.B. and Breck, D.W. (1956) Crystalline zeolites. II. Crystal structure of synthetic zeolite, type A. *J. Am. Chem. Soc.*, **78**, 5972–5977.
21 Milton, R.M. (1989) Molecular sieve science and technology: a historical perspective, in *Zeolite Synthesis, ACS Symposium Series 398* (eds M.L. Occelli, and H.E. Robson), American Chemical Society, Washington, D.C., pp. 1–10.
22 Davis, S. and Inoguchi, Y. (2009) *CEH Marketing Research Report: Zeolites*, SRI Consulting.
23 Mumpton, F.A. (1984) Zeolite exploration: the early days, in *Proc. 6th Intl. Zeolite Conf., Reno, USA, 1983* (eds D. Olson, and A. Bisio), Butterworth, Guilford, Surrey, UK, pp. 68–82.
24 Torii, K. (1978) Natural zeolites: utilization of natural zeolite in Japan, in *Natural Zeolites, Occurrence, Properties, Use* (eds L.B. Sand, and F.A. Mumpton), Pergamon Press, New York, pp. 441–450.
25 Yu, J. (2007) Synthesis of zeolites, in *Introduction to Zeolite Science and Practice*, 3rd edn (eds J. Cejka, H. Van Bekkum, A. Corma, F. Schuth), Stud. Surf. Sci. Catal., vol. **168**, Elsevier, Amsterdam, pp. 39–103.
26 Wilson, S.T., Lok, B.M., Messina, C.A., Cannan, T.R., and Flanigen, E.M. (1982) Aluminophosphate molecular sieves: a new class of microporous crystalline inorganic solids. *J. Am.Chem. Soc.*, **104**, 1146–1147.
27 Flanigen, E.M., Lok, B.M., Patton, R.L., and Wilson, S.T. (1987) Aluminophosphate molecular sieves and the periodic table, in *New Developments in Zeolite Science and Technology, Proc. 7th Intl. Zeolite Conf., Tokyo, 1986* (eds Y. Murakami, A. Ijima, and J.W. Ward) Elsevier, Amsterdam, pp. 103–112.
28 Szostak, R. (1998) *Molecular Sieves, Principles of Synthesis and Identification*, 2nd edn, Blackie Academic & Professional, London, pp. 208–244.
29 Davis, M.E., Saldarriaga, C., Montes, C., Garces, J., and Crowder, C. (1988) A molecular sieve with eigthteen-membered rings. *Nature*, **331**, 698–699.
30 Lok, B.M., Messina, C.A., Patton, R.L., Gajek, R.T., Cannan, T.R., and Flanigen, E.M. (1984) Silicoaluminophosphate molecular sieves: another new class of microporous crystalline inorganic solids. *J. Am. Chem. Soc.*, **106**, 6092–6093.
31 Derouane, E.G., Maistriau, L., Gabelica, Z., Tuel, A., Nagy, J.B., and Von Ballmoos, R. (1989) Synthesis and characterization of the very large pore molecular sieve MCM-9. *Appl. Catal.*, **51**, L13–L20.
32 Martens, J.A., Mertens, M., Grobet, P.J., and Jacobs, P.A. (1988) Synthesis and characterization of silicon-rich SAPO-5, in *Innovation Zeolite Mater. Sci.* (eds P.J. Grobet, W.J. Mortier, E.F. Vansant, and G. Schulz Ekloff), Stud. Surf. Sci. Catal. **37**, Elsevier, Amsterdam, pp. 97–105.
33 Van Nordstrand, R.A., Santilli, D.S., and Zones, S.I. (1988) An all-silica molecular

sieve that is isostructural with aluminum phosphate(AlPO4-5), in *Perspect. Mol. Sieve Sci., ACS Symp. Ser. 368* (eds W.H. Flank, and T.E. Whyte, Jr.), American Chemical Society, Washington, DC, pp. 236–245.

34. Bennett, J.M., and Marcus, B.K. (1988) The crystal structures of several metal aluminophosphate molecular sieves, in *Innovation Zeolite Mater. Sci.* (eds P.J. Grobet, W.J. Mortier, E.F. Vansant, and G. Schulz Ekloff), Stud. Surf. Sci. Catal., vol. 37, Elsevier, Amsterdam, pp. 269–279.

35. Wilson, S.T. (2001) Phosphate-based molecular sieves: novel synthetic approaches to new structures and compositions, in *Introduction to Zeolite Science and Practice*, 2nd Revised edn (eds H. Van Bekkum, E.M. Flanigen, P.A. Jacobs, and J.C. Jensen), Stud. Surf. Sci. Cat., vol. 137, Elsevier, Amsterdam, pp. 229–260; Wilson, S.T. (2007) Phosphate-based molecular sieves: new structures, synthetic approaches, and applications, in *Introduction to Zeolite Science and Practice*, 3rd Revised edn (eds J. Cejka, H. Van Bekkum, A. Corma, and F. Schuth), Stud. Surf. Sci. Catal., vol. 168, Elsevier, Amsterdam, pp. 105–135.

36. Petersen, O.V., Giester, G., Brandstatter, F., and Niedermayr, G. (2002) Nabesite, Na2BeSi4O10·4H2O, a new mineral species from the Ilímaussaq alkaline complex, South Greenland. *Can. Mineral.*, **40**, 173–181.

37. Ueda, S., Koizumi, M., Baerlocher, C., McCusker, L.B., and Meier, W.M. (1986) Preprints of poster papers. *Int. Zeolite Conf.*, **7**, 23.

38. Cheetham, T., Fjellvag, H., Gier, T.E., Kongshaug, K.O., Lillerud, K.P., and Stucky, G.D. (2001) Very open microporous materials: from concept to reality, in *Zeolites and Mesoporous Materials at the Dawn of the 21st Century* (eds A. Galarneau, F. Di Renzo, F. Fajula, and J. Vedrine), Stud. Surf. Sci. Catal., vol. 135, Elsevier, Amsterdam, pp. 788–795.

39. Roehrig, C., Gies, H., and Marler, B. (1994) Rietveld refinement of the crystal structure of the synthetic porous zincosilicate VPI-7. *Zeolites*, **14**, 498–503.

40. Freyhardt, C.C., Lobo, R.F., Khodabandeh, S., Lewis, J.E., Jr., Tsapatsis, M., Yoshikawa, M., Camblor, M.A., Pan, M., and Helmkamp, M.M. (1996) VPI-8: a high-silica molecular sieve with a novel "pinwheel" building unit and its implications for the synthesis of extra-large pore molecular sieves. *J. Am. Chem. Soc.*, **118** (31), 7299–7310.

41. McCusker, L.B., Grosse-Kunstleve, R.W., Baerlocher, C., Yoshikawa, M., and Davis, M.E. (1996) Synthesis optimization and structure analysis of the zincosilicate molecular sieve VPI-9. *Microporous Mater.*, **6**, 295–309.

42. Takewaki, T., Beck, L.W., and Davis, M.E. (1999) Zincosilicate CIT-6: a precursor to a family of *BEA-type molecular sieves. *J. Phys. Chem. B*, **103** (14), 2674–2679.

43. Harvey, G., and Meier, W.M. (1989) The synthesis of beryllophosphate zeolites, in *Zeolites: Facts, Figures, Future* (eds P.A. Jacobs, and R.A. van Santen), Stud. Surf. Sci. Catal., vol. **49A**, Elsevier, Amsterdam, pp. 411–420.

44. Wang, J., Feng, S., and Xu, R. (1989) Synthesis and characterization of zeolitic microporous alumino-borates, in *Zeolites: Facts, Figures, Future* (eds P.A. Jacobs, and R.A. van Santen), Stud. Surf. Sci. Catal., vol. 49A, Elsevier, Amsterdam, pp. 143–150.

45. Yang, G., Li, L., Chen, J., and Xu, R. (1989) Synthesis and structure of a novel aluminoarsenate with an open framework. *J. Chem. Soc. Chem. Commun.*, **1989**, 810.

46. Chen, J., and Xu, R. (1989) Syntheses and characterization of two novel inclusion compounds: (AlAsO4:0.2(CH3)4NOH:0.3H2O) and (GaAsO4:0.2(CH3)4NOH:0.1H2O.) *J. Solid State Chem.*, **80**, 149–151.

47. Martens, J.A., and Jacobs, P.A. (1999) Phosphate-based zeolites and molecular sieves, in *Catalysis and Zeolites, Fundamentals and Applications* (eds J. Weitkamp, and L. Puppe), Springer-Verlag, Berlin, pp. 53–80.

48. Yamagishi, Y., Namba, S., and Vashima, T. (1989) Preparation and acidic properties of antimonosilicate with MFI structure, in *Zeolites: Facts, Figures,*

Future, (eds P.A. Jacobs, and R.A. van Santen), Stud. Surf. Sci. Catal., vol. 49A, Elsevier, Amsterdam, pp. 459–467.

49 Gabelica, Z., and Guth, J.L. (1989) Germanium-rich MFI zeolites: the first example of an extended framework substitution of silicon by another tetravalent element, in *Zeolites: Facts, Figures, Future*, (eds P.A. Jacobs, and R.A. van Santen), Stud. Surf. Sci. Catal., vol. 49A, Elsevier, Amsterdam, pp. 421–430.

50 Peacor, D.R., Rouse, R.C., and Ahn, J.H. (1987) Crystal structure of tiptopite, a framework beryllophosphate isotypic with basic cancrinite. *Am. Mineral.*, 72, 816–820.

51 Rouse, R.C., Peacor, D.R., and Merlino, S. (1989) Crystal structure of pahasapaite, a beryllophosphate mineral with a distorted zeolite rho framework. *Am. Mineral.*, 74, 1195–1202.

52 Bedard, R.L., Wilson, S.T., Vail, L.D., Bennett, J.M., and Flanigen, E.M. (1989) The next generation: synthesis, characterization, and structure of metal sulfide-based microporous solids, in *Zeolites: Facts, Figures, Future* (eds P.A. Jacobs, and R.A. van Santen), Stud. Surf. Sci. Catal., vol. 49A, Elsevier, Amsterdam, pp. 375–387.

53 Scott, R.W.J., MacLachlan, M., and Ozin, G.A. (1999) *Curr. Opin. Solid State Mater. Sci.*, 4, 113–121; Bowes, C.L., and Ozin, G.A. (1996) Self-assembling frameworks: beyond microporous oxides. *Adv. Mater.*, 8, 13–28.

54 Estermann, M., McCusker, L.B., Baerlocher, C., Merrouche, A., and Kessler, H. (1991) A synthetic gallophosphate molecular sieve with a 20-tetrahedral-atom pore opening. *Nature*, 352, 320–323.

55 Freyhardt, C.C., Tsapatsis, M., Lobo, R.F., Balkus, K.J., Jr., and Davis, M.E. (1996) A high-silica zeolite with a 14-tetrahedral-atom pore opening. *Nature*, 381, 295–298; Lobo, R.F., Tsapatsis, M., Freyhardt, C.C., Khodabandeh, S., Wagner, P., Chen, C.-Y., Balkus, K.J., Jr., Zones, S.I., and Davis, M.E. (1996) *J. Am. Chem. Soc.*, 119 (36), 8474–8484.

56 Wagner, P., Yoshikawa, M., Lovallo, M., Tsuji, K., Tsapatsis, M., and Davis, M.E. (1997) CIT-5: a high-silica zeolite with 14-ring pores. *Chem. Commun.*, 1997, 2179–2180.

57 Bu, X., Feng, P., and Stucky, G.D. (1997) *Science*, 278 (5346), 2080–2085; Feng, P., Bu, X., and Stucky, G.D. (1997) Hydrothermal syntheses and structural characterization of zeolite analogue compounds based on cobalt phosphate. *Nature*, 388, 735–741; Feng, P., Bu, X., Gier, T.E., and Stucky, G.D. (1998) Amine-directed syntheses and crystal structures of phosphate-based zeolite analogs. *Microp. Mesoporous Mater.*, 23 (3–4), 221–229.

58 Gier, T.E., and Stucky, G.D. (1991) Low-temperature synthesis of hydrated zinco(beryllo)-phosphate and arsenate molecular sieves. *Nature*, 349, 508–510.

59 Puppe, L., and Weisser, J. (1984) Crystalline aluminosilicate PSH-3 and its process of preparation. US Patent 4,439,409; Leonowicz, M.E., Lawton, J.A., Lawton, S.L., and Rubin, M.K. (1994) MCM-22: a molecular sieve with two independent multidimensional channel systems. *Science*, 264, 1910; Lawton, S.L., Fung, A.S., Kennedy, G.J., Alemany, L.B., Chang, C.D., Hatzikos, G.H., Lissy, D.N., Rubin, M.K., Timken, H.-K.-C., Steuernagel, S., and Woessner, D.E. (1996) Zeolite MCM-49: a three-dimensional MCM-22 analog synthesized by in situ crystallization. *J. Phys. Chem.*, 100, 3788–3798.

60 Delprato, F., Delmotte, L., Guth, J.L., and Huve, L. (1990) Synthesis of new silica-rich cubic and hexagonal faujasites using crown-ether based supramolecules as templates. *Zeolites*, 10, 546.

61 Pluth, J.J., and Smith, J.V. (1990) Crystal structure of boggsite, a new high-silica zeolite with the first three-dimensional channel system bounded by both 12- and 10-rings. *Am. Mineral.*, 75, 501.

62 Boggs, R.C., Howard, D.G., Smith, J.V., and Klein, G.L. (1993) Tschernichite, a new zeolite from Goble, Columbia County, Oregon. *Am. Mineral.*, 78, 822.

63 Alberti, A., Vezzalini, G., Galli, E., and Quartieri, S. (1996) The crystal structure

64 Galli, E., Quartieri, S., Vezzalini, G., Alberti, A., and Franzini, M. (1997) Terranovaite from Antarctica: a new "pentasil" zeolite. *Am. Mineral.*, **82**, 423.

of gottardiite, a new natural zeolite. *Eur. J. Mineral.*, **8**, 69–75.

65 Vezzalini, G., Quartieri, S., Galli, E., Alberti, A., Cruciani, G., and Kvick, A. (1997) Crystal structure of the zeolite mutinaite, the natural analog of ZSM-5. *Zeolites*, **19**, 323.

66 Menshikov, Y.P. (1984) Perlialit, K9Na(Ca, Sr)[Al12Si24O72]:15H2O – a new potassium zeolite. *Zap. Vses. Mineral. O-va*, **113**, 607; Artioli, G., and Kvick, A. (1990) Synchrotron x-ray Rietveld study of perlialite, the natural counterpart of synthetic zeolite-L. *Eur. J. Mineral.*, **2**, 749.

67 Effenberger, H., Giester, G., Krause, W., and Bernhardt, H.-J. (1998) Tschortnerite, a copper-bearing zeolite from the Bellberg volcano, Eifel, Germany. *Am. Mineral.*, **83**, 607.

68 Chapman, D.M., and Roe, A.L. (1990) Synthesis, characterization and crystal chemistry of microporous titanium-silicate materials. *Zeolites*, **10**, 730.

69 Kuznicki, S.M., Trush, K.A., Allen, F.M., Levine, S.M., Hamil, M.M., Hayhurst, D.T., and Mansom, M. (1992) Synthesis and adsorptive properties of titanium silicate molecular sieves, in *Synthesis of Microporous Materials, Molecular Sieves*, vol. 1 (eds M.L. Ocelli, and H.E. Robson), Van Nostrand Reinhold, New York, pp. 427–453.

70 Kresge, C.T., Leonowicz, M.E., Roth, W.J., Vartuli, J.C., and Beck, J.S. (1992) Ordered mesoporous molecular sieves synthesized by a liquid-crystal template mechanism. *Nature*, **359**, 710–712; Beck, J.S., Vartuli, J.C., Roth, W.J., Leonowicz, M.E., Kresge, C.T., Schmitt, K.D., Chu, C.T.W., Olson, D.H., Sheppard, E.W., McCullen, S.B., Higgins, J.B., and Schlenker, J.L. (1992) A new family of mesoporous molecular sieves prepared with liquid crystal templates. *J. Am. Chem. Soc.*, **114**, 10834.

71 Inagaki, S., Fukashima, Y., and Kuroda, K. (1993) Synthesis of highly ordered mesoporous materials from a layered polysilicate. *J. Chem. Soc. Chem. Commun.*, **1993**, 680; Inagaki, S., Fukashima, Y., and Kuroda, K. (1994) Synthesis and characterization of highly ordered mesoporous material; FSM-16, from a layered polysilicate, in *Zeolites and Related Microporous Materials: State of the Art 1994* (eds J. Weitkamp, H.G. Karge, H. Pfeifer, and W. Holderich), Stud. Surf. Sci. Catal., vol. 84, Elsevier, Amsterdam, pp. 125–132; Inagaki, S., Fukashima, Y., Akada, A., Kurauchi, T., Kuroda, K., and Kato, C. (1993) New silica–alumina with nano-scale pores prepared from kanemite, in *Proceedings from the Ninth International Zeolite Conference, Montreal 1992* (eds R. von Ballmoos, J.B. Higgins, and M.M.J. Treacy), Butterworth–Heinemann, London, pp. 305–311; Yanagisawa, T., Shimizu, T., Kuroda, K., and Kato, C. (1990) The preparation of alkyltrimethyl-ammonium–kanemite complexes and their conversion to microporous materials. *Bull. Chem. Soc. Jpn.*, **63**, 988–992.

72 Chen, C.-Y., Xiao, S.-Q., and Davis, M.E. (1995) Studies on ordered mesoporous materials III. Comparison of MCM-41 to mesoporous materials derived from kanemite. *Microporous Mater.*, **4**, 1–20.

73 Zhao, D., and Wan, Y. (2007) The synthesis of mesoporous molecular sieves, in *Introduction to Zeolite Science and Practice*, 3rd Revised edn (eds J. Cejka, H. Van Bekkum, A. Corma, and F. Schuth), Stud. Surf. Sci. Catal., vol. 168, Elsevier, Amsterdam, pp. 241–300.

74 Vartuli, J., and Degnan, T., Jr. (2007) Applications of mesoporous molecular sieves in catalysis and separations, in *Introduction to Zeolite Science and Practice*, 3rd Revised edn (eds J. Cejka, H. Van Bekkum, A. Corma, and F. Schuth), Stud. Surf. Sci. Catal., vol. 168, Elsevier, Amsterdam, pp. 837–854.

75 Di Renzo, F., Cambon, H., and Dutarte, R. (1997) A 28-year-old synthesis of micelle-templated mesoporous silica. *Microporous Mater.*, **10**, 283–286; Chiola, V., Ritsko, J.E., and Vanderpool, C.D. (1971) Process for producing low-bulk density silica. US Patent 3,556,725, assigned to Sylvania Electric Products, Inc.

76 Burton, A., and Zones, S. (2007) Organic molecules in zeolite synthesis: their preparation and structure-directing effects, in *Introduction to Zeolite Science and Practice*, 3rd Revised edn (eds J. Cejka, H. Van Bekkum, A. Corma, and F. Schuth), Stud. Surf. Sci. Catal., vol. 168, Elsevier, Amsterdam, pp. 137–179.

77 Lewis, G.J., Miller, M.A., Moscoso, J.G., Wilson, B.A., Knight, L.M., and Wilson, S.T. (2004) Experimental charge density matching approach to zeolite synthesis, in *Recent Advances in the Science and Technology of Zeolites and Related Materials, Proceedings of the 14th International Zeolite Conference* (eds E.W.J. van Steen, I.M. Claeys, L.H. Callanan, and C.T. O'Connor), Stud. Surf. Sci. Catal., vol. 154A, Elsevier, Amsterdam, pp. 364–372.

78 Lewis, G.J., Jan, D.Y., Mezza, B.J., Moscoso, J.G., Miller, M.A., Wilson, B.A., and Wilson, S.T. (2004) UZM-4: a stable Si-rich form of the BPH framework type, in *Recent Advances in the Science and Technology of Zeolites and Related Materials, Proceedings of the 14th International Zeolite Conference* (eds E.W.J. van Steen, I.M. Claeys, L.H. Callanan, and C.T. O'Connor), Stud. Surf. Sci. Catal., vol. 154A, Elsevier, Amsterdam, pp. 118–125.

79 Miller, M.A., Lewis, G.J., Moscoso, J.G., Koster, S., Modica, F., Gatter, M.G., and Nemeth, L.T. (2007) Synthesis and catalytic activity of UZM-12, in *From Zeolites to Porous MOF Materials – The 40th Anniversary of International Zeolite Conference* (eds R. Xu, Z. Gao, J. Chen, and W. Yan), Stud. Surf. Sci. Catal., vol. 170A, Elsevier, Amsterdam, pp. 487–492.

80 Blackwell, C.S., Broach, R.W., Gatter, M.G., Holmgren, J.S., Jan, D.-Y., Lewis, G.J., Mezza, B.J., Mezza, T.M., Miller, M.A., Moscoso, J.G., Patton, R.L., Rohde, L.M., Schoonover, M.W., Sinkler, W., Wilson, B.A., and Wilson, S.T. (2003) *Angew. Chem. Int. Ed.*, **42** (15), 1737–1740.

81 Zones, S.I., Hwang, S.-J., Elomari, S., Ogino, I., Davis, M.E., and Burton, A.W. (2005) The fluoride-based route to all-silica molecular sieves; a strategy for synthesis of new materials based upon close-packing of guest-host products. *C. R. Chimie*, **8**, 267–282.

82 Corma, A., Rey, F., Rius, J., Sabater, M.J., and Valencia, S. (2004) Supramolecular self-assembled molecules as organic directing agent for synthesis of zeolites. *Nature*, **431**, 287–290.

83 Paillaud, J.-L., Harbuzaru, B., Patarin, J., and Bats, N. (2004) Extra-large-pore zeolites with two-dimensional channels formed by 14 and 12 rings. *Science*, **304**, 990.

84 Corma, A., Diaz-Cabanas, M.J., Rey, F., Nicolopoulus, S., and Boulahya, K. (2004) ITQ-15: the first ultralarge pore zeolite with a bi-directional pore system formed by intersecting 14- and 12-ring channels, and its catalytic implications. *Chem. Commun.*, **2004**, 1356.

85 Ferey, G. (2007) Hybrid porous solids, in *Introduction to Zeolite Science and Practice*, 3rd Revised edn, (eds J. Cejka, H. Van Bekkum, A. Corma, and F. Schuth), Stud. Surf. Sci. Cat., vol. 168, Elsevier, Amsterdam, pp. 327–374; Muller, U., Schubert, M.M., and Yaghi, O.M. (2008) Chemistry and applications of porous metal-organic frameworks, in *Handbook of Heterogeneous Catalysis*, Wiley-VCH Verlag GmbH, Weinheim, pp. 247–262.

86 Breck, D.W., Eversole, W.G., and Milton, R.M. (1956) New synthetic crystalline zeolites. *J. Am. Chem. Soc.*, **78**, 2338.

87 Villaescusa, L., and Camblor, M. (2003) The fluoride route to new zeolites. *Recent Res. Dev. Chem.*, **1**, 93–141.

88 Ozin, G.A., Kuperman, A., and Stein, A. (1989) Advanced zeolite, material science. *Angew. Chem. Int. Ed. Engl.*, **28**, 359–376.

89 Wang, Z., Mitra, A., Wang, H., Huang, L., and Yan, Y. (2001) Pure-silica zeolite low-k dielectric thin films. *Adv. Mater.*, **13**, 746; Wang, Z., Mitra, A., Wang, H., Huang, L., and Yan, Y. (2001) Pure silica zeolite films as low-k dielectrics by spin-on of nanoparticle suspension. *Adv. Mater.*, **13**, 1463.

90 Bedard, R.L., and Flanigen, E.M. (1991) Nanoscale engineered ceramics from zeolites: creating the ideal precursor for high-quality cordierite, in *Synthesis/*

Characterization and Novel Applications of Molecular Sieve Materials (eds R.L. Bedard, T. Bein, M.E. Davis, J. Garcia, V.A. Maroni, and G.D. Stucky), Materials Research Society, Pittsburgh, *Mater. Res. Soc. Symp. Proc.*, **233**, 219–224.

91 International Zeolite Association (1977) *Newsletter 1*, Aug. 22.

92 Flanigen, E.M. (1980) Molecular sieve zeolite technology – the first twenty-five years, in *Proc. 5th Int. Conf. on Zeolites, Naples, Italy, June, 1980* (ed. L.V.C. Rees), Heyden, London, pp. 760–780.

Further Reading

Breck, D.W. (1974) *Zeolite Molecular Sieves, Structure, Chemistry and Use*, John Wiley & Sons, New York; reprinted (1984) by Krieger, Malabar, Florida.

Barrer, R.M. (1978) *Zeolites and Clay Minerals as Sorbents and Molecular Sieves*, Academic Press, London.

Barrer, R.M. (1982) *Hydrothermal Chemistry of Zeolites*, Academic Press, London.

Szostak, R. (1992) *Handbook of Molecular Sieves*, Van Nostrand Reinhold, New York.

Szostak, R. (1998) *Molecular Sieves, Principles of Synthesis and Identification*, 2nd Edn, Blackie Academic & Professional, London.

Van Bekkum, H., Flanigen, E.M. and Jansen, J.C. (eds) (1991) *Introduction to zeolite science and practice*, Stud. Surf. Surf. Sci. Catal., **58**, Elsevier, Amsterdam.

Van Bekkum, H., Flanigen, E.M., Jacobs, P.A. (eds) (2001) *Introduction to zeolite science and practice*, 2nd edn, Stud. Surf. Sci. Catal., **137**, Elsevier, Amsterdam.

Cejka, J., Van Bekkum, H., Corma, A., and Schuth, F. (eds) (2007), *Introduction to zeolite science and practice*, 3rd Edn, Stud. Surf. Sci. Catal., **168**, Elsevier, Amsterdam.

Xu, R., Pang, W., Huo, O., and Chen, J. (2007) *Chemistry of Zeolites and Related Porous Materials, Synthesis and Structure*, John Wiley & Sons (Asia) Pte. Ltd.

2
Zeolite Types and Structures
Robert W. Broach

2.1
Introduction

Molecular sieves are porous solids with pores of the size of molecular dimensions. Traditionally, zeolites are considered to be a subset of the molecular sieve family with structures based on an infinitely extending three-dimensional, four-connected framework of AlO_4 and SiO_4 tetrahedra linked to each other by the sharing of oxygen atoms. However, the usage of the term "zeolite" has evolved to include tetrahedral oxide structures with framework atoms other than Al and Si, but where zeolitic properties are exhibited (e.g., channels, voids, reversible hydration–dehydration). The framework structure of a zeolite defines intracrystalline voids and channels occupied by cations (to balance the net negative charge for each AlO_4 in the framework), water molecules and perhaps other guest species.

IUPAC provides detailed guidelines [1] for specifying the chemical formula for zeolites. In the simplest form, a general formula for most zeolites can be given as:

$$|M_x(H_2O)_y|[Al_x Si_{(t-x)} O_{2t}] - \text{IZA}$$

where the guest species are listed between the braces, | |, and the host framework is listed between the brackets, [ ]. M represents a univalent charge-balancing cation, x is the number of framework Al atoms in the unit cell, y is the number of adsorbed water molecules, t is the total number of framework tetrahedral atoms in the unit cell (Al + Si) and **IZA** is the code for the framework type assigned by the Structure Commission of the International Zeolite Association [2] (see Section 2.2). For $x = 0$ the framework is purely siliceous and no cations are needed to balance framework charge, and when multivalent cations are present the number of total cations is less than x.

2.2
Building Units for Zeolite Frameworks

The basic building unit (BBU) for the framework in a zeolite is a TO_4 tetrahedron where the central T-atom is typically Si or Al and the peripheral atoms are O. It is sometimes convenient to discuss composite building units (CBUs) formed by combining several BBUs. For example, eight tetrahedra can be connected by sharing vertices to form a cube, or double four-ring (*d4r*). Three common ways of representing a CBU are shown in Figure 2.1. A common standard view (Figure 2.1a) represents T and O atoms as spheres and is useful when it is important to show details of the O-atom positions. Tetrahedral representations (Figure 2.1b) show TO_4 tetrahedra with shared corners and still give some feel for the O-atom positions but simplify the views. The framework representation (Figure 2.1c) displays T–T linkages with the T-atom at the vertices and with oxygen atoms assumed to be positioned near the center of each edge. This representation is most useful for simplifying complex structures and displaying the overall framework topology, but does not give a good picture of the O-atom positions.

Polyhedral CBUs are described using common names (such as sodalite cage), three letter codes (such as *d4r* or *cub*), or the descriptors $[n1^{m1}\ n2^{m2}\ ...]$ where $m1$ is the number of $n1$-rings, $m2$ is the number of $n2$-rings, etc. For example, the *d4r* CBU has the descriptor $[4^6]$ because it is defined by six equivalent four-rings. Examples of some polyhedral CBUs found in known zeolite framework types are shown in Figure 2.2. BBUs can also be connected in infinite chains forming chain CBUs, some of which are shown in Figure 2.3. Other common building units are given at the IZA Structure Commission web site [4] and a compilation of building units found in known structures up to 1999 is given by J.V. Smith [3].

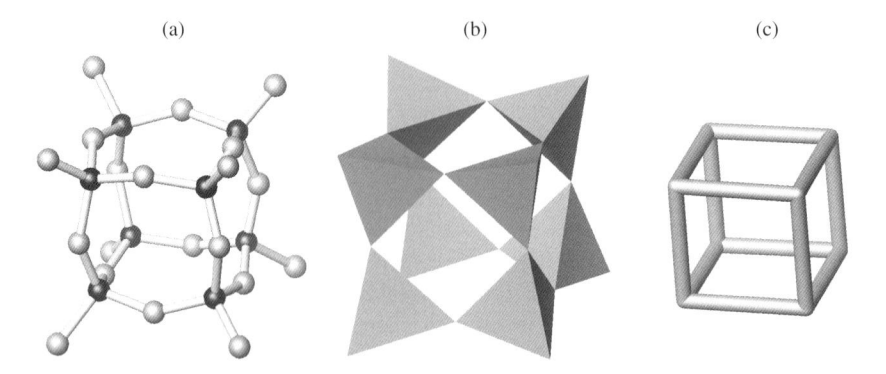

Figure 2.1 Common ways of representing the d4r composite building unit formed from eight tetrahedra in the shape of a cube. The standard view (a) represents T and O atoms as spheres. Tetrahedral representations (b) show TO4 tetrahedra with shared corners. The framework representation (c) displays T–T linkages with the T-atom at the vertices and with oxygen atoms assumed to be positioned near the center of each edge.

[4⁶]
double 4-ring (*d4r*)
or cube (*cub*)

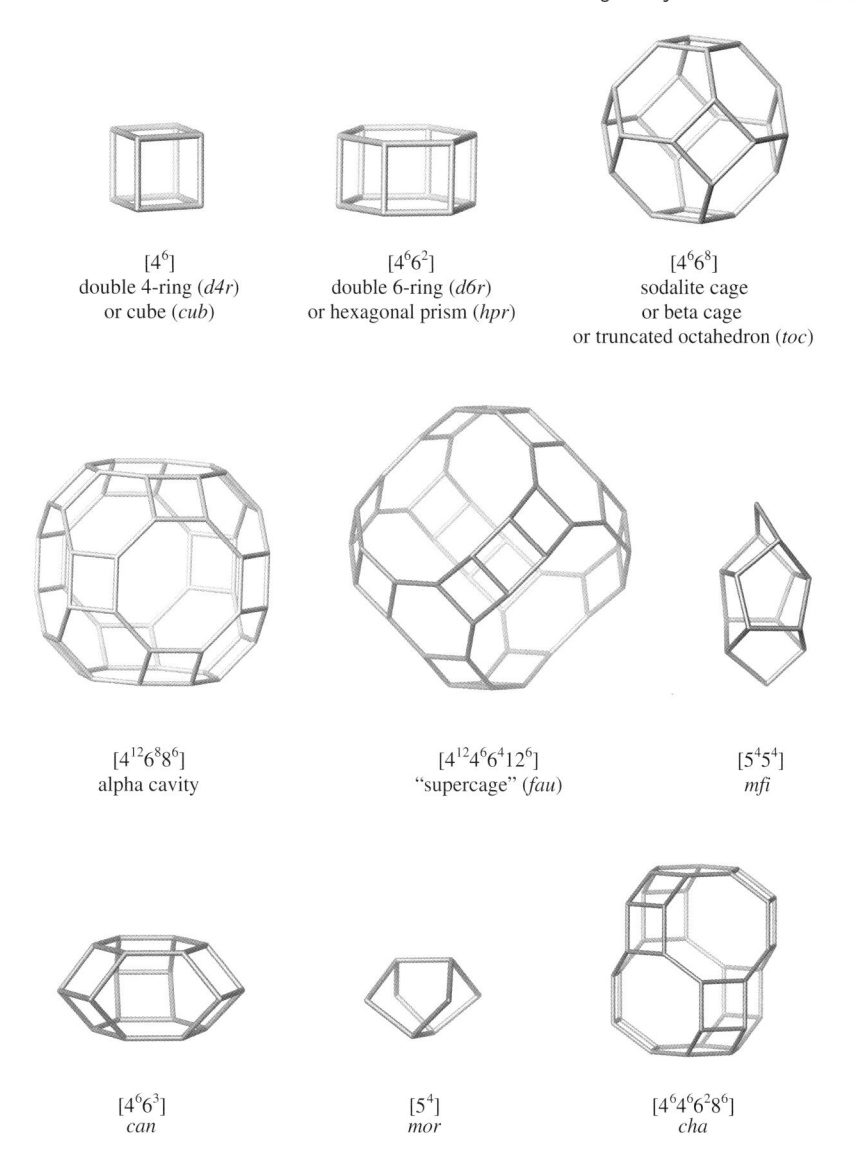

[4⁶6²]
double 6-ring (*d6r*)
or hexagonal prism (*hpr*)

[4⁶6⁸]
sodalite cage
or beta cage
or truncated octahedron (*toc*)

[4¹²6⁸8⁶]
alpha cavity

[4¹²4⁶6⁴12⁶]
"supercage" (*fau*)

[5⁴5⁴]
mfi

[4⁶6³]
can

[5⁴]
mor

[4⁶4⁶6²8⁶]
cha

Figure 2.2 Some examples of polyhedral composite building units with their corresponding pore symbols and common names. The nodes are tetrahedrally coordinated atoms such as Si or Al. Bridging oxygen atoms have been left out for clarity.

The three-dimensional framework structure of a zeolite is formed by linking BBUs in an infinite repeating lattice. The framework structure for zeolite type A (framework code LTA) is shown in Figure 2.4. Figure 2.4a shows the T-atom connectivity. Figure 2.4b is the same view with all rings of size 6 and smaller filled in

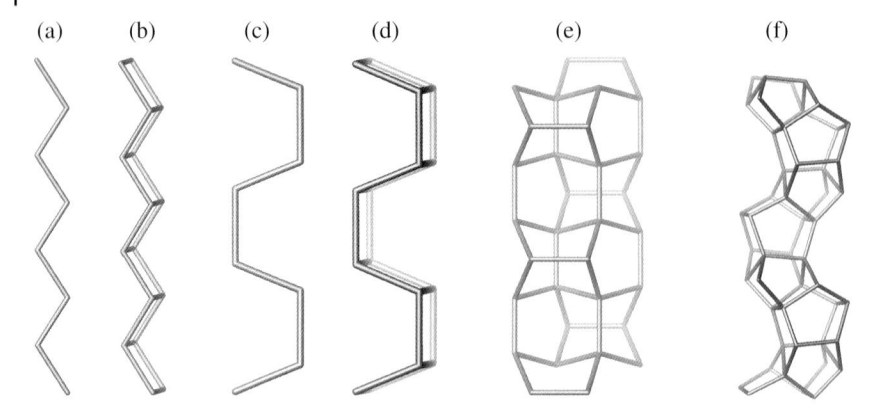

Figure 2.3 Some examples of chain composite building units. (a) Single chain (zig-zag) and (b) double chain (double zig-zag) with periodicities of two. (c) Single chain (crankshaft) and (d) double chain (double crankshaft) with periodicities of four. (e) Complex chain (designated *mod* by Smith [Ref [3]]) found in the MOR framework type. (f) Complex pentasil chain (designated *kgl* by Smith [Ref [3]]) found in the MFI framework type. The nodes are tetrahedrally coordinated atoms such as Si or Al. Bridging oxygen atoms have been left out for clarity.

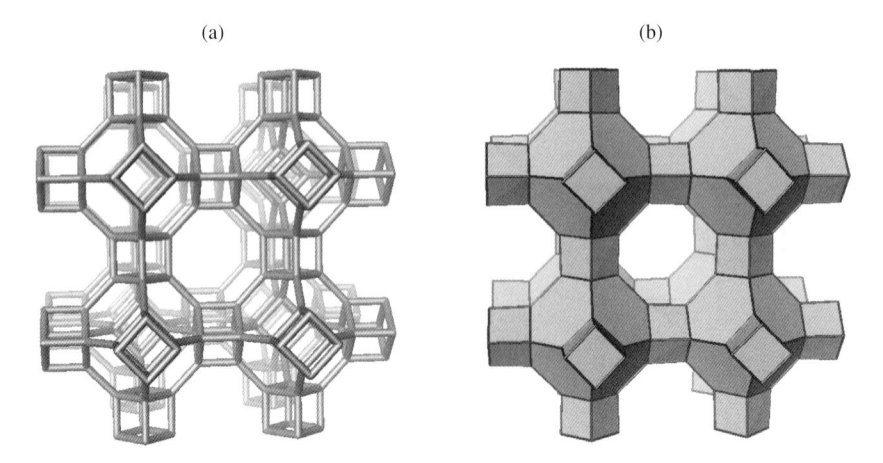

Figure 2.4 (a) Skeletal diagram of the LTA framework structure and (b) polyhedral representation emphasizing the volume in the framework accessible to molecules large than H_2O.

to emphasize the pore volume accessible to molecules larger than water. Notice that the LTA framework type contains within its structure several different polyhedral CBUs, including the double four-ring (*d4r*), the beta cage and the alpha cavity (all shown in Figure 2.2).

2.3
Zeolite Framework Types

Zeolites can be classified by considering the topology of the framework without regard to chemical composition (for example, by considering a framework consisting of SiO_2 only) and choosing the highest possible symmetry for the topology. All zeolites having the same topology constitute a zeolite framework type. Currently there are 191 known framework types recognized by the Structure Commission of the International Zeolite Association [2, 4] and assigned a three letter code. Details of these framework types are available at the Structure Commission web site [4]. New structure types continually are being discovered by researchers and as shown in Figure 2.5 new codes have been approved every year since 1992.

Not all frameworks built from tetrahedra as described above are considered to be zeolites. Dense phases are not considered to be zeolites, only those phases with some porosity. Generally, materials with pores accessible by windows defined by six T-atoms or less (six-rings) are not considered to be zeolites. In fact, the boundary between zeolites and dense phases is somewhat nebulous. IUPAC defines [1] zeolites as a subset of microporous or mesoporous materials containing voids arranged in an ordered manner and with a free volume larger than a 0.25 nm diameter sphere. The Structure Commission of the International Zeolite Association uses the criterion of framework density (T-atoms per 1000 Å^3) with the maximum framework density for zeolites ranging from 19 to 21.

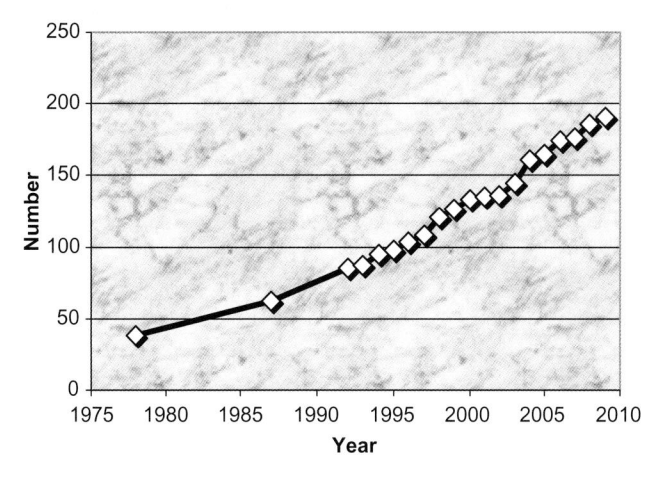

Figure 2.5 Cumulative number of framework types by year approved by the IZA Structure Commission and given a three letter code (assignment of dates before 1992 is uncertain).

2.4
Pores, Channels, Cages and Cavities

The *n*-rings (where n is the number of T atoms in the ring) defining the face of a polyhedral CBU are called windows or *pores*. Polyhedra whose faces are no larger than six-rings are called *cages*, because the faces are too narrow to pass molecules larger than H_2O. Polyhedra with a least one face larger than a six-ring are called *cavities*. Pores that are infinitely extended in one dimension and are large enough to allow diffusion of guest species (i.e., larger than six-rings) are called *channels*. Framework types can contain one-, two-, or three-dimensional channels. To illustrate, the LTA structure type (Figure 2.4) contains two types of cages, the *d4r* [4^6] and the beta cage [$4^6 6^8$] and one type of cavity, the alpha cavity [$4^{12} 6^8 8^6$]. In addition it contains a three-dimensional intersecting channel system with eight-ring pores.

An important characteristic of zeolites is the effective width of the channels which determines the accessibility of the channels to guest species. The effective width is limited by the smallest free aperture along the channel. For the LTA structure type this is the eight-ring window (Figure 2.6), which has a maximum free aperture of about 0.43 nm. The aperture is calculated by subtracting the diameter of an oxygen ion in silicate structures (0.27 nm) from the interatomic distance of two opposing oxygen atoms across the ring. Free apertures for channels limited by several ring sizes are shown in Table 2.1.

Knowing the framework type of a material, the size of molecules that can be adsorbed can be estimated. Kinetic diameters for various molecules [5–9] are given in Table 2.2. Thus neopentane (kinetic diameter of 0.62 nm) is expected to be adsorbed by NaX zeolite (FAU structure type) which has channels defined by 12-

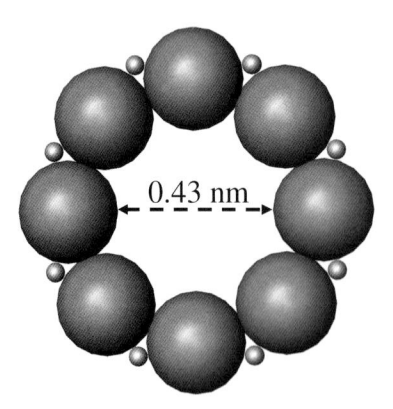

Figure 2.6 Eight-ring pore in LTA framework types. O-atoms and T-atoms are represented are large (0.135 nm radius) and small (0.026 nm radius) spheres respestively. The effective pore width is shown as the interatomic distance between two opposing O-atoms minus the diameter of the O-atom (0.27 nm).

Table 2.1 Apertures for different ring sizes.

T-atoms in ring	Maximum free aperture (nm)	Typical free apertures (nm)
4	0.16	–
5	0.15	–
6	0.28	–
8	0.43	0.30–0.45
10	0.63	0.45–0.60
12	0.80	0.60–0.80

Table 2.2 Kinetic diameters for various molecules.

Molecule	Kinetic diameter (nm)	Molecule	Kinetic diameter (nm)	Molecule	Kinetic diameter (nm)
He	0.26	N_2	0.36	2-Methylbutane	0.50
NH_3	0.26	CO	0.38	2-Methylpentane	0.50
Ne	0.28	CH_4	0.38	3-Methylpentane	0.50
H_2O	0.27	C_2H_4	0.39	SF_6	0.55
H_2	0.29	Xe	0.40	p-Xylene	0.57
NO	0.32	Cyclopropane	0.42	Benzene	0.59
Cl_2	0.32	Propane	0.43	Toluene	0.59
HCl	0.32	n-Butane	0.43	CCl_4	0.59
CO_2	0.33	n-Pentane	0.43	Cyclohexane	0.60
C_2H_2	0.33	n-Hexane	0.43	Ethylbenzene	0.60
N_2O	0.33	n-Heptane	0.43	Neopentane	0.62
Methanol	0.34	n-Nonane	0.43	2,2-DMB	0.62
Ar	0.34	n-Decane	0.43	o-Xylene	0.63
O2	0.35	1-hexene	0.43	m-Xylene	0.63
Br_2	0.35	CF_2Cl_2	0.44	Mesitylene	0.77
HBr	0.35	Propylene	0.45	(C2F5)2NC3F7	0.77
Kr	0.36	Butene-1	0.45	$(C_2H_5)_3N$	0.78
SO_2	0.36	CF_4	0.47	$(C_4H_9)_3N$	0.81
H_2S	0.36	i-Butane	0.50	$(C_4F_9)_3N$	1.02
CS_2	0.36				

ring pores (maximum aperture 0.80 nm), but should not be adsorbed by chabazite (CHA structure type) which has eight-ring pores (maximum aperture 0.43 nm). The estimates are only approximate since the apparent sizes of adsorbate molecules and pores can vary with conditions. Some adsorbate molecules are more flexible than others and can distort while passing through pores. Also, the maximum free aperture given in Table 2.1 is for a symmetrical ring, but the rings can be distorted to varying degrees in real materials reducing the diameter to

Table 2.3 Framework types with larger than 12-rings.

FTC	Type material	FD_{Si}	SBU	CBU	Rings	Dim
AET	AlPO-8	18.2	6	afi bog	14 6 4	1
CFI	CIT-5	16.8	5-[1,1,1]	mtt cas ton	14 6 5 4	1
-CLO	Cloverite	11.1	4-4 or 4	d4r clo lta	20 8 6 4	3
DON	UTD-1F	17.1	5-3	afi mel	14 6 5 4	1
ETR	ECR-34	15.4	4		18 8 6 4	3
OSO	OSB-1	13.3	Spiro-5 and 3-1 (1:1)	lov	14 8 3	3
SFH	SSZ-53	16.5	5-3	afi cas mtw	14 6 5 4	1
SFN	SSZ-59	16.6	5-3	jbw cas mtw	14 6 5 4	1
UTL	IM-12	15.6	Combinations only	d4r cas fer non ton	14 12 6 5 4	2
VFI	VPI-5	14.5	18 or 6 or 4-2	afi bog	18 6 4	1

values less than the maximum value. Some typical values for the larger ring sizes are indicated in the table. It is necessary to know details of the crystalline structure including the atomic positions of the T-atoms and oxygen atoms in order to more correctly predict adsorption properties. In addition, temperature and thermal vibration can affect the apertures and apparent size can change with temperature. Also, as shown later, guest species can reduce the apparent apertures in some zeolites.

It is easy to search the IZA database of zeolite structures for framework types with particular properties such as ring size using the advanced search options. For example, searching the database for frameworks with larger than 12-rings gives the result shown in Table 2.3. Shown in the table are the framework type code (FTC), the material for which the code was granted (Type Material), the framework density for the hypothetical all silica version of the zeolite (FD_{Si}), secondary building units (SBUs) and common composite building units (CBUs) found in the structure, the sizes of rings in the structure (Rings), and the dimension of the channel system for the largest ring size (Dim). This list could be the start for selecting candidates for adsorption of molecules as large as $(C_2H_5)_3N$. Secondary building units (SBUs) are component units with up to 16 T-atoms that are derived assuming that the entire framework is made up of only one type of unit only. For further details and a complete list of SBUs see the compendium by van Koningsveld [10].

2.5
Materials Versus Framework Types

The framework type codes accepted by the IZA Structure Commission gives a convenient way to discuss general structural features of materials but it is important to note that the properties of real materials can vary greatly. Many different materials of each structure type have been prepared, each varying in details of their structures and properties. Their properties are determined not only by the structure type as given by the IZA three letter code, but also by their chemical and physical properties.

One way the chemical properties of a zeolite can be modified is by replacing some or all of the tetrahedral framework atoms with other atom types. An all-silica framework is neutral and requires no counterbalancing cations. Substitution of some of the Si^{4+} cations in the framework by Al^{3+} leaves a net negative charge which is balanced by extra framework cations. Other cations besides Al and Si that can be incorporated into the frameworks include B, Al, P, Fe, Zn, Ga and Ge. It was recognized as early as 1971 that P^{5+} cations could be substituted into zeolite frameworks [11] and later that aluminophosphate materials, $AlPO_4$, could be prepared with no silicon in the framework [12]. This new class of aluminophosphate materials has grown to over 50 members, some of which have aluminosilicate analogs, but some of which are completely new framework types with no aluminosilicate analog. There is evidence that at least 16 other elements (Li, Be, B, Mg, Si, Ti, V, Cr, Mn, Fe, Co, Ni, Zn, Ga, Ge, As) can be incorporated into $AlPO_4$ frameworks.

Another way to modify zeolite properties is to change the nature of the non-framework cations. Some examples of how cation exchange can affect properties are noted in Section 2.6. For catalytic applications it is often desirable to create active acidic sites. This can be done by exchanging the cations with NH_4^+ and then calcining the zeolite, removing NH_3 and leaving behind protons attached to framework oxygen atoms.

2.6
Structures of Commercially Significant Zeolites

Commercially significant zeolites include the synthetic zeolites type A (LTA), X (FAU), Y (FAU), L (LTL), mordenite (MOR), ZSM-5 (MFI), beta (*BEA/BEC), MCM-22 (MTW), zeolites F (EDI) and W (MER) and the natural zeolites mordenite (MOR), chabazite (CHA), erionite (ERI) and clinoptilolite (HEU). Details of the structures of some of these are given in this section. Tables in each section lists the type material (the common name for the material for which the three letter code was established), the chemical formula representative of the unit cell contents for the type material, the space group and lattice parameters, the pore structure and known mineral and synthetic forms.

To provide a common basis for research on three widely used industrial zeolites, NIST has issued reference materials for zeolite Y (RM 8850), zeolite A (RM 8851) and ammonium ZSM-5 zeolite (RM 8852). Reference and information values are provided for major and trace element content, key atomic ratios, enthalpy of formation, unit cell parameters and particle size distributions.

2.6.1
Linde Type A (LTA)

The LTA framework type (Table 2.4) can be built by linking sodalite cages through double four-rings (Figure 2.7). This creates an alpha cavity accessible to molecules larger than water via a three-dimensional eight-ring channel system. Figure 2.8a shows the shape of the channel system in LTA and Figure 2.8b shows the cages in the structure not accessible to molecules larger than water (generated using the 3D drawing option at the IZA Structure Commission web site [4]). Thus, the large area in the alpha cavity is accessible through the eight-ring pores, but the smaller area in the sodalite cages is not accessible.

LTA is commonly synthesized with a Si/Al ratio of 1, which allows for strict alternation of the Si and Al atoms in the structure and a doubling of the unit cell

Table 2.4 LTA structure type

Type material	Linde type A (zeolite A) [13, 14]
Chemical formula	$\|Na_{12}(H_2O)_{27}\|_8 [Al_{12}Si_{12}O_{48}]_8$–**LTA**
Space group	Pseudo cell: cubic, Pm3m, $a = 12.30$ Å. Super cell (8× volume of pseudo cell): cubic, Fm-3c, $a = 24.61$ Å
Pore structure	Three-dimensional eight-ring
Mineral forms	Not known
Synthetic forms	Alpha, ITQ-29, Linde type A, 3A, 4A, 5A, LZ-215, N-A, SAPO-42, ZK-21, ZK-22, ZK-4, UZM-9

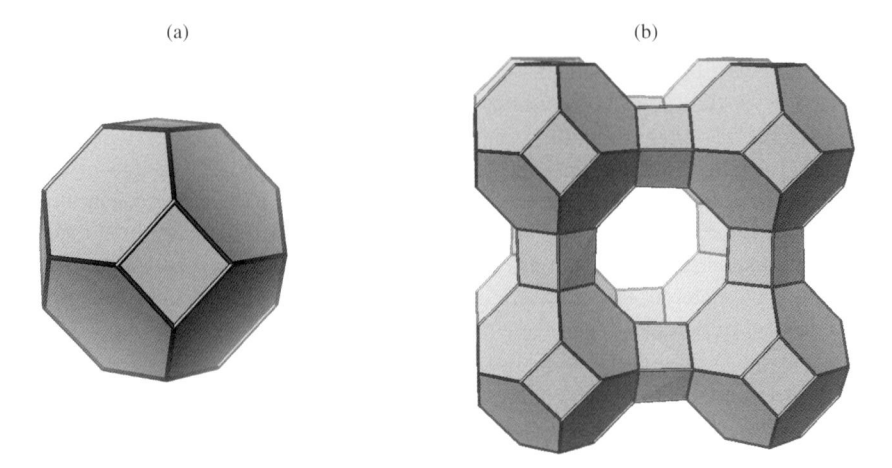

Figure 2.7 (a) Sodalite cage and (b) LTA framework formed by linking sodalite cages through double four-rings.

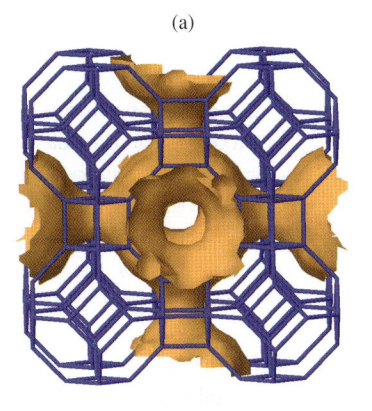

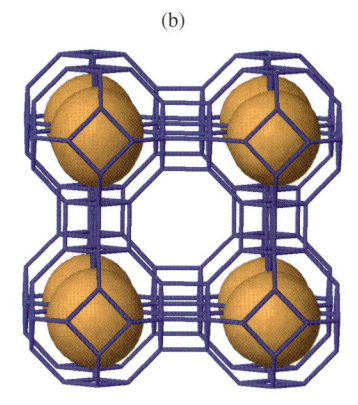

2.6 Structures of Commercially Significant Zeolites

Figure 2.8 (a) Channel system and (b) cages in LTA.

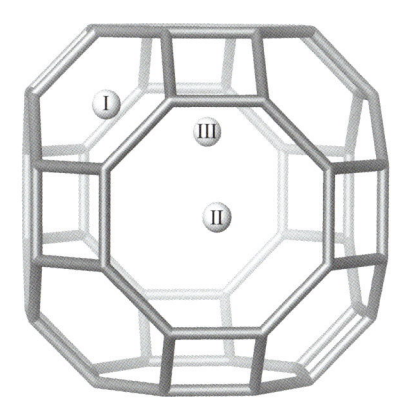

Figure 2.9 Cation sites in LTA. Site I is centered on the six-rings and displaced into the alpha cavity. Site II is near the center of the eight-rings and close to the eight-ring planes. Site III is centered on the four-rings and displaced into the alpha cavity.

lattice parameter leading to a super cell with eight times the volume of the smaller cell. Si/Al ratios generally vary from 1 to about 6; and a pure silica form was reported [15].

Primary cation sites (Figure 2.9) are: site I (centered on the six-rings and displaced into the alpha cavity), site II (near the center of the eight-rings and close to the eight-ring planes) and site III (centered on the four-rings and displaced into the alpha cavity). Sites I, II and III can accommodate 8, 3 and 12 cations per alpha cavity, respectively, and cations tend to prefer sites in the order I, then II, then III. Since site II is near the center of the eight-ring, cations which occupy site II will affect diffusion. A fully Na-exchanged LTA zeolite with a Si/Al ratio of 1 has 12 cations per alpha cavity, which is enough to fully occupy sites I and II. The site II cations significantly alter accessibility into the alpha cavity. Replacing some of the monovalent Na cations with divalent cations such as Ca^{2+} reduces the number of cations in site II and frees-up access to the cavities [16, 17]. Thus, exchanging the sodium cations in a 4A zeolite with calcium ions (5A) can increase the effective aperture size from about 0.4 to 0.5 nm. Conversely, exchanging the sodium cations with potassium (3A) can reduce the aperture size to about 0.3 nm.

2.6.2
Faujasite (FAU)

The framework for FAU type zeolite (Table 2.5) can be built (Figure 2.10) by linking sodalite cages through double six-rings (instead of double four-rings as for LTA). This creates a large cavity in FAU called the "supercage" (which should really be called a supercavity) accessible by a three-dimensional 12-ring pore system (Figure 2.11).

Table 2.5 FAU structure type

Type material	Faujasite [18, 19]
Chemical formula	$|(Ca,Mg,Na)_{29}(H2O)_{240}|\ [Al_{58}Si_{134}O_{384}]$ –**FAU**
Space group	Cubic, Fd-3m, a = 24.74 Å
Pore structure	Three-dimensional 12-ring
Mineral forms	Faujasite
Synthetic forms	Beryllophosphate X, Li-LSX, LZ-210, SAPO-37, siliceous Na-Y, zeolite X (Linde X), zeolite Y (Linde Y), zincophospate X

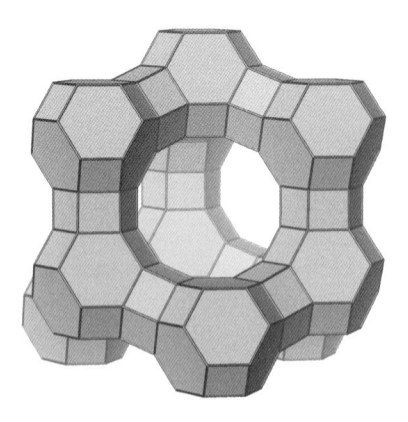

Figure 2.10 Framework structure for FAU zeolite formed by linking sodalite cages through double six-rings.

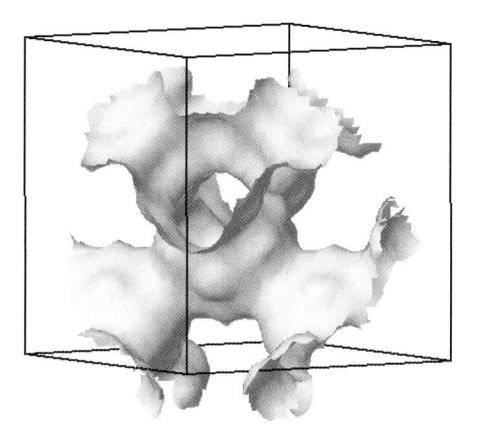

Figure 2.11 The channel system in FAU is three-dimensional with 12-ring pores opening into large supercages.

FAU type zeolites exchanged with many different cations (Na^+, K^+, Ba^{2+}, Cu^{2+}, Ni^{2+}, Li^+, Rb^+, Sr^{2+}, Cs^+, etc.) have been extensively studied. The unit cell contents of hydrated FAU type zeolite can be represented as $|M_x(H_2O)_y|$ $[Al_xSi_{192-x}O_{384}]$ – **FAU**, where x is the number of Al atoms per unit cell and M is a monovalent cation (or one-half of a divalent cation, etc.). The number of Al atoms per cell can vary from 96 to less than 4 (Si/Al ratios of 1 to more than 50). Zeolite X refers to zeolites with between 96 and 77 Al atoms per cell (Si/Al ratios between 1 and 1.5) and Zeolite Y refers to zeolites with less than 76 Al atoms per cell (Si/Al ratios higher than 1.5).

Primary cation sites (Figure 2.12) are site 1 in (or near) the center of the D6R, site 1' inside the sodalite cage on the face of the six-ring of the D6R, site 2' in the sodalite cage on the six-ring connecting to the supercage, site 2 in the supercage on the face of the six-ring and site 3 in the supercage near the four-rings. The important cation sites for interaction with adsorbed molecules are sites 2 and 3, since the other sites are buried in cages accessible only through six-rings. A recent review by Frising and Leflaive [20] gives details of the cation distributions reported in the literature through mid-2007. The FAU unit cell contains 16 *d4r*, eight sodalite cages and eight supercages. Sites 1, 1', 2', 2 and 3 can accommodate 16, 32, 32, 32 and 96 cations, respectively. For dehydrated Na-exchanged zeolites, the cations tend to occupy sites which maximize their interaction with framework oxygen atoms and minimize cation–cation electrostatic repulsion. This means sites 1 and 2 tend to be occupied first, accommodating up to 48 cations. When more than 48 cations are present, site 1' is progressively favored over site 1, up to 64 cations per cell. Above 64, site 3 in the supercage begins to be occupied. The addition of water or other adsorbates can modify the cation site distribution, since the cations may be able to find more favorable interactions with the adsorbates. For example, the addition of water tends to move Na cations from site 1 to site 1'

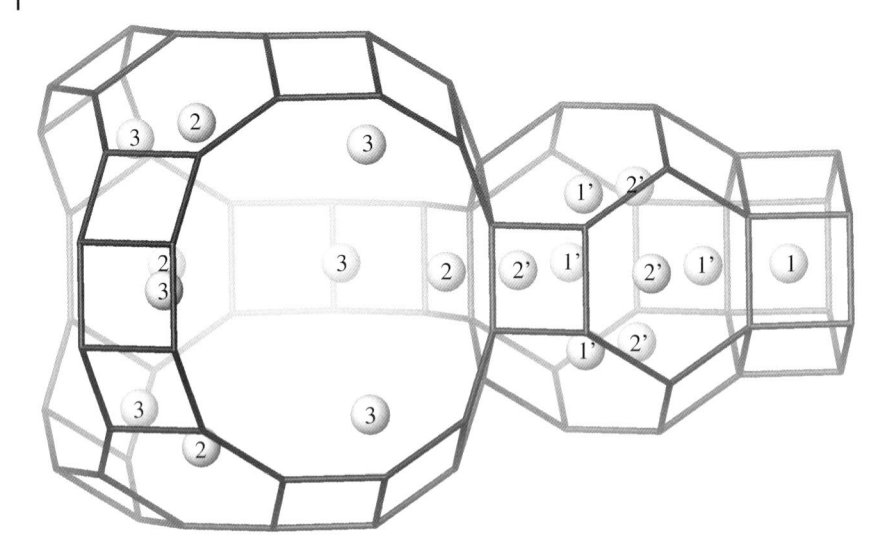

Figure 2.12 Cation sites in FAU. Site 1 is in (or near) the center of the D6R, site 1' is inside the sodalite cage on the face of the six-ring of the D6R, site 2' is in the sodalite cage on the six-ring connecting to the supercage, site 2 is in the supercage on the face of the six-ring and site 3 is in the supercage near the four-rings.

Table 2.6 MOR structure type

Type material	Mordenite [21]
Chemical formula	$\|Na_8^+(H_2O)_{24}\|[Al_8Si_{40}O_{96}]$ – **MOR**
Space group	Orthorhombic, Cmcm, $a = 18.1$ Å, $b = 20.5$ Å, $c = 7.5$ Å
Pore structure	One-dimensional 12-ring, one-dimensional eight-ring
Mineral forms	Maricopaite, mordenite
Synthetic forms	Ca-Q, LZ-211, Na-D

where the cations can favorably interact with both framework oxygens and water molecules. With other cation types, distributions among the sites can be different than with sodium because of differences in cation size and charge.

2.6.3
Mordenite (MOR)

The MOR framework (Table 2.6, Figure 2.13) is built from chains of five-membered rings linked laterally to form the three-dimensional structure. The framework defines a one-dimensional 12-ring channel system connected in one direction by eight-ring channels (Figure 2.14).

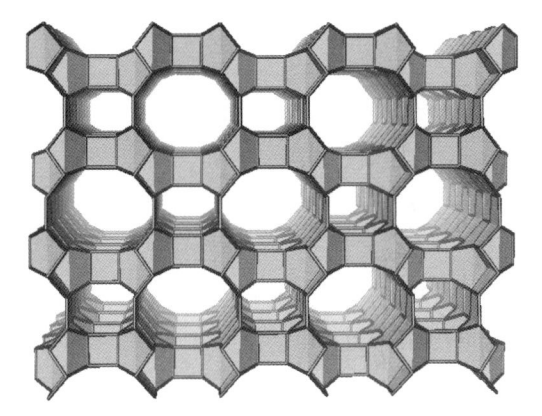

Figure 2.13 Framework structure for MOR viewed down the 12-ring one-dimensional channels.

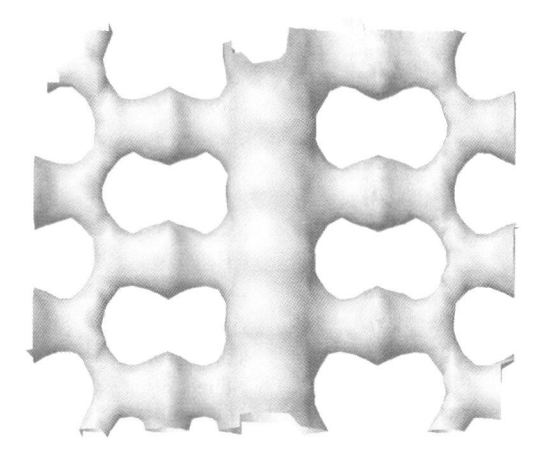

Figure 2.14 The channel system in MOR viewed normal to the 12-rings showing the connectivity between 12-rings via secondary eight-ring channels.

Natural mordenite is highly siliceous and exists with a nearly constant Si/Al ratio of 5, while synthetic mordenites have been made with Si/Al ratios from about 4 to 12 [22]. The primary cation sites [23] are indicated in Figure 2.15. Sites A, B and C are in the eight-ring channels. Only Sites D and E are in the 12-ring channel. Site D is on the face of an eight-ring inside the 12-ring channel and Site E is on the face of a six-ring. In dehydrated samples, cations tend to favor site A, but distribute among all the sites. In hydrated samples, site A is still favored, but other cations tend to move into the 12-ring channels where they can coordinate to water.

For most applications, which take advantage of the large 12-ring channels, the channel system can be considered to be one-dimensional. Since the only com-

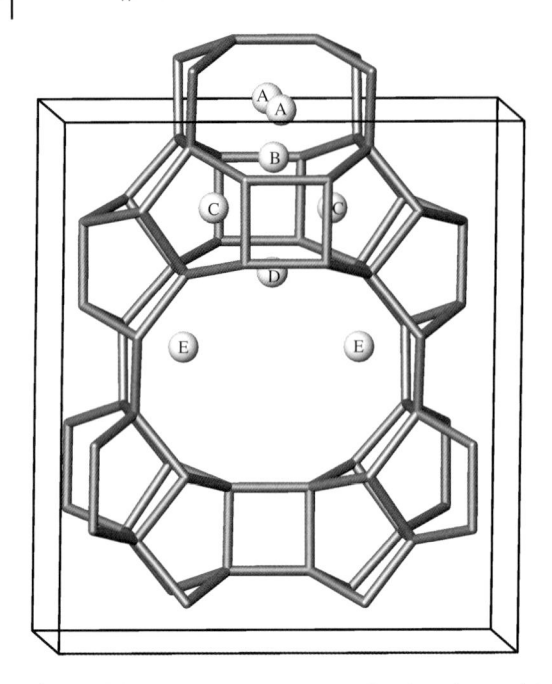

Figure 2.15 Primary cation sites in MOR. Sites A, B and C are in the eight-ring channels. Only Sites D and E are in the 12-ring channel. Site D is on the face of an eight-ring inside the 12-ring channel and Site E is on the face of a six-ring.

munication between the large channels is through smaller eight-rings, this limits diffusion of large molecules to one-dimension. Any blockage of a 12-ring channel may block the entire channel. Thus, morphology may have an affect on adsorption properties. For example, needle-like crystals with the 12-ring channels aligned along the needle axis may be undesirable. Many natural mordenites have severely restricted diffusion and were historically called small-port mordenites. The reason for the restrictions is still not completely understood and has been attributed to structural defects in the framework [24–26]. Synthetic mordenites can be prepared as the large-port or small-port variety depending on synthesis conditions [27] and small-port mordenites can be converted to large-port forms by acid treatments and calcination [28, 29].

2.6.4
Chabazite (CHA)

The CHA framework structure (Table 2.7, Figure 2.16) can be built by linking layers of *d6r* cages through tilted four-rings. The framework defines a large cavity, called the *cha* cavity, accessible through a three-dimensional eight-ring pore system (Figure 2.17).

2.6 Structures of Commercially Significant Zeolites

Table 2.7 CHA structure type

Type material	Chabazite [30, 31]
Chemical formula	$|Ca_6^{2+}(H_2O)_{40}|[Al_{12}Si_{24}O_{72}]$ – **CHA**
Space group	Rhombohedral, R-3m, a = 9.42 Å, α = 94.47°
Pore structure	Three-dimensional eight-ring
Mineral forms	Chabazite, wilhendersonite
Synthetic forms	AlPO-34, CoAPO-44, CoAPO-47, DAF-5, GaPO-34, Linde D, Linde R, LZ-218, MeAPO-47, MeAPSO-47, (Ni(deta)2)-UT-6, Phi, SAPO-34, SAPO-47, UiO-21, ZK-14, ZYT-6

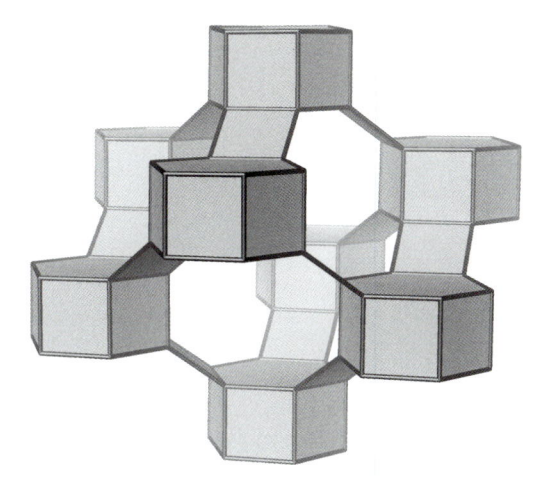

Figure 2.16 CHA framework structure built from *d6r* units linked by four-rings. The stacking sequence of six-ring units as shown is AABBCCAA.

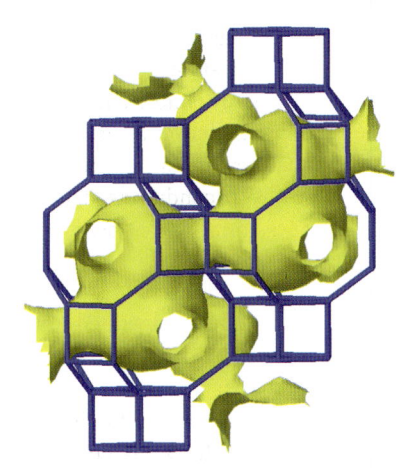

Figure 2.17 Channel system in CHA. The three-dimensional eight-ring pore system opens into large *cha* cavities.

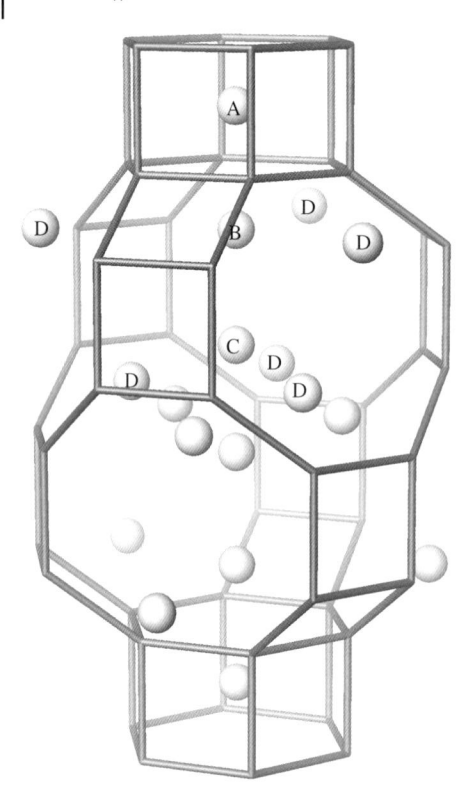

Figure 2.18 Cation sites in CHA. Site A is in the *d6r*. All other sites are in the large *cha* cavity.

CHA frameworks have been found with Si/Al ratios from 1 [32] to infinity [33]. Principle cation sites are: in the center of the d6r (Figure 2.18a), in the *cha* cavity on the face of the six-ring of the *d6r* in the large cage (Figure 2.18b), in the *cha* cavity along the three-fold axis (Figure 2.18c) and in or near the eight-ring window (Figure 2.18d). Small cations tend to prefer sites a and b in dehydrated samples and as for other zeolites, cations tend to migrate to the large cage on hydration to maximize coordination to water. Large cations such as Cs^+ fit nicely in the eight-ring windows.

The CHA framework is a member of a rich, diverse family of zeolites known as the ABC-6 family. All members of the ABC-6 family consist of layers of six-rings or double six-rings, arranged in a hexagonal array, interconnected by tilted four-rings [34–37]. The three possible six-ring layers are designated A, B and C. The stacking sequence of layers determines the framework type. An A layer can be followed by an A, B, or C layer, but a sequence of three or more identical layers (e.g., AAA) is not likely because of the severe strain that would be imposed on bond angles. The known members of this family are listed in Table 2.8. In real crystals stacking sequences may not be perfect and faulting can arise where the

Table 2.8 Known members of the ABC-6 family of frameworks.

Layers	Code	Name	Stacking sequence
2	CAN	Cancrinite	AB
3	SOD	Sodalite	ABC
3	OFF	Offretite	AAB
4	LOS	Losod	ABAC
4	GME	Gmelinite	AABB
6	LIO	Liottite	ABACAC
6	ERI	Erionite	AABAAC
6	EAB	TMA-E(AB)	AABCCB
6	CHA	Chabazite	AABBCC
8	AFG	Afganite	ABABACAC
8	AFX	SAPO-56	AABBCCBB
9	LEV	Levyne	AABCCABBC
10	FRA	Franzinite	ABCABACABC
12	SAT	STA-2	AABABBCBCCAC
12	AFT	AlPO-52	AABBCCBBAACC

Table 2.9 MFI structure type

Type material	ZSM-5 [41–43]
Chemical formula	$\lvert Na_n^+ (H_2O)_{16}\rvert[Al_nSi_{96-n}O_{192}] -$ **MFI**, $n < 27$
Space group	Orthorhombic, Pnma, $a = 20.07$ Å, $b = 19.92$ Å, $c = 13.42$ Å
Pore structure	Three-dimensional ten-ring
Mineral forms	Encilite, mutinaite
Synthetic forms	AMS-1B, AZ-1, Bor-C, Boralite C, FZ-1, LZ-105, NU-4, NU-5, Silicalite, TS-1, TSZ, TSZ-III, TZ-01, USC-4, USI-108, ZBH, ZKQ-1B, ZMQ-TB, ZSM-5

stacking sequence is disrupted. The complete behavior has been observed, from essentially no faulting (e.g., in AlPO-52 [38]) to moderate faulting (e.g., in LZ-276 [39]) and to faulting at a 50% probability level that gives maximum randomness (e.g., in babelite [40]).

2.6.5
ZSM-5 (MFI)

The MFI framework (Table 2.9, Figure 2.19) can be built from five-rings and contains cavities interconnected by a straight ten-ring channel systems and a zigzag ten-ring channel system (Figure 2.20). Primary cation sites (Figure 2.21)

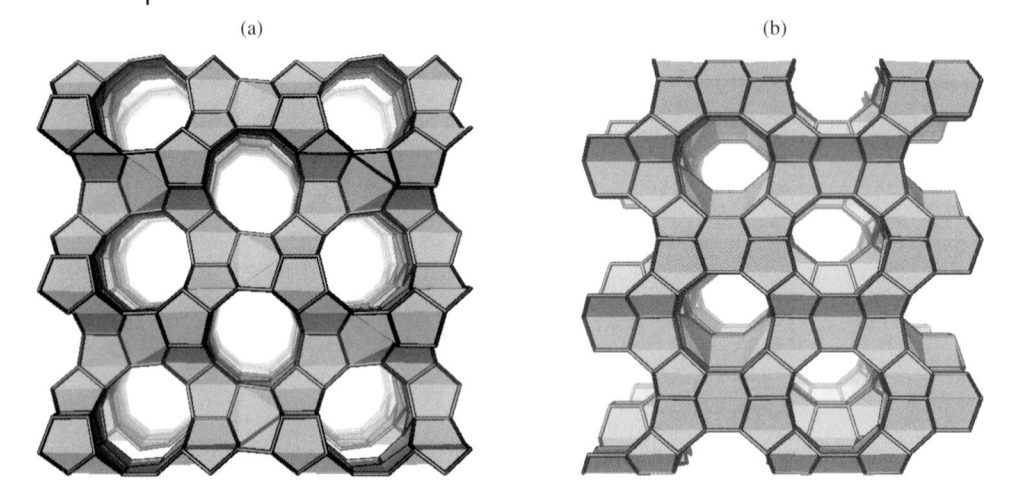

Figure 2.19 MFI framework views: (a) down the straight ten-ring channel system, (b) down the zig-zag ten-ring channel system.

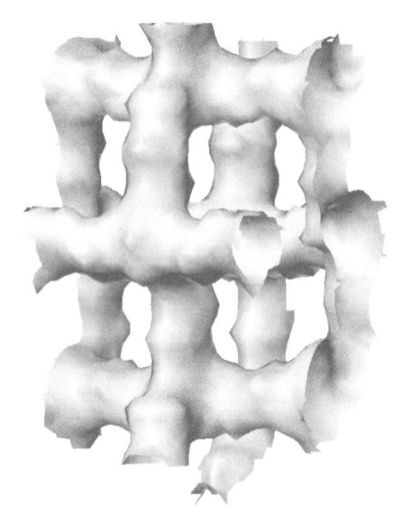

Figure 2.20 Diagram of the channel system in MFI showing the connectivity between the straight ten-ring channels (horizontal) and the zig-zag ten-ring channels (vertical).

are site 1 near the intersection of the two ten-ring channel systems, site 2 in the straight ten-ring channel, and site 3 in the zigzag ten-ring channel [44].

MFI-type zeolites are highly siliceous with Si/Al ratios from about 10 to infinity; generally, the higher the Si/Al ratio the easier the synthesis. MFI zeolites are reported to be hydrophobic and organophilic and for that reason useful for removing organics from water streams and stable for separations and catalysis in the presence of water.

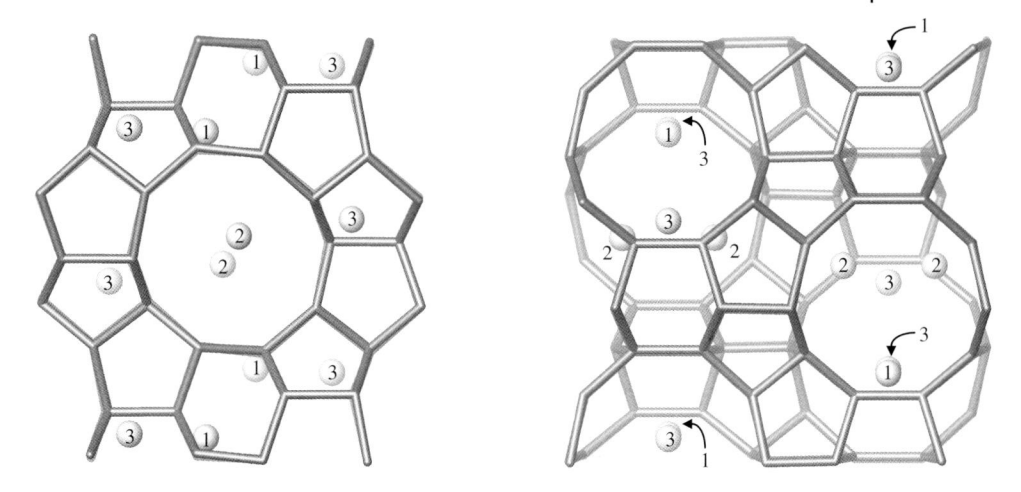

Figure 2.21 Cation sites in MFI. Site 1 is in near the intersection of the two ten-ring channel systems. Site 2 is in the straight ten-ring channel. Site 3 is in the zigzag ten-ring channel.

Table 2.10 LTL structure type

Type material	Linde type L [50]
Chemical formula	$|K_6^+ Na_3^+ (H_2O)_{21}|[Al_9Si_{27}O_{72}]$ – **LTL**
Space group	Hexagonal, P6/mmm: $a = 18.40$ Å, $c = 7.52$ Å
Pore structure	One-dimensional 12-ring
Mineral forms	Perlialite
Synthetic forms	Gallosilicate L, (K,Ba)-G, L, Linde type L, LZ-212

Since MFI zeolites are highly siliceous the number of cations is small. However, since all the cation sites are in the MFI channels, all changes to cation number and type can affect the adsorption properties. The adsorption capacity of alkanes in MFI has been shown to increase with decreasing non-framework cation density (increasing Si/Al ratio) [45] and the linear/branched separation selectivity increases with increasing non-framework cation density [46–48]. Molecular simulations indicate that for a given cation, the adsorption of alkanes in MFI increases with decreasing the nonframework cation concentration and, for a given Si/Al ratio, the adsorption of alkanes in MFI increases with decreasing atomic weight of the non-framework cation [49].

2.6.6
Linde Type L (LTL)

The LTL framework (Table 2.10) can be built entirely from the cancrinite cage (*can*) shown in Figure 2.2 by connecting all T-atoms of the axial six-rings along the c-axis direction forming alternating *can* and *d6r* cages along the *c*-axis, and by

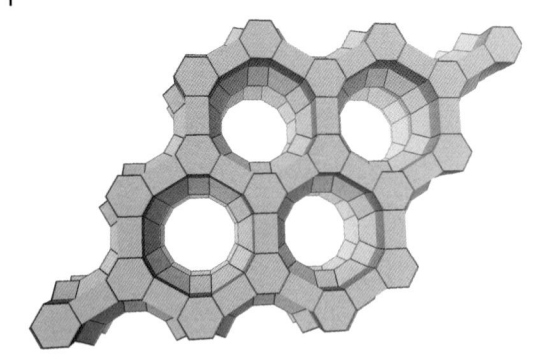

Figure 2.22 LTL framework viewed down the one-dimensional 12-ring channels.

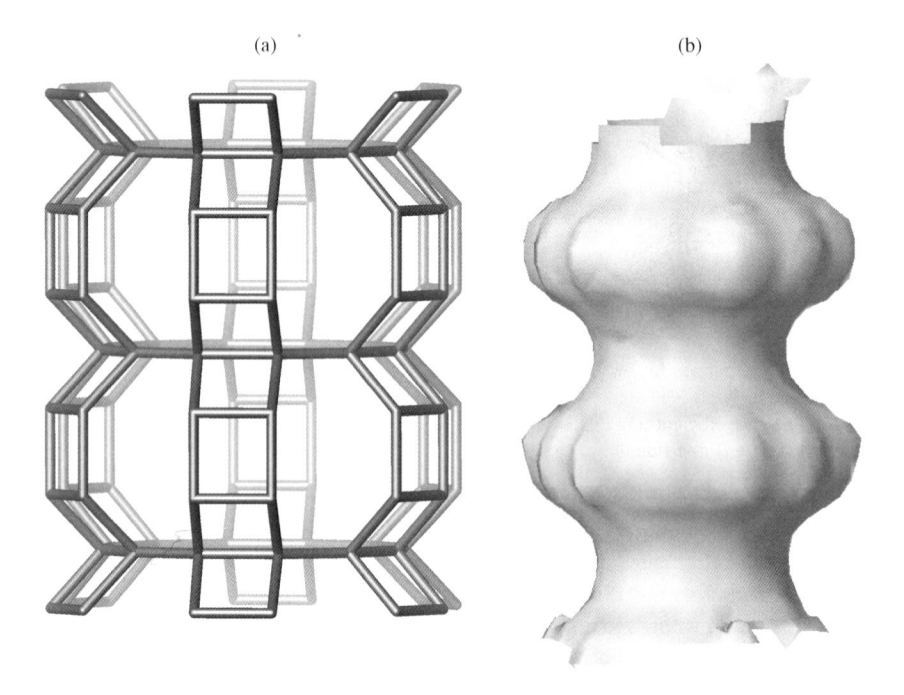

Figure 2.23 (a) Details of the T-atom connectivity comprising the LTL 12-ring channel system. (b) Diagram of the resulting channels.

connecting the equatorial T-atoms of the equatorial six-rings in the *a*- and *b*-axis directions forming rings of alternating *can* cages and highly distorted eight-rings. The resulting framework (Figure 2.22) contains a one-dimensional 12-ring channel system connected by a three-dimensional eight-ring channel system. However, the eight-ring apertures are so highly distorted that they do not allow diffusion, and the framework can be considered to allow only one-dimensional diffusion along the 12-ring channels (Figure 2.23).

Figure 2.24 Cation sites in LTL. Site A is in the *d6r* cage, site B is in the *can* cage, site C is in the puckered eight-ring connecting *can* cages and site D is inside the 12-ring channels near the eight-ring. Only cations in site D are easily exchanged and accessible to interactions with adsorbates.

Si/Al ratios for the LTL structure type vary from about 1 to 6. Cation sites in LTL are shown in Figure 2.24. Site A is in the *d6r* cage, site B is in the *can* cage, site C is in the puckered eight-ring connecting *can* cages and site D is inside the 12-ring channels near the eight-ring. Cations in sites A, B and C are buried in cages, not easily cation exchanged and not accessible to interactions with adsorbates. Only cations in site D are easily exchanged and accessible to interactions with adsorbates.

2.6.7
Beta Polymorphs *BEA and BEC

*BEA and BEC (Table 2.11) refer to two end members of a disordered family of zeolites commonly called beta [52, 53]. *BEA refers to beta polymorph A, an end member which has not been synthesized yet in its pure state. BEC refers to

Table 2.11 *BEA and BEC structure types

Type material	Beta polymorph A [51, 52]
Chemical formula	$|Na_7^+|[Al_7Si_{57}O_{128}]$ – *BEA
Space group	Tetragonal, $P4_122$; $a = 12.661$ Å, $c = 26.406$ Å
Pore structure	Three-dimensional 12-ring
Mineral forms	Tschernichite
Synthetic forms	Beta, Al-rich beta, CIT-6

Type material	FOS-5 (beta polymorph C) [53]
Chemical formula	$|(C_3H_9N)_{48} (H_2O)_{36}| [Ge_{256}O_{512}]$ –**BEC**
Space group	Tetragonal, $I4_1/amd$; $a = 25.990$ Å, $c = 27.271$ Å
Pore structure	Three-dimensional 12-ring
Mineral forms	Tschernichite
Synthetic forms	FOS-5 (beta polymorph C), ITQ-14, ITQ-17

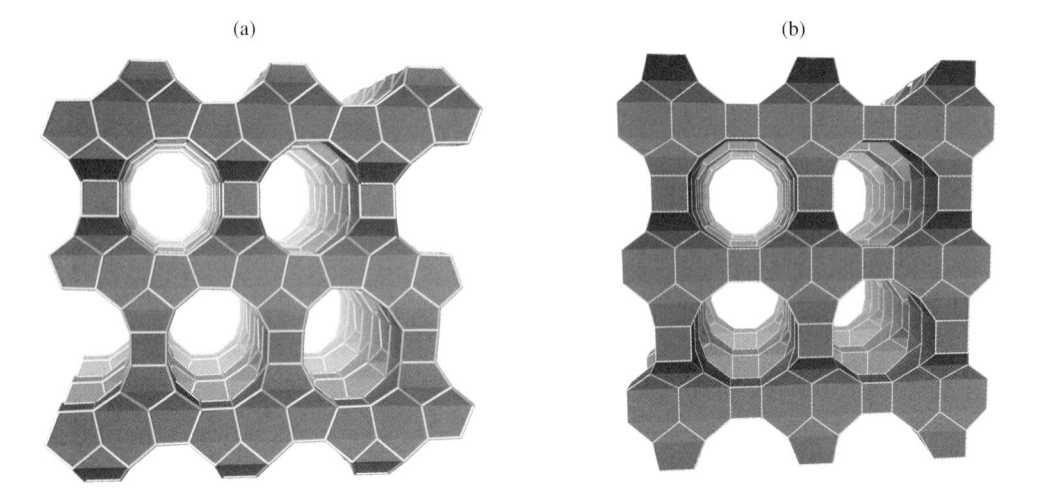

(a) (b)

Figure 2.25 Framework structures for two polymorphs of beta zeolite: (a) *BEA and (b) BEC. All beta polymorphs contain three-dimensional 12-ring channels.

polymorph C, another end member which has been synthesized [54], but only as a Ge polytype. Details of the nature of the disorder are given at the IZA-SC web site [4]. All of the polymorphs have three-dimensional 12-ring channel systems. The framework structures and channel systems for *BEA and BEC are shown in Figures 2.25 and 2.26.

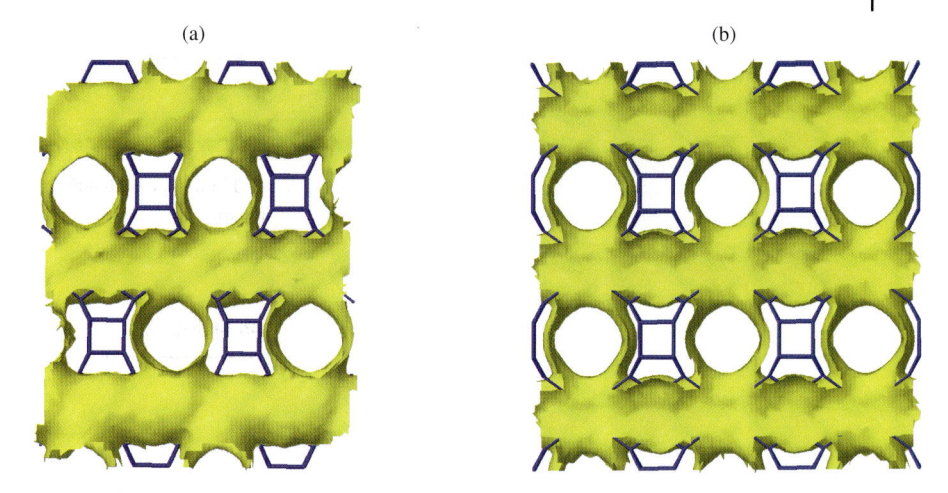

Figure 2.26 Comparison of channel systems in (a)*BEA and (b) BEC.

Table 2.12 MWW structure type

Type material	MCM-22 [56]
Chemical formula	$\lvert H_{2.4}^{+}Na_{3.1}^{+}\rvert[Al_{0.4}B_{5.1}Si_{66.5}O_{144}]$ – **MWW**
Space group	Hexagonal, P6/mmm; $a = 14.208$ Å, $c = 24.945$ Å
Pore structure	Two-dimensional ten-ring
Mineral forms	Not known
Synthetic forms	ERB-1, ITQ-1, MCM-22, PSH-3, SSZ-25

The beta family are high-silica frameworks and are synthesized with Si/Al ratios of about 5 to infinity, although tschernichite with the beta framework is found in nature with a Si/Al ratios of 3–4. Most of the synthesized materials are highly disordered, usually as intergrowths of polymorphs A and B (a monoclinic form). The high degree of intergrowth can lead to a large number of defects [54] giving beta unique acid properties [55]. Because of the high degree of disorder in beta zeolites details of the cations sites are not known.

2.6.8
MCM-22 (MWW)

The MWW framework structure (Table 2.12, Figure 2.27) contains two independent 2-dimensional channel systems accessible by ten-ring pores [57]. One is

52 | *2 Zeolite Types and Structures*

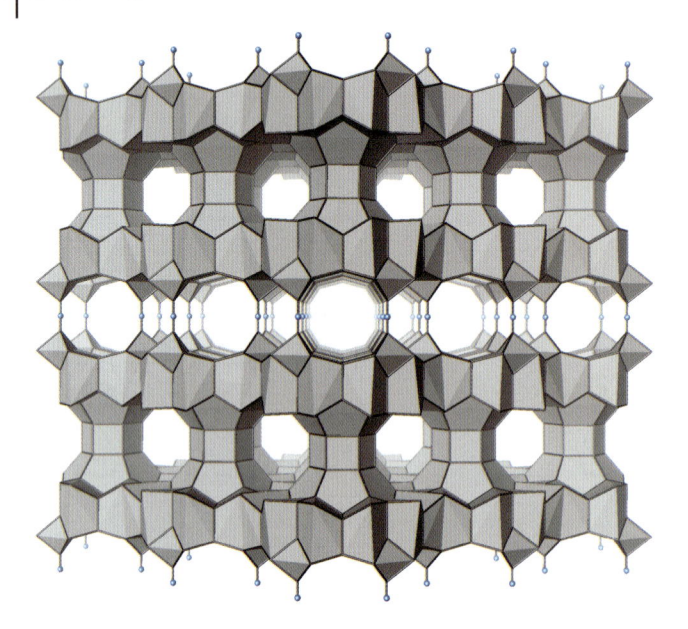

Figure 2.27 MWW framework structure which contains two independent two-dimensional channel systems accessible by ten-ring pores.

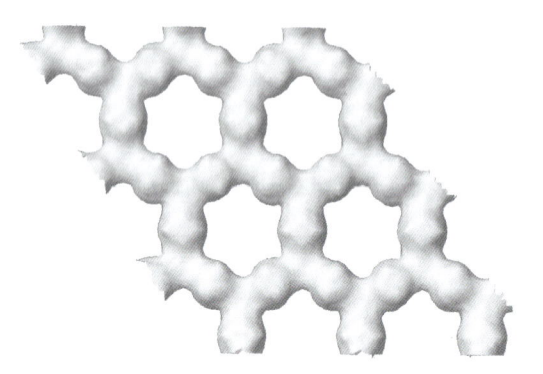

Figure 2.28 View of the ten-ring channel system in MWW defined by two-dimensional zig-zag channels.

defined by two-dimensional zig-zag channels with ten-ring pores (Figure 2.28). The other is defined by two-dimensional straight channels with ten-ring pores that open into large cavities with inner free diameters of 0.71 nm and inner free heights of 1.82 nm (Figure 2.29).

Although access to the interior of the MWW framework is through ten-ring pores, some adsorption and catalytic studies demonstrate some

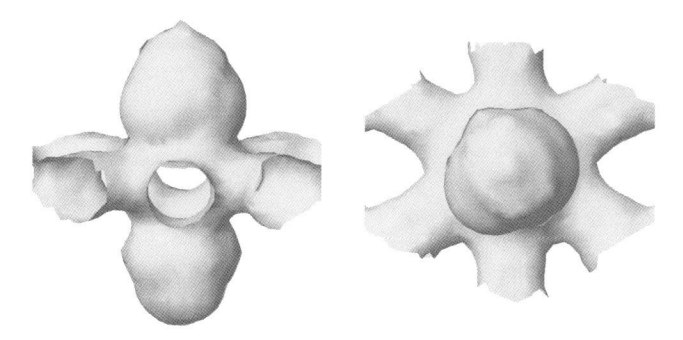

Figure 2.29 Two views of the ten-ring channel system in MWW defined by two-dimensional straight channels with ten-ring pores that open into large cavities with inner free diameters of 0.71 nm and inner free heights of 1.82 nm.

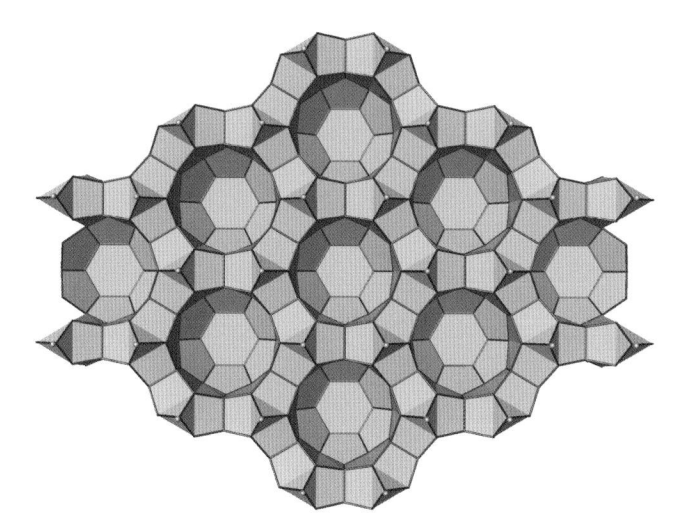

Figure 2.30 View showing 12-ring pockets thought to be present at the surface of MWW samples which crystallize as very thin plates.

characteristics of a 12-ring zeolite [57, 58]. High resolution electron microscopy and X-ray diffraction studies show that MWW samples tend to crystallize as very thin plates with a high concentration of external pockets with 12-ring openings [59]. A diagram of the surface showing 12-ring pockets is shown in Figure 2.30.

MWW-type frameworks have been synthesized with Si/Al ratios of about 10 to infinity [59, 60]. Cation sites [61] are shown in Figure 2.31.

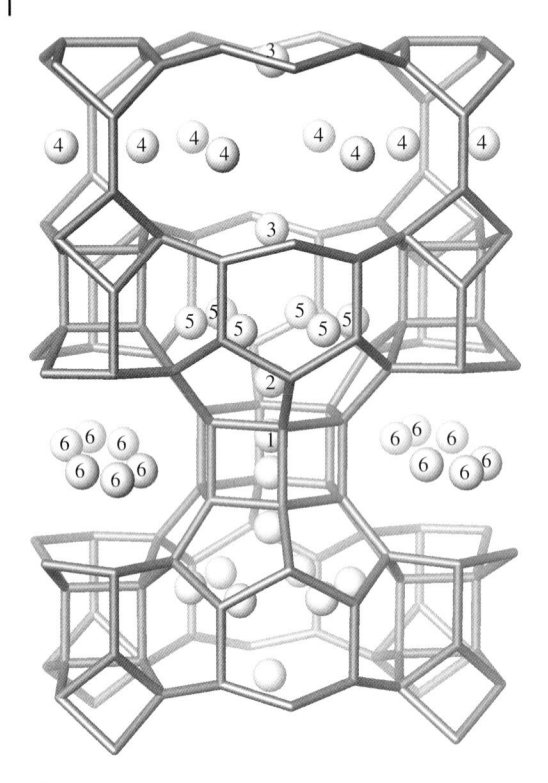

Figure 2.31 MWW cation sites.

2.7
Hypothetical Zeolite Frameworks

The fact that many more frameworks are theoretically possible than have been observed has been recognized and researchers today have generated millions of theoretical frameworks that would seem to be feasible. Joe Smith began enumerating theoretical frameworks in the 1960s by considering conversions and connectivity of building units and generated over 1000 theoretical frameworks [62–64]. Since then, others have used numerical methods for generating millions of hypothetical frameworks [65–71]. Treacy *et al.* find hypothetical frameworks [67, 72] by performing a symmetry-constrained combinatorial search over all possible permutations of bonds between atoms sitting in all possible combinations of crystallographic sites for each space group. The number of generated frameworks by year using this method dwarfs the number of known frameworks (Figure 2.32). The current limitation in enumeration of hypothetical frameworks seems to be computational speed for generating frameworks and algorithms for analysis of the huge amount of data generated. Many of the frameworks appear to have interesting properties such as large channel sizes and would be interesting to synthesize. A major ques-

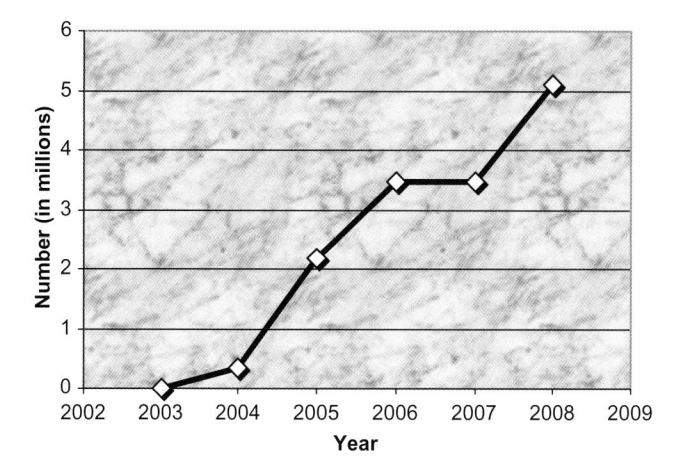

Figure 2.32 Number of hypothetical frameworks contained in the Database of Hypothetical Frameworks [Ref [66]], by year. The number of hypothetical frameworks dwarfs the number of natural and synthetic frameworks.

tion that remains to be answered is why nature and synthetic chemists have only managed to produce a tiny fraction of the possible zeolitic framework types.

Acknowledgments

Framework and polyhedral figures were drawn using ATOMS V6.2 by Shape Software. Channel system diagrams were generated using the 3D drawing option at the IZA Structure Commission web site [4].

References

1 McCusker, L.B., Liebau, F., and Engelhardt, G. (2003) Nomenclature of structural and compositional characteristics of ordered microporous and mesoporous materials with inorganic hosts (IUPAC recommendations 2001). *Micropor. Mesopor. Mat.*, **58** (11), 3–13.
2 Baerlocher, C., McCusker, L.B., and Olson, D.H. (2007) *Atlas of Zeolite Framework Types*, 6th Revised edn, Elsevier, Amsterdam.
3 Smith, J.V. (ed.) (2000) *Microporous and Other Framework Materials with Zeolite-Type Structures, Subvolume A: Tetrahedral Frameworks of Zeolites, Clathrates, and Related Materials*, 331 pp., Springer-Verlag, Berlin, Germany.
4 Baerlocher, C. and McCusker, L.B. (2009) Database of Zeolite Structures, http://www.iza-structure.org/databases/ (accessed June 2009).
5 Breck, D.W. (1974) *Zeolite Molecular Sieves: Structure, Chemistry and Use*, John Wiley & Sons, Inc., New York.
6 Long, Y., Xu, T., Sun, Y., and Dong, W. (1998) Adsorption behavior on defect structure of mesoporous molecular sieve MCM-41. *Langmuir*, **14** (21), 6173–6178.
7 Chen, N.Y., Degnan, T.F., and Smith, C.M. (1994) *Molecular Transport and*

Reactions in Zeolites, VCH, New York, p. 134.

8 Harrison, I.D., Leach, H.F., and Whan, D.A. (1987) Some sorptive and catalytic properties of zeolite Nu-10. *Zeolites*, **7** (1), 28–34.

9 Burggraaf, A.J., Vroon, Z.A.E.P., Keizer, K., and Verweij, H. (1998) Permeation of single gases in thin zeolite MFI membranes. *J. Membr. Sci.*, **144** (1–2), 77–86.

10 van Koningsveld, H. (2007) *Compendium of Zeolite Framework Types. Building Schemes and Type Characteristics*, Elsevier, Amsterdam.

11 Flanigen, E.M. and Grose, R.W. (1971) Phosphorus substitution in zeolite frameworks. *Adv. Chem. Ser.*, **101** (Molecular Sieve Zeolites-I), 76–101.

12 Wilson, S.T., Lok, B.M., Messina, C.A.,Cannan, T.R., and Flanigen, E.M. (1982) Aluminophosphate molecular sieves: a new class of microporous crystalline inorganic solids. *J. Am. Chem. Soc.*, **104**, 1146–1147.

13 Gramlich, V. and Meier, W.M. (1971) The crystal structure of hydrated NaA: a detailed refinement of a pseudosymmetric zeolite structure. *Z. Kristallogr.*, **133**, 134–149.

14 Reed, T.B. and Breck, D.W. (1956) Crystalline zeolites. II. Crystal structures of synthetic zeolite, type A. *J. Am. Chem. Soc.*, **78**, 5972–5977.

15 Corma, A., Rey, F., Rius, J., Sabater, M.J., and Valencia, S. (2004) Supramolecular self-assembled molecules as organic directing agent for synthesis of zeolites. *Nature*, **431**, 287–290.

16 Garcia-Sanchez, A., Garcia-Perez, E., Dubbeldam, D., Krishna, R., and Calero, S. (2007) A simulation study of alkanes in Linde type A zeolites. *Adsorp. Sci. Technol.*, **25**, 417–427.

17 Breck, D.W., Eversole, W.G., Milton, R.M., Reed, T.B., and Thomas, T.L. (1956) Crystalline zeolites. I. Properties of a new synthetic zeolite, type A. *J. Am. Chem. Soc.*, **78**, 5963–5971.

18 Baur, W.H. (1964) On the cation and water positions in faujasite. *Am. Mineral.*, **49**, 697–704.

19 Bergerhoff, G., Baur, W.H., Nowacki, and Über, W. (1958) die Kristallstrukturen des Faujasits. *N. Jb. Miner. Mh*, 193–200.

20 Frising, T. and Leflaive, P. (2008) Extraframework cation distributions in X and Y faujasite zeolites: a review. *Micropor. Mesopor. Mat.*, **114**, 27–63.

21 Meier, W.M. (1961) The crystal structure of mordenite (ptilolite). *Z. Kristallogr.*, **115**, 439–450.

22 Jacobs, P.A. and Martens, G.A. (1987) *Synthesis in High-Silica Aluminosilicate Zeolites*, Elsevier, Amsterdam, p. 321.

23 Mortier, W.J., Pluth, J.J., and Smith, J.V. (1975) Positions of cations and molecules in zeolites with the mordenite-type framework. I. Dehydrated calcium-exchanged ptilolite. *Mat. Res. Bull.*, **10** (10), 1037–1045.

24 Raatz, F., Marcilly, C., and Freund, E. (1985) Comparison between small port and large port mordenites. *Zeolites*, **5** (5), 329–333.

25 Van Geem, P.C., Scholle, K.F.M.G.J., Van der Velden, G.P.M., and Veeman, W.S. (1988) Study of the transformation of small-port into large-port mordenite by magic-angle spinning NMR and infrared spectroscopy. *J. Phys. Chem.*, **92** (6), 1585–1589.

26 Moreno, S., and Poncelet, G. (1997) Dealumination of small- and large-port mordenites: a comparative study. *Microporous Mat.*, **12** (4/6), 197–222.

27 Sand, L.B. (1968) Synthesis of large-port and small-port mordenites. *Mol. Sieves, Pap. Conf.*, **1967**, 71–77.

28 Freund, E., Marcilly, C., and Raatz, F. (1982) Pore opening of a small-port mordenite by air-calcination. *J. Chem. Soc., Chem. Commun.*, **5**, 309–310.

29 Stach, H., Jaenchen, J., Jerschkewitz, H.G., Lohse, U., Parlitz, B., Zibrowius, B., and Hunger, M. (1992) Mordenite acidity: dependence on the silicon/aluminum ratio and the framework aluminum topology. 1. Sample preparation and physicochemical characterization. *J. Phys. Chem.*, **96** (21), 8473–8479.

30 Dent, L.S. and Smith, J.V. (1958) Crystal structure of chabazite, a molecular sieve. *Nature*, **181**, 1794–1796.

31 Smith, J.V., Rinaldi, F., and Dent Glasser, L.S. (1963) Crystal structures with a chabazite framework. II. Hydrated

Ca-chabazite at room temperature. *Acta Crystallogr.*, **16**, 45–53.

32. Tillmanns, E., Fischer, R.X., and Baur, W.H. (1984) Chabazite-type framework in the new zeolite willhendersonite, KCaAl3Si3O12·5H2O. *Neues Jahrb. Mineral. Monats.*, **12**, 547–558.

33. Diaz-Cabanas, M.J., Barrett, P.A., and Camblor, M.A. (1998) Synthesis and structure of pure Si O2 chabazite: the Si O2 polymorph with the lowest framework density. *Chem. Commun.*, **1998**, 1881–1882.

34. Kokotailo, G.T. and Lawton, S.L. (1964) Possible structures related to gmelinite. *Nature*, **203** (4945), 621–623.

35. Smith, J.V. and Bennett, J.M. (1981) Enumeration of 4-connected 3-dimensional nets and classification of framework silicates: the infinite set of ABC-6 nets; the Archimedean and. sigma.-related nets. *Am. Mineral.*, **66** (7–8), 777–788.

36. Akporiaye, D.E. (1992) Structural relations of zeolite frameworks: ABC-6 and other frameworks derived from hexagonal 4.6.12 and (4.6.8)1(4.8.12)1 3-connected nets. *Zeolites*, **12** (2), 197–201.

37. Plevert, J., Kirchner, R.M., and Broach, R.W. (1999) Faulting effects in the CHA-GME group of ABC-6 materials. Proceedings of the International Zeolite Conference, 12th, Baltimore, July 5–10, 1998. Meeting Date 1998, 4, pp. 2445–2452.

38. McGuire, N.K., Bateman, C.A., Blackwell, C.S., Wilson, S.T., and Kirchner, R.M. (1995) Structure refinement, electron microscopy, and solid-state magic angle spinning nuclear magnetic resonance characterization of AlPO4-52: an aluminophosphate with a large cage. *Zeolites*, **15** (5), 460–469.

39. Skeels, G.W., Sears, M., Bateman, C.A., McGuire, N.K., Flanigen, E.M., Kumar, M., and Kirchner, R.M. (1999) Synthesis and characterization of phi-type zeolites LZ-276 and LZ-277: faulted members of the ABC-D6R family of zeolites. *Micropor. Mesopor. Mat.*, **30** (2–3), 335–346.

40. Szostak, R. and Lillerud, K.P. (1994) Babelite: the random member of the ABC-D6R family of zeolites. *J. Chem. Soc. Chem. Commun.*, **20**, 2357–2358.

41. Kokotailo, G.T., Lawton, S.L., Olson, D.H., and Meier, W.M. (1978) Structure of synthetic zeolite ZSM-5". *Nature*, **272**, 437–438.

42. Olson, D.H., Kokotailo, G.T., Lawton, S.L., and Meier, W.M. (1981) Crystal structure and structure-related properties of ZSM-5. *J. Phys. Chem.*, **85**, 2238–2243.

43. van Koningsveld, H., van Bekkum, H., and Jansen, J.C. (1987) On the location and disorder of the tetrapropylammonium (TPA) ion in zeolite ZSM-5 with improved framework accuracy. *Acta Crystallogr.*, **B43**, 127–132.

44. Mentzen, B.F. (2007) Crystallographic determination of the positions of the monovalent H, Li, Na, K, Rb, and Tl cations in fully dehydrated MFI Type Zeolites. *J. Phys. Chem. C*, **111** (51), 18932–18941.

45. Calleja, G., Pau, J., and Calles, J.A. (1998) Pure and multicomponent adsorption equilibrium of carbon dioxide, ethylene, and propane on ZSM-5 zeolites with different Si/Al ratios. *J. Chem. Eng. Data*, **43** (6), 994–1003.

46. Vroon, Z.A.E.P., Keizer, K., Gilde, M.J., Verweij, H., and Burggraaf, A.J. (1996) Transport properties of alkanes through ceramic thin zeolite MFI membranes. *J. Membr. Sci.*, **113** (2), 293–300.

47. Tuan, V.A., Falconer, J.L., and Noble, R.D. (2000) Isomorphous substitution of Al, Fe, B, and Ge into MFI-zeolite membranes. *Micropor. Mesopor. Mat.*, **41** (1–3), 269–280.

48. Ciavarella, P., Moueddeb, H., Miachon, S., Fiaty, K., and Dalmon, J.-A. (2000) Experimental study and numerical simulation of hydrogen/isobutane permeation and separation using MFI-zeolite membrane reactor. *Catal. Today*, **56** (1–3), 253–264.

49. Beerdsen, E., Dubbeldam, D., Smit, B., Vlugt, T.J.H., and Calero, S. (2003) Simulating the effect of nonframework cations on the adsorption of alkanes in MFI -type zeolites. *J. Phys. Chem. B*, **107** (44), 12088–12096.

50. Barrer, R.M. and Villiger, H. (1969) The crystal structure of the synthetic zeolite L. *Z. Kristallogr.*, **128**, 352–370.

51 Higgins, J.B., LaPierre, R.B., Schlenker, J.L., Rohrman, A.C., Wood, J.D., Kerr, G.T., and Rohrbaugh, W.J. (1988) The framework topology of zeolite beta. *Zeolites*, **8**, 446–452.

52 Newsam, J.M., Treacy, M.M.J., Koetsier, W.T., and de Gruyter, C.B. (1988) Structural characterization of zeolite beta. *Proc. R. Soc. Lond. A*, **420**, 375–405.

53 Conradsson, T., Dadachov, M.S., and Zou, X.D. (2000) Synthesis and structure of (Me3N)6[Ge32O64](H2O)4.5, a thermally stable novel zeotype with 3D interconnected 12-ring channels. *Micropor. Mesopor. Mat.*, **41**, 183–191.

54 Wright, P.A., Zhou, W., Pérez-Pariente, J., and Arranz, M. (2005) Observation of growth defects in zeolite beta. *J. Am. Chem. Soc.*, **127** (2), 494–495.

55 Jansen, C., Creyghton, E.J., Njo, S.L., van Koningsveld, H., and van Bekkum, H. (1997) On the remarkable behaviour of zeolite Beta in acid catalysis. *Catal. Today*, **38** (2), 205–212.

56 Leonowicz, M.E., Lawton, J.A., Lawton, S.L., and Rubin, M.K. (1994) MCM-22: a molecular sieve with two independent multidimensional channel systems. *Science*, **264**, 1910–1913.

57 Russo, P.A., Carrott Ribeiro, M.M.L., Carrott, P.J.M., Matias, P., Lopes, J.M., Guisnet, M., and Ribeiro, F.R. (2009) Characterisation by adsorption of various organic vapours of the porosity of fresh and coked H- MCM–22 zeolites. *Micropor. Mesopor. Mat.*, **118** (1–3), 473–479.

58 Lawton, S.L., Leonowicz, M.E., Partridge, R.D., Chu, P., and Rubin, M.K. (1998) Twelve-ring pockets on the external surface of MCM-22 crystals. *Micropor. Mesopor. Mat.*, **23**, 473–479.

59 Pawlesa, J., Bejblova, M., Sommer, L., Bouzga, A.M., Stocker, M., and Cejka, J. (2007) Synthesis, modification and characterization of MWW framework topology materials. Stud. Surf. Sci. Catal., vol. 170A, Elsevier, Amsterdam, (From Zeolites to Porous MOF Materials), pp. 610–615.

60 Camblor, M.A., Corma, A., Diaz-Cabanas, M.-J., and Baerlocher, C. (1998) Synthesis and structural characterization of MWW type zeolite ITQ-1, the pure silica analog of MCM-22 and SSZ-25. *J. Phys. Chem. B*, **102** (1), 44–51.

61 Polato, C.M.S., Henriques, C.A., Perez, C.A., and Monteiro, J.L.F. (2004) Alkali cations exchange in MCM-22. Stud. Surf. Sci. Catal., vol. 154B, Elsevier, Amsterdam, (Recent Advances in the Science and Technology of Zeolites and Related Materials), pp. 1912–1919.

62 Smith, J.V. and Rinaldi, F. (1962) Framework structures formed from parallel four- and eight-membered rings. *Mineral. Mag.*, **33** (258), 202–212.

63 Smith, J.V. (1977) Enumeration of 4-connected 3-dimensional nets and classification of framework silicates. I. Perpendicular linkage from simple hexagonal net. *Am. Mineral.*, **62** (7–8), 703–709.

64 Han, S. and Smith, J.V. (1999) Enumeration of four-connected three-dimensional nets. III. Conversion of edges of three-connected two-dimensional nets into saw chains. *Acta Crystallogr. A*, **A55** (2 pt 2), 360–382.

65 Foster, M.D. and Treacy, M.M.J. (2009) A Database of Hypothetical Zeolite Structures, http://www.hypotheticalzeolites.net (accessed June 2009).

66 Treacy, M.M.J., Foster, M.D., and Rivin, I. (2008) Towards a catalogue of designer zeolites. *Turning Points Solid State Mat. Surf. Sci.*, 208–220.

67 Treacy, M.M.J., Rivin, I., Balkovsky, E., Randall, K.H., and Foster, M.D. (2004) Enumeration of periodic tetrahedral frameworks. II. Polynodal graphs. *Micropor. Mesopor. Mat.*, **74**, 121–132.

68 Treacy, M.M.J., Randall, K.H., Rao, S., Perry, J.A., and Chadi, D.J. (1997) Enumeration of periodic tetrahedral frameworks. *Z. Kristallogr.*, **212**, 768–791.

69 Li, Y., Yu, J., Liu, J., Yan, W., Xu, R., and Xu, Y. (2003) Design of zeolite frameworks with defined pore geometry through constrained assembly of atoms. *Chem. Mat.*, **15**, 2780–2785; see also http://mezeopor.jlu.edu.cn/ (accessed June 2009).

70 Le Bail, A. (2005) Inorganic structure prediction with GRINSP. *J. Appl. Cryst.*, **38**, 389–395; see also http://sdpd.univ-lemans.fr/cod/pcod/index.html (accessed June 2009).

71 Falcioni, M., and Deem, M.W. (1999) A biased Monte Carlo shceme for zeolite structure solution. *J. Chem. Phys.*, **110** (3), 1754–1766.

72 Treacy, M.M.J., Rao, S., and Rivin, I.A., Combinatorial method for generating new zeolite frameworks, in *Proceedings of the 9th International Zeolite Conference Montreal 1992* (eds R. Von Ballmoos, J.B. Higgins, and M.M.J. Treacy), Butterworth-Heinemann, Stoneham, Massachusetts, pp. 381–388.

3
Synthesis of Zeolites and Manufacture of Zeolitic Catalysts and Adsorbents

Robert L. Bedard

3.1
Introduction

This chapter outlines general procedures for synthesizing and scaling-up zeolites and aluminophosphate molecular sieves and conditions to avoid during reagent mixing and digestion. Typical synthetic procedures are described for both types of materials and important synthetic parameters are introduced that must be considered and controlled. Important chemical and engineering issues relating to scale-up are outlined including raw materials, reaction mixture make-up, stirring, heating methods and solid–liquid separation.

Forming operations are described from both a chemical and ceramic engineering perspective, with emphasis on design of structures with desired mass transfer and strength characteristics. Specific forming methods that are discussed include extrusion, bead forming and spray-drying. Newer forms will also be introduced.

Catalyst and adsorbent finishing procedures are outlined including preparation of binder-free forms, zeolite stabilization procedures, ion exchange and metal loading procedures, and drying and firing operations.

Finally, new developments in zeolite catalyst and adsorbent manufacture will be outlined in the form of a survey of recent open and patent literature. Patents are extensively cited as reference materials for this chapter; however no effort is made to identify specific manufacturing techniques actually used by any particular company to manufacture zeolites or any catalytic or adsorbent products. This is because there is no way to determine whether manufacturing patents are practiced as is or whether further refinements have been accomplished subsequent to the filing of the patents that may be held by specific companies as trade secrets.

The health, safety and environmental (HS&E) aspects of manufacturing zeolites and the ultimate catalyst and adsorbent materials derived from them are the most important issues with respect to a maintaining a sustainable zeolite business. A recent paper treats some of these HS&E issues in a cursory way [1]. Treatment of this subject is beyond the scope of this chapter, but the unique HS&E issues

Zeolites in Industrial Separation and Catalysis. Edited by Santi Kulprathipanja
Copyright © 2010 WILEY-VCH Verlag GmbH & Co. KGaA, Weinheim
ISBN: 978-3-527-32505-4

pertaining to each product and manufacturing situation needs to be carefully studied and mapped out to assure, at the minimum, full compliance with applicable occupational safety regulations and environmental laws.

3.2
Synthesis of Zeolites and Aluminophosphate Molecular Sieves

It is a daunting task to summarize the synthetic knowledge available to the scientist and engineer on the hydrothermal synthesis of zeolites; and insufficient space is available here. Numerous high quality books and monographs as well as publications from the International Zeolite Association have treated the important aspects of zeolite synthesis [2–7]. Historical background is provided in Chapter 1. Significantly more information is available in the patent literature, although the experimental detail given in patents sometimes gives insufficient information to either effectively reproduce the work or determine if the invention is a significant advance. It is accurate to say that more manufacturing information is present in the patent literature than the open literature, which makes careful study of these legal documents a necessity when a complete picture of this field is desired. Synthetic recipes are predominantly given for small bench-scale syntheses, although there are a number of patents that list larger-scale procedures [8].

Synthetic zeolites and other molecular sieves are important products to a number of companies in the catalysis and adsorption areas; and numerous applications, both emerging and well-established, are encouraging the industrial synthesis of the materials. There are currently no more than a few dozen crystalline microporous structures that are widely manufactured for commercial use, in comparison to the hundreds of structures that have been made in the laboratory. See Chapter 2 for details on zeolite structures. The highest volume zeolites manufactured are two of the earliest-discovered materials: zeolite A (used extensively as ion exchangers in powdered laundry detergents) and zeolite Y (used in catalytic cracking of gas oil).

3.2.1
Hydrothermal Synthesis – The Key to Metastable Phases

Hydrothermal synthesis refers to the formation of materials in a primarily aqueous solvent. The water solubilizes the components of the reagent mixture to varying degrees, develops the concentration and pH-dependent speciation of each framework component and aids in ultimate stabilization of the crystalline microporosity by coordinating with charge-balancing cations in the final product and by void filling part of the resulting microporosity. The water solvent is cheap and readily available, easy to recycle and not an issue with respect to release of non-contaminated quantities into the environment. For this reason nearly all manufacturing of zeolites and other oxide molecular sieves is carried out in water; and even in the case of the manufacture of emerging microporous materials, such as metal organic frameworks, the economic and most environmentally sound first choice for reaction solvent will likely be water.

With the landmark discovery of synthetic zeolites at Union Carbide Corporation in the early 1950s, a new area of science and technology was born [9–11]. Today millions of kilos of synthetic zeolites are manufactured at a number of catalyst and adsorbent companies throughout the world using synthesis recipes that are modifications of those used by the early pioneers. A significant effort had already been made prior to Milton and Breck's discoveries, notably by the group of Richard Barrer, to understand hydrothermal chemistry and its relationship to the synthesis of metastable zeolite phases [12–16]. This early work largely involved high temperatures and pressures and relatively low-reactivity precursors, which limited the types of zeolite frameworks that could be formed. The key discovery of using reactive gels as aluminosilicate precursors and digesting at lower temperatures had not been anticipated prior to Robert Milton's and Donald Breck's work starting in the late 1940s at Union Carbide.

Similarly, reactive oxide mixtures are also used to synthesize aluminophosphate molecular sieves, usually starting from phosphoric acid along with the addition of alumina and silica sources analogous to those used in zeolite synthesis with a notable exception; alkylammonium salts and amines were ultilized in structure-directing and space filling to the exclusion of alkali hydroxide solutions and alkali metal salts.

Shortly after their discovery, work began at Union Carbide to scale-up the syntheses of zeolites A and X, and later zeolite Y to yield plant-compatible formulations suitable for large-scale manufacture. It became apparent as manufacturers were attempting these that a number of factors can complicate scale-up efforts [17]. One of the most challenging complications is the fact that crystallization fields of different zeolites can overlap significantly, causing a particular gel composition to form several zeolites depending on the crystallization parameters [2]. Overlapping crystallization fields also occur in aluminophosphate molecular sieves [18]. Several aluminophosphate-based molecular sieves have also been scaled up and similar complications with respect to the larger scale can be expected in these syntheses.

3.2.2
Typical Zeolite Syntheses

A number of reagents containing oxide components are used in zeolite manufacture [19]. Silica is provided by addition of sodium or other alkali silicate solutions, precipitated, colloidal, or fumed silica, or tetraalkylorthosilicate (alkyl = methyl, ethyl) and certain mineral silicates such as clays and kaolin. Alumina is provided as sodium aluminate, aluminum sulfate solution, hydrous aluminum oxides such as pseudo boehmite, aluminum nitrate, or aluminum alkoxides. Additional alkali is added as hydroxide or as halide salts, while organic amines and/or alkylammonium species are added as neat amines, solutions of hydroxides, or halide salts.

Technical grade reagents that are frequently used in manufacturing generally are not as pure as those used in laboratory syntheses, nor is there as much quality control in terms of actual composition of reagent solutions or freshness of solution reagents such as alkali hydroxide, silicate and aluminate solutions. Caustic solutions can age rapidly if storage precautions are not taken. Absorption of CO_2 can result in pH and reactivity changes in hydroxide solutions or precipitation of hydroxyl-carbonates or aluminum hydroxides from aluminate solutions. Unreactive impurities such as quartz or cristobalite can be present in silicate reagents, which may facilitate nucleation of impurity phases. High levels of iron impurities in aluminate or silicate can affect the catalytic reactivity or optoelectronic properties (in the case of certain emerging applications) of manufactured zeolites.

Care must be taken to account for each component in the formulation when specific oxide reagent ratios are needed to produce a desired zeolite phase. Water stoichiometry in a zeolite gel can be critical, not only to produce the desired phase but also because of its effect on crystal size and product morphology [20]. Many reagents bring significant additional water beyond the free water added as solvent which cannot be ignored in reagent calculations. Reagent interactions must be accounted for, particularly where acid base reactions take place as the reaction mixture is formulated, since the final mixture pH is an important factor in speciation of the solution components. Hydroxide ion is understood to fulfill a mineralizing and complexing role in zeolite crystallization; and its concentration can have dramatic affects on reaction rates [21, 22]. The presence of alcohols as a result of hydrolysis of alkoxide reagents may, under some circumstances, cause problems with nucleation or crystallization, potentially requiring procedures that mitigate the problem, such as alcohol removal via distillation prior to digestion.

The heats of reaction/solution of some reagents as hydrolysis/dissolution takes place can cause substantial elevation in slurry/solution temperatures, particularly at a large scale where heat transfer and radiative cooling are not nearly as efficient as it is in small laboratory vessels. Other reagents, such as certain sodium aluminates and particularly reagents that are not freshly prepared, may need elevated temperatures for full dissolution in water. These hot or very warm solutions can adversely affect early nucleation conditions in some zeolite syntheses. Hot reagent solutions and mixtures are sometimes cooled prior to their addition to other reagents to better control the early reactions and speciation of aluminosilicate and silicate precursors.

In a typical aluminosilicate zeolite synthesis, sodium aluminate is dissolved in water along with some fraction of the additional sodium hydroxide that is needed in the reactant mixture. Separately, sodium silicate is mixed with the remainder of the sodium hydroxide. The two solutions are combined using the required mix order and agitation level, resulting in the initial zeolite gel. In some cases this initial gel is aged at an intermediate temperature for a time to allow evolution of the gel chemistry and perhaps initial nucleation of the system [2, 3, 10, 23]. The reaction mixture is then digested at a higher temperature, usually between 50 and 200 °C for a prescribed time until the desired level of product crystallinity is reached. Over-digestion of some zeolite reaction mixtures is possible, for instance converting the desired phases such as zeolite A or X to zeolite P (gismondine type) [24, 25]. Of course, to call the above procedure a "typical synthesis" is an over-simplification, since many approaches and variations to zeolite synthesis have been developed over the years, including at the manufacturing scale.

There are a number of patents where "seed" or nucleation solutions are prepared and added to the zeolite reaction mixture, starting with work outlined by W.R. Grace in several patents [26–28]. It is not clear which if any of these nucleation sols have a crystalline structure or how they effect the nucleation kinetics of the crystallization process. Some of these nucleation sols can facilitate the crystallization of more than one zeolite structure. One example of nucleation sol can be used to produce A-, X-, or Y-type zeolites [29]. Utilizing these nucleation sols can complicate the process of gel formulation, resulting in multi-step procedures of dissolving, mixing and digesting multiple precursor mixtures and carefully controlling the addition of each of the mixtures to generate the final gel. However, the use of nucleation enhancing sols appears to allow better product morphological control as well as faster crystallization, which can provide significant cost benefit as well as more uniform product properties. Another apparent advantage of nucleation solutions is the ability in some cases of preparing stoichiometric zeolite gels, which reduces waste silicate in the manufacturing. The number of patents in this area underscores the apparent utility of nucleation sols in manufacturing [30–33].

A number of zeolite syntheses result in incomplete conversion of all of the gel components to solid zeolite. In many cases the reactant Si/Al ratio is different than that of the product (usually higher), resulting in a silicate solution remaining behind. At the industrial scale this silicate solution is often recycled to minimize waste and raw material cost [34–36].

3.2.3
Important Synthesis Parameters – Zeolites

The order of mixing the reagents into the final slurry or gel can be critical. One order of mixing may give a coarse slurry while a reversed order of mixing of the same reagents may give a thick gel. Complex reagent addition procedures have sometimes been developed, including addition of supplementary reagents after crystallization to obtain a second crystallization of unreacted gel components [37]. The rheology of the resulting mixture may have a bearing on the ultimate nucleation conditions as well as on the power needed to stir the reaction mixture. If insufficient power is used for stirring or inappropriate shear conditions are used an undesired phase may nucleate during the digestion process or the product mixture may be at least partially amorphous. Some reaction mixtures give a different phase under high shear versus lower shear mixing [38–40].

Inadequate stirring power may also give rise to inhomogeneous regions of the reaction gel both in terms of composition and temperature, leading to non-uniform nucleation conditions in different regions of the gel which can lead to unwanted product mixtures or particle size and/or morphological variations [41, 42]. The presence of hot surfaces coupled with inadequate mixing can give rise to hot spots, which could result in the nucleation of denser phase impurities or amorphous glassy gel in the product. Because of the critical effects of stirring on the outcome of zeolite syntheses, a detailed understanding of agitation scale-up factors is paramount [43].

The method of heating a reaction mixture can also be a critical factor. Obviously, large kettles and their contents can be more difficult to heat as quickly as laboratory-sized reactors. Varying engineering solutions can be brought to bear to mitigate these problems. These solutions include steam sparging, kettles with steam-heated walls and the pumping of reagent mixtures through heat exchangers [44, 45]. This heat transfer problem can also arise in the cooling of reaction mixtures after crystallization, where products may remain in hot solutions for longer than desired due to unfavorable cooling rates of large-scale equipment. Gel heat-up/cool-down rate factors combine with agitation issues to create challenges in zeolite manufacturing scale-up. Seeding with crystalline zeolite is sometimes used in zeolite synthesis to mitigate other scale-up difficulties [46–52], although in certain cases a zeolite with a different structure than the seed material can crystallize, perhaps by heterogeneous nucleation [53].

3.2.4
Typical Aluminophosphate Syntheses

Many reagents used in aluminophosphate, or AlPO, synthesis are similar to those used in zeolite synthesis. Alumina can be provided as aluminum sulfate solution, hydrous aluminum oxides such as pseudo boehmite, aluminum nitrate, or aluminum alkoxides. Alkali aluminates and alkali reagents in general are not used in aluminophosphate manufacturing to avoid formation of dense alkali aluminophosphate phases. Phosphorous is almost always added as phosphoric acid, but can also be provided as phosphorous oxides. Silica is provided by precipitated, colloidal, or fumed silica, or tetraethylorthosilicate. The elemental richness of the aluminophosphate-based molecular sieves can include numerous elements from main group, alkaline earth and transition metals, including magnesium, zinc, cobalt, manganese, gallium, germanium, etc. These elements can be added to reaction mixtures as oxides, hydrous oxides, or soluble nitrate, sulfate, or halide salts. Some studies have added transition metals as sparingly soluble carbonates [54].

There is limited patent literature available on manufacturing techniques for aluminophosphates. Although many patents describe AlPO synthesis, most described examples are small-scale preparations. The fact that at least two catalytic applications have been commercialized for SAPO molecular sieves indicates that they have been scaled-up to large quantities [55, 56]. A large-scale preparation of SAPO-34 is described in a recent patent [57].

In a typical aluminophosphate synthesis, 85% phosphoric acid is diluted with a predetermined amount of water and then a calculated amount of hydrated aluminum oxide, usually a pseudoboehmite, AlOOH, is added and stirred until homogeneous. A predetermined amount of tetraalkylammonium hydroxide solution is added to the slurry and the mixture is once again stirred until homogeneous. The Al/P atomic ratio in the reaction mixture is usually close to 1, although in certain cases some benefit with respect to purity of products can be derived by a slight excess of phosphorus since unreacted phosphate is usually in a soluble form while unreacted aluminum oxide is often present as a sparingly soluble precipitate. The pH conditions of aluminophosphate synthesis is usually significantly lower than that of zeolite synthesis (pH 3–10), although variation in pH to higher values may facilitate synthesis of silicon-containing frameworks, known as SAPOs.

3.2.5
Important Synthesis Parameters – Aluminophosphates

The synthesis of aluminophosphates has been actively studied for almost 30 years, and an extensive open and patent literature has developed during that time. However, most of the manufacturing information is likely still trade secret to the companies who have been involved.

Heat-up rate effects have been investigated with respect to microwave synthesis of AlPO phases, however there are few publications concerning the heat-up effects in conventional heating [58]. There has also been at least one study of pH and H_2O level on aluminophosphate crystallization [59]. A recent paper attempts to study the unique crystallization process of several aluminophosphate molecular sieve compositions [60].

3.2.6
Dewatering, Filtration and Washing of Molecular Sieve Products

An important aspect of molecular sieve manufacturing is the separation of the crystalline product from the mother liquor and subsequent washing of the material to prepare it for further processing. Solid–liquid separation can be accomplished in a number of ways usually related either to filtration or centrifugation. Equipment that is used include belt and drum filters, plate and frame filter presses, basket, screen, bowl and other types of centrifuges. Washing is usually done on the filtration equipment although removal of the filter cake and reslurrying can be beneficial if the process does not increase the cost unnecessarily. Inorganic and organic flocculants similar to those used in the water treatment industry may also be used to facilitate filtration and dewatering of the zeolite product [61]. More recently, microfiltration or ultrafiltration of zeolite slurries, specifically for purification and concentration of micro- or nanocrystalline zeolites, has been described [62, 63].

Zeolites can be over-washed during workup, resulting in decationization, also known as cation hydrolysis [64]. Decationization involves replacement of alkali cations with H_3O^+ cations, which also generates hydroxide anions in the surrounding solution. Decationization is usually more of a problem with low-silica zeolites that are most often used as adsorbent products, in which the cation composition of the final product is often critical to the adsorption application. Additional problems can arise because of increased acidity from the protonic sites generated by decationization. Acid sensitive molecules, such as olefins and certain aromatics, can be inadvertently subjected to catalytic transformations during purification processes if the adsorbent materials were decationized during manufacture. Decationization can be reversed in many instances by treatment of the zeolite with dilute alkali solutions at some point in the manufacturing process.

3.3
Forming Zeolite Powders into Usable Shapes

Many procedures that are used in zeolite catalyst and adsorbent manufacture are common to other manufactured products. The unit operations and equipment that are used can therefore be found throughout the catalyst and adsorbent manufacturing industries. There are a number of books and monographs that outline equipment and manufacturing procedures that are relevant to zeolite catalysts and adsorbents [65–68].

3.3.1
Chemical Engineering Considerations in Zeolite Forming

The necessity of forming zeolite powders into larger particles or other structures stems from a combination of pressure drop, reactor/adsorber design and mass transfer considerations. For an adsorption or catalytic process to be productive, the molecules of interest need to diffuse to adsorption/catalytic sites as quickly as possible, while some trade-off may be necessary in cases of shape- or size-selective reactions. A schematic diagram of the principal resistances to mass transfer in a packed-bed zeolite adsorbent or catalyst system is shown in Figure 3.1 [69].

The speed of provision of the feed molecules to the adsorption/catalytic sites must be balanced with engineering issues such as pressure drop in a reactor/adsorber, so the particle size and pore structure of engineered forms must be optimized for each application. A hierarchy of diffusion mechanisms interplays in processes using formed zeolites. Micropore, molecular, Knudsen and surface diffusion mechanisms are all more or less operative, and the rate limiting diffusion mechanism in each case is directly affected by synthesis and post-synthesis manufacturing processes. Additional details are provided in Chapter 9.

Zeolite particles are incorporated into a number of different engineered forms, including small spherical particles for fluidized bed applications and small granules for powdered detergents. Larger forms include extruded pellets with various cross-sectional shapes and beads made by bead-forming processes.

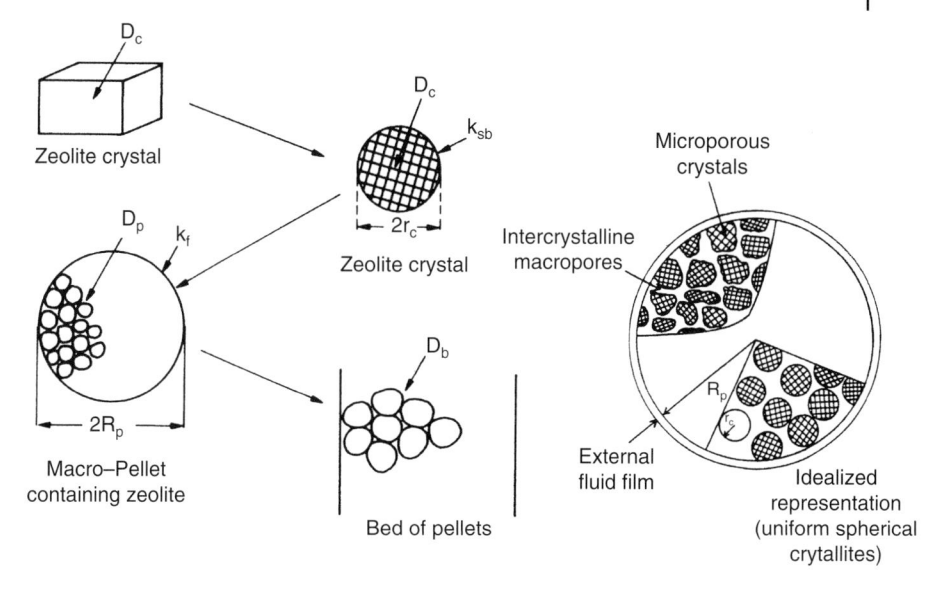

Figure 3.1 Principal mass transfer resistances in a packed bed of zeolite beads [69].

Zeolite catalysts and adsorbents have also been incorporated into monolithic contactors by several routes, including extruded zeolite/binder composites [70], wash-coated ceramic monoliths [71] and corrugated thin-sheet monoliths [72].

3.3.2
Ceramic Engineering Considerations in Zeolite Forming

Many engineered forms, particularly those that are clay, alumina, silica, or aluminophosphate-bound composites, are in essence specialized ceramic articles, albeit low-density ones. The forming operation is facilitated by a working knowledge of the surface chemistry of the various components of the final composite. For instance, a detailed understanding of the surface charge of the various particles involved as a function of pH, adsorbed ions and other factors can dramatically reduce the amount of trial and error involved in developing a successful formulation [73, 74].

The bonding forces that hold a zeolite catalyst or adsorbent particle together are created by high-temperature treatment of hydrated oxides and hydroxylated zeolite surfaces and are presumed to be hydroxide and perhaps oxide bridges between the zeolite and binder components. Under the high temperatures of calcination, binder components can undergo phase transformations to crystalline forms that sometimes are stronger than the original amorphous hydrous oxide or hydroxide. Forming unit operations are similar to those used in the ceramics industry, including extrusion, bead forming and slurry casting in cases where zeolite powders are coated on surfaces.

3.3.3
Bound Zeolite Forms

Extrusion, bead forming and spray-drying are the most common methods for bonding of zeolite powders. Binders can include amorphous aluminosilicates, kaolin and montmorillonite type clays[75, 76], alumina[77–79], silica[80, 81], titania, zirconia[82] and amorphous aluminophosphate [83]. The zeolite powder–binder composite used in extrusion or bead forming is usually formulated using a kneading machine called a wet muller. Sometimes extrusion/forming aids are added to facilitate a smooth extrusion or bead forming operation [84]. Dissolvable or combustible void-forming agents have also been formulated into bound forms to produce additional porosity to facilitate mass transfer in the final zeolite form [85, 86]. Sol-gel oil dropping is also used as a means to bind zeolites into strong, attrition-resistant forms [87].

Spray drying requires formulation of a relatively stable, usually aqueous dispersion of zeolite and colloidal binder particles, which are then fed to the spray drier [88, 89]. Organic dispersing agents may be added to the spray drier feed slurry to stabilize the dispersion while the slurry is fed to the drier.

Organic polymers and resins have also been used for zeolite binding. An early example is the use polyurethane in the formation of vibration-resistant zeolite porous bodies for refrigerant drying [90]. Organic binders such as cellulose acetate and other cellulose-based polymers have also used to mitigate problems with binder dissolution in aqueous phase separations [91, 92]. Latex has also been used as a water-stable organic binder [93]. More recently, thermoplastic resins, such as polyethylene have also been used as binders for zeolites [94].

3.3.4
Other Zeolite Forms – Colloids, Sheets, Films and Fibers

In recent years zeolites have been utilized in non-conventional forms which emphasize certain chemical or engineering characteristics of the materials. Novel forms include thin sheets made by coating processes and paper-making technology and zeolite-containing fibers.

Thin zeolite sheets offer improved mass transfer for possible rapid cycle adsorption processes. One of the first studies of zeolite as fillers in paper-making was issued to NCR in 1955, although at that time the term "zeolite" was more often used to describe any ion exchanger whether or not it was actually a crystalline microporous zeolite [95]. A later patent described the incorporation of microporous zeolite powders in paper sheets [96]. More recently a number of patents described zeolite-containing papers in adsorption processes [97–99].

There are also reports of thin zeolite coatings on various surfaces other than ceramic monoliths. These coatings facilitate mass transfer for several applications including adsorption, catalysis and heat exchange. Heat exchange applications include coated dessicant heat exchange wheels and coated tube-shell heat exchangers for HVAC and refrigeration applications [100–102]. Zeolite-coated screens have been applied to evaporative emissions control, adsorptive separations and catalytically active surfaces [103, 104].

Zeolites have been incorporated into and attached to fibers of various compositions for a number of applications. The types of fibers that have been used include glass [105], polymers by addition to the monomer prior to polymerization [106], optical fibers [107], cellulose pulp [108] and self-supported hollow zeolite fibers by a templating route [109].

3.4
Finishing: Post-Forming Manufacturing of Zeolite Catalysts and Adsorbents

Most zeolite-containing products require multiple manufacturing steps, including further chemical transformation, modification, ion exchange, impregnation, forming, drying and calcination, prior to their application as adsorbents or catalysts. Selected manipulations are detailed in the next sections.

3.4.1
Post-Forming Crystallization

A series of zeolite products have appeared in the patent literature in which all or part of the crystallization process has been accomplished after the material was formed into its final shape. Binderless zeolite products fall into this category, and a number of precursors to this binderless technology has appeared in the older patent literature related to manufacturing zeolites from clays and other mineral sources. An early patent assigned to Minerals and Chemicals Phillip Corp., a precursor company to Engelhard, described the forming of dehydrated kaolin clay and alkali hydroxide into spray-dried, extruded, beaded, or other aggregates, followed by digesting the resulting forms in dilute alkali solutions, forming a strong shaped article composed essentially entirely of aggregated zeolite particles [110]. The method can provide substantial cost savings in raw materials and manufacturing time, although not all zeolites can necessarily be made by this pre-form method. Union Carbide filed on similar concepts shortly after Minerals and Chemicals Phillip [111].

Although the use of kaolin and other disilicate clays had been anticipated for making zeolites earlier, the use of reactive dehydrated clays and the pre-forming of the material into forms were not suggested [112]. A later patent by UOP taught the synthesis of mixed zeolites A and X by digesting pre-formed spheres of silica and alumina [113]. W.R. Grace incorporated their nucleation center technology to the perform idea by using them to facilitate conversion of preformed metakaolin bodies into zeolites X and Y [114]. Microspherical Y zeolite catalysts aimed at catalytic cracking were also prepared by W.R. Grace from spray-dried metakaolin [115]. A number of other companies have added to the art in this area, indicating significant application of this concept [116–118]. Additional development of this concept involves binding a zeolite powder made by conventional methods with a reactive dehydrated clay and then converting the clay to more of the zeolite, thereby generating a "binderless" form [119]. Zeolite monoliths have also been produced by the pre-form method [120].

3.4.2
Stabilization and Chemical Modification of Zeolites

Many zeolites cannot be directly synthesized in a composition that optimizes their catalytic properties. Improvements in these properties related to acidity, stability and porosity can be made by a variety of post-synthetic or secondary synthetic treatments. The two major goals for secondary synthesis treatments are generation of strongly acidic protonic sites and increase of the Si/Al ratio of the zeolite. Simple acid exchange of zeolites, particularly lower SiO_2 compositions such as A, X and Y, drastically destabilizes their frameworks because of the tendency of aluminum cations to increase their coordination and thereby migrate out of the zeolite framework. It is known that mere hydration of these zeolites after acid exchange results in structure collapse [121]. Many studies have attempted to mitigate dealumination by carrying out ammonium exchange followed by heating to dehydrate the zeolite and generate the protonic form.

Dealumination processes are usually used in conjunction with production of the acid form of zeolite Y for many catalytic applications, and zeolites A and X are in most cases no longer used in acid catalytic applications because the high amount of aluminum in their frameworks makes them difficult to stabilize using various dealumination techniques. Successful dealumination and at least partial annealing of defects, resulting from movement of silicon cations to the aluminum vacancies, can be assessed by measurement of the reduction of the unit cell size of the zeolite. This unit cell reduction is a consequence of the relative ionic radii of Al^{3+} (0.54 Å) and Si^{4+} (0.40 Å).

There are a myriad of ways to combine acid exchange, ammonium exchange, calcination and steaming treatments to create rich varieties of hydrothermally stable acidic zeolite materials that can be incorporated into catalysts [122–130]. Some of the original patents covering these methods are outlined in Table 3.1. The original ultrastable Y zeolite patent shows the effect of the multiple ion exchange and steaming treatments on the Si/Al ratio and unit cell size [123].

Various fluoride-containing reactants have also been used to dealuminate, dealuminate/reinsert silica and/or modify the acidity of zeolites for catalytic applications [132–137].

An additional way to effect the shape/size selectivity of zeolite catalysts was developed which entailed deposition of various silicon-containing deposits on the external surfaces of zeolites to create an additional diffusion barrier [131, 138].

Many of these treatments can be straightforwardly scaled-up to industrial quantities by adding steam to calcination furnace atmospheres and carrying out various forms of solution ion exchange or chemical treatment and/or acid extractions in suitable powder and granule contacting equipment. The products vary in stability and catalytic properties because the treated materials are no longer freshly crystal-

Table 3.1 Stabilization/dealumination treatments of zeolites.

United States patent	Date	Description	Reference
3257310	1966	Steam stabilization	[122]
3293192	1966	Ultrastable Y–ammonium exchange → calcination → ammonium exchange → calcination	[123]
3374057	1968	Ammonium exchange in the presence of aluminum salts	[124]
3383169	1968	Ammonium exchange with added acid	[125]
3402996	1968	Cation exchange → calcination cation redistribution → followed by cation exchange	[126]
3449070	1969	Acid or ammonium exchange then calcination	[127]
3506400	1970	Steaming followed by acid extraction	[121]
3929672	1975	Partial ammonium exchange → steaming → additional ammonium exchange	[128]
4503023	1985	Silicon substitution for aluminum via hexafluorosilicate treatment	[131]
5013699	1991	Ammonium exchange → steaming → low pH ammonium exchange	[129]
5242677	1993	Ammonium exchange → steaming → aluminum exchange → steaming → acid extraction	[130]

lized, pure zeolites [139]. Impurities such as non-framework aluminum, defects such as tetrahedral hydroxyl nests, etched crystal surfaces, compositional gradients and porosity in the mesopore range are all generated in varying amounts depending on the types of treatment employed and their sequence. Variation in manufacturing equipment design is also expected to give rise to differences in secondary synthesis products from one supplier to another.

There have also been a number of reports of realumination treatments of zeolites which purport to reinsert aluminum into sites in the zeolite framework. The first report was a vapor phase reaction of Silicalite with $AlCl_3$ [140]. Another method of realumination has been proposed as a recrystallization of the surface region of the zeolite in the presence of a basic alkali aluminate solution in a lower Si/Al ratio and then diffusion of the aluminum into the interior of the crystals by a hydrolysis process [141]. In another realumination treatment, zeolite containing non-framework aluminum is treated with aqueous ammonium hydroxide to cause reinsertion of the aluminum into defect sites such as hydroxyl nests. Some of these reports imply that the aluminum distributions, and therefore the acidity characteristics, are different depending on the nature of realumination treatment [142].

3.4.3
Ion Exchange and Impregnation

Often zeolite products must be ion exchanged into alkali metal, alkaline earth, transition element, or rare earth forms to generate the desired adsorbent materials for a given separation or purification. Catalytically active metals can also be introduced into zeolites by ion exchange. Most ion exchange processes are carried out on an engineered form of the zeolite to facilitate solid–liquid separation and further processing and also to minimize possible losses of expensive metal components that have been ion exchanged into the zeolite. Ion exchange selectivity series have been published in a number of reports, including seminal works by Breck, Barrer and Sherry [143–145].

It is usually critical in the case of ion exchange to have a well-developed ion exchange isotherm for the ions of interest in the zeolite product that is to be manufactured, particularly if a narrow range in cation stoichiometry is needed for the application. The isotherm enables the design of the ion exchange manufacturing process in terms of concentrations and temperatures needed as well as the efficiency of salt usage and throughput rate. Although the aforementioned seminal papers feature a number of ion exchange isotherms, it may be necessary to develop one for the specific manufacturing process being developed. Efficient ion exchange processes are important particularly when expensive noble metal is being loaded into the zeolite.

Impregnation of zeolite products is generally understood to be a process that is different from ion exchange. In impregnation, a zeolite-containing product is contacted with a concentrated metal salt solution which is then evaporated to dryness or drained off then evaporated to dryness. Most impregnation solutions introduce the metal as an anionic complex which has higher affinity for the binders in the zeolite engineered forms, such as alumina, than for the zeolite powder contained in the engineered form. This affinity is for at least two reasons, the main one being that there is usually considerably more external surface area associated with the binder in a bound zeolite form than with external surface of the zeolite. The anionic complex of the metals have little tendency to enter the zeolite pores, however surface reactions of the anionic species on the binders can generate cations that can then enter the zeolites by ion exchange.

Another reason that anions do not have a high affinity for zeolite surfaces is because of the negative surface charge generally associated with template-free zeolites [73]. Impregnation solutions of cations or cationic complexes can also be used in the case where there is low ion exchange capacity in the zeolite being used. In this case the metal distributes itself both inside the zeolite and on the binder. There may also be cases in which an anionic complex is used in impregnation to purposely avoid its exchange into the zeolite, for instance if there is a desire not to generate zeolitic acid sites which arise upon reduction of exchanged metal ions.

3.4.4
Drying and Firing

Drying and firing processes are essential parts of all zeolite product manufacturing. Given the considerations above regarding hydrothermal and thermal stability, it is clear the precise control of calcination temperature and atmospheres are important manufacturing considerations, particularly since a number of the dealumination and stabilization treatments outlined above require calcination or steam calcination as a step in the manufacturing process. Another important function of the calcination step of manufacture is in setting the binder to form a strong engineered form that will be attrition-resistant in applications such as fluidized-bed, high-vibration environments and in other continuously regenerated, moving-bed applications.

By definition, calcination provides oxides and a volatile reaction product [146]. In the case of calcination of zeolite powders the volatile product is water, which can cause hydrothermal damage to low stability zeolites if the calcination process or equipment is not well designed. In the case of setting a binder in a zeolite-containing engineered form, hydrothermal damage must again be considered, and finally when calcining impregnated or ion-exchanged zeolite beads or extrudates, volatile reaction products such as SO_2, CO_2, NO_X and HCl can cause additional damage or unforeseen changes in the products if possible high-temperature gas–solid chemistry is not considered.

3.5
Selected New Developments in Catalyst and Adsorbent Manufacture

Zeolite and molecular sieve science is still a very active area of investigation and numerous new developments are continually appearing. This section is topical but not comprehensive, covering some new areas that should be of interest to the reader.

Zeolite and Molecular Sieve Nanoparticles: Significant exploration and development has occurred in recent years in the area of nanoparticle zeolites and other molecular sieves. Principal justifications for this effort include the expected improved accessibility of inter-zeolite catalytic and adsorption sites to feed molecules. Another active area of research has been the possible use of nano-zeolites, particularly high SiO_2 compositions, as components of thin, low k dielectric layers in microelectronic circuits. Although a few companies and academic groups continue to work in this area it is not clear whether the technological problems, including strength and defect control, will be solved in the near future.

Early reports include a patent from W.R. Grace describing nano-sized zeolite X and Y, another from Mobil for nano-sized MFI catalysts and a third describing the preparation of nano-sized zeolite L [147–149]. The explosion of nanotechnology research in the past few decades has led to an enormous number of additional patents and literature papers in nano-sized zeolites and AlPO-based materials. Commercial production of nano-zeolites is in its infancy. One company, Clariant, is offering nano-zeolite L with 30–60 nm particle size in up to 100 kg quantities. Nano-zeolites have also been investigated as seed layers in zeolite membranes and in mixed matrix membranes, both of which are briefly outlined below. Solid–liquid separation is likely an issue with nanozeolites unless the particles are agglomerated; however, several targeted applications, such as low k dielectric films, require completely dispersed particles. Highly dispersed zeolite particles are probably better provided as colloidal slurries rather than as dry powders to avoid the solid–liquid separation problem and prevent the formation of hard aggregates in dry powders.

Inorganic Zeolite Membranes: Synthetic zeolite membranes were first described in a 1960 patent by Union Oil, where the zeolite membranes were included as part of a fluid diffusion fractionation process; however preparation of the membranes was not outlined in this early patent [150]. In a 1939 report, zeolite electrochemical membranes were made by cutting and polishing natural chabasite and gmelinite zeolite samples [151]. These mechanically fabricated membranes were necessarily thick (1 mm) to provide structural integrity. Zeolite membranes on inorganic supports and their preparation were first described in a 1987 patent, assigned to Hiroshi Suzuki [152]. Extensive development efforts have been carried out in this area by Mobil and Exxon for possible application in petrochemical separations. The first few Mobil patents described self-supporting membranes made only with zeolite particles [153, 154]. Membrane manufacture has evolved after years of development, with requirements in terms of flux and structural integrity driving the further development of supported membranes [155]. Supports vary from macroporous sintered metal to porous alumina or silica. Efforts have been made to address cracks and defects in the zeolite membrane layers to increase membrane selectivity [156, 157]. Recent reports describe mesoporous transition layers between the macroporous supports and thin oriented zeolite layers [158]. Commercial zeolite separation membranes are manufactured by Mitsui and NGK [159–161]. A detailed presentation of the area of zeolite membranes is given in Chapter 10.

Mixed Matrix Membranes: Mixed matrix membranes are composites of organic polymers and zeolite particles. The first report of fabrication by dry pressing of zeolite-containing polyethylene, acrylic and polystyrene mixed matrix membranes was by Barrer in 1960 [162]. These early membranes were rather thick and were used to study membrane potentials in ionic solutions. The Union Oil patent referenced above also anticipates certain kinds of mixed matrix membranes without outlining preparation methods [150]. The first functional mixed-matrix separation membranes were reported as zeolite–cellulose acetate composites by UOP [163]. Zeolite–polymer composites have also been made by infiltrating A- and X-type zeolite pellets with epoxy resins followed by curing [164]. Mixed matrix membranes made by a film-casting technique produced zeolite–silicone membranes in another report [165]. A detailed treatment of mixed matrix membranes is given in Chapter 11.

Microwave Synthesis of Zeolites and Molecular Sieves: The use of microwaves holds promise for efficiency improvements in zeolite synthesis due to the rapid heating possible when using microwave radiation [166]. The first report of microwave synthesis of zeolites was by Mobil Oil in 1988, which broadly claimed the synthesis of zeolite materials in the presence of a microwave-sympathetic material, such as water or other protic component [167]. A number of reports have appeared since, including synthesis of zeolites Y, ZSM-5 [168] and metalloaluminophosphate-type materials, such as MAPO-5 [169]. There have also been extensive investigations in using microwaves for zeolite membrane synthesis. Recent reviews discuss the progress in microwave zeolite synthesis [170, 171].

Continuous and Semi-Continuous Zeolite Synthesis: The first report of continuous synthesis, rather than the usual batch-wise synthesis of zeolites, was from Mobil Oil in 1963 [172]. Another similar patent was from Alcoa Aluminum in 1969 [173]. A 1969 paper presented a mathematical crystallization model for the continuous crystallization of zeolite A [174]. The successful implementation of continuous synthesis of zeolites must accommodate the relatively slow crystallization rates with the reactor design to allow sufficient residence time at the necessary digestion temperature. A recent patent publication describes continuous zeolite synthesis using microwave heating, which couples the often significant advantages of faster zeolite crystallization under microwave radiation with a continuous synthesis, dewatering and work-up process [175].

References

1 Casci, J.L. (2005) Zeolite molecular sieves: preparation and scale-up. *Micropor. Mesopor. Mater.*, **82**, 217–226.
2 Breck, D.W. (1974) *Zeolite Molecular Sieves*, John Wiley & Sons, Inc., New York.
3 Barrer, R.M. (1982) *Hydrothermal Chemistry of Zeolites*, Academic Press, London.
4 Szostak, R. (1989) *Molecular Sieves Principals of Synthesis and Identification*, Van Nostrand Reinhold, New York.
5 van Bekkum, H., Flanigen, E.M., Jacobs, P.A., and Jansen, J.C. (eds.) (2001) *Introduction to Zeolite Science and Practice*, Studies in Surface Science and Catalysis, vol. 137, Elsevier Science Publishers B V, Amsterdam.
6 Robson, H. (ed.) (2001) *Verified Syntheses of Zeolitic Materials*, Elsevier, Science Publishers B V, Amsterdam.
7 Xu, R., Pang, W., Yu, J., Huo, Q., and Chen, J. (2007) *Chemistry of Zeolites and Related Porous Materials: Synthesis and Structure*, John Wiley & Sons (Asia) Pte Ltd.
8 Hinchey, R.J. and Weber, W.W. (1982) Zeolite LZ-200. US Patent 4,348,369.
9 Milton, R.M. (1959) Molecular sieve adsorbents. US Patent 2,882,243.
10 Breck, D.W. (1964) Crystalline zeolite Y. US Patent 3,130,007.
11 Breck, D.W., Eversole, W.G., Milton, R.M., Reed, T.B., and Thomas, T.L. (1956) Crystalline zeolites. I. The properties of a new synthetic zeolite, type A. *J. Am. Chem. Soc.*, **78** (23), 5963–5971.
12 Barrer, R.M. (1948) Synthesis of a zeolitic mineral with chabazite-like sorptive properties. *J. Chem. Soc.*, 127–132.
13 Barrer, R.M. and White, E.A.D. (1951) The hydrothermul chemistry of silicates. Part I. Synthetic lithium aluminusilicates. *J. Chem. Soc.*, 1267–1278.

14 Barrer, R.M. and White, E.A.D. (1952) The hydrothermal chemistry of silicates. Part II. Synthetic crystalline sodium aluminosilicates. *J. Chem. Soc.*, 1561–1571.

15 Barrer, R.M., Baynham, J.M., and McCallum, N. (1953) Hydrothermal Chemistry of Silicates. Part V. Compounds structurally related to Analcite. *J. Chem. Soc.*, 4035–4041.

16 Barrer, R.M. and Baynham, J.W. (1956) The hydrothermal chemistry of the silicates. Part VII synthetic potassium aluminosilicates. *J. Chem. Soc.*, 2882–2891.

17 Schwochow, F. and Heinze, G. (1973) Production of synthetic zeolites of faujasite structure. US Patent 3,720,756.

18 Wilson, S.T. (2001) Phosphate-based molecular sieves, in *Introduction to Zeolite Science and Practice*, Studies in Surface Science and Catalysis, vol. 137, Elsevier Science Publishers B V, Amsterdam, pp. 229–297.

19 Kühl, G. (1998) Source Materials for Zeolite Synthesis. *Micropor. Mesopor. Mater.*, **22** (3), 515–516.

20 Yu, J. (2007) Synthesis of zeolites, in *Introduction to Zeolite Zeolite Molecular Sieves*, Studies in Surface Science and Catalysis, vol. 168, Elsevier Science Publishers B V, Amsterdam, p. 52.

21 Barrrer, R.M. (1982) *Hydrothermal Chemistry of Zeolites*, Academic Press, London, p. 154.

22 Yu, J. (2007) Synthesis of zeolites, in *Introduction to Zeolite Zeolite Molecular Sieves*, Studies in Surface Science and Catalysis, vol. 168, Elsevier Science Publishers B V, Amsterdam, p. 51.

23 Ginter, D.M., Bell, A.T., and Radke, C.J. (1992) The effects of gel aging on the synthesis of NaY zeolite from colloidal silica. *Zeolites*, **12** (6), 742–749.

24 Breck, D.W. (1974) *Zeolite Molecular Sieves*, John Wiley & Sons, Inc., New York, pp. 276 and 386.

25 Weber, W.W. (1975) Zeolite Y synthesis. US Patent 3,920,798.

26 McDaniel, C.V., Maher, P.K., and Pilato, J.P. (1974) Preparation of zeolites. US Patent 3,808,326.

27 McDaniel, C.V. and Duecker, H.C. (1971) Process for preparing high silica faujasite. US Patent 3,574,538.

28 Elliot, C.H. and McDaniel, C.V. (1972) Process for preparing high silica faujasite. US Patent 3,639,099.

29 Vaughan, D.E.W., Edwards, G.C., and Barrett, M.G. (1982) Preparation of zeolites. US Patent 4,340,573.

30 Albers, E.W. and Vaughan, D.E.W. (1976) Method for producing open framework zeolites. US Patent 3,947,482.

31 McDaniel, C.V., Maher, P.K., and Pilato, J.M. (1979) Preparation of zeolites. US Patent 4,166,099.

32 Vaughan, D.E.W., Edwards, G.C., and Barrett, M.G. (1979) Synthesis of type Y zeolties. US Patent 4,178,352.

33 Strack, H. and Kleinschmit, P. (1986) Process for the production of a seed mixture for faujasite synthesis. US Patent 4,608,236.

34 Milton, R.M. (1959) Molecular sieve adsorbents. US Patent 2,882,244.

35 Weber, W.W. (1975) Process for preparing zeolite Y. US Patent 3,898,319.

36 Elliott, C.H. (1979) Preparation of zeolite. US Patent 4,164,551.

37 Micco, D.J. and Hinchey, R.J. (2003) Method for making zeolites and zeolite mixtures having enhanced cation exchange properties. US Patent 6,641,796.

38 Hu, P.C. and Liimatta, E.W. (1996) Process for preparation of zeolite "X". US Patent 5,487,882.

39 Hanif, N., Anderson, M.W., Alfredsson, V., and Terasaki, O. (2000) The effect of stirring on the synthesis of intergrowths of zeolite Y polymorphs. *Phys. Chem. Chem. Phys.*, **2**, 3349–3357.

40 Marrot, B., Bebon, C., Colson, D., and Klein, J.P. (2001) Influence of the shear rate during the synthesis of zeolites. *Cryst. Res. Technol.*, **36** (3), 269–281.

41 Coker, E.N., Dixon, A.G., Thompson, R.W., and Sacco, A. (1995) Zeolite synthesis in unstirred batch reactors II. Effect of non-uniform pre-mixing on the crystallization of zeolites A and X. *Micropor. Mater.*, **3**, 637–646.

42 Ding, L., Zheng, Y., Zhang, Z., Ring, Z., and Chen, J. (2006) Effect of agitation on the synthesis of zeolite beta and its synthesis mechanism in absence

43 Nienow, A.W. (1992) *The Suspension of Solid Particles. Mixing in the Process Industries,* 2nd edn, Butterworth-Heinemann, Oxford, pp. 364–411.
44 Mattox, W.J. and Arey, W.F. (1959) Alkylation of aromatics. US Patent 2,904,607.
45 Weber, H. (1962) Process for production of sodium zeolite A. US Patent 3,058,805.
46 Kerr, G.T. (1967) Process for making crystalline zeolites. US Patent 3,321,272.
47 Kerr, G.T. (1966) Chemistry of crystalline aluminosilicates I. Factors affecting the formation of zeolite A. *J. Phys. Chem.,* **70** (4), 1047–1050.
48 Kacirek, H. and Lechert, H. (1975) Growth of the zeolite type NaY. *J. Phys. Chem.,* **79** (15), 1589–1593.
49 Vaughan, D.E.W. and Edwards, G.C. (1976) Synthetic ferrierite synthesis. US Patent 3,966,883.
50 Bronić, J., Subotić, B., and Škreblin, M. (1999) Investigation of the influence of seeding on the crystallization of zeolite A in the membrane-type reactor micropor. *Mesopor. Mater.,* **28**, 73–82.
51 Whittam, T.V. (1981) Zeolite synthesis. US Patent 4,275,047.
52 Grose, R.W. and Flanigen, E.M. (1981) Novel zeolite compositions and processes for preparing and using same US Patent 4,257,885.
53 Rouleau, L., Lacombe, S., Alario, F., Merlen, E., Kolenda, F., and Magne-Drisch, J. (2002) Process for preparing A zeolite with structure type EUO. US Patent 6,342,200.
54 Bu, X., Gier, T.E., Feng, P., and Stucky, G.D. (1998) Cobalt phosphate based zeolite structures with the edingtonite framework topology. *Chem. Mater.,* **10**, 2546–2551.
55 UOP (2008) UOP's MTO technology licensed to Eurochem for New Nigerian Petrochemicals plant focus on catalysis, *Chem. Week,* 4 February.
56 Vora, B., Chen, J.Q., Bozzano, A., Glover, B., and Barger, P. (2009) Various routes to methane utilization – SAPO-34 catalysis offers the best option. *Catal. Today,* **141**, 77–83.
57 Barger, P.T., Wilson, S.T., and Holmgren, J.S. (1992) Metal aluminophosphate catalyst for converting methanol to light olefins. US Patent 5,126,308.
58 Jhung, S.H., Lee, J.H., and Chang, J. (2008) Crystal size control of transition metal ion-incorporated aluminophosphate molecular sieves: effect of ramping rate in the syntheses. *Micropor. Mesopor. Mater.,* **112**, 178–186.
59 Chen, C. and Jehng, J. (2003) Effect of synthesis pH and H2O molar ratio on the structure and morphology of aluminum phosphate (AlPO-5) molecular sieves. *Catal. Lett.,* **85** (1–2), 73–80.
60 Beale, A.M. and Weckhuysen, B.M. (2007) Understanding the crystallisation processes leading to the formation of microporous aluminophosphates, in *Zeolites to Porous Materials – The 40th Anniversary of International Zeolite Conference,* Studies in Surface Science and Catalysis, vol. 170, Elsevier Science Publishers B V, Amsterdam, pp. 748–755.
61 Chang, Y., Martens, L.R., and Vaughn, S.N. (2006) Molecular sieve catalyst composition, its making and use in conversion processes. US Patent 7,122,500.
62 Potter, M.J. (1999) Microfiltration of zeolites. US Patent 5,919,721.
63 Clem, K.R., Martens, L.R.M., Vaughn, S.N., Stafford, P.R., Kress, J.W., and Mertens, M.M. (2003) Preparation of molecular sieve catalysts microfiltration. US Patent 6,521,562.
64 Sherry, H.S. (2003) *Ion Exchange, Handbook of Zeolite Science and Technology,* Marcel Dekker, New York, 1008.
65 Stiles, A.B. (1987) *Catalyst Supports and Supported Catalysts,* Butterworths, Boston.
66 Stiles, A.B. and Koch, T.A. (1995) *Catalyst Manufacture,* 2nd edn, Marcel Dekker, New York.
67 Satterfield, C.N. (1991) *Heterogeneous Catalysis in Industrial Practice,* 2nd edn, McGraw-Hill, Inc.
68 LePage, J.F. (1987) *Applied Heterogeneous Catalysis: Design, Manufacture, Use of Solid Catalysts,* TECHNIP.

69. Ruthven, D.M. and Post, M.F.M. (2001) Diffusion in zeolite molecular sieves, in *Introduction to Zeolite Science and Practice*, Studies in Surface Science and Catalysis, vol. 137, Elsevier Science Publishers B V, Amsterdam, p. 533.
70. Liepa, A.L. and Japikse, C.H. (1977) Beverage carbonation device. US Patent 4,007,134.
71. Marques, P., Remy, C., and Sorensen, C.M. (2005) Monolithic zeolite coated structures and a method of manufacture. US Patent 6,936,561.
72. Norback, P.G. (1974) Moisture exchanger for gaseous media. US Patent 3,807,149.
73. Mäurer, T., Müller, S.P., and Kraushaar-Czarnetzki, B. (2001) Aggregation and peptization behavior of zeolite crystals in sols and suspensions. *Ind. Eng. Chem. Res.*, **40**, 2573–2579.
74. Kuzniatsova, T., Kim, Y., Shqau, K., Dutta, P.K., and Verweij, H. (2007) Zeta potential measurements of zeolite Y: application in homogeneous deposition of particle coatings micropor. *Mesopor. Mater.*, **103**, 102–107.
75. Mitchell, W.J. and Moore, W.F. (1961) Bonded molecular sieves. US Patent 2,973,327.
76. Breck, D.W. and Acara, N.A. (1965) Crystalline zeolite L. US Patent 3,216,789.
77. Van Dyke, R.E. and Trainer, R.P. (1958) Solid contact material. US Patent 2,865,867.
78. Young, D.A. and Mickelson, G.A. (1971) Alumina-bonded catalysts. US Patent 3,557,024.
79. Mitsche, R.T., Kuntz, H.L., and Hayes, J.C. (1977) Method of manufacturing an extruded catalyst composition. US Patent 4,046,713.
80. Gladrow, E.M. and Smith, W.M. (1967) Catalyst composition of a crystalline aluminosilicate and a binder. US Patent 3,326,818.
81. Absil, R.P.L., Angevine, P.J., Herbst, J.A., Klocke, D.J., McWilliams, J.P., Han, S., and Shihabi, D.S. (1991) Method for preparing a zeolite catalyst bound with a refractory oxide of low acidity. US Patent 5,053,374.
82. Marler, D.O. (1993) Catalysts bound with low acidity refractory oxide. US Patent 5,182,242.
83. Sharma, S.B., Gurevich, S.V., Riley, B.D., and Rosinski, G.A. (2000) Selective xylenes isomerization and ethylbenzene conversion. US Patent 6,143,941.
84. Gordina, N.E., Prokof'ev, V.Y., and Il'in, A.P. (2005) Extrusion molding of sorbents based on synthesized zeolite. *Glass Ceram.*, **62** (9–10), 282–286.
85. Flank, W.H., Fethke, W.P., and Marte, J. (1989) Process for preparing molecular sieve bodies. US Patent 4,818,508.
86. Howell, P.A. and Acara, N.A. (1964) Process for producing molecular sieve bodies. US Patent 3,119,660.
87. Takahashi, K., Kumata, F., Nozaki, H., Inoue, S., and Makabe, T. (1995) Process for producing spherical zeolite catalyst and apparatus for producing the same, US Patent 5,464,593.
88. Gladrow, E.M., Ruhnke, E.V., and Kimberlin, C.N. (1962) Process for preparing attrition-resistant adsorbents. US Patent 3,055,841.
89. Lim, J., Brady, M., and Stamires, D. (1982) Attrition resistant zeolite containing catalyst. US Patent 4,333,857.
90. Kuhn, K. and Boedecker, M. (1972) Shaped body for filtering and drying of liquids and gases and process of making the same. US Patent 3,687,297.
91. Kulprathipanja, S. (1981) Technique to reduce the zeolite molecular sieve solubility in an aqueous system. US Patent 4,248,737.
92. Kulprathipanja, S. and deRosset, A.J. (1983) Separatory process using organic bound adsorbents. US Patent 4,421,567.
93. Chao, C.C., Sherman, J.D., and Barkhausen, C.H. (1989) Latex polymer bonded crystalline molecular sieves. US Patent 4,822,492.
94. Chu, J., Pryor, J.N., and Welsh, W.A. (1999) Desiccation using polymer-bound desiccant beads. US Patent 5,879,764.
95. Marra, W.H. and Stinchfield, J.C. (1955) Paper coating compositions comprising an adhesive, an alkali metal silicate and

an attapulgite or zeolite material. US Patent 2,699,432.
96 Crowley, R.P. (1966) Method of preparing adsorbent filter paper containing crystalline zeolite particles, and paper thereof. US Patent 3,266,973.
97 Macriss, R.A., Rush, W.F., and Weil, S.A. (1974) Desiccant system for an open cycle air-conditioning system. US Patent 3,844,737.
98 Belding, W.A., Janke, S., Holeman, W.D., and Delmas, M.P.F. (1996) Enthalpy wheel. US Patent 5,542,968.
99 Dunne, S.R. (2000) Apparatus for use in sorption cooling processes. US Patent 6,102,107.
100 Hemstreet, R.A. (1967) Adsorbent-coated thermal panels. US Patent 3,338,034.
101 Fischer, J.C. (1988) High efficiency sensible and latent heat exchange media with selected transfer for a total energy recovery wheel. US Patent 4,769,053.
102 Dunne, S.R. and Behan, A.S. (1996) Adsorbent composites for sorption cooling process and apparatus. US Patent 5,518,977.
103 Hoke, J.B., Buelow, M.T., and Kauffman, J.J. (2006) Coated screen adsorption unit for controlling evaporative hydrocarbon emissions. US Patent Application 20060272508.
104 Caesar, P.D. (1987) Method for preparing structured catalytic solids. US Patent 4,692,423.
105 Louis, B., Tezel, C., Kiwi-Minsker, L., and Renken, A. (2001) Synthesis of structured filamentous zeolite materials via ZSM-5 coatings of glass fibrous supports. Catal. Today, 69, 365–370.
106 Kubota, K., Katoh, T., Hirata, M., and Hayashi, K. (1993) Dyed synthetic fiber comprising silver-substituted zeolite and copper compound, and process for preparing same US Patent 5,180,402.
107 Pradhan, A.R., Macnaughtan, M.A., and Raftery, D. (2002) Preparation of zeolites supported on optical microfibers. Chem. Mater., 14, 3022–3027.
108 Larsen, G., Vu, D., and Marquez-Sanchez, M. (2004) Stable zeolite/cellulose composite materials and method of preparation. US Patent 6,814,759.
109 Wang, Y.J., Tang, Y., Wang, X.D., Yang, W.L., and Gao, Z. (2000) Fabrication of hollow zeolite fibers through layer-by-layer Adsorption Method. Chem. Lett., 1344–1345.
110 Haden, W.L. and Dzierzanowski, F.J. (1961) Method for making synthetic zeolite material. US Patent 2,992,068.
111 Taggart, L.R.L. and Rabaud, G.L. (1964) Process for producing molecular sieve bodies. US Patent 3,119,659.
112 Sensel, E.E. (1961) Process for production of type A zeolite. US Patent 3,009,776.
113 Michalko, E. (1967) Method of preparing high rate zeolitic molecular sieve particles. US Patent 3,348,911.
114 Rundell, C.A. and McDaniel, C.V. (1973) Process for preparing zeolitic bodies having high strength characteristics. US Patent 3,777,006.
115 Haden, W.L. and Dzierzanowski, F.J. (1970) Microspherical zeolitic molecular sieve composite catalyst and preparation thereof. US Patent 3,506,594.
116 Rollmann, L.D. (1976) Manufacture of crystalline aluminosilicate zeolites. US Patent 3,939,246.
117 Arika, J., Aimoto, M., and Miyazaki, H. (1986) Process for preparation of high-silica faujasite type zeolite. US Patent 4,587,115.
118 Murrell, L.L., Overbeek, R.A., Chang, Y., Van der Puil, N., and Yeh, C.Y. (1999) Method for making molecular sieves and novel molecular sieve compositions. US Patent 6,004,527.
119 Hildebrandt, D.E. (1983) Method of producing binderless zeolite extrudates. US Patent 4,381,256.
120 Pryor, J.N. and Chi, C.W. (1984) Preparation of binderless 3a adsorbents. US Patent 4,424,144.
121 Ebery, P.E., Laurent, S.M., and Robson, H.E. (1970) High silica crystalline zeolites and process for their preparation. US Patent 3,506,400.
122 Plank, C.J. and Rosinski, E.J. (1966) Steam activated catalyst. US Patent 3,257,310.
123 Maher, P.K. and McDaniel, C.V. (1966) Zeolite Z-14US and method of preparation thereof. US Patent 3,293,192.

124 McDaniel, C.V. and Maher, P.K. (1968) Process for ion exchanging crystalline zeolites with nitrogenous bases. US Patent 3,374,057.

125 Young, D.A. (1968) Process for the preparation of crystalline ammonium zeolites. US Patent 3,383,169.

126 Maher, P.K. and McDaniel, C.V. (1968) Ion exchange of crystalline zeolites. US Patent 3,402,996.

127 McDaniel, C.V. and Maher, P.K. (1969) Stabilized zeolites. US Patent 3,449,070.

128 Ward, J.W. (1975) Ammonia-stable Y zeolite compositions. US Patent 3,929,672.

129 Vassilakis, J.G. and Best, D.F. (1991) Novel zeolite compositions derived from zeolite Y. US Patent 5,013,699.

130 Cooper, D.A., Denkewicz, R.P., and Hertzenberg, E.P. (1993) Stable zeolite of low unit cell constant and method of making same. US Patent 5,242,677.

131 Shively, J.H. and Archer, E.D. (1972) Selected adsorption with a silanized crystalline alumino-silicate. US Patent 3,658,696.

132 Sato, M. and Otani, S. (1971) Method of the disproportionation of toluene. US Patent 3,553,277.

133 Voorhies, A. and Kimberlin, C.N. (1971) Hydrocarbon conversion catalyst. US Patent 3,630,965.

134 Wilson, W.B. (1972) Fluoride-containing crystalline alumino-silicates. US Patent 3,702,312.

135 Elliott, C.H. (1974) Synthetic fluoride containing zeolite systems. US Patent 3,839,539.

136 Kokotailo, G.T., Rohrman, A.C., and Sawruk, S. (1983) Modification of catalytic activity of synthetic zeolites. US Patent 4,414,189.

137 Breck, D.W. and Skeels, G.W. (1985) Silicon substituted zeolite compositions and process for preparing same US Patent 4,503,023.

138 Allen, P.T. and Drinkard, B.M. (1972) Separation of mixtures with modified zeolites. US Patent 3,698,157.

139 Szostak, R. (2001) *Introduction to Zeolite Science and Practice*, Studies in Surface Science and Catalysis, vol.137, Elsevier Science Publishers B V, Amsterdam, p. 270.

140 Anderson, M.W., Klinowski, J., and Xinsheng, L. (1984) Alumination of Highly Siliceous Zeolites. *J. Chem. Soc. Chem. Commun.*, 1596–1597.

141 Caglione, A.J., Cannan, T.R., Greenlay, N., and Hinchey, R.J. (1994) Process for preparing low silica forms of zeolites having the faujasite type structure. US Patent 5,366,720.

142 Narayana, M. and Murray, B.D. (1992) Process for realuminating zeolites. US Patent 5,118,482.

143 Barrer, R.M. and Klinowski, J. (1974) Ion-exchange selectivity and electrolyte concentration. *J. Chem. Soc. Faraday*, 2080–2091.

144 Sherry, H.S. (1966) The ion-exchange properties of zeolites. I. Univalent ion exchange in synthetic faujasite. *J. Phys. Chem.*, **70** (4), 1158–1168.

145 Sherry, H.S. (1968) The ion-exchange properties of zeolites. IV. Alkaline earth ion exchange in the synthetic zeolites linde X and Y. *J. Phys. Chem.*, **72** (12), 4086–4094.

146 Kingery, W.D., Bowen, H.K., and Uhlmann, D.R. (1976) *Introduction to Ceramics*, 2nd edn, John Wiley & Sons, Inc., NewYork, pp. 414–420.

147 Maher, P.K. and Scherzer, J. (1970) Method of preparing microcrystalline faujasite-type zeolite. US Patent 3,516,786.

148 Plank, C.J., Rosinski, E.J., and Schwartz, A.B. (1975) Hydrocarbon conversion. US Patent 3,926,782.

149 Vaughan, D.E.W. and Strohmeier, K.G. (1994) Process for preparing LTL nano-crystalline zeolite compositions. US Patent 5,318,766.

150 Fleck, R.N. and Wight, C.G. (1960) Fluid diffusion fractionation. US Patent 2,924,630.

151 Marshall, C.E. (1939) The use of zeolitic membrane electrodes. *J. Phys. Chem.*, **43** (9), 1155–1164.

152 Suzuki, H. (1987) Composite membrane having a surface layer of an ultrathin film of cage-shaped zeolite and processes for production thereof. US Patent 4,699,892.

153 Haag, W.O. and Tsikoyiannis, J.G. (1991) Membrane composed of a pure molecular sieve. US Patent 5,019,263.

154 Haag, W.O., Tsikoyiannis, J.G., and Valuocsik, E.W. (1992) Synthesis of a membrane composed of a pure molecular sieve. US Patent 5,100,596.
155 Barri, S.A.I., Bratton, G.J., and de Villiers Naylor, T. (1994) Process for the production of a membrane. US Patent 5,362,522.
156 Geus, E.R., Jansen, J.C., Jaspers, B.C., Schoonman, J., and Van Bekkum, H. (1995) Inorganic composite membrane comprising molecular sieve crystals. US Patent 5,429,743.
157 Geus, E.R., Jansen, J.C., Jaspers, B.C., Schoonman, J., and Van Bekkum, H. (1998) Inorganic composite membrane comprising molecular sieve crystals. US Patent 5,753,121.
158 Choi, J., Ghosh, S., Lai, Z., and Tsapatsis, M. (2006) Uniformly a-oriented MFI zeolite films by secondary growth. *Angew. Chem. Int. Ed.*, **45** (7), 1154–1158.
159 Okamoto, K., Kita, H., Kondo, M., Miyake, N., and Matsuo, Y. (1996) Membrane for liquid mixture separation. US Patent 5,554,286.
160 Morigami, Y., Kondo, M., Abe, J., Kita, H., and Okamoto, K. (2001) The first large-scale pervaporation plant using tubular-type module with zeolite NaA membrane separ. *Purif. Tech.*, **25**, 251–260.
161 Nakayama, K., Suzuki, K., Yoshida, M., Yajima, K., and Tomita, T. (2006) Method for preparing DDR type zeolite membrane, DDR type zeolite membrane, and composite DDR type zeolite membrane, and method for preparation thereof. US Patent 7,014,680.
162 Barrer, R.M. and James, S.D. (1960) Electrochemistry of crystal—polymer membranes. Part I. Resistance measurements. *J. Phys. Chem.*, **64** (4), 417–421.
163 Kulprathipanja, S., Neuzil, R.W., and Li, N.N. (1988) Separation of fluids by means of mixed matrix membranes. US Patent 4,740,219.
164 Demertzis, M. and Evmiridis, N.P. (1986) Potentials of ion-exchanged synthetic zeolite-polymer membranes. *J. Chem. Soc. Faraday Trans. I*, **82**, 3647–3655.
165 te Hennepe, H.J.C., Bargeman, D., Mulder, M.H.V., and Smolders, C.A. (1987) Zeolite-filled silicone rubber membranes part 1. Membrane preparation and pervaporation results. *J. Membrane Sci.*, **35**, 39–55.
166 Mingos, D.M.P. and Baghurst, D.R. (1991) Applications of microwave dielectric heating effects to synthetic problems in chemistry. *Chem. Soc. Rev.*, **20**, 1–47.
167 Chu, P., Dwyer, F.G., and Vartuli, J.C. (1988) Crystallization method employing microwave radiation. US Patent 4,778,666.
168 Arafat, A., Jansen, J.C., Ebaid, A.R., and van Bekkum, H. (1993) Microwave preparation of zeolite Y and ZSM-5. *Zeolites*, **13**, 162–165.
169 Cresswell, S.L., Parsonage, J.R., Riby, P.G., and Thomas, M.J.K. (1995) Rapid synthesis of magnesium aluminophosphate-5 by microwave dielectric heating. *J. Chem. Soc. Dalton Trans.*, 2315–2316.
170 Thompsett, G.A., Conner, W.C., and Yngvesson, K.S. (2006) Microwave synthesis of nanoporous materials. *Chemphyschem*, **7** (2), 292–319.
171 Li, Y. and Yang, W.J. (2008) Microwave synthesis of zeolite membranes: a review. *Membrane Sci.*, **316**, 3–17.
172 Frilette, V.J. and Kerr, G.T. (1963) Process for making crystalline zeolites. US Patent 3,071,434.
173 Hirsh, W. (1969) Production of crystalline zeolites. US Patent 3,425,800.
174 Lui, S. (1969) Continuous process of zeolite A. *Crystallization. Chem. Eng. Sci.*, **24**, 57–64.
175 Park, S., Kim, D.S., Chang, J., and Kim, J. (2001) Continuous process and apparatus for preparing inorganic materials employing microwave. US Patent Application 20010054549.

4
Zeolite Characterization

Steven A. Bradley, Robert W. Broach, Thomas M. Mezza, Sesh Prabhakar, and Wharton Sinkler

4.1
Introduction

4.1.1
Importance of Characterization

Characterization is an integral tool for the development of new zeolites and for the development and commercialization of zeolitic catalysts and adsorbents. Single techniques are not sufficient as they rarely provide full details of the system. A combination of selective characterization techniques is required. As suggested by Deka [1] even a single acidity characterization method may be insufficient to provide the necessary detailed information to understand the zeolite acid sites. Thus according to Deka the combination of different experimental techniques is required to shorten the time of development for a new catalyst.

Haensel and Haensel [2] describe the criticality of the role of catalyst characterization in process and catalyst development. They conclude that catalyst characterization is the cornerstone for the science of catalysis and for industrial progress. They emphasize that the characterization must have a purpose and be targeted for understanding and solving a specific problem in order to achieve commercial success.

The purpose of this chapter is to provide general guidance for the various characterization techniques that are most often employed in the development and commercialization of new zeolites, catalysts and adsorbents. Each of these techniques can be a volume unto itself. Thus we only briefly describe the technique but emphasize the information that can be obtained by the particular method and the utility of that information for characterizing zeolites, zeolitic catalysts and zeolitic adsorbents. The limitations of the various techniques are included to assist the novice. For a more in depth understanding of the various techniques, references are given throughout the chapter.

A good general reference is *Zeolite Molecular Sieves – Structure, Chemistry and Use* by Breck [3]. The *Molecular Sieves – Science and Technology* volumes have excellent reviews of various characterization techniques. *Characterization 1* (volume 4) [4] and *Characterization 2* (volume 5) [5] cover general characterization techniques. Included in these two volumes but not discussed in this chapter are: the application of UV/VIS spectroscopy, photoelectron spectroscopy, chemical analyses of zeolites, pore size analysis using probe molecules and techniques to characterize coke on a spent zeolitic catalyst. *Acidity and Basicity* (volume 6) [6] details the application of NMR, microcalorimetry and test reactions for acid site characterization. *Adsorption and Diffusion* (volume 7) [7] describes the measurement of diffusion under non-equilibrium conditions and by pulse field gradient NMR as well as spectroscopic methods for measuring adsorption and adsorption kinetics.

In this chapter we start with the methodology for the identification of the structure of a newly invented zeolite. To accomplish this task a multi-technique approach was necessary and is used as an example for the necessity of employing complementary techniques to solve a problem. If such a methodology had not been pursued, the structure would not have been correctly elucidated.

Diffraction is a central technique for characterizing the structure of a zeolite. X-ray diffraction is commonly used to identify the presence and quantitatively determine the amount of crystalline zeolite present. Electron microscopy is useful for structural and morphological analysis and for chemical analyses at specific locations. The spectroscopic techniques of IR and NMR provide insight into acid sites and framework structure. Additional understanding of the acid sites can be achieved by various techniques described in the *Physical/Chemical* section along with the application of porosimetry and thermal analysis. Throughout this chapter examples are given for various techniques characterizing the same ammonium exchanged and steamed Y-zeolites.

4.2
Multi-Technique Methodology

4.2.1
Identification of the Structure of a Newly Invented Zeolite

Low Si/Al ratio zeolites are attractive because they can offer high acid site density for catalytic applications and greater ion-exchange capacity and compositional diversity for separation processes. The zeolite UZM-5 (framework code UFI) was synthesized at low Si/Al ratios (Si/Al < 10) using combinations of two of the most common organic templating agents: the tetramethylammonium (TMA^+) and tetraethylammonium (TEA^+) cations. The strategy used in the synthesis was the charge density mismatch (CDM) approach [8], where a stable homogeneous reaction mixture is created and then at the desired time an agent is added to induce crystallization.

Unfortunately, the X-ray diffraction patterns for different UZM-5 preparations showed broad peaks (due to the very small crystallite size and anisotropic morphology) and were not of sufficient quality for *ab initio* structure solution. However, indexing the patterns did provide the crystal system (tetragonal) and approximate lattice parameters ($a = 12.3$ Å, $c = 28.5$ Å). In order to solve the structure for UZM-5 a broad range of analytical techniques was needed. Adsorption measurements showed a diminished uptake of i-C_4H_{10} versus n-C_4H_{10}, suggesting an eight-ring or small ten-ring channel system. A McBain–Bakr gravimetric apparatus was used to determine adsorption capacities. At room temperature and $P/P_0 = 0.3$, UZM-5 adsorbed 0.22 cm^3 n-C_4H_{10} per gram and 0.050 cm^3 i-C_4H_{10} per gram. The kinetic diameters of n-C_4H_{10} and i-C_4H_{10} are 4.3 Å and 5.0 Å, respectively. Also, peaks in the infrared absorption spectrum at 1216 cm^{-1} and 545 cm^{-1} implied the presence of five-rings, while a band at 569 cm^{-1} suggested the possibility of double four- or double six- rings in the structure [9]. Electron diffraction was used to confirm the unit cell from X-ray diffraction and to narrow the range of possible space groups to five body-centered tetragonal groups. The information from high-resolution transmission electron microscopy in Figure 4.1 was key to the structure solution,

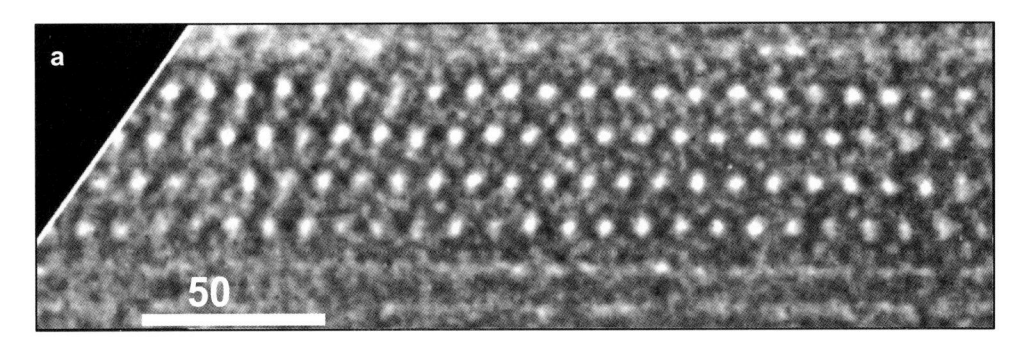

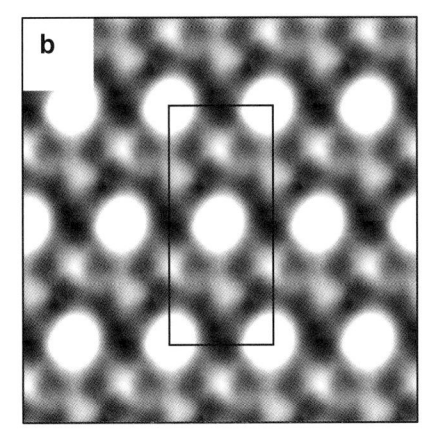

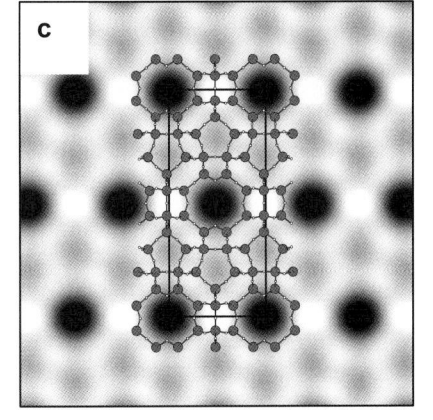

Figure 4.1 (a) Raw high-resolution TEM image of UZM-5 zeolite in [100] orientation. The image is a detail of a thin area of a single plate, viewed edge-on. (b) Result of real-space averaging of ten unit cells. Unit cells were selected using self-similarity and intensities within the unit cells were averaged to reduce noise and emphasize systematic high-resolution features. (c) Same as (b) but after applying c2mm plane group symmetry consistent with the UZM-5 I4/mmm space group. Image has been contrast-inverted. Overlay is a ball and stick model of the UFI framework structure showing agreement with the experimental image details.

Table 4.1 Distance least-squares optimized atomic coordinates for UZM-5.

Atom label	Wyckoff notation	X	Y	Z
Si1	16n	0.1776	0.0000	0.0556
Si2	32o	0.3784	0.1909	0.2013
Si3	16n	0.3727	0.0000	0.1281
O1	16n	0.2941	0.0000	0.0827
O2	8i	0.1993	0.0000	0.0000
O3	16m	0.1085	0.1085	0.0698
O4	16k	0.3757	0.1243	0.2500
O5	32o	0.3483	0.1085	0.1590
O6	16n	0.5000	0.2410	0.1931
O7	16m	0.2899	0.2899	0.2030
O8	8g	0.5000	0.0000	0.1119

Space group I4/mmm; optimized cell constants: a = 12.332 Å, c = 29.096 Å.
Reference constants: $\langle T\text{-}O \rangle$ = 1.639 Å, $\langle O\text{-}O \rangle$ = 2.676 Å, $\langle T\text{-}T \rangle$ = 3.126 Å (assuming Si:Al = 7.0, Si-O = 1.623 Å, Al-O = 1.748 Å, $\langle O\text{-}T\text{-}O \rangle$ = 109.5° and $\langle T\text{-}O\text{-}T \rangle$ = 145°).
Final convergence gave R_{DLS} = 0.0017 σ = 0.008
$R_{DLS} = \{\Sigma[w \times (D_o - D)]^2 / \Sigma(w \times D_o)^2\}^{1/2}$
$\sigma_{DLS} = \{\Sigma[w \times (D_o - D)]^2 / (M - NV)\}^{1/2}$

as it pinpointed the locations of the channels and double four-rings. Figure 4.1a shows the original high-resolution TEM image of a plate-like crystal along a <100> zone axis direction. The image was processed in order to reduce noise using a real-space averaging of ten unit cells of the original image, as shown in Figure 4.1b. The figure shows light areas indicating positions of low atomic density (e.g., the main pore system), and dark areas indicating positions of high projected atomic density (e.g., the double four-rings). A trial model of the structure consistent with the TEM images and the analytical data mentioned above was built in the space group *I4/mmm*. The atomic coordinates were optimized by distance least squares (DLS) refinement [10]. The constraints used, the atomic coordinates and the residuals are given in Table 4.1. The DLS refinement technique is commonly used particularly in the zeolite community for the evaluation and preliminary refinement of hypothetical structures. The calculated diffraction pattern based on the DLS coordinates agrees well with the experimental diffraction pattern as shown in Figure 4.2. The UZM-5 framework is confirmed by agreement between the model and high-resolution TEM images along several crystallographic directions (e.g., see Figure 4.1c). Rietveld refinement of the X-ray powder data also confirms the structure.

The structure of UZM-5 consists of alternating layers along the c-axis of two different composite building units or polyhedra, designated *cub* and *cle* (Figure 4.3) [11]. Alternatively, the structure can be visualized as being built from two larger composite building units, designated *grc* and *wbc* (Figure 4.4a). The *grc* polyhedron, found in several other zeolites, including LTA, has usually been called an alpha cage but according to recent IUPAC recommendations is more properly

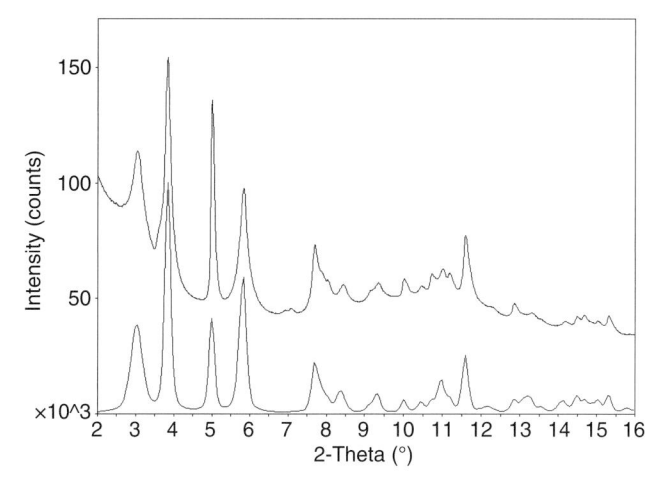

Figure 4.2 Comparison of the experimental diffraction pattern for a calcined never rehydrated UZM-5 (top) with a simulation using DLS coordinates (bottom). For the observed pattern, synchrotron wavelength = 0.7527 Å, space group = I4/mmm, a = 12.193 ± 0.003 Å, c = 28.403 ± 0.017 Å. The simulation was done using Accelrys materials studio ver. 2.1.5 and pseudo-voigt profiles with anisotropic crystallite size broadening (coherence lengths: $L_a = L_b = 125$ Å, $L_c = 50$ Å).

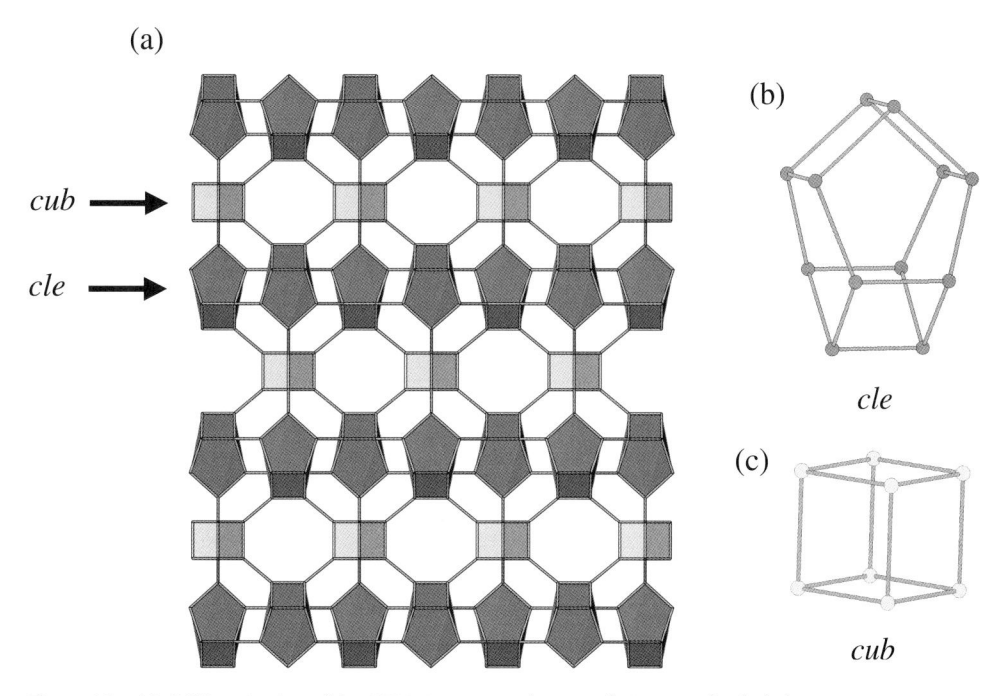

Figure 4.3 (a) (100) projection of the UZM-5 structure showing *cle* (gray) and *cub* (light gray) polyhedra. (b) *cle* polyhedron (c) *cub* polyhedron (double four-ring).

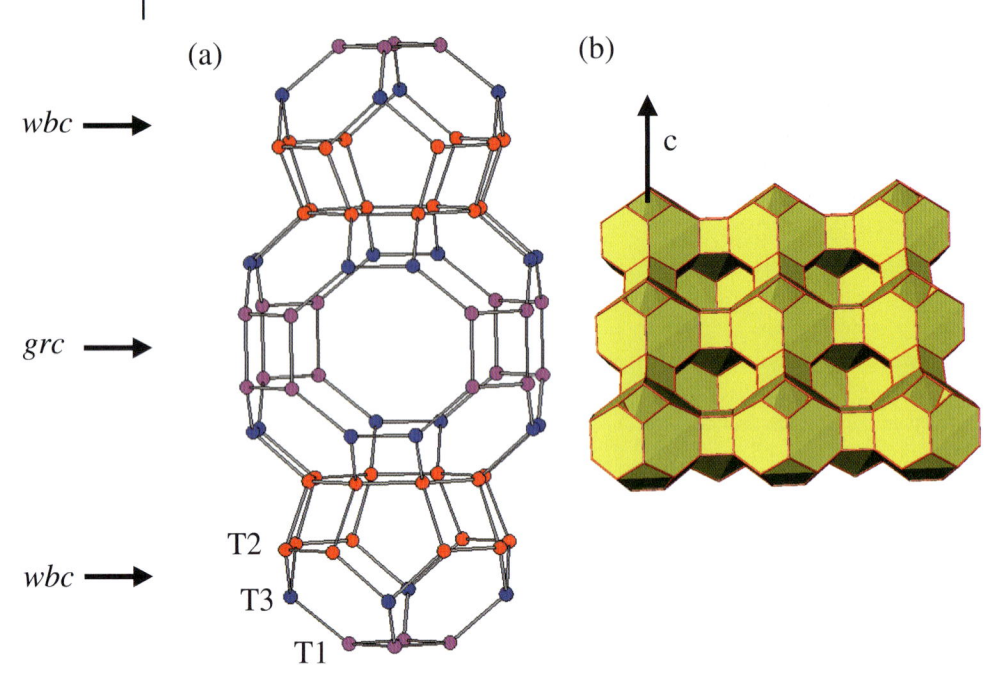

Figure 4.4 (a) View of the alpha (*grc*) cavity bicapped in the *c*-axis direction by *wbc* cavities forming pockets accessible though eight-rings on opposite sides of each alpha cavity. The three unique T-atoms in the structure are color-coded. (b) The predominant (001) surface contains 16-ring cups with an eight-ring bottom. These cups are essentially the bottom halves of alpha cavities sitting over *wbc* cavities.

called an alpha cavity [12]. The *wbc* polyhedron has not previously been found in any zeolite to our knowledge. Both of these cavities are accessible for adsorption of molecules larger than water through eight-ring windows. In UZM-5, the alpha cavities are fused through eight-rings along the *a*- and *b*-axis directions to form a two-dimensional eight-ring channel system. In the *c* direction *wbc* cavities are fused with the eight-rings above and below each alpha cavity (Figure 4.4a). The three-dimensional structure is formed by linking the bicapped alpha cavities in the *c*-direction, creating a *cub* cage at the apices. The UZM-5 framework is projected on a high-resolution TEM image along the ⟨100⟩ direction in Figure 4.1c. The body-centered eight-ring pore system and the five-ring channels in the model fit the features of the high-resolution image.

The layers of fused alpha cavities in UZM-5 are similar to those found in zeolites of framework type LTA; both structures have the same 12.3 Å lattice repeat distance along *a* and *b*. In LTA-type zeolites successive layers of alpha cavities are connected through eight-ring windows allowing three-dimensional diffusion. However, in UZM-5 successive alpha cavity layers are separated by intervening layers of *wbc* cavities which block diffusion along the *c*-axis direction. The resulting eight-ring pore system is consistent with the observed adsorption of n-C_4H_{10} and the relative exclusion of i-C_4H_{10}. In addition, the greatly favored crystal growth of UZM-5 in the a and b directions results in a fine platelike morphology with many sites on the (001) surface. This surface formed by the layer of the *wbc* cavities consists of 16-ring cups that are essentially the bottom halves of alpha cavities (Figure 4.4b). These cups are likely the sites where important catalytic reactions such as xylene isomerization and benzene alkylation occur, both of which involve molecules too large to fit in the eight-ring pores [13]. The presence of these surface cups is reminiscent of the MWW structure [14].

Given the complex nature of the crystal structure and small crystal size with an anisotropic morphology of UZM-5, the normal X-ray diffraction patterns were not sufficient to deduce an unambiguous structure. Thus a multi-technique approach was required to successfully solve the structure, to explain the adsorption properties and by analogy to the structure of other zeolites in order to assess potential applications.

4.3
X-Ray Powder Diffraction Characterization of Zeolitic Systems

4.3.1
Interpretation of Powder Diffraction Data for Zeolites

X-Ray powder diffraction is a powerful tool for characterization of zeolites. The basic experiment is relatively easy to perform and can be done in most labs on standard diffractometers; and the data obtained is easy to analyze for many applications. Powder diffraction can be used to determine whether a new zeolite has been synthesized, whether a desired zeolite has been made or whether a crystallization process has completed. As noted in Section 4.2, X-ray powder diffraction can be an integral tool in determining the details of the structure of a newly synthesized zeolite. In addition, it is a critical characterization technique that can be routinely used, for example, to identify contaminants present in a synthesis, to determine how much zeolite has been bound into a catalyst or adsorbent pellet, or to ascertain if heat treatment alters the zeolite structure. Of the techniques described in this chapter, powder diffraction is probably the most commonly used. Additional details can be found elsewhere [15–19].

Consider the example of a powder pattern for zeolite X (FAU structure type) shown in Figure 4.5 which gives diffracted intensity on the vertical axis as a function of diffraction angle (two-theta) on the horizontal axis. The three basic things to look for in a diffraction pattern are peak positions, peak intensities and peak shapes. The peak positions of a crystalline material are determined solely by the size and shape of the unit cell. Since all zeolite framework types have unique unit cells, the peak positions will be a fingerprint for each zeolite type. Peak intensities are determined by the positions and types of atoms in the unit cell. So, variations in intensities give information about the chemical composition and siting of atoms in the unit cells. Peak shapes and widths give an idea of the quality of the crystal-

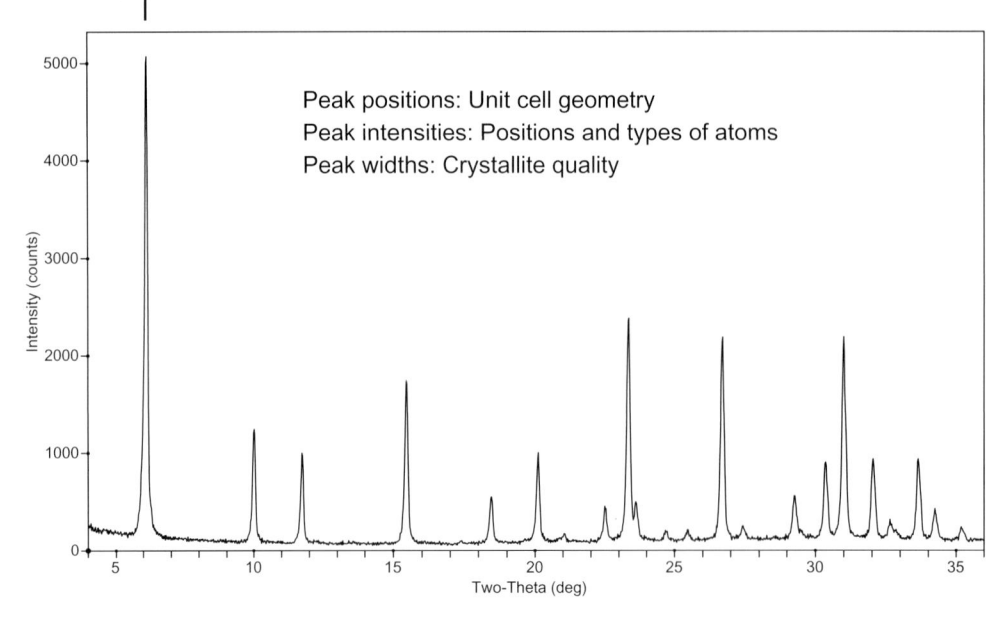

Figure 4.5 X-Ray powder diffraction pattern for zeolite X (FAU).

lites. In general small crystallite size and high crystallite strain lead to peak broadening. In the following sections some of the basic methods for using powder diffraction will be described followed by a brief discussion of more advanced techniques.

4.3.2
Phase Identification and Quantification

One application of powder diffraction is phase identification. Since zeolites of the same structure type give similar powder patterns, the powder pattern can be used as a fingerprint to identify the zeolite type. Furthermore, when multiple phases are present, the powder pattern is a superposition of the patterns for each of the separate phases and the relative overall intensities of the peaks is related to the amount of each phase. Thus patterns from mixtures of phases can be analyzed to determine the identity and relative amount of each phase.

This approach is commonly employed for bound zeolites. In this manner the relative amount of crystalline zeolite present can be determined. As usually implemented, the method provides a number that is the ratio of intensities for peaks in the diffraction pattern of the sample of interest to the intensity of the same peaks in the pattern of a reference zeolite. Since the intensity ratio is often expressed as a percentage, it is commonly referred to as percent zeolite crystallinity. The better terminology is relative amount of crystalline zeolite compared to a specific reference.

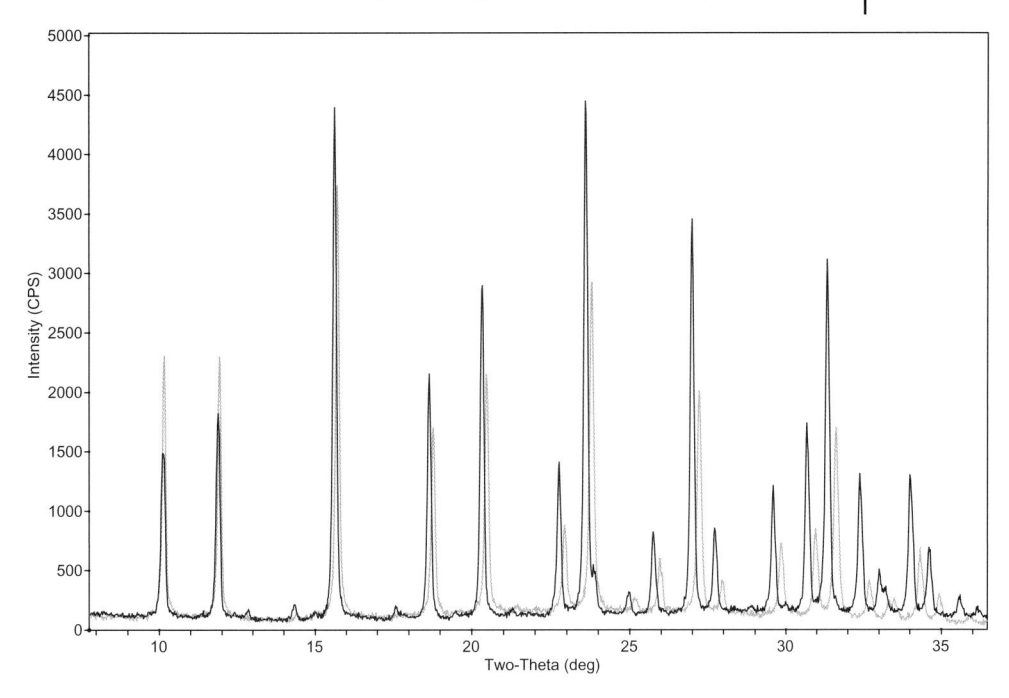

Figure 4.6 Affect on steaming dealumination on the X-ray diffraction pattern for a Y-zeolite (FAU).

It is important to note that the selection of the reference zeolite is critical since, even for a given zeolite structure, there can be significant changes in the peak intensities because of different amounts of cations or type of cations that can be exchanged in the zeolite (Figure 4.6), whether the zeolite pores are filled with guest molecules and on the nature of other guest species. A reference zeolite should be selected that is very similar to the zeolite sample in question. The reference zeolite should also be highly crystalline since the amount present in the sample will be compared to it. This sometimes creates an anomaly when the sample zeolite has a relative amount of crystalline zeolite that is greater than unity. This only means that the reference zeolite was not as crystalline as the sample being analyzed. Since water in the pores often suppresses the low-angle intensities, pretreatment conditions must be standardized and moisture change during analysis must be minimized. Many labs initially dry the sample in a drying oven and then equilibrate it at a constant humidity. The instrument may also be kept at a similar humidity level to minimize changes in moisture content. Other laboratories analyze the dry sample under moisture-free conditions. Since the reference sample must also be periodically run for comparison, some laboratories use a less moisture sensitive reference as a standard.

Typically several diffraction peaks are used for comparison. ASTM method D3906 for faujasite materials uses a comparison of eight different peaks. Although the method was developed for FCC catalysts, it is also applicable to other FAU-type

4 Zeolite Characterization

adsorbents. ASTM method D5357 for LTA zeolites uses the six strongest peaks. If there is suspicion that one or more of the peaks to be used has interference from other crystalline species in the sample, it should not be included in the analysis. When the amount of zeolite present is extremely low, the single most intense peak can be used for comparison.

Changes in peak intensities can be used to monitor treatment processes. For example, an overall decrease in peak intensity may indicate destruction of zeolite. Other changes in peak intensities may indicate the treatment has modified the atomic positions. Often, a general increase in low-angle intensity relative to high angle intensity indicates removal of material such as water from the pores. Presence of coke in the pore structure can also modify the XRD pattern. Non-uniform changes may indicate changes in cation content or cation location. Comparing these two patterns also demonstrates the necessity as mentioned previously with respect to using a reference zeolite that resembles the zeolite of interest for determining relative zeolite content.

4.3.3
Unit Cell Size Determination

Changes in peak positions indicate changes in the unit cell size or shape. For example, shifts of peaks to smaller d-spacings (higher two-theta) might indicate dealumination of the zeolite, since on average Al-O distances are longer than Si-O distances (see Figure 4.7). Such dealumination can occur during manufacture,

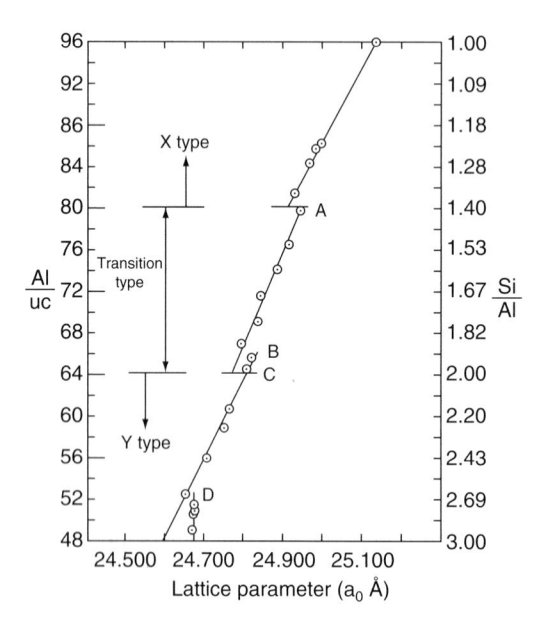

Figure 4.7 The number of Al atoms per unit cell and the Si:Al ratio versus a_0 for hydrated sodium faujasites (from Dempsey [21]).

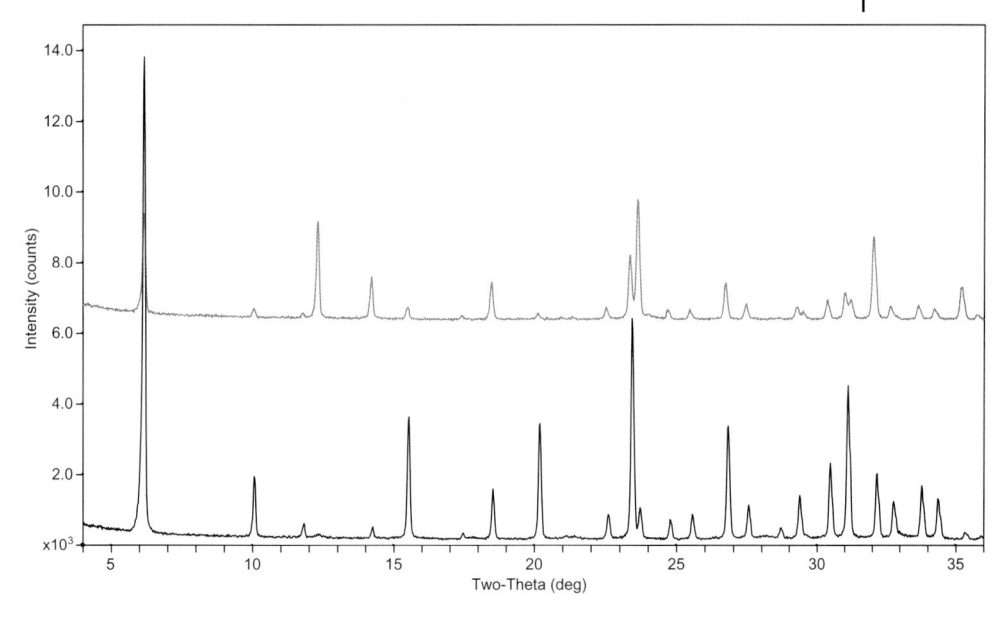

Figure 4.8 X-Ray powder patterns for Na-X (bottom) and Ba-X (top) zeolites.

service, regeneration, or other treatments. For hydrated Na-exchanged faujasite zeolites the unit cell size, a_o, increases from about 24.6 to 25.1 Å as the number of Al atoms in the unit cell increases from about 48 to 96 [20]. Since there is a close to linear relationship between the aluminum atom density in number/unit cell and the unit cell size (Figure 4.8), the Si/Al ratio of the zeolite can be obtained [20, 21]. This is quite useful for bound systems that have either Al or Si in the binder or matrix from other sources because the use of chemical analyses does not provide the Si/Al specifically of the zeolite. Similar relationships can be developed for the presence of other cations [20] and for other zeolites.

4.3.4
Crystallite Size

Average crystallite sizes can be estimated from peak widths (e.g., see [22]) using the Scherrer equation:

$$L_{hkl} = K \times \lambda / (\beta_{hkl} \times \cos \theta_{hkl}) \qquad (4.1)$$

where L_{hkl} is the crystallite length in the direction given by the Miller indices hkl, K is a constant which depends on the method of measuring the peak width (commonly set to 0.9), λ is the wavelength, β_{hkl} is the width of the hkl peak corrected for instrumental broadening and θ_{hkl} is one-half of the diffraction angle. For in-house diffractometers estimates are only accurate for crystallites smaller than about 1000 Å. This may also provide crystallite shape information when the peak broadening is found to be different in different crystallographic directions. Such crystallite size determinations may or may not be comparable to those obtained by scanning electron microscope (SEM), as noted in Section 4.4.2.1.

Unusual broadening and appearance of intensity in apparently non-Bragg positions may indicate the presence of faulting and/or intergrowths. Some examples of pairs of zeolite types that can lead to faulting are AEI/CHA and FAU/EMT. More complex faulting involving more than two end-members can occur (see the IZA Structure Commission web site for more details). The simulation of these types of patterns is complicated but can be done with the software program DIFFaX [23]. For example, Treacy et al. [24] used DIFFaX to simulate the full powder pattern profiles for FAU/EMT intergrowth materials and to then estimate the faulting probabilities. They showed that the twin faults from TEM did not typically occur randomly but tended to cluster into segregated blocks of intergrown hexagonal and cubic material.

4.3.5
Rietveld Refinement

It is not hard to extrapolate from the above information that because the powder pattern is determined by the unit cell size and shape and the positions and types of atoms in the unit cell, that it might be possible to work backwards and determine the crystalline structure for an unknown material from its powder diffraction pattern. Those familiar with single-crystal diffraction know that this is usually a routine procedure given a suitably large single crystal (greater than about 100 μm on edge) on modern in house single-crystal diffractometers. Unfortunately industrial zeolite powders usually have crystals that are much too small for single-crystal diffraction. So when a powder diffraction pattern for a zeolite shows that it is a new material, usually powder methods must be used. The fundamental problem with powder data is that there is strong overlap of diffraction peaks from different reflections and complete deconvolution of separate intensities for each of the peaks is not possible. The difficulty then lies in how to partition the intensities among all of the overlapping peaks. Various methods have been developed to do this and research is still continuing [25–30]. Once intensities have been partitioned, single-crystal-like methods sometimes can be used to solve the structure. Once a model has been developed for a structure, the Rietveld refinement is used to provide confidence that the model is correct.

Rietveld refinement [25, 26] is a method of whole pattern refinement, where a calculated diffraction pattern for a structure model is a least-squares fit to an observed diffraction pattern. Originally, it was used as a means of verifying proposed structure models. For zeolites, Rietveld refinement is still used for the same purpose and provides details of the structure including atomic positions of framework atoms and cation sittings. Data with accurate intensities and well-resolved peaks are needed for the most accurate work, and so often a synchrotron source is used for data collection since it can provide higher intensity and peak resolution than an in-house diffractometer. However, modern in-house diffractometers often provide good enough data for some refinements.

Increasingly, Rietveld refinement is also used as a more accurate way for determining the other types of information discussed above, including phase identification, phase quantization, unit cell size determination and crystallite size analysis. The advantage of the Rietveld method for these applications is that it uses the entire diffraction pattern instead of just a few diffraction peaks. Details of the methods used are given in several good books and references [27–30].

4.4 Electron Microscopy Characterization of Zeolitic Systems

4.4.1 Importance of Electron Microscopy for Characterizing Zeolites

Electron microscopy has contributed widely to many aspects of current understanding of zeolitic materials. These aspects may broadly be classified as structural, morphological and compositional. Associated with these three aspects are basic applications of three dominant uses of electron-beam based materials characterization: diffraction and high-resolution TEM imaging (for bulk and defect structure), conventional imaging using TEM and SEM (for morphology) and energy-dispersive X-ray spectroscopy (EDS; for compositional zeolite characterization).

Since about the late 1950s, electron microscopy has been used for structural and morphological characterization of zeolites. TEM was used in early zeolite studies for routine morphological examination and unit cell determination [31]. Although most early structural work on zeolites relied heavily on X-ray diffraction, electron microscopy and diffraction were applied in a number of important cases. Thus evidence from electron diffraction was applied beginning in the late 1950s, for example to distinguish offretite from erionite structures [32, 33] and to show the presence of superlattice intensities related to Si-Al ordering in A-zeolite [34]. Diffraction contrast imaging was used in 1972 by Kokotailo *et al.* to study stacking defects in erionite [35]. In this same period of early electron microscope instrumentation, work on zeolites also played a role in the development of direct structural imaging with TEM. The first demonstration of direct imaging with TEM of a crystal lattice spacing was performed in 1958 by Menter [36], who obtained images showing the 14.7 Å <111> lattice plane of faujasite. As the range of instrumentation increased, with improvements in TEM resolution, the advent of the scanning electron microscope (SEM) and the development of EDS, electron microscopy increasingly became a routine tool to address a broad range of aspects of zeolite science.

Starting in the early 1980s, improved instrumentation for high-resolution TEM made it possible to study details of zeolite structure directly by means of TEM imaging [37–40]. With the advent of the commercial scanning electron microscope (SEM) in the 1960s, the SEM rapidly became a routine tool for imaging zeolite morphologies, for example [41]. The development of EDS beginning in the late 1960s [42] led to a further broadening of the applicability of electron microscopy to the study of local chemical composition, for example, changes of chemical composition towards the exterior of a zeolite crystal, or variations in composition from particle to particle.

While the range of techniques available in electron microscopy are widely used and have contributed greatly to understanding of zeolite morphology, structure and local chemical composition, it is important to recognize some limitations and aspects of zeolites for which electron microscopy is not as well-suited. One of the technologically most important aspects of zeolites is their acidity, provided by hydroxyl groups associated with an electronic defect due to the presence of substitutional cations in the framework. Because of the weak scattering of electrons by a hydroxyl group and the typically small atomic number differences of the substitutional cations, electron microscopy is limited in its ability to detect and characterize acidity in zeolites. While there may be some latent possibilities due to the sensitivity of low-angle electron scattering to charge distribution [43], in general, structural phenomena without long-range order, and associated with light elements will always be a challenge for electron microscopy. Thus the effect of dealumination of a zeolite would be very difficult to characterize directly using electron microscopy. Fortunately the use of NMR, FTIR and many analyses using probe molecules are ideally suited for the study of phenomena such as dealumination or acidity in zeolites. Thus an approach using a combination of disparate characterization techniques presently available can provide a thorough fundamental understanding of virtually all zeolite properties.

4.4.2
Scanning Electron Microscopy

4.4.2.1 Morphological Characterization

SEM is broadly applicable to the morphological study of zeolites. For a typical zeolite powder sample, little effort is required for sample preparation and interpretation of morphological information from the images is usually straightforward. There are several important advantages of SEM for morphological characterization. First, SEM can be applied without special sample preparation to larger particle morphologies, above a few tenths of a micron, which in TEM are not accessible due to the requirement that the electron beam penetrate through the sample. SEM morphological images of zeolites are usually intuitively interpretable due to the sensitivity of the detected signal (secondary electrons) to the local surface orientation, thus producing a "shadow" effect reflecting the surface topology. The fact that SEM secondary images are effectively surface images means that there is no ambiguity as to whether or not a feature in an image resides on the particle surface or is embedded within the bulk. This is not true for TEM in which the image represents a projection through the particle, and this can add to the difficulty of assessing particle morphologies with TEM images. One challenge for zeolites is that due to the low density and low atomic number the beam penetration can be fairly large, leading to delocalization of the secondary electron emission and degrading the sensitivity to the smallest surface features. This makes it desir-

Figure 4.9 SEM image of Y zeolite taken with field emission SEM, operated at 0.5 kV. Sample prep was by dispersing powder in isopropyl alcohol and placing a droplet of the suspended powder on a standard TEM grid. Prior to examining in SEM a light chromium coating was applied to the sample to prevent charging.

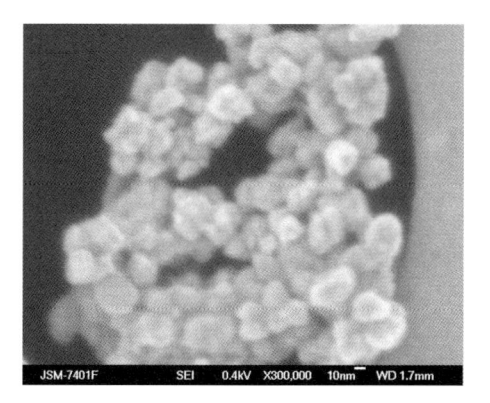

Figure 4.10 SEM image of nanocrystalline zeolite taken with field emission SEM operated at 0.5 kV. Sample prep was by dispersing powder in isopropyl alcohol and placing a droplet of the suspended powder on a standard TEM grid. Sample is uncoated.

able to use lower voltages which can reduce the penetration depth and increase the surface sensitivity of the images. Recent developments in instrumentation, particularly field emission SEMs, have significantly improved SEM imaging capabilities at low voltage, permitting routine observation of surface details in the size range of a few nanometers, at accelerating voltages of 1 kV or below. Figures 4.9 and 4.10 show field-emission SEM images of a conventional Y zeolite and a nanocrystalline zeolite, both taken at low accelerating voltage. In the case of the Y zeolite, a fine chromium coating was applied to reduce sample charging, whereas this was not necessary in the case of the nanocrystalline zeolite due to the small morphology and low accelerating voltage. Note that in Figure 4.10 identifiable features of the morphology are imaged, which are in the range of about 5 nm or less in size.

Most applications of zeolites require a feed or adsorbate to access the interiors of the zeolitic micropores. In many instances the rate at which this occurs will tend to increase with the density of pore openings, which for example in a 3D pore system depends in the simplest approximation on $1/d^2$ where d is the crystal diameter. Thus, for zeolite applications in catalysis and adsorption the crystal morphology is often a dominant factor in the material's performance [44]. Crystal morphology is often related to the activity in catalytic applications, smaller crystals often showing enhanced activity. Morphology may also affect the dwell time of a molecule within the zeolite crystal, which may determine the extent of side reactions. Finally, morphology may impact forming processes used to make commercial catalysts. For this reason, SEM plays a vital role in zeolite science and development, providing important understanding for applications as well as feedback for tailoring synthesis conditions to achieve a desired morphology.

4.4.2.2 Compositional Characterization

Compositional characterization by energy dispersive X-ray spectroscopy (EDS) is a basic tool in SEM as well as TEM. Useful introductory texts providing an overview of EDS in SEM and TEM respectively can be found in references [45, 46]. Using EDS it is straightforward, for example, to qualitatively identify an impurity in a sample based on its composition. Due to the local nature of the analysis in the case of SEM (typically a few microns cubed are analyzed, depending on the accelerating voltage and X-ray line energy), it allows the analyst to determine the composition of any observed morphological heterogeneities. Aside from routine applications such as identification of impurities which are critical for development of materials, there are two classes of applications which advance understanding of zeolite properties and synthesis. The first of these is the detection of compositional gradients within a zeolite crystal or agglomerate and the second is the detection of particle to particle variation through automated acquisition of large numbers of single-particle EDS analyses. Both of these advanced applications are applicable to zeolite systems which are fully developed (i.e., in which impurities or other issues have already been dealt with). Examples of both of these are given below.

In order to detect compositional gradients within a single particle, it is generally necessary to obtain a cross-section of a single crystal or particle. For particles smaller than several microns in diameter, this is not practical using standard mounting and polishing techniques because of problems with particles pulling out of the acrylic or other mounting medium. Thus in order to detect compositional gradients within a zeolite crystal of typically micron-sized proportions, it is preferable to use a microtome with diamond knife in order to expose the crystal cross-section. Furthermore, in order to reduce the interaction volume to a small fraction of the total crystal, it is necessary to obtain a thin section in the same thickness range as a TEM sample. Fortunately, this can readily be done for zeolites using a microtome and for more details the reader is referred to Section 4.3.1 on TEM sample preparation. Use of modern EDS equipment and software permits acquisition of a "spectrum image" [47], which for each sample position or pixel

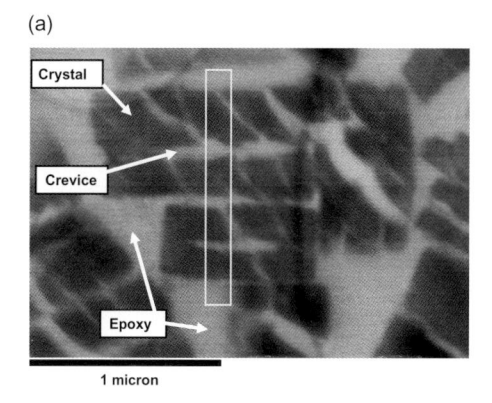

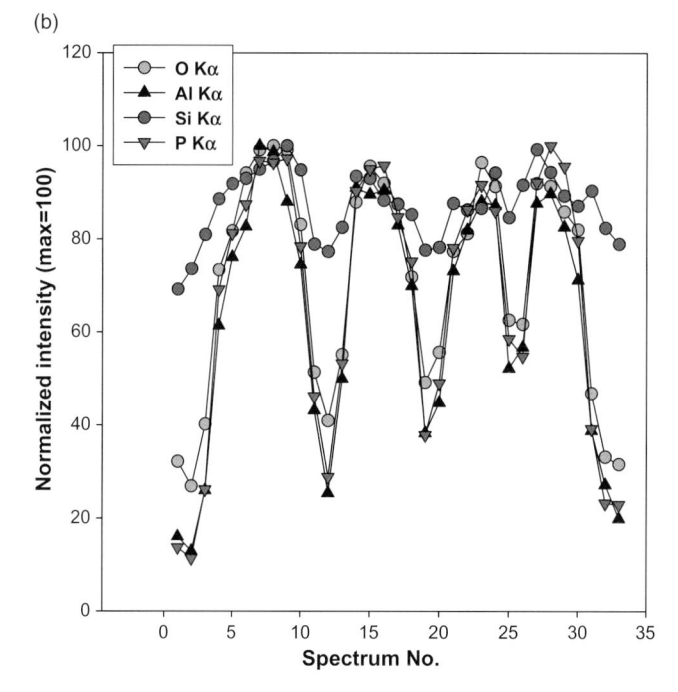

Figure 4.11 (a) SEM image of thin-sectioned silico-aluminophosphate (SAPO) crystal showing platelike morphology of adhering plates. Box indicates location of X-ray map. (b) Plot of peak areas from 32 spectra extracted from X-ray map. Each spectrum is a horizontal strip of the map shown in Figure 4.11a, with 35 nm distance between them. Peak areas have been scaled so that the maximum in the set equals 100. Whereas O, Al and P vary together, there is excess Si at the surfaces and within crevices.

within an image records the EDS spectrum (counts per energy channel). An example of SEM spectrum imaging applied to compositional heterogeneity within a silico-aluminophosphate (SAPO) crystal is shown in Figure 4.11a, indicating that the surfaces of the crystals are Si-enriched. In order to prevent the sample from drifting during acquisition, software-driven sample-drift correction was used. The type of individual-particle EDS analysis shown in Figure 4.11b can provide improved understanding of the sequences of events in zeolite or molecular sieve synthesis or subsequent treatments. It also may support efforts to tailor the composition gradients to optimize performance due to their impact on the spatial distribution of acid sites.

An additional advanced application of EDS in the SEM is the determination of compositional variation among particles in a powder. This general approach to powder characterization by SEM was first used by Byers et al. in 1971 [48], and may be of particular relevance to zeolite science and technology in determining the influence of synthesis parameters on compositional heterogeneities in a zeolite powder. Because of the need to analyze large numbers of individual particles (typically at least a few hundred) this is also a software-driven application. For preparing samples, the powder can be dispersed in a solvent such as isopropyl alcohol and a droplet placed on a bulk carbon substrate. For zeolite powders this typically results in high-contrast secondary electron images which can be used for automated threshholding to obtain the locations and outlines of the particles. Figure 4.12 shows an example of an application of this type of study to a problem of zeolite synthesis, in which a silica source was slowly added to an alumina source in synthesis solution and aged under stirring at different temperatures (room temperature, 75 °C) for the same amount of time, followed by separation and drying of the solids. The aim was to determine whether different makeup procedures might affect the degree of variation in the particle-to-particle compositional variation. The SEM study utilized an automated routine to locate the particles in an initial image, then scan within the particle outline during the EDS spectrum acquisition (20 kV beam energy). The compositional histograms shown in Figure 4.12a (room temperature aging) and Figure 4.12b (75 °C aging) show different widths, indicating that the slow addition of silica nutrient combined with low temperature aging can result in greater particle-to-particle heterogeneity than occurs by aging at elevated temperature. Both the asymmetry of the distribution at 75 °C as well as the overall shifted composition are indications of a greater extent of reaction of the silica and alumina, as well as possible depletion of alumina source prior to addition of the final silica.

There are numerous experimental factors which need to be carefully controlled in conducting studies of particle-to-particle compositional variations. First, it is important to be sure that the preparation of the sample for SEM has not produced large agglomerates of solids. In cases where agglomeration occurs on evaporation of the solvent, one possible approach may be to suspend in a camphor–naphthalene flux above the 32 °C melting point, followed by rapid solidification on the substrate and then by sublimation of the camphor–naphthalene in vacuum [49]. In addition to potential pitfalls in the sample preparation, it is important to be aware of systematic artifacts in the compositions resulting from the EDS analyses, particularly if the particles vary significantly in size (e.g., see [50]). This can be controlled up to a point by rejecting particles outside of

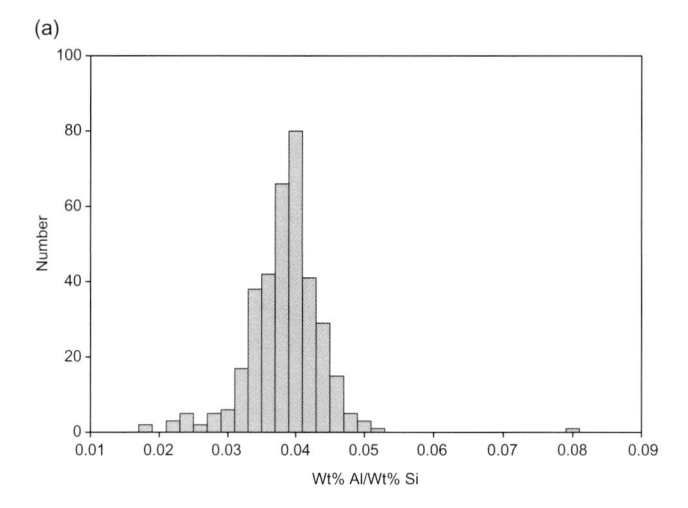

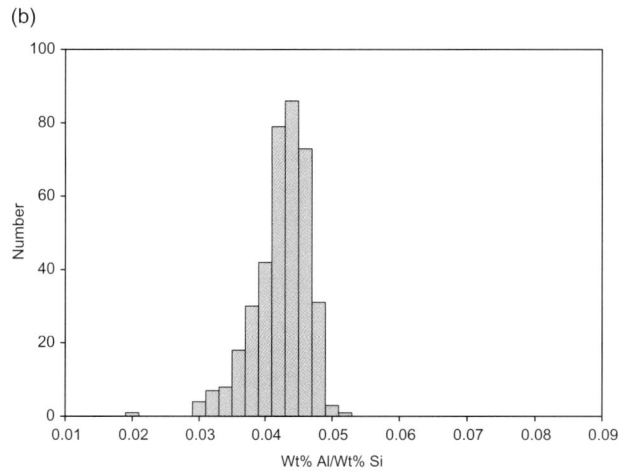

Figure 4.12 Histogram from automated SEM-EDS particle analysis of 362 zeolite precursor particles. Particles were reacted from silica and alumina sources slowly combined and aged at (a) room temperature and (b) 75 °C.

defined size limits etc. However in general comparing the compositional distribution of two powders with different size distributions must be approached with extreme caution. Finally, the histograms obtained contain broadening effects due to uncertainty in quantitation (e.g., random errors related to noise in the spectra). This will add a broadening on top of the true compositional spread and might in some cases overwhelm the compositional spread. The broadening effect due to random experimental errors such as signal to noise in the spectra may be assessed if a suitable standard can be found with negligible inter-particle compositional variation.

4.4.3
Transmission Electron Microscopy

4.4.3.1 Sample Preparation

In contrast to SEM, in which the detectors used to produce an image are generally situated on the side of the sample closest to the electron source, both TEM or STEM imaging use detectors located below the sample (or generally on the far side of the electron source). Thus in spite of the significantly higher voltages (usually 100–400 kV) in TEM or STEM, it is often necessary to resort to some means of reducing the sample thickness prior to examination to permit electron transmission, usually about 100 nm thickness. In the case of zeolites, the need for this depends on the morphology of the crystals (or more generally particles), as well as on the type of data sought (e.g., qualitative diffraction spots, high-resolution images, quantitative EDS spectra). For zeolite powders with particles below a few hundred nanometers one may be able to do much useful analysis by simply dispersing the particles in a solvent and placing a droplet of the suspension on a carbon support grid (e.g., "holey carbon"). In cases of nanocrystalline zeolites, such as that pictured in Figure 4.10, this would be the preparation of choice for all applications (diffraction, imaging and EDS analysis). For larger morphologies, for example, the Y zeolite shown in Figure 4.9, most TEM applications except possibly qualitative electron diffraction spot patterns would be severely hampered by the large thickness, and it will be necessary to find some means of sectioning the crystals to obtain a smaller and/or more uniform thickness. In special cases of highly platelike crystals, sectioning is also needed in order to be able to obtain thin samples oriented on zone axes contained within the plane of the plates.

In general for materials science applications involving dense materials, ion beam thinning or mechanical (tripod) polishing are the most common means of producing samples. Fortunately these difficult and time-consuming thinning techniques are not suitable for zeolites, which are easily damaged by ion beams and being powders do not lend themselves easily to mechanical polishing. Rather, due to their softness and low density, zeolites behave quite well in thin sectioning techniques developed for biological materials, specifically using microtomy with the help of a diamond knife. With this approach it is usually possible to obtain ~70–100 nm thick sections of randomly oriented zeolite crystals [51].

The first step in microtomy is to embed the powder in one of a variety of media, which can be either an epoxy [21] or pure sulfur [52]. There is a range of epoxies suitable for microtomy of zeolites, but generally the requirements are that it be hard and that it infiltrate or adequately wet the zeolite powder before curing. Both LR white and Epon-type epoxies (if mixed properly) meet the hardness requirement. The use of Epon enhances the wetting of silicious or carbon-containing zeolites, but it contains low levels of chlorine which may cause overlaps in EDS and will make it impossible to analyze for Cl. Typically a small pinch of powder is blended with two drops of epoxy in the tip of an embedding capsule. Mixing can be done with a clean toothpick, combined if needed with a cycle or two of evacuation in order to improve penetration. After blending, the capsule is topped off and cured. In the case of sulfur embedding, a small droplet of molten sulfur can be blended rapidly on an inert substrate on a hot plate with a pinch of zeolite powder. This is then solidified on the tip of a spatula and the solid granule is re-melted on the tip of a cylindrical tapered base of suitable size to be mounted in the microtome's chuck. With either epoxy or sulfur, thin sections are cut with a high-quality diamond knife using thickness settings in the range of about 50–100 nm. These are collected on holey carbon grids and in the case of sulfur embedding the sulfur is sublimed off in an auxiliary pumping station before introducing into the TEM. If using sulfur, this allows one to obtain a section which is devoid of embedding medium, at the cost of some residual sulfur in the EDS spectra. Both epoxy and sulfur embedded thin sections can produce excellent samples of quite uniform thickness for compositional analysis and high-resolution imaging. The epoxies can present some challenges in high-resolution imaging due to their instability and tendency to deform under intense electron bombardment. Because zeolites are essentially brittle materials, sectioning does not result in perfectly uniform thickness but instead tends to shave away segments of parallel chips of material [53].

4.4.3.2 Structural Characterization

The crystal structures of zeolites largely determine their potential as catalysts and adsorbents. The pore dimensions dictate what molecules can access the interiors of the pores, as well as the product shape selectivity in catalytic applications. As indicated in Section 4.4.1, application of high-resolution TEM in zeolite structure solutions and other structure characterizations has become fairly routine and has contributed greatly to structure solutions (e.g., refs. [38–40, 54–57], see review in [58]). Due to developments over the past quarter century, conventional TEM and field-emission STEM instruments are capable of achieving resolutions better than 0.2 nm, and are thus capable of revealing structural features of zeolites on the atomic scale. One of the primary challenges of working with zeolites in the TEM is beam sensitivity [38, 59], which in severe cases may make it virtually impossible to acquire images at magnifications sufficient to record details of the zeolite structure. Treacy and Newsam [59] found that the beam sensitivity increases with the degree of hydration and also with the framework charge, and they proposed a model to explain their observations based on generation of hydroxyl radicals due to beam ionization. The problem of beam sensitivity can be addressed by using low-dose methods and may be reduced by using cryogenic sample cooling and/or higher accelerating voltages. It is also reduced by dealumination of zeolites [60]. In the case of bulk structure studies low-dose methods combined with processing of images by real-space averaging can significantly improve the quality of high-resolution structural details [57, 61]. Figure 4.1 (see Section 4.2) shows HR-TEM work on the zeolite UZM-5 (UFI framework type). As shown in Figure 4.1b and c, the raw HR-TEM image was processed using a real-space averaging of the image intensities of ten unit cells. As described in Section 4.2, the image in Figure 4.1 provided the basis for solving the UFI structure using model building, in combination with other techniques [57] such as electron and X-ray diffraction, FTIR, and the gravimetric adsorption technique described in Section 4.7.3.

In addition to TEM imaging, electron diffraction has also routinely been used to aid in symmetry determination. Although full space group determination using convergent beam diffraction is generally not feasible due to beam sensitivity of zeolites, spot patterns can generally resolve the crystal Laue symmetry. Recently, the combination of direct methods with electron diffraction intensities is increasingly being used for bulk structure solutions. In this case, measured electron diffraction intensities are used to provide more complete structural information, going beyond the auxiliary use of electron diffraction geometry and systematic extinctions for symmetry information. Since the first demonstration of the solution of MCM-22 by Nicolopoulos *et al.* in 1997 [62], a number of zeolite structural studies based on diffraction and direct methods have been published [63, 64], including some *ab initio* solutions of unknown structures [65, 66].

Most of the applications of electron diffraction intensities for structure analysis rely on a kinematical approximation and thus do not account for the effects of dynamical multiple diffraction. The use of intensities which may be strongly perturbed by multiple scattering results in many cases in poor or misleading structure indications in the direct methods results. One approach which can be shown to reduce dynamical effects somewhat is to use precession electron diffraction (PED) [67] which involves conical rotation of the incident beam about a zone axis direction and thus avoids the strongly dynamical direct zone axis orientation. Although the intensities collected with this technique are still significantly perturbed by dynamical effects [68, 69] results obtained by this approach for zeolites are encouraging [70–72].

Recently, improvements in resolution of both TEM and STEM instruments has been achieved by the development of advanced electron optics for correcting high order aberrations. These developments have extended the resolution on current state of the art aberration corrected instruments to below 0.1 nm. While this is a relatively new field, it holds significant prospects for obtaining improved structural information in high-resolution TEM or STEM images of zeolites. In both cases beam sensitivity is still a concern and may place a greater limitation on the results than does the instrumental point resolution. However, recent results using aberration-corrected STEM on clathrates [73], an equally beam-sensitive family of materials, suggests that improved resolution in STEM images of zeolites might also be feasible using aberration corrected STEM. The application of aberration TEM might also be expected to yield similar improvement, possibly with less impact of beam sensitivity.

4.4.3.3 Morphological Characterization

The use of TEM for morphological characterization has increasingly been supplanted as improvements in SEM, particularly combined with the use of lower voltage, steadily decrease the size range for which TEM has clear benefits. Nevertheless, for some important cases of zeolite morphologies, TEM may still be the

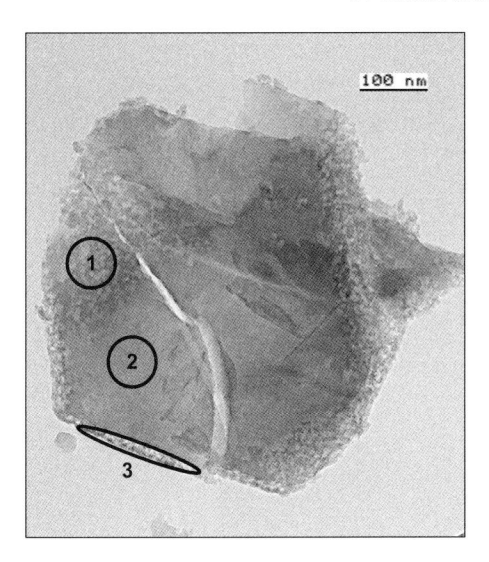

Figure 4.13 Sectioned crystal of dealuminated Y zeolite showing mesoporous region at edge. EDS analyses were performed using TEM mode at marked locations. The results show molar Si/Al ratios of approximately 7.5 ± 0.1, 7.1 ± 0.2 and 13.0 ± 0.3. Condenser astigmatism coils were used to produce an elliptically-distorted beam shape for acquiring spectrum 3. The crack traversing the crystal is from thin sectioning.

method of choice. One case in point is the UZM-5 zeolite described in the previous section, which crystallizes as rosette-shaped particles comprised of many very fine platelets (typically only on the order of 100 Å thick). Because of the very fine morphology and the greater difficulty of controlling sample charging in the SEM than in TEM, the latter can be very effectively used for morphological characterization. Another case would be extremely fine-grained zeolites such as shown in Figure 4.10. In this case a TEM image may be preferable because it can reveal lattice planes which contain supplementary information about the extent of individual crystals, whether the particles are single crystals, etc. A final factor which may influence the choice of TEM for morphological work is that it provides insight into internal mesoporosity due to the fact that the image represents a projection of image effects through the thickness of the particles (provided these are small enough to permit adequate transmission of intensity). Examples of this are shown in Figures 4.13 and 4.14 which are respectively images of a Y zeolite crystal showing mesoporosity caused by dealuminating treatments, and a particle of zeolite (crystallized as a colony of multiple, mutually aligned nanocrystals).

The fact that TEM images for adequately thin objects are effectively projection images has allowed the recent development of TEM-based tomography, and its successful application to molecular sieves and other catalytic materials [74–77]. This technique uses image tilt series (single or dual axis) comprising on the order of 100–200 images, which are then aligned and processed to produce a tomogram containing a three-dimensional reconstruction of the sample topology

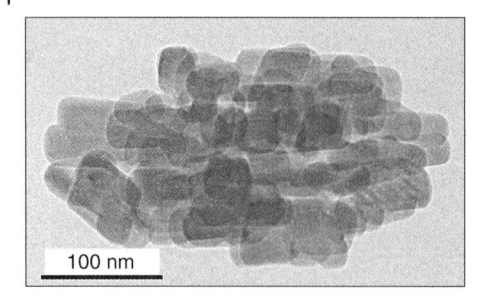

Figure 4.14 Image of cluster of oriented zeolite crystals. Because the TEM image shows a projection through the entire thickness of the cluster, it can be seen that the crystals terminate within the cluster leaving internal mesoporosity.

at a resolution on the order of one to several nanometers [78]. One limitation for zeolites is due to the issue of beam sensitivity. For intermediate resolutions (e.g., imaging of mesoporosity such as shown in Figure 4.14) it may not be necessary that the crystallinity be preserved. However any significant distortion of the morphology in the course of acquiring the image series would degrade the tomogram quality and resolution or prevent successful alignment and computation of the tomogram. The range of applicability to zeolites and other catalytic materials is nevertheless impressive and this technique is likely to have an increasing presence in the science and technology of zeolites and other mesoporous catalytic materials.

4.4.3.4 Compositional Characterization

If a thin section is used, either TEM or STEM or SEM can be used for local chemical characterization of zeolites using an approach not significantly different from what was described in Section 4.4.2.1 for Figure 4.11. In particular, it is the use of a thin-sectioned sample more than the choice of instrument which makes possible the compositional characterization at scales finer than an individual crystal. The choice of STEM over TEM allows the added convenience of selecting for analysis a location within a stored image, or collecting a spectrum image. When operating in TEM mode the location for analysis is selected by placing the beam on a feature of interest and collecting for a suitable time to provide a low-noise spectrum. The ability to analyze irregularly shaped locations is somewhat constrained in TEM mode. However by using condenser lens and astigmatism coils it is possible to analyze thin elongated areas of roughly elliptical shape. An example of this is shown in Figure 4.13, which concerns a surface-leached Y zeolite, and indicates by EDS a dealumination of the exterior regions of the crystal, to a depth of about 20 nm.

Provided that a suitably thin sample is used and also that the beam energy is sufficient to neglect large changes in ionization cross-section with depth in the sample, it is appropriate to use a Cliff–Lorimer approach to spectrum quantification regardless of what type of instrument is used. The main differentiator with respect to instrumentation will be the detector solid angle, which is typically on the order of 0.1–0.2 steradians for TEM, STEM and in-lens SEM instruments and closer to 0.01 steradians for most standard SEM instruments.

Along with the advantages of flexibility in the areas analyzed and the possibility of mapping or spectral imaging, scanning microscopy can have drawbacks for STEM EDS of zeolites, which arises from the very fine intense beam used. If this beam is left for even milliseconds in a fixed location it can cause compositional shifts due to preferential migration of one or more elements away from the beam [59, 60], again an effect of beam sensitivity. The possibility of migration of one or more elements should be carefully considered in evaluating compositional data acquired using extensive electron-beam exposure, such as that shown in Figure 4.11. The presence of artifacts can be checked for by acquiring spectra in TEM mode or with a defocused probe, and checking for systematic disagreement in the elemental analyses. Another factor that can affect STEM-mode EDS work is sample drift. As discussed in Section 4.4.2.2, software-based sample drift correction using image cross correlation is now available from a number of instrument vendors and if used properly can largely remove this issue.

4.4.3.5 STEM Application to Metals in Zeolites and Coke Analysis

Characterizing the size, location and chemistry of small metal particles that provide the metal functionality in a catalyst is important for both the development of a catalyst as well as identifying whether unusual performance in service is caused by the metal function. The advantage of using microscopy for the characterization of the metal clusters is that it is site-specific.

High-angle annular dark-field or Z-contrast imaging in the STEM is extremely useful for characterizing small metal clusters. With this technique the high angle annular dark-field detector can distinguish between metal clusters with high atomic numbers and the zeolite or catalyst support with low atomic number constituents. A small coherent electron probe is scanned across the sample. At small scattering angle, mass thickness and coherent diffraction contrast are the predominant contrast mechanisms in STEM bright-field imaging. At larger scattering angles, the coherent scattering is progressively replaced by incoherently generated thermal diffuse scattering, which then becomes dependent on Z, the atomic number. At sufficiently large angles, the scattering approaches Z^2 dependence, which is the unscreened Rutherford scattering. To image the small metal clusters in a zeolite, a focused incident probe smaller than 0.5 nm is necessary with a probe approaching 0.2 nm or smaller being preferred. As with TEM to minimize beam damage to the zeolite, using low-dose methods and cryogenic sample cooling are often necessary.

Rice *et al.* [79] were the first to demonstrate small metal clusters randomly located throughout the channels of a zeolite L crystal. Numerous channel-blocking

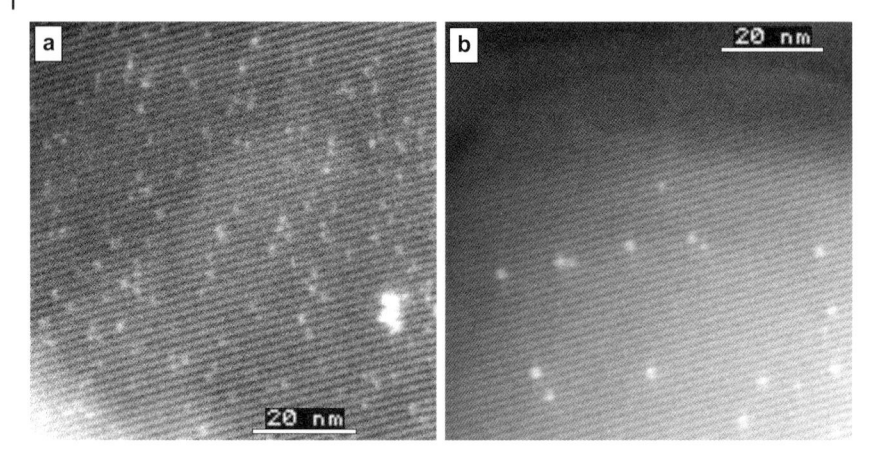

Figure 4.15 (a) Pt clusters for fresh catalyst are about 1 nm in size and found in the channels of the Y-zeolite. (b) After time under reaction conditions the clusters have agglomerated to about 2 nm in size.

clusters were observed, which suggested a deactivation mechanism. Such clusters were not visible in the bright-field or TEM imaging mode. At approximately the same time Pan et al. [80] imaged and performed electron microdiffraction of Pt clusters in two forms of Y-zeolite.

For a zeolitic catalyst where Pt, Pd or other transition metal might be present to provide metal activity, STEM can be used to determine whether the metal is agglomerated and to what extent, whether the metal is in the zeolite or present on the geometric exterior or whether the metal is associated with the zeolite or binder. As an example of the utility of the technique, Figure 4.15 shows the growth of Pt clusters for fresh and spent faujasite zeolite catalyst. After time under reaction conditions, the Pt clusters have grown from 1 nm to 2 nm. The clusters have remained in the channels of the faujasite. Pt agglomeration can be concluded as the deactivation mechanism.

The combination of STEM with analytical techniques is a powerful way for probing at the nano-scale. With analytical electron microscopy, concentrations from particles as small as 1 nm^3 can be determined. EDS is the most common approach but electron energy loss (EELS) is very useful for determining the presence and electronic state of first row transition metals and for low atomic number elements. EELS involves analyzing the energy distribution of initially monoenergetic electrons after they have interacted with a specimen. EELS [and the related imaging techniques of energy-filtered TEM (EFTEM) and EELS spectral imaging] is most useful for characterizing spent catalysts. Figure 4.16 is an example where EELS identified the coating on the zeolite to be carbonaceous material that had grown while the catalyst was under reaction conditions and thus is the probable deactivation mechanism for this reaction. Variations in the C_K edge can be used to differentiate carbonaceous species.

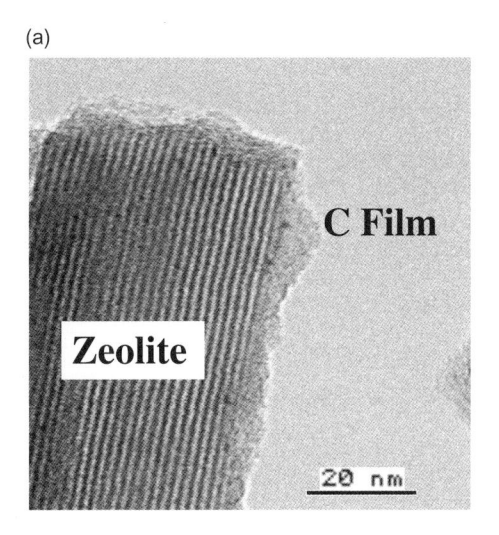

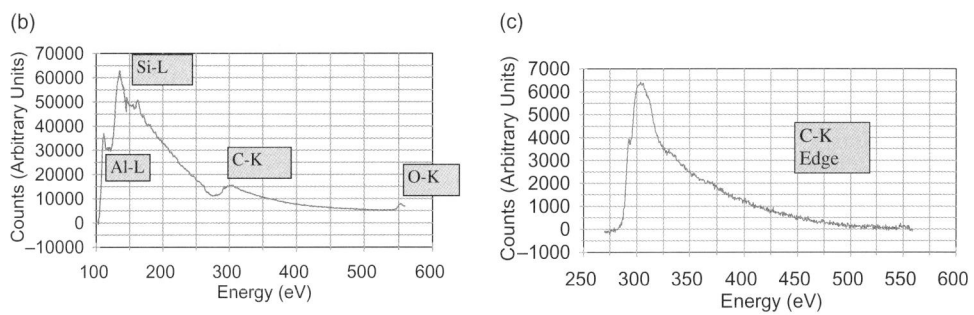

Figure 4.16 (a) Bright-field image of film on spent zeolite catalyst showing surface carbon buildup. (b) EELS specturm of zeolite and carbon. Identification of film is not obvious. (c) EELS spectrum from film showing it to be amorphous carbon.

4.5
Infrared Spectroscopy Characterization of Zeolitic Systems

4.5.1
Introduction to Infrared Spectroscopy

Infrared spectroscopy has proven to be a very informative and powerful technique for the characterization of zeolitic materials. Most infrared spectrometers measure the absorption of radiation in the mid-infrared region of the electromagnetic spectrum (4000–400 cm^{-1} or 2.5–25 μm). In this region of the spectrum, absorption is due to various vibrational modes in the sample. Analysis of these vibrational absorption bands provides information about the chemical species present. This includes information about the structure of the zeolite as well as other functional groups that may be present as a result of how the material was synthesized and treated (e.g., ion exchanged, calcined).

Many excellent reviews on the application of infrared spectroscopy for characterization of silicoaluminates have been written and some are referenced here if a more thorough coverage of this topic is desired [81–89]. The purpose of this section is not to cover infrared spectroscopy of zeolites in great detail, but to give the reader an overview of the utility of the technique and the types of information that can be obtained. References are provided if more in-depth information is desired.

4.5.2
Modes of Measurement

Infrared spectra of zeolitic samples can be measured in several different modes. These include transmission, diffuse reflectance, attenuated total internal reflection (ATR) and emission. Transmission and diffuse reflectance are by far the most widely used of these techniques. In the transmission mode, the sample is placed directly in the infrared beam of the instrument and the light passing through or transmitted is measured by the detector. This transmitted signal (T) is ratioed to the open beam (no sample) signal (T_0) to get the transmission spectrum of the sample. The transmission spectrum is converted to an absorbance spectrum:

$$A = -\log(T/T_0) \tag{4.2}$$

The primary advantage of the transmission mode is that measured absorbance is directly proportional to the concentration (C) of the absorber according to the well-known Beer–Lambert law:

$$A = abC \tag{4.3}$$

where a is the IR absorptivity of the component and b is the thickness of the sample. For a more precise measure of concentration, IR absorbance band areas are typically measured using instrument software so the Beer–Lambert law becomes:

$$A_{int} = \varepsilon_{int} \times b \times C \tag{4.4}$$

where A_{int} is the integrated absorbance band area and ε_{int} is the apparent integrated absorption intensity. In many cases, sufficient information is obtained by comparing the integrated absorbance band intensities between samples. In cases where actual concentration information is desired, then it is necessary to determine the apparent integrated absorption intensity for the sample. This can be difficult and usually requires that the concentration of the component of interest be determined by another method such as gravimetrically, volumetrically, or by another spectroscopic technique. For transmission spectroscopy, samples are usually pressed into pellets or mounted in a support so they can be placed directly in the IR beam. Zeolites can have some strong absorption bands and significant scattering of the IR radiation, which is dependent on particle size, can occur. This reduces the amount of radiation that reaches the detector. Therefore, samples for use in transmission experiments must be very thin (~0.1 mm) or diluted in a non-IR-absorbing matrix such as potassium bromide.

In diffuse reflectance spectroscopy, the sample is usually in powdered form and is placed in a sample cup. The sample cup is placed in an accessory that typically uses a combination of flat and elliptical mirrors to direct the IR beam onto the sample and collect the light that is reflected off the sample. The mirrors are aligned such that primarily only the diffusely reflected light is collected and directed to the detector. As in the transmission measurements, the reflected signal (R) is ratioed to a background signal (R_0) without the sample present. The background signal is obtained using either a non-IR-absorbing sample (e.g., potassium bromide, potassium chloride) or a mirror. The result is a reflectance spectrum. This reflectance spectrum can be related to concentration by converting it to an absorbance-like spectrum by using either the Kubelka–Munk or the $\log(1/R)$ relationships:

$$KM = (1-R)^2/(2R) \tag{4.5}$$

where KM is Kubelka–Munk units, and R is measured reflectance value [90].

The primary advantage of the diffuse reflectance measurement mode is that an IR spectrum can be measured on a powdered sample which can greatly simply sample preparation compared to the transmission mode. Unfortunately, quantitative measurements using the diffuse reflectance mode are more difficult because the amount of sample contributing to the reflected signal is not known. The amount of diffusely reflected light depends on many factors, including the index of refraction of the sample, particle size, packing density and surface roughness. Several approaches have been used to circumvent these issues. One way is to develop and follow a rigorous sample preparation routine in which the sample is ground, sieved and carefully packed into the sample cup. Another approach is to mix in an internal standard or reference material such a KCN that has a distinct, strong IR absorbance band around $2200\,cm^{-1}$ that does not overlap the bands of interest in the sample. Spectra can then be normalized to the absorbance of the reference compound. In studies where the samples are modifications of the same material, a common absorbance band that is not expected to vary much between samples can be used for normalization.

Infrared spectroscopy can be used to obtain a great deal of information about zeolitic materials. As mentioned earlier, analysis of the resulting absorbance bands can be used to get information about the structure of the zeolite and other functional groups present due to the synthesis and subsequent treatments. In addition, infrared spectroscopy can be combined with adsorption of weak acid and base probe molecules to obtain information about the acidity and basicity of the material. Other probe molecules such as carbon monoxide and nitric oxide can be used to get information about the oxidation state, dispersion and location of metals on metal-loaded zeolites.

4.5.3
Framework IR

"Framework" IR refers to measuring the infrared spectrum of a zeolite to obtain information about the structural units present. The spectral region between 1400 and 400 cm^{-1} is where vibrational (lattice) bands that contain this structural information appear. A zeolite consists of a framework of tetrahedra (primary building unit) that contain silicon and aluminum atoms linked by oxygen atoms. Zeolites also contain secondary building units (SBUs), for example, single rings and double rings. Various structures are formed depending on how these primary and secondary units are packed together. Flanigen, Khatami and Szymanski were the first to systematically study the use of mid-IR spectroscopy to identify structural units in zeolite frameworks [91]. From their work, they were able to assign regions of IR vibrational bands in zeolites to various structural units. These assignments are shown in Table 4.2.

They concluded that the infrared spectrum contained vibrational modes from both structure insensitive internal tetrahedra and structure sensitive external linkages. The exact frequency of these bands depends on the structure of the zeolite as well as its silicon to aluminum ratio (Si/Al). A typical framework IR spectrum for a Y zeolite sample is shown in Figure 4.17. The accepted band assignments and frequency ranges are shown on the figure.

The most frequently used sample preparation method for measuring framework IR spectra is the KBr wafer technique. KBr does not have any interfering absorption bands. This involves dispersing finely ground zeolite powder into high purity KBr powder and compressing this mixture in a hydraulic press to make a clear glassy pellet. A dilution of 1:500 to 1:1000 is typically required to obtain good quality spectra since the zeolite internal tetrahedron vibrations (1250–950 cm^{-1}), common to all zeolites, are very strong. Without dilution, these absorption bands are so highly absorbing that they cannot be accurately measured by typically available commercial infrared spectrometers.

Table 4.2 Zeolite vibrational band assignments as per Flanigen et al. [91].

Vibrational mode	Frequency (cm^{-1})
Internal tetrahedra	
Asymmetric stretch	1250–950
Symmetric stretch	720–650
T-O bend	500–420
External linkages	
Double rings	650–500
Pore openings	420–300
Symmetric stretch	820–750
Asymmetric stretch	1150–1050 (shoulder)

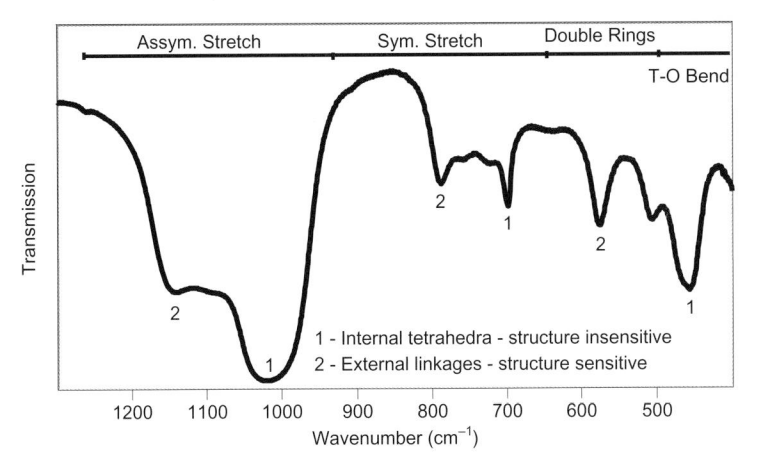

Figure 4.17 Framework spectrum of FAU (Si/Al$_2$ = 5.5) with band assignments as per Flanigen et al. [91].

The utility of measuring lattice vibrations for obtaining information about zeolites has been widely demonstrated. Applications include determining the structure of zeolites by the identification of the structural units present, measuring changes in the framework Si/Al within materials with the same zeolite structure and tracking the formation of zeolite during synthesis.

4.5.3.1 Zeolite Structure

Infrared spectroscopy has been used to help solve or determine the structure of zeolites. The technique is particularly useful for identifying the presence of double four- and six-rings as well as five-membered pentasil rings. In the structural characterization of beta zeolite, Newsam and coworkers used a variety of techniques including IR, electron microscopy (TEM), X-ray diffraction (XRD) and sorption data to solve the stacked, faulted structure [57]. The presence of IR absorption bands at 1232 and 560 cm^{-1} indicated that the structure contained five-member pentasil building units.

Van Bekkum and coworkers extensively studied the IR spectra of zeolites with and without five-member pentasil ring structural units. They concluded that the presence of pentasil structural units in a zeolite will give rise to two vibrational bands around 1230 and 550 cm^{-1}. They proposed that the bands around 1230 and 550 cm^{-1} are related to five-member ring chains and blocks, respectively [92].

As mentioned in Section 4.2 earlier in this chapter, infrared spectroscopy was used to provide information about the structural units present in UZM-5. The framework IR spectrum of a UZM-5 sample is shown in Figure 4.18. The characteristic vibrational bands for double four-rings (D4R) and pentasil rings (S5) are present. This provided some valuable information about the types of linkage units present and combined with data from other techniques such as XRD and TEM allowed the structure of UZM-5 to be solved.

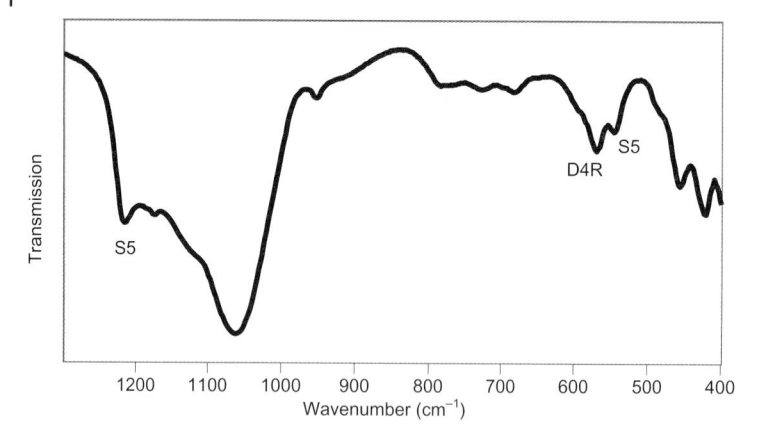

Figure 4.18 Framework spectrum of UZM-5 with structural groups identified. S5 – single five-ring, D4R – double four-ring.

4.5.3.2 Tracking Zeolite Framework Si/Al

Some vibrational modes of zeolites are sensitive to the amount of aluminum in the framework [93]. The substitution of aluminum for silicon atoms in the zeolite framework changes the T–O–T bond angles (where T is a tetrahedral atom that can be either Si or Al). This is primarily due to the smaller size and different charge density of the aluminum atoms compared to silicon. This results in a shift in frequency for vibrational modes in the zeolite due to external linkages. The T–O–T asymmetric (1100–980 cm^{-1}) and symmetric (800–600 cm^{-1}) stretching modes as well as structural unit vibrations like double four- and double six-rings exhibit a shift to lower frequencies as the aluminum content of the framework is increased. Figure 4.19 shows this relationship for a faujasite-type framework.

This technique has been used to measure changes in amount of framework aluminum in a given zeolite due to changes in the synthesis conditions or by post treatments such as ion exchange, steaming and calcination. Steaming and calcination usually result in the loss of aluminum from the framework. Such changes can affect both the catalytic and adsorptive performance. The dependence of the adsorptive properties of zeolites on their framework structure and aluminum content is discussed in more detail in Section 6.5.11. The loss of framework aluminum can be detected in infrared measurements by the shift of the structural vibrational bands to higher frequency whereas this loss may not be reflected in bulk elemental analyses for silicon and aluminum. Correlation plots of asymmetric T–O–T stretching frequency and framework aluminum content have been developed for several zeolites. The amount of aluminum in the framework has to be determined by an independent technique. ^{29}Si and ^{27}Al NMR are commonly used for this determination (see Sections 4.6.2.1 and 4.6.2.2 for more detail on these techniques).

In a study on acid-leached mordenites, Vansant and coworkers found a linear correlation between the frequency of a skeletal vibration mode and the framework

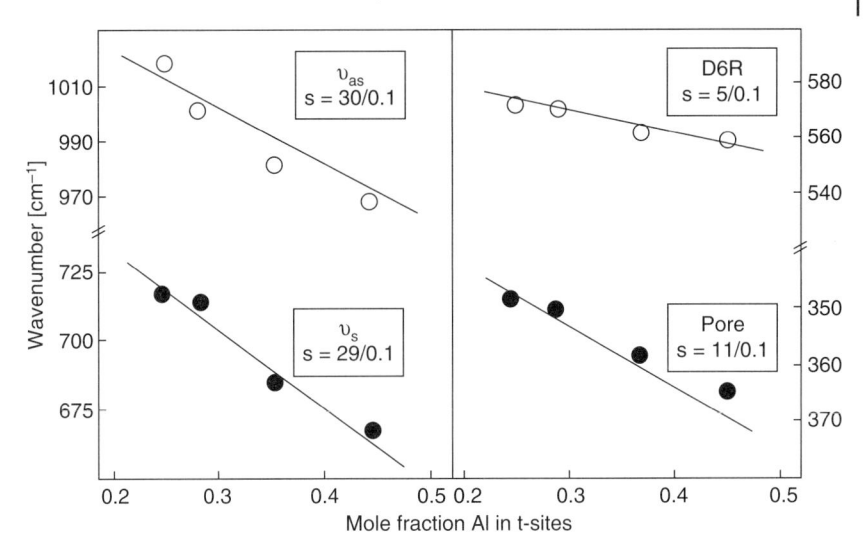

Figure 4.19 Frequency dependence on mole fraction of framework aluminum for various vibrational modes of X and Y zeolites. D6R means double six-ring, ν_{as} and ν_s are the assymmetric and symmetric stretches, respectively. s is the slope of the correlation lines (cm^{-1}/0.1 mol fraction Al). Adapted from Flanigen [91], with permission from ACS.

aluminum content as measured by ^{27}Al MAS NMR [94]. They assigned an absorption band that varied between 680 and 620 cm^{-1} to a structural Al–O vibration. Samples with different levels of framework aluminum were made by acid-leaching the starting sodium mordenite under various conditions. They found a good linear correlation between both the Al–O and the asymmetric T–O–T stretching frequencies with framework Al content over a range of two to eight aluminum atoms per unit cell.

Measuring the shift of structural vibrational frequencies can be a fast and easy way to measure changes in framework aluminum content. However, there are some limitations. When comparing samples, the zeolites must all be in the same form, i.e., as-synthesized, calcined, dehydrated, hydrated, or ion-exchanged. The presence of organic template molecules, cations and water in the zeolite pores can affect the pore size and T–O bonding. This can change both the T–O–T bond angles and bond strengths. The result is a shift of the vibrational frequencies which are not entirely dependent of framework aluminum content.

Another limitation is that sensitivity of the technique for detecting changes in the aluminum content of the framework varies depending on the Si/Al content of the sample. In low aluminum content zeolites (Si/Al$_2$ > 60), there is very little aluminum in the framework. Since there are very few aluminum atoms per unit cell to begin with, a loss of some of these aluminum atoms from the framework will have a very small effect on the overall structural T–O–T bond angles. This effect is shown in Figure 4.20 for a series of NH$_4^+$ form MFI zeolites where the number of framework aluminum atoms per unit cell varies between 0.7 and 8.0. The

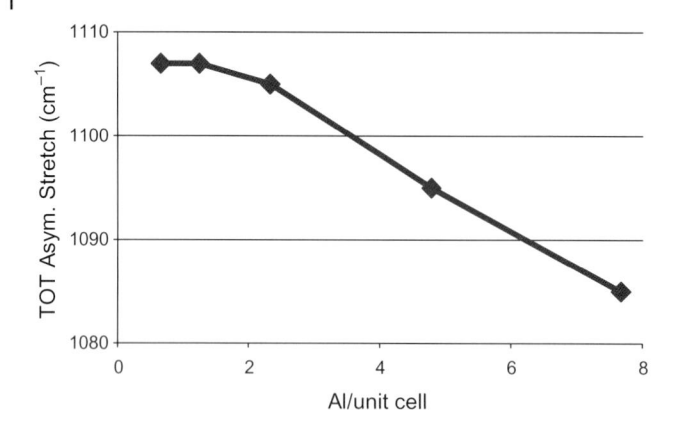

Figure 4.20 Dependence of the asymmetric T–O–T stretch of NH_4^+-MFI zeolite samples on number of aluminum atoms per unit cell.

T–O–T asymmetric frequency varies linearly over the range of two to eight atoms per unit cell. Below two aluminum atoms per unit cell, the stretching frequency is not very sensitive to further removal of aluminum from the framework.

4.5.3.3 Zeolite Synthesis

Framework infrared has also been used to look for the formation of a zeolite during synthesis. Since many of the secondary building units can be detected in the infrared spectrum, it is possible to see zeolite formation at very early stages in the synthesis. In fact, zeolite formation can be detected in the infrared before crystallinity is observed by X-ray diffraction. Most of the reported work has been done by sampling the zeolite synthesis at various stages, isolating the solids and measuring infrared spectra of the dried samples.

Flanigen monitored the changes in the IR spectra that occur during the synthesis of NaX zeolite from a sodium aluminosilicate gel. The appearance of absorption bands due to the formation of structural units in the zeolite as the crystallization of NaX proceeded were observed [93]. In particular, the growth of a band around 575 cm^{-1} indicated the formation of double six-rings which is one of the structural sub-units of X zeolite.

The IR spectra of samples from a hydrothermal LTA synthesis from an organic free sodium aluminate gel are shown in Figure 4.21. The starting gel has broad bands centered around 1100 and 500 cm^{-1} that correspond to T–O–T (where T is a tetrahedrally coordinated Si or Al atom) stretching and T–O bending modes. These bands shift as more aluminum is incorporated into the solid over the first hour. These bands continue to shift and sharpen with the appearance of a band around 580 cm^{-1} after 2 h. This band is due to the formation of double four-rings (D4R) which is one of the structural linkage units in the LTA structure. The XRD pattern of the 2 h sample shows very weak reflections due to LTA indicating that crystallites are forming that are large enough to be detected by XRD. The intensity

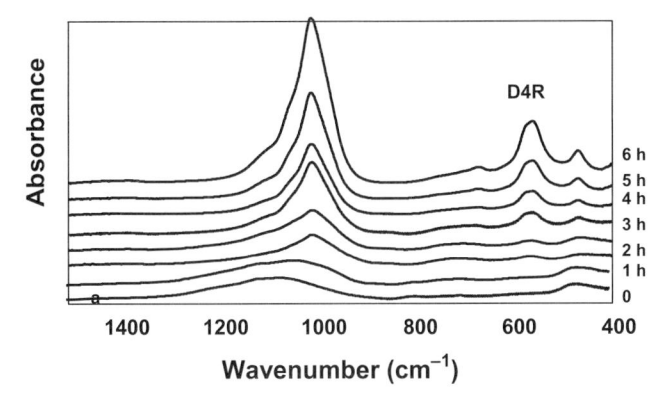

Figure 4.21 IR spectra of solids isolated at various times during a LTA synthesis. D4R refers to a double four-ring structural linkage.

of the bands increase with longer synthesis time until the LTA is fully formed after about 6 h.

Vedrine and coworkers studied vibrational bands for different types of zeolites with different particle sizes [95]. They concluded that during the synthesis of ZSM-type zeolite that the presence of vibrational bands at 550 and 450 cm^{-1} indicate that a ZSM-type zeolite may have formed. Absence of the 550 cm^{-1} band indicates that such a structure has not formed. The 550 cm^{-1} band is characteristic of five-member pentasil rings which are a structural unit of ZSM-type zeolites.

4.5.4
Methods Requiring Sample Pretreatment

The IR spectra in the work described above were measured on the zeolite samples "as-received", usually diluted in a non-absorbing matrix such as KBr, without any prior treatment. This works well if information in the structural or "framework" region of the spectrum is desired. However, if information about the hydroxyl groups, acid/base properties, dispersion of metals, or reactivity of the zeolite is of interest, then some sort of pretreatment step is required. The pretreatment step is used to at minimum, remove adsorbed water from the zeolite sample since water has strong absorption bands in the OH stretch regions (3600–3300 cm^{-1}) and the H–O–H bending region (~1640 cm^{-1}). The presence of adsorbed water makes it difficult or impossible to observe the discrete OH groups in a zeolite due to hydrogen bonding and it also affects the ability to probe its acid/base properties. Another reason for pretreatment at elevated temperature is to convert the zeolite to its hydrogen form, which is the form most commonly used in catalysis. Therefore, most researchers employ a sample pretreatment step before measuring IR spectra of zeolitic samples.

The most common method for preparing samples for pretreatment and subsequent measurement of IR spectra is the self-supporting pellet technique. In this method, the sample is ground to a fine powder and pressed into a thin, self-supporting disk using a commercially available hydraulic sample press and pellet die set. The resulting pellets are typically 5–20 mg/cm^2 and about 0.1–0.2 mm thick. Other ways of preparing samples for IR analysis include pressing the zeolite powder into a fine metal grid [96] and spray-drying a powder slurry onto a support [97]. Making measurements in the diffuse reflectance mode has the advantage of minimal sample preparation since powdered sample is loaded directly into the sampling cup.

The prepared sample is then placed into a holder and loaded into an apparatus that allows for exposure to elevated temperature under the desired atmosphere with windows that allow the IR beam to enter and exit for making the IR measurement. There are a wide variety of "home-built" and commercially available IR treatment chambers or cells available. The two most widely used configurations are those that separate the sample treatment and IR measurement sections and those that allow treatment of the sample while it is in the IR beam. In addition, some IR chambers are designed for use in vacuum and others are designed for gas flow at either atmospheric pressure or elevated pressures. Each design has its advantages and disadvantages. Therefore, the operating conditions, the type of measurement desired, and required flexibility will determine the best chamber design for a given application. This is why there are so many "home-built" designs reported in the IR literature.

4.5.5
Hydroxyl IR

Whenever an aluminum (+3 formal charge) atom is substituted for a silicon (+4 formal charge) atom in a zeolite framework, there is a net charge of −1 on the framework. This charge is balanced by cations. For zeolites used in catalysis, it is usually desirable to have them in the acid or proton form. The protons are part of the bridging hydroxyl groups (Si–OH–Al) in the zeolite framework and give the material Brönsted acidity. IR spectroscopy can be used to directly observe these hydroxyl groups by measuring the O–H stretching modes associated with them. However, in order to be able to see these discrete O–H stretching modes, it is necessary to at least dehydrate the zeolite as described above. Figure 4.22 shows the IR spectra of a self-supported pellet of a ammonia exchanged Y zeolite sample before and after treatment at various temperatures in helium flow. The spectrum of the "as-pressed" material shows very strong and broad absorbance bands due to the presence of adsorbed water and ammonium ion. These absorbance of these bands is so high (>6) that they cannot be measured by a typical IR instrument and therefore are truncated on the absorbance axis. Due to these broad absorption bands, it is virtually impossible to obtain any information about the different types hydroxyl groups that are present in the zeolite. As the pretreatment temperature is increased, water is desorbed, and hydrogen bonding between the surface hydroxyl groups decreases. After treatment at 300 °C, sufficient dehydroxylation has occurred that allows the various O–H stretching vibrational bands to be distinguished. The accepted assignments of the primary O–H stretching modes are:

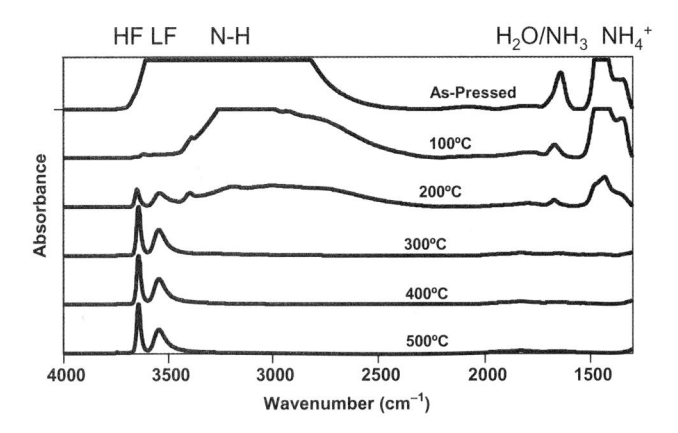

Figure 4.22 IR spectra of NH_4^+-FAU zeolite (Si/Al_2 = 5.5) after 1 h treatment in flowing He at various temperatures. Spectra collected at 25 °C. HF–OH is in supercages, LF–OH in sodalite cages.

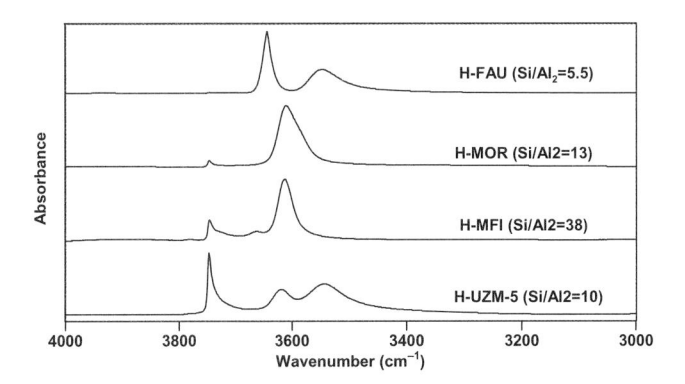

Figure 4.23 IR spectra showing hydroxyl region of various H-form zeolites. Samples pretreated for 2 h at 500 °C, spectra recorded at 25 °C.

Si–OH or external silanol around 3745 cm^{-1} (very weak) and the high field (HF) Si–OH–Al in the supercages (~3645 cm^{-1}) and low field (LF) hydroxyls in the sodalite cages (~3545 cm^{-1}) [98].

In addition, there are some weaker broad bands that are typically observed between 1900 and 1500 cm^{-1} that have been assigned to overtones of framework vibrations. The IR spectra in Figure 4.22 are truncated at 1300 cm^{-1} because the absorbance of the sample is too high to measure at lower frequencies (<1200 cm^{-1}). This is due to the very strong T–O–T stretching vibrations of the zeolite as mentioned in the previous section on framework IR measurements.

Information about the number and types of hydroxyl groups present in a particular zeolite can be obtained from its IR spectrum. Si–OH–Al groups in different locations in a zeolite structure can have different stretching frequencies. Figure 4.23 shows the hydroxyl IR spectra for several different zeolites. Both the FAU and

UZM-5 samples show two distinct Si–OH–Al bands which are due to OH groups vibrating in the different size cages of these zeolites. Depending on the structure of the zeolite, the spectrum can have multiple Si–OH–Al vibrational bands and they can vary in position.

The frequency of the OH stretch is dependent on the size and shape of the rings or cages in which the OH group is located as well as the number of aluminum atoms per unit cell (framework Si/Al). Jacobs and Mortier studied the IR spectra of many different hydrogen form zeolites in an attempt to rationalize the difference in stretching frequencies for lattice hydroxyl groups [99]. They found that the frequency of the lattice OH groups could be rationalized in terms of two concepts, depending on the size of the rings. For zeolites with larger than eight-membered rings, the frequency of the OH stretch is linearly correlated with the Sanderson intermediate electronegativity (see Section 6.3.1 for a more detailed discussion). The Sanderson electronegativity is a measure of the electron-accepting ability of the zeolite framework and is dependent on the chemical composition such framework Si/Al and the cations present. For hydroxyl groups vibrating in six- or eight-membered rings, there is a shift to lower frequencies (bathochromic) due to electrostatic interactions with nearest oxygen atoms. The strength of this interaction is inversely proportional to the squared distance between them. Therefore, the relative strength of the acidic hydroxyl cannot be directly correlated to its vibrational frequency (i.e., lower frequency → stronger acidity).

The hydroxyl spectra of zeolites can be much more complex than shown in Figure 4.23. Depending on how it was prepared and especially on post-synthesis treatments, dealumination of the zeolite framework can result in the creation of many additional OH containing species. These include extra-framework aluminum (Al–OH), internal silanols (Si–OH), silanol nests (Si–OH — HO–Si), and hydrogen-bonded hydroxyl groups. The presence of internal silanols and silanol nests was demonstrated by Woolery and coworkers for a series of highly siliceous MFI zeolites (Si/Al_2 = 70–26 000) [100]. As the amount of aluminum in the framework is reduced the band due to the presence of the bridging acidic hydroxyl groups disappears and is replaced by a broad band centered around 3500 cm^{-1}. This broad band has been attributed to hydrogen bonding between the Si–OH groups located in silanol nests (Figure 4.24).

Calcined and steamed FAU samples also have complex hydroxyl IR spectra. Figure 4.25 shows the difference between an ammonium ion-exchanged FAU before and after steaming and calcination. The very simple, easily interpretable hydroxyl spectrum of the ammonium exchanged FAU sample is transformed into a complex series of overlapping hydroxyl bands due to contributions from framework and non-framework aluminum atoms in the zeolite resulting from the hydrothermal treatment conditions [101].

Information about the zeolite crystallite size can also be inferred from the IR spectrum. Changes in the shape and slope of the baseline of the spectrum between 3600 and 2000 cm^{-1} is associated with light scattering from different crystal sizes and shapes. In general, the scattering increases with increased crystal size [102].

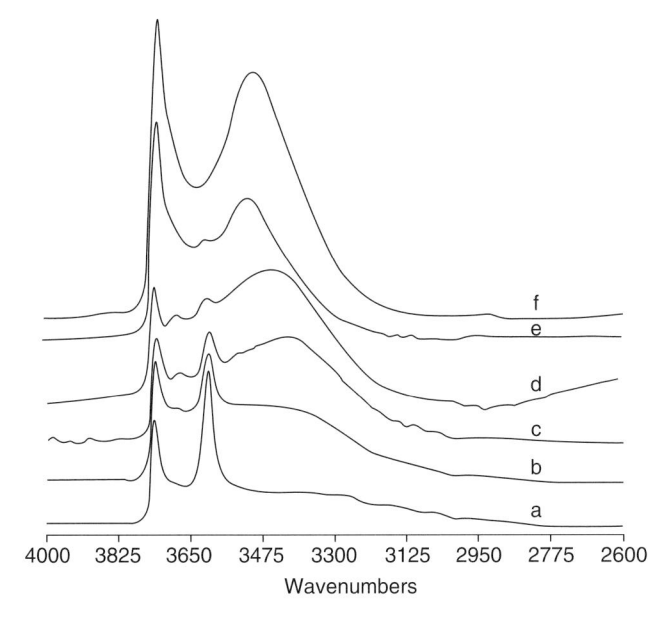

Figure 4.24 FTIR spectra of HZSM-5 with varying Si/Al$_2$ ratios: (a) 70, (b) 140, (c) 200, (d) 500, (e) 600, (f) 26 000. From Woolery, et al. [101], with permission from Elsevier

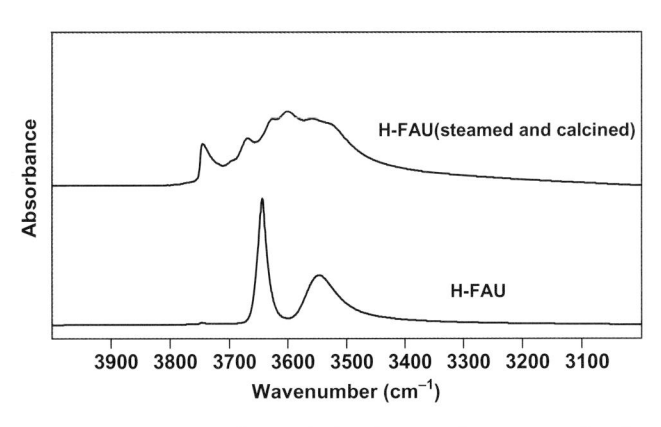

Figure 4.25 IR spectra showing hydroxyl region of H-FAU samples showing the effect of steaming and calcination.

4.5.6
Acidity

One of the important properties that make zeolites useful for catalysis is their acidity – particularly, their Brönsted acidity. The strength, location and accessibility of these sites will change depending on the structure of the zeolite and its composition (Si/Al). Therefore, it is important to be able to measure these differences in order to develop an understanding of the performance/structure/composition relationships for a given process chemistry. Although hydroxyl IR spectra can provide information about the framework and extra-framework aluminum content of the zeolite (Brönsted and Lewis acidity), there is little strength and accessibility information. One way to obtain information about acidity is by adsorption of probe molecules. In these experiments, the zeolite sample is exposed to a weakly basic probe and the IR absorption spectrum is measured. From analysis of the absorption bands due to the interaction of the adsorbed probe molecule with the zeolite it is possible to determine the number of accessible Brönsted and Lewis acid sites. There are a large number of probe molecules of different sizes and basicities that can be used, depending on the exact information that one is trying to obtain. It is important to realize that there is no one absolute acidity number/strength scale that exists for solid acid materials. This is why there are so many different techniques available for the measurement of acidity. Each of these acidity methods provides a measure of the acidity of the sample. Depending on the technique used, the conditions under which is it done and the particular probe molecule used, different results can be obtained. Although different, these results are complimentary and can help to provide more thorough understanding of the acidity of the material.

4.5.6.1 General Theory

As mentioned above, an acidic zeolite can provide both protonic (Brönsted) and aprotonic (Lewis) sites. The Brönsted sites are typically structural or surface hydroxyl groups and the Lewis sites can be charge compensating cations or arise from extra-framework aluminum atoms. A basic (proton acceptor) molecule B will react with surface hydroxyl groups (OH_s) via hydrogen bonding

$$OH_s + B = OH_s\text{---}B \qquad (4.6)$$

The strength of this interaction is determined by the relative proton affinities of the oxygen atom and the basic molecule. This interaction weakens the O–H bond which results in a bathochromic (lower frequency) shift in the O–H stretching vibration. In addition, this hydrogen bonding results in significant broadening of the band due to coupling with the O—B stretching mode and a large increase in the integrated absorbance [86, 103]. If the interaction is strong enough, the base molecule will be protonated

$$OH_s + B \rightarrow O_s^- + HB \qquad (4.7)$$

and the O–H vibration will no longer be observed. Also, as the proton affinity of the probe molecule increases, the vibrational spectrum becomes more complex due to the appearance of additional bands due to combination and overtone modes. This is described in the Bellamy–Hallam–Williams (BHW) theory which is explained in more detail by Zechhina [103].

For aprotonic or Lewis acid sites (L), the base can form an adduct

$$L + B = L:B \qquad (4.8)$$

Table 4.3 Criteria for selection of IR probe molecules (adapted from Knözinger and Huber [86]).

1. The interaction of the probe and acid/base site should have a unequivocal spectral response.
2. The probe molecule should interact selectively with acidic or basic sites.
3. Frequency shifts due to interaction must be measurable with sufficient accuracy to distinguish them from those of the unperturbed frequencies.
4. Molar absorption coefficients should be large enough to allow high detection sensitivity.
5. Molar absorption coefficients of the vibrational modes of interest should be experimentally available to allow quantitative measurements.
6. The probe molecule should provide high specificity to allow discrimination of similar sites that may have only small differences in strength.
7. The size of the probe molecule should be selected so that it is accessible to the sites of interest, for example, internal versus external acid sites in a zeolite.
8. The reactivity of the probe molecule under experimental conditions should be low to avoid reactions other than simple acid-base interactions.

For this case, the primary change that is observable in the IR spectrum is due to changes in the vibrational frequencies of the probe molecule due to modifications in bond energies. This can lead to changes in bond force constants and the normal mode frequencies of the probe molecule. In some cases, where the symmetry of the molecule is perturbed, un-allowed vibrational modes in the unperturbed molecule can be come allowed and therefore observed. A good example of this effect is with the adsorption of homonuclear diatomic molecules, such as N_2 and H_2 (see Section 4.5.6.8).

4.5.6.2 Selection of Probe Molecules

Essentially, any molecule could potentially be used as a probe molecule. However, practically, a useful probe molecule should meet certain minimum criteria. Such criteria for the selection of an "ideal" probe molecule were first developed by Paukshtis and Yurchenko [104]. These have been refined by several research groups and summarized by Knözinger and Huber [86]. The criteria are summarized in Table 4.3. Based on these criteria, small, weakly interacting probe molecules are recommended for zeolites particularly when you want to probe sites within the pore structure.

As mentioned above, many different probe molecules have been used to measure the acidity of zeolites. These molecules vary over a wide range of proton affinities, size (kinetic diameter) and shape. Table 4.4 lists these properties for several of the more commonly used probes.

4.5.6.3 Quantitation of Sites

Quantitation of the number of acid sites present can be obtained from the integrated band areas for a particular vibrational band of the probe molecule and its molar absorption or molar extinction coefficient using the Lambert–Beer law as previous described in Section 4.5.2. In order to determine the molar extinction

Table 4.4 Properties of common basic probe molecules.

Probe molecule	Proton affinity (PA; kJ mol^{-1})[a]	Kinetic diameter (Å)	Geometric shape
N_2	494	3.76	Linear
CH_4	544	3.8	Pyramidal
CO	594	3.64	Linear
NH_3	854	3.0	Trigonal
Pyridine	930	5.7	Planar
2,6-Dimethylpyridine (lutidine)	963	6.7	Planar
2,4,6-Trimethylpyridine (collidine)	975–980[b]	7.4	Planar
2,6-Di-t-butyl-pyridine	983	7.9	Planar

a) From [140].
b) Estimated.

Table 4.5 Selected experimentally determined IR extinction coefficients for surface species and adsorbates on zeolites. (Adapted from Karge and Geidel [87]).

Species	Zeolite type	Extinction coefficient (cm mmol^{-1})	Remarks	Band position (cm^{-1})
OH group	FAU	3.1	HF Brönsted site	3625
OH group	FAU	3.1	LF Brönsted site	3550
OH group	MOR	8.5	Brönsted site in main channels	3612
OH group	MOR	1.55	Brönsted site in side pockets	3584
NH_3 : L	MCM-41	1.98	Ammonia on Lewis site	1620
$NH_4^+ \to B$	MCM-41	1.47	Ammonia on Brönsted site	1450
$CO \to B$	MOR	2.7	CO bound to HF and LF Brönsted sites	2177 (HF) 2169 (LF)
$PyrH^+ \to B$	FAU	3.0	Pyr on Brönsted site. Results from different groups/samples	1540
		1.8		1542
		1.3		1540
	MOR	1.8	Pyr on Brönsted site.	1540
	MFI	1.3	Pyr on Brönsted site.	1540
Pyr : L	FAU	1.5	Pyr on Lewis site	1450
	MFI	1.5	Pyr on Lewis site	1450
	MOR	1.5	Pyr on Lewis site	1450

coefficient, the amount of probe molecule adsorbed on the sample must be known. This is usually done by combining the IR measurement with other volumetric, gravimetric, or spectroscopic techniques. Table 4.5 shows some of the extinction coefficients reported by other researchers. Exact measurements of molar absorp-

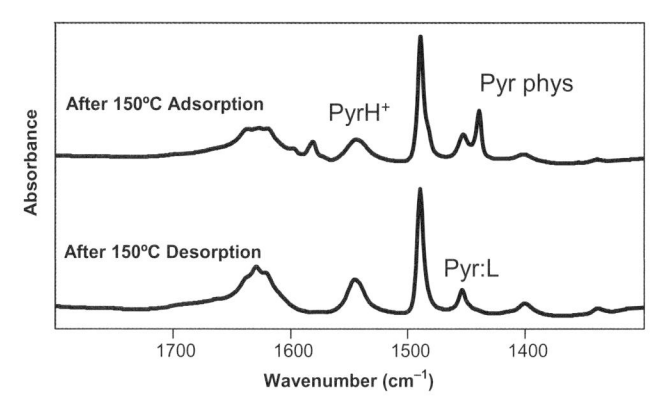

Figure 4.26 IR spectra showing absorbance bands due to pyridine adsorption on a H-FAU sample. PyrH⁺ is protonated pyridine on Brönsted sites, Pyr:L is pyridine coordinated to Lewis acid sites, Pyr phys is physisorbed pyridine. All spectra recorded at 25 °C.

tion coefficients can be extremely difficult and reliable values are available only in specific cases. In addition, by examination of the coefficients listed Table 4.5, the value can be material-dependent. Furthermore, different values have been measured on the same zeolite by different research groups. This could be due to differences in the methodologies used to determine the extinction coefficients or due to working with different Si/Al zeolite samples. Therefore, for true quantitative determinations, the molar absorption coefficient should be determined for a particular probe on a given zeolite. In most situations, it is not necessary to determine the exact number of acid sites, since relative differences in number and type are sufficient.

4.5.6.4 Pyridine Adsorption

One of the first probe molecules used in combination with IR to measure acidity is pyridine [105–107]. Pyridine has been and continues to be one of the most widely used probes in investigations of surface acidity. Figure 4.26 shows the IR spectrum of pyridine adsorbed on a FAU sample after adsorption and subsequent desorption at 150 °C. Pyridine is able to clearly distinguish between Brönsted and Lewis acidity due to distinct ring mode vibrational bands. The 1540 cm^{-1} band is due to the 19b ring mode vibration of the adsorbed pyridinium ion (protonated pyridine) and provides a measure of the number of accessible Brönsted sites in the zeolite. The coordinated pyridine band (~1450 cm^{-1}) is due to the 19b ring vibration of the adsorbed non-protonated pyridine molecule. The position of this band is sensitive to the electron acceptor strength of the site. Physisorbed pyridine (weakly adsorbed) has an absorbance band around 1440 cm^{-1}. Pyridine adsorption is usually followed with a desorption step at 150 °C to remove the physisorbed pyridine from the sample.

For weak acceptors like charge balancing cations (Na⁺, K⁺, etc.), the interaction with the lone pair on the nitrogen is weak and results in a absorbance band around 1440 cm^{-1}. For stronger acceptors like extra-framework aluminum atoms (Lewis

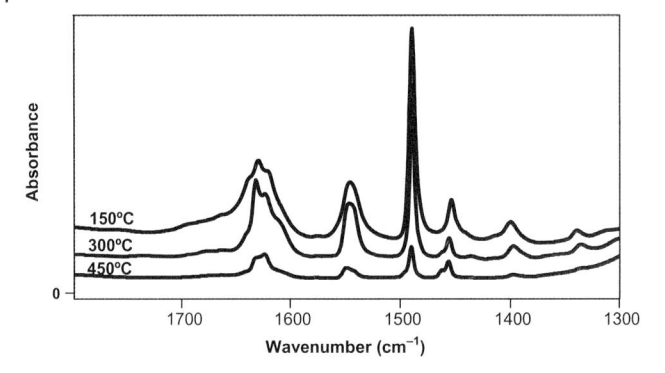

Figure 4.27 IR spectra showing absorbance bands due to pyridine adsorption on a H-FAU sample after desorption at 150, 300 and 450 °C. All spectra recorded at 25 °C.

acid sites), this band is around 1450 cm^{-1}. In fact, the position of this band has been shown to be linearly correlated to the electrostatic coulombic field of the cations in the zeolite [108]. The strength of distribution of the Brönsted acid sites cannot be determined directly from the adsorption spectrum of the pyridium ion since it is formed by any acid site with a strength sufficient to protonate pyridine. However, if the pyridine is desorbed at different temperatures, acid site strength distribution information can be obtained by assuming that stronger sites will retain pyridine at higher temperatures. This is the basic assumption that is made when using any of the thermal desorption techniques such as ammonia temperature programmed desorption (TPD). Pyridine adsorption on a FAU sample after desorption at several temperatures is shown in Figure 4.27. Integration of the pyridine bands after each desorption step provides the amount of pyridine remaining on the sample. The desorption of pyridine can be done either stepwise as in this example or in the TPD mode.

A convenient way to describe the differences in the acid site strength distributions between materials is in terms of "weak", "moderate" and "strong" sites. These values are calculated from the Brönsted and Lewis pyridine band areas measured at each of the three desorption temperatures. In this case, the total number of sites is proportional to the band area after 150 °C desorption ($Area_{150}$). The weak sites are those that are removed by 300 °C desorption ($Area_{150}$–$Area_{300}$) and the moderate sites are those that remain after 300 °C desorption but are removed by 450 °C desorption ($Area_{300}$–$Area_{450}$). Finally, the strong sites are those that remain after 450 °C desorption ($Area_{450}$). The difference in acid site strength distributions can be calculated as described above and conveniently displayed in a bar chart format. This method of comparing the acid site strength distributions between samples is shown in Figure 4.28 for the ammonium exchanged and the steamed and calcined FAU samples from Figure 4.24. The distributions for both the Brönsted and the Lewis sites as measured by pyridine adsorption/desorption IR are shown. From these plots, the main effects of steaming and calcination on the FAU sample are a loss of about one-third of the total Brönsted sites and a

Brönsted Acid Site Distribution

Lewis Acid Site Distribution

■ Weak ▨ Moderate ▧ Strong ☰ Total

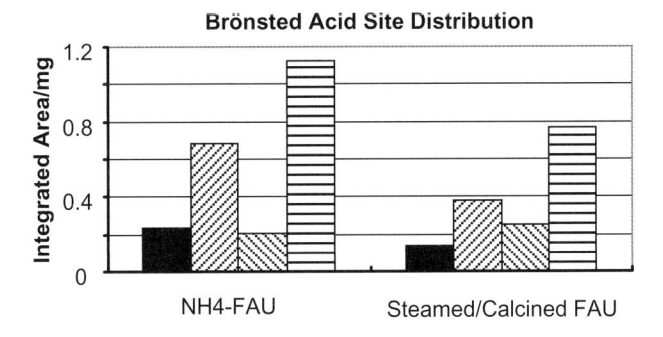

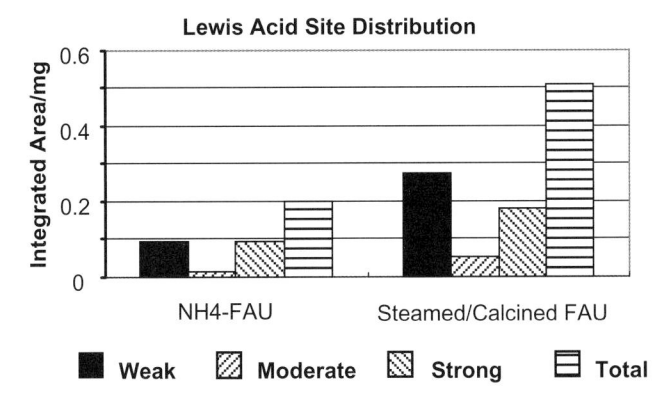

Figure 4.28 Effect of steaming and calcination on Brönsted and Lewis acid site strength distributions of a FAU-type zeolite as determined by pyridine adsorption/desorption IR.

factor of 2.5 increase in total Lewis sites. Examination of the weak, moderate, and strong contributions show that this hydrothermally treated sample has a slight increase in strong Brönsted acid sites and significantly lower weak and moderate sites. In terms of Lewis acidity, the steamed sample has an increase in medium and strong sites and a threefold increase in weak sites. These changes in acid site type and strength can be attributed to framework dealumination caused by the hydrothermal treatment conditions. The loss of aluminum from the framework is also evident by the shift to higher frequency ($1018 \rightarrow 1050\,cm^{-1}$) in the asymmetric T–O–T stretch measured by framework IR, as discussed in Section 4.5.3.2. The comparison of areas as described above does provide quantitative information about the relative changes in acidity between the samples since the area is direction proportional to the concentration (Beer–Lambert law, discussed in Section 4.5.2.) It most cases, this relative, but quantitative comparison between samples is sufficient to provide information about how various treatments or modifications have altered acid site distributions. Since extinction coefficients can change with zeolite type (Table 4.5), these comparisons are best for samples of the same zeolite type. Therefore, caution should be used when comparing data from samples with different zeolite structures.

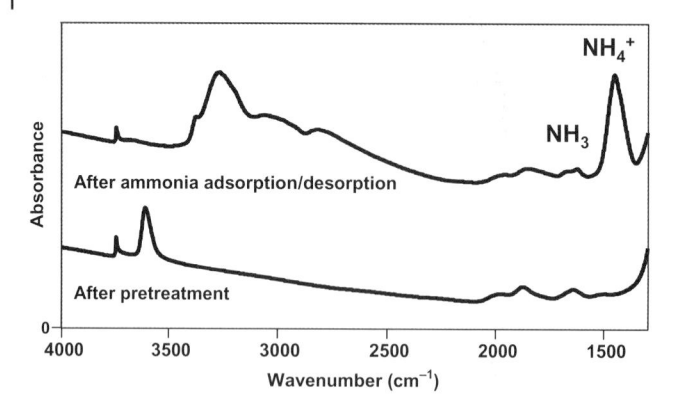

Figure 4.29 IR spectra of H-MOR sample after pretreatment at 500 °C and after adsorption/desorption of ammonia at 150 °C. All spectra recorded at 25 °C.

One limitation of the use of pyridine as a probe molecule in zeolites is its rather large size. Depending on the particular zeolite structure, the pore openings of the zeolite cavities may be too small to allow pyridine access. Therefore, pyridine cannot be used to measure acid sites in zeolites with pore openings that have six-member rings like sodalite cages and even eight-member rings like ferrierite [109]. However pyridine can be used on these types of materials to provide information about acid sites that are external to the pore system of the zeolite, as discussed later in Section 4.5.6.9.

4.5.6.5 Ammonia

Ammonia is another relatively strong basic molecule (PA = 854 kJ mol^{-1}) that is commonly used to probe acid sites in zeolites. Like pyridine, ammonia has distinct vibrational absoption bands when it is protonated by Brönsted acid sites and coordinated to Lewis acid sites, as shown in Figure 4.29. Brönsted acids sites can protonate ammonia to form an ammonium ion ([NH$_4$-B]$^+$) which has absorbance bands around 1450 and 3130 cm^{-1}. The absorbance bands for coordinated form of ammonia (NH$_3$:L) appear around 1630 and 3330 cm^{-1}. The main advantage of ammonia is that its smaller kinetic diameter (3.0 Å) allows it to probe acid sites in the cages of smaller pore zeolites (e.g., eight-member rings) that are not accessible to other larger probes like pyridine (5.7 Å). There are other changes that are evident in the IR spectrum of the H-MOR sample due to the adsorption of the basic probe molecule. The bridging acidic hydroxyl band around 3610 cm^{-1} is completely attenuated by the ammonia since it has reacted to form a surface NH$_4^+$ species. N-H stretching bands and other broad bands due to combination and overtone bands are observed [103]. From examination of the hydroxyl region, it is evident that ammonia has reacted with most or all of the acidic bridging hydroxyl groups in the MOR sample. In addition, there is very little reaction with the silanol groups which are considered to be non-acidic. The same type of experiments described for pyridine in Section 4.5.6.4 can also be performed with ammonia to obtain similar information about acid site accessibility, strength and type in zeolites.

Some researchers have successfully combined ammonia IR and gas adsorption/desorption (ammonia TPD) measurements to provide quantitative Brönsted/Lewis acid site strength distributions on zeolitic materials [110–112]. The ammonia TPD technique is described in more detail in Section 4.7.4. This technique measures the total amount of ammonia desorbed from the sample as a function of time (also temperature due to linear heating ramp). Since it is measuring desorbed ammonia, it cannot be used to distinguish on what type of site the ammonia came from. However, the ammonia IR adsorption/desorption IR data provides the Brönsted/Lewis distribution information. Combining these two complimentary techniques allows the determination of quantitative acid site strength distributions with Brönsted/Lewis speciation. Zhang and coworkers used the combined techniques to characterize the acidity on Y, mordenite and ZSM-12 zeolites [111]. An erionite-rich zeolite T was examined by Martin and coworkers [112].

4.5.6.6 Low-Temperature Acidity Probes

Although the use of weak base probes like pyridine and ammonia can provide important information about sample acidity, they do have some drawbacks. Even though they are considered weak bases, they are strong enough to be protonated by the Brönsted sites present in zeolites and this reaction is essentially irreversible under ambient conditions. The same protonated base species is formed regardless of the strength of the acid group. Therefore, no strength distribution information is available by simple adsorption measurements. This goes against one of the properties of an ideal probe molecule as discussed in Section 4.5.6.2 the probe molecule should provide high specificity to allow discrimination of similar sites that may have only small differences in strength. To obtain the strength information, the probe molecule is usually desorbed at multiple temperatures. This works in many cases, but increasing temperature can result in probe reaction, decomposition, or a change in the sample itself which can lead to erroneous results.

The adsorption of very weakly basic probes like carbon monoxide, nitrogen and even hydrogen at low temperature has been demonstrated to be a very useful technique for measuring acidity in zeolites. Use of these probes at low temperature (~77 K, the temperature of liquid nitrogen) has several advantages over stronger bases. These molecules are non-reactive with the acid sites at room temperature, only weakly interact at low temperature, the adsorption process is reversible and they meet essentially all of the criteria for ideal infrared probe molecules (see Section 4.5.6.2). Therefore, when the probe is added in controlled doses, it interacts with the strongest sites first and eventually interact with the weaker sites. For Brönsted acid sites, this interaction is not strong enough to protonate the probe molecule, but results in hydrogen-bonding with the O–H group as shown below:

$$Z-O-H + B = Z-O\ldots H\ldots B \qquad (4.9)$$

where Z–O is the oxygen atom in the zeolite framework, H is the acidic proton and B is the probe molecule. The hydrogen bonding with the probe molecule

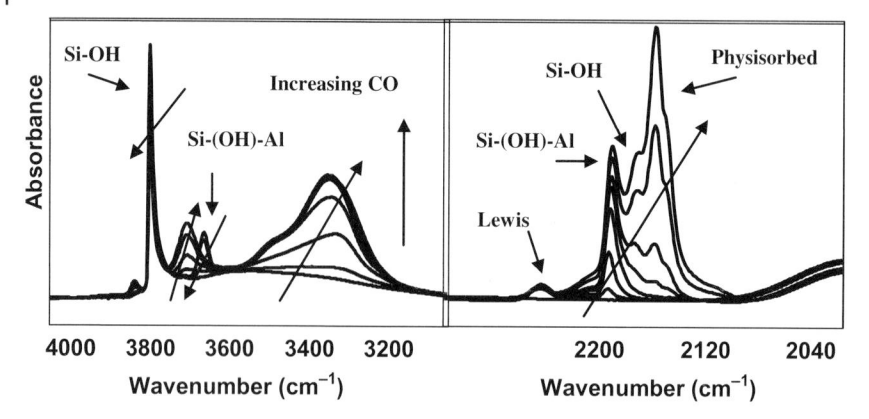

Figure 4.30 Overlay of IR spectra during CO adsorption at 90 K on H-BEA zeolite. The diagonal arrows indicate whether the band is increasing or decreasing with increasing CO coverage.

weakens the O–H bond which causes its absorption band to shift to lower frequency, broaden, with a large increase in its extinction coefficient. This is typical of the effect of hydrogen-bonding on IR spectra [86, 104]. The magnitude of the shift in the hydroxyl band is proportional to the acid strength of the proton.

4.5.6.7 Carbon Monoxide

Carbon monoxide is the most commonly used low-temperature IR probe. With a proton affinity of 594 kJ mol^{-1} it is a significantly weaker base than pyridine (PA = 930 kJ mol^{-1}). IR spectra of multiple doses of CO during adsorption on a H-BEA zeolite at 90 K are shown in Figure 4.30. When adsorbed at low temperature, changes in two different regions of the IR spectrum are observed. One is the shifting and broadening of the hydroxyl groups of the zeolite due to hydrogen-bonding with CO and the other is the appearance of C-O stretching bands in the 2300–2000 cm^{-1} region due to the adsorbed CO. CO interacts first with the strongly acidic bridging hydroxyl groups and then begins to interact with the much weaker silanol groups. Difference spectra are typically used to analyze the data. Figure 4.31 shows the difference spectrum for the H-BEA sample in which the spectrum before CO addition is subtracted from the spectrum after CO adsorption. The effects of CO adsorption are clearly visible showing the disappearance of the bridging hydroxyl band at 3610 cm^{-1} and the growth of the shifted band at 3295 cm^{-1} (315 cm^{-1}). The partially attenuated silanol band at 3745 cm^{-1} is shifted to 3650 cm^{-1} (95 cm^{-1}). In the CO stretching region, multiple overlapping bands are observed due to the perturbation of the CO bond from its gas phase value of 2143 cm^{-1} through interaction with the various acid sites present in the zeolite. At higher CO coverages, an absorbance band around 2140 cm^{-1} due to physisorbed CO is observed. Although the shifts in the CO stretching frequency due to interaction with acid sites are not as large as those for the OH bonds, it is possible to distinguish CO adsorption on silanols versus bridging acidic OH groups. CO interacting with strong Lewis sites is characterized by absorption bands ≥2200 cm^{-1}.

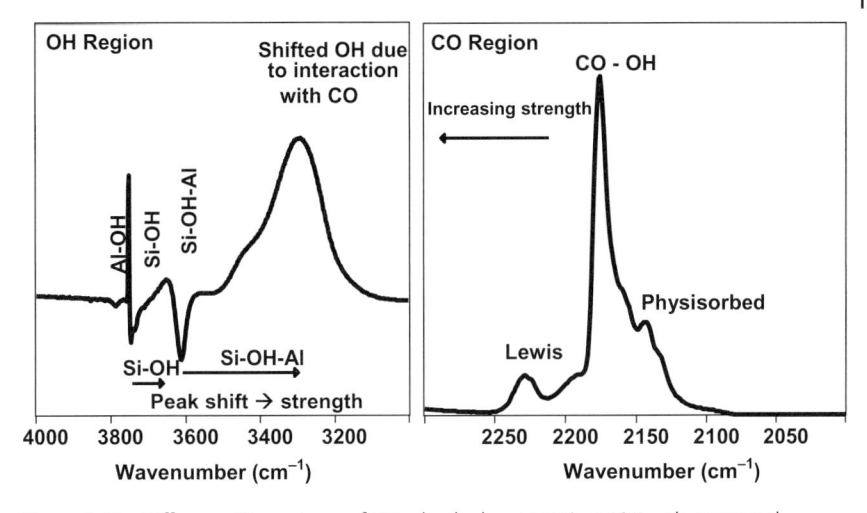

4.5 Infrared Spectroscopy Characterization of Zeolitic Systems | 133

Figure 4.31 Difference IR spectrum of CO adsorbed on H-BEA at 90 K with pretreated spectrum (before CO addition) subtracted.

Table 4.6 Hydroxyl band shifts measured by low-temperature CO adsorption for various zeolites.

Sample ID	Si/Al$_2$	Hydroxyl peak shift (cm^{-1})
H-MOR	21	95 [Si-OH] 330 [Si-OH-Al]
H-MOR	13	310 [Si-OH-Al]
H-FAU	5.1	100 [Si-OH] 293 [Si-OH-Al]
H-FAU	60	320 [Si-OH-Al]
H-BEA	26.4	85 [Si-OH] 345 [Si-OH-Al]
H-MFI	38	95 [Si-OH] 305 [Si-OH-Al]

By measuring the shifts of the various hydroxyl bands of the zeolite, a direct measure of the relative acid site strengths can be made without the need for thermal desorption. Table 4.6 lists the measured hydroxyl band shifts for a variety of hydroxyl groups on different zeolites using low temperature CO adsorption. This data indicates that there is indeed a difference in the intrinsic acid strength of the bridging hydroxyl groups in different zeolites as well as in the same zeolite structure with different framework aluminum content.

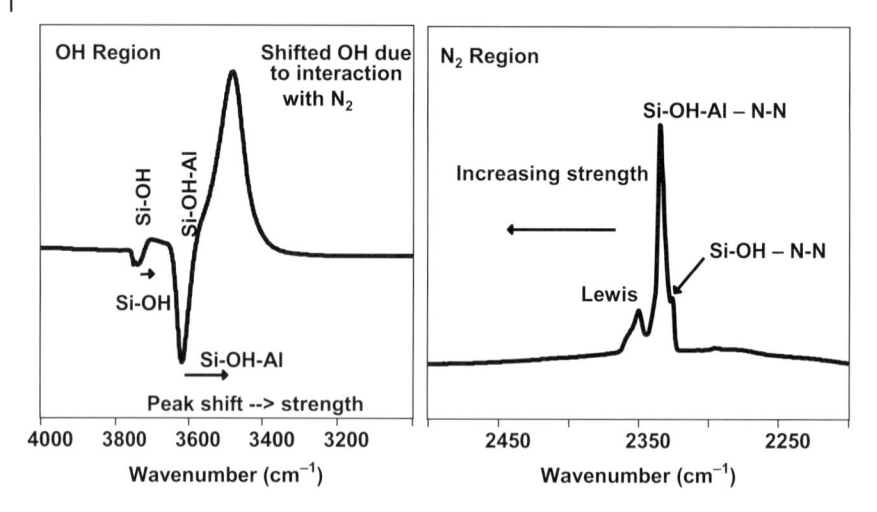

Figure 4.32 Difference IR spectrum of N_2 adsorbed on H-MOR at 95 K with presented spectrum (before N_2 addition) subtracted.

4.5.6.8 Nitrogen (N_2)

Di-nitrogen (PA = 494 kJ mol^{-1}) is isoelectronic with CO and is another probe molecule that is being used more frequently for low temperature adsorption experiments to measure acidity on zeolites. Since it is a weaker base than CO it can be even more specific for measuring differences in the strength between strong Brönsted sites and is completely unreactive at low temperatures. Another advantage to using nitrogen as a probe is that being a homo-nuclear diatomic molecule, it is not infrared active in the gas phase. When it adsorbs or interacts with protonic or aprotic sites on a zeolite, the N-N stretching mode becomes IR active due to a reduction in molecular symmetry caused by the anisotropic environment. The resulting IR active N-N stretching bands appear around the gas phase Raman shift at 2332 cm^{-1}. Depending on the strength of the acid site, the absorption band is shifted to higher frequency (blue shift). Figure 4.32 shows the IR difference spectrum of a mordenite sample showing the effect of N_2 adsorption. The measured hydroxyl band shifts are smaller for nitrogen adsorption compared to carbon monoxide due to the lower proton affinity of N_2. The bridging acidic hydroxyl group at 3620 cm^{-1} is shifted to 3478 cm^{-1} (142 cm^{-1}). The silanol band is only shifted 45 cm^{-1}. In the N_2 stretching region, IR active bands appear due to N_2 interaction with the various acid sites present. Stronger sites cause larger shifts in the N_2 vibrational frequency. Several research groups have advocated the use of N_2 as an acidity probe [113–115].

4.5.6.9 Measurement of External Acidity

Larger probe molecules such as alkyl substituted pyridines and hindered amines have been used to probe only the external acidity of zeolites. The large kinetic diameter of these molecules prevents them from entering the pores of the zeolite and therefore they can only interact with acid sites that are external to the zeolite cages. For nano-zeolites and other zeolites with high surface area and/or surface pockets such as MCM-22, ITQ-2, and MTW, and MCM-56 the surface acidity can dominate and effect catalytic properties [116, 117]. In many cases it is desirable minimize external acidity since the presence of these sites can lead to non-selective reactions that would not occur in the zeolite pores. Many studies on the role and modification of surface acidity on selectivity of ZSM-5 zeolite catalysts have been reported [118–121]. In other cases, for reactions such as toluene disproportionation that involve bulky intermediates, surface acidity is desirable. Wu and coworkers concluded that toluene disproportionation over MCM-22 occurs primarily in the MWW cages and subsequent p-xylene isomerization occurs in the ten-member ring 2D channels and the crystal exterior [122]. Studies with alkyl substituted pyridines such as 2,6-dimethyl pyridine (Lutidene) [123], 2,4,6-trimethyl pyridine (Collidine)[124] and 2,6-di-t-butyl pyridine [125] have been reported. Since the lone pair of the nitrogen atom in the aromatic ring is sterically hindered by the alkyl groups, these molecules react primarily with only Brönsted sites. A comparison of the kinetic diameters and proton affinities of these alkyl pyridines is listed in Table 4.4.

These large probe molecules can be combined with smaller probe molecules to determine the amount of surface acidity versus total acidity. By choosing a smaller probe molecule that can access the zeolite cages, the total acidity (external + internal) can be determined. A second experiment using a probe that cannot enter the zeolite pores provides the external only acid site information. Nesterenko and coworkers reported on a novel systematic methodology for the characterization and quantitation of acid sites on a series of dealuminated mordenites by using a variety of probes with different kinetic diameters combined with subsequent low-temperature CO adsorption [126]. They used the adsorption of pyridine, collidine and 2,6-di-t-butylpyridine to react with accessible acid sites and then performed a low temperature CO adsorption experiment to probe the remaining sites. Using this approach, they were able to obtain a more detailed characterization about changes in the accessibility of acid sites in mordenites. They found that dealuminum resulted in partial destruction of the side pockets which made them accessible to pyridine and further dealumination results in formation of mesopores that makes the zeolite crystals completely accessible to the more bulky lutidene, collidine and 2,6-di-t-butylpyridine molecules.

4.5.6.10 Other Probes

The acidity probes discussed above are the most commonly used. However, the use of many different probes has been reported in the literature. This list includes nitriles, alkanes, amines, water, di-hydrogen, deuterium, isotopically labeled molecules, benzene, etc. Probe molecules can also be used to measure basicity on zeolites. In this case, weakly acidic molecules such as CO_2, pyrrole, acetic acid and halogenated light paraffins have been used. Space does not permit discussion of these in any detail, but information about these probes and their applications can be found in the following references [87, 127–130].

4.5.7
In Situ/In Operando Studies

The use of probe molecules for measuring acidity on zeolites can provide a great deal of insight into the materials performance as a catalyst or adsorbent. By combining the acidity data with performance data from micro-reactor or pilot plant testing, it is possible to gain insight into how of the strength, location and number of acid sites in a material affects its performance. This information provides leads on how to change the zeolites properties to improve performance. Although this approach can be very useful, it does not allow one to see what is actually happening during reaction. Infrared spectroscopy can also be used to study an adsorbent or catalyst under operating conditions. *In operando* spectroscopy is a relatively new methodology that combines spectroscopic characterization of the catalytic material under reaction conditions with simultaneous measurement of catalyst activity and selectivity. This can provide molecular level information about the dynamic states of catalytically active sites and about the surface intermediates present during reaction. The design of the spectroscopic reaction cell is critical because it must be able to generate catalytic conditions so performance data equivalent to a conventional catalytic reactor can be obtained while allowing simultaneous spectroscopic measurements. Bañares described the difference between *in situ* and *in operando* experiments and discussed the important considerations that need to be addressed when putting together equipment for *in operando* experiments [131]. Fu and Ding followed the alkylation of benzene with propylene over MCM-22 [132]. They found that propylbenzene was formed only when benzene was pre-adsorbed on the acid sites of the MCM-22. Introduction of propylene displaced some of the benzene and the activated propylene molecules then reacted with the benzene to form propylbenzene.

Lercher and coworkers studied xylene isomerization on surface modified HZSM-5 zeolites [133]. They used time-resolved *in situ* IR spectroscopy to monitor the concentration of reactants and product inside the pores of the zeolite. Massiani and coworkers studied the conversion of xylene over mordenites [134]. They used *in operando* IR to characterize the adsorbed surface species and evolution of the active sites as the reaction proceeded.

4.5.8
Characterization of Metal-Loaded Zeolites

Zeolites can be ion-exchanged with cations or impregnated with various metals to modify their performance for use in applications such as separations, adsorption and catalysis. For example, faujasite zeolites exchanged with Na, Li, K, Ca, Rb, Cs, Mg, Sr, Ba, Cr, Mn, Fe, Co, Ni, Cu, Zn, Ru, Pd, Ag, Cd, In, Pt, Tl, Pb, La, Ce, Nd, Gd, Dy and Yb have been made and studied due to their use in separation and catalysis [135]. The ability to determine the distributions of these cations in the zeolitic structure is one of the key parameters needed in understanding adsorption mechanisms and molecular selectivities. Little has compiled an excellent reference source for infrared spectra of various adsorbent molecules on a variety of different materials and metals [136]. Although this reference was published over 40 years ago, it is still a very good starting place to help determine the best probe molecule for your particular application.

4.5.8.1 Cation Exchange for Adsorption/Separation

For adsorbents, an acidic zeolite is not typically desired. The acidity is usually removed by exchange of the protons with a group I or group II cation. These cations can then act as adsorption sites for nucleophilic molecules. The number of cations (degree of exchange), their location, and accessibility can be important parameters when it comes to a materials performance in an adsorption process. If the zeolite is not completely cation exchanged, there can be residual Brönsted acidity which can lead to undesirable reactions and loss of adsorption selectivity. Testing for residual acidity can be done by using one or more of the acidity characterization techniques discussed in Section 4.5.6. A fully exchanged zeolite should not have any measurable Brönsted acid sites. The location of the cations can also be important and are affected by the degree of exchange and the water content of zeolite. The same type of molecules used to probe acidity can be used to provide information about the siting of cations in zeolites. Busca and coworkers used low-temperature CO adsorption IR to compare K-LTA (3A), Na-LTA (4A) and Ca-LTA (5A) [137]. They found that, for K-LTA, weak potassium carbonyls were formed at the external surface. There was very little penetration of the CO into the channels. On the Na-LTA sample, Na^+ carbonyls formed in the zeolite cavities. They found that a small number of CO molecules bond to more than one cation, forming a strongly adsorbed species. Ca^{+2}-$(CO)_2$ *gem*-dicarbonyls were observed on Ca-LTA and a small number of CO molecules strongly adsorbed to more than one cation, as in the Na-LTA sample. The presence of strongly adsorbed CO and the ability to form dicarbonyls can be related to the materials performance in CO separations and its regenerability.

Martra and coworkers studied low-temperature CO and room-temperature CO_2 adsorption on Na- and Ba-exchanged X and Y zeolites [138]. They were able to determine from the CO adsorption IR data that Ba^{+2} ions located in the S_{II} sites in both X and Y zeolites exhibit higher Lewis acid strength compared to Na^+ ions in the same sites. Furthermore, the Lewis acid strength of both Ba^{+2} and Na^+ ions was dependent on the framework aluminum content of the zeolite. Using CO_2 as a probe, they found that substituting Na^+ with Ba^{+2} caused a reduction in the basicity of the framework oxygens in proximity to the cations.

Co-adsorption of benzene and ammonia was used to study the competition between cations and the 12-member rings of Cs(Na)-EMT zeolite by Su and Norberg [139]. They used changes in two C-H out of plane vibrations of benzene to indicate its reversible migration between the cations and the 12-member rings. Co-adsorption of ammonia displaced the benzene from the cations into the zeolite cages. These results illustrate how the chemical and adsorptive properties of the cages in zeolites can be modified by use of co-adsorption of a second molecule, ion-exchange and isomorphous substitution in the framework.

4.5.8.2 Metal-Loading for Catalysis

Metals have been loaded into zeolites by ion exchange and by impregnation for use in catalysis. Noble metals such as Pt and Pd have been added to provide a hydrogenation/dehydrogenation function. This hydrogenation function can be used alone or combined with the acidity of the zeolite to obtain a bi-functional catalyst for use in aromatization, isomerization, cracking and polymerization reactions. Other non-noble metals such as W, Mo, Co and Ni have been used in hydrocracking catalysts for processing heavy feeds into lighter components. In general, it is desirable to have the metals in a highly dispersed state to maximize catalyst activity. Infrared spectroscopy using probe molecules such as CO and NO has been successfully used to characterize these materials.

CO is an excellent probe molecule for probing the electronic environment of metals atoms either supported or exchanged in zeolites. Hadjiivanov and Vayssilov have published an extensive review of the characteristics and use of CO as a probe molecule for infrared spectroscopy [80]. The oxidation and coordination state of the metal atoms can be determined by the spectral features, stability and other characteristics of the metal-carbonyls that are formed. Depending on the electronic environment of the metal atoms, the vibrational frequency of the C–O bond can shift. When a CO molecule reacts with a metal atom, the metal can back-donate electron density into the anti-bonding pi-orbital. This weakens the C–O bond which results in a shift to lower vibrational frequencies (bathochromic) compared to the unperturbed gas phase CO value (2143 cm^{-1}) [62]. These carbonyls form and are stable at room temperature and low CO partial pressures, so low temperature capabilities are not necessary to make these measurements.

The maximum frequency of the C–O vibrational band and the overall band shape reflect the average electronic environment of the platinum atoms. Changes in band intensity, frequency and shape depend on the cluster size, exposed crystallographic planes, location (inside versus outside zeolite pores) and oxidation state of the metal. The number of publications for CO adsorption on zeolite supported noble metals are too numerous to list here, but a few examples that illustrate the type of information that can be obtained are included below.

4.5.8.3 Noble Metal-Loading for Catalysis

Kubelkova and coworkers studied the interaction of CO with platinum in a NaX zeolite [141]. They decomposed $Pt(NH_3)_4^+$ on the zeolite under various conditions and observed distinct CO vibrational spectra for PtO-CO, $Pt^{\delta+}$-CO, Pt^{+2}-CO and $[Pt(CO)_2]^+$ complexes after CO adsorption at room temperature. Subsequent reduction with hydrogen resulted in characteristic particle size distributions with distinct CO vibrational frequencies depending on whether the clusters (~1.2 nm) were inside the supercages or outside the cages (~4.0 nm).

Sachtler and coworkers showed evidence of CO adsorption on monoatomic Pt^0 with a CO vibrational frequency at 2123 cm^{-1} in H-MOR zeolite [142]. This band is shifted to significantly higher frequencies than the ~2070 cm^{-1} band that is typically observed for CO on multiatomic clusters. They proposed that these electron-deficient Pt atoms are interacting with protons in the zeolite forming [Pt-H$_z$] adducts in the mordenite.

The effect of crystallite size and shape of K-L zeolite on the dispersion of Pt was examined by a variety of techniques by Resasco and coworkers [143]. They obtained multiple overlapping CO bands on these samples and were able to assign them to Pt clusters located inside the zeolite pores (<2050 cm^{-1}), near the pore mouth (2050–2075 cm^{-1}) and outside the pores (>2075 cm^{-1}). They were able to correlate high n-octane aromatization activity with the K-L zeolite samples with short channels where most of the Pt is inside the pores.

Bischoff and coworkers showed that there is a clear particle size dependent shift of the linearly absorbed CO stretching vibrational frequency on Pt supported in a zeolite [144]. They observed shift from 2055 cm^{-1} for CO chemisorbed on 1–2 nm particles to 2070 cm^{-1} for CO on 4–5 nm particles. They correlated the frequency shift with the occurrence of higher indexed crystal planes for the smaller particles.

4.5.8.4 Non-Noble Metal-Loading for Catalysis

Non-noble metals such as Ni, Co, Mo, W, Fe, Ag and Cu have been added to zeolites for use in catalysis. In addition to CO, nitric oxide (NO) has been shown to be a good adsorbate for probing the electronic environment of these metals. When NO chemisorbs on these metals, it can form mononitrosyl (M-NO) and dinitrosyl species (ON-M-NO). The monontrosyl species has a single absorption band and the dinitrosyl species has two bands due to asymmetric and symmetric vibrational modes of the (ON-M-NO) moiety. Again, there have been many studies reported in the literature on the use of NO and/or CO adsorption on non-noble metals supported on zeolites and they are too numerous to list here. Several examples have been selected and summarized to provide the reader with the type of information that can be provided by this method.

Ni species in faujasite were studied by CO adsorption IR by Kebulkova and coworkers [145]. They found that partially reduced Ni in dealuminated Y zeolite forms bicarbonyl complexes that vibrate at 2100 and 2140 cm^{-1}. These bicarbonyls are transformed into monocarbonyls which vibrate at 2118 cm^{-1} upon removal of gas phase CO. Big clusters of Ni^0 on the outer surface of Ni-NaX zeolite was redispersed by CO at ambient temperatures by formation of Ni(CO)$_4$ which is mobile and can be trapped inside the zeolite cavities. This nickel carbonyl species can then be decomposed into polynuclear carbonyls with characteristic bands between 2070 and 1812 cm^{-1} and metallic nickel clusters. Hadjiivanov and coworkers studied the effect of preparation method on the state of nickel ions in betra zeolite using CO and NO adsorption combined with XRD and temperature-programmed reduction (TPR) [146]. They found that incorporation of Ni by ion exchange in beta zeolite (Si/Al = 11) did not cause any structural changes in the zeolite. Two different types of Ni^{2+} sites with high coordinative unsaturation were detected by CO and NO adsorption. Some of these Ni^{2+} cations were easily reduced to Ni$^+$ by CO. In contrast, Ni exchanged into de-aluminated beta zeolite (Si/Al > 1300) changed the unit cell parameters of the zeolite indicating incorporation of the Ni into the framework. Very little CO and NO was adsorbed on the Ni^{2+} in these samples since they are more coordinately saturated or inaccessible compared to the aluminum containing beta sample. This result has implications on the reactivity of the Ni^{2+} in catalytic reactions such as selective reduction on nitric oxides [147].

The adsorption of NO and CO has been used to characterize the properties of Co in Co-exchanged zeolites [148–151]. NO is a selective probe for Co^{3+} and CO is selective for Co^{2+} species. Datka and coworkers used the combination of CO and NO adsorption IR to quantitatively determine the concentration of Co^{3+} as an oxide and the Co^{2+} present as in exchange positions, oxide-like clusters, and cobalt oxide in a series of Co-exchanged ZSM-5 and ferrierite (FER) zeolites [151]. They established conditions under which the CO and NO would react selectively with the various types of sites and established absorption coefficients for the quantitative calculations. Differences in the distributions of the various forms of Co species were found to be dependent on both the structure and framework Si/Al of the zeolite.

4.6
NMR Characterization of Zeolitic Systems

4.6.1
Introduction to NMR

To design an active and stable zeolite catalyst we need knowledge of the local framework structure, nature, location and distribution of active sites. Magic angle spinning (MAS) NMR spectroscopy has been widely employed for the investigation of local structural elucidation of zeolites. The most commonly investigated nuclei present in zeolites and related microporous materials are ^{29}Si, ^{27}Al, ^{31}P, ^{1}H, ^{17}O, ^{23}Na and $^{6,7}Li$ [1–9]. Other nuclei like ^{13}C, ^{25}Mg, ^{11}B, ^{47}Ti, ^{51}V, ^{95}Mo, ^{119}Sn, ^{133}Cs, ^{205}Tl, etc. have also been investigated. This technique is very powerful when single crystals are not available or when the samples lack sufficient long range order as in amorphous or partially crystalline materials that might result in steaming or calcining procedures in an industrial environment. Also for aluminosilicates Si and Al have similar X-ray scattering powers, which poses difficulty in X-ray structural determination. MAS NMR provides complementary information to X-ray diffraction. The most important parameter we measure by NMR is the chemical shift and the chemical shift ranges for many of the nuclei are well-established. The chemical shifts are found to be sensitive to local structural coordination, bond lengths, bond angles, hybridization, next nearest neighbors, etc. ^{27}Al NMR is routinely used to identify the proportions of framework and extra-framework Al upon a variety of treatments to zeolites (for e.g., dealumination, realumination, etc). ^{29}Si NMR is used to identify the structural species and framework Si/Al ratio. The chemical shift ranges for ^{29}Si, ^{31}P and ^{1}H are given in Figures 4.33–4.35. ^{17}O NMR identifies the structural environments around bridging oxygen. In the following

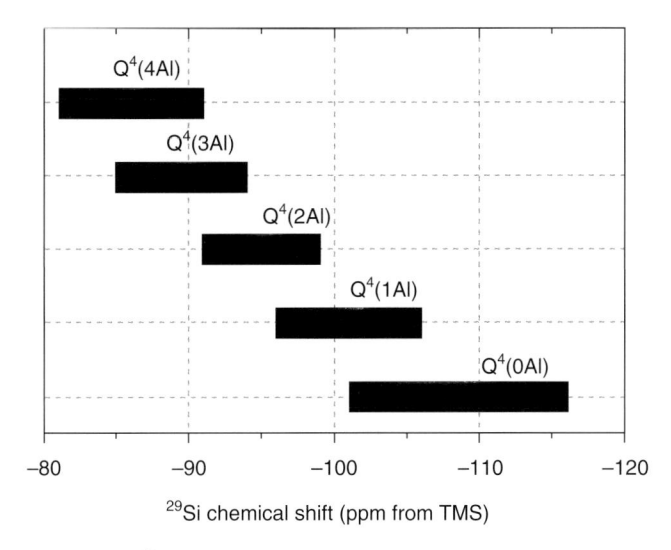

Figure 4.33 ^{29}Si MAS NMR chemical shift ranges for aluminosilicates. This figure is redrawn based on references [152–155]. Here Q^4 is the tetrahedral silicon connected to n aluminum atoms ($n = 0$–4) via oxygen bridges.

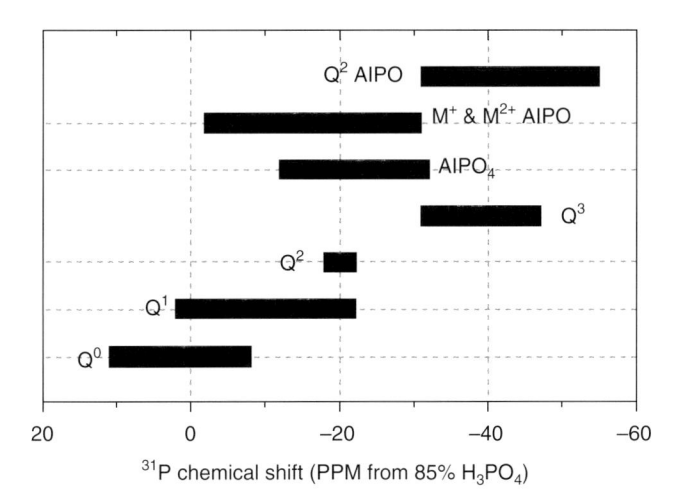

Figure 4.34 ^{31}P MAS NMR chemical shift ranges for crystalline phosphates and aluminophosphates. This figure is redrawn based on references [155, 156]. Here Q^n is the tetrahedral phosphorous connected to n bridging oxygens ($n = 0$–3), the remaining oxygen is double bonded. In AlPO$_4$ all the oxygens are bridging oxygens. M$^+$ and M^{2+} are the mono- and divalent cations, respectively.

we present a qualitative overview of nuclear spin interactions of interest, followed by applications of MAS NMR techniques to zeolites. For a detailed literature on this subject the reader is referred to excellent books and reviews that have been published already on this topic [152–160].

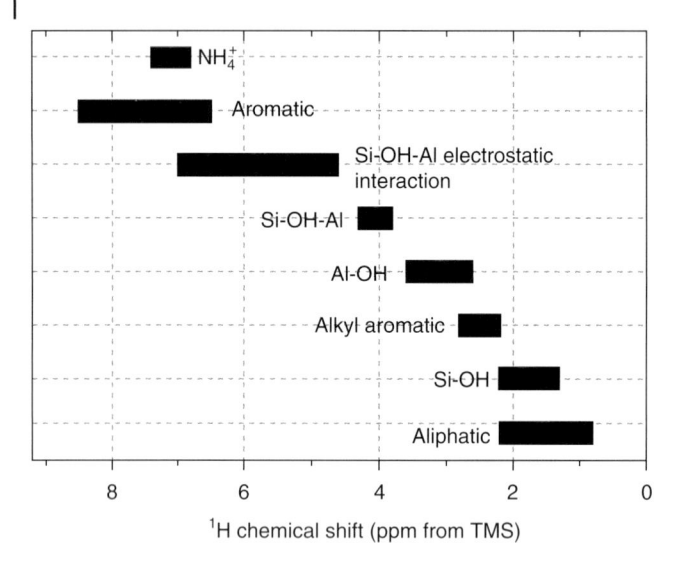

Figure 4.35 ^{1}H MAS NMR chemical shift ranges for H-forms of zeolites. This figure is drawn based on the data from references [155, 157].

4.6.1.1 Spin-Half Nuclei

The major interactions that broaden the NMR spectra of spin-half nuclei are magnetic dipolar interactions between nuclei (homo- and heteronuclear interactions) and anisotropic chemical shift interactions [161–163]. For a less abundant nucleus like Si (natural abundance 4.7%), the homonuclear interactions are negligible. Both these interactions are described by equations containing the geometric term $(3\text{Cos}^2\theta-1)$, where θ is the angle between internuclear vector connecting the nuclei and the external magnetic field [164, 165]. Since a typical powder sample contains crystallites oriented in all possible directions, we need to spin the sample rapidly at the magic angle (54° 44′) to average the geometric term to zero. The spinning speed should be sufficiently fast (≥static linewidth) to narrow the spectrum and if the spinning speed is lower than the linewidth, sidebands spaced at multiples of the spinning speed occurs on either side of isotropic peak.

4.6.1.2 Cross Polarization

This technique involves transfer of polarization from one NMR active nucleus to another [166–168]. Traditionally cross polarization (CP) was employed to transfer polarization from a more abundant nucleus (I) to a less abundant nucleus (S) for two reasons: to enhance the signal intensity and to reduce the time needed to acquire spectrum of the less abundant nuclei [168]. Thus CP relies on the magnetization of I nuclei which is large compared to S nuclei. The short spin-lattice relaxation time of the most abundant nuclei (usually proton) compared to the long spin-lattice relaxation time of the less abundant nuclei, allows faster signal averaging (e.g., ^{29}Si or ^{13}C). CP is not quantitative as the intensity of S nuclei closer to I nuclei are selectively enhanced. Nowadays CP has been extended to other pairs of nuclei (e.g., ^{19}F → ^{29}Si [169], ^{27}Al → ^{31}P [170, 171], ^{27}Al → ^{29}Si [172], ^{23}Na → ^{31}P [173]) to obtain connectivity information about these nuclei. Theoretical and experimental aspects of cross polarization have been discussed extensively [174, 175]. ^{1}H → ^{29}Si CPMAS has been widely applied to zeolites to identify the presence of silanol species [176]. CP was shown to be very useful in understanding probe molecules adsorbed on surfaces and their dynamics [177]. Mobile and immobile phases can also be easily distinguished by CP [178].

4.6.1.3 Quadrupolar Nuclei

Nuclei with spin greater than $I = 1/2$ have an electric quadrupole moment due to non-spherical charge distribution of the nucleus. In these systems the nuclear electric quadrupole moment interacts with electric field gradient at the nuclear site. For half-integer nuclei like ^{27}Al ($I = 5/2$) the central transition ($I = +0.5$ to $I = -1/2$), which is not affected by first order quadrupolar interaction, is detected [179, 180]. Magic angle spinning NMR averages only up to the first order quadrupolar interactions, and when the second order quadrupolar interactions are substantial, significant shifting and broadening of signals occur. Also second order quadrupolar interactions are inversely proportional to the applied magnetic field. Hence the peak position measured by MAS NMR varies with the applied magnetic field, and second order averaging is required to measure the true isotropic chemical shift. The second order term contains more complex geometric terms and hence no single angle rotation can average the second order quadrupolar interactions. These second order interactions are averaged by spinning the samples at two angles simultaneously (double rotor spinning [DOR] [181–183]) or sequentially (dynamic angle spinning [DAS] [183–187]). Another method of averaging these interactions is utilizing multiple quantum transitions and magic angle spinning (MQMAS) [188–190]. Of these three techniques, MQMAS became popular as it can be implemented in existing commercial MAS probes, whereas DAS or DOR requires specialized probes. MQMAS is essentially a 2D NMR technique; one axis corresponds to regular MAS dimension and the other axis corresponds to isotropic dimension. The information obtained from MQMAS is used for the quantitative simulation of MAS spectrum to get site populations. Satellite transition MAS (STMAS) [191–193], which correlates single quantum satellite transitions to central transition, has also been developed as complimentary technique to MQMAS. Nutation NMR [194, 195] is also used to resolve peaks based on their quadrupolar parameters (quadrupole coupling constant, asymmetry parameter). It is also a 2D technique as one axis corresponds to MAS dimension and the other axis is related to quadrupolar parameters and radiofrequency field strength.

4.6.1.4 Dipolar Recoupling

Since dipolar interaction is a distance-dependent interaction, the heteronuclear interaction between spins I and S can be exploited to get information about the distance between these nuclei. Heteronuclear dipolar interactions that are averaged by magic angle spinning can be "reintroduced" by suitable dephasing pulses that are synchronized with the sample spinning. It is a difference spectroscopy

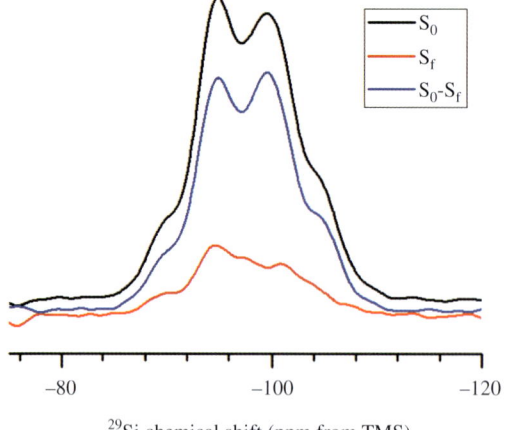

Figure 4.36 ^{29}Si{^{27}Al}TRAPDOR NMR of a typical NH$_4$Y zeolite. The top spectrum, S_o, is the ^{29}Si spin-echo spectrum, the middle spectrum is the ^{29}Si spectrum with ^{27}Al dephasing pulses turned on. The bottom spectrum is the difference between S_o and S_f, that is solely due to ^{29}Si close in space to ^{27}Al. The more pronounced peak around −95 ppm in the bottom spectrum is due to silicon connected to two Al atoms [Q^4(2Al)].

where I spin-echo MAS spectrum (S_f) acquired with rotor-synchronized dephasing pulses on the S spins is subtracted from the spectrum (S_o) obtained without dephasing pulses. The dephasing pulses "interrupt" the averaging of dipolar interaction in every rotor cycle. The difference spectrum is solely due to I-S dipolar coupling. If both I and S spins are spin-half, the experiment is called *rotational echo double resonance* (REDOR) [196–198]. *Transferred-echo double resonance* (TEDOR) [199, 200] is a closely related technique. If the S spin is quadrupolar *transfer of population in double resonance* (TRAPDOR) [201–203] or *rotational echo adiabatic passage double resonance* (REAPDOR) [204, 205] is employed. An example of a dipolar recoupling experiment applied to a zeolite sample is shown in Figure 4.36.

4.6.1.5 Pulsed Field Gradient NMR–diffusion

Pulsed field gradient (PFG) NMR has been widely used to measure diffusion coefficients of hydrocarbons located inside zeolites [206–208]. This technique is based on the spatial dependence of resonance frequencies in the presence of magnetic field gradients. Usually ^{1}H diffusion coefficient is determined from these measurements, although other nuclei can also be employed. This experiment is based on spin-echo sequence. In the simplest case, the spin-echo sequence contains two pulses ($\pi/2$, π) separated by time τ and acquisition of signal after time 2τ. After the $\pi/2$ pulse, the nuclear spins evolve for time τ due to magnetic field inhomogenity, heteronuclear dipolar and chemical shift interactions, and the evolution is reversed by a π pulse, and after another time, τ the spins refocus and an echo (maximum signal, neglecting transverse relaxation) forms. PFG NMR involves application of magnetic field gradients on both sides of the π pulse. Diffusion of molecules during the τ interval, results in attenuation of echo intensity. The decay of the echo amplitude is followed as a function of magnetic field gradient strength, and diffusion coefficient is calculated from echo signal decay. In practice transverse relaxation, magnetic susceptibility etc complicates the situation, and accordingly techniques were developed to address the issue. Broad range of time (10 ms to several seconds) and length scales (<1 μm to several microns) are available. Larger gradient strength allows measurement at smaller diffusion times and diffusion coefficients varying from 10^{-6} to $10^{-14}\,m^2\,s^{-1}$ have been measured by this method.

4.6.2
Applications

As mentioned earlier, MAS NMR is a very powerful local structural probe that is sensitive to coordination number, and next nearest neighbors, and provides complimentary information to diffraction techniques. Lack of Si, Al ordering in zeolites poses challenges to structural determination by diffraction techniques and NMR provides a powerful complimentary tool for structural analysis. Zeolites can be tailored to a specific application by modification of the structure by processes such as ion exchange, calcination, steaming, acid wash etc and ^{29}Si and ^{27}Al NMR is routinely employed to follow the structural changes occurring during these modifications. Thermal stability of zeolites, which is an important criteria for an industrial catalyst, is also evaluated by following the structural changes obtained through ^{29}Si and ^{27}Al NMR. Also ^{27}Al, ^{29}Si, ^{17}O and ^{1}H MAS NMR provides a means to quantify framework Al, Si/Al ratio and acidity. Much work is devoted on ultrastable Y zeolites to understand the zeolite structure and catalytic performance. The charge compensating cations have also been investigated. Typical MAS NMR experiment involves minimal sample preparation. Usually powdered samples are packed in a zirconia rotor and dry air or nitrogen is used to spin the sample located inside the NMR probe. Below only a few representative examples are given to illustrate the power of this technique.

4.6.2.1 ^{29}Si NMR

^{29}Si NMR is used to directly probe the local structure of zeolites and each resonance in the spectrum is due to crystallographically inequivalent Si atom. The line shapes and peak positions are very sensitive to subtle changes in the zeolites lattice structure. The chemical shift for silicon in zeolites is well-established and range from −60 to −120 ppm (chemical shifts are usually referenced with respect to tetra methyl silane [TMS]). ^{29}Si NMR shows distinct resonances depending on the number of Si–O–Si bonds around each silicon. The silicon species are represented by Q^n (where Q is the tetrahedral silicon, n is the number of Si–O–Si bonds or bridging oxygens). Thus Q^4 is the tetrahedral silicon connected to four neighboring silicons via oxygen bridges, Q^2 has two bridging and two non-bridging oxygens and Q^0 is the isolated silicate anion (SiO_4^{4-}). Substitution of one of the silicons in

Q⁴ by Al results in a downfield shift (less negative, high frequency) of 5–8 ppm [153, 158]. The silicon–aluminum environments in this case are represented by Q⁴(nAl) where n is the number of Si–O–Al bonds and can range from 0 to 4. Thus ^{29}Si MAS NMR resolves five distinguishable resonances in aluminosilicates. Extra framework silicon can be estimated from the −110 peak. Based on the NMR investigation of large number of aluminosilicates, a variety of empirical relationships between ^{29}Si NMR chemical shifts and Si–O bond lengths, Si–O–Si and Si–O–Al bond angles, hybridizations, electronegativities of the surrounding groups have been established [146, 153, 158].

^{29}Si NMR provides quantitative information about the framework composition, and framework Si/Al ratio is an important parameter used to tune the catalyst property. Zeolite acidity is directly related to the amount of framework Al. Framework Si/Al ratio can directly be obtained from just ^{29}Si NMR alone. Si/Al ratio can be calculated from ^{29}Si NMR intensities if the resonances due to different Q⁴(nAl) species are well-resolved using Eq. (4.10), assuming there is no Al–O–Al bonds present:

$$\frac{\text{Si}}{\text{Al}} = \frac{\sum_{n=0}^{4} I_{Q^4(nAl)}}{\sum_{n=0}^{4} \frac{n}{4} I_{Q^4(nAl)}} \quad (4.10)$$

Here $I_{Q4(nAl)}$ is the intensity of Q⁴(nAl) species. Thus the ^{29}Si NMR intensity of each Si–O–Al bond is equivalent to one-quarter Al atom, as each Al atom is surrounded by four other Si atoms. Figure 4.37 illustrates ^{29}Si NMR of Y zeolite sample and

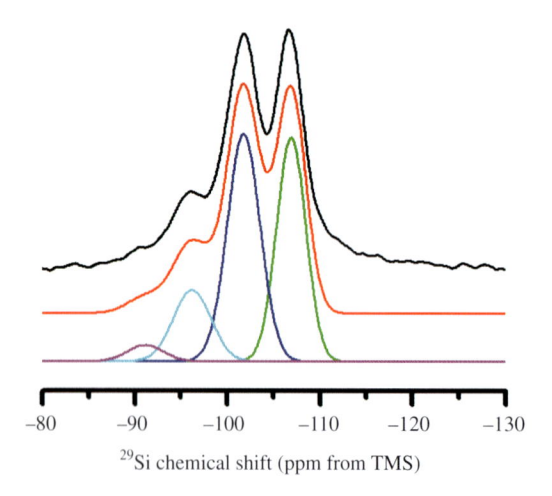

Figure 4.37 ^{29}Si MAS NMR of a calcined Y zeolite. The top spectrum is the experimental ^{29}Si MAS NMR spectrum and the middle spectrum is based on computer simulation using Gaussian components shown in the bottom spectrum. The Si/Al ratio calculated for this material using the equation shown in the text is 4.5.

their individual population of $Q^4(nAl)$ species obtained from computer simulation. Si/Al ratios of synthetic faujasites obtained from NMR agree quite well with that obtained from X-ray fluorescence [209]. Comparison of bulk versus framework Si/Al obtained from NMR allows us to calculate extent of framework damage to post-synthesis modifications. The Si/Al ratios from this measurement are not reliable if there are significant amounts of silanol groups (Q^n) present, as there is a significant overlap of resonances due to Si–OH groups and Si–O–Al groups. However the presence of different Q^n species due to Si–OH groups can be identified with the ^{29}Si cross polarization technique. Using the peak positions and widths obtained from CP it is possible to fit the ^{29}Si NMR data and obtain corrected Si/Al ratios. Dealumination of zeolites results in highly siliceous materials thereby enhancing the thermal stability. It is difficult to measure Si/Al ratios for highly siliceous materials, as the intensity of peaks due to Si–O–Al groups is very low.

4.6.2.2 ^{27}Al NMR

^{27}Al MAS NMR has been demonstrated to be an invaluable tool for the zeolite scientist. It provides a simple and direct way to quantify the proportions of Al in four [Al(4)], five [Al(5)] and six [Al(6)] coordinations. Quantitative determination of these species is an important issue in catalysis, and major effort is devoted on this topic. As mentioned already, for Al only the central transition (+half to –half "selective excitation") is detected. The central transition is unaffected by first order quadrupolar interaction, but the presence of second order effects causes broadening and complicates the quantitation of the Al species. Usually hydrated samples and short radiofrequency pulses are employed for quantitative determination of framework and extra framework aluminum species. It is uncertain whether hydration changes the coordination of Al species. Certain extra framework Al can have very large quadrupolar interactions resulting in very broad lines ("NMR invisible") [155, 202]. Unlike ^{29}Si NMR, ^{27}Al has a short relaxation time due to its quadrupolar nature, and the ^{27}Al NMR spectrum with good signal to noise can be obtained in a relatively short time.

Acidity of the zeolite framework can be tailored by dealumination or realumination that results in modification of framework to extra framework Al ratio. ^{27}Al MAS NMR is extensively used to quantify the population of these Al species. An example of ^{27}Al NMR spectral changes upon calcination, steaming or acid wash is shown in Figure 4.38. In fresh zeolite, Al is present only as Al(4) species. Calcination results in the conversion of 25% of Al(4) to Al(6) species (Figure 4.38, bottom spectrum). Steaming the zeolites results (Figure 4.38, middle) in generation of Al(5) in addition to Al(6). The sharp peak in the octahedral region is due hexa aquo aluminum species. Steaming followed by acid wash (Figure 4.38, top spectrum) results in increase in Al(5) and narrowing of the Al(6) peak. The intensity extra framwork aluminum species provide a direct measure of the extent of dealumination. The presence of extra framwork aluminum results in pore blocking resulting in reduction in the availability of acid sites, directly affecting the catalytic performance of the zeolite. ^{29}Si NMR provides complementary information and the corresponding ^{29}Si NMR is shown in Figure 4.38b. ^{29}Si NMR of the base zeolite shows

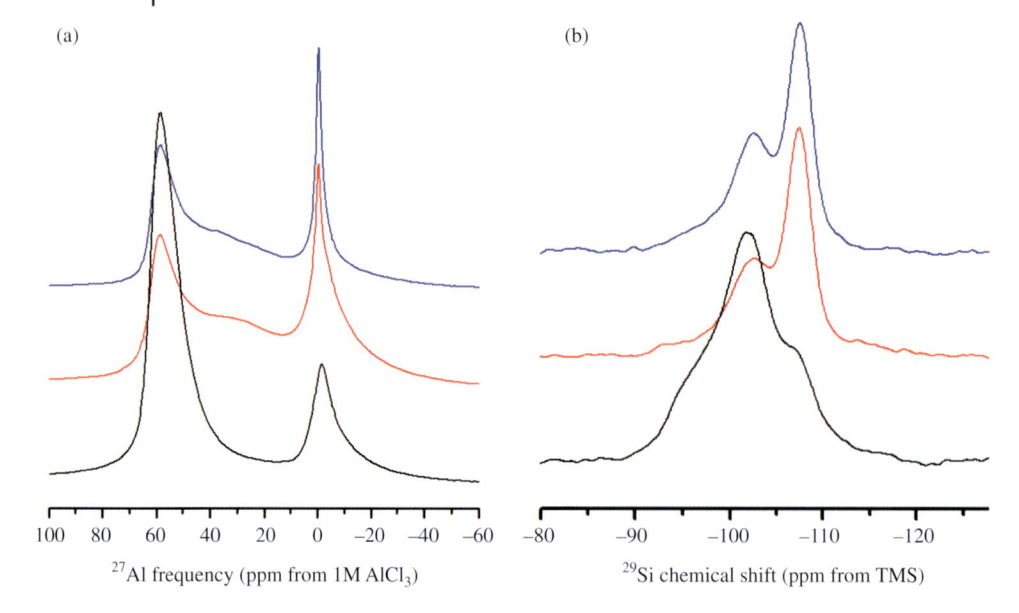

Figure 4.38 (a) ^{27}Al MAS NMR of Y zeolites subjected to calcination (bottom spectrum), steaming and calcination (middle spectrum) and steaming and acid wash (top). The peak at ~60 ppm is the tetrahedral Al, the peak around 0 ppm is the octahedral Al, the peak in the middle region is due to distorted tetrahedral Al. ^{27}Al NMR clearly shows the generation of extra framework Al upon various modifications. The sharp peak close to ~0 ppm is due to hexa aquo aluminum species. (b) ^{29}Si MAS NMR of Y zeolites subjected to calcination (bottom spectrum), steaming and calcination (middle spectrum) and steaming and acid wash (top spectrum). Steaming and calcination results in increase of Q^4(0Al) species at the expense of other species clearly demonstrating the dealumination of the material.

Q^4(0Al), Q^4(1Al), Q^4(2Al) and Q^4(3Al) species. Calcination (Figure 4.38, bottom spectrum) results in increase in Q^4(0Al), Q^4(1Al) species at the expense of Q^4(2Al), Q^4(3Al), clearly demonstrating dealumination of the zeolite. The framework Si/Al ratio calculated from ^{29}Si NMR at this stage is 3.8. Steaming and acid wash results in further increase in Q^4(0Al), Q^4(1Al) species (Figure 4.36 middle, top), resulting in Si/Al ratio of 6.6. Framework Si/Al ratio can also be calculated using Al NMR intensities and bulk Si/Al ratio obtained from chemical analysis. But care should be taken in interpretation of the data, as some of the aluminum may be "invisible." The nature of extra framework Al can be elucidated from ^{1}H → ^{27}Al CP MAS NMR experiments [210, 211]. MQMAS is a powerful technique to distinguish subtle structural differences that are masked by regular MAS due to second order quadrupole broadening. For example, Figure 4.39 demonstrates that MQMAS is capable of distinguishing Al(5) from distorted Al(4) environments in steamed MTW powder. Omegna et al. [212] employing high magnetic fields and high spinning speeds, identified at least six Al species that are present as Al(4), Al(5), or Al(6) in ultrastabilized Y. Conversion of Al(5) and Al(6) coordination (so-called flexible Al)

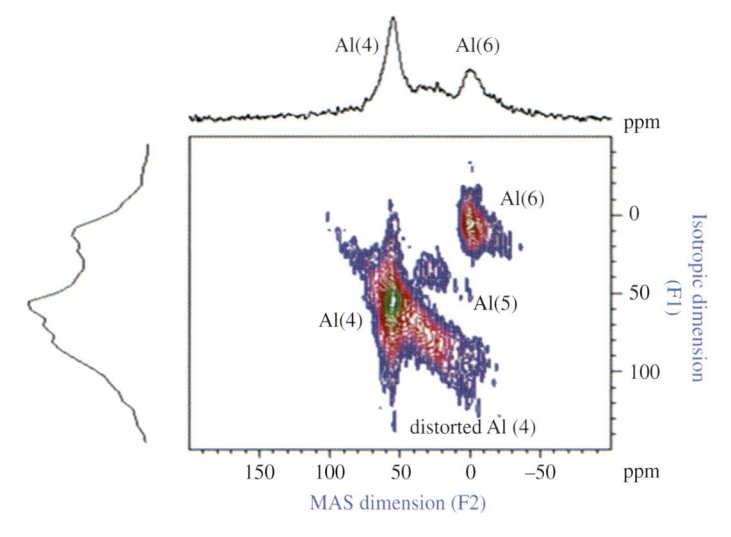

Figure 4.39 ^{27}Al MQMAS NMR of a steamed MTW zeolite powder. The spectrum was collected at a ^{27}Al frequency of 104.2639 MHz and a sample spinning speed of 32 kHz. The top spectrum is the projection along the MAS dimension and the left spectrum is the projection along the isotropic dimension. MQMAS clearly shows that the middle region in the 1D MAS spectrum is due to combination of five coordinate Al and distorted four coordinate aluminums.

to Al(4) coordination upon ammonia adsorption has also been demonstrated. ^{27}Al nutation NMR [194, 195] has also been shown to resolve Al environments that are overlapping in regular MAS NMR based on differing quadrupolar parameters of the Al species.

4.6.2.3 ^{31}P NMR

^{31}P nucleus is a spin-half nucleus with 100% abundance and is hence attractive for NMR. ^{31}P MAS NMR has shown to be very sensitive to local coordination, bond angle, bond distances, degree of hybridization, etc. [156]. The chemical shifts for phosphorus in ortho-, pyro- and meta-phosphate-type environments are quite distinct, and the chemical shift ranges are given in Figure 4.34. ^{31}P NMR has been widely employed to investigate the local structural differences in a variety of aluminophosphate (AlPO), silicoaluminophospahte (SAPO), P-modified zeolites and other microporous materials [155]. In addition heteronuclear correlation between P and Al are also demonstrated on variety of microporous materials [155, 172, 200].

The choice of probe molecule for acidity determination is largely dictated by the sensitivity of the nuclei, and the high sensitivity of ^{31}P nuclei has been exploited to quantify acidity in zeolites. The use of phosphorus containing bases as probe molecules to measure solid acidity circumvents insensitivity problems associated with ^{13}C (1.1% abundance) and ^{15}N (0.4% abundance) in bases such as ammonia,

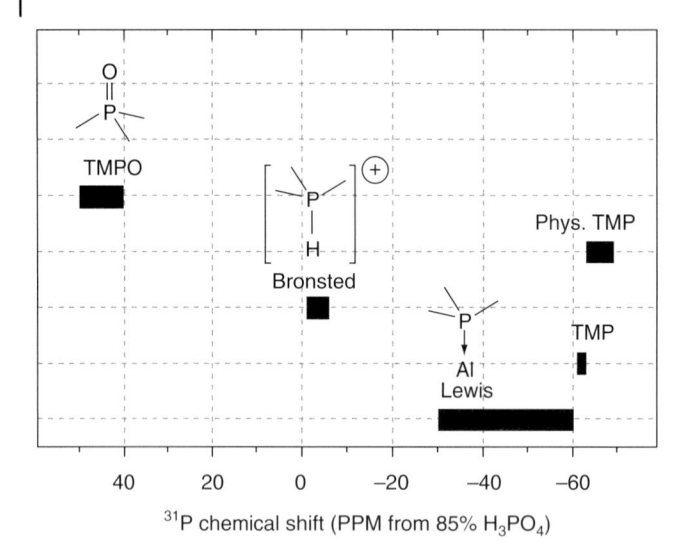

Figure 4.40 ^{31}P MAS NMR chemical shift ranges for Brönsted and Lewis acid sites in trimethyl phosphine (TMP) loaded zeolites.

butylamines and pyridine. Catalytic activity of zeolites depends on the strength, concentration and location of acid sites in zeolites. ^{31}P MAS NMR is extensively used to investigate the number, type and location (internal or external) of Brönsted and Lewis acid sites present zeolites. Typical phosphorus-based probe molecules employed are trimethyl phosphine (TMP) and trimethylphosphine oxide (TMPO) [213]. TMP is a small molecule (kinetic diameter 5.5 Å) that forms complexes with Brönsted acid sites and framework and non framework Lewis sites. Figure 4.40 shows the chemical shift ranges observed for Lewis, Bronsted sites in TMP-loaded zeolites. Larger molecules like triphenyl phosphine (kinetic diameter ~10.2 Å) is used to characterize external acidity [214]. Adjacent Brönsted acid pairs were investigated by using alkyl bridged diphosphine probe molecules [215, 216]. By using [1,n] Bis(diphenyphosphino) alkanes, where the two P are separated by an alkyl chain with n varying from 1 to 6 (distance between the two phosphorus is ~3 to 9 Å), existence of a variety of adjacent acid sites have been demonstrated in zeolite Y with different Si/Al ratios.

4.6.2.4 ^{1}H NMR

^{1}H NMR provides a direct way to quantify the number and type of OH groups present in zeolites. Brönsted acid sites and terminal OH groups were identified in a number of zeolites. Usually the proton resonance due to adsorbed water dominates the spectrum and the zeolite need to be pretreated in helium at high temperatures (300–450 °C) to remove the adsorbed water. Broad-line ^{1}H NMR has been widely employed to measure the distance between proton and aluminum in zeolites [155, 159]. High-speed magic angle spinning is required to average the strong ^{1}H–^{1}H homonuclear interactions. High-speed ^{1}H MAS NMR is shown to be sensitive to terminal OH groups and bridging framework OH groups (Brönsted sites). Thus ^{1}H NMR provides a direct measure of Brönsted acids and does not suffer from the limitations of other spectroscopic techniques based on probe molecules, which are highly dependent on their size and basicity. The chemical shift ranges for different species are given in Figure 4.35. ^{1}H chemical shifts are shown to increase with increase in mean electronegativity of the zeolite framework [157]. If the bridging OH group is involved in hydrogen bonding with the zeolite framework, the ^{1}H peak position shifts downfield (high frequencies), and has been demonstrated in dehydrated HY or HZSM-5. ^{1}H-^{27}Al TRAPDOR was useful in identifying "invisible" Al present in HY [202].

4.6.2.5 ^{17}O and Other Nuclei

^{17}O NMR is a promising tool for identifying the structural environments around oxygen in zeolites. However due to the low abundance of ^{17}O isotope (0.04%) the application of ^{17}O NMR is limited to ^{17}O-enriched zeolites only. The oxygen environment of highly siliceous Y was investigated by DOR and DAS NMR. ^{17}O DOR spectrum collected at above 14 T resolves all four crystallographically different oxygen species [217]. In the case of low silica zeolite A and LSX, Fredue et al. [218], based on ^{17}O MQMAS and DOR, found correlation between isotropic ^{17}O chemical shifts of the oxygen sites and Si–O–Al bond angle. Hydration of zeolites is shown to cause a downfield shift of 8 ppm from dehydrated zeolites. ^{17}O NMR has also been utilized to resolve Brönsted acid sites present in supercages and sodalite cages in HY [219]. Well-resolved ^{17}O DOR peaks were reported for Li-LSX zeolites, but individual peaks were not resolved for Na-, K and Cs-LSX samples. DOR and MQMAS NMR showed similar sets of resonances for mono- and divalent cation-exchanged LTA and LSA [220].

The shape selectivity of zeolites is influenced by the location and distribution of charge-compensating cations. The charge-compensating ions other than protons are all quadrupolar. ^{6}Li and ^{7}Li NMR spectra of dehydrated LiX-1.0 identified three crystallographically distinct sites [221]. In the case NaX with Si/Al ratio of 1.23, six distinct sodium sites were identified using fast ^{23}Na NMR, DOR and nutation techniques [222]. ^{23}Na MQMAS has been extensively studied for zeolites X and Y [155]. Other cations like Cs and La in zeolites have also been investigated [155, 158].

4.6.2.6 Diffusion of Hydrocarbons in Zeolites

Understanding the adsorption, diffusivities and transport limitations of hydrocarbons inside zeolites is important for tailoring zeolites for desired applications. Knowledge about diffusion coefficients of hydrocarbons inside the micropores of zeolites is important in discriminating whether the transport process is micropore or macropore controlled. For example, if the diffusion rate is slow inside zeolite micropores, one can modify the post-synthesis treatment of zeolites such as calcination, steaming or acid leaching to create mesopores to enhance intracrystalline diffusion rates [223]. The connectivity of micro- and mesopores then becomes an important issue and can be studied by choosing probe molecules that can penetrate only mesopores or both. PFG NMR has been employed widely to measure diffusivities and molecular displacements of guest molecules in several zeolites. In addition to the measurement of diffusion coefficient, PFG NMR provides clues about the nature of diffusion and barriers to transport [223, 224]. Commercial NMR probes that are capable of creating large magnetic gradients (up to 3500 Gauss cm^{-1}) are available. This allows measurement of molecular displacement to about less than a micron. Intracrystalline diffusivities (i.e., diffusivities measured for means square displacements that are smaller than the zeolite crystal size) for n-octane and 1, 3, 5 triisopropyl benzene loaded Y and USY have been determined [223], and from that study it was concluded that mesopores in USY does not play a significant role in intracrystalline diffusivity. Diffusion of methane and butane in MFI was measured as a function of temperature and diffusion times [224]. The variation of diffusion coefficients with molecular displacement at low temperatures was attributed to existence of transport barriers at the intersection of MFI crystals or its intergrowth components [224]. Thus PFG NMR is useful for studying intracrystalline and intercrystalline or long range diffusion and also to understand external surface barriers that may limit diffusion into the zeolite. Catalytic performance of FCC catalysts has been shown to improve with increase in intraparticle diffusion [208].

4.7
Physical/Chemical Characterization

4.7.1
Nitrogen Physisorption

The micro-, meso- and macro-sized pores of a zeolite impact the catalytic and separation properties. Based on IUPAC terminology micropores are defined by pore sizes smaller than 2 nm, mesopores are between 2 and 50 nm and pores greater than 50 nm are referred to as macropores.

One of the most widely used methods for determining the pore size and surface area of zeolites is nitrogen physisorption. From the shape of the nitrogen adsorption and desorption isotherm the presence and shape of the mesopores can be deduced. As shown in Figure 4.41 a faujasite without mesopores have a type I isotherm since the micropores fill and empty reversibly, while the presence of mesopores results in a combination of type I and IV isotherms. The existence of a hysteresis loop in the isotherms indicates the presence of mesopores while the shape of this hysteresis loop is related to their geometric shape.

Although there are several methods for analysis of nitrogen physisorption data, the most commonly used is BET surface area. Because for microporous materials the boundary conditions for multilayer adsorption are not fulfilled, the calculated BET surface area has no physical meaning. Such data should be considered proportional to the total micropore volume rather than the specific surface area. The t-plot method can be used to calculate the micropore volume and the mesopore

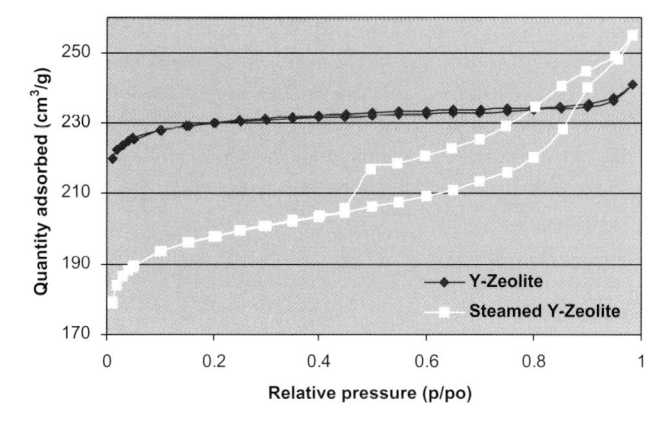

Figure 4.41 Isotherm for ammonium exchanged Y-zeolite showing type I and steamed Y-zeolite showing a type IV isotherm.

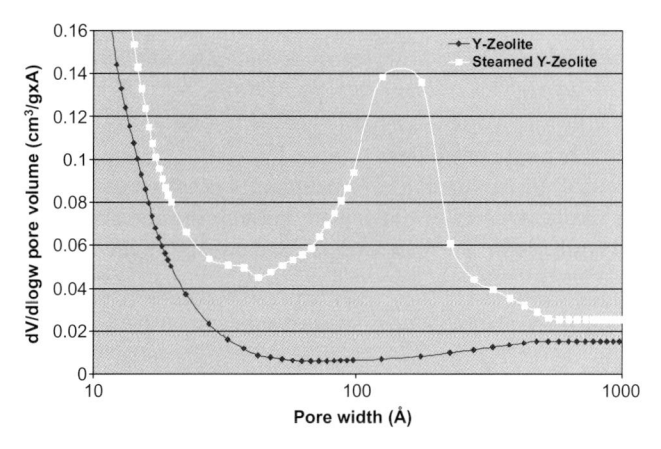

Figure 4.42 Calculated mesopore size distribution for the steamed Y-zeolite based on the BET adsorption. The ammonium exchanged Y-zeolite has no mesopores.

area. The micropore volume is determined from the intercept of the t-plot while the mesopore area is calculated from the slope of the t-plot. The micropore volume is relatable to the amount of zeolite present. Loss of structure of the zeolite from processing can be determined by this method and often the micropore volume technique is more sensitive to minor structural degradation than is the X-ray diffraction measurement. If some zeolite micropores are filled or blocked, the adsorption method yields a lower result. Although the adsorption method allows the nitrogen to equilibrate, a highly blocked three-dimensional zeolite such as a heavily coked Y-faujasite can yield an abnormally low micropore volume. This effect is most dramatic in the case for zeolites with one-dimensional channels. Examination of the mesopore volume can provide insight into the effect of various steaming treatments if it is assumed that the crystal size has remained unchanged. Figure 4.42 compares the calculated mesopore size distribution for an ammonium

154 | *4 Zeolite Characterization*

exchanged and steamed Y-zeolite. Because water must be removed from the zeolite pores before the adsorption of nitrogen, the heating procedure must be carefully considered to prevent structural damage to the zeolite such as from self-steaming. Greg and Sing [225], Lowell and Shields [226], Webb and Orr [227] and Breck [3] provide insights into the pore volume of various zeolites. ASTM method D-4365 for determining micropore volume of a catalyst provides a procedure for performing these analyses.

4.7.2
Thermal and Mechanical Analyses

Good thermal and hydrothermal stability are important parameters for several reasons. For some zeolites, the thermal removal of the templating agent can destroy the framework structure. When a zeolite is bound to form a catalyst, the zeolite must be able to withstand the calcination temperature. Finally, the preservation of good activity and selectivity during high-temperature regeneration, as occurs with FCC catalysts is a critical necessity. For a Y-zeolite the stability can be controlled by the silica to alumina ratio, degree of crystallinity, ionic exchange form (sodium vs rare earth) and many other factors. Brown and Gallagher [228] is an excellent reference for thermal analysis of materials and catalysts with Pal-Borbely [229] specifically focused on zeolites.

Loss of crystalline structure as a function of temperature can also be measured *ex situ* by performing the heat treatment and then determining the change in zeolite content by X-ray diffraction. Such an analysis is shown in Figure 4.43 for

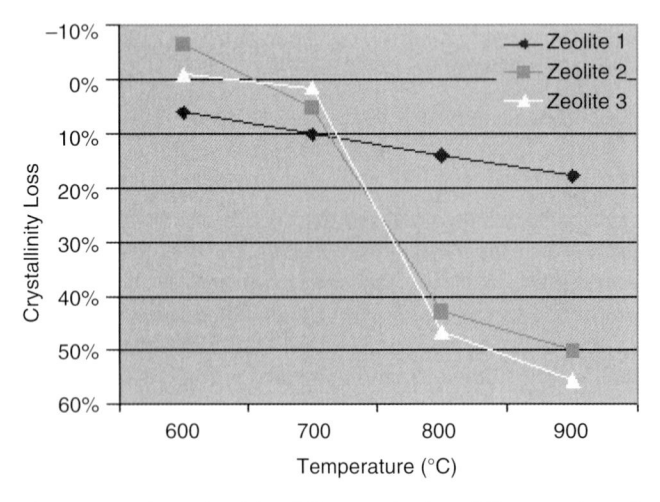

Figure 4.43 Thermal stability differences for three different zeolites all treated at various temperatures for 1 h in air. Zeolites 2 and 3 have very poor high temperature thermal stability when compared to zeolite 1.

(a)

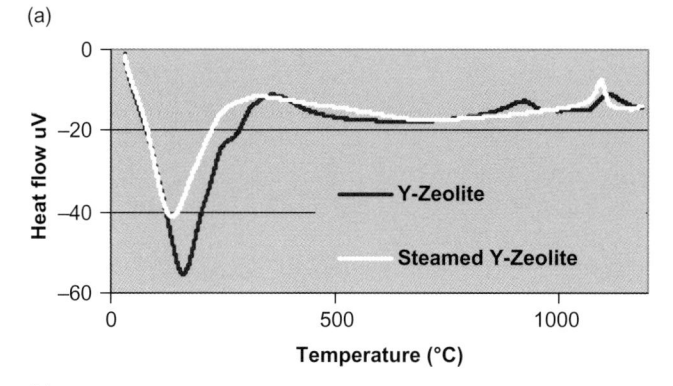

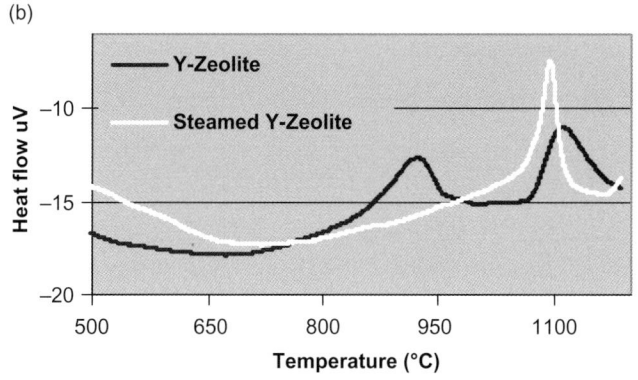

(b)

Figure 4.44 (a) DTA trace showing a lower temperature for the collapse of crystal structure for the unsteamed ammonium exchanged Y-zeolite compared to the steamed Y-zeolite; (b) is an enlargement of the DTA trace.

three different zeolites. With this approach a commercial elevated temperature exposure can be better simulated. Alternatively, high-temperature X-ray diffraction can be used to monitor changes in zeolite structure *in situ*.

Thermal measurements such as DSC and DTA can be used to determine the crystal collapse temperature. The presence of the exothermic peak is associated with the lattice collapse. As shown in Figure 4.44 for a steamed and unsteamed faujasite, the thermal stability improves with increasing silica/alumina framework.

Thermogravimetric or TGA analysis can be useful for determining temperature of decomposition of the templating agent or the temperature of burn out of coke for a spent catalyst. Figure 4.45 demonstrates the burn character for a spent FCC catalyst where the start of the decomposition of the C is at about 380 °C. This analysis was performed in 5% O_2 but other gases such as water or nitrogen can be used. The incorporation of a mass spectrometer can provide insight into the species that are evolved. However, for hydrocarbon species the analysis can become quite complex to interpret.

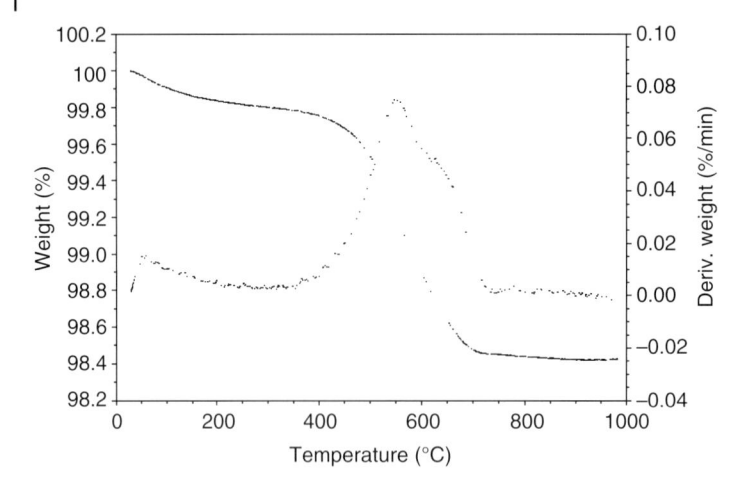

Figure 4.45 TGA curve showing carbon burn for spent FCC catalyst heated at 10 °C/min in 0.5% O_2.

Some zeolites such as $AlPO_4$-5 (AFI) have a negative coefficient of thermal expansion over certain temperature ranges [230, 231]. For a zeolite formed into a pellet, one method to verify the presence of a negative coefficient of thermal expansion is by thermal mechanical analysis (TMA). For just the zeolite, powder X-ray diffraction at various temperatures can be used. Such an analysis can be of importance for identifying pellet strength or vessel containment issues.

Of the various mechanical properties of a formed catalyst containing zeolite, attrition resistance is probably the most critical. This is particularly the case for FCC catalysts because of the impact on the addition rate of fresh catalyst, particulate emissions of fines and overall catalyst flow in the reactor and regenerator. Most attrition methods are a relative determination by means of air jet attrition with samples in the 10 to 180 μm size range. For example the ASTM D5757 method attrites a humidified sample of powder with three high velocity jets of humidified air. The fines are continuously removed from the attrition zone by elucidation into a fines collection assembly. The relative attrition index is calculated from the elutriated fines removed at a specific time interval.

Thermogravimetry can be used to measure the amount of water [232] or other molecule adsorbed on a zeolite. DSC can be utilized to study the thermal effects during adsorption and desorption of water [233] because the peak area under the heat flow time curve is related to the sorption heat.

4.7.3
Adsorption Capacity

There are two methods for determining the adsorption isotherm. The volumetric approach determines the quantity of gas present in the system by measurement of the pressure, volume and temperature. After exposing the dehydrated or activated zeolite to a quantity of gas in a closed system, the quantity adsorbed is determined from the pressure, temperature and volume at equilibrium. An alternative approach is the gravimetric method, which measures the amount of gas adsorbed by weighing the sample in a closed system on a quartz spring type balance. This technique requires a buoyancy correction that involves the determination of the volume occupied by the sample. A good description of these techniques is provided by Orr and Dallavalle [234].

4.7.4
Acid Sites

The acid function of a zeolite is unique and is an important parameter when incorporated into a catalyst. A specific zeolite consists of a well-defined, discrete number of acid sites. The acidity of zeolites is difficult to characterize because these materials contain both Lewis and Brönsted acid sites and tend to have a heterogeneous distribution of acid sites. There are many excellent review papers detailing the interactions, reactivity and characterization of zeolite acid sites [235–240]. As noted already, pyridine IR (Section 4.5.6.4) and magic angle spinning P-NMR (Section 4.6.2.3) are excellent techniques for probing and measuring Brönsted and Lewis acid sites. The Hammet titration approach determines acid strength by measuring the degree of protonation of basic indicators in acid solutions by measuring an effective pK_a. Although adapted to solid acids, it has minimal utility and is rarely used for characterizing acid strength of zeolites. There are many other techniques that provide greater insight into the zeolite/catalyst and can provide information as to the strength distribution of acid sites.

Adsorption microcalorimetry for measuring acid strength distribution is widely used and is based on the assumption that the acid strength is directly related to the heat of adsorption. Several important factors for consideration include the adsorption temperature, the pretreatment temperature and conditions and the selection of a probe molecule. A variety of probe molecules such as ammonia or pyridine have been employed to elucidate various insights. The shape of the differential heat curve versus coverage demonstrates the number, reactivity and distribution of surface sites, which can be used to evaluate changes in framework aluminum or zeolite structure as a function of different treatment conditions. More details of this technique are given by Gravelle [241], Auroux [242, 243] and Dumesic [244].

Temperature programmed desorption (TPD) is another methodology for measuring acid strength distributions [235, 236, 245]. With the TPD technique a sample is heated at a constant rate and the quantity of material desorbed at each temperature is measured. The data can be converted into a desorption energy distribution which is related to the acid strength distribution. It is assumed that the desorption energy is primarily from the energy to reprotonate the acid so that a stronger acid requires a higher energy to reprotonate for a Brönsted acid site. A variety of probe molecules such as ammonia, propylamine and various alcohols can be used.

Ammonia TPD is very simple and versatile. The use of propylamine as a probe molecule is starting to gain some popularity since it decomposes at the acid site to form ammonia and propene directly. This eliminates issues with surface adsorption observed with ammonia. The conversion of the TPD data into acid strength distribution can be influenced by the heating rate and can be subjective based on the selection of desorption temperatures for categorizing acid strength. Since basic molecules can adsorb on both Brönsted and Lewis acid sites, the TPD data may not necessarily be relevant for the specific catalytic reaction of interest because of the inability to distinguish between Bronsted and Lewis acid sites.

Using a test reaction for comparison of activity or acid strength can provide insight into the reaction mechanism or effects of various processing steps [246]. Selection of the probe molecule and test conditions must be done with caution so as to prevent ambiguity of the results. It is often desired to select a reaction that has a single pathway and to operate under conditions where a low conversion can be maintained. Cumene cracking has been frequently employed [247]. Many prefer to use alkane cracking as a model reactant for probing Brönsted acid sites with *n*-heptane [248, 249], *n*-hexane [250] and hexadecane [251] being the alkanes of choice. For isomerization reactions, *m*-xylene isomerization [249] has been used. Selection of the appropriate probe molecule is a function of many parameters such as size of the probe molecule and accessibility to the desired acid sites as well as the type of reaction to be modeled. The associated factors that may lead to variable results when performing such studies using ethylbenzene disproportionation as the probe molecule is detailed by Cotterman *et al.* [251].

4.8
Conclusions

4.8.1
Future Characterization Directions

As noted throughout this chapter, a single characterization technique cannot provide a full understanding or explain catalyst or adsorbent performance. The selection of an appropriate methodology requires: (i) knowledge of the problem, (ii) the information that can be obtained from the various techniques and (iii) recognizing the limitations of the chosen approach. This overview of zeolite characterization has emphasized examples from an industrial perspective.

Characterization techniques continue to develop and will impact their application to zeolitic systems. Aberration corrected electron microscopes are now being used to improve our understanding of catalysts and other nano-materials and will do the same for zeolites. For example, individual Pt atoms dispersed on a catalyst support are now able to be imaged in the STEM mode [252]. The application of this technique for imaging the location of rare-earth or other high atomic number cations in a zeolite would be expected to follow. Combining this with tomography it should be possible to identify the location of nanometer-sized metal clusters in the zeolite channels. The beam sensitivity of zeolites as well as the mobility of the metal atoms under the electron beam makes this a particularly challenging and difficult experiment.

Many of the characterization techniques described in this chapter require ambient or vacuum conditions, which may or may not be translatable to operational conditions. *In situ* or *in operando* characterization avoids such issues and can provide insight and information under more realistic conditions. Such approaches are becoming more common in X-ray adsorption spectroscopy (XAS) methods of XANES and EXAFS, in NMR and in transmission electron microscopy where environmental instruments and cells are becoming common. *In situ* MAS NMR has been used to characterize reaction intermediates, organic deposits, surface complexes and the nature of transition state and reaction pathways. The formation of alkoxy species on zeolites upon adsorption of olefins or alcohols have been observed by ^{13}C *in situ* and *ex situ* NMR [253]. Sensitivity enhancement techniques play an important role in the progress of this area. *In operando* infrared and RAMAN is becoming more widely used. *In situ* RAMAN spectroscopy has been used to online monitor synthesis of zeolites in pressurized reactors [254]. Such techniques will become commonplace.

Another approach is to perform *ex situ* reactions and insert the sample into a high vacuum system without exposure to ambient conditions. Incorporating N_2 glove boxes or reactor systems with X-ray photoelectron spectroscopy (XPS) sample handling can also provide information that is closer to operational conditions. In a similar manner *ex situ* reactions and sample handling are starting to be applied to electron microscopy studies. Commercially available sample transfer systems will accelerate the application of this methodology.

Theoretical calculations and simulations using *ab initio* and density function theory (DFT) methodologies are also seeing increasing use. Combining these theoretical calculations with spectroscopic data can assist in the interpretation of the observed spectral features and an improved understanding of how a probe molecule interacts with the various types of sites in zeolitic systems.

In the near term diffusion measurements should become more widely used as measurements of diffusion under non-equilibrium conditions and by pulse field gradient NMR become refined. This will require the development of useful correlations that can be extended for adsorbent and catalyst development as opposed to making individual measurements on zeolitic systems. As a result, the development of more standardized procedures, that is, probe molecule, conditions and supplementary data requirements such as particle size distributions will enhance these analyses. Combination of microscopic such as pulse field gradient NMR and macroscopic such as zero length column measurements for initial studies will become more common place. The application of these techniques will be essential for investigating the affect of various processing treatments required for making the adsorbent or catalyst. For example the impact of the binder in blocking access to the zeolite can be analyzed.

References

1 Deka, R.C. (1998) Acidity in zeolites and their characterization by different spectroscopic methods. *Ind. J. Chem. Technol.*, **5**, 109–123.

2 Haensel, V. and Haensel, H.S. (1989) The role of catalyst characterization in process development. *ACS Symp. Ser.*, **411**, 2–11.

3 Breck, D.W. (1974) *Zeolite Molecular Sieves*, John Wiley and Sons, Inc., New York.

4 Karge, H.G. and Weitkamp, J. (eds) (2004) *Molecular Sieves – Science and Technology, Characterization 1*, vol. 4, Springer, Berlin.

5 Karge, H.G., and Weitkamp, J. (eds) (2007) *Molecular Sieves – Science and Technology, Characterization 2*, vol. 5, Springer, Berlin.

6 Karge, H.G. and Weitkamp, J. (eds) (2008) *Molecular Sieves – Science and Technology, Acidity and Basicity*, vol. 6, Springer, Berlin.

7 Karge, H.G. and Weitkamp, J. (eds) (2008) *Molecular Sieves – Science and Technology, Adsorption and Diffusion*, vol. 7, Springer, Berlin.

8 Miller, M.A., Moscoso, J.G., Koster, S.C., Gatter, M.G., and Lewis, G.J. (2007) Synthesis and characterization of the 12-ring zeolites UZM-4 (BPH) and UZM-22 (MEI) via the charge density mismatch approach in the Choline-Li_2O-SrO-Al_2O_3-SiO_2 system, Stud. Surf. Sci. Catal., vol 170A, Elsevier, Amsterdam, pp. 347–354.

9 Flanigen, E.M. (1976) *Zeolite Chemistry and Catalysis. ACS Monograph*, **171**, 80–117; Jansen, J.C., van der Gaag, F.J., and van Bekkum, H. (1984) Identification of ZSM-type and other 5 ring containing zeolites by i.r. spectroscopy. *Zeolites*, **4**, 369–372.

10 Baerlocher, C., Hepp, A., and Meier, W.M. (1977) *DLS-76, A Program for the Simulation of Crystal Structures by Geometric Refinement*, Institute of Crystallography and Petrography, ETH, Zurich, Switzerland.

11 Smith, J.V. (2000) *Tetrahedral Frameworks of Zeolites, Clathrates, and Related Materials*, Microporous and Other Framework Materials with Zeolithe-Type Structures, vol. A (eds W.H. Baur, R.X. Fischer), Springer, Berlin.

12 McCusker, L.B., Liebau, F., and Engelhardt, G. (2001) Nomenclature of structural and compositional characteristics of ordered microporous and mesoporous materials with inorganic hosts. *Pure Appl. Chem.*, **73**, 381–394.

13 Jan, D.-Y., Lewis, G.J., Moscoso, J.G., and Miller, M.A. (2002) Aromatic alkylation process using ZSM-5 and UZM-6 aluminosilicates. US Patent 6,388,157; Jan, D.-Y., Lewis, G.J., Moscoso, J.G., and Miller, M.A. (2002) Xylene isomerizaton process using UZM-5 and UZM-6 zeolites. US Patent 6,388,159.

14 Leonowicz, M.E., Lawton, L.A., Lawton, S.L., and Rubin, M.K. (1994) MCM-22: a molecular Sieve with two independent multidimensional channel systems. *Science*, **264**, 1910–1913.

15 van Koningsveld, H. and Bennett, J.M. (1999) Zeolite structure determination from X-ray diffraction. *Mol. Sieves*, **2**, 1–29.

16 Kokotailo, G.T. and Fyfe, C.A. (1995) Zeolite structure analysis with powder X-ray diffraction and solid-state NMR techniques. *Rigaku J.*, **12**, 3–10.

17 McCusker, L.B. (1991) Zeolite crystallography, structure determination in the absence of conventional single-crystal data. *Acta Cryst.*, **A47**, 297–313.

18 Baerlocher, C. and McCusker, L.B. (1994) Practical aspects of powder diffraction data analysis, Stud. Surf. Sci. Catal., vol. 85, Elsevier, Amsterdam, pp. 391–428.

19 Rohrbaugh, W.J. and Wu, E.L. (1989) Factors affecting X-ray diffraction characteristics of catalyst materials. *ACS Symp. Ser.*, **411**, 279–302.

20 Sohn, J.R., DeCanio, S.J., Lunsford, J.H., and O'Donnell, D.J. (1986) Determination of framework aluminum content in dealuminated Y-type zeolites: a comparison based on unit cell size and wavenumber of i.r. bands. *Zeolites*, **6**, 225–227.

21 Dempsey, E., Kuhl, G.H., and Olson, D.H. (1969) Variation of the lattice parameter with aluminum content in synthetic faujasites. Evidence for ordering of the framework ions. *J. Phys. Chem.*, **73**, 387–390.

22 Klug, H.P. and Alexander, L.E. (1974) *X-Ray Diffraction Procedures for Polycrystalline and Amorphous Materials*, John Wiley and Sons, Inc., New York.

23 Treacy, M.M.J., Newsam, J.M., and Deem, M.W. (1991) A general recursion method for calculating diffracted intensities from crystals containing planar faults. *Proc. R. Soc. Lond.*, **A443**, 499–520.

24 Treacy, M.M.J., Vaughn, D.E.W., Strohmaier, K.G., and Newsam, J.M. (1996) Intergrowth segregation in FAU-EMT zeolite materials. *Proc. R. Soc. Lond. A*, **452**, 813–840.

25 Rietveld, H.M. (1967) Line profiles of neutron powder-diffraction peaks for structure refinement. *Acta Crystallogr.*, **22** (1), 151–152.

26 Rietveld, H.M. (1969) Profile refinement method for nuclear and magnetic structures. *J. Appl. Crystallogr.*, **2** (Pt 2), 65–71.

27 Young, R.A. (ed.) (1993) The rietveld method. *Int. Union Crystallogr. Monogr. Crystallogr.*, **5**, 298.

28 McCusker, L.B., Von Dreele, R.B., Cox, D.E., Louer, D., and Scardi, P. (1999) Rietveld refinement guidelines. *J. Appl. Crystallogr.*, **32** (1), 36–50.

29 Baerlocher, C. and McCusker, L.B. (2007) New advances in zeolite structure analysis, Stud. Surf. Sci. Catal., vol. 170A, Elsevier, Amsterdam, pp. 657–665.

30 David, W.I.F., Shankland, K., McCusker, L.B., and Baerlocher, C. (eds) (2002) Structure determination from powder diffraction data. *Int. Union Crystallogr. Monogr. Crystallogr.*, **13**, 337.

31 Gard, J.A., Barrer, R.M., and Baynham, J. (1955) The hydrothermal chemistry of silicates. Part VI. A lamellar habit in synthetic feldspar. *J. Chem. Soc.*, 2480–2481.

32 Staples, L.W. and Gard, J.A. (1959) The fibrous zeolite erionite; its occurrence, unit cell, and structure. *Mineral. Mag.*, **32**, 261–281.

33 Bennett, J.M. and Gard, J.A. (1967) Non-identity of the zeolites erionite and offretite. *Nature*, **214**, 1005–1006.

34 Gard, J.A. (1961) *Proceedings of the Second European Regional Conference on Electron Microscopy, Delft*, vol. 1, Nederlandse Vereniging voor Electronenmicroscopie, pp. 203–206.

35 Kokotailo, G.T., Sawruk, S., and Lawton, S.L. (1972) Direct observation of stacking faults in the zeolite erionite. *American Mineralogist*, **57**, 439–444.

36 Menter, J. (1958) *Adv. Mod. Phys.*, **7**, 9.

37 Sanders, J.V. (1978) *Physics of Materials* (eds B. Borland, L. Clareborough, and A.W. Moore), University of Melbourne Press, Melbourne, p. 244.

38 Bursill, L.A., Lodge, E.A., and Thomas, J.M. (1980) Zeolitic structures as revealed by high-resolution electron microscopy. *Nature*, **286**, 1–13.

39 Thomas, J.M., Millward, G.R., and Ramdas, S. (1981) New methods for the structural characterization of shape-selective zeolites. *Faraday Discuss. Chem. Soc.*, **72**, 5–52.

40 Terasaki, O., Thomas, J.M., and Millward, G.R. (1984) Imaging the structures of zeolite L and synthetic mazzite. *Proc. R. Soc. Lond. A*, **395**, 153–164 K.

41 Khatami, H., Flanigen, E.M., and Mumbach, N.R. (1973) , in *Third International Conference on Molecular Sieves*, (ed. J.B. Uytterhoeven), Leuven University Press, Zurich, Switzerland.

42 Fitzgerald, R., Keil, K., and Heinrich, K.F.J. (1968) Solid-state energy-dispersion spectrometer for electron-microprobe X-ray analysis. *Science*, **159**, 528–530.

43 Zuo, J.M., Kim, M., O'Keefe, M., and Spence, J.C.H. (1999) Direct observation of d-orbital holes and Cu-Cu bonding in Cu_2O. *Nature*, **401**, 49–52.

44 Haag, W.O., Lago, R.M., and Weisz, P.B. (1981) Transport and reactivity of hydrocarbon molecules in a shape-selective zeolite. *Faraday Discuss. Chem. Soc.*, **72**, 317–330.

45 Goldstein, J.I., Newbury, D.E., Joy, D.C., Lyman, C.E., Echlin, P., Lifshin,

E., Sawyer, L., and Michael, J.R. (2003) *Scanning Electron Microscopy and X-Ray Microanalysis*, Kluwer Academic, New York.
46 Williams, D.B. and Carter, C.B. (1996) *Transmission Electron Microscopy*, Plenum Press, New York.
47 Hunt, J.A., and Williams, D.B. (1992) The current state of spectrum imaging, in *50th Annual Meeting of the Electron Microscopy Society of America* (eds G.W. Bailey, J. Bentley, and J.A. Small), San Francisco Press, San Francisco, pp. 1200–1201.
48 Byers, R.L., Davis, J.W., White, E.W., and McMillan, R.E. (1971) A computerized method for size characterization of atmospheric aerosols by the scanning electron microscope. *Environ. Sci. Technol.*, **5**, 517–521.
49 Thaulow, N. and White, E.W. (1971) General method for dispersing and disaggregating particulate samples for quantitative SEM and optical microscope studies, Technical Report 5 (US Department of Commerce Technical Report AD733457), Office of Naval Research, Metallurgical Program.
50 Armstrong, J.T. (1991), in *Electron Probe Quantitation* (eds K.F.J. Heinrich, and D.E. Newbury), Plenum Press, New York, p. 261.
51 Csencsits, R. and Gronsky, R. (1988), in *Specimen Preparation for Transmission Electron Microscopy of Materials MRS Symp. Proc.*, vol. 115 (eds J.C. Bravman, R.M. Anderson, and M.L. McDonald), Materials Research Society, Pittsburg, PA, pp. 103–108.
52 Hugo, R.C. and Cady, S.L. (2004) Preparation of geological and biological TEM specimens by embedding in sulfur. *Microsc. Today*, **12**, 28–30.
53 Lynch, J., Raatz, F., and Dufresne, P. (1987) Characterisation of the textural properties of dealuminated HY forms. *Zeolites*, **7**, 333–340.
54 Newsam, J.M., Treacy, M.M.J., Koetsier, W.T., and de Gruyter, C.B. (1988) Structural characterization of zeolite beta. *Proc. R. Soc. Lond. A*, **420**, 375–440.
55 Lobo, R., Pan, M., Chan, I., Li, H.-X., Medrud, R.C., Zones, S.I., Crozier, P.A., and Davis, M.E. (1993) SSZ-26 and SSZ-33: Two molecular sieves with intersecting 10- and 12-ring pores. *Science*, **262**, 1543–1546.
56 Lobo, R. and Davis, M.E. (1995) CIT-1: a new molecular sieve with intersecting pores bounded by 10- and 12-rings. *J. Am Chem. Soc.*, **117**, 3766–3779.
57 Blackwell, C.S., Broach, R.W., Gatter, M.G., Holmgren, J.S., Jan, D.-Y., Lewis, G.J., Mezza, B.J., Mezza, T.M., Miller, M.A., Moscoso, J.G., Patton, R.L., Rohde, L.M., Schoonover, M.W., Sinkler, W., Wilson, B.A., and Wilson, S.T. (2003) Open framework materials synthesized in the TMA+/TEA+ mixed template system: the new low Si/Al ratio zeolites UZM-4 and UZM-5. *Angew. Chem. Int. Ed.*, **42**, 1737–1740.
58 Pan, M. (1996) High resolution electron microscopy of zeolites. *Micron*, **27**, 219–238.
59 Treacy, M.M.J. and Newsam, J.M. (1987) Electron beam sensitivity of zeolite L. *Ultramicroscopy*, **23**, 411–420.
60 Bradley, S.A. and Targos, W.M. (1991) Analytical electron microscopy of catalytic materials. *Ultramicroscopy*, **37**, 210–215.
61 Pan, M. and Crozier, P.A. (1993) Quantitative imaging and diffraction of zeolites using a slow-scan CCD camera. *Ultramicroscopy*, **52**, 487–498.
62 Nicolopoulos, S., Gonzalez-Calbet, J.M., Vallet-Regi, M., Camblor, M.A., Corell, C., Corma, A., and Diaz-Cabanas, M.J. (1997) Use of electron microscopy and microdiffraction for zeolite framework comparison. *J. Am. Chem. Soc.*, **119**, 11000–11005.
63 Dorset, D.L. and Kennedy, G.J. (2005) Crystal structure of MCM-70: a microporous material with high framework density. *J. Phys. Chem. B*, **109**, 13891–13898.
64 Gilmore, C.J., Dong, W., and Dorset, D.L. (2008) Solving the crystal structures of zeolites using electron diffraction data. I. The use of potential-density histograms. *Acta Crystallogr. A*, **A64**, 284–294.
65 Wagner, P., Terasaki, O., Ritsch, S., Nery, J.G., Zones, S.I., Davis, M.E., and Hiraga, K. (1999) Electron diffraction

structure solution of a nanocrystalline zeolite at atomic resolution. *J. Phys. Chem.*, **B103**, 8245–8250.

66 Dorset, D.L., Strohmaier, K.G., Kliewer, C.E., Corma, A., Diaz-Cabanas, M.J., Rey, F., and Gilmore, C.J. (2008) Crystal structure of ITQ-26, a 3D framework with extra-large pores. *Chem. Mater.*, **20**, 5325–5331.

67 Vincent, R. and Midgley, P.A. (1994) Double conical beam-rocking system for measurement of integrated electron diffraction intensities. *Ultramicroscopy*, **53**, 271–282.

68 Own, C.S., Marks, L.D., and Sinkler, W. (2006) Precession electron diffraction 1: multislice simulation. *Acta Crystallogr. A*, **A62**, 434–443.

69 Sinkler, W., Own, C.S., and Marks, L.D. (2007) Application of a 2-beam model for improving the structure factors from precession electron diffraction intensities. *Ultramicroscopy*, **107**, 543–550.

70 Dorset, D.L., Gilmore, C.J., Jorda, J.L., and Nicolopoulos, S. (2007) Direct electron crystallographic determination of zeolite zonal structures. *Ultramicroscopy*, **107**, 462–473.

71 Own, C.S., Sinkler, W., and Marks, L.D. (2006) Rapid structure determination of a metal oxide from pseudo-kinematical electron diffraction data. *Ultramicroscopy*, **106**, 114–122.

72 Ciston, J., Own, C.S., and Marks, L.D. (2008) Cone-angle dependence of Ab-initio structure solutions using precession electron diffraction. *AIP Conf. Proc.*, **999**, 53–65.

73 Neiner, D., Okamoto, N.L., Condron, C.L., Ramasse, Q.M., Yu, P., Browning, N.D., and Kauzlarich, S.M. (2007) Hydrogen encapsulation in a silicon clathrate type I structure: $Na_{5.5}(H_2)_{2.15}Si_{46}$: synthesis and characterization. *J. Am. Chem. Soc.*, **129**, 13857–13862.

74 Koster, A.J., Ziese, U., Verkleij, A.J., Janssen, A.H., de Graaf, J., Geus, J.W., and de Jong, K.P. (2000) Development and application of 3-dimensional transmission electron microscopy (3D-TEM) for the characterization of metal-zeolite catalyst systems, Stud. Surf. Sci. Catal., vol. 130, Elsevier, Amsterdam, pp. 329–334.

75 de Jong, K.P. (2006) Support materials and characterization tools for nanostructured catalysts. *Oil Gas Sci. Technol.*, **61**, 527–534.

76 de Jong, K.P., Koster, A.J., Janssen, A.H., and Ziese, U. (2005) Electron tomography of molecular sieves, Stud. Surf. Sci. Catal., vol. 157, Elsevier, Amsterdam, pp. 225–242.

77 Ziese, U., de Jong, K.P., and Koster, A.J. (2004) Electron tomography: a tool for 3D structural probing of heterogeneous catalysts at the nanometer scale. *Appl. Catal. A*, **260**, 71–74.

78 Mastronarde, D.N. (1997) Dual-axis tomography: an approach with alignment methods that preserve resolution. *J. Struct. Biol.*, **120**, 343–352.

79 Rice, S.B., Disko, M.M., and Treacy, M.M.J. (1990) On the imaging of Pt atoms in zeolite frameworks. *Ultramicroscopy*, **34**, 108–118.

80 Pan, M., Cowley, J.M., and Chan, I.Y. (1990) Study of highly dispersed Pt in Y-zeolites by STEM and electron microdiffraction. *Ultramicroscopy*, **34**, 93–101.

81 Hadjiivanov, K.I. and Vayssilov, G.N. (2002) Characterization of oxide surfaces and zeolites by carbon monoxide as an IR probe molecule. *Adv. Catal.*, **47**, 307–511.

82 Coluccia, S., Marchese, L., and Martra, G. (2000) Molecular probes for the characterization of adsorption sites in micro-and mesoporous materials, in *Photofunctional Zeolites* (ed. M. Anpo), NOVA Science, New York, pp. 39–74.

83 Wakabayashi, F. and Domen, K.A. (1997) new method for characterizing solid surface acidity–an infrared spectroscopic method using probe molecules such as N2 and rare gases. *Catal. Surv. Jpn*, **1** (2), 181–193.

84 Kustov, L.M. (1997) New trends in IR-spectroscopic characterization of acid and basic sites in zeolites and oxide catalysts. *Topics in Catalysis*, **4**, (1, 2, Acidity in Aluminas, Amorphous and Crystalline Silico-Aluminas), 131–144.

85 Zecchina, A., Otero Arean, C. (1996) Diatomic molecular probes for mid-IR

studies of zeolites. *Chem. Soc. Rev.*, **25** (3), 187–197.
86 Knözinger, H. and Huber, S. (1998) IR spectroscopy of small and weakly interacting molecular probes for acidic and basic zeolites. *J. Chem. Soc. Faraday Trans.*, **94** (15), 2047–2059.
87 Karge, H.G. and Geidel, G. (2004) Vibrational spectroscopy. *Mol. Sieves*, **4**, 1–200.
88 Busca, G. (1996) The use of vibrational spectroscopies in studies of heterogeneous catalysis by metal oxides: an introduction. *Catal. Today*, **27**, 323–352.
89 Zecchina, A., Spoto, G., and Bordiga, S. (2005) Probing the acid sites in confined spaces of microporous materials by vibrational spectroscopy. *Phys. Chem. Chem. Phys.*, **7**, 1627–1642.
90 Fuller, M.P. and Griffiths, P.R. (1978) Diffuse reflectance measurements by Fourier-transform infrared spectroscopy. *Anal. Chem.*, **50** (11), 1906–1910.
91 Flanigen, E.M., Khatami, H., and Szymanski, H.A. (1971) Infrared structural studies on zeolite frameworks. *Adv. Chem. Ser.*, **101**, 201–228.
92 Jansen, J.C., van der Gaag, F.J., and van Bekkum, H. (1984) Identification of ZSM-type and other 5-ring containing zeolites by i.r. spectroscopy. *Zeolites*, **4** (4), 369–372.
93 Flanigen, E.M. (1976) Structural analysis by infrared spectroscopy. *ACS Monogr.*, **171**, 80–117; Rabo, J. (ed.) (1976) *Zeolite Chemistry and Catalysis*, American Chemical Society, Washington, D.C.
94 Goovaerts, F., Vansant, E.F., De Hulsters, P., and Gelan, J. (1989) Stuctural vibrations of acid-leached mordenites. *J. Chem. Soc. Faraday Trans. 1*, **85** (11), 3687–3694.
95 Coudurier, G., Naccache, C., and Vedrine, J.C. (1982) Uses of IR. spectroscopy in identifying ZSM zeolite structure. *J. Chem. Soc., Chem. Commun.*, **24**, 1413–1415.
96 Ballinger, T.H., Wong, J.C.S., and Yates, J.T., Jr. (1992) Transmission infrared spectroscopy of high area solid surfaces. A useful method for sample preparation. *Langmuir*, **8**, 1676–1678.
97 Beebe, T.P., Gelin, P., and Yates, J.T., Jr. (1984) Infrared spectroscopic observations of surface bonding in physical adsorption: the physical adsorption of CO on SiO_2 surfaces. *Surf. Sci.*, **148**, 526–550.
98 Jacobs, P.A. and Uytterhoevin, J.B. (1973) Assignment of the hydroxyl bands in the infrared spectra of zeolites X and Y Part 1. Na-H Zeolites. *J. Chem. Soc. Faraday Trans. I*, **69**, 359–372.
99 Jacobs, P.A. and Mortier, W.J. (1982) An attempt to rationalize stretching frequencies of lattice hydroxyl groups in hydrogen-zeolites. *Zeolites*, **2** (3), 226–230.
100 Woolery, G.L., Alemany, L.B., Dessau, R.M., and Chester, A.W. (1986) Spectroscopic evidence for the presence of internal silanols in highly siliceous ZSM-5. *Zeolites*, **6** (1), 14–16.
101 Jacobs, P.A. and Uytterhoevin, J.B. (1973) Assignment of the hydroxyl bands in the infrared spectra of zeolites X and Y part 2. After different treatments. *J. Chem. Soc. Faraday Trans. I*, **69**, 373–386.
102 Amaroli, T., Simon, L.J., Digne, M., Montanari, T., Bevilacqua, M., Valtchev, V., Patarin, J., and Busca, G. (2006) Effects of crystal size and Si/Al ratio on the surface properties of H-ZSM-5 zeolites. *Appl. Catal. A*, **306**, 78–84.
103 Zecchina, A., Marchese, L., Bordiga, C., Pazè, C., and Gianotti, E. (1997) Vibrational spectroscopy of NH4+ ions in zeolitic materials: An IR study. *J. Phys. Chem. B*, **101**, 10128–10135.
104 Paukshtis, E.A. and Yurchenko, E.N. (1983) Study of the acid-base properties of heterogeneous catalysts by infrared spectroscopy. *Russ. Chem. Rev*, **52**, 242–258.
105 Liengme, B.V. and Hall, W.K. (1966) Studies of hydrogen held by solids. *Trans. Farady Soc.*, **62**, 3229–3243.
106 Richardson, R.L. and Benson, S.W. (1957) A study of the surface acidity of cracking catalyst. *J. Phys. Chem.*, **61**, 405–411.
107 Parry, E.P. (1963) An infrared study of pyridine adsorbed on acidic solids. Characterization of surface acidity. *J. Catal.*, **2**, 371–379.

108 Ward, J.W. (1968) A spectroscopic study of the surface of zeolite Y: the adsorption of pyridine. *J. Colloid Interface Sci.*, **28**, 269–278.

109 Pieterse, J.A.Z., Veefkind-Reyes, S., Seshan, K., Domokos, L., and Lercher, J.A. (1999) On the accessibility of acid sites in ferrierite for pyridine. *J. Catal.*, **187**, 518–520.

110 Lok, B.M., Marcus, K.K., and Angell, C.L. (1986) Characterization of zeolite acidity. II. Measurement of zeolite acidity by ammonia temperature programmed desorption and FTIR spectroscopy techniques. *Zeolites*, **6**, 185–194.

111 Zhang, W.Z., Burckle, E.C., and Smirniotis, P.G. (1999) Characterization of the acidity of ultrastable Y, mordenite, and ZSM-12 via NH_3-stepwise temperature programmed desorption and Fourier transform infrared spectroscopy. *Micropor. Mesopor. Mater.*, **33**, 173–185.

112 Martin, A., Wolf, U., Berndt, H., and Lucke, B. (1993) Nature, strength, and concentration of acid sites of erioite-rich zeolite T determined by ammonia-i.r. spectroscopy. *Zeolites*, **13**, 309–314.

113 Zverev, S.M., Smirov, K.S., and Tsyganenko, A.A. (1988) IR spectroscopic study of low-temperature adsorption of molecular nitrogen on the surface of oxides. *Kinet. Katal.*, **29**, 1251–1257.

114 Wakabayashi, F., Kondo, J., Domen, K., and Hirose, C. (1993) Dinitrogen as a probe of acid sites of zeolites. *Catal. Lett.*, **21**, 257–264.

115 Geobaldo, F., Lamberti, C., Ricchiardi, G., Bordiga, S., Zecchina, A., Palomino, G.T., and Areán, C.O. (1995) N_2 adsorption at 77 K on H-mordenite and alkali-metal-exchanged mordenites: an IR study. *J. Phys. Chem.*, **99**, 11167–11177.

116 Onida, B., Berello, L., Geobaldo, F., and Garrone, E. (2003) IR study of the acidity of ITQ-2, an "all-surface" zeolitic system. *J. Catal.*, **241**, 191–199.

117 Juttu, G.G. and Lobo, R.F. (2000) Characterization and catalytic properties of MCM-56 and MCM-22 zeolites. *Micropor. Mesopor. Mater.*, **40**, 9–23.

118 Bhat, Y.S., Das, J., Rao, K.V., and Halgeri, A.B. (1996) Inactivation of external surface of ZSM-5: zeolite morphology, crystal size, and catalytic activity. *J. Catal.*, **159**, 368–374.

119 Wang, I., Ay, C-L., Lee, B-J., and Chen, M-H. (1989) Para-selectivity of dialkylbenzenes over modified HZSM-5 by vapour phase deposition of silica. *Appl. Catal.*, **54**, 257–266.

120 Kürschhner, U., Jerschkewitz, H.-G., Schreier, E., and Völter, J. (1990) Shape selectivity of hydrothermally treated H-ZSM-5 in toluene disproportionation and xylene isomerization. *Appl. Catal.*, **57**, 167–177.

121 Zheng, S., Heydenrych, H.R., Jentys, A., and Lercher, J.A. (2002) Influence of surface modification on the acid site distribution of HZSM-5. *J. Phys. Chem. B*, **106**, 9552–9558.

122 Wu, P., Komatsu, T., and Yashima, T. (1998) Selective formation of p-xylene with disproportionation of toluene over MCM-22 catalysts. *Micropor. Mesopor. Mater.*, **22**, 343–356.

123 Bensisi, H.A. (1973) Determination of proton acidity of solid catalysts by chromatographic adsorption of sterically hindered amines. *J. Catal.*, **28** (1), 176–178.

124 Thibault-Starzyk, F., Vimont, A., and Gilson, J.P. (2001) 2D-COS IR study of coking in xylene isomerisation on H-MFI zeolite. *Catal. Today*, **70**, 227–241.

125 Corma, A., Fornes, V., Forni, L., Marquez, F., Martinez-Triguero, J., and Moscotti, D. (1998) 2,6-Di-tert-butylpyridine as a probe molecule to measure external acidity of zeolites. *J. Catal.*, **179** (2), 451–458.

126 Nesterenko, N.S., Thibault-Starzyk, F., Montouillout, V., Yuschenko, V.V., Fernandez, C., Gilson, J.-P., Fajula, F., and Ivanova, I.I. (2004) Accessibility of the acid sites in dealuminated small-port mordenites studied by FTIR of co-adsorbed alkylpyridines and CO. *Micropor. Mesopor. Mater.*, **71** (1–3), 157–166.

127 LaValley, J.C. (1996) Infrared spectrometric studies of the surface basicity of

metal oxides and zeolites using adsorbed probe molecules. *Catal. Today*, **27**, 377–401.

128 Gribov, E.N., Cocina, D., Spoto, G., Bordiga, S., Ricchiardi, G., and Zecchina, A. (2006) Vibrational and thermodynamic properties of Ar, N_2, O_2, H_2, and CO adsorbed and condensed on (H,Na)-Y zeolite cages as studied by variable temperature IR spectroscopy. *Phys. Chem. Chem. Phys.*, **8**, 1186–1196.

129 Lercher, J.A., Grundling, C., and Eder-Mirth, G. (1996) Infrared studies of the surface acidity of oxides and zeolites using adsorbed probe molecules. *Catal. Today*, **27**, 353–376.

130 Makarova, M.A., Ojo, A.F., Karim, K., Hunger, M., and Dwyer, J. (1994) FTIR study of weak hydrogen bonding of Brönsted hydroxyls in zeolites and aluminophosphates. *J. Phys. Chem.*, **98**, 3619–3623.

131 Banares, M.A. (2005) Operando methodology: combination of in situ spectroscopy and simultaneous activity measurements under catalytic reaction conditions. *Catal. Today*, **100**, 71–77.

132 Fu, J. and Ding, C. (2005) Study on alkylation of benzene with propylene over MCM-22 zeolite catalyst by in situ IR. *Catal. Commun.*, **6**, 770–776.

133 Zheng, S., Jentys, A., and Lercher, J.A. (2006) Xylene isomerization with surface-modified HZSM-5 zeolite catalysts: an in situ IR study. *J. Catal.*, **241** (2), 304–311.

134 Marie, O., Thibault-Starzyk, F., and Massiani, P.J. (2005) Conversion of xylene over mordenites: an operando infrared spectroscopy study of the effect of Na+. *J. Catal.*, **230** (1), 28–37.

135 Frising, T. and Leflaive, P. (2008) Extraframework cation distributions in X and Y faujasite zeolites: a review. *Micropor. Mesopor. Mater.*, **114**, 27–63.

136 Little, L.H. (1966) *Infrared Spectra of Adsorbed Species*, Academic Press, New York.

137 Montanari, T., Salla, I., and Busca, G. (2008) Adsorption of CO on LTA zeolite adsorbents: an IR investigation. *Micropor. Mesopor. Mater.*, **109**, 216–222.

138 Martra, G., Ocule, R., Marchses, L., Centi, G., and Coluccia, S. (2002) Alkali and alkaline-earth exchanged faujasites: strength of lewis base and acid centres and cation site occupancy in Na- and BaY and Na- and BaX zeolites. *Catal. Today*, **73**, 83–93.

139 Su, B-L. and Norberg, V. (1998) Migration of adsorbed benzene molecules from cations to 12R windows in the large cages of Cs(Na)-EMT zeolite upon coadsorption of NH_3. *Langmuir*, **14**, 2353–2360.

140 Lide, D.R. (ed.) (2008) Proton affinities, in *CRC Handbook of Chemistry and Physics* (Internet Version), CRC Press, Boca Raton.

141 Bischoff, H., Jaeger, N.I., Schultz-Ekloff, G., and Kubelkova, L. (1993) Interaction of CO with platinum species in zeolites. Evidence for platinum carbonyls in partially reduced zeolite PtNaX. *J. Mol. Catal.*, **80**, 95–103.

142 Zholobendo, V.L., Lei, G-D., Carvill, B.B., Lerner, B.A., and Sachtler, W.M.H. (1994) Identification of isolated Pt atoms n H-mordenite. *J. Chem. Soc. Faraday Trans.*, **90**, 233–238.

143 Trakarnroek, S., Jongpatiwut, S., Rirksomboon, T., Osuwan, S., and Resasco, D.E. (2006) n-octane aromatization over Pt/KL of varying morphology and channel lengths. *Appl. Catal. A*, **313**, 189–199.

144 Bischoff, H., Jaeger, N.I., and Schultz-Ekloff, G. (1990) FTIR study of the particle size dependent chemisorption of CO on Pt dispersed within a faujasite matrix. *Z. Phys. Chemie*, **217**, 1093–1101.

145 Kebulkova, L., Nováková, J., Jaeger, N.I., and Schultz-Ekloff, G. (1993) Characterization of nickel species at $Ni/\gamma\text{-}Al_2O_3$ and Ni/faujasite catalysts by carbon monoxide adsorption. *Appl. Catal. A*, **95**, 87–101.

146 Penkova, A., Dzwigaj, S., Kefirov, R., Hadjiivanov, K., and Che, M. (2007) Effect of preparation method on the state of BEA zeolites. A study by Fourier transform Ir spectroscopy of adsorbed CO and NO, temperature programmed reductions, and x-ray

diffraction. *J. Phys. Chem. C*, **111**, 8623–8631.

147 Tang, J., Zhang, T., Ma, L., Li, L., Zhao, J., Zheng, M., and Lin, L. (2001) Microwave discharge-assisted NO reduction by CH_4 over Co/HZSM-5 and Ni/HSM-5 under O_2 excess. *Catal. Lett.*, **73**, 193–197.

148 Zhu, C.Y., Lee, C.W., and Chong, P.J. (1996) FT i.r. study of NO adsorption over CoZSM-5. *Zeolites*, **17**, 483–488.

149 Góra-Marek, K., Gil, B., Śliwa, M., and Datka, J. (2007) An IR spectroscopy study of Co sites in zeolites CoZSM-5. *Appl. Catal. A*, **330**, 33–42.

150 Gil, B., Janas, J., Wloch, E., Olejniczak, Z., Datka, J., and Sulikowski, B. (2008) The influence of the initial acidity of HFER on the status of Co species and catalytic performance of CoFER and InCoFER in CH_4-SCR-NO. *Catal. Today*, **137**, 174–178.

151 Góra-Marek, K., Gil, B., and Datka, J. (2009) Quantitative IR studies of the concentration of Co^{+2} and Co^{+3} sites in zeolites. *Appl. Catal. A*, **353**, 117–122.

152 Fyfe, C.A. (1984) *Solid State NMR for Chemists*, C.F.C. Press, Guelph.

153 Klinowski, J. (1984) Nuclear magnetic resonance studies of zeolites. *Progr. NMR Spectrosc.*, **16**, 237–309.

154 Klinowski, J. (1993) Applications of solid-state NMR for the study of molecular sieves. *Anal. Chim. Acta*, **283**, 929–965.

155 Mackenzie, K.J.D. and Smith, M.E. (2002) Multinuclear solid-state NMR of inorganic materials. *Pergamon Mater. Ser.*, **6**, 23–105.

156 Kirkpatrick, R.J. and Brow, R.K. (1995) Nuclear magnetic resonance investigation of the structures of phosphate and phosphate-containing glasses: a review. *Solid State Nucl. Magn. Reson.*, **5**, 9–21.

157 Hunger, M. and Brunner, E. (2004) NMR Spectroscopy. *Mol. Sieves*, **4**, 201–293.

158 Engelhardt, G. and Michel, D. (1987) *High-Resolution Solid-State NMR of Silicates and Zeolites*, John Wiley & Sons, Ltd, Chichester.

159 Thomas, J.M. and Klinowski, J. (1985) The study of aluminosilicate and related catalysts by high-resolution solid state NMR spectroscopy. *Appl. Catal.*, **33**, 199–374.

160 Klinowski, J. (1991) Solid-state NMR studies of molecular sieve catalysts. *Chem. Rev.*, **91**, 1459–1479.

161 Slichter, C.P. (1990) *Principles of Magnetic Resonance*, Springer-Verlag, New York.

162 Mehring, M. (1983) *Principles of High-Resolution NMR in Solids*, Springer-Verlag, Berlin.

163 Gerstein, B.C. and Dybowski, C.R. (1985) *Transient Techniques in NMR of Solids*, Academic Press Inc., New York, NY.

164 Andrew, E.R., Bradbury, A., and Eades, R.G. (1959) Removal of dipolar broadening of nuclear magnetic resonance spectra of solids by specimen rotation. *Nature*, **183**, 1802.

165 Lowe, I.J. (1959) Free induction decays of rotating solids. *Phys. Rev. Lett.*, **2**, 285–287.

166 Pines, A., Gibby, M.G., and Waugh, J.S. (1973) Proton-enhanced NMR of dilute spins in solids. *J. Chem. Phys.*, **59**, 569–590.

167 Hartmann, S.R. and Hahn, E.L. (1962) Nuclear double resonance in the rotating frame. *Phys. Rev.*, **128**, 2042–2053.

168 Schaefer, J. and Stejskal, E.O. (1976) Carbon-13 nuclear magnetic resonance of polymers spinning at the magic angle. *J. Am. Chem. Soc.*, **98**, 1031–1032.

169 Bertani, P., Raya, J., Reinheimer, P., Gougeon, R., Delmotte, L., and Hirschinger, J. (1999) $^{19}F/^{29}Si$ distance determination in fluoride-containing octadecasil by hartmann-hahn cross-polarization under fast magic-angle spinning. *Solid State Nucl. Magn. Reson.*, **13**, 219–229.

170 Fyfe, C.A., Grondey, H., Mueller, K.T., Wong-Moon, K.C., and Markus, T. (1992) Coherence transfer involving quadrupolar nuclei in solids: ^{27}Al (^{31}P cross-polarization NMR in the molecular sieve VPI-5. *J. Am. Chem. Soc.*, **114**, 5876–5878.

171 Egan, J.M., Wenslow, R.M., and Mueller, K.T. (2000) Mapping aluminum/phosphorus connectivities in

172 Fyfe, C.A., Wong-Moon, K.C., Huang, Y., Grondey, H., and Mueller, K.T. (1995) Dipolar-based ^{27}Al - ^{29}Si solid-state NMR connectivity experiments in zeolite molecular sieve frameworks. *J. Phys. Chem.*, **99**, 8707–8716.

173 Prabakar, S., Wenslow, R.M., and Mueller, K.T. (2000) Structural properties of sodium phosphate glasses from ^{23}Na → ^{31}P cross-polarization NMR. *J. Non-Cryst. Solids*, **263**, 82–93.

174 Vega, A.J. (1992) MAS NMR spin locking of half-integer quadrupolar nuclei. *J. Magn. Reson.*, **96**, 50–68.

175 De Paul, S.M., Ernst, M., Shore, J.S., Stebbins, J.F., and Pines, A. (1997) Cross-polarization from quadrupolar nuclei to silicon using low-radio-frequency amplitudes during magic-angle spinning. *J. Phys. Chem. B*, **101**, 3240–3249.

176 Engelhardt, G., Loshe, U., Samoson, A., Magi, M., Tarmak, M., and Lippmaa, E. (1982) High resolution ^{29}Si N.M.R. of dealuminated and ultrastabel Y-zeolites. *Zeolites*, **2**, 59–62.

177 Rauscher, H.J., Michel, D., Deininger, D., and Geschke, D. (1980) Carbon-13 NMR study of pyridinium ion formation in zeolites. *J. Mol. Catal.*, **9**, 369–379.

178 Lyerla, J., Yannoni, C.S., and Fyfe, C.A. (1982) Chemical applications of variable-temperature CPMAS NMR spectroscopy in solids. *Acc. Chem. Res.*, **15**, 208–216.

179 Samoson, A. and Lippmaa, E. (1983) Excitation phenomena and line intensities in high-resolution NMR powder spectra of half-integer quadrupolar nuclei. *Phys. Rev. B*, **28**, 6567–6570.

180 Man, P.P., Klinowski, J., Trokiner, T., Zanni, H., and Papon, P. (1988) Selective and non-selective NMR excitation of quadrupolar nuclei in the solid state. *Chem. Phys. Lett.*, **151**, 143–150.

181 Samoson, A., Lippmaa, E., and Pines, A. (1988) High resolution solid-state N.M.R. averaging of second-order effects by means of a double-rotor. *Mol. Phys.*, **65**, 1013–1018.

182 Samoson, A., and Lippmaa, E. (1989) Synchronized double-rotation NMR spectroscopy. *Solid State Magn. Reson.*, **84**, 410–416.

183 Samoson, A. and Pines, A. (1989) Double rotor for solid-state NMR. *Rev. Sci. Instrum.*, **60**, 3239–3241.

184 Chmelka, B.F., Mueller, K.T., Pines, A., Stebbins, J., Wu, Y., and Zwanziger, J.W. (1989) O-17 NMR in solids by dynamic-angle spinning and double rotation. *Nature*, **339**, 42–44.

185 Mueller, K.T., Sun, B.Q., Chingas, G.C., Zwanziger, J.W., Terao, T., and Pines, A. (1990) Dynamic-angle spinning of quadrupolar nuclei. *J. Magn. Reson.*, **86**, 470–487.

186 Llor, A. and Virlet, J. (1988) Towards high-resolution NMR of more nuclei in solids: sample spinning with time-dependent spinner axis angle. *Chem. Phys. Lett.*, **152**, 248–253.

187 Mueller, K.T., Wooten, E.W., and Pines, A. (1991) Pure-absorption-phase dynamic-angle spinning. *J. Magn. Reson.*, **92**, 620–627.

188 Frydman, L. and Harwood, J.S. (1995) Isotropic spectra of half-integer quadrupolar spins from bidimensional magic-angle spinning NMR. *J. Am. Chem. Soc.*, **17**, 5367–5368.

189 Medek, A., Harwood, J.S., and Frydman, L. (1995) Multiple-quantum magic-angle spinning NMR: a new method for the study of quadrupolar nuclei in solids. *J. Am. Chem. Soc.*, **117**, 12779–12787.

190 Goldbourt, A. and Madhu, P.K. (2005) Multiple-quantum magic-angle spinning: high-resolution NMR spectroscopy of half-integer spin quadrupolar nuclei. *Annu. Rep. NMR Spectrosc.*, **54**, 81–153.

191 Gan, Z. (2000) Isotropic NMR spectra of half-integer quadrupolar nuclei using satellite transitions and magic-angle spinning. *J. Am. Chem. Soc.*, **122**, 3242–3243.

192 Gan, Z. (2001) Satellite transition magic-angle spinning nuclear magnetic resonance spectroscopy of half-integer

quadrupolar nuclei. *J. Chem. Phys.*, **114**, 10845–10853.

193 Ashbrook, S.E. and Wimperis, S. (2002) High-resolution NMR spectroscopy of quadrupolar nuclei in solids: satellite-transition MAS with self-compensation for magic-angle misset. *J. Am. Chem. Soc.*, **124**, 11602.

194 Hamdan, H. and Klinowski, J. (1989) ^{27}Al quadrupole nutation NMR studies of amorphous aluminosilicates. *Chem. Phys. Lett.*, **158**, 447–452.

195 Man, P.P. and Klinowski, J. (1988) Quadrupole nutation ^{27}Al NMR studies of isomorphous substitution of aluminum in the framework of zeolite Y. *Chem. Phys. Lett.*, **147**, 581–584.

196 Gullion, T. and Schaefer, J. (1989) Rotational-echo double-resonance NMR. *J. Magn. Reson*, **81**, 196–200.

197 Gullion, T. and Schaefer, J. (1989) Detection of weak heteronuclear dipolar coupling by rotational-echo double-resonance nuclear magnetic resonance. *Adv. Magn. Reson.*, **13**, 57–83.

198 Gullion, T. (1997) Measurement of heteronuclear dipolar interactions by rotational-echo, double-resonance nuclear magnetic resonance. *Magn. Reson. Rev.*, **17**, 83–131.

199 Hing, A.W., Vega, S., and Schaefer, J. (1992) Transferred-echo double-resonance NMR. *J. Magn. Reson.*, **96**, 205–209.

200 Fyfe, C.A., Mueller, K.T., Grondey, H., and Wong-Moon, K.C. (1992) Dipolar dephasing between quadrupolar and spin-1/2 nuclei. REDOR and TEDOR NMR experiments on VPI-5. *Chem. Phys. Lett.*, **199**, 198–204.

201 Grey, C.P., Veeman, W.S., and Vega, A.J. (1993) Rotational echo ^{14}N/^{13}C/^{1}H triple resonance solid-state nuclear magnetic resonance: a probe of ^{13}C-^{14}N internuclear distances. *J. Chem. Phys.*, **98**, 7711–7724.

202 Grey, C.P. and Vega, A.J. (1995) Determination of the Quadrupole Coupling Constant of the Invisible Aluminum Spins in Zeolite HY with ^{1}H/^{27}Al NMR TRAPDOR NMR. *J. Am. Chem. Soc.*, **117**, 8232–8242.

203 van Eck, E.R.H., Janssen, R., Maas, W.E.J.R., and Veeman, W.S. (1990) A novel application of nuclear spin-echo double-resonance to aluminophosphates and aluminosilicates. *Chem. Phys. Lett.*, **174**, 428–432.

204 Gullion, T. (1995) Measurement of dipolar interactions between spin-1/2 and quadrupolar nuclei by rotational-echo, adiabatic-passage, double-resonance NMR. *Chem. Phys. Lett.*, **246**, 325–330.

205 Gullion, T. (1995) Detecting ^{13}C-^{17}O dipolar interactions by rotational-echo, adiabatic-passage, double-resonance NMR. *J. Magn. Reson.*, **A117**, 326–329.

206 Kärger, J. and Ruthven, D.M. (1992) *Diffusion in Zeolites and Other Microporous Solids*, John Wiley & Sons, New York.

207 Stallmach, F. and Galvosas, P. (2007) Spin Echo NMR Diffusion Studies. *Annu. Rep. NMR Spectrosc.*, **61**, 51–131.

208 Kärger, J. and Vasenkov, S. (2005) Quantitation of diffusion in zeolite catalysts. *Micropor. Mesopor. Mater.*, **85**, 195–206.

209 Klinowski, J., Ramdas, S., Thomas, J.M., Fyfe, C.A., and Hartman, J.S. (1982) A re-examination of Si, Al ordering in zeolites Na-X and Na-Y. *Chem. Soc. Faraday Trans. II*, **78**, 1025–1050.

210 Rocha, J., Carr, S.W., and Klinowski, J. (1991) ^{27}Al quadrupole nutation and ^{1}H – ^{27}Al cross-polarization solid-state NMR studies of ultrastable zeolite Y with fast magic-angle spinning. *Chem. Phys Lett.*, **187**, 401–408.

211 Rocha, J. and Klinowski, J. (1991) ^{27}Al solid-state NMR spectra of ultrastable zeolite Y with fast magic-angle spinning and ^{1}H-^{27}Al cross-polarization. *J. Chem. Soc. Chem. Commun.*, 1121–1122.

212 Omegna, A., van Bokhoven, J.A., and Prins, R. (2003) Flexible aluminum coordination in alumino-silicates. Structure of zeolite H-USY and amorphous silica-alumina. *J. Phys. Chem. B*, **107**, 8854–8860.

213 Lunsford, J.H. (1997) Characterization of acidity in zeolites and related oxides using trimethylphosphine as a probe. *Top. Catal.*, **4**, 91–98.

214 Lashdaf, M., Nieminen, V., Tiitta, M., Venäläinen, T., Osterholm, H., and

Krause, O. (2004) Role of acidity in hydrogenation of cinnamaldehyde on platinum beta zeolite. *Micropor. Mesopor. Mater.*, **75**, 149–158.

215 Peng, L., Chupas, P.J., and Grey, C.P. (2004) Measuring Brønsted acid densites in zeolite HY with diphosphine molecules and solid state NMR spectroscopy. *J. Am. Chem. Soc.*, **126**, 12254–12555.

216 Peng, L. and Grey, C.P. (2008) Diphosphine probe molecules and solid-state NMR investigations of proximity between acidic sites in zeolite HY. *Micropor. Mesopor. Mater.*, **116**, 277–283.

217 Bull, L.M., Cheetham, A.K., Anupold, T., Reinhold, A., Samoson, A., Sauer, J., Bussemer, B., Lee, Y., Gan, Z., Shore, J., Pines, A., and Dupree, R.A. (1998) High-resolution ^{17}O NMR study of siliceous zeolite faujasite. *J. Am. Chem. Soc.*, **120**, 3510–3511.

218 Freude, D., Loeser, T., Michel, D., Pingel, U., and Prochnow, D. (2001) ^{17}O NMR studies of low silicate zeolites1. *Solid State Nucl. Magn. Reson.*, **20**, 46–60.

219 Peng, L., Huo, H., Liu, L., and Grey, C.P. (2007) ^{17}O magic angle spinning NMR studies of Brønsted sites in zeolites HY and HZSM-5. *J. Am. Chem. Soc.*, **129**, 335–346.

220 Readman, J.E., Grey, C.P., Ziliox, M., Lucy, M., Bull, L.M., and Samoson, A. (2004) Comparison of the ^{17}O NMR spectra of Zeolites LTA and LSX. *Solid State Nucl. Magn. Reson.*, **26**, 153–159.

221 Feuerstein, M. and Lobo, R.F. (1998) Characterization of Li cations in zeolite LiX by solid-state NMR spectroscopy and neutron diffraction. *Chem. Mater.*, **10**, 2197–2204.

222 Feuerstein, M., Hunger, M., Engelhardt, G., and Amoureux, J.P. (1996) Characterisation of sodium cations in dehydrated zeolite NaX by ^{23}Na NMR. *Solid State Nucl. Magn. Reson.*, **7**, 95–103.

223 Kortunov, P., Vasenkov, S., Kärger, J., Valiullin, R., Gottschalk, P., Fé Elía, M., Perez, M., Stöcker, M., Drescher, B., McElhiney, G., Berger, C., Gläser, R., and Weitkamp, J. (2005) The role of mesopores in intracrystalline transport in USY zeolite: PFG NMR diffusion study on various length scales. *J. Am. Chem. Soc.*, **127**, 13055–13059.

224 Vasenkov, S., Böhlmann, W., Galvosas, P., Geier, O., Liu, H., and Kärger, J. (2001) PFG NMR study of diffusion in MFI-Type zeolites: evidence of the existence of intracrystalline transport barriers. *J. Phys. Chem. B*, **105**, 5922–5927.

225 Gregg, S.J. and Sing, K.S.W. (1982) *Adsorption, Surface Area and Porosity*, Academic Press, New York.

226 Lowell, S. and Shields, J.E. (1979) *Powder Surface Area and Porosity*, Chapman and Hall, New York.

227 Webb, P.A. and Orr, C. (1997) *Analytical Methods in Fine Particle Technology*, Micromeritics Instrument Corporation, Norcross, GA.

228 Brown, M.E. and Gallagher, P.K. (2003) *Applications to Inorganic and Miscellaneous Materials*, Handbook of Thermal Analysis and Calorimetry, vol. 2, Elsevier, Amsterdam.

229 Pal-Borbley, G. (2007) Thermal analysis of zeolites. *Molec. Sieves Sci. Technol.*, **5**, 67–101.

230 Lightfoot, P., Woodcock, D.A., Maple, M.J., Villaescusa, L.A., and Wright, P.A. (2001) The widespread occurrence of negative thermal expansion in zeolites. *J. Mater. Chem.*, **11**, 212–216.

231 Tschaufeser, P. and Parker, S.C. (1995) Thermal expansion behavior of zeolites and AlPO$_{4S}$. *J. Phys. Chem.*, **99**, 10609–10615.

232 Li, C-Y. and Rees, L.V.C. (1986) Thermogravimetric studies of faujasite with different Si/Al ratios. *Zeolites*, **6**, 217–220.

233 Muller, J.C.M., Hakvoort, G., and Jansen, J.C. (1998) DSC and TG study of water adsorption and desorption on zeolite NaA. *J. Therm. Anal. Calorim.*, **53**, 449–466.

234 Orr, C., and Dallavalle, J.M. (1959) *Fine Particle Measurement, Size, Surface and Pore Volume*, Macmillan, New York.

235 Gorte, R.J. (1999) What do we know about the acidity of solid acids? *Catal. Lett.*, **62**, 1–13.

236 Farneth, W.E. and Gorte, R.J. (1995) Methods for characterizing zeolite acidity. *Chem. Rev.*, **95**, 615–635.

237 Gorte, R.J. and White, D. (1997) Interactions of chemical species with acid sites in zeolites. *Top. Catal.*, **4**, 57–69.

238 Corma, A. (1995) Inorganic Solid acids and their use in acid catalyzed hydrocarbon reactions. *Chem. Rev.*, **95**, 559–614.

239 Rabo, J.A. and Gajda, G.J. (1989) Acid function in zeolites: recent progress. *Catal. Rev. Sci. Eng.*, **31**, 385–430.

240 Weitkamp, J. (2000) Zeolites and catalysis. *Solid State Ionics*, **131**, 175–188.

241 Gravelle, P.C. (1977) Calorimetry in adsorption and heterogeneous catalysis studies. *Cat. Rev. Sci. Eng.*, **16**, 37–110.

242 Auroux, A. (1997) Acidity characterization by microcalorimetry and relationship with reactivity. *Top. Catal.*, **4**, 71–89.

243 Auroux, A. (2002) Microcalorimetry methods to study the acidity and reactivity of zeolites. Pillared clays and mesoporous materials. *Top. Catal.*, **19**, 205–213.

244 Cardona-Martinez, N. and Dumesic, J.A. (1992) Applications of adsorption microcalorimetry to the study of heterogeneous catalysis. *Adv. Catal.*, **38**, 149–244.

245 Katada, N., Igi, H., Kim, J.H., and Niwa, M. (1997) Determination of the acidic properties of zeolite by theoretical analysis of temperature programmed desorption of ammonia based on adsorption equilibrium. *J. Phys. Chem. B*, **101**, 5969–5977.

246 Lercher, J.A., Jentys, A., and Brait, A. (2008) Catalytic test reactions for probing the acidity and basicity of zeolites. *Mol. Sieves*, **6**, 153–212.

247 Liu, Y., Zhang, W., and Pinnavaia, T.J. (2000) Steam-stable aluminosilicate mesostructures assembled from zeolite type Y seeds. *J. Am. Chem. Soc.*, **122**, 8791–8792.

248 Marques, J.P., Gener, L., Lopes, J.M., Ramoa Ribeiro, F., and Guisnet, M. (2005) n-Heptane cracking on dealuminated HBEA zeolites. *Catal. Today*, **107**, 726–733.

249 Vasques, M.H., Ramoa Ribeiro, F., Gnep, N., and Guisnet, M. (1989) Effects of steaming on the shape selectivity and on the acidity of HZSM-5. *React. Kinet. Catal. Lett.*, **38**, 301–306.

250 Babitz, S.M., Williams, B.A., Miller, J.T., Snurr, R.Q., Haag, W.O., and Kung, H.H. (1999) Monomolecular cracking of n-hexane on Y, MOR and ZSM-5 zeolites. *Appl. Catal. A*, **179**, 71–86.

251 Cotterman, R.L., Hickson, D.A., and Shatlock, M.P. (1989) Relationship between structure and catalytic performance of dealuminated Y zeolite. *ACS Symp. Ser.*, **411**, 24–39.

252 Blom, D.A., Bradley, S.A., Sinkler, W., and Allard, L.F. (2006) Observation of Pt atoms, clusters and rafts on oxide supports, by sub-Ångstrom Z-contrast imaging in an aberration-corrected STEM/TEM. *Proc. Microsc. Microanal.*, **12**, 50–51.

253 Hunger, M. and Wang, W. (2006) Characterization of solid catalysts in the functioning state by Nuclear Magnetic Resonance spectroscopy. *Adv. Catal.*, **50**, 149–225.

254 Fan, F., Feng, Z., Li, G., Sun, K., Ying, P., and Li, C. (2008) In situ UV raman spectroscopic studies on the synthesis mechanism of zeolite X. *Chem. Eur. J.*, **14**, 5125–5129.

5
Overview in Zeolites Adsorptive Separation
Santi Kulprathipanja and Robert B. James

5.1
Introduction

Adsorption is the process whereby molecules of a gas or liquid species adhere to a solid surface. If a certain species A has a greater affinity for the solid surface than another species B in the mixture the preferentially adsorbed species can in principle be separated from the other molecules in the gas or liquid mixture. If the solid adsorbent is to be reused the adsorbed species must be desorbed from the solid. In gas phase adsorption the adsorbed material is most often removed by changing the temperature and/or the pressure of the system along with a carrier or sweeper gas. For liquid systems a chemical desorbent must be found that preferentially displaces the desired product species from the solid. The desorbent must itself be easily separated from the product in another separation step, usually distillation.

5.2
Industrial Adsorptive Separation

5.2.1
Gas Separation

Activated carbons have been used for many years to remove impurities from gases and from water. Zeolites also have been commonly employed in gas separation processes to remove water (drying) and for producing oxygen-enriched air by exploiting the different affinities between zeolites and water, oxygen and nitrogen. Recent work with gas separations is aimed at natural gas cleanup and CO_2 capture. Chapter 9 details industrial gas separations. Chapters 10 and 11 explore adsorptive gas separations using zeolitic and zeolitic–polymer membranes in detail.

5.2.2
Liquid Separation

Industrial adsorption separation processes for liquids are most successful when the species involved have very close boiling points, making distillation expensive or even impossible or are thermally sensitive at convenient distillation temperatures.

Industrial applications of zeolitic separation processes fall into the following general application categories:

- Aromatics and derivatives separation (e.g., xylene isomers, *p*-diethylbenzene, cymene);
- Non-aromatic hydrocarbons separation (e.g., olefin/paraffin, *n*-paraffin/non-*n*-paraffin);
- Carbohydrates and organic acids separation (e.g., fructose/glucose, citric acid, fatty acids);
- Fine chemical and pharmaceuticals separation (e.g., enantiomers, vitamins);
- Trace impurity removal (e.g., sulfur compounds, nitrogen compounds, oxygen compounds, iodide, aromatics, metals).

5.2.2.1 Aromatics

Para-xylene is an industrially important petrochemical. It is the precursor chemical for polyester and polyethylene terephthalate. It usually is found in mixtures containing all three isomers of xylene (*ortho*-, *meta*-, *para*-) as well as ethylbenzene. The isomers are very difficult to separate from each other by conventional distillation because the boiling points are very close. Certain zeolites or mol sieves can be used to preferentially adsorb one isomer from a mixture. Suitable desorbents exist which have boiling points much higher or lower than the xylene and displace the adsorbed species. The boiling point difference then allows easy recovery of the xylene isomer from the desorbent by distillation. Because of the basic electronic structure of the benzene ring, adsorptive separations can be used to separate the isomers of families of substituted aromatics as well as substituted naphthalenes.

5.2.2.2 Non-Aromatic Hydrocarbons

Other adsorptive separations of industrial importance are olefin from paraffin and *n*-paraffin from branched paraffin. Adsorptive separation is attractive for these systems because the initial separation does not depend on boiling points but rather on molecule type. As an example, natural or processed naphtha boiling range material is made up of a wide range of chemical species. The normal straight chain or normal paraffin is a much better feedstock for feeding ethylene plants or steam crackers than the branched hydrocarbon. The branched hydrocarbon is a much better feedstock for catalytic reforming to produce aromatics and high octane gasoline. Zeolites can be used to adsorb the C5–C10 normal paraffins from a naphtha mixture. The separation allows optimum use of each feedstock type in downstream processing and achieves an overall more valuable product yield from the naphtha feed.

5.2.2.3 Non-Petrochemicals

A good example of using adsorptive separation in a non-refining/petrochemical application is the separation of fructose from an aqueous solution of mixed sugars. This process allows the production of high concentration fructose which has a much higher sweetness to calorie ratio than simple glucose or sucrose. As in fine chemical and pharmaceutical applications we can often use adsorption when distillation is not possible or feasible or when the material is thermally sensitive.

5.2.2.4 Trace Component Removal

Industrial examples of adsorbent separations shown above are examples of bulk separation into two products. The basic principles behind trace impurity removal or purification by liquid phase adsorption are similar to the principles of bulk liquid phase adsorption in that both systems involve the interaction between the adsorbate (removed species) and the adsorbent. However, the interaction for bulk liquid separation involves more physical adsorption, while the trace impurity removal often involves chemical adsorption. The formation and breakages of the bonds between the adsorbate and adsorbent in bulk liquid adsorption is weak and reversible. This is indicated by the heat of adsorption which is <2–3 times the latent heat of evaporation. This allows desorption or recovery of the adsorbate from the adsorbent after the adsorption step. The adsorbent selectivity between the two adsorbates to be separated can be as low as 1.2 for bulk liquid adsorptive separation. In contrast, with trace impurity removal, the formation and breakages of the bonds between the adsorbate and the adsorbent is strong and occasionally irreversible because the heat of adsorption is >2–3 times the latent heat of evaporation. The adsorbent selectivity between the impurities to be removed and the bulk components in the feed is usually several times higher than the adsorbent selectivity for bulk liquid adsorptive separation.

In the majority of impurity removal processes, the adsorbent functions both as a catalyst and as an adsorbent (catalyst/adsorbent). The impurity removal process often involves two steps. First, the impurities react with the catalyst/adsorbent under specified conditions. After the reaction, the reaction products are adsorbed by the catalyst/adsorbent. Because this is a chemical adsorption process, a severe regeneration condition, or desorption, of the adsorbed impurities from the catalyst/adsorbent is required. This can be done either by burning off the impurities at an elevated temperature or by using a very polar desorbent such as water to desorb the impurities from the catalyst/adsorbent. Applications to specific impurities are covered in the followings section. The majority of industrial applications involve the removal of species containing hetero atoms from bulk chemical products as purification steps.

5.3
R&D Adsorptive Separation

When selecting candidate zeolitic materials for effecting an adsorptive separation the researcher faces an enormous number of possible combinations of materials and desorbents. In the absence of any established algorithm for selecting materials the researcher is forced to rely on analogy and published experience to make choices for experimentation.

The sections below survey the literature and report literature separation work on specific separation applications.

5.3.1
Aromatic Hydrocarbon Separation

Table 5.1 summarizes work done on aromatic and substituted aromatic and other ring systems.

5.3.2
Non-Aromatic Hydrocarbon Separation

Table 5.2 summarizes work done on paraffin and olefin and isomer separations systems.

5.3.3
Carbohydrate Separation

Table 5.3 contains reports and surveys work done on carbohydrate and fatty acid and organic chemical separations.

5.3.4
Pharmaceutical Separation

Table 5.4 contains survey results for pharmaceutical and fine chemical separations work.

5.3.5
Trace Impurities Removal

5.3.5.1 Sulfur Removal
Sulfur and organo-sulfur species are known process contaminants in petroleum refining as well as environmental contaminants in fuels. The types of compounds found in the process stream are dependent on the boiling range of the process streams, which may be gasoline, kerosene, diesel and fuel oil. Typical organo-sulfur compounds found in gasoline are mercaptans (RSH), organic sulfides (R-S-R), organic disulfides (R-SS-R), carbon disulfides (S-C-S), and thiophene. In

Table 5.1 Survey of liquid separations using crystalline materials: aromatic applications.

Separation	Adsorbent	Desorbent	Reference
Alkyltetralin from alkyltetralin, linear alkylbenzene	NaY	t-Butylbenzene	[1, 2]
m-Chlorobenzotrifluoride from m-, p-chlorobenzotrifluoride	NaY		[3]
Chloroethylbenzene from chlorobenzene, dichlorotoluene	NaSrX		[4]
p-Chloroethylbenzene from o-, m-, p-cholroethylbenzene	AgBaX	3,4-Dichlorotoluene	[5]
m-Chloroethylbenzene from o-, m-, p-cholroethylbenzene	NaX, KPbNaY	p-Xylene, dichloro-toluene	[5, 6]
2,4-Chlorofluorotoluene from 2,4-, 3,4-, 2,5-chlorofluorotoluene	NaY		[7]
o-Chloronitrobenzene from o-, m-, p-chloronitrobenzene	KX, KY, CoY, CaY	Nitrobenzene/ toluene	[8]
p-Chloronitrobenzene from p-, o-chloronitrobenzene	H-ZSM5		[9]
m-Chloro-n-propylbenzene from o-, m-, p-chloro-n-propylbenzene, o-, m-, p-chloroisopropylbenzene	CsPbKY, 20%LiY	o-Xylene, o-chloro-toluene	[10]
m-Chlorophenol from m-, p-, o-chlorophenol	NaY		[11]
Coumarone, indene from coumarone, indene, coal tar distillate	NaY	Toluene	[12]
m-Cresol from m-, p-cresol	H-ZSM5	Compressed propane	[13, 14]
m-Cresol from m-cresol, 2,6-xylenol	NaX		[15]
p-, m-Cresol from p-, m-, o-cresol, 2-ethylphenol, 2,6-dimethylphenol	KBaX (5 wt%–water)	1-Pentanol	[16]
p-Cresol from xylenol (cresol isomers)	KBaX, NaX	Pentanol	[15, 17, 18]
m-Cymene from p-, m-cymene	NaX	Toluene	[19]
p-Cymene from p-cymene, m-cymene	KBaX	Toluene	[20]
m-Dichlorobenzene (rejective) from m-, p-, o-dichlorobenzene	CaKY, KPbY, KY	3,4-Dichlorotoluene, m-xylene	[21–24]
m-Dichlorobenzene from m-, p-, o-dichlorobenzene	LiX, NaKX	Toluene	[25, 26]
m-Dichlorobenzene from m-, p-, o-dichlorobenzene	Silicalite-1	Chlorobenzene	[27]
m-Dichlorobenzene from m-, p-dichlorobenzene	ZSM-5		[13]

Table 5.1 Continued

Separation	Adsorbent	Desorbent	Reference
p-Dichlorobenzene from p-, m-, o-dichlorobenzene	BaY	p-Xylene	[28]
p-Dichlorobenzene from p-, m-dichlorobenzene	HZSM-5, MgZSM-5	4-Chlorotoluene	[29, 30]
p-Dichlorobenzene from p-, o-dichlorobenzene	Silicalite-1	Chlorobenzene	[31]
3,5-Dichlorobromobenzene from 3,5-, 2,4-, 2,6-dichlorobromobenzene, 1,5-dichloro-2,4-dibromobenzene, m-dichlorobenzene	NaX		[32]
3,5-Dichlorocumene from 3,5-dichlorocumene, 2,4-dichlorocumene, 2,5-dichlorocumene	HKY	m-Xylene	[33]
2,4-Dichlorofluorobenzene from 3,4-, 2,4-, 2,5-dichlorofluorobenzene	BaAgX		[34]
2,4-Dichlorophenol from 2,4-, 2,6-dichlorophenol	KL, Ca/Li/NiX		[35]
2,4-Dichlorotoluene from 2,3-, 2,4-, 2,5-, 2,6-, 3,4-dichlorotoluene	SrNaX	m-Xylene	[36]
2,5-Dichlorotoluene from 2,3-, 2,4-, 2,5-, 2,6-, 3,4-dichlorotoluene	AgNaY, CsKX	1-Hexanol	[37]
2,6-Dichlorotoluene from 2,3-, 2,4-, 2,5-, 2,6-dichlorotoluene	AgNaX, AgMgX	Toluene, xylenes, chlorotoluenes	[38, 39]
2,5-Dichlorotoluene from 2,3-, 2,4-, 2,6-, 2,5-dichlorotoluene	KL	Chlorobenzene/n-heptane	[40]
2,6-Dichlorotoluene from 2,3-, 2,4-, 2,6-, 2,5-dichlorotoluene	HAgNaX	o-Xylene	[41]
2,6-Dichlorotoluene from 2,6-, 2,4- 2,5-dichlorotoluene	NaX, CaX	Isopropylnaphthalene	[42, 43]
2,6-Dichlorotoluene from 2,6-, 2,4- 2,5-dichlorotoluene	SrNaX	m-xylene	[6]
3,5-Dichlorotoluene from isomers	NaX		[44]
m-Diethylbenzene (rejective) from m-, o-, p-diethylbenzene	AgKY		[45]
o-Diethylbenzene from diethylbenzene isomers	NaY		[46]
p-Diethylbenzene from diethylbenzene isomers	KY, modified ZSM-5		[47, 48]

Table 5.1 Continued

Separation	Adsorbent	Desorbent	Reference
p-Diethylbenzene from o-, m-, p-diethylbenzene	Silicalite-1	Compressed CO_2	[49]
p-, m- Diethylbenzene from diethylbenzene isomers	KBaX	Xylene mixture	[50]
2,6-Diethylnaphthalene from 2,6-, 2,7-, 1,6-, 1,7-, 1,3-diethylnaphthalene isomers and others (naphthalene ethylation products)	KY	p-Diethylbenzene	[51]
2,6-Diethylnaphthalene from 2,6-, 2,7-, 1,3-diethylnaphthalene	KBaY	Toluene	[52]
2,6-Diethylnaphthalene from 2,6-diethylnaphthalene, 2,6-naphthalenedicarboxylic acid	KY	n-Heptane, p-diethylbenzene	[53]
3,5-Diethyltoluene, 2,6-diethyltoluene from diethyltoluene isomers	KBaX	Toluene	[54]
3,5-Diethyltoluene from 3,5-, 2,4-, 2,3-, 2,6-, 2,5-diethyltoluene isomers	KX	p-Diethylbenzene/ iso-octane	[55]
p-Diisopropenylbenzene from p-diisopropenylbenzene, p-isopropenylisopropylbenzene	NaY		[56]
4,4′-Diisopropylbiphenyl from 3,5′-, 3,3′-, 3,4′-, 4,4′-diisopropylbiphenyl	Na-mordenite	Toluene	[57]
2,6-Diisopropylnaphthalene from 2,6-, 2,7-, 1,3-, 1,5-, 1-4-, 1,6-, 1,7-diisopropylnaphthalene	ZSM-5	Xylenes, p-diethylbenzene	[58]
2,6-Diisopropylnaphthalene from 2,7-, 2,6-diisopropylnaphthalene	NaY	Diethylbenzene	[59]
2,7-Diisopropylnapththalene from 1,5-, 1,4-, 1,6-, 2,3-, 2,7-, 2,6-diisopropylnaphthalene	KY	Diethylbenzene, toluene	[60, 61]
2,7-Diisopropylnaphthalene from 2,7-, 2,6-diisopropylnaphthalene	KSrY, BaY	Toluene, diethylbenzene	[59, 62]
1,5-Dimethylnaphthalene from 1,5-, 2,6-, 2,7-dimethylnaphthalene	L type	o-Xylene/n-decane	[63]
1,7-Dimethylnaphthalene from 1,7-, 2,7-dimethylnaphthalene	L type		[64]
2,6-Dimethylnaphthalene from 2,6-, 1,4-, 1,3-dimethylnaphthalene	KX, SrX	Toluene	[65, 66]
2,6-Dimethylnaphthalene from 2,6-, 1,6-, 1,5-, 1,3-, 1,7-, 2,7-dimethyl-naphthalene	KY	m-Xylene	[67]

Table 5.1 Continued

Separation	Adsorbent	Desorbent	Reference
2,6-Dimethylnaphthalene from 2,6-, 1,6-, 1,5-dimethylnaphthalene	KX, KY, NaY	Toluene	[68]
2,6-Dimethylnaphthalene from 2,6-, 2,7-dimethylnaphthalene	KY	o-Xylene	[69]
2,6-Dimethylnaphthalene from isomers	KX/activated carbon	Monochlorobenzene	[70]
2,7-Dimethylnaphthalene from 2,6-, 2,7-dimethylnaphthalene	KY	Isooctane	[71]
α,β-, β,β-Dimethylnaphthalene from α,α-, α,β-, β,β-dimethylnaphthalene	Na-mordenite/HY		[72]
2,4-Dimethylphenol from 2,4-, 2,5-dimethylphenol	SrX	1-Hexanol	[73]
2,6-Dimethylphenol from 2,6-dimethylphenol, p- or m-cresol	CaX, BaX		[15]
3,5-Dimethylphenol from 3,5-, 2,4-, 2,3-dimethylphenol	KY	Hexanol	[74]
2,3-Dinitrotoluene from 2,3-, 2,4-, 2,6-, 3,4-dinitrotoluene	NaY	Propyl acetate	[75]
2,4-Dinitrotoluene from 2,4-, 2,6-dinitrotoluene	KX	C_3–C_5 alcohol, ketone ester, or nitrobenzene	[76]
m-divinylbenzene from m-, p-divinylbenzene	NaY		[77]
Durene from durene, isodurene	Na-beta prehnitene		[78]
Durene from isodurene and other unknowns	LiX	1,2,3-Trimethyl-benzene	[79]
Ethylbenzene from ethylbenzene, p-, m-, o-xylene	Beta zeolite, gallium beta zeolite	p-Diethylbenzene	[80, 81]
Ethylbenzene (extractive) from ethylbenzene, p-, m-, o-xylene	CsX (SiO_2/$Al_2O_3 < 2.2$), CsX, KX, RbX	Toluene, m-diisopropylbenzene/dodecane	[82–86]
Ethylbenzene (rejective) from ethylbenzene, p-, m-, o-xylene	SrKX, NaY	Toluene	[87–89]
1-Ethylnaphthalene from 1-, 2-ethylnaphthalene	HY		[90]
m-Ethylphenol from m-, p-ethylphenol	KNaX		[91]
p-Ethylphenol and m-ethylphenol from p-ethylphenol, m-ethylphenol, butyl alcohol	KPbY		[92]

Table 5.1 Continued

Separation	Adsorbent	Desorbent	Reference
p-Ethylphenol from p-, m-ethylphenol	KY, KBaX, KBaY	Pentanol, butyl alcohol	[93–95]
o-Ethyltoluene from o-, m-, p-ethyltoluene	NaX	Toluene	[96]
p-Ethyltoluene from p-, m-, o-ethyltoluene	KY, ferrierite	Tetralin	[97, 98]
p-Ethyltoluene from p-, m-, pseudocumene, hemimellitene	Cs-beta	Toluene	[99]
Iodonaphthalene from iodonaphthalene, 2,6-, 2,7-diiodonaphthalene	KX		[100]
4-Isopropylbiphenyl from 4-, 3-isopropylbiphenyl	KY	Diethylbenzene	[101]
4-Isopropyldibenzofuran from 1-, 2-, 3-, 4-isopropyldibenzofuran	NaX		[102]
2,6-Isopropylmethylnaphthalene from 1,4-, 2,5-, 2,7-, 2,6-isopropylmethylnaphthalene	KBaY, NaY		[103]
2-Methylnaphthalene from 2-, 1-methylnaphthalene	NaY (modified by CVD)		[104]
m-Nitrobenzaldehyde from m-, o-, p-nitrobenzaldehyde	LiX	Acetonitrile, benzene	[105]
m-Nitrobenzaldehyde from m-, o-, p-nitrobenzaldehyde	NaY	Nitrobenzene	[106]
p-Nitrobenzaldehyde from p-, m-, o-nitrobenzaldehyde	FeZSM-5		[107]
o-Nitrotoluene from m-, o-, p-nitrotoluene	KX	Nitrobenzene	[108]
Phenanthrene from phenanthrene, anthracene	NaX	Benzene	[109]
m-Tertiarybutylphenol from m-, o-, p-tertiarybutylphenol	AgNaX		[110]
1,3,5-Trimethylbenzene, n-decane from 1,3,5-trimethylbenzene, 3,3,5-trimethyl-haptane, 2,6-dimethyloctane, 2-methylnonone, n-decane	Silicalite-1	n-Heptane/isooctane	[111]
Trimethylbenzenes from xylenes	Na-dealuminated Y	Toluene	[112]

Table 5.1 Continued

Separation	Adsorbent	Desorbent	Reference
2,4-Toluenediamine from 2,4-,2,6-toluenediamine	Na/Ca/Li/MgY	Toluene	[113]
2,6-Toluenediamine from 2,4-,2,6-toluenediamine	BaX	Methyl alcohol	[114]
2,6-Toluenediamine from 2,4-,2,6-toluenediamine	KY	Toluene	[20]
p-Toluidine from p-, m-, o-toluidine	NiX, NiY, CoY, KBaY	Aniline, 3-methylpyridine	[13, 115]
1,2,4-Trichlorobenzene from 1,2,4-, 1,2,3-trichlorobenzene	H-ZSM-5		[116]
1,2,4-Trimethylbenzene from 1,2,4-, 1,3,5-trimethylbenzene	NaY		[104]
2,3,5-Trimethylcumene from 2,3,5-, 2,4,5-, 2,3,6-trimethylcumene	KX	Toluene/isooctane	[117, 118]
m-Xylene from m-, o-, p-xylene	NaY, NaLiY	Indan, toluene	[119–123]
m-Xylene (rejective) from m-, o-, p-xylene, benzene, C_9 aromatics	H-beta	Toluene	[124]
m-Xylene from m-, p-xylene	ZSM-5		[30]
o-Xylene from o-, m-xylene	$AlPO_4$-8		[125]
o-Xylene from o-, m-, p-xylene, ethylbenzene	AgX	p-Diethylbenzene	[126]
o-Xylene from o-, m-, p-xylene, ethylbenzene	IM-12	Touene	[127]
p-Xylene (extractive) from p-, m-, o-xylene, ethylbenzene	BaX, KBaX, BaSrX, pyrolyzed BaX	p-Diethylbenzene, diethyltoluene, toluene, p-diisopropylbenzene, indan, diisopropyl ether, methyl tertiary-butyl ether	[128–138]
p-Xylene (extractive) from p-, m-, o-xylene, ethylbenzene	KBaX, KBaY	1,2-Difluorobenzene, 1,3-difluorobenzene	[139, 140]
p-Xylene (extractive) from p-, m-, o-xylene, ethylbenzene	Cs-SSC-25	Benzene	[141]
p-Xylene (extractive) from p-, m-, o-xylene, ethylbenzene	Silicalite-1		[142]
p-Xylene (rejective) from p-, o-, m-xylene, ethylbenzene	NiY	p-Diethylbenzene	[143]

Table 5.2 Survey of liquid separations using crystalline materials: non-aromatic hydrocarbon applications.

Separation	Adsorbent	Desorbent	Reference
Aromatics (benzene, toluene, m-xylene, mesitylene) from olefin (octene)	NaX, NaY		[144]
Benzene from benzene, octane	NaX, NaY		[144]
1,3-Butadiene from 1,3-butadiene, 1-butene	AgY		[145]
1-Butene from 1-butene, cis- and trans-2-butene	ALPO-34/Si-CHA		[146]
1-Butene from 1-butene, isobutene, cis- and trans-2-butene	KX	1-Hexene/cyclohexane	[147, 148]
Trans-2-butene from cis-2-butene	Silicalite-1	Pentene/i-pentene	[149]
Cis-3-heptene from cis-3-heptene, trans-3-heptene	AgY	n-Pentane/diethyl ether	[150]
Hydrocarbons (aromatic) from liquid paraffins	NaX, CaX	n-Heptane, ethyl alcohol	[151]
Hydrocarbons (multibranched chain) from singly branched and straight-chain hydrocarbons	KBaX	Isobutane	[152]
Indan from hydrocarbons	NaX		[153]
Isobutene from n-C_4 hydrocarbons	Silicalite-1	1-Pentene	[154]
2-Methylpentane from 2-methylpentane, 2,2-dimethylbutane	KH-mordenite, Sr/Ba, HZSM-5, Na-EU-1 (Si/Al = 24), Na-NU-87 (Si/Al = 16)		[155, 156]
Monomethyl paraffins (C_{10}–C_{15}) from monomethyl paraffins, di-isoparaffins, di-isoolefins, aromatics	Silicalite-1	Cyclohexane/n-hexane	[146, 157]
Naphthalene from coal tar distillate	LiX	1,2,3-Trimethylbenzene	[158]
n-Octane, isooctane from n-hexadecane	CaA (5A)		[159]
1-Octene, 1-decene from 1-octene, 1-decene, octane, decane	CuY		[160]

Table 5.2 Continued

Separation	Adsorbent	Desorbent	Reference
Olefin from paraffin	NaX (caustic treated)	Light olefin	[161]
Olefins (hexane, octane, dodecene) from paraffin (heptane)	NaX, NaY		[144]
Olefins (normal) from C_5–C_8 branched olefins	Silicalite-1, ZSM-5	C_3–C_8 ketones	[162]
Olefins (n-C_6) from mixtures containing C_6 branched olefins and cyclic hydrocarbons	Silicalite-1	Pentene, 1-butene	[163]
Oxygenates from paraffin stream (Fischer Tropsch product)	NaX, CaA		[164]
n-Paraffins (C_5–C_8), mono-branched paraffins (C_5–C_8), multi-branched paraffins (C_5–C_8) from hydrocarbons (C_5–C_8)	CaA, ZSM-5, ZSM-11, ZSM-23, ferrierite	i-Pentane	[165]
n-Paraffins (C_5–C_{10}) from straight-run gasoline fraction	CaA		[166]
n-Paraffins (C_{10}–C_{14}) from non-n-paraffins (C_{10}–C_{14})	CaA, MgA, silicalite-1	n-Pentane/i-octane	[167, 168]
n-Paraffins (C_{10}–C_{35}) from non-n-paraffins (C_{10}–C_{35})	CaA (5A)	n-Octane/i-octane	[169]
n-Paraffins (C_{20}–C_{30}) from refinery slack wax	CaA (5A)	Steam	[170]
n-Paraffins from petroleum fractions	CaA	Ammonia	[171]
Paraffins (liquid) from diesel fuel fractions	Tseosorb 5 AM	Ammonia	[172]
Paraffins (liquid) from kerosene, diesel fuel	MgA	Steam	[173]
n-Pentane, cis-2-pentene, 1-pentene from n-pentane, cis-2-pentene, 1-pentene, 2-methyl-2-butene, 2-methyl-1-butene, 3-methyl-1-butene, cyclopentene	Silicalite-1	Methylcyclohexane	[174]
Propane from propane, propene	Silicalite-1		[175]

Table 5.3 Survey of liquid separations using crystalline materials: carbohydrates, fatty acids and oxygenated organics applications.

Separation	Adsorbent	Desorbent	Reference
Arachidonate esters, linolenate esters from $C_{15}H_{31}CO_2Me$, $C_{15}H_{29}CO_2Me$, $C_{17}H_{35}CO_2Me$, $C_{17}H_{35}CO_2Me$, $C_{17}H_{33}CO_2Me$, $C_{17}H_{31}CO_2Me$, arachidonate esters, linolenate esters	NH_4Y	Chloroform	[176]
Diglycerides from triglycerides	K-omega	Ketone/n-aliphatic hydrocarbon	[177]
Fatty acids (oleic and linoleic acids) from resin acids	Silicalite-1	Acetone, methyl ethyl ketone, 3-pentanone	[178]
Fatty acids (palmitic, oleic, linoleic acids) from unsaponifiables	Silicalite-1	Acetone	[179]
Fatty acids, saturated (palmitic and stearic acids)	Silicalite-1	Acetone	[180]
Fatty acid methyl ester (FAME), resin acid (RA) from esterified tall oil (FAME, RA and free fatty acid)	NaX	Petroleum naphtha	[181]
Trans-unsaturated fatty acid from its *cis*- and *trans*-isomeric mixture	Na-ZSM-5		[182]
Methyl esters of fatty acid from methyl esters of resin acid	Silicalite-1	Cyclohexane	[183]
Methyl linoleate, methyl oleate from methyl linoleate, methyl oleate, methyl esters of saturated fatty acids	KY	Toluene	[184]
Monoglycerides from monoglycerides, diglycerides, triglycerides	KX	Methyl ethyl ketone	[185]
Oleic acid from oleic acid, diolein, triolein	$AlPO_4$-11	2-Heptanone	[186]
Oleic acid from oleic acid, linoleic acid	Silicalite-1	Methyl ethyl ketone/ propionic acid	[187, 188]
Oleic acid from oleic acid, linoleic acid	Steam-treated silicalite	Methyl ethyl ketone/ acetic aid	[189]
1,3-Palmitoyl-2-oleyl triglyceride from 1-palmitoyl-2,3-dioleyl triglyceride	AgX	Ethyl acetate/n-heptane	[190]

Table 5.3 Continued

Separation	Adsorbent	Desorbent	Reference
Saturated-, unsaturated fatty acids, triglyceride from interesterification products	silicalite-1	2-Heptanone/acetone	[191]
Acetaldehyde from C_3–C_{15} hydrocarbons	NaA		[192]
Acetic acid, butyric acid, formic acid, propionic acid from acetic acid, butyric acid, formic acid, propionic acid, other oxygenated aliphatic compounds	Silicalite-1		[193]
Aniline from water	H-beta, Cu-beta ($SiO_2/Al_2O_3 = 75:1$), H-ZSM5, Cu-ZSM5 ($SiO2/Al2O3 = 80:1$)		[194, 195]
Butanol from fermentation broth	silicalite-1		[196]
Epsilon-caprolactam from epsilon caprolactam, 4-ethyl-2-pyrrolidinone, 5-methyl-2-piperdinone, 3-ethyl-2-pyrrolidinone, 3-methyl-2-piperdinone	Silicalite-1	Methanol, acetonitrile	[197, 198]
Cyclohexanone from cyclohexanone, cyclohexanol	Ca/Ba/SrX, Li/Na/BaY		[199]
Ethyl alcohol from water	silicalite-1		[200]
Ethylaniline from ethylaniline, diethylaniline	NaY	Methyl alcohol	[201]
Ethylene glycol from ethylene glycol, propylene glycol, *tert*-butyl alcohol, water	NH_4Y, 4A, 5A, NaX	Methyl alcohol/water	[202]
Indole from indole, 1-methylnaphthalene, 3-methylindole	AgY	Acetonitrile	[203]
Lactic acid from lactic acid, citric acid	ZSM-5, dealuminated Y		[204]
Malic acid from malic acid, fumaric acid, maleic acid	Silicalite-1	Water/acetone	[205]
Methyl tertiary butyl ether from methyl tertiary butyl ether, methanol, acetone, tertiary butyl alcohol, methyl ethyl ketone, diisobutylene	NaX		[206]

Table 5.3 Continued

Separation	Adsorbent	Desorbent	Reference
Myristic acid from myristic acid, lauric acid	Silicalite-1	Methyl acetate, acetone, 3-pentanone	[207]
Phenol from water	siliceous beta zeolite (Si/Al = ∞)		[208]
Phenylalanine from phenylalanine, tyrosine, water	NaZSM-5		[209]
Gamma-picoline from gamma-picoline, alpha-picoline, beta-picoline	KX	Pyridine	[210]
Beta-pinene from alfa-pinene	KX	Hexene	[211]
Water from water, ethanol	NaA		[212, 213]
Arabinose from arabinose, xylose, galactose, mannose, glucose, color bodies	CaY	Water	[214, 215]
Arabinose from arabinose, Maltin 150, glucose, xylose, galactose, mannose	NH$_4$X	Water	[216]
L-Arabinose from L-arabinose, galactose, glucose, mannose, xylose	BaX	Water	[217]
Fructose from fructose, glucose	BaX	Water	[218]
Fructose from glucose, sucrose, and salts	Organically bound CaY, esterified CaY	Water	[219–223]
L-Fructose from L-fructose, L-mannose, L-psicose	CaY	Water	[224]
Glucose from fructose	KX, CaY	Water	[225, 226]
Inositol from inositol, glucose, fructose, sucrose, sorbitol	BaX	Water	[227]
Isomaltose from isomaltose, glucose	Dealuminated-beta zeolite	Water/ethanol	[228]
Lactulose from lactulose, lactose	KY, BaY	Water	[229, 230]
Maltose from maltose, glucose, polysaccharides	Dealuminated Y	Water	[231]
Mannose from glucose	Dealuminated Y (H$^+$, Ca^{2+}, Sr^{2+}, NH$_4^+$)	Water	[156]
D-Mannose from D-mannose, glucose	BaY	Water	[232]
Sucrose from raffinose, betaine, salts	LZ-Y20	Water/ethanol	[233]

Table 5.4 Survey of liquid separations using crystalline materials: pharmaceutical and fine chemical applications.

Separation	Adsorbent	Desorbent	Reference
L-Arginine from L-arginine, L-phenylalanine, water	H/Na-Y (Si/Al = 2.5)		[234]
Chlorophyll-a from chlorophyll-a, β-carotene	MCM-41	Methanol	[235]
Ethyl eicosapentanoate from ethyl eicosapentanoate, ethyl eicosanoate, ethyl eicosatetraenoate	KY		[236]
Fulvic acids from fulvic acids, water	Hexadecyl trimethyl ammonium modified clinoptilolite	Ethanol	[237]
2,6-Lutidine (rejective) from 2,6-lutidine, 2,4-lutidine, α-picoline, β-picoline,	AgNa-5.1Y	2,4-Xylidine	[23]
Mannitol from mannitol, galactitol, sorbitol	BaX	Water	[238]
Methylelaidate from methylelaidate, methyloleate	Na-ZSM-5 (Si/Al = 77.5)	n-Hexane	[239]
Methylparaben from methylparaben, wintergreen oil			[240]
Tetralone from tetralone, alpha-naphthol, tetralin	NaY	Methyl acetate	[241]

kerosene and diesel, substituted thiophenes such as benzothiophenes, alkylthiophenes, alylbenzothiophenes and alkyldibenzothiophenes are also found. Physical adsorption as well as hydrocatalytic adsorbents is used extensively in the refining industry to remove sulfur compounds. Table 5.5 surveys some sulfur species impurity separations

5.3.5.2 Oxygenate Removal

Current interest in synthetic fuels production by Fischer–Tropsch (FT) reactions have created a need for removal of byproduct oxygenates, formed by the FT reaction. The oxygenates consist of primary and internal alcohols, aldehydes, ketones, esters and carboxylic acids. The hydrocarbon products derived from the FT reaction range from methane to high molecular weight paraffin waxes containing more than 50 carbon atoms.

There are a number of industrial applications for paraffins, olefins and mixtures in the C9–C16 range that are produced in the FT process. Among such uses is as a precursor to linear alkylbenzene (LAB), which is used to produce linear

Table 5.5 Survey of liquid separations using crystalline materials: trace impurities removal applications.

Separation	Adsorbent	Desorbent	Reference
Acetonitrile from light aromatics (C_7)	NaX, H-ZSM-5, H-MCM-22		[242]
Alkyl iodines from alkyl iodines, acetic acid	AgY (SiO_2/Al_2O_3 = 10) with zirconia or silica binder		[243]
Arsenate and arsenite from ground water	Zr-clinoptilolite		[228]
Benzothiophene from benzothiophene, naphthalene	NaY	p-Xylene	[244–246]
Benzothiophene, dibenzothiophene and 4,6-dimethyldibenzothiophene from diesel fuel	NiY		[247]
p-Dichlorobenzene from p-, o-dichlorobenzene, water (simulated waste water)	HZSM-5	Ethanol	[9]
Dioxane from light aromatics (C_7)	NaX, H-ZSM-5, H-MCM-22		[248]
Iodide from acetic acid	AgY (SiO_2/Al_2O_3 =12), Ag-mordenite		[249–252]
Metals – Rh(III), Ir(III), Os(III), Co(II), Ni(II), Mo(V), V(III), Fe(III), Ti(IV), Zn(IV) from acetic acid	HY (SiO_2/Al_2O_3 =12)		[252]
Naphthalene, benzothiophene from naphthalene, benzothiophene, isooctane	high-silica zeolite (Tosoh Corporation, SiO_2/Al_2O_3= 6.2)		[239]
Nitrogen compounds from diesel fuel	CuX, CuY		[253]
Sulfur compounds from hydrocarbons (gasoline)	NiX, NiY, MoX, CeY		[254, 255]
Sulfur compounds from hydrocarbons (diesel oil)	CeY, NiY, CuY, FeY		[255, 256]
Sulfur compounds from hydrocarbons (jet fuel)	CuY		[257]
Sulfur (total) compounds, mercaptans, naphthenic acids from diesel fuel	natural clinoptilolite		[258]
Thiophene and thiophene compounds from liquid fuel	AgY, Cu(I)-Y/activated carbon, Cu(I)-SBA-15		[259–262]

alkylbenzene sulfonate (LAS). LAS is a surfactant commonly used in the detergent industry. Therefore, a large portion of the syncrude is preferably hydroprocessed into usable products. The presence of oxygenates can cause a number of problems with processing the syncrude, including adversely impacting hydroprocessing catalyst activity necessitating an increase in the hydroprocessing severity. Alumina, silica, 13X molecular sieves and 5A molecular sieve adsorbents are used to remove oxygenates from the FT product [166, 263].

5.3.5.3 Nitrogenate Removal

As with organo-sulfur compounds, the types of organo-nitrogen species found in refinery process streams depends on the boiling range of the hydrocarbon stream. Organo-nitrogen compounds include, but are not limited to: acetonitrile, pyridine, aniline, indoles, uinoline, carbazoles, acridines and benzocarbazoles. Basic organo-nitrogen compounds have often been described as the strongest inhibitors of the hydrodesulfurization process for ultra-low sulfur diesel production [264].

Adsorbents such as H-ZSM-5, H-MCM-22 and Na-X can be used to removing nitrogen containing contaminants such as acetonitrile, propionitrile and acrylonitrile from toluene [242].

5.3.5.4 Iodide Removal

In the chemical industry, iodine and/or iodine compounds are often used as catalysts and/or catalytic promoters for the production of value-added organic chemicals. As with other catalytic reactions, the catalyst or promoter must be removed from the products after completing the reaction. However, removing trace amounts of organic iodide contaminates from the product by conventional distillation techniques is difficult primarily due to the fact that iodine compounds are unstable and split off into various boiling ranges.

Ag exchanged zeolite is used to remove iodine compounds. More recently Ag-LZ-210™, an Ag-exchanged zeolite-Y adsorbent developed by UOP and having a high silica/alumina ratio (Si/Al > 5), has been used commercially to remove iodide from acetic acid streams [250–252].

5.3.5.5 Aromatic Removal

Trace aromatics removal from linear paraffins in the C_{10}–C_{15} range is an important step in producing linear alkylbenzene (LAB) which in turn is used to make linear alkylbenzene sulfonate (LAS) an important constituent of detergents. High purity linear paraffins are required to produce superior detergent properties. For this application, MgY and NaX adsorbents are reported to be effective adsorbents in removing aromatics from C_{10}–C_{14} n-paraffins [265–267].

5.3.5.6 Metal Removal

Removal of trace metals by adsorption is feasible in polar liquid streams, such as aqueous streams. One example is the removal of trace metal contaminants such as Rh^{+3}, Ir^{+3}, Os^{+3}, Co^{+2}, Ni^{+2}, Mo^{+5}, V^{+3}, F^{+3}, Ti^{+4} and Zn^{+4} from an acetic acid product stream. These metals need to be removed before the iodide removal step (see iodide removal section) to protect the activity of the iodide removal adsorbent, Ag-LZ-210. If metal contaminants are present, Ag^{+1} on the Ag-LZ-210 zeolite is exchanged by trace metals in the acetic acid. This causes a decrease in the ability of the adsorbent to remove the iodide. To protect the capacity of the Ag-LZ-210 adsorbent for adsorption of iodide, either H-LZ-210 or a cationic exchange resin is used [252].

Another application for adsorption of metal impurities is in the nuclear power industry. Radioactive cesium is one of the compounds that is difficult to remove from radioactive waste. This is because ordinary resins and zeolites do not effectively adsorb radioactive cesium. In 1997, IONSIV IE-911™ crystalline silicotitanate (CST) ion exchangers were developed and effectively used to clean up radioactive wastes in the Melton Valley tanks at Oak Ridge [268, 269]. CST was discovered [270] by researchers at Sandia National Laboratories and Texas A&M University, with commercial manufacture carried out by UOP.

5.4
Summary Review of Zeolites in Adsorptive Separation

The most commonly employed crystalline materials for liquid adsorptive separations are zeolite-based structured materials. Depending on the specific components and their structural framework, crystalline materials can be zeolites (silica, alumina), silicalite (silica) or AlPO-based molecular sieves (alumina, phosphorus oxide). Faujasites (X, Y) and other zeolites (A, ZSM-5, beta, mordenite, etc.) are the most popular materials. This is due to their narrow pore size distribution and the ability to tune or adjust their physicochemical properties, particularly their acidic–basic properties, by the ion exchange of cations, changing the SiO_2/Al_2O_3 ratio and varying the water content. These techniques are described and discussed in Chapter 2. By adjusting the properties almost an infinite number of zeolite materials and desorbent combinations can be studied.

The survey summaries show that zeolite adsorbents are most often employed for non-aqueous systems. This is because the material generally used as a binder to fabricate an agglomerated zeolite, is a clay comprising silicon dioxide and aluminum oxide which tends to dissolve in water. This dissolution results in negative changes in physical characteristics of the adsorbent as well as silicon contamination of the solution which manifests itself as turbidity in the product.

A proven solution to the binder problem is to use water insoluble organic polymer binders instead of clay. For example cellulose acetates and cellulose ethers binders are successfully employed to make commercial zeolitic adsorbents for sugar separation in aqueous solutions [154, 205, 218–223, 225–226, 231–232, 238]. This technique allows the use of zeolite adsorbents in aqueous separation processes.

Silicalite is another crystalline material whose use in liquid adsorptive separation is found in the surveys. Applications for silicalite are noted in categories 1 and 2 below with the main applications falling into category 2 due to the small pore size aperture of its MFI structure and low electrostatic field compared to those of aluminosilicates. Examples of the separation of species with a lower electron capability from others with a higher capability include:

1) The separation of alcohol from fermentation broth [196, 200] and saturated from unsaturated species [175, 187–189, 193, 205].
2) The separation of hydrocarbons in accordance with their molecular sizes [31, 49, 142, 146, 149, 154, 157, 162, 163, 167, 178, 183, 207].

The surveys show that advances in adsorptive separations come not only from exploring new pairings of established materials and desorbents but also through modification of a zeolite's physicochemical characteristics and synthesis of novel zeolitic materials. Much work is devoted to seeking advancements by these last two means. This work will inevitably lead to the development of new applications and establishment of new industrial liquid adsorptive separation processes.

Acknowledgments

Parts of this chapter are reprinted from the chapter "Liquid separations" by Santi Kulprathipanja and James A. Johnson, in *Handbook of Porous Solids*, edited by Ferdi Schuth, Kenneth S.W. Sing and Jens Weitkamp (2002), Wiley–VCH, with permission.

References

1 Aoki, S., Kato, T., Taniguchi, H., Morotomi, H., and Ono, M. (1994) Method of removing alkyl tetralin coexisting in alkylbenzene. Jpn Patent 5221885.
2 Aoki, S., Kato, T., Taniguchi, H., Morotomi, H., and Ono, M. (1995) Method of removing alkyltetralins from an alkylbenzene mixture. U.S. Patent 5,414,191.
3 Miwa, K., Nagaoka, Y., and Inoue, T. (1983) Separation of chlorobenzotrifluoride isomers. Jpn Patent 58131922.
4 Iwayama, K., Yamakawa, S., Kato, M., and Okino, H. (1999) Use of an adsorbent for separating halogenated aromatic compounds and method of separation using it. Eur. Patent 948,988.
5 Minomiya, E. and Miyata, S. (2001) Method for isomerizing halogenated ethylbenzene and method for separating halogenated ethylbenzene isomers. U.S. Patent 6,235,952.
6 Iwayama, K., Yamakawa, S., Kato, M., and Okino, H. (2001) Adsorbent for separating halogenated aromatic compounds and separation method. U.S. Patent 6,320,087.
7 Suzuki, M., Maeda, S., and Tada, K. (1994) Method for separating 2-chloro-4-fluorotoluene. Jpn Patent 5320079.
8 Priegnitz, J.W. (1981) Process for the separation of ortho-chloronitrobenzene. U.S. Patent 4,240,986.
9 Guo, Z., Zheng, S., Zheng, Z., Jiang, F., Hu, W., and Ni, L. (2005) Selective adsorption of p-chloronitrobenzene from aqueous mixture of p-chloronitrobenzene and o-chloronitrobenzene using HZSM-5 zeolite. *Water Res.*, **39**, 1174.

10 Nakatani, J., Minomiya, E., Inohara, M., Iwayama, K., and Kato, T. (2002) Method for producing aromatic compounds having alkyl group with at least three carbon atoms. U.S. Patent 6,462,248.
11 Shigehara, I., Yoshizawa, H., and Maeda, M. (1988) Separation of monohalogenated phenol. Jpn Patent 62273929.
12 Zinnen, H.A. (1991) Separation of coumarone from indene. U.S. Patent 4,992,621.
13 Yue, Y.H., Tang, Y., Liu, Y., and Gao, Z. (1996) Chemical liquid deposition zeolites with controlled pore-opening size and shape-selective separation of isomers. *Ind. Eng. Chem. Res.*, **35**, 430.
14 Lee, K.R. and Tan, C.S. (2000) Separation of m- and p-cresols in compressed propane using modified HZSM-5 pellets. *Ind. Eng. Chem. Res.*, **39**, 1035.
15 Raychoudhuri, A. and Gaikar, V.G. (1995) Adsorptive separations of 2,6-xylenol/cresol mixtures with zeolites. *Sep. Technol.*, **5**, 91.
16 Zinnen, H.A. (1994) Adsorptive separation of cresol isomers. Eur. Patent 587,949.
17 Neuzil, R.W. (1983) Process for the separation of cresol from xylenol. U.S. Patent 4,386,225.
18 Neuzil, R.W. and Rosback, D.H. (1976) Process for the separation of cresol isomers. U.S. Patent 3,969,422.
19 Nixon, J.R. (1973) p-and m-cymene separation. Ger. Patent DE 2,239,423.
20 Matsunaga, K., Pponda, H., and Tada, K. (1995) Separation of p-toluidine. Jpn Patent 06247907.
21 Kanai, T., Kimura, M., and Noguchi, Y. (1993) Separation of dichlorobenzene isomer. Jpn Patent 04330025.
22 Ishiyama, K. and Watanabe, M. (1999) Separation of dichlorobenzene isomer. Jpn Patent 11180911.
23 Watanabe, M., Yoshikawa, M., and Iwayama, K. (2001) Adsorbent for aromatic isomers and production of aromatic isomers. U.S. Patent 6,262,322.
24 Kato, M., Kato, Y., and Okada, K. (2006) Method for producing aromatic compound isomers. U.S. Patent 7,091,389.
25 McCulloch, B. and Gatter, M.G. (1991) Process for extracting meta-dichlorobenzene from isomer mixtures with mixed alkali metal exchanged X zeolite adsorbents. U.S. Patent 4,996,380.
26 Miwa, K., Watanabe, M., and Tada, K. (1987) Separation of dichlorobenzene isomer. Jpn Patent 61268636.
27 Guo, G., Long, Y., and Sun, Y. (2001) Hydrophobic silicalite method for liquid-phase selective adsorption, separation and mixing of dichlorobenzene. C.N. Patent 1,315,217.
28 Yokota, Z., Wakamatsu, K., and Shimura, M. (1998) Separation of paradichlorobenzene. Jpn Patent 10017504.
29 Pentling, U., Buysch, H.J., Puppe, L., Pies, M., and Paul, H.I. (1995) Process for the separation of mixtures of m- and p-dichlorobenzene. Ger. Patent DE 4,325,484.
30 Pentling, U., Buysch, H.J., Puppe, L., Roehik, K., Grosser, R., and Paul, H.I. (1994) Process for separating mixtures of m- and p-dichlorobenzene. German Patent DE 4,218,841.
31 Guo, G. and Long, Y. (2001) Static equilibrium studies on separation of dichlorobenzene isomers on binder-free hydrophobic adsorbent of MFI type zeolite. *Sep. Purif. Technol.*, **24**, 507.
32 Yamada, B., Kimura, M., and Noguchi, Y. (1990) Separation of bromodichlorobenzene isomer. Jpn Patent 01213243.
33 Bunji, Y., Kimura, M., and Noguchi, Y. (1991) Separation of dichlorocumene isomer. Jpn Patent 02229122.
34 Suzuki, M. and Tada, K. (1992) Separation of 2,4-dichlorofluorobenzene. Jpn Patent 04013639.
35 Zinnen, H.A. and Issa, K.C. (1992) Separation of dichlorophenol isomers with zeolite adsorbents. U.S. Patent 5,118,876.
36 Kanai, T., Ichanagi, H., and Noguchi, Y. (1993) Method for separating 2,4-dichlorotoluene. Jpn Patent 05070383.
37 Yoshikawa, T., Watanabe, M., and Iwayama, K. (2004) Method for

separating dichlorophenol. Jpn Patent 2004238391.
38 Maeda, M. and Imada, H. (1997) Separation of 2,4-dichlorotoluene or 2,6-dichlorotoluene. Jpn Patent 09316015.
39 Miwa, K., Nagaoka, Y., Inoue, T., and Tada, K. (1988) Adsorptive separation process. U.S. Patent 4,774,371.
40 McCulloch, B. and Oroskar, A.R. (1990) Process for extracting 2,5-dichlorotoluene from isomer mixtures with sodium-L zeolite adsorbents. U.S. Patent 4,922,040.
41 Imada, Y. and Maeda, S. (1997) Separation of 2,4-dichlorotoluene or 2,6-dichlorotoluene. Jpn Patent 09316014.
42 Shirato, Y., Shimokawa, K., and Shimura, M. (1991) Separation of 2,6-and 2,5-dichlorotoluene. Jpn Patent 03020232.
43 Wambach, R., Hartung, S., Reiss, G., and Puppe, L. (1981) Separation process. U.S. Patent 4,254,062.
44 Miwa, K., Nagaoka, Y., and Tada, K. (1985) Separation of 3,5-dichlorotoluene. Jpn Patent 60013727.
45 Miwa, K., Nagaoka, Y., Tada, K., and Inoue, T. (1985) Separation of diethylbenzene isomer. Jpn Patent 60025941.
46 Choi, J.T., Kim, C.G., Kim, C.Y., Kim, M.J., Kwak, B.S., and Lee, S.J. (2003) Process for separating 1,2-diethylbenzene from amixture of diethylbenzene isomers. K.R. Patent 2003-046,548.
47 Xiao, G. and Wu, P. (1997) *Shiyou Huagong*, **26**, 509.
48 Xiao, G., Zou, X., and Wu, P. (2002) *Shiyou Huagong*, **31**, 616.
49 Tan, C. and Chiang, S. (1994) Separation of diethylbenzene isomers on silicalite in the presence of high pressure carbon dioxide and propane. U.S. Patent 5,336,837.
50 Choi, S., Jang, S.U., Kim, S.J., Kim, Y.S., Lee, S.J., and Son, J.Y. (2003) Method for separating 1,4-diethylbenzene and1,3-diethylbenzene from diethylbenzene isomer mixture. K.R. Patent 2003-069,482.
51 Takeuchi, G., Kariu, K., and Shiroshita, M. (1994) Method for separating 2,6-diethylnaphthalene. U.S. Patent 5,300,721.
52 Takeuchi, H., Kario, K., and Shiroshita, M. (1991) Separation of 2,6-diethylnaphthalene. Jpn Patent 03204826.
53 Yoshimizu, M. and Takeuchi, H. (1992) Production of 2,6-napthalenedicarboxylic acid. Jpn Patent 03258746.
54 Zinnen, H.A. (1990) Chromatographic separation process for recovering individual diethyltoluene isomers. U.S. Patent 4,940,548.
55 Zinnen, H.A. (1991) Extractive chromatographic separation process for recovering 3,5-diethyltoluene. U.S. Patent 5,019,271.
56 Senzaki, T., Takayama, T., and Hara, T. (1998) Separation of isopropenylbenzenes. Jpn Patent 10182512.
57 Morita, M., Takeuchi, H., Kario, K., and Kamei, T. (1990) Separation of 4,4-dialkylbiphenyl. Jpn Patent 01249729.
58 Fellmann, J.D., Wentrcek, P.R., and Kilner, P.H. (1991) Selective sorption of dialkylated multinuclear aromatic compounds. W.O. Patent 9,101,286.
59 Ootani, S., Taniguchi, H., Shirato, Y., and Shimura, M. (1993) Separation of 2, 6-or 2,7-diisopropylnaphthalene. Jpn Patent 05039230.
60 Shirato, Y., Shimura, M., Shimokawa, K., Fukui, Y., Tachibana, Y., Tate, K., and Taniguchi, H. (1991) Separation of 2,7-diisopropylnaphthalene. Jpn Patent 02172931.
61 Barder, T. (1991) Separation of 2,7 diisopropylnaphthalene from a feed mixture comprising various diisopropylnaphthalene isomers with a zeolite adsorbent. U.S. Patent 5,012,039.
62 Shirato, Y., Shimokawa, K., Kaita, J., Fukui, Y., and Hirohama, S. (1989) Adsorption and separation of 2,6-diisopropylnaphthalene. Jpn Patent 63243040.
63 Verduijn, J., Janssen, M., De, G.C., Koetsier, W., and Van, O.C. (1992) Process for separating dimethylnaphthalene isomers with zeolite L agglomerates. U.S. Patent 5,146,040.
64 Nagao, S. and Ogawa, H. (2006) Method of separating Dimethylnaphtha-

lene isomers. W.O. Patent 2006-109,667.
65 Taniguchi, K. and Fujimoto, T. (1988) Method for separating dimethylnaphthalene isomer. Jpn Patent 62240632.
66 Hobbs, S.H. and Barder, T.J. (1989) Two-stage adsorptive separation process for purifying 2, 6. Dimethylnaphthalene. U.S. Patent 4,835,334.
67 Munson, C., Bigot, P., and He, Z.A. (2000) Method for producing 2,6-DMN from mixed dimethylnaphthalenes by crystallization, adsorption and isomerization. U.S. Patent 6,057,487.
68 Kraikul, N., Rangsunvigit, P., and Kulprathipanja, S. (2006) Study on the adsorption of 1,5-1, 6- and 2,6-dimethylnaphthalene on a series of alkaline and alkaline earth ion-exchanged faujasite zeolites. *Adsorption*, 12, 317.
69 Nakao, N., Yamamoto, K., and Motoyuki, M. (2004) Method for concentrating 2,6-dimethylnaphthalene. U.S. Patent 6,706,939.
70 Barger, P.T., Barder, T.J., Lin, D.Y., and Hobbs, S.H. (1991) Continuous process for the production of 2,6-dimethylnaphthalene. U.S. Patent 5,004,853.
71 Motoyuki, M., Yamamoto, K., Sapre, A.V., McWilliams, J.P., and Donnelly, S.P. (2000) Process for preparing 2,6-dialkylnaphthalene. U.S. Patent 6,121,501.
72 Nagao, S. and Ogawa, H. (2006) Method of separating Dimethylnaphthalene isomers. W.O. Patent 2006-068,174.
73 Morimoto, M., Miyata, A., and Ishikawa, M. (1997) Separation of xylenol isomer. Jpn Patent 09151147.
74 Shirato, Y., Shimokawa, K., and Shimura, M. (1991) Separation of 3, 5-xylenol. Jpn Patent 03020234.
75 Zinnen, H.A. and Franczyk, T.S. (1988) Process for separating the minor isomers of dinitrotoluene. U.S. Patent 4,717,778.
76 Zinnen, H.A. and Franczyk, T.S. (1987) Process for separating isomers of dinitrotoluene. U.S. Patent 4,642,397.
77 Senzaki, T., Noguchi, K., and Shiraishi, K. (1998) Separation of divinylbenzenes. Jpn Patent 10182511.
78 Rosenfeld, D.D. and Daniel, L.G. (1988) Process for the separation of C10 aromatic isomers. U.S. Patent 4,743,708.
79 Kulprathipanja, S., Kuhnle, K.K., and Patton, M.S. (1993) Process for the separation of C.sub.10 aromatic isomers. U.S. Patent 5,223,589.
80 Barthomeuf, D.M. (1986) Process for separating ethylbenzene from xylenes by selective adsorption on a Beta zeolite. U.S. Patent 4,584,424.
81 Barthomeuf, D.M. and Daniel, L.G. (1989) Process for separating ethylbenzene from xylenes by selective adsorption on a gallium beta zeolite (ATD-35). U.S. Patent 4,751,346.
82 Exxon Research and Engineering Co. (1977) Zeolite adsorbent for separating hydrocarbon mixtures. Neth. Patent 7602348.
83 Kulprathipanja, S. (1995) Process for adsorptive separation of ethylbenzene from aromatic hydrocarbons. U.S. Patent 5,453,560.
84 Barthomeuf, D.M. (1986) Process for separating ethylbenzene from xylenes by selective adsorption on a cesium substituted X zeolite. U.S. Patent 4,593,149.
85 Barthomeuf, D.M. (1986) Process for separating ethylbenzene from xylenes by selective adsorption on a rubidium substituted X zeolite. U.S. Patent 4,613,725.
86 Neuzil, R.W. and Rosback, D.H. (1976) Process for the separation of ethylbenzene by selective adsorption on a zeolitic adsorbent. U.S. Patent 3,943,182.
87 Neuzil, R.W. and Rosback, D.H. (1977) Separation of ethylbenzene with an adsorbent comprising Sr and K exchanged X or Y zeolite. U.S. Patent 3,998,901.
88 Neuzil, R.W. and Rosback, D.H. (1977) Process for the separation of ethylbenzene. U.S. Patent 4,028,428.
89 Neuzil, R.W. (1976) Process for the separation of ethylbenzene. U.S. Patent 3,997,619.
90 Yoshimizu, M., Yamaguchi, T., and Fujii, H. (1993) Separation of 2-ethylnaphthalene. Jpn Patent 05163169.

91 Inoe, S., Mifuji, Y., and Nagae, S. (1993) Adsorbent excellent in selectivity to m-ethylphenol. Jpn Patent 05253476.
92 Suzuki, M., Morimoto, M., and Ishikawa, M. (1996) Separation of alkylphenol isomer. Jpn Patent 08245457.
93 Waters, J.A. (1998) Recovery of para-ethylphenol. U.S. Patent 5,744,654.
94 Matsumoto, M., Hakozaki, T., Idai, T., Shimura, M., Yokota, Y., and Shiroto, Y. (1996) U.K. Patent 2,291,056.
95 Suzuki, M., Morimoto, M., and Ishikawa, M. (1996) Separation of ethylphenol isomer. Jpn Patent 08291096.
96 Neuzil, R.W. (1985) Separation of ortho bi-alkyl substituted monocyclic aromatic isomers. U.S. Patent 4,482,777.
97 Olken, M.M., Lee, G.S.J., and Garces, J.M. (1991) Separation of ethyltoluene isomers. U.S. Patent 4,996,388.
98 Zinnen, H.A. (1991) Zeolitic para-ethyltoluene separation with tetralin heavy desorbent. U.S. Patent 4,956,522.
99 Barthomeuf, D.M. and Rosenfeld, D.D. (1986) Process for the separation of C.sub.9 aromatic isomers. U.S. Patent 4,554,398.
100 Rule, M. and Browning, H.L., Jr. (1989) Selective adsorption/separation of diiodonaphthalenes. U.S. Patent 4,814,526.
101 Aoki, S. (1997) Separation of monoisopropylbiphenyl isomer. Jpn Patent 09067275.
102 Kito, T. and Takeuchi, H. (1991) Separation of alkyldiphenylene oxides. Jpn Patent 02243683.
103 Shirato, Y., Shimura, M., Shimokawa, K., and Hirohama, S. (1989) Production of 2,6-methylisopropylnaphthalene. Jpn Patent 63243043.
104 Yue, Y., Tang, Y., Kan, Y., and Gao, Z. (1996) Pore size control of NaY zeolite by chemical liquid deposition. *Huaxue Xuebao*, **54**, 591.
105 Zinnen, H.A. and Franczyk, T.S. (1988) Separation of nitrobenzaldehyde isomers. U.S. Patent 4,714,783.
106 Kimura, M. and Noguchi, Y. (1987) Method of separating nitrobenzaldehyde isomer. Jpn Patent 61130259.
107 Wimmer, P., Buysch, H.J., Grosser, R., and Puppe, L. (1992) Process for separating mixtures of nitrobenzaldehyde isomers. German Patent DE 4033613.
108 Zinnen, H.A. (1986) Process for the separation of ortho-nitrotoluene. U.S. Patent 4,620,047.
109 Aoki, S. and Ono, M. (1997) Jpn Patent 09067276.
110 Yamada, B., Kimura, M., and Noguchi, Y. (1990) Separation of t-butylphenol isomer. Jpn Patent 01211541.
111 Kulprathipanja, S., Marinangeli, R.E., Sohn, S.W., Fritsch, T.R., and Lawson, R.J. (2003) Process for producing phenyl-alkanes using an adsorptive separation section. U.S. Patent 6,617,481.
112 Bricker, M.L., McGonegal, C.P., and Zinnen, H.A. (2004) Process for separating alkylaromatic hydrocarbons. U.S. Patent 6,673,979.
113 Zinnen, H.A. (1989) Process for separating a mixture comprising 2,6-toluene diisocyanate and 2,4-toluene diisocyanate. Eur. Patent 324215.
114 Zinnen, H.A. (1987) Process for separating isomers of toluenediamine. U.S. Patent 4,633,018.
115 Priegnitz, J.W. and Zinnen, H.A. (1985) Process for separating isomers of toluidine. U.S. Patent 4,480,129.
116 Otomo, K., Yamaguchi, M., Ito, M., and Tokunaga, H. (1990) Method for separating a trihalogenobenzene isomer. Eur. Patent 324041.
117 UOP Inc. (1977) Separation of tetraalkylsubstituted aromatic hydrocarbon isomers. Jpn Patent 51143624.
118 Campbell, D.R. and Priegnitz, J.W. (1975) Separation of tetra-alkyl substituted aromatic hydrocarbon isomers. U.S. Patent 3,864,416.
119 Kulprathipanja, S. (1999) Process for adsorptive separation of metaxylene from xylene mixtures. U.S. Patent 5,900,523.
120 Ichikawa, Y. and Iwata, K. (1977) Jpn Patent 52000933.
121 Shioda, T., Yasuda, H., Saito, N., Asatari, H., and Furuichi, H. (1995)

Separation of m-xylene. Jpn Patent 07033689.
122 Neuzil, R.W. (1982) Process for the separation of meta-xylene. U.S. Patent 4,326,092.
123 Kulprathipanja, S. (1995) Process for the adsorptive separation of metaxylene from aromatic hydrocarbons. U.S. Patent 5,382,747.
124 Zinnen, H.A. and McGonegal, C.P. (2000) Process for separating meta-xylene. U.S. Patent 6,137,024.
125 Lucena, S.M.P., Cavalcante, C.L., Jr., and Pereira, J.A.F.R. (2006) Ortho-selectivity in aluminophosphate molecular sieves: a molecular simulation study. *Adsorption*, **12**, 423.
126 Antos, G.J. and Flint, N.J. (1985) Separation of ortho-xylene. U.S. Patent 4,529,828.
127 Leflaive, P., Dubreuil, A.C., Caullet, P., Patarin, J., and Paillaud, J.L. (2006) Process for separation by selective adsorption on a solid containing a zeolite with a crystalline structure analogous to IM-12. F.R. Patent 2,877,237.
128 Neuzil, R.W. (1976) Process for separating para-xylene. U.S. Patent 3,997,620.
129 Kulprathipanja, S. (1984) Separation of bi-alkyl substituted monocyclic aromatic isomers with pyrolyzed adsorbent. U.S. Patent 4,423,279.
130 Zinnen, H.A. (1990) Zeolitic para-xylene separation with diethyltoluene heavy desorbent. U.S. Patent 4,864,069.
131 Furlan, L.T., Chaves, B.C., and Santana, C.C. (1992) Separation of liquid mixtures of p-xylenes and o-xylenes in X zeolites: the role of water content on the adsorbent selectivity. *Ind. Eng. Chem. Res.*, **31**, 1780.
132 Zinnen, H.A. (1993) Zeolitic para-xylene separation with indan and indan derivatives as heavy desorbent. U.S. Patent 5,159,131.
133 Kulprathipanja, S. (1998) Adsorptive separation of para-xylene using isopropylbenzene desorbent. U.S. Patent 5,849,981.
134 Cain, J.J. (2002) Process for separation of para-xylene using an ether as desorbent. E.P. Patent 1,165,471.
135 Plee, D. and Methivier, A. (2002) Agglomerated zeolite adsorbents, process for their preparation, and their use for adsorbing paraxylene from aromatic C8 fractions. U.S. Patent 6,410,815.
136 Roeseler, C.M., Kulprathipanja, S., and Rekoske, J.E. (2004) Adsorptive separation process for recovery of para-xylene. U.S. Patent 6,706,938.
137 Plee, D. and Methivier, A. (2005) Agglomerated zeolitic adsorbents, method for obtaining same uses thereof. U.S. Patent 6,884,918.
138 Leflaive, P., Wolff, L., Hotier, G., and Methivier, A. (2005) Process for co-production of paraxylene, metaxylene and/or orthoxylene. U.S. Patent 6,841,714.
139 Neuzil, R.W. and Antos, G.J. (1994) Use of a fluro-aromatic desorbent in a process for adsoptive separation of para-alkylaromatic hydrocarbons. Can. Patent 1322176.
140 Neuzil, R.W. and Antos, G.J. (1990) Separation of para-xylene. U.S. Patent 4,886,929.
141 Cain, J.J. (2001) Adsorption process for paraxylene purifacation using Cs SSZ-25 adsorbent with benzene desorbent. U.S. Patent 6,281,406.
142 Guo, G. and Long, Y. (2001) Process for separating p-xylene with hydrophobic silicic zeolite by selective adsorption. C.N. Patent 1,280,977.
143 Zinnen, H.A. and Fergin, R.L. (1990) Rejective separation of para-xylene from xylene isomers and ethylbenzene with zeolites. U.S. Patent 4,940,830.
144 Daems, I., Leflaive, P., Methivier, A., Baron, G.V., and Denayer, J.F.M. (2006) Influence of Si:Al-ratio of faujasites on the adsorption of alkanes, alkenes and aromatics. *Micropor. Mesopor. Mater.*, **96**, 149.
145 Takahashi, A., Yang, R.T., Munson, C.L., and Chinn, D. (2001) Cu(I)–Y-zeolite as a superior adsorbent for diene/olefin separation. *Ind. Eng. Chem. Res.*, **40**, 3979.
146 Kulprathipanja, S. (2001) Process for monomethyl acyclic hydrocarbon adsorptive separation. U.S. Patent 6,252,127.

147 Priegnitz, J.W. (1973) The selective separation of Butene-1 from a hydrocarbon mixture employing zeolites X and Y. U.S. Patent 3,723,561.

148 Neuzil, R.W. and Fergin, R.L. (1979) Desorbent for separation of butene-1 from a C4 hydrocarbon mixture using zeolite X. U.S. Patent 4,119,678.

149 Kulprathipanja, S. (1984) Separation of trans- and cis-olefins. U.S. Patent 4,433,195.

150 UOP Inc. (1978) Jpn Patent 52027123.

151 Owaysi, F.A. and Al-Ameeri, R.S. (1986) Purification of liquid paraffins. Eur. Patent 164,905.

152 Neuzil, R.W. (1972) Selectively adsorbing multibranched paraffins. U.S. Patent 3,706,813.

153 Park, J.Y.G., Kulprathipanja, S., and Haizmann, R.S. (1999) Zeolitic reforming with selective feed-species adjustment. U.S. Patent 5,922,923.

154 Kulprathipanja, S. (1991) Process for separating glucose and mannose with dealuminated Y zeolites. U.S. Patent 5,000,794.

155 Namba, S., Yoshimura, A., and Yashima, T. (1979) Separation of 2-methylpentane and 2,2-dimethylbutane by means of shape-selective adsorption into modified H-mordenite. *Chem. Lett.*, **7**, 759.

156 Ducreux, O. and Jolimaitre, E. (2004) Process combining hydroisomerisation and separation using a zeolitic adsorbent with a mixed structure for the production of high octane number gasolines. U.S. Patent 6,809,228.

157 Sohn, S., Kulprathipanja, S., and Rekoske, J.E. (2003) Monomethyl paraffin adsorptive separation process. U.S. Patent 6,670,519.

158 Kulprathipanja, S., Kuhnle, K., Patton, M.S., and Fergin, R.L. (1993) Process for separating naphthalene from substituted benzene hydrocarbons. U.S. Patent 5,177,300.

159 Yusubov, F.V., Zeinalov, R.I., Ibragimov, V.S., and Babaev, R.K. (2006) Liquid-phase adsorption of binary mixtures of alkanes. *Chem. Technol. Fuels Oils*, **42**, 293.

160 Rosback, D.H. (1973) Adsorbing olefins wih a copper-exchanged type Y zeolite. U.S. Patent 3,720,604.

161 Rosback, D.H. and Neuzil, R.W. (1977) Olefin separation process. U.S. Patent 4,036,744.

162 Funk, G.A., Lansbarkis, J.R., Oroskar, A.R., and McCulloch, B. (1993) Process for separating normal olefins from non-normal olefins. U.S. Patent 5,220,102.

163 Kulprathipanja, S. and Neuzil, R.W. (1985) Process for separating C.sub.6 olefin hydrocarbons. U.S. Patent 4,486,618.

164 Kulprathipanja, S., Priegnitz, J.W., Sohn, S.W., Glover, B.K., and Vora, B.V. (2006) Process for removal of oxygenates from a paraffin stream. U.S. Patent 7,102,044.

165 Ragil, K., Bailly, M., Jullian, S., and Clause, O. (2002) Process for chromatographic separation of a C5-C8 feed or an intermediate feed into three effluents, respectively rich in straight chain, mono-branched and multi-branched paraffins. U.S. Patent 6,353,144.

166 Gadzhiev-Shengeliya, M.K., Areshidze, G.Kh., Datashvili, T.K., and Bezarashvili, G.S. (2005) Influence of conditions of the process on adsorption dewaxing of straight-run gasoline. *Pet. Chem.*, **45**, 404.

167 Kulprathipanja, S. and Neuzil, R.W. (1983) Process for separating normal paraffins using silicalite adsorbent. U.S. Patent 4,367,364.

168 Borisova, L.V. and Mirskii, Y.V. (1972) *Tr. Grozn. Neft. Nauch-Issled. Inst.*, **25**, 90.

169 Kulprathipanja, S. (1991) Adsorptive separation process for the purification of heavy normal paraffins with non-normal hydrocarbon pre-pulse stream. U.S. Patent 4,992,618.

170 Tomas-Alonso, F., Angosto, L.A. Olmos, and Munecas Vidal, M.A. (2004) Selective separation of normal paraffins from slack wax using the molecular sieve adsorption technique. *Sep. Sci. Technol.*, **39**, 1577.

171 Asher, W.J., Campbell, M.L., Epperly, W.R., and Robertson, J.L. (1969) Desorb n-paraffins with ammonia. *Hydrocarbon Process.*, **48**, 134.

172 Bolotov, L.T., Pereverzev, A.N., Filippova, T.F., and Borisova, L.V.

(1981) Feedstocks for the adsorptive recovery of liquid paraffins. *Khim. Tekhnol. Topl. Masel.*, **12**, 5.

173 Mirskii, Y.V., Votlokhin, Y.Z., Breshchenko, E.M., Leont'ev, A.S., Remova, M.M., Basin, B.Y., Makar'ev, S.V., Zamanov, V.V., and Khadzhiev, S.N. (1978) *Adsorbenty, Ikh Poluch., Svoistva Primen., Tr. Vses. Soveshch. Adsorbentam*, **4**, 175–179.

174 McCulloch, B. and Lansbarkis, J.R. (1994) Process for separating normal olefins from non-normal olefins. U.S. Patent 5,276,246.

175 Padin, J. and Yang, R.T. (2000) New sorbents for olefin/paraffin separations by adsorption via π-complexation: Synthesis and effects of substrates. *Chem. Eng. Sci.*, **55**, 2607.

176 Arai, M. and Fukuda, H. (1989) Concentration and separation of unsaturated fatty acid esters. Jpn Patent 63264555.

177 Zinnen, H.A. (1989) Process for separating di- and triglycerides. U.S. Patent 4,770,819.

178 Cleary, M.T., Kulprathipanja, S., and Neuzil, R.W. (1985) Process for separating fatty acids from rosin acids. U.S. Patent 4,522,761.

179 Cleary, M.T., Kulprathipanja, S., and Neuzil, R.W. (1985) Process for separating fatty acids from unsaponifiables. U.S. Patent 4,519,952.

180 Cleary, M.T., Kulprathipanja, S., and Neuzil, R.W. (1985) Process for separating fatty acids. U.S. Patent 4,524,029.

181 Üstün, G. (1996) Separation of fatty acid methyl esters from tall oil by selective adsorption. *JAOCS*, **73**, 203.

182 Jacobs, P.A., Maes, P., Paulussen, S.J., Tielen, M., Van Steenkiste, D.F.E., and Van Looveren, L.K. (2001) Elimination of trans-unsaturated fatty acid compounds by selective adsorption with zeolites. U.S. Patent 6,229,032.

183 Cleary, M.T., Kulprathipanja, S., and Neuzil, R.W. (1982) Process for separating esters of fatty and rosin acids. U.S. Patent 4,329,280.

184 deRosset, A.J. (1980) Two-stage process for separating mixed fatty-acid esters. U.S. Patent 4,213,913.

185 Zinnen, H.A. (1989) Process for separating mono-, di- and triglycerides. U.S. Patent 4,797,233.

186 Zinnen, H.A. (1992) Process for separating fatty acids and triglycerides. U.S. Patent 5,102,582.

187 Cleary, M.T., Kulprathipanja, S., and Neuzil, R.W. (1985) Process for separating oleic acid from linoleic acid. U.S. Patent 4,529,551.

188 Cleary, M.T., Kulprathipanja, S., and Neuzil, R.W. (1985) Process for separating oleic acid from linoleic acid. U.S. Patent 4,511,514.

189 Patton, R.L., McCulloch, B., and Nickl, P.K. (2000) Chromatographic separation of fatty acids using ultrahydrophobic silicalite. U.S. Patent 6,013,186.

190 Ou, J.D. (1991) Process for separating triglycerides and regenerating absorbent used in said separation process. U.S. Patent 4,961,881.

191 Zinnen, H.A. (1993) Process for separating fatty acids and. Triglycerides. U.S. Patent 5,225,580.

192 Mueller, U., Weiss, R., Diehl, K., Sandrick, G., and Sauvage, L. (1994) Removal of acetaldehyde from hydrocarbons. German Patent DE 4,226,302.

193 Groszek, A.J. (1985) Process for separating a carboxylic acid of 1–8 carbon atoms from a mixture thereof with water and/or one or more other oxygenated aliphatic compounds. Eur. Patent 132,049.

194 O'Brien, J., Curtin, T., and O'Dwyer, T.F. (2005) An investigation into the adsorption of aniline from aqueous solution using H-beta zeolites and copper-exchanged beta xeolites. *Adsorpt. Sci. Technol.*, **23**, 255.

195 O'Brien, J., Curtin, T., and O'Dwyer, T.F. (2004) Adsorption of aniline from aqueous solution using copper-exchanged ZSM-5 and unmodified H-ZSM-5. *Adsorpt. Sci. Technol.*, **22**, 743.

196 Qureshi, N., Hughes, S., Maddox, I.S., and Cotta, M.A. (2005) Energy efficient recovery of butanol from model solutions and fermentation broth by adsorption. *Bioprocess Biosyst. Eng.*, **27**, 215.

197 Miller, J.F. and Keller, G.E. II (1999) Separation of epsion caprolactam from isomers. W.O. Patent 9,965,873.
198 Miller, J.F. (2000) Separation processes. U.S. Patent 6,045,703.
199 Chao, C.C. and Sherman, J.D. (1982) Bulk cyclohexanol/cyclohexanone separation by selective adsorption on zeolitic molecular sieves. U.S. Patent 4,283,560.
200 Farhadpour, F.A., Bono, A., and Tuzun, U. (1984) Separation of alcohol-water mixtures by liquid phase adsorption. Monogr. Eur. Brew. Conv., 9, 203.
201 Buysch, H.J., Huellmann, M., and Puppe, L. (1989) Process for separating aniline derivatives. EP Patent 300,285.
202 Chou, Y.C.T. (1986) Process for separation of ethylene glycol and propylene glycol on selected zeolites. U.S. Patent 4,588,847.
203 Chemical Industry, A. and Japan, C., L. (1984) Novel process for separation of indole. Jpn Patent 59013758.
204 Sextl, E., Kiss, A., Kinz, H., Schaefer-Treffenfeldt, W., Yonsel, S., and Stockhammer, S. (1995) Process for isolating hydroxymonocarboxylic and tricarboxylic acids. U.S. Patent 5,676,838.
205 Kulprathipanja, S. (1990) Process for the adsorptive separation of hydroxy paraffinic dicarboxylic acids from olefinic dicarboxylic acids. U.S. Patent 4,902,829.
206 Karas, L.J., Candela, L.M., and Kahn, A.P. (2006) Purification of methyl tertiary butyl ether. U.S. Patent 7,022,885.
207 Cleary, M.T. (1986) Process for separating saturated fatty acids from each other. U.S. Patent 4,578,223.
208 Khalid, M., Joly, G., Renaud, A., and Magnoux, P. (2004) Removal of phenol from water by adsorption using zeolites. Ind. Eng. Chem. Res., 43, 5275.
209 Titus, E., Kalkar, A.K., and Gaikar, V.G. (2003) Equilibrium studies of adsorption of amino acids on NaZSM-5 zeolite. Colloids Surf., A, 223, 55.
210 Zinnen, H.A. (1986) Separation of picoline isomers. U.S. Patent 4,594,423.
211 deRosset, A.J. and Neuzil, R.W. (1974) Process for the separation of pinene isomers. U.S. Patent 3,851,006.
212 Carmo, M.J. and Gubulin, J.C. (1997) Ethanol-water separation in the PSA process. Braz. J. Chem. Eng., 14, 217.
213 Al-Asheh, S., Banat, F., and Al-Lagtah, N. (2004) Separation of ethanol–water mixtures using molecular sievesand biobased adsobents. Chem. Eng. Res. Des., 82, 855.
214 Chang, C.H. (1987) Process for separating arabinose from a pentose/hexose mixture. U.S. Patent 4,664,718.
215 Chang, C.H. (1991) Process for separating arabinose. C.A. Patent 1,288,095.
216 Kulprathipanja, S. (1990) Process for separating arabinose from a mixture of other aldoses. U.S. Patent 4,857,642.
217 Sherman, J.D. and Chao, C.C. (1984) Separation of arabinose by selective adsorption on zeolitic molecular sieves. U.S. Patent 4,516,566.
218 Neuzil, R.W. and Priegnitz, J.W. (1977) Separating a ketose from an aldose by selective adsprption. U.S. Patent 4,024,331.
219 Kulprathipanja, S. and Neuzil, R.W. (1982) Technique to reduce the zeolite molecular sieve solubility in an aqueous system. U.S. Patent 4,333,769.
220 Kulprathipanja, S. (1981) Technique to reduce the zeolite molecular sieve solubility in an aqueous system. U.S. Patent 4,248,737.
221 Kulprathipanja, S. (1981) Cellulose acetate butyrate bound zeolite adsorbents. U.S. Patent 4,295,994.
222 Kulprathipanja, S. and deRosset, A.J. (1983) Separatory process using organic bound adsorbents. U.S. Patent 4,421,567.
223 Kulprathipanja, S. and Neuzil, R.W. (1981) Esterified aluminosilicate adsorbent as for resolution of sugar components. U.S. Patent 4,287,001.
224 Chang, C.H. (1990) Process for separating ketoses from alkaline-or pyridine-catalyzed isomerization products. U.S. Patent 4,880,920.
225 Neuzil, R.W. and Priegnitz, J.W. (1982) Process for separating glucose from fructose by selective adsorption. U.S. Patent 4,349,668.
226 Heper, M., Türker, L., and Kincal, N.S. (2007) Sodium, ammonium, calcium,

and magnesium forms of zeolite Y for the adsorption of glucose and fructose from aqueous solutions. *J. Colloid Interface Sci.*, **306**, 11.
227 Chao, C.C. and Sherman, J.D. (1985) Bulk separation of inositol and sorbitol by selective adsorption on zeolitic molecular sieves. U.S. Patent 4,482,761.
228 Berensmeier, S. and Buchholz, K. (2004) Separation of isomaltose from high sugar concentrated enzyme reaction mixture by dealuminated beta-zeolite. *Sep. Purif. Technol.*, **38**, 129.
229 Chao, C.C. and Sherman, J.D. (1983) Bulk lactulose/lactose separation by selective adsorption on zeolitic molecular sieves. U.S. Patent 4,394,178.
230 Odawara, H. (1977) Process for preparation of hexafluoropropanone-2. Jpn Patent 52071409.
231 Goodman, W.H. (1988) Process for separating maltose from mixtures of maltose, glucose and other saccharides. U.S. Patent 4,707,190.
232 Sherman, J.D. and Chao, C.C. (1984) Separation of mannose by selective adsorption on zeolitic molecular sieves. U.S. Patent 4,471,114.
233 Kulprathipanja, S., Neuzil, R.W., and Landis, A.M. (1984) Separation of sucrose from thick juice. U.S. Patent 4,475,954.
234 Krohn, J.E. and Tsapatsis, M. (2006) Phenylalanine and arginine adsorption in zeolites X, Y, and beta. *Langmuir*, **22**, 9350.
235 Mehraban, Z. and Farzaneh, F. (2006) MCM-41 as selective separator of chlorophyll-a from b-carotene and Chlorophyll-a mixture. *Micropor. Mesopor. Mater.*, **88**, 84.
236 Shirato, Y., Shimokawa, K., Shimura, M., and Hirohama, S. (1991) Separation of eicosapentaenoic acid ester and/or docosahexaenoic acid ester. Jpn Patent 02258742.
237 Wang, S., Gong, W., Liu, X., Gao, B., and Yue, Q. (2006) Removal of fulvic acids using the surfactant modified zeolite in a fixed-bed reactor. *Sep. Purif. Technol.*, **51**, 367.
238 Sherman, J.D. and Chao, C.C. (1984) Bulk separation of polyhydric alcohols by selective adsorption on zeolitic molecular sieves. U.S. Patent 4,456,774.
239 Furuya, E., Sato, K., Kataoka, T., Horiguchi, T., and Otake, Y. (2004) Amount of aromatic compounds adsorbed on inorganic adsorbents. *Sep. Purif. Technol.*, **39**, 73.
240 deRosset, A.J., Priegnitz, J.W., and Landis, A.M. (1984) Process for the separation of methylparaben from wintergreen oil. U.S. Patent 4,408,065.
241 Zinnen, H.A. and Johnson, S.P. (1991) Purification of alpha naphthol-containing feedstock by adsorption. U.S. Patent 4,962,273.
242 Palkhiwala, A.G., Lin, Y.H., Perlmutter, D.D., and Olson, D.H. (1999) Liquid phase separation of polar hydrocarbons from light aromatics using zeolites. *Adsorption*, **5**, 399–407.
243 Kulprathipanja, S., Sherman, J., Napolitano, A., and Markovs, J. (2002) Method for treating a liquid stream contaminated with an iodine-containing compound using a cation-exchanged zeolite. U.S. Patent 6,380,428.
244 Senzaki, T., Noguchi, K., Shimoura, Y., and Matsumoto, T. (1997) Separation and recovery of benzothiopene and naphthalene. Jpn Patent 09151139.
245 Senzaki, T. and Noguchi, K. (1998) Separation and recovery of benzothiopene and naphthalene. Jpn Patent 10265416.
246 Senzaki, T., Noguchi, K., Imamura, T., and Takayama, T. (2000) Separation of benzothiopene and naphthalene. Jpn Patent 2000191556.
247 Bhandari, V.M., Ko, C.H., Park, J.G., Han, S., Cho, S., and Kim, J. (2006) Desulfurization of diesel using ion-exchanged zeolites. *Chem. Eng. Sci.*, **61**, 2599.
248 Jeanneret, J.J. (1996) UOP parex process, in *Handbook of Petroleum Refining Process*, 2nd edn (ed. R.A,, Meyers), McGraw-Hill, New York, pp. 2.47–2.54.
249 Kulprathipanja, S., Sherman, J., Napolitano, A., and Markovs, J. (2002) Method for treating a liquid stream contaminated with an iodine-containing compound using a cation-exchanged zeolite. U.S. Patent 6,380,428.

250 Kulprathipanja, S., Spehlmann, B.C., Willis, R.R., Sherman, J.D., and Leet, W.A. (1999) Method for treating an organic liquid contaminated with an iodide compound. U.S. Patent 5,962,735.

251 Kulprathipanja, S. and Spehlmann, B.C. (2001) Direct removal of trace ionic iodide from acetic acid. Proceedings of the 13th International Zeolite Conference, Montpellier, France, 8–13 July.

252 Kulprathipanja, S., Vora, B.V., and Leet, W.A. (2003) Combination pretreatment/ adsorption for treating a liquid stream contaminated with an iodine-containing compound. U.S. Patent 6,506,935.

253 Hernandez-Maldonado, A.J., and Yang, R.T. (2004) Denitrogenation of transportation fuels by zeolites at ambient temperature and pressure. *Angew. Chem. Int. Ed.*, **43**, 1004.

254 Kulprathipanja, S., Nemeth, L.T., and Holmgren, J.S. (1998) Process for removing sulfur compounds from hydrocarbon streams. U.S. Patent 5,807,475.

255 Xue, M., Chitrakar, R., Sakane, K., Hirotsu, T., Ooi, K., Yoshimura, Y., Toba, M., and Feng, Q. (2006) Preparation of cerium-loaded Y-zeolites for removal of organic sulfur compounds from hydrodesulfurizated gasoline and diesel oil. *J. Colloid Interface Sci.*, **298**, 535.

256 Bhandari, V.M., Ko, C.H., Park, J.G., Han, S., Cho, S., and Kim, J. (2006) Desulfurization of diesel using ion-exchanged zeolites. *Chem. Eng. Sci.*, **61**, 2599.

257 Hernandez-Maldonado, A.J., Yang, R.T., and Cannella, W. (2004) Desulfurization of commercial jet fuel by adsorption via -complexation with vapor phase exchanged (VPIE) Cu(I)-Y zeolites. *Ind. Eng. Chem. Res.*, **43**, 6142.

258 Benashvili, E.M., Uchaneishvili, T.G., Alibegashvili, M.S., Cherkezishvili, N.I., and Kvitaishvili, K.E. (1985) Adsorptive separation of hetero compounds from petroleum fractions by natural zeolites. *Soobshch. Akad. Nauk Gruz. SSR*, **118**, 537.

259 Yang, R.T., Yang, F.H., Takahashi, A., and Hernandez-Maldonado, A.J. (2006) Selective sorbents for purification of hydrocarbons. U.S. Patent 7,053,256.

260 Yang, R.T., Yang, F.H., Takahashi, A., and Hernandez-Maldonado, A.J. (2006) Selective sorbents for purification of hydrocartons. U.S. Patent 7,148,389.

261 Yang, F.H., Hernandez-Maldonado, A.J., and Yang, R.T. (2004) Selective adsorption of organosulfur compounds from transportation fuels by p-complexation. *Sep. Sci. Technol.*, **39**, 1717.

262 Dai, W., Zhou, Y., Sun, S.L.i, W.L.i, W.S.u, Y., and Zhou, L. (2006) Thiophene capture with complex adsorbent SBA-15/Cu(I). *Ind. Eng. Chem. Res.*, **45**, 7892.

263 DeWet, J.P., Jacobson, P., and Janen, W. (2004) Extraction of oxygenates from a hydrocarbon stream. W. O. 2004/080927.

264 Grigis, M.J. and Gates, B.C. (1991) Reactivities, reaction networks, and kinetics in high-pressure catalytic hydroprocessing. *Ind. Eng. Chem. Res.*, **30**, 2021.

265 Dickson, C.T., Fitzke, J.R., and Becker, C.L. (1992) Process control of process for purification of linear paraffins. U.S. Patent 5,109,139.

266 Dickson, C.T., Fitzke, J.R., and Becker, C.L. (1992) Recycle for process for purification of linear paraffins. U.S. Patent 5,171,923.

267 Schrelner, J.L., Britton, R.A., Dickson, C.T., and Pehler, F.A. (1993) Purification of a hydrocarbon feedstock using a zeolite adsorbent. U.S. Patent 5,220,099.

268 Lee, D.D., Taylor, J.F., Jr., Walker, P.A., and Hendrickson, D.W. (1997) Cesium-removal flow studies using ion exchange. *Environ. Prog.*, **16**, 251.

269 Sherman, J.D. (1999) Synthetic zeolites and other microporous oxide molecular sieves. *Proc. Natl. Acad. Sci. U. S. A.*, **96**, 3471.

270 Anthony, R.G., Dosch, R.G., Gu, D., and Philip, C.V. (1994) Use of silicotitanates for removing cesium and strontium from defense waste. *Ind. Eng. Chem. Res.*, **33**, 2702.

6
Aspects of Mechanisms, Processes, and Requirements for Zeolite Separation

Santi Kulprathipanja

6.1
Introduction

As documented in Chapter 5, zeolites are very powerful adsorbents used to separate many products from industrial process steams. In many cases, adsorption is the only separation tool when other conventional separation techniques such as distillation, extraction, membranes, crystallization and absorption are not applicable. For example, adsorption is the only process that can separate a mixture of C_{10}–C_{14} olefins from a mixture of C_{10}–C_{14} hydrocarbons. It has also been found that in certain processes, adsorption has many technological and economical advantages over conventional processes. This was seen, for example, when the separation of *m*-xylene from other C_8-aromatics by the HF–BF$_3$ extraction process was replaced by adsorption using the UOP MX Sorbex™ process. Although zeolite separations have many advantages, there are some disadvantages such as complexity in the separation chemistry and the need to recover and recycle desorbents.

This chapter addresses the fundamentals of zeolite separation, starting with: (i) impacts of adsorptive separation, a description of liquid phase adsorption, (ii) tools for adsorption development such as isotherms, pulse and breakthrough tests and (iii) requirements for appropriate zeolite characteristics in adsorption. Finally, speculative adsorption mechanisms are discussed. It is the author's intention that this chapter functions as a bridge to connect the readers to Chapters 7 and 8, *Liquid Industrial Aromatics Adsorptive Separation* and *Liquid Industrial Non-Aromatics Adsorptive Separation*, respectively. The industrial mode of operation, the UOP Sorbex™ technology, is described in Chapters 7 and 8.

6.2
Impacts of Adsorptive Separation Versus Other Separation Processes

Separation processes play a significant role in industry. These processes include: distillation, absorption, extraction, adsorption, crystallization and membrane separation. The most utilized and well-known technique is distillation, which is used in 90–95% of all processes in the chemical industry [1, 2]. However, due to technological and economic limitations, distillation is not always feasible. For the remaining applications other techniques must be explored. Zeolite adsorption is one of the more desirable options because of its extensive and flexible applications. Adsorption technologies have a wide range of applications throughout a wide range of industries, including the chemical, petroleum and allied industries. Zeolite separation as it applies to industrial applications is part of the focus of this book.

An example of a technologically difficult separation is the separation of C_5 and C_6 linear paraffins from their branched and cyclic paraffins when upgrading gasoline octane. This separation cannot be accomplished by distillation because of the nearly indistinguishable boiling points of the feed components. As shown in Figure 6.1, there are boiling point differences among the linear, branched and cyclic paraffins for both C_5 and C_6. However, because the boiling points increase between C_5 and C_6 in a broad boiling mixture, the boiling points of various C_5 and C_6 paraffins overlap, making recovery of the normal paraffings as a class by distillation infeasible. It is true that *n*-pentane can be fractionated from iso-pentane, but the problem of recovering the *n*-hexane would remain.

In this particular case, the adsorption process can be used to overcome the distillation limitation. This is demonstrated in Figure 6.2, which represents the relative adsorption of C_5 and C_6 linear, branched and cyclic paraffins from the liquid phase of the 5A adsorbent used in the UOP Gasoline Molex™ process, licensed by UOP. In this process, only linear paraffins can enter the pores of 5A zeolite, while branched and cyclic paraffins are completely excluded due to their large kinetic diameters. Also, the selectivity for linear paraffins with respect to other types of paraffins is infinite. Consequently, the separation of linear paraffins from branched and cyclic paraffins becomes possible.

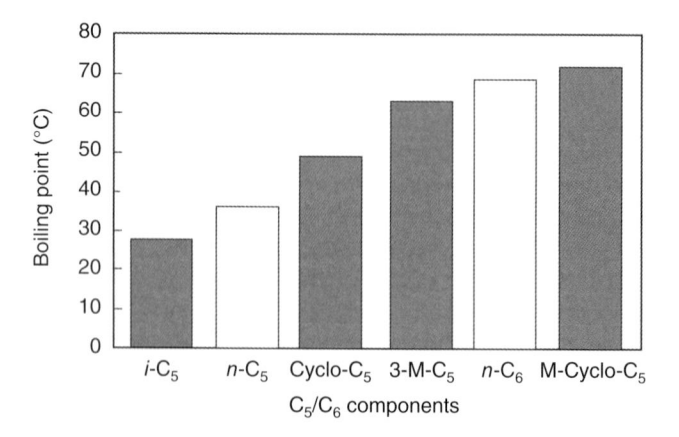

Figure 6.1 C_5/C_6 boiling point distribution.

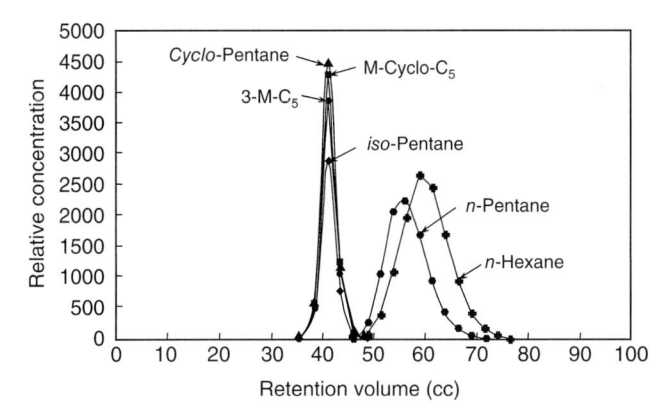

Figure 6.2 Chromatographic separation of n-paraffin from non-n-paraffin.

Table 6.1 Boiling point (°C) of C_{10}–C_{14} olefin/paraffin.

	C_{10}	C_{11}	C_{12}	C_{13}	C_{14}
Olefin	166–173	192–193	213	232–233	251
Paraffin	174	196	216	234	252–254

The commercial liquid adsorptive separation process of C_{10}–C_{14} n-olefins from C_{10}–C_{14} n-paraffins is another unique example of how zeolite adsorption can be applied. As shown in Table 6.1, distillation is not an option to separate C_{10}–C_{14} olefins from C_{10}–C_{14} paraffins because of their close boiling points. In this case, the UOP Olex™ process using NaX adsorbent is used to separate C_{10}–C_{14} olefins from C_{10}–C_{14} paraffins.

Until late 1990s, purified m-xylene was produced predominantly by the HF/BF$_3$ process developed by Mitsubishi Gas Chemical Co. The separation is based on the complex formation between m-xylene and solvent HF/BF$_3$. However, concerns about the process operation, environment, metallurgy and safety render the process commercially unattractive due to its use of HF/BF$_3$. These concerns led to many developments in the adsorptive separation process for m-xylene separation [3–8]. The UOP MX Sorbex process, developed by UOP and commercialized in 1998, already accounts for more than 70% of the world's m-xylene capacity. A 95% m-xylene recovery with 99.5% purity can be achieved by the MX Sorbex™ process.

Traditionally, citric acid is recovered from fermentation broth via a lime and sulfuric acid process. In this process, filtered fermentation broth is treated with lime to precipitate calcium citrate. This precipitate is washed and acidified with sulfuric acid to convert the calcium citrate to solid calcium sulfate (gypsum) and

Table 6.2 Analysis of citric acid product by adsorption.

Compound	Feed	Product specification	Sorbex pilot plant product
Citric acid purity, wt%	85.0	99.5+	99.5+
Citric acid, wt%	50	50	50
Total nitrogen, wt%	0.5	0.04	Not detectable
Readily carbonizable substance	32	0.90	<0.30
Cations, wt%	0.5	0.04	<0.01
Anions, wt%	1.2	0.09	<0.03
APHA color	100 000	300	<50

citric acid. The solid and liquid are separated and the citric acid solution is taken through a series of crystallizations with intermediate carbon treatment and ion exchange to remove the remaining impurities. This conventional process is costly, requires extensive solids handling and is environmentally objectionable. The disposal of one pound of gypsum (1 lb = 0.45 kg) per pound of citric acid produced is an increasingly expensive problem for citric acid manufacturers. In addition to the lime and sulfuric acid processes, solvent extraction can also be used commercially to recover citric acid. In this process, citric acid is extracted from its broth with a mixture of trilaurylamine, n-octanol and a C_{10} or C_{11} isoparaffin. With the addition of heat, the acid from the solvent phase is subsequently recovered from the solvent phase into water. Again, this is a complex and expensive process because it generates a substantial amount of waste for disposal.

Recognizing the need for a more economically and environmentally friendly citric acid recovery process, an adsorptive separation process to recover citric acid from fermentation broth was developed by UOP [9–14] using resin adsorbents. No waste gypsum is generated with the adsorption technique. The citric acid product recovered from the Sorbex pilot plant either met or exceeded all specifications, including that for readily carbonizable substances. An analysis of the citric acid product generated from a commercially prepared fermentation broth is shown in Table 6.2, along with typical production specifications. The example sited here is not related to zeolite separation. It is intent to demonstrate the impact of adsorption to other separation processes.

6.3
Liquid Phase Adsorption

Liquid phase zeolitic separation includes two main events: adsorption and desorption. Adsorption of an adsorbate (liquid component being adsorbed by solid) onto zeolitic adsorbent is dictated by the characteristics of the adsorbate-adsorbent interaction. A zeolitic adsorbent is a crystalline porous solid having particular characteristics (see Chapter 2). When immersed in a liquid mixture, the porous solid has the capability to adsorb specific species from the liquid. At equilibrium, the composition of material inside the pores differs from that of the liquid surrounding the porous solid. The amount of adsorbate that can be taken up by the adsorbent(i.e., the adsorbent capacity) is dependent on the nature of the adsorbate-adsorbent interaction, surface area and pore volume of the adsorbent. In general, a higher adsorbent capacity is seen with higher adsorbent selectivity, higher adsorbent surface area and higher pore volume. This phenomenon creates a higher concentration profile of the more selective component in the pores of the adsorbent than in the surrounding liquid. The process of adsorbing the adsorbate into the zeolitic adsorbent is known as the adsorption step.

To desorb the adsorbate from the zeolitic adsorbent, a desorbent is added. A desorbent is a suitable liquid that is capable of displacing or desorbing the adsorbate from the selective pores of the adsorbent. The process of recovering or desorbing the adsorbate from the adsorbent is known as the desorption step. In the chromatographic liquid adsorptive separation process, the adsorption and desorption processes must occur simultaneously. After the desorption step, both the rejected product (product with lower selectivity, resulting in less adsorption by adsorbent) and the extracted product (product with higher selectivity, resulting in strong adsorption by adsorbent) contain desorbent. In general, the desorbent is recovered by fractionation or evaporation and recycled back into the system.

6.3.1
Sanderson's Model of Intermediate Electronegativity

One of the parameters in the broad class of liquid adsorption mechanisms is the interaction between the acidic and basic sites of the adsorbent and the adsorbate. The acidity of zeolitic adsorbent is normally affected by the zeolite SiO_2/Al_2O_3 molar ratio, the ionic radii and the valence of the cations exchanged into the zeolite. In this contribution, Sanderson's model of intermediate electronegativity of zeolitic adsorbent acidity (S_{int}) can be calculated as a representation of the strength of the adsorbent acidity based on the following equation:

$$S_{int} = \left(\prod_i S_i^{n_i} \right)^{1/(\Sigma n_i)} \tag{6.1}$$

where S_i is the Sanderson's electronegativity of the atom and n_i is the stoichiometry of the atom in a unit cell of zeolite [15]. For the ion-exchanged zeolites with a formula of $Na_pM_q(SiO_2)_r(AlO_2)_s$, the S_{int} can be calculated by:

$$S_{int} = \left(S_{Na}^p S_M^q S_{Si}^r S_{Al}^s S_O^{2(r+s)} \right)^{1/[p+q+r+s+2(r+s)]} \tag{6.2}$$

where S_{Na}, S_M, S_{Si}, S_{Al} and S_O represent the electronegativity of Na, M, Si, Al, and O, respectively, and p, q, r and s represent the total number of atoms of the corresponding element in the unit cell. Sanderson's intermediate electronegativity (S_{int}) of the alkaline and alkaline earth exchanged faujasite zeolites are calculated using Eq. (6.1) and summarized in Table 6.3. Table 6.3, shows S_{int} increases as:

Table 6.3. The Sanderson's intermediate electronegativity (S_{int}) of the zeolite [19].

Adsorbent	Chemical composition	Ionic radius (Å)	S_{int}
LiX	$Li_{20}Na_{64}(AlO_2)_{84}(SiO_2)_{108}$	0.76	3.284
NaX	$Na_{84}(AlO_2)_{84}(SiO_2)_{108}$	1.02	3.278
KX	$K_{82}Na_2(AlO_2)_{84}(SiO_2)_{108}$	1.38	3.076
RbX	$Rb_{71}Na_{13}(AlO_2)_{84}(SiO_2)_{108}$	1.52	3.052
CsX	$Cs_{73}Na_{11}(AlO_2)_{84}(SiO_2)_{108}$	1.67	2.962
MgX	$Mg_{30}Na_{24}(AlO_2)_{84}(SiO_2)_{108}$	0.72	3.765
CaX	$Ca_{40}Na_4(AlO_2)_{84}(SiO_2)_{108}$	0.99	3.754
SrX	$Sr_{41}Na_2(AlO_2)_{84}(SiO_2)_{108}$	1.12	3.732
BaX	$Ba_{41}Na_2(AlO_2)_{84}(SiO_2)_{108}$	1.35	3.657
LiY	$Li_{22}Na_{31}(AlO_2)_{53}(SiO_2)_{139}$	0.76	3.587
NaY	$Na_{53}(AlO_2)_{53}(SiO_2)_{139}$	1.02	3.580
KY	$K_{51}Na_2(AlO_2)_{53}(SiO_2)_{139}$	1.38	3.435
RbY	$Rb_{47}Na_6(AlO_2)_{53}(SiO_2)_{139}$	1.52	3.409
CsY	$Cs_{49}Na_4(AlO_2)_{53}(SiO_2)_{139}$	1.67	3.334
MgY	$Mg_{20}Na_{13}(AlO_2)_{53}(SiO_2)_{139}$	0.72	3.907
CaY	$Ca_{23}Na_7(AlO_2)_{53}(SiO_2)_{139}$	0.99	3.890
SrY	$Sr_{24}Na_5(AlO_2)_{53}(SiO_2)_{139}$	1.12	3.883
BaY	$Ba_{24}Na_5(AlO_2)_{53}(SiO_2)_{139}$	1.35	3.836

(i) ionic radii decreases from Rb to Li and from Ba to Mg, (ii) SiO_2/Al_2O_3 increases from X-zeolite (SiO_2/Al_2O_3 of 2.5) to Y-zeolite (SiO_2/Al_2O_3 of 5.0) and (iii) the valence of exchanged cations on the same SiO_2/Al_2O_3 ratio zeolite increases such as from Li to Mg. The higher S_{int} indicates high electro-accepting ability and strong adsorbent acidity [16–18].

6.4 Modes of Operation

As pointed out later in this chapter, zeolite separation mechanisms are highly complex due to the interaction between the zeolitic adsorbent, adsorbate and desorbent. In order to develop an adsorptive separation process, an understanding of the zeolite adsorbent, adsorbate and desorbent interaction is required. This has led to the development of appropriate techniques such as fundamental adsorption isotherms, pulse testing and breakthroughs. Isotherms, pulse testing and breakthroughs are suitable tools for zeolite adsorbent and desorbent screening in adsorption process development. The tools also allow scientists and engineers to optimize initial operating parameters such as temperature, liquid flow circulation, adsorbent moisture, etc. However, isotherms, pulse testing and breakthroughs do not offer sufficient information for scaling-up the commercial unit. The scale-up information can be obtained from the Sorbex technology given in Chapters 7 and 8.

6.4.1
Adsorption Isotherms

Typically, adsorption isotherms are generated using a batch experiment at a fixed temperature and a fixed feed composition. These experiments include exposing a known amount of adsorbent to a known concentration of adsorbate at a constant temperature. Once equilibrium is established, the net adsorbate concentration in the liquid is measured. This process is repeated at multiple adsorbate concentrations and temperatures. A plot of adsorbate loading (g adsorbate/g adsorbent) versus adsorbate concentration reveals the adsorption isotherm with the shape of the isotherm determining the suitability of a particular adsorbent for a particular system [20].

6.4.2
Pulse Test Procedure

When developing a liquid phase adsorptive separation process, a laboratory pulse test is typically used as a tool to search for a suitable adsorbent and desorbent combination for a particular separation. The properties of the suitable adsorbent, such as type of zeolite, exchange cation and adsorbent water content, are a critical part of the study. The desorbent, temperature and liquid flow circulation are also critical parameters that can be obtained from the pulse test. The pulse test is not only a critical tool for developing the equilibrium-selective adsorption process; it is also an essential tool for other separation process developments such as rate-selective adsorption, shape-selective adsorption, ion exchange and reactive adsorption.

A pulse test procedure [6] begins with an injection of a small pulse of the feed mixture to be separated into a desorbent stream flowing through a packed adsorbent column at a fixed flow rate and temperature. The on-line column effluent composition is then determined as a function of time or volume of desorbent passed by gas or liquid chromatography. Particularly important is the sequence and time when each of the feed components exit the packed adsorbent column because these characteristics describe the specific adsorbate and adsorbent interactions. By determining the interactions using the pulse test, the separation process can be optimized.

An example of results from a pulse test separating arbitrary components A and B along with a tracer is shown in Figure 6.3. An inert tracer is selected that will not be adsorbed by the system being studied. Each component elution time or volume, shown in Figure 6.3, is taken at the maximum peak height or concentration of a specific component. Furthermore, the net retention volume of each component is measured based on the maximum peak height or concentration of the tracer as the zero origin. Because the net retention volume of any component is ideally proportional to its distribution coefficient (its concentration in the adsorbed phase divided by its concentration in the un-adsorbed phase), the adsorbent selectivity for the more strongly adsorbed component B over component A can

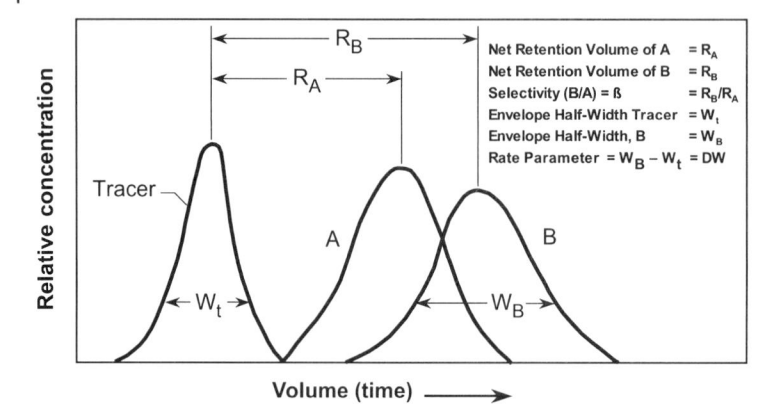

Figure 6.3 Schematic pulse test: two components and paraffin tracer.

Legend in figure:
- Net Retention Volume of A = R_A
- Net Retention Volume of B = R_B
- Selectivity (B/A) = ß = R_B/R_A
- Envelope Half-Width Tracer = W_t
- Envelope Half-Width, B = W_B
- Rate Parameter = $W_B - W_t$ = DW

be calculated from the ratio of the net retention volumes of component B to component A. Component B is more strongly adsorbed in the adsorbent packed bed since it elutes after component A.

6.4.3 Breakthrough Procedure

The adsorbent and adsorbate interaction obtained using the pulse test is at a diluted feed condition. So, the interaction information might not fully represent the actual interaction since the commercial feed concentration is normally much higher. To be more representative, the breakthrough technique is introduced.

The breakthrough procedure is similar to the pulse test procedure except a large amount of high feed concentration is used. The breakthrough procedure can be described as follows:

1) Introduce solvent into a packed adsorbent column at a fixed flow rate, pressure and temperature.

2) Introduce a feed mixture to be separated until the effluent reaches the feed composition. The feed breakthrough comprises the components to be separated, the tracer and the desorbent.

3) Introduce solvent until no component in the feed mixture is detected in the effluents.

A plot of column effluent composition as a function of time or volume of solvent and feed reveals the adsorption and desorption behaviors of the components in the feed mixture and illustrates whether the adsorption system is suitable for separating the components. Particularly important is the sequence and time when each of the feed components exit the packed adsorbent column because these characteristics describe the specific adsorbate and adsorbent interactions. From the breakthrough plot, zeolitic adsorbent selectivity, capacity, mass transfer rate and desorbent strength can be calculated.

6.5
Zeolite Separation Processes

The degree of liquid phase adsorption is almost infinite due to the number of ways of modifying the zeolite characteristics. Key zeolitic adsorbent characteristics include framework structure, zeolite particle sizes, chemical composition, binder, counter exchange ion and water content. These variables are carefully modified to selectively adsorb one particular component over others. The adsorbed component can then be removed or desorbed from the adsorbent using a suitable solvent functioning as a desorbent. To optimize separation by liquid phase adsorption, two opposing forces must be balanced: the adsorptive force of the adsorbent to the adsorbate and the desorptive force of the desorbent to the adsorbate. Ideally, the adsorbent should have a lower selectivity for the desorbent than the adsorbed component. However, the adsorbent should have a higher selectivity for the desorbent than the rejected components in the mixture. This is to ensure that the desorbent does not utilize most of the adsorbent capacity and at the same time, has the ability to desorb the adsorbate from the adsorbent.

There are many adsorption mechanisms that zeolite separation processes can account for. This depends on many factors such as zeolite structures, zeolite pore sizes, zeolite Si/Al ratios, exchanged cations and the physical and chemical properties of adsorbates and desorbents. In general, the mechanisms present in zeolite adsorption can be classified as following:

1) Equilibrium-selective adsorption;
2) Rates-selective adsorption;
3) Shape-selective adsorption;
4) Ion-exchange;
5) Reactive adsorption.

6.5.1
Equilibrium-Selective Adsorption

The foundation of equilibrium-selective adsorption is based on differences in the equilibrium selectivity of the various adsorbates with the adsorbent. While all the adsorbates have access to the adsorbent sites, the specific adsorbate is selectively adsorbed based on differences in the adsorbate–adsorbent interaction. This in turn results in higher adsorbent selectivity for one component than the others. One important parameter that affects the equilibrium-selective adsorption mechanism is the interaction between the acidic sites of the zeolite and basic sites of the adsorbate. Specific physical properties of zeolites, such as framework structure, choice of exchanged metal cations, SiO_2/Al_2O_3 ratio and water content can be manipulated to influence the acidity of zeolites, which in turn affects separation performance. The degree of separation is characterized by the zeolite selectivity and capacity as well as the choice of desorbent and operating conditions such as temperature, feed and desorbent flow rates. Therefore, to achieve a meaningful separation using the equilibrium-selective adsorption mechanism, the following zeolite characteristics and operating conditions are to be considered:

1) Zeolite framework structure;
2) Metal cation exchanged in zeolite;
3) Zeolite SiO_2/Al_2O_3 molar ratio;
4) Moisture content in zeolite;
5) Characteristic of desorbent;
6) Operating temperature.

These variables are described in the following sections.

6.5.1.1 Zeolite Framework Structure

One of the most significant variables affecting zeolite adsorption properties is the framework structure. Each framework type (e.g., FAU, LTA, MOR) has its own unique topology, cage type (alpha, beta), channel system (one-, two-, three-dimensional), free apertures, preferred cation locations, preferred water adsorption sites and kinetic pore diameter. Some zeolite characteristics are shown in Table 6.4. More detailed information on zeolite framework structures can be found in Breck's book entitled *Zeolite Molecular Sieves* [21] and in Chapter 2.

The variety of the different framework structures result in different adsorbent characteristics: acid strength, size of molecule adsorbed, adsorption/desorption rate of different molecules, capacity and stability. As a result, these differences characterize the adsorbent's selectivity to a specific molecule and adsorbent–adsorbate interactions. Take for example, the difference in selectivity of BaY and Ba-Mordenite [24] to p-xylene (PX), m-xylene (MX) and o-xylene (OX):

BaY selectivity: PX > OX > MX
Ba-Mordenite selectivity: PX > MX > OX

Ba-Modenite's selectivity to MX is higher than OX, but the opposite is true for BaY. This reversal in selectivity is a result of differences in adsorbent framework characteristics: mordenite has higher acid strength compared to Y zeolite. Adsorption and desorption rates of xylenes are expected to be faster in BaY compared to Ba-Mordenite because Mordenite is a one-dimensional channel system while Y zeolite is a three-dimensional channel. With the reason stated, a three-dimensional channel Zeolite is the preferred mass separating agent of choice compared to one- or two-dimensional channels for the liquid adsorption separation.

6.5.1.2 Metal Cation Exchanged in Zeolite

One of the parameters in the broad class of equilibrium-selective adsorption mechanisms is the interaction between the acidic and basic sites of the adsorbent and the adsorbate. Zeolites can be ion-exchanged with a variety of metal cations

Table 6.4 Typical properties of common zeolites [21–23].

Zeolite type	Channel system	Pore openings (Å; hydrated form)	Typical SiO_2/Al_2O_3 mole ratio	Theoretical ion exchange capacity (meq/g; Na form, anhydrous)
Analcime	One-dimensional	2.6	4	4.9
Chabazite	Three-dimensional	3.7 × 4.2 and 2.6	4	4.9
Clinoptilolite	NK	4.0 × 5.5, 4.4 × 7.2 and 4.1 × 4.7	10	2.6
Erionite	Three-dimensional	3.6 × 5.2	6	3.8
Ferrierite	Two-dimensional	4.3 × 5.5 and 3.4 × 4.8	11	2.4
Phillipsite	Three-dimensional	4.2 × 4.4, 2.8 × 4.8 and 3.3	4.4	4.7
Zeolite A	Three-dimensional	4.2 into alfa-cage; 2.2 into beta-cage	2	7.0
Zeolite L	One-dimensional	7.1	6	3.8
Mordenite	Two-dimensional	2.9 × 5.7	10	2.6
Zeolite Omega	One-dimensional	7.5	7	3.4
Silicalite-1	Three-dimensional	(5.7–5.8) × (5.1–5.2)	50	0.63
Zeolite X	Three-dimensional	7.4 into supercage; 2.2 into beta-cage	2.5	6.4
Zeolite Y	Three-dimensional	7.4 into supercage; 2.2 into beta-cage	4.8	4.4

to alter their acidity and other properties. There is a strong correlation between the total acidity of a zeolite (the sum of both Brönsted and Lewis acids) and the ionic radius of the cation as well as the valence charge of the exchanged cation [18, 25, 26]. Exchanged cations with lower ionic radii have higher zeolite acidity. The correlation between zeolite acidity and ionic radius or exchanged cation valence is measured by titration with n-butylamine and a Hamett indicator, methyl red. Zeolite acidity increases for monovalent exchanged cations from Cs < K < Na < Li and increases from Ba < Sr < Ca < Mg and Ba < Pb < La < Ca < Zn < Co for divalent exchanged cations. For cations of similar ionic radii, divalent cations have higher zeolite acidity than monovalent cations. For example, Ba-zeolite has a higher acidity than K-zeolite, and Ca-zeolite has a higher acidity than Na-zeolite and Li-zeolite. Furthermore, strong acidic Bronsted properties can be observed in hydrated zeolites due to polarization of adsorbed water in the strong electrostatic field between the exchanged cations and the AlO_4^- anions [26], For hydrated zeolites, higher moisture content results in higher zeolite acidity.

Table 6.5 Base strengths of aromatic hydrocarbons relative to HF.

Compound	Relative basicity at 0.1 M
Benzene	0.09
Toluene	0.63
p-Xylene	1.0
o-Xylene	1.1
m-Xylene	26
Durene (1,2,4,5-tetramethylbenzene)	140
Isodurene (1,2,3,5-tetramethylbenzene)	16 000

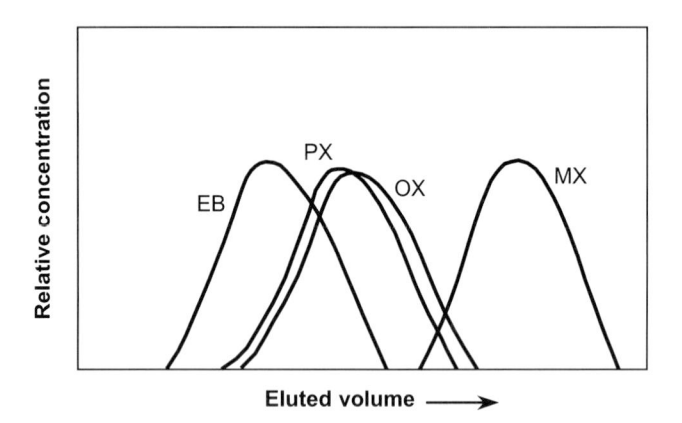

Figure 6.4 Separation of m-xylene from C_8-aromatics using NaY adsorbent.

Utilization of zeolite acidic strength in C_8-aromatics (xylens) systems is illustrated next. In the presence of strong acids, xylene isomers have varying basicity (Table 6.5), with m-xylene being the most basic and p-xylene the least basic among the C_8-aromatics [27]. Based on the basicity of the xylenes, the acidity of each zeolite can be properly adjusted to selectively adsorb m-xylene or p-xylene. As demonstrated in Figure 6.4, a more acidic zeolite such as NaY will selectively adsorb m-xylene from other C_8-aromatics [28, 29]. In contrast, Figure 6.5 shows that a weaker acidic zeolite such as KY will selectively adsorb p-xylene from other C_8-aromatics [30, 31]. In both systems, toluene was used as the desorbent.

Acid–base interactions between zeolitic adsorbents and adsorbates do not always correctly predict the trend of adsorbent selectivity. This is illustrated by the adsorptive separation of durene from isodurene. Pulse test experiments indicated that the adsorbent selectivity for durene/isodurene increases from KX < NaX < LiX, shown in Table 6.6 [32]. Because isodurene is a stronger base than durene (Table 6.5), one would expect that the results for adsorbent selectivity

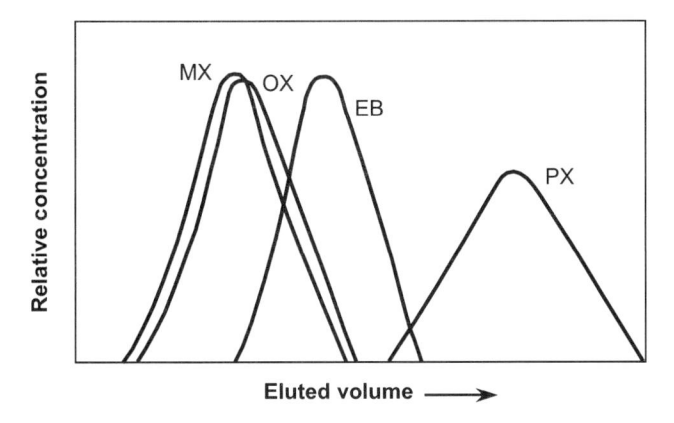

Figure 6.5 Separation of p-xylene from C_8-aromatics using KY adsorbent.

Table 6.6 Durene/isodurene selectivity as a function of X-zeolite exchanged cation.

Cation	Durene/isodurene selectivity
K	0.59
Na	0.97
Li	3.50

shown in Table 6.6 would be the reverse order (LiX < NaX < KX). The counterintuitive trend may be explained by steric hindrances caused by the differences in ionic radii of the cations, which could play a role in the adsorbent and adsorbate interaction.

The separation of p-xylene from m-xylene and o-xylene using X- and Y-zeolites exchanged with monovalent cations (Li, Na, K, Rb, Cs) and divalent cations (Mg, Ca, Sr, Ba) were investigated by Suntornpun [33] and Limsamutchaikul [34], respectively, using pulse tests. Results of the study of the PX/MX and PX/OX selectivities were tabulated against Sanderson's model of intermediate electronegativity (S_{int}), as shown in Table 6.7. The results show some correlation between selectivity versus S_{int} that, within the same class of zeolite (X or Y) and valence of exchanged cations (mono- or divalent), lower S_{int} adsorbents give higher PX selectivity. For example, BaX (S_{int} of 3.657) and BaY (S_{int} of 3.836) have a PX/MX selectivity of 2.25 and 1.49, respectively. However, as one can also see from Table 6.7, acid–base interactions cannot be solely used to explain the adsorption mechanism for the whole range of cations studied. Again, differences in ionic radii of the cations and other unidentified factors complicate the adsorbate and adsorbent interaction mechanism.

Table 6.7 Correlation of exchanged X- and Y-zeolite S_{int} values to PX/MX, PX/OX selectivity.

Adsorbent	S_{int}	PX/MX	PX/OX
LiX	3.284	0.86	0.76
NaX	3.278	0.93	0.76
KX	3.076	1.35	1.04
RbX	3.052	1.22	0.96
CsX	2.962	1.16	0.89
MgX	3.765	0.91	0.87
CaX	3.754	0.61	0.52
SrX	3.732	0.67	0.91
BaX	3.657	2.25	2.00
LiY	3.587	0.67	0.88
NaY	3.580	0.60	1.02
KY	3.435	2.16	1.84
RbY	3.409	1.53	1.35
CsY	3.334	1.43	1.34
MgY	3.907	1.03	0.89
CaY	3.890	0.82	0.70
SrY	3.883	0.98	0.84
BaY	3.836	1.49	1.26

6.5.1.3 Zeolite SiO_2/Al_2O_3 Molar Ratio

As mentioned earlier, multiple factors determine zeolite acidity, including the SiO_2/Al_2O_3 molar ratio. Zeolite acidity increases in strength as the molar ratio of SiO_2/Al_2O_3 decreases [25] due to the increase in AlO_4^- sites, which strengthens the electro-static field in the zeolite and increases the number of acid sites. However, the wide array of cage and channel arrangements and electrochemical properties that result from various crystalline structures and different SiO_2/Al_2O_3 ratios also affect zeolite acid strength. In certain conditions, a high density of AlO_4^- in the zeolite framework could actually lower the acid strength of the adsorbent. The reverse in acid strength can be explained by the dipolar repulsion of the AlO_4^- groups outweighing the increase in polarizability. [18, 35]. For the reasons stated, an increase in zeolite acidity as the molar ratio of SiO_2/Al_2O_3 increases is normally observed. Manipulating both the exchanged cations and the SiO_2/Al_2O_3 ratio offers a great flexibility in tailoring adsorbents for a specific application. However, the more variables that are altered, the more difficult the adsorption behavior becomes to predict.

In addition to the zeolite SiO_2/Al_2O_3 molar ratio, the previous section noted that the ionic radii and the valence of the cations exchanged into the zeolite also affect the total acidity of the zeolite. Exchanged cations with lower ionic radii and higher valence have been demonstrated experimentally to give higher zeolite acidity [17, 25]. The general trend of zeolite acidity, as affected by SiO_2/Al_2O_3 molar ratio, ionic radii and valence of exchanged cation, can be calculated using the Sanderson's

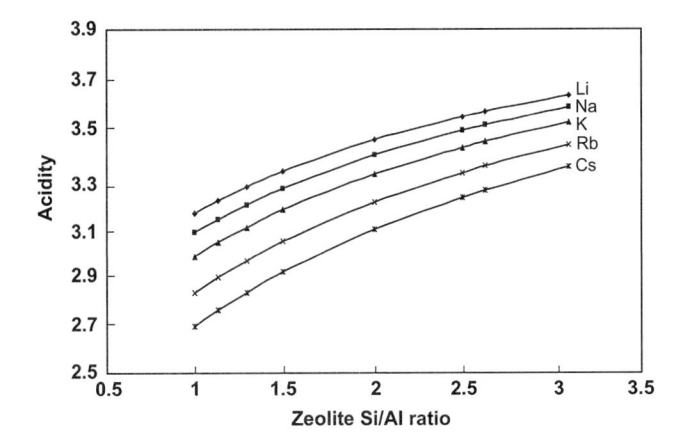

Figure 6.6 Acidity (sint) versus Si/Al ratio for mono-valence cation exchanged zeolites.

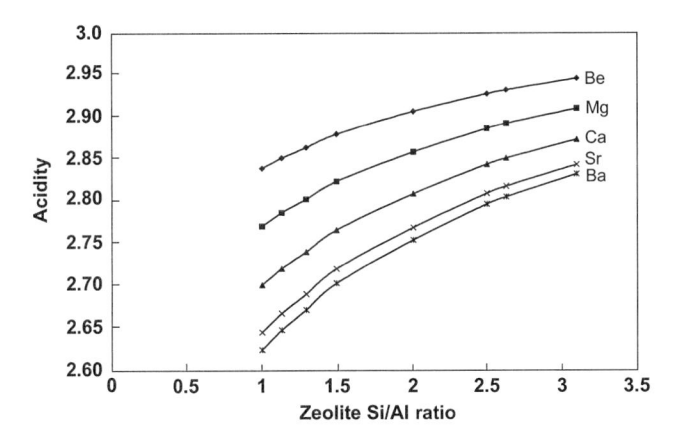

Figure 6.7 Acidity (sint) versus Si/Al ratio for di-valence cation exchanged zeolites.

model of intermediate electronegativity (S_{int}) with the ion-exchanged zeolites with a formula of $M_q(SiO_2)_r(AlO_2)_s$ [36]. The relationship between S_{int} and SiO_2/Al_2O_3 zeolite molar ratio for exchanged monovalent cations and divalent cations are shown in Figures 6.6 and 6.7, respectively. The plots show that increasing the SiO_2/Al_2O_3 molar ratio increases the S_{int}, which suggests that, within the limits of Sanderson's model of intermediate electronegativity calculation, the dipolar repulsion outweighs the increase in polarizability. The plots also confirm that increasing the S_{int}, decreases the cationic radii.

The adsorptive separation of durene and isodurene is used here to illustrate the effect of zeolite acidity. As shown in Figure 6.8, LiX is more selective towards durene than isodurene [32]. However, LiY is more selective towards isodurene

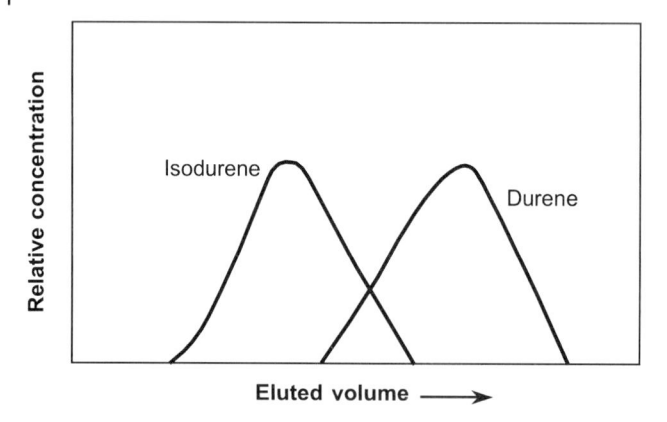

Figure 6.8 Separation of durene and isodurene using LiX adsorbent.

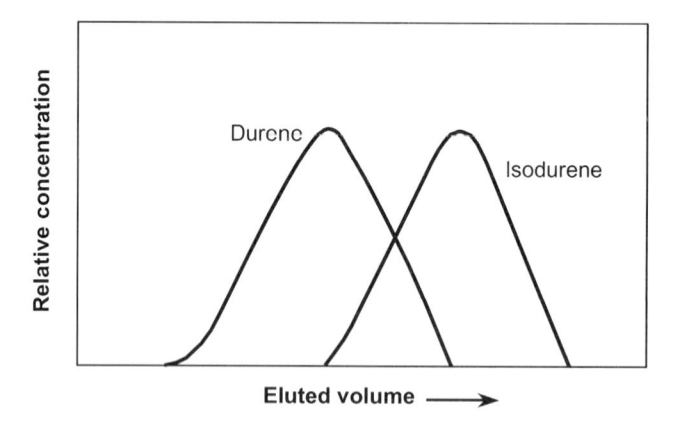

Figure 6.9 Separation of isodurene and durene using LiY adsorbent.

than durene (Figure 6.9). These results are consistent to what is expected since zeolite Y (Si/Al of 2.5) is more acidic than zeolite X (Si/Al of 1.25).

Figure 6.10 shows separation of fructose and glucose using adsorbents CaY, CaX and KX [37]. The combination of the zeolite framework structure and exchanged cations are instrumental to this intricate mechanism as seen with the separation of fructose and glucose. Ca^{2+} forms complexes with fructose and, thus, CaY shows a high selectivity for fructose over glucose. In contrast, CaX does not exhibit high selectivity for fructose. However, KX performs good separation for fructose. No simple explanation can be offered for this elaborate system.

6.5.1.4 Moisture Content in Zeolite

Adsorbed water molecules on a zeolite adsorbent are polarizable due to a strong electrostatic field between the exchanged cations and alumina framework [26].

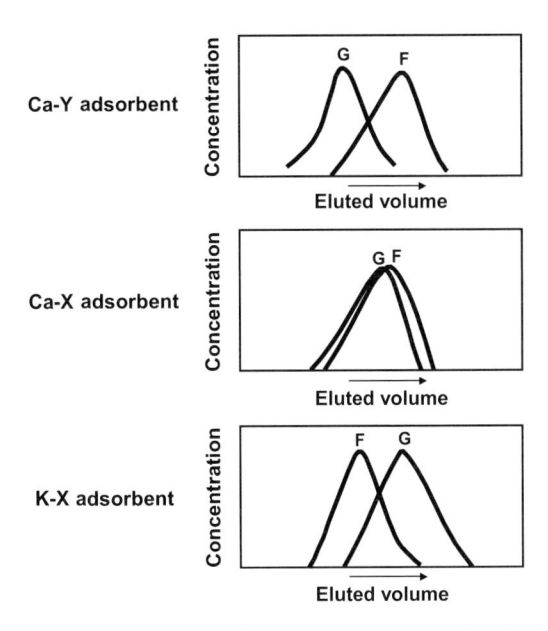

Figure 6.10 Fructose/glucose separation on zeolite adsorbents.

Hence, water molecules enhance the acidic properties of the zeolite's Brönsted acids. Adsorbate–adsorbent interactions and, therefore, adsorbent selectivity and adsorbate mass transfer rates are altered due to water polarization. When developing an adsorbent to be used in a commercial adsorptive separation process, the water content of the adsorbent is adjusted to balance adsorbent selectivity and component mass transfer rate.

6.5.1.5 Characteristics of the Desorbent

In liquid phase adsorption, some particular components of the feed steam are selectively adsorbed or extracted by a solid zeolitic adsorbent. At the same time, other components of the feed stream are rejected by the adsorbent. At equilibrium, the liquid composition within the zeolite pores differs from that of the liquid surrounding the zeolite. In the process, a second liquid component, the desorbent, is also introduced into the system. The function of the desorbent is to desorb and recover the extracted feed components from the adsorbent. In order for the desorbent to perform well in the process, a suitable interactive force between the desorbent and the extracted components to the adsorbent is required. If the selectivity is too high, it requires high desorbent volume to desorb the extracted components from the adsorbent. If the selectivity is too low, the desorbent tends to compete with extracted components for capacity of adsorbent.

A basic guideline for choosing a desorbent is to match the chemical properties of the extracted components and the desorbent, along with appropriate selection of boiling point differences to allow recovery from the feed components after the

Table 6.8 Separation of p-xylene using KY and BaX adsorbents with various desorbents [38, 39] by pulse tests.

Adsorbent	Desorbent	PX net retention volume (ml)	PX/EB selectivity	PX/MX selectivity	PX/OX selectivity	PX stage time (s)
BaX	Phenyldecane	107	1.84	3.02	2.83	22.2
KY	Phenyldecane	No PX desorbent				
KY	Diphenylmethane	10.5	1.68	3.08	1.89	29.7
BaX	Diphenylmethane	0	1	1	1	
BaX	1,3-Diisopropylbenzene	No PX desorbent				
BaX	1,4-Diisopropylbenzene	56.4	2.28	3.45	3.34	19.8
KY	1,4-Diisopropylbenzene	No PX desorbent				
BaX	1,3,5-Triethylbenzene	77.6	1.3	3.0	2.32	43.9
KY	1,3,5-Triethylbenzene	61.4	2.33	7.58	5.23	40.8
BaX	5-Tertiarybutyl-m-xylene	73.2	1.94	3.28	2.87	23.4
KY	5-Tertiarybutyl-m-xylene	47.1	2.05	3.72	3.37	54.5

adsorption section. The following examples illustrate this concept: (i) for a n-paraffin separation, a n-paraffin desorbent is preferable, (ii) for an aromatic separation, an aromatic desorbent is preferable, (iii) for a highly polar adsorbate, a desorbent should be selected from a class of alcohol, ketone, acid, or water.

Desorbent features are illustrated in Table 6.8 using C_8-aromatic adsorbates with BaX and KY adsorbents. The results in Table 6.8 further emphasize the desorbent characteristic requirement mentioned above. For instance, phenyldecane is a suitable desorbent for PX separation using BaX adsorbent. However, phenyldecane is too weak to desorb PX from KY adsorbents. In contrast, diphenylmethane offers good separation of PX with KY adsorbent but not with BaX. With BaX adsorbent, PX is separated from other C_8-aromatics using 1,4-diisopropylbenzene but not with other isomers of diisopropylbenzene, such as 1,3-diisopropylbenzene.

6.5.1.6 Operating Temperature

Another critical variable in liquid phase adsorptive separation is the operating temperature. Liquid phase adsorption must be operated at a temperature that optimizes selectivity and mass transfer rates. Generally, selectivity increases and transfer rates decrease at lower temperatures.

6.5.1.7 Operating Pressure
In general, operating pressure does not affect liquid phase adsorption. However, sufficient pressure must be applied to maintain the system in the liquid phase during the entire process.

6.5.2
Rate-Selective Adsorption

Although most of the commercial adsorptive separation processes are operated under the selective-equilibrium adsorption mechanism, adsorptive separation may also be based on diffusion rates through a permeable barrier which are designated as "rate-selective adsorption" processes. In some instances there may be a combination of equilibriums as well as rate selective adsorption. A rate-selective adsorption process yields good separation when the diffusion rates of the feed components through the permeable barrier differ by a wide margin.

Examples of rate-selective adsorption are demonstrated using silicalite adsorbent for separation of C_{10}–C_{14} n-paraffins from non-n-paraffins [40, 41] and C_{10}–C_{14} mono-methyl-paraffins from non-n-paraffins [42–45]. Silicalite is a ten-ringed zeolite with a pore opening of 5.4 Å × 5.7 Å [22]. In the case of n-paraffins/non-n-paraffins separation [40, 41], n-paraffins enter the pores of silicalite freely, but non-n-paraffins such as aromatics, naphthenes and iso-paraffins diffuse into the pores more slowly. However, the diffusion rates of both normal n-paraffins and non-n-paraffins increase with temperature. So, one would expect to see minimal separation of n-paraffins from non-n-paraffins at high temperatures but high separation at lower temperature.

For the rate-selective separation of C_{10}–C_{14} mono-methyl-paraffins from non-n-paraffins [42–45], diffusion simulations were carried out using the Solids Diffusion module in the Accelrys Insight II molecular modeling package [44]. The modeling results from the diffusion simulations of four paraffins of varying carbon numbers in silicalite are summarized in Table 6.9.

Table 6.9 Energetics and predictions from modeling molecular diffusion in silicalite.

Organic molecules	Energy barrier (kcal/mol)	Maximum energy (kcal)	Diffusion prediction
n-Decane	2	−31	Fast
2-Methylnonane	8	−29	Moderate
2,6-Dimethyloctane	15	−17	Slow
3,3,5-Trimethylheptane	45	+35	Excluded

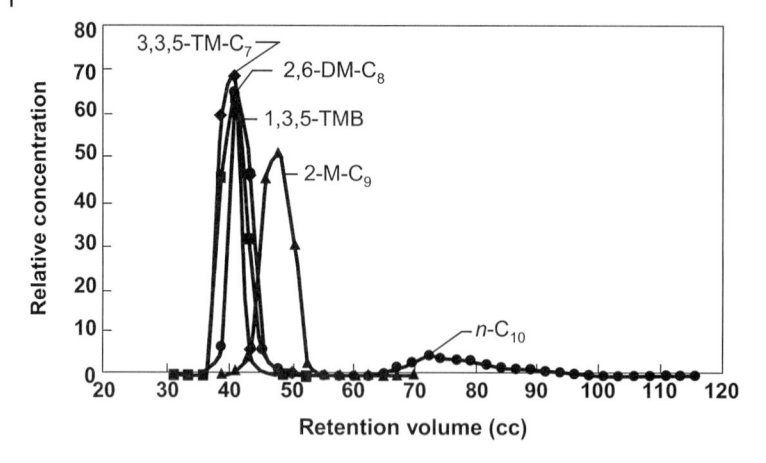

Figure 6.11 Chromatographic separation of mono-methyl paraffin.

Note that the predicted diffusivity of the molecules, based on the magnitude of the energy barrier, agrees with the experimental data (see Table 6.9). The retention volumes and hence diffusivities increase in the following order: 3,3,5-trimethylheptane < 2,6-dimethyloctane < 2-methylnonane < n-decane. The increase in retention volumes of the paraffins is consistent with the decrease in barriers energies. Both the calculations and pulse test experiments indicate that the tri-substituted paraffin, 3,3,5-trimethylheptane, is too large to fit in silicalite pores. The di-substituted paraffin, 2,6-dimethyloctane, has limited diffusivity and is mostly excluded from the pores. The mono-substituted paraffin, 2-methylnonane has good diffusion but is slower than the n-paraffin, n-decane. These results suggest that, to a great extent, normal and mono-methyl paraffins can be separated from di-/tri-substituted paraffins, naphthenes and aromatics using silicalite adsorbent. Laboratory pulse tests have, in fact, demonstrated these separations, as shown in Figure 6.11. Separation of mono-methylparaffin from depleted kerosene (removal of n-paraffins) was also demonstrated in a simulated moving bed (SMB) pilot plant with >90% mono-methylparaffins purity at >70% recovery [44, 45].

Another example of rate-selective adsorption is the separation of diisopropylbenzene isomers using a silicalite adsorbent. Figure 6.12 shows the adsorption rates of 1,3-diisopropylbenzene and 1.4-di-isopropylbenzene into silicalite adsorbent. In particular, it illustrates the more rapid adsorption of 1,4-di-isopropylbenzene compared to 1,3-di-isopropylbenzene.

6.5.3
Shape-Selective Adsorption

Shape-selective adsorption, also known as molecular sieving, is a process that separates molecules based on inclusion or exclusion from specific zeolite pores. In contrast, the equilibrium- and rate-selective mechanisms are based on adsorb-

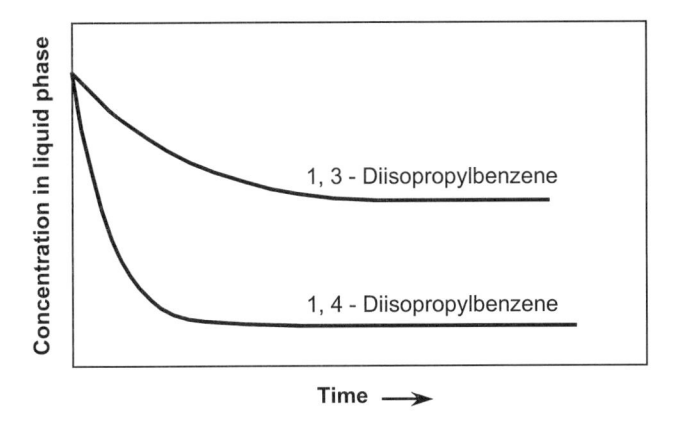

Figure 6.12 Rate-selective adsorption of diisopropylbenzene isomers on Silicalite adsorbent.

ate-adsorbent interactions and molecular diffusion rates through zeolite pores. Thus, for shape-selective adsorption, the theoretical selectivity between molecules that can and cannot enter a specific zeolite pore can be infinite. However, in most cases, adsorbate–adsorbent and/or adsorbate–adsorbate interactions are also involved in the shape-selective mechanism.

Figure 6.2 illustrates the separation of n-$C_{5/6}$ and non-n-$C_{5/6}$ in CaA molecular sieves or 5A. The separation mechanism is obvious when the kinetic diameter of the molecules and molecular sieve pore size opening are compared. n-$C_{5/6}$ have kinetic diameters of less than 4.4 Å which can diffuse freely into the 4.7 Å pores of the CaA molecular sieve, while non-n-$C_{5/6}$ have kinetic diameters of 6.2 Å. A commercial example of shape-selective adsorption is the UOP Molex process, which uses CaA molecular sieves to separate C_{10}–C_{14} n-paraffins from non-n-paraffins (aromatics, branched, naphthenes).

6.5.4
Ion Exchange

Zeolitic adsorbents are composed of a large number of ionic (or potentially ionic) sites. Zeolites are crystalline, hydrated aluminosilicates containing most commonly Na^+, K^+ and H^+ cations. For most zeolites, the aluminosilicate structure is a three-dimensional open framework of AlO_4 and SiO_4 tetrahedrals linked to each other by oxygen molecules. The framework contains channels and interconnected voids occupied by cations and water molecules. The cations are mobile and can be exchanged with other cations to varying degrees. In ion exchange separation, cations in liquid are reversibly exchanged with cations in a solid adsorbent. More specifically, cations are interchanged with other cations without changing the structure of zeolites. At equilibrium, electroneutrality in both zeolite and liquid phases is maintained. Liquid separation based on ion exchange is important in industrial applications. Critical factors to consider in developing an ion exchange separation process are adsorbent capacity, selectivity and kinetics. These factors are described in the following sections.

6.5.4.1 Ion Exchange Capacity

Ion exchange capacity is a measure of the quantity of cations adsorbed or removed by the zeolitic adsorbent. As described in Section 6.5.1.3, the total capacity of a zeolite is a function of its SiO_2/Al_2O_3 mole ratio. Theoretical ion exchanged capacities of some common zeolites are calculated and summarized in Table 6.4.

6.5.4.2 Ion Exchange Selectivity

In general, zeolites have higher ion exchange selectivity for higher-charged cations. For cations having the same valence, the ion exchange selectivity often depends on the hydrated ionic radius. This is seen from the zeolite ion exchange selectivity decreasing with ionic radii [23].

$$Cs^+ > Rb^+ > K^+ > Na^+ > Li^+$$
$$Ba^{++} > Sr^{++} > Ca^{++} > Mg^{++}$$

6.5.4.3 Kinetics

Ion exchange involves the formation and breakage of bonds between ions in solution and exchange sites in a zeolitic adsorbent. The reaction equilibrium of the ion exchange process depends most significantly on contact time, operating temperature and ionic concentration.

6.5.5
Reactive Adsorption

In conventional chemical process industry designs, a chemical reactor is typically sequenced with a downstream separator. The reaction is carried out initially to convert raw materials into value added products which are then isolated and recovered in the separator. The operating conditions of the reactor and separator can be varied to achieve product yield and purity. However, the variations of operating conditions to achieve optimum performance are subject to prevailing constraints. In many cases, recycle streams are incorporated into the process to reprocess un-reacted raw materials or intermediate by-products back through the reactor and separator to increase overall process yields. To overcome some of the sequential process constraints, reactive separation is an alternative choice. The reactive separations, such as the reactive adsorption process, combine the unit operations of reaction and zeolite adsorption into a single process operation with simultaneous reaction and separation. In combining sequential processing steps into an integrated processing, one can eliminate one or more recycle streams which are associated with optimizing performance of the original sequential process configuration. Besides eliminating some of the recycling streams, the integration may lead to the design of a separation process which cannot be achieved with separate reactor and separator process flow elements. Other advantages of reactive adsorption over the sequential reaction and adsorption include reducing energy and capital costs, overcoming the equilibrium-limited conversion, operating at less severe conditions and increasing reaction efficiency. More detailed information on reactive adsorption can be found in a book entitled *Reactive Separation Process* [46].

Developments in reactive separation date back to applications of simple chemical treatments including the use of acidic clays for the removal of olefins in hydrocarbon streams, acid-catalyzed polymerization over clay beds and the use of solid KOH to remove sulfur from various hydrocarbon streams [47]. More recent developments have followed on the heels of the discovery and development of better-engineered synthetic adsorbents and catalysts for applications in the food, biotechnology, pharmaceutical, chemical, refining, environmental and nuclear industries. For example, Ag-substituted molecular sieves and ion exchange resins have been used to remove trace iodides by reaction and precipitation from vapor and liquid streams to facilitate safe operation of nuclear reactors [48] and to purify acetic acid produced by methanol carbonylation [49–54]. Many reactive adsorption processes developed to date utilize traditional fixed bed and fluidized/moving bed absorber designs. However, much of the recent development efforts have been focused on improving purification processes by reactive chromatographic methods and simulated moving bed technologies.

6.6
Summary

Adsorptive separation is a powerful technology in industrial separations. In many cases, adsorption is the only technology available to separate products from industrial process streams when other conventional separation tools fail, such as distillation, absorption, membrane, crystallization and extraction. It is also demonstrated that zeolites are unique as an adsorbent in adsorptive separation processes. This is because zeolites are crystalline solids that are composed of many framework structures. Zeolites also have uniform pore openings, ion exchange ability and a variety of chemical compositions and crystal particle sizes. With the features mentioned, the degree of zeolite adsorption is almost infinite. It is also noted that because of the unique characteristics of zeolites, such as various pore openings, chemical compositions and structures, many adsorption mechanisms are in existence and are practiced commercially.

Acknowledgments

Parts of this chapter are reprinted from the chapter "Liquid separations" by Santi Kulprathipanja and James A. Johnson, in *Handbook of Porous Solids*, edited by Ferdi Schuth, Kenneth S.W. Sing and Jens Weitkamp (2002), Wiley–VCH, with permission.

References

1 Humphrey, J.L. and Keller, G.E., II (1997) *Separation Process Technology*, Mcgraw-Hill, USA.
2 Humphrey, J.L., Sebert, A.F., Koort, R.A. (1991) Separation Technologies – Advances and Priorities, U.S. DOE Final Report, Contract No. DE-AC07-90ID12920-1, 3 Feb.
3 Kulprathipanja, S. (1999) Process for adsorptive separation of metaxylene. U.S. Patent 5,900,523.
4 Ichikawa, Y. and Iwata, K. (1977) Separation of m-xylene. Japanese Patent 52,000,933.
5 Shioda, T., Yasuda, H., Saito, N., Asatari, H., and Furuichi, H. (1995) Method for Separation of m-Xylene. Japanese Patent 07,033,689.
6 Neuzil, R.W. (1982) Separation of m-xylene. U.S. Patent 4,326,092.
7 Kulprathipanja, S. (1995) Process for the adsorptive separation of metaxylene from aromatic hydrocarbons. U.S. Patent 5,382,747.
8 Yue, Y.H., Tang, Y., Liu, Y., and Gao, Z. (1996) Chemical liquid deposition zeolites with controlled pore opening size and shape-selective separation of isomers. *Ind. Eng. Chem. Res.*, **35** (2), 430–433.
9 Kulprathipanja, S. (1988) Separation of citric acid from fermentation broth with a neutral polymeric adsorbent. U.S. Patent 4,720,579.
10 Kulprathipanja, S., Oroskar, A.R., and Priegnitz, J.W. (1989) Separation of citric acid from fermentation broth with a weakly basic anionic exchange resin adsorbent. U.S. Patent 4,851,573.
11 Kulprathipanja, S. and Oroskar, A.R. (1991) Separation of an organic acid from a fermentation broth with an anionic polymeric adsorbent. U.S. Patent 5,068,419.
12 Kulprathipanja, S. (1989) Separation of citric acid from fermentation broth with a strongly basic anionic exchange resin adsorbent. U.S. Patent 4,851,574.
13 Kulprathipanja, S. and Oroskar, A.R. (1991) Separation of lactic acid from fermentation broth with an anionic polymeric absorbent. U.S. Patent 5,068,418.
14 Kulprathipanja, S. (2007) Separation of citric acid from gluconic acid in fermentation broth using a weakly or strongly basic anionic exchange resin adsorbent. U.S. Patent 7,241,918 B1.
15 Dixit, L. and Roa, T.S.R. (1996) Polarizability model of acidity of zeolites. *Zeolites*, **16**, 287–293.
16 Mortier, W.J. (1978) Zeolite electronegativity related to physicochemical properties. *J. Catal.*, **55**, 138–145.
17 Barthomeuf, D. (1991) Acidity and basicity in zeolites, in *Catalysis and Adsorption by Zeolites* (eds G. Ohhlmann et al.), Elsevier Science, Amsterdam, pp.157–169.
18 Barthomeuf, D. (1996) Basic zeolites: characterization and uses in adsorption and catalysis. *Cat. Rev. Sci. Eng.*, **34**, 521–612.
19 Kraikul, N., Rangsunvigit, P., and Kulprathipanja, S. (2006) Study on the adsorption of 1,5-, 1,6- and 2,6-dimethyl-naphthalene on a series of alkaline and alkaline earth ion-exchanged faujasite zeolites. *AdsorptionI*, **12** (5–6), 317–327.
20 Ruthven, D.M. (1984) *Principles of Adsorption and Adsorption Processes*, John Wiley & Sons, Inc., USA.
21 Break, D.W. (1973) *Zeolite Molecular Sieves*, John Wiley & Sons, Inc., USA.
22 Flanigen, E.M., Benneth, J.M., Gross, R.W., Cohen, J.P., Patten, R.I., Kirchner, R.M., and Smith, J.V. (1978) Silicalite, a new hydrophobic crystalline silica molecular sieve. *Nature*, **271**. 512.
23 Sherman, J.D. (1984) Ion exchange separations with molecular sieve zeolites, in *Zeolites: Science and Technology* (eds G. Ohhlmann, F.R. Ribeiro, A.E. Rodrigues, L.D. Rollmann, C. Naccache), NATO Scientific Affairs Division/Martinus Nijhoff, The Hague, pp.583–623
24 Namba, S., Kanai, Y., Shoji, H., and Yashima, T. (1984) Separation of p-isomers from disubstituted benzenes by means of shape-selective adsorption on mordenite and ZSM-5 zeolites. *Zeolite*, **4**, 77–80.

25 Seko, M., Miyake, T., and Inada, K. (1979) Economical *p*-xylene and ethylbenzene separated from mixed xylene. *Ind. Eng. Chem. Prod. Res. Dev.*, **18** (4), 263–268.

26 Ward, J.W. (1968) The nature of active sites on zeolites. III. The alkali and alkaline earth ion-exchanged forms. *J. Catal.*, **10**, 34–46.

27 Kilpatrick, M. and Luborsky, I.E. (1953) The base strengths of aromatic hydrocarbons relative to hydrofluoric acid in anhydrous hydrofluoric acid as the solvent. *J. Am. Chem. Soc.*, **75**, 577.

28 Kulprathipanja, S. (1995) Process for the adsorptive separation of metaxylene from aromatic hydrocarbons. U.S. Patent 5,382,747.

29 Kulprathipanja, S. (1983) Separation of bi-alkyl-substituted monocyclic aromatic isomers with pyrolyzed adsorbent. U.S. Patent 4,423,279.

30 Shimura, M., Wakamatsu, S., and Shirato, Y. (1996) Separation of *p*-xylene by using zeolitic adsorbents. Japanese Patent 08,217,700.

31 Zinnen, H.A. (1989) Zeolitic *p*-xylene separation with tetralin heavy desorbent. U.S. Patent 4,886,930.

32 Kulprathipanja, S., Kuhnle, K.K., and Patton, M.S. (1993) Process for separating durene from substituted benzene hydrocarbons. U.S. Patent 5,223,589.

33 Suntornpun, R. (2002) Acid-base interaction between C_8 aromatics and X and Y zeolites. M.S. Thesis in Petrochemical Technology. The Petroleum and Petrochemical College, Chulalongkorn University, Bangkok, Thailand.

34 Limsamutchaikul, S. (2003) C_8 aromatics adsorption: effects of zeolite acidity, M.S. Thesis in Petrochemical Technology. The Petroleum and Petrochemical College, Chulalongkorn University, Bangkok, Thailand.

35 Rabo, J.A. and Gajda, G.J. (1989–1990) Acid function in zeolites: recent process. *Catal. Rev. Sci. Eng.*, **31** (4), 385–430.

36 Heidler, R., Janssens, G.O.A., Mortier, W.J., and Schoonheydt, R.A. (1996) Charge sensitivity analysis of intrinsic basicity of faujasite-type zeolites using the electronegativity equalization method (EEM). *J. Phys. Chem.*, **100** (50), 19728–19734.

37 Johnson, J.A. and Kulprathipanja, S. (1989) Proceedings of The International Conference on Recent Development in Petrochemical and Polymer Technologies, Petroleum and Petrochemical College, Chulalongkorn University, Bangkok, Thailand, Dec. 12–16.

38 Kulprathipanja, S. (1996) Adsorptive separation of para-xylene with high boiling desorbents. U.S. Patent 5,495,061.

39 Kulprathipanja, S. (1998) Adsorptive separation of para-xylene using isopropylbenzene desorbent. U.S. Patent 5,849,981.

40 Kulprathipanja, S. and Neuzil, R.W. (1984) Low temperature process for separating hydrocarbons. U.S. Patent 4,455,444.

41 Kulprathipanja, S. (2001) Monomethyl paraffin adsorptive separation process. U.S. Patent 6,222,088 B1.

42 Kulprathipanja, S. (2001) Process for monomethyl acyclic hydrocarbon adsorptive separation. U.S. Patent 6,252,127 B1.

43 Sohn, S.W., Kulprathipanja, S., and Rekoske, J.E. (2003) Monomethyl paraffin adsorptive separation process. U.S. Patent 6,670,519 B1.

44 Kulprathipanja, S., Rekoske, R.E., Gatter, M.G., and Sohn, S.W. (2003) Proceedings of the Third Pacific Basin Conference on Adsorption Science and Technology, Kyongju, Korea, May 25–29.

45 Kulprathipanja, S. (1991) Adsorptive separation process for the purification of heavy normal paraffins with non-normal hydrocarbon pre-pulse stream. U.S. Patent 4,992,618.

46 Kulprathipanja, S. (2002) *Reactive Separation Process*, Taylor & Francis, New York, USA.

47 Bland, W.F. and Davidson, R.L. (1967) *Petroleum Processing Handbook*, McGraw-Hill, New York, USA.

48 Pence, D.T. and Macek, W.J. (1970) Silver Zeolite: Iodide Adsorption Studies. The U.S. Atomic Energy Commission, Idaho Operations Office, Under Contract #AT(10-1)-1230, Nov.

49 Hilton, C.B. (1986) Removal of iodide compounds from nonaqueous organic media. U.S. Patent 4,615,806.

50 Kulprathipanja, S., Spehlmann, B.C., Willis, R.R., Sherman, J.D., and Leet, W.A. (1999) Method for treating an organic liquid contaminated with an iodide compound. U.S. Patent 5,962,735.

51 Kulprathipanja, S., Vora, B.V., and Li, Y. (1999) Method for treating a liquid stream contaminated with an iodide-containing compound using a solid absorbent comprising a metal phthalocyanine. U.S. Patent 6,007,724.

52 Kulprathipanja, S., Lewis, G.J., and Willis, R.R. (2001) Method for treating a liquid stream contaminated with an iodide-containing compound using a cation-exchanged crystalline manganese phosphate. U.S. Patent 6,190,562 B1.

53 Kulprathipanja, S., Sherman, J.D., Napolitano, A., and Markovs, J. (2002) Method for treating a liquid stream contaminated with an iodide-containing compound using a cation-exchanged zeolite. U.S. Patent 6,380,428 B1.

54 Kulprathipanja, S., Vora, B.V., and Leet, W.A. (2003) Combination pretreatment/adsoption for treating a liquid stream contaminated with an iodine-containing compound. U.S. Patent 6,506,935 B1.

7
Liquid Industrial Aromatics Adsorbent Separation
Stanley J. Frey

7.1
Introduction

Zeolites have been used in the industrial adsorptive purification of aromatic petrochemicals since the early 1970s. The application of zeolites to aromatic adsorptive purification and extraction is a particularly suitable fit because of three major factors. The first is the inherent difficulty involved in separating certain aromatic components by distillation. Petrochemical production requires individual components be obtained in very high purity, often in excess of 99.5%. While distillation is the most popular method of separation in the petrochemical industry, it is not well suited for the final step of producing high purity single component streams from close boiling multi-component aromatics-rich mixtures.

The second factor that enables the use of zeolites in the separation and purification of aromatics is the development of naphtha reforming and naphtha steam cracking technologies for the high volume production of aromatic streams that are quite free of heteroatom contaminants and contain primarily the simplest aromatics. Typically a molecule's polarity is the chemical characteristic that most strongly determines its affinity for the zeolite. Sulfur, nitrogen or oxygen-containing hydrocarbons in the source streams to zeolitic separation processes strongly adsorb on the zeolite if they are included in the feed stream, reducing the adsorptive capacity for the desired aromatic compounds to impractical levels [1]. Before the 1950s the primary source of the aromatics for petrochemicals was derived from coal via coke ovens which produced low yields of liquids boiling at <200 °C. This material, known as "coal tar," contains desirable aromatics along with many heteroatom contaminants [2]. With the advent of technologies that generate large volumes of relatively contaminant-free aromatic mixtures from naphtha, especially hydrotreating, catalytic reforming and to a lesser extent pyrolysis gasoline from naphtha steam cracking, the way was made easier (albeit unknowingly) to apply zeolites that might otherwise be susceptible to poisoning by polar hydrocarbons.

The third factor that makes zeolites particularly suitable for simple aromatic separation is the subtle difference in adsorptive affinity between the various aromatics onto the zeolite. Zeolites, especially the eight-member ring (e.g., ZSM-5) and 12-member ring (e.g., X and Y zeolites) varieties, have physical characteristics that allow for discrimination between various aromatic isomers in the commercially interesting eight-carbon number group known as mixed xylenes. (Note: strictly speaking ethylbenzene is not a xylene but is always part of mixed xylene streams produced by catalytic reforming or naphtha cracking, along with the three true xylenes. For convenience and in accord with conventional industrial lexicon we refer to the stream of the four C8 aromatics as "mixed xylenes" throughout this chapter.) For continuous simulated moving bed applications the 12-member ring zeolites have found much more application to aromatic separations than the eight-member ring zeolites, as all of the isomers in mixed xylenes can enter the X and Y faujasite pores [3]. However, the eight-member rings have also been investigated [4].

The liquid phase adsorption processes for aromatics extraction are made economically relevant by the large world demand for aromatic petrochemicals. The global per annum production rates of the highest capacity aromatic petrochemicals derived from reformate or pygas for the recent past are shown in Table 7.1.

Benzene and *para*-xylene are the most sought after components from reformate and pygas, followed by *ortho*-xylene and *meta*-xylene. While there is petrochemical demand for toluene and ethylbenzene, the consumption of these cannot be discussed in the same way as the other four. Toluene is used in such a large quantity in gasoline blending that its demand as a petrochemical pales in comparison. Ethylbenzene from reformate and pygas is typically dealkylated to make benzene or isomerized to make xylenes. On-purpose production of petrochemical ethylbenzene (via ethylene alkylation of benzene) is primarily for use as an intermediate in the production of another petrochemical, styrene monomer. Ethylbenzene plants are typically built close coupled with styrene plants.

The large demand for benzene is due to its use as a starting material in the production of polystyrene, acrylonitrile styrene butadiene rubber, nylons, polycarbonates and linear alkyl benzene detergent. All of these final chemical products that are suitable to form into consumer goods have multiple chemical transformations in various industrial processes to obtain them from benzene. Because the production of benzene does not involve a liquid adsorptive process on a zeolite, these processes are not described here but can be found in other sources. However, it is important to note that benzene is typically a large byproduct from an aromatics

Table 7.1 Approximate world demand of high-volume aromatic petrochemicals.

	10^6 t/year (2005–2008)
Benzene	40
p-Xylene	26
o-Xylene	5
m-Xylene	0.4

complex producing p-xylene, and its economics are important in the profitability of any aromatics complex.

Each of the isomers of the aromatics containing eight carbons has petrochemical use. Because of their utility, the industrial counter-current adsorptive separation technique trademarked as the UOP Sorbex™ processes by UOP LLC has been applied to producing three of the four close-boiling isomers; o-xylene can be economically separated by distillation. The Sorbex processes are the dominant technology for industrial counter-current adsorptive separation. However, other technologies by Axens (Eluxyl technology) [5] and by Toray [6] do exist. The continuous countercurrent adsorptive separation processes (otherwise known as simulated moving bed; SMB) for the separation of p-xylene, m-xylene and ethylbenzene in addition to that of extraction of 2,6 dimethyl naphthalene are now discussed.

7.2
Major Industrial Processes

7.2.1
p-Xylene

7.2.1.1 Industrial Uses and Demand

Since the early 1970s p-xylene has grown to become a large volume petrochemical. It is used primarily for the production of polyester fibers, films and resins, such as PET (polyethylene terephthalate) [7]. Demand for p-xylene has increased tenfold since 1970 to about 26×10^6 t/year. Almost all of this additional production has been by the UOP Parex™ process as shown in Figure 7.1. A baseline production of p-xylene is maintained by crystallization based sites that existed before the SMB adsorptive separation technology was established [8].

Virtually all of the p-xylene demand comes directly from the demand for polyester. Since 1995 a significant fraction of the demand for p-xylene has been driven

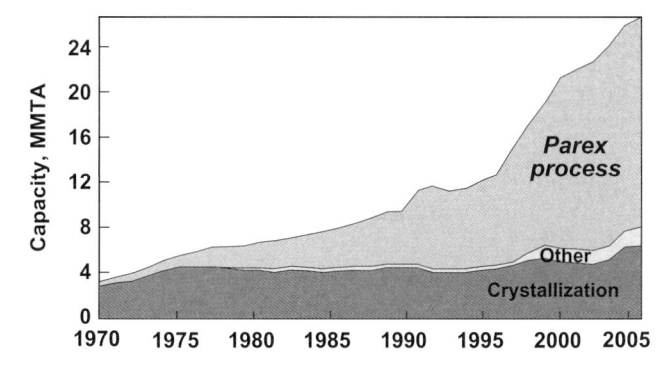

Figure 7.1 Worldwide p-xylene production capacity.

by use of PET for packaging of carbonated beverages and bottled water. PET resin has gas barrier properties making it particularly well suited for these applications [9]. Other major uses of polyester are in the production of textile fibers and engineering thermoplastics. This broad applicability in polyester utilization is due to its impact resistance and superior performance as a gas barrier material.

Prior to polymerization, p-xylene is first oxidized to terephthalic acid (TA) or dimethyl terephtalate (DMT). These diacid or dimethyl ester monomers are then polymerized via a condensation reaction with ethylene glycol to form the polyester. Prior to the development of a method to purify TA to make purified terephtalic acid (PTA, >99% pure) by the Mid-Century Corporation in the 1950s [10], DMT was the primary way to obtain the purified dicarboxylate. The Amoco Oil Company, now part of BP International, made several improvements to the PTA process since its inception [11]. Since the advent of the availability of PTA, it has become the monomer of choice over DMT. PTA avoids the complications of including methanol to enable purification and handling the methanol evolved during the polymerization to polyester.

7.2.1.2 Method of Production

The production of p-xylene begins with petroleum naphtha, as does the production of the other mixed xylene components, benzene and toluene. Naphtha is chemically transformed to the desired petrochemical components and the individual components are recovered at required purity in what is known in the industry as an "aromatics complex" [12]. A generic aromatics complex flow scheme is shown in Figure 7.2. It is useful to briefly review the general flow scheme of this complex for subsequent discussion of the liquid adsorptive processes. The process blocks

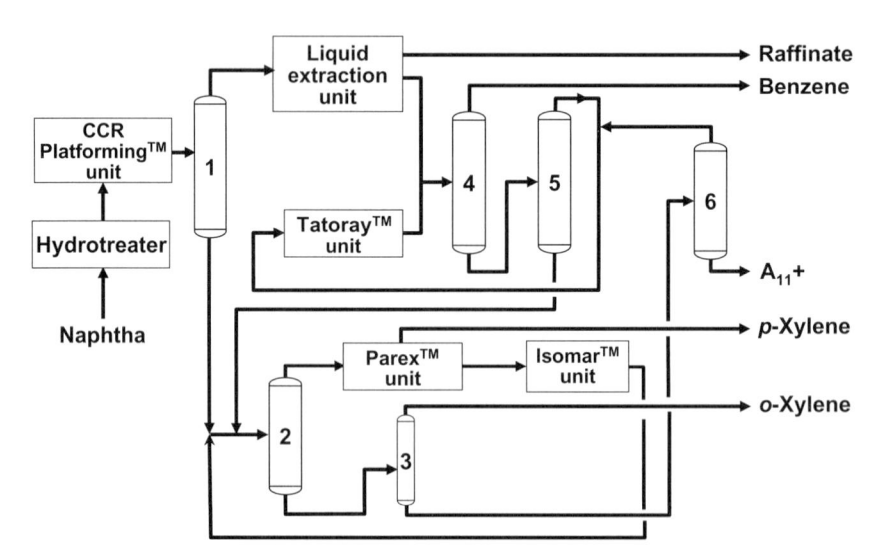

Figure 7.2 Typical aromatics complex with UOP technology.

in Figure 7.2 are labeled according to the corresponding UOP process units. The columns shown are distillation columns.

Naphtha enters the aromatics complex and is hydrotreated to remove sulfur, oxygen and nitrogen-containing organic compounds [13]. The naphthenes and paraffins in the treated naphtha are then reformed to aromatics in the UOP CCR Platforming™ process unit, producing a C5+ stream referred to as reformate [14]. The reformate is then sent to Column 1 where toluene and lighter boilers (A7−) are split from mixed xylene and heavier components (A8+). The A8+ components go to Column 2 where they are sent along with an A8 recycle stream from the UOP Isomar™ process unit and an A8+ stream from the UOP Tatoray™ process unit [15, 16]. A portion of the o-xylene and all the A9+ leaves the bottom of Column 2. The mixed xylenes go overhead and on to the UOP Parex™ process unit. The Parex unit extracts the p-xylene in the stream at 99.7–99.9% purity at a recovery in excess of 97%. The other stream exiting the Parex unit contains the unextracted mixed xylene components. This Parex raffinate stream is sent to the Isomar unit where the mixed xylenes are reacted in the presence of hydrogen over a zeolitic catalyst to an equilibrium mixture of mixed xylenes containing about 22% p-xylene. The equilibrated xylenes are recycled to Column 2 to remove trace A9+ by-product formed in the xylene isomerization reaction. Some o-xylene is recovered in Column 3 from the bottoms stream of Column 2 to produce a >98.5% purity o-xylene product. The A9+ from Column 3 is rerun in Column 4 so that the A11+ components are removed before sending the A9–A10 components to the Tatoray unit. Meanwhile, the A7− aromatics from the Column 1 overhead are extracted by extractive distillation using Sulfolane solvent in the UOP Sulfolane™ process unit [17]. The non-aromatics in the C5–C8 range are rejected as raffinate for use as gasoline or feed to a naphtha steam cracker for ethylene and propylene production. The extracted benzene and toluene are sent to Column 4 along with A6+ produced in the Tatoray unit. A >99.9% benzene product is produced from the overhead of Column 4. The bottoms of Column 4 are sent to Column 5 where the toluene is taken overhead and returned to the Tatoray unit to be transalkylated over a zeolitic catalyst in the presence of hydrogen with the A9–A10 material from Column 6, to produce more benzene and mixed xylenes. The A8+ material from the bottom of Column 5 is sent to Column 2 for recovery of the mixed xylenes produced in the Tatoray unit.

The aromatics complex converts approximately 75% of the feed naphtha to petrochemical aromatics with the vast majority of the remainder being exported as raffinate and some hydrogen. With a modern aromatics complex flowscheme, a little over half of the mixed xylenes are produced in the Tatoray unit while the rest are produced in the CCR Platforming unit directly from the naphtha reforming. Having reviewed the framework of an aromatics complex we are now in a better position to understand the context of the continuous countercurrent liquid adsorptive Parex process which produces the primary aromatics complex product, p-xylene.

Combined mixed xylenes from the CCR Platforming and Tatoray units make up the fresh feed to the Parex unit and Isomar unit loop. A typical composition

Table 7.2 Mixed xylene composition comparison for A_8-producing technologies.

	Reformate mixed xylenes (wt%)	Tatoray mixed xylenes (wt%)
Ethylbenzene	17.5	1.5
p-Xylene	18.5	24.0
m-Xylene	39.5	52.5
o-Xylene	24.0	22.0
Non-aromatics	0.5	0.0

Table 7.3 Mixed xylene composition comparison for A_8 isomerization technologies.

	Mixed xylenes from Isomar ethylbenzene dealkylation catalyst (wt%)	Mixed xylenes from Isomar ethylbenzene isomerization catalyst (wt%)
Ethylbenzene	1.5	8.0
p-Xylene	23.6	19.2
m-Xylene	52.7	45.8
o Xylene	22.2	21.0
Non-aromatics	0.0	6.0

for both of these streams is shown in Table 7.2. The Tatoray unit mixed xylenes derived using modern catalysts typically have an ethylbenzene content around 1%, while the reformate mixed xylenes are much higher in ethylbenzene. As seen in Figure 7.2, the feed to the Parex unit is a mixture of fresh mixed xylenes from the reformer and transalkylation units and recycled Parex raffinate that has been chemically re-equilibrated in the Isomar unit. The recycled mixed xylenes from the Isomar unit typically amount to about 75% of the Parex feed and can vary significantly in concentration depending upon the Isomar catalyst type, as shown in Table 7.3. The selection of Isomar catalyst is usually guided by the degree of desirability of benzene as a by-product. When benzene is desired then it is advantageous to convert the ethylbenzene to benzene at high conversion in the Isomar unit.. Alternatively when minimization of naphtha feed consumption for p-xylene production is the goal, then ethylbenzene to xylene isomerization at a lower per reactor pass conversion is preferred.

The rather low concentration of the desired p-xylene component in the Parex unit feed means a large fraction of the feed stock contains other A8 components that are competing for adsorption sites in the adsorbent zeolite cages. Due to this typically lean feed, a significant hike in the Parex unit capacity can be obtained by even a small increase in the composition of the p-xylene. Techniques to increase the p-xylene feed concentration include greater dealkylation of the ethylbenzene in the Isomar unit by converting from an ethylbenzene isomerization catalyst to

Table 7.4 Examples of capacity expansion in UOP Parex technology.

Location	Original startup	Revamped p-xylene (KMTA)	Increase (%)	Revamped startup
Far East	1987	230	70	1997
Far East	1990	600	33	1997
Far East	1990	230	70	1997
Far East	1980	700	88	1998
Americas	1979	230	109	1998
Far East	1979	625	46	1999
Europe	1981	165	74	2001
Americas	1981	290	49	2001
Far East	1998	350	17	2001
Far East	1995	65	18	2002
SE Asia	1996	542	74	2004
India	1999	1620	35	2004
Far East	1997	540	35	2005
SE Asia	2000	521	25	2005
SE Asia	2001	463	60	2008

a higher activity ethylbenzene dealkylation catalyst, producing a greater fraction of the mixed xylenes via transalkylation (which are lower in ethylbenzene relative to reformate), and using p-xylene-selective toluene disproportion technology to generate an A8 stream with p-xylene concentrations in excess of 85% [18]. A survey of a few Parex units shown in Table 7.4 indicates that the original capacity of a Parex unit has more than doubled in the same adsorbent chamber volume (much of the increase is due to improvements in adsorbent:desorbent combinations and adsorbent chamber design). Not only does an increase in the p-xylene composition of the fresh feed allow for an increase in the Parex unit's capacity, but it also greatly reduces the operating costs by reducing the size of the recycle stream required to convert the residual A8 isomers to p-xylene at a rate limited by the 22% p-xylene mixed xylene equilibrium.

7.2.1.3 Characteristics of Zeolitic Adsorptive Process

Equipment The Parex unit uses Sorbex technology to produce the effect of the simulated counter current moving bed of solid adsorbent relative to the flow of liquid [19]. The equipment used in the adsorption section of a Parex unit is shown in Figure 7.3. The primary equipment pieces are the two adsorbent chambers, liquid flow distributors, piping and pumps directing the circulating flow from the bottom of one chamber to the top of the other, 24 bedlines connecting each distributor grid to the rotary valve, the rotary valve to step the net streams around the 24 distributor grids and a control system to regulate the flows, pressures and rotary valve movement in the adsorption section.

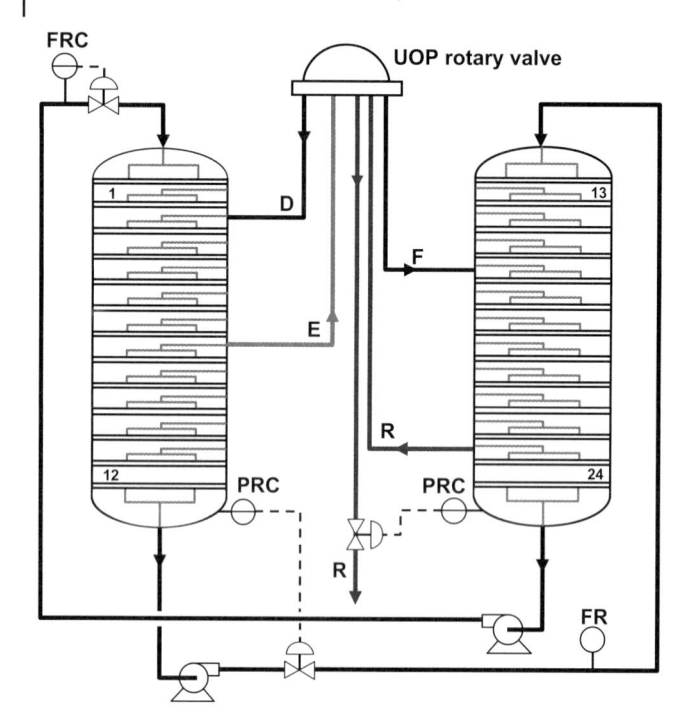

Figure 7.3 UOP Parex unit adsorbent section.

The two adsorbent chambers contain the zeolitic adsorbent, the liquid xylenes and p-diethylbenzene desorbent. Proper loading of the adsorbent into the large diameter vessels in industrial production plants is of critical importance to maximize adsorbent mass in the fixed vessel volume and not generate low and high density areas within the adsorbent bed. Density inconsistencies could adversely affect liquid flow distribution and thereby have a detrimental effect on the performance of the process. Adsorbent loading methods are a matter of proprietary know how of the technology licensors. However, Seko has published a paper on the practical matters involved in an actual problem case [20].

For practical reasons, the adsorbent is divided between two chambers. The liquid flow distributor grids separate the adsorbent into 24 distinct beds. The distributor grids are designed to fully mix the incoming desorbent and feed streams with the circulating liquid before contacting the adsorbent, ensure radial and circumferential mixing of the circulating liquid and to allow for the even withdrawal of the extract and raffinate products [21–23]. The distributor grid design ensures maintenance of ideal plug flow [24]. Twenty-four beds were chosen as an economic balance between the ideal of continuous movement of the net streams around the chamber and adequate frequency of radial and circumferential mixing, versus capital cost of additional distributor grids and related equipment.

The UOP rotary valve has been used in hundreds of Sorbex units across a variety of applications. The purpose of the rotary valve is to simply move the inlet and outlet ports of the net streams (feed, desorbent, raffinate, extract) around the 24 beds in stepwise fashion, creating a semi-continuous countercurrent flow of adsorbent relative to the entry and exit points of the net streams to and from the adsorbent chambers [25]. The rotary valve consists of a rotor plate pressed against a stator plate inside a pressure vessel that indexes the net desorbent, feed, extract and raffinate streams around the adsorbent chambers [26]. An alternative method of moving the inlet and outlet streams around the adsorbent chambers is used in other technologies where multiple automatic on-off valves at each distributor grid inlet are employed [5].

The method of connecting the chambers via the pumps and piping is done according to certain design criteria established to be critical for efficient production. The piping is designed according to specialized criteria to ensure plug flow which, in turn, ensures that the ultimate product purity is not compromised [27].

The raffinate and extract streams leave the UOP Parex unit adsorption section via the UOP rotary valve and are respectively routed to the raffinate and extract columns for separation of the mixed xylene components from the PDEB as shown in Figure 7.4. Because the desorbent has a higher boiling point than the mixed xylenes, the desorbent exits the bottom of the distillation column and is pumped back to the adsorbent chamber section. The mixed xylene raffinate stream is taken as a side cut from the raffinate column to remove water. The extract *p*-xylene stream is taken from the top of the extract column and routed to a finishing column where any toluene that was in the Parex feed is removed. The *p*-xylene product exits the bottom of the finishing column. The adsorbent has some selectivity for toluene as well as *p*-xylene.

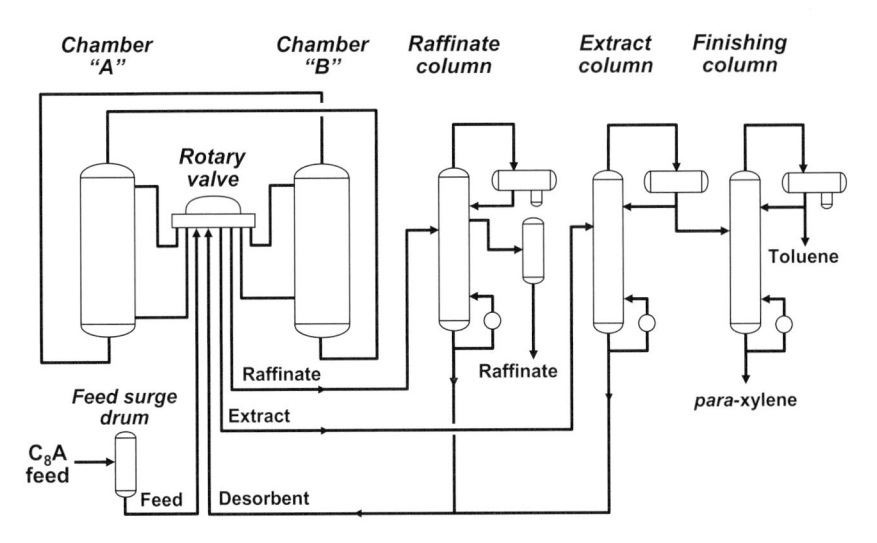

Figure 7.4 UOP Parex process.

Zeolite/Desorbent Combination The desorbent used in the UOP Parex unit is *p*-diethylbenzene (PDEB) [28]. It has been found to have approximately the same affinity for the faujasite zeolite as does *p*-xylene, balancing the amount of desorbent required for *p*-xylene desorption while not excluding the *p*-xylene from adsorbing in the adsorption zone.

PDEB boils at 351 °F (177 °C) while the highest-boiling xylene, *o*-xylene, boils at 291 °F (144 °C) so there is substantial relative volatility to perform the distillative separation economically. Using a desorbent that has a higher boiling point than the feed stock typically results in the most economical process because the desorbent to feed ratio in commercial processes using UOP Sorbex technology is typically higher than unity. Less feed than desorbent results in less material boiling overhead in the raffinate and extract distillation columns at equivalent feed:desorbent relative volatility, resulting in lower reboiler heat requirements.

As mentioned earlier, faujasite X and Y zeolites are favored for aromatics adsorption. The large pore structure promotes fast mass transfer rates into and out of the zeolite cage. Varying the silica:alumina ratio of the base zeolite affects final adsorbent performance. While Pentasil zeolites such as ZSM-5 might be more selective, the mass transfer rates in liquid phase multi-component mixtures are usually relatively slow. It has been noted that a faujasite cage can contain typically up to three mixed xylene molecules. The nature in which these molecules pack into a zeolite cage has a profound effect on zeolite capacity and selectivity. Modification of X and Y zeolites with various ion exchanged alkali earths and alkali metals affect the packing of the mixed xylenes in the cages and the adsorption phenomena [29]. Varying the amounts and combinations of various exchanged ions produces different aromatic component selectivities and capacities.

Adsorption phenomena for aromatics on the X and Y zeolites is quite different than that relevant to the highly shape selective systems used for normal/iso paraffin separations. In aromatics systems with wide-pore zeolites, all molecules adsorb to some finite extent. This means that adsorbent selectivity can actually translate quite practically to UOP Parex unit capacity. Adsorbents with increased selectivity for *p*-xylene in the same zeolitic pore volume means more of the zeolite is occupied with *p*-xylene rather than the other xylenes, increasing the amount of *p*-xylene that can be recovered for a given amount of circulated adsorbent. These selectivity changes are typically affected by modifications in exchanged ion quantity and type, Si:Al ratio and operating conditions under which the separation takes place. The large effect that the multitude of possible zeolitic modifications has on UOP Parex unit capacity provides a great incentive for the continuous development of new adsorbents.

The adsorbent typically consists of the selectively adsorbing zeolite and a binder. The binding material is required to make particles of the zeolites of a large enough size to allow for practical use in a commercial application. The particle formation method and resulting characteristics of the binder/zeolite combination is of critical importance, especially as it relates to mass transfer through the particle and the characteristics of the zeolite [30]. The practical aspects of forming and treating the adsorbent particles are a matter of intellectual property of adsorbent manufacturers. However, a general outline of the steps involved is given by Ruthven [31].

Critical Variables The critical operating variables in a constructed UOP Parex unit are the individual adsorbent chamber zone ratio settings. The adsorbent chamber zone ratio settings are easily manipulated during operation. However, given that the process operates at steady-state, few adjustments actually have to be made on a day-to-day basis. Settling on the optimal zone ratio settings is usually accomplished in the first few weeks of operation, with minor adjustments made periodically thereafter to respond to varying conditions such as feed stock composition. Operating parameters do affect the condition of the zeolite, and certain key ones are critical to adsorbent selectivity and capacity.

Critical variables that affect the separation that cannot be changed in a constructed UOP Parex unit include the style and type of exchanged ions on the adsorbent. The location of the large exchanged ions (e.g., Ba, Cs, Na, K) affects the hydrocarbon packing in the zeolite cages, hydrocarbon access to the zeolite pores and strength of the zeolite–hydrocarbon interactions [32]. Obviously all of these effects have a significant impact on the performance of the chromatographic separation of the *p*-xylene.

The simulated moving bed operational mode involves four distinct functional zones, the adsorption, purification, desorption and buffer zones. These zones are described in detail in other parts of this book. We now examine the function of each zone as it applies to *p*-xylene adsorption and which can be extrapolated to the other aromatics separations.

The faujasite zeolite in the UOP Parex process has some finite affinity for all the aromatic species in the mixed xylene feed, indicated by the fact that selectivities between the components are typically less than five. Because the adsorbent has the tendency to adsorb all aromatic species in the feed to some extent, the fundamental variable dictating the adsorption zone operation is the ratio of zeolitic selective pore volume circulated past the feedpoint by the stepping action of the rotary valve per the volume of aromatics conveyed to the adsorption chambers. Typically this ratio is set to obtain a certain target recovery of *p*-xylene.

At a given amount of selective pore volume circulation, *p*-xylene is adsorbed by an adsorbent that is highly selective for *p*-xylene over the other mixed xylene components. A more selective adsorbent effectively gives the adsorbent more capacity to carry *p*-xylene out of the adsorption zone into the purification zone. Increasing the volumetric flow of *p*-xylene in the fresh feed by concentrating the *p*-xylene in the feed increases the required selective pore volume circulation per aromatic fresh feed by a ratio that is somewhat less than one-to-one. Because the required rise in adsorbent circulation is less than the increase in the volumetric rate of *p*-xylene in the feed, the overall Parex unit capacity is effectively increased for feeds that are more concentrated in *p*-xylene. Finally, the desorbent that enters the adsorption zone from the purification zone has a significant impact on the amount of *p*-xylene that can be adsorbed for a given selective pore volume circulation per aromatic fresh feed ratio. Because the desorbent is selected and the adsorbent designed to provide about equal selectivities for the desorbent as for the desired p-xylene, the desorbent is the strongest competitor for the zeolitic selective pores. Therefore, all desorbent entering the adsorption zone essentially takes available selective pore volume from p-xylene. The nature of flow in simulated moving bed technology, however, necessitates some desorbent flow into the adsorption zone from the purification zone to effect purification. Again, here the selectivity of the adsorbent for the p-xylene over the other mixed xylene components is important because a higher selectivity will allow purification at lower desorbent rates in the purification zone resulting in less desorbent flowing into the adsorption zone.

The purification zone inflows are adsorbent flowing in from the adsorption zone containing the adsorbed p-xylene and, to a lesser extent, other feed components and a liquid flow of desorbent and purified p-xylene from the desorption zone on the opposite end of the zone at the extract point. The fundamental operating variable of the purification zone is the ratio of the net liquid flow rate through the purification zone to the rate of circulating zeolitic selective pore volume. This ratio is the primary control point to maintain purity in an operating UOP Parex unit. The required value of this ratio is affected by the feed p-xylene concentration and adsorbent selectivity for p-xylene. Adsorbents more selective for p-xylene require less desorbent to enter the purification zone (and consequently enter the adsorption zone) because the p-xylene itself is in high enough concentration to displace the other mixed xylene components. As mentioned in the previous paragraph, the operation of the purification zone has a significant impact on the adsorption zone primarily due to the resultant desorbent and refluxed p-xylene that flows from the purification zone to the adsorption zone.

The desorption zone is the simplest of all the SMB zones in that it basically contains only two components – the desired component (p-xylene) and the desorbent (PDEB). The desorbent zone is run such that essentially all of the p-xylene is displaced from the adsorbent by the time the desorbent inlet stream reaches any particular bed of adsorbent. This is accomplished by saturating the adsorbent with PDEB through the desorption zone. The fundamental variable in determining the flow rates through the desorption zone is the ratio of the net flow of desorbent into the desorption zone to the rate of circulating selective zeolitic pore volume. The required amount of desorbent is influenced by the adsorbent selectivity for PDEB versus p-xylene. Adsorbents highly selective for PDEB over p-xylene require fewer theoretical stages to remove the p-xylene in the desorption zone. However, higher PDEB selectivity versus p-xylene reduces the efficiency of the adsorption zone where an adsorbent highly selective for PDEB would reduce the ultimate p-xylene capacity due to PDEB occupying more of the selective zeolitic pores than would p-xylene. The need for a balanced selectivity between the p-xylene and PDEB can be seen in the tension between the optimal operating conditions for the desorption and adsorption zones.

The fourth and final zone in SMB technology is known as the buffer zone which separates the outgoing raffinate stream containing the undesired C8s from the desorbent stream. The desorbent stream must be free of all A8 contaminants entering the desorption zone because impurities in the material in the desorption zone affect the composition of the extract. Any C8 contaminants that might enter the desorption zone would flow out the extract stream with the *p*-xylene and distill with the *p*-xylene in the extract column, resulting in contamination of the product.

7.2.2
m-Xylene

7.2.2.1 Industrial Uses and Demand

Industrial production of *m*-xylene was first accomplished by Mitsubishi Gas Chemical in the early 1980s using an HF/BF_3 ionic liquid [33, 34]. Since then all new *m*-xylene has been produced via the UOP MX Sorbex™ process which, like the Parex process, utilizes SMB principles [35–38]. The liquid phase, adsorptive process is favored for its operational, maintenance and environmental advantages over the previous technology [39]. About 300 000 t/year of *m*-xylene capacity was constructed in the late 1990s using the MX Sorbex process. It is expected that another 300 000 t/year of capacity will be added between 2005 and 2011. Currently world *m*-xylene demand is about 600 000 metric tons per annum or about 2% that of *p*-xylene. The growth in demand of *m*-xylene is closely linked to the demand growth of *p*-xylene, as it is used as a co-monomer in the production of PET-based packaging resins with very low oxygen permeability. Analogous to *p*-xylene processing, *m*-xylene is first oxidized to IPA (purified isophthalic acid) prior to co-polymerization with PTA and ethylene glycol to produce the enhanced gas barrier resins. The required *m*-xylene purity to make acceptable IPA is minimally 99.5%. However, restrictions on the maximum allowable concentration of certain contaminants such as *o*-xylene, *p*-xylene and ethylbenzene are often mandated by IPA manufacturing. These added purity specifications on the *m*-xylene provide operational benefits, especially in environmental control, for IPA manufacturing.

Significant progress is being made in using *m*-xylene as the primary monomer in other resins with low oxygen permeability such as *meta*-xylene diamine [40–42]. A breakthrough in a polymer application where *m*-xylene is the primary component could greatly increase the demand of *m*-xylene [43].

7.2.2.2 Method of Production

The industrial production of *m*-xylene is very similar to that of *p*-xylene. In fact, most of the production of *m*-xylene is done in facilities where a much larger quantity of *p*-xylene is produced. Figure 7.5 is a typical flow diagram for an aromatics complex where *m*-xylene is produced. It is quite like the flow diagram for the production of *p*-xylene except that a fraction of the Parex unit raffinate, containing typically over 60% *m*-xylene, is used as fresh feed to the MX Sorbex unit for *m*-xylene extraction. Because the required *m*-xylene production is typically much lower than that of *p*-xylene and the MX Sorbex fresh feed stream is three times more concentrated than the Parex unit fresh feed stream, the feed stream to the

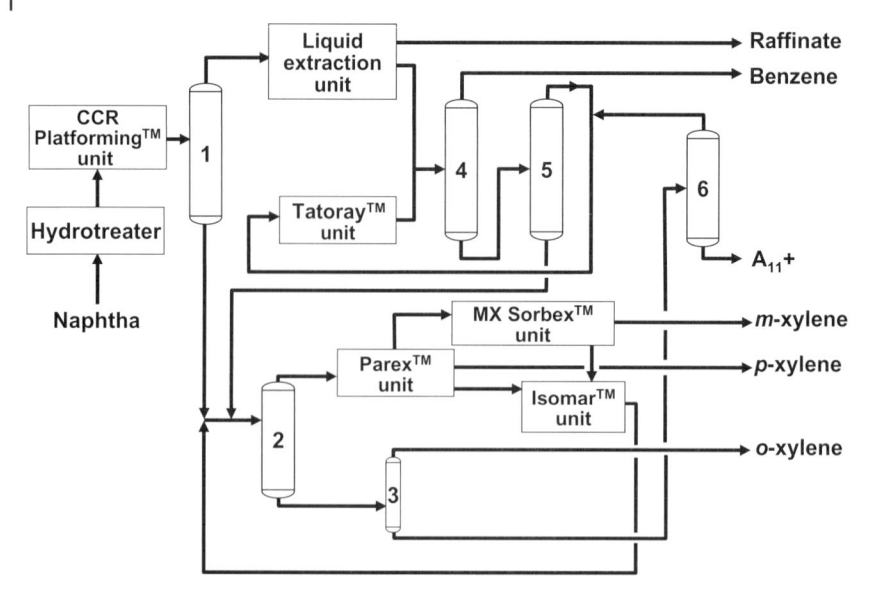

Figure 7.5 Typical aromatics complex with UOP MX sorbex.

MX Sorbex unit is on the order of one-tenth the size of the Parex unit raffinate stream within the same aromatics complex. This flow configuration where *m*-xylene capacity is added to an existing aromatics complex has been commonly adopted. A few *m*-xylene producers do not also produce *p*-xylene at the same site. In such cases, flow configurations have included once through *m*-xylene extraction of a mixed xylene stream or an MX Sorbex unit and Isomar unit integrated within a loop, with the same flow configuration as is conventionally used for a Parex unit and Isomar unit loop.

The Parex and MX Sorbex processes are quite similar with regard to the adsorbent section mechanics so all of the discussion about the functional zones in the section for the Parex process applies to the MX Sorbex process also. The MX Sorbex process produces *m*-xylene at 99.5–99.8% purity at a recovery in excess of 95%. The major differences between the two technologies are the choice of adsorbent and desorbent.

7.2.2.3 Characteristics of Zeolitic Adsorptive Process

Equipment The flow diagram of an MX Sorbex unit is exactly like that of a Parex unit around the adsorbent chambers. Only the flows from the distillation columns are different due to the light desorbent used.

Zeolite/Desorbent Combination A light desorbent, that is a desorbent that boils lighter than the mixed xylene feed, is used in the MX Sorbex process. This means that the energy demand of the distillation columns per unit feed for the MX Sorbex

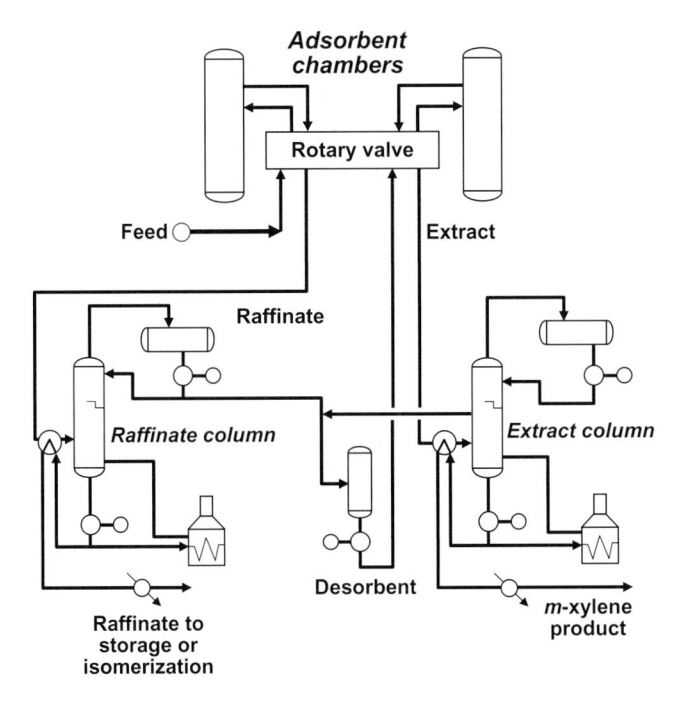

Figure 7.6 MX Sorbex process.

process would be higher than that of a Parex unit however heat integration within the MX Sorbex unit and the fact that the feed is more concentrated in the extracted component, m-xylene, mean that utility cost per product is actually less for the MX Sorbex process than for the Parex process. A typical unit-wide flow diagram for the MX Sorbex process is shown in Figure 7.6.

The adsorbent is an alkali exchanged faujasite bound into a bead. Because m-xylene is by far the most basic of the four mixed xylene isomers, the selectivity of the zeolite is tailored for this chemical trait.

7.3
Other Significant Processes

The Parex, Toray Aromax and Axens Eluxyl processes are the three adsorptive liquid technologies for the separation and purification of p-xylene practiced on a large scale today. The MX Sorbex process is the only liquid adsorptive process for the separation and purification of m-xylene practiced on an industrial scale. We now consider a few other liquid adsorptive applications using Sorbex technology for aromatics separation that have commercial promise but have not found wide application.

7.3.1
2,6-Dimethylnaphthalene

7.3.1.1 Industrial Uses and Demand
The di-aromatic 2,6 dimethylnaphthalene (26DMN) is of particular interest for the production of polyethylene naphthalate (PEN). PEN has superior oxygen barrier and temperature resistance properties relative to PET, making it a top prospect for applications in beer bottling and reusable plastic bottle applications. Because of its superior properties to neat PET (PEN can also be incorporated into PET resin) it has a good possibility of growing in demand.

7.3.1.2 Method of Production
PEN is produced from 26DMN analogously to how PET is produced from *p*-xylene. The 26DMN is oxidized to dimethyl 2,6 naphthalene dicarboxylate and then polymerized with ethylene glycol to produce PEN. The 26DMN can be synthesized by a number of methods mentioned by Lillwitz [44].

An SMB-type process has been disclosed to extract the 26DMN using liquid phase adsorption in a simulated moving bed two stage system [45, 46]. The source of the 26DMN is not specified but could be by one of the methods outlined by Lillwitz. Currently almost all separation of 26DMN is by crystallization.

7.3.2
Ethylbenzene

7.3.2.1 Industrial Uses and Demand
Ethylbenzene is a high volume petrochemical used as the feed stock for the production of styrene via dehydrogenation. Ethylbenzene is currently made by ethylene alkylation of benzene and can be purified to 99.9%. Ethylbenzene and styrene plants are usually built in a single location. There is very little merchant sale of ethylbenzene, and styrene production is about 30×10^6 t/year. For selective adsorption to be economically competitive on this scale, streams with sufficiently high concentration and volume of ethylbenzene would be required. Hence, although technology has been available for ethylbenzene extraction from mixed xylenes, potential commercial opportunities are limited to niche applications.

7.3.2.2 Method of Production
The UOP Ebex™ process has been available for license since the 1970s. This process is a rejective simulated moving bed process where the ethylbenzene is the least adsorbed member of the mixed xylenes and is recovered in high purity in the raffinate stream [47]. Other liquid phase simulated moving bed concepts selective for ethylbenzene have been considered. These would ostensibly require less adsorbent circulation per unit feed because ethylbenzene is typically at <20% concentration in mixed xylenes [48, 49]. A process is disclosed by Broughton [50] that produces a pure *m*-xylene stream along with a pure ethylbenzene stream.

7.3.3
p-Cresol

7.3.3.1 Industrial Uses and Demand
p-Cresol is used primarily for the production of 2,6-di-tert-butyl-p-cresol otherwise known as butylated hydroxyl toluene (BHT). BHT demand is on the order of 5,000 t/year and is used as an antioxidant and stabilizer in plastics and resins (http://www.mindfully.org/Plastic/BHT-Butylated-Hydroxytoluene-DBPC.htm) [51].

7.3.3.2 Method of Production: Adsorbent–Desorbent
A simulated moving bed system has been proposed for the production of p-cresol from mixtures of cresol isomers even derived from coal tar [52]. Neuzil et al. give details of the development of the adsorbent and desorbent system reviewing balancing mass transfer issues with selectivity [53]. The desorbent for the cresol system is 1-pentanol. For these liquid adsorptive systems where highly polar molecules are adsorbed and desorbed with polar desorbents, the tolerance of the system for trace polar contaminants is higher because the feed and desorbent can more easily exchange with them on the surface of the zeolites. Any contaminants that come with the feed tend to not permanently stay on the adsorbent, as in completely non-polar systems seen in the other aromatic applications.

7.4
Summary

The use of Sorbex technology in aromatics production has proved to be the largest application of bulk chemical production by liquid adsorption. The Parex process, first applied commercially in the early 1970s, has been continuously improved through the years, going through five generations of adsorbent and four generations of desorbent. The complexity of the fundamentals of the adsorbent synthesis encompasses different ion selection or combinations thereof, degree of ion exchange, zeolite type, variation in silica:alumina ratio, particle-forming method, binder selection and other parameters. Add to that different methods to enrich feed stocks, selection of desorbent and continuous improvements in the adsorbent chamber process equipment and design and it is not surprising that simulated moving bed technology continues to see consistent gains in productivity even after being practiced commercially for almost 40 years by dozens of operating companies.

Expansion of Sorbex technology to the production of m-xylene shows how the process concept can be used for multiple applications in separations that cannot be performed by other means. One can expect that, as demand for new, difficult to separate aromatics increases, the simulated moving bed liquid adsorption processes can provide a means for production.

References

1 Methivier, A. (1998) Influence of oxygenated contaminants on the separation of C8 aromatics by adsorption of faujasite zeolites. *Ind. Eng. Chem. Res.*, **37**, 604–608.

2 Wittcoff, H. and Reuben, B. (1996) *Industrial Organic Chemicals*, John Wiley & Sons, Inc., New York, pp. 348–350.

3 Ruthven, D. (1984) *Principles of Adsorption and Adsorption Processes*, John Wiley & Sons, Inc., New York, pp. 9–19.

4 Guo, G.-Q., Chen, H., and Ying-Cai, L. (2000) Separation of p-xylene from C8 aromatics on binder-free hydrophobic adsorbent of MFI zeolite. I. Studies on static equilibrium. *Micropor. Mesopor. Mater.*, **39**, 149–161.

5 Ash, G., Barth, K., Hotier, G., Mank, L., and Renard, P. (1994) Eluxyl: a new paraxylene separation process. *Rev. Inst. Fr. Petrole*, **49** (5), 541–549.

6 Otani, S., Sato, M., Kanaoka, M., Akita, S., Ogawa, D., and Tsuchiya, Y. (1977) Development of p-xylene manufacturing processes utilizing a new continuous adsorption technology. *Am. Soc. Mech. Eng., Appl. Mech. Div.*, **1**, 550–556.

7 Wyeth, N. (1986) *PET – A Global Perspective*, High Performance Polymers: Their Origin and Development, Proceedings of the Symposium. Presented at the 91st Meeting of the American Chemical Society, Elsevier, New York, pp. 417–423.

8 Spiller, C. (1958) Paraxylene purification system. US Patent 2,866,833.

9 Wyeth, N. (1988) Inventing the PET bottle. *Res. Technol. Manag.*, **31** (4), 53–55.

10 Saffer, A. and Barker, R. (1958) Production of aryloxy compounds. US Patent 2,833,816.

11 Lammers, G. (1965) Process for the recovery of paraxylene. US Patent 3,177,265.

12 Meyers, R. (ed.) (2003) *Handbook of Petroleum Refining Processes*, 3rd edn, McGraw-Hill Companies, Inc., New York, pp. 2.3–2.11.

13 Meyers, R. (ed.) (2003) *Handbook of Petroleum Refining Processes*, 3rd edn, McGraw-Hill Companies, Inc., New York, pp. 8.31–8.41.

14 Meyers, R. (ed.) (2003) *Handbook of Petroleum Refining Processes*, 3rd edn, McGraw-Hill Companies, Inc., New York, pp. 4.3–4.31.

15 Meyers, R. (ed.) (2003) *Handbook of Petroleum Refining Processes*, 3rd edn, McGraw-Hill Companies, Inc., New York, pp. 2.39–2.46.

16 Meyers, R. (ed.) (2003) *Handbook of Petroleum Refining Processes*, 3rd edn, McGraw-Hill Companies, Inc., New York, pp. 2.55–2.63.

17 Meyers, R. (ed.) (2003) *Handbook of Petroleum Refining Processes*, 3rd edn, McGraw-Hill Companies, Inc., New York, pp. 2.13–2.23.

18 UOP LLC (2009) UOP PX-Plus Technical Sheet, http://uop.com/objects/53%20 PX%20Plus.pdf.

19 Broughton, D., Neuzil, R., Pharis, J., and Brearley, C. (1970) The parex process for recovering paraxylene. *Chem. Eng. Prog.*, **66** (9), 70–75.

20 Seko, M., Takeuchi, H., and Inada, T. (1981) Scale-up from chromatographic separation of p-xylene and ethylbenzene. *Ind. Eng. Chem., Prod. Res. Dev.*, **21** (4), 656–661.

21 Broughton, D. (1965) Fluid distributing means for packed chambers. US Patent 3,214,247.

22 Carson, D. (1974) Flow distribution apparatus. US Patent 3,789,989.

23 Harter, I., Darmancier, D., and Renard, P. (2000) Single-phase fluid distributor-mixer-extractor for beds of granular solids. US Patent 6,024,871.

24 Frey, S., Sechrist, P., and Kauff, D. (2008) Flow distribution apparatus. US Patent 7,343,551.

25 Broughton, D. and Gerhold, C. (1961) Continuous sorption process employing fixed bed of sorbent and moving inlets and outlets. US Patent 2,985,589.

26 Carson, D. and Purse, F. (1962) Rotary valve. US Patent 3,040,777.

27 Migliorini, C., Mazzotti, M., and Morbidelli, M. (1999) Simulated

moving-bed unit with extra-column dead volume. *AIChE J.*, **45** (7), 1411–1421.
28. UOP D-1000 Technical Sheet, http://uop.com/objects/D_1000_Desorbent.pdf.
29. Neuzil, R. (1971) Aromatic hydrocarbon separation by adsorption. US Patent 3,558,730.
30. Raksh, J., Tyaqi, B., Badheka, Y., Choudary, V., and Bhat, T. (2003) Effect of clay binder on sorption and catalytic properties of zeolite pellets. *Ind. Eng. Chem. Res.*, **42** (14), 3263–3272.
31. Ruthven, D. (1984) *Principles of Adsorption and Adsorption Processes*, John Wiley & Sons, Inc., New York, p. 19.
32. Furlan, L., Beatriz, C., and Santana, C. (1992) Separation of liquid mixtures of p-xylene and o-xylene in X zeolites: the role of water content on the adsorbent selectivity. *Ind. Eng. Chem. Res.*, **31** (7), 1780–1784.
33. Igarashi, Y. and Ueno, T. (1968) A new xylene separation. *Proc. Am. Chem. Soc., Div. Pet. Chem., Preprints*, **13** (4), A79–A83.
34. McCaulay, D., Shoemaker, B., and Lien, A. (1950) Hydrogen fluoride-boron trifluoride extraction of xylene isomers. *Ind. Eng. Chem.*, **42** (10), 2103–2107.
35. M-xylene gets a market boost. (1999) *Chem. Eng.*, **106** (12), 96.
36. Neuzil, R. (1982) Process for the separation of meta-xylene. US Patent 4,326,092.
37. Kulprathipanja, S. (1995) Process for the adsorptive separation of metaxylene from aromatic hydrocarbons. US Patent 5,382,747.
38. Kulprathipanja, S. (1999) Process for adsorptive separation of metaxylene from xylene mixtures. US Patent 5,900,523.
39. Business update: plants and projects: Mitsubishi gas chemical chooses UOP's MX sorbex technology. (2007) *Chem. Eng. Prog.*, **103** (12), 20–21.
40. Harada, M. (1988) New barrier resin for food packaging. *Plast. Eng.*, **44** (1), 27–29.
41. Zimmerman, D. (2001) Flexible Packaging–An Overview. Technical Association of the Pulp and Paper Industry Press–Polymers, Laminations and Coatings Conference, pp. 233–238
42. Toft, N. and Postoaca, I. (2002) Laminated packaging material, method of manufacturing of said laminated material and packaging containers produced therefrom. US Patent 6,436,547.
43. Business update: plants and projects: mitsubishi to bolster nylon supply in NA. (2002) *Chem. Eng. Prog.*, **98** (9), 18.
44. Lillwitz, L. (2001) Production of dimethyl-2,6-naphthalenedicarboxylate: precursor to polyethylene naphthalate. *Appl. Catal. A Gen.*, **221**, 337–358.
45. Hobbs, S. and Barder, T. (1989) Two-stage adsorptive separation process for purifying 2,6 dimethylnaphthalene. US Patent 4,835,334.
46. Barger, P., Barder, T., Lin, D., and Hobbs, S. (1991) Continuous process for the production of 2,6-dimethylnaphthalene. US Patent 5,004,853.
47. deRosset, A. (1975) Process for the separation of ethylbenzene. US Patent 3,917,734.
48. Neuzil, R. and Rosback, D. (1976) Process for the separation of ethylbenzene by selective adsorption on a zeolitic adsorbent. US Patent 3,943,182.
49. Kulprathipanja, S. (1995) Process for adsorptive separation of ethylbenzene from aromatic hydrocarbons. US Patent 5,453,560.
50. Broughton, D., (1981) Production of pure M-xylene and pure ethyl benzene from a mixture of C8 aromatic isomers. US Patent 4,306,107.
51. Davassy, B., Shanbhag, G., Lefebvre, F., and Halligudi, S. (2004) Alkylation of p-cresol with tert-butanol catalyzed by heteropoly acid supported on zirconia catalyst. *J. Mol. Catal. A Chem.*, **210**, 125–130.
52. Neuzil, R., and Rosback, D. (1976) Process for the separation of cresol isomers. US Patent 3,969,422.
53. Neuzil, R., Rosback, D., Jensen, R., Teague, J., deRosset, A. (1980) An energy-saving separation scheme. *Chemtech*, **10** (8), 498–503.

8
Liquid Industrial Non-Aromatics Adsorptive Separations
Stephen W. Sohn

8.1
Introduction

This chapter reviews the adsorptive separations of various classes of non-aromatic hydrocarbons. It covers three different normal paraffin molecular weight separations from feedstocks that range from naphtha to kerosene, the separation of mono-methyl paraffins from kerosene and the separation of mono-olefins both from a mixed C_4 stream and from a kerosene stream. In addition, we also review the separation of olefins from a C_{10-16} stream and review simple carbohydrate separations and various acid separations.

8.2
Normal Paraffin Separations

The first family of non-aromatic liquid adsorptive separation processes that employ UOP Sorbex™ technology are the UOP Molex™ processes [1]. The separation is based more on size exclusion rather than selectivity. The family of Molex processes is capable of separating normal paraffins that range from C_5 to C_{18} from their corresponding mixture of normal and non-normal hydrocarbons. As explained in earlier chapters, Sorbex technology consists of a liquid-phase selective adsorption onto a proprietary solid adsorbent using a technique developed by UOP for separating a component or group of components. The Sorbex technique simulates countercurrent contact of the adsorbent and the liquid phase, without requiring movement of the solid bed. The first large-scale simulated moving bed process unit was commercialized in 1961 and in a Molex process unit for the separation of C_{10-13} n-paraffins from kerosene. This process (in addition to other Molex separations) is examined in detail in this chapter.

In this section we will review the following Molex family processes: the gasoline Molex, MaxEne and detergent Molex processes. A brief introduction to the Molex process family starts at the lower molecular weight range: the gasoline Molex process. A gasoline Molex process is capable of separating normal C_{5-6} paraffins from light naphtha. It is well known that normal paraffins found in naphtha have low octane values. Extracting and subsequently isomerizing the normal paraffins into non-normals (via the UOP Penex™ process) upgrades the naphtha's octane value for blending into the refiner's gasoline pool. The next member of the Molex process family is the MaxEne™ process: the MaxEne process is similar to the gasoline Molex process in that it processes naphtha; but the MaxEne process extends the molecular weight range of the naphtha feedstock from C_{5-6} to C_{5-10}. This broader molecular weight feedstock is classified as "full range" naphtha. Heavier normal paraffins are difficult to isomerize in a reformer (for increased octane value) but they are very desirable components of feedstock for naphtha crackers. The MaxEne process extracts the normal paraffins (extract) from full-range naphtha to produce an excellent naphtha cracker feedstock for ethylene production while simultaneously producing normal paraffin depleted raffinate, which is an excellent reformer feedstock for motor fuel production. The last member of the Molex process family we review is the detergent Molex process. A detergent Molex unit processes the heaviest normal paraffin feedstock and is capable of separating C_{10-13} n-paraffins from kerosene and up to C_{18} from gas oil. It is desirable to isolate C_{10-13} normal paraffins because they are one of the components used in creating the detergent precursor; linear alkylbenzene (LAB).

Each Molex process employs a unique set of process operating conditions, process configuration and desorbent. The specific process details for each of the three n-paraffin separation process are revealed in this chapter, but before we review these details, we first discuss the important adsorbent and desorbent performance characteristics that are common to all.

8.2.1
Characteristics of Adsorbent for Normal Paraffin Extraction

The criteria for adsorbent selection are covered by the following five main performance characteristics: selectivity, capacity, mass transfer rate and long-term stability and zeolite type [2]. A brief description of each follows.

8.2.1.1 Selectivity
Selectivity is a relative term and is defined in the Molex process as the adsorbent's preference for desired component (in this case, normal paraffins) over the undesired feed components (cyclic paraffins, iso-paraffins, aromatics) while employing a particular desorbent. One can easily determine an adsorbent and desorbent combination selectivity using a pulse test screening apparatus. This apparatus consists of a known volume of adsorbent placed in a fixed bed. A stream of desorbent is then passed over the bed to fill the pore and interstitial volume of the bed. A known quantity of feed is introduced to the feed at the top of the adsorbent bed and passed across the column as a "pulse" of feed. This pulse of feed is then pushed through the adsorbent bed using a known desorbent flow rate. Effluent from the column is monitored for the various feed components and the concentrations of each component noted (with respect to time) as they elude from the

column. Based on the relative elution position of the various feed components (via analytical results) and the known desorbent rate, one can calculate the adsorbent's selectivity (for a given desorbent system) of adsorbed components relative to non-adsorbed components. Selectivity is an important consideration for both the feed components and desorbent components. Ultimately the selectivity of the desired feed component (in this case normal paraffins) must be at least equivalent to the selectivity of the desorbent for a successful Molex process.

In order to compare the performance of various adsorbent and desorbent systems for the same feedstock on must use the selectivity beta ratio. Equation (8.1) depicts the selectivity beta ratio of an adsorbent/desorbent pair is defined as the ratio of the two components in the adsorbed phase divided by the ratio of the same two components in the unabsorbed phase at equilibrium conditions.

$$\text{Selectivity (beta)} = (\text{wt\% C}/\text{wt\% D})_A / (\text{wt\% C}/\text{wt\% D})_U \qquad (8.1)$$

Where C and D are two components of the feed represented in percent weight and the subscripts "A" and "U" represent the adsorbed and the non-adsorbed phases, respectively. Equilibrium conditions are achieved when the feed passing over a bed of adsorbent does not change composition and there is no net transfer of material occurring between the non-adsorbed and adsorbed phases. Relative selectivity can be expressed not only for one feed compound as compared to another but can also be expressed between any feed mixture component and the desorbent material.

Where selectivity of two feed components approaches 1.0, there is no preferential adsorption of one component by the adsorbent with respect to the other; they are both adsorbed to about the same degree with respect to teach other. As beta becomes less than or greater than 1.0, there is a preferential adsorption by the adsorbent for one component with respect to the other. When comparing the selectivity of an adsorbent for component C over component D, beta > 1.0 indicates preferential adsorption of component C within the adsorbent. Beta < 1.0 indicates that component D is preferentially adsorbed leaving an non-adsorbed phase richer in component C and an adsorbed phase richer in component D.

In the case of the Molex process, the pulse test separation between the desired normal paraffins and the non-normal feed components produces discrete and separate peaks. This high degree of separation means a Molex unit can achieve high degrees of purity and recovery but the ultimate purity and recovery are dictated by non-ideal conditions such as back mixing or flow mal-distributions.

8.2.1.2 Capacity

The second adsorbent characteristic is capacity. Capacity is defined as the quantity of desired component for example normal paraffins adsorbed from the feed while the adsorbent is exposed to the feed. Capacity is reported either as a weight or volume of the desired component retained by the adsorbent per volume or weight of adsorbent. It is desirable to have capacity values as great as physically practical. Capacity, just like selectivity, is measured using the pulse test apparatus.

8.2.1.3 Mass Transfer Rate

The third adsorbent characteristic is mass transfer rate. In the case of the Molex process, mass transfer rate is defined as the time necessary to achieve either: (i) adsorption of feed normals and simultaneous displacement of desorbent normals from the selective pores or (ii) feed normals displacement from the selective pores of the adsorbent by desorbent normals. The mass transfer rate is an important adsorbent characteristic because the rate of exchange fixes the amount of adsorbent required to recover the desired quantity of feed normal paraffins. It also fixes the quantity of desorbent required to recover the extracted n-paraffins from the adsorbent's selective volume. Fast exchange rates result in the most cost-effective separation process because it results in less desorbent normals circulated through the process and recovered from the extract stream and less adsorbent inventory required to achieve the desired feed normal paraffin recovery [3]. The combination of both results in a reduction in the process operating costs. It is well known that mass transfer exchange rates are temperature-dependent. Therefore a testing matrix using the pulse test apparatus allows one to determine the optimum operating process temperature for a given feed, desorbent and adsorbent combination.

8.2.1.4 Stability

The fourth adsorbent characteristic is stability. Long-term stability is defined as the retention of the fresh adsorbent's capacity and performance even after processing feed for a defined time period set by process economics. Stability is measured by monitoring the combination of both the capacity and/or mass transfer rate for normal paraffins. In order for the Molex separation process to be economical, an adsorbent must demonstrate both capacity and mass transfer rate retention when exposed to contaminants present in the feed. Typically, these contaminants are classified as polar materials that are strongly held or attracted to specific sites of the adsorbent. These contaminants are either strongly adsorbed or reacted with the adsorbent resulting in some performance degradation. In some cases a contaminant occupies adsorbent's selective volume and in other cases they may partially block the pore opening to the adsorbent's selective volume. Both situations result in performance deterioration. In these situations, hydro-processing pretreatment of the feedstock is necessary to remove these contaminants to achieve the desired stability target. The extent of pretreatment is determined by the economics of cost of pretreatment compared to the cost of adsorbent replacement.

8.2.1.5 Zeolite Types

The fifth and final adsorbent characteristic is zeolite type. The adsorbent used in the Molex process is a proprietary and is a particularly effective adsorbent for normal paraffin separation [4, 5] and has achieved purity and recovery targets for the Molex processes. A sampling of various molecules (and their corresponding dimensions) that Molex can easily separate is listed in Table 8.1. As discussed in Chapter 6, a zeolites's pore structure is dependent on its silica:aluminum ratio and the proprietary Molex adsorbent possess a uniform repeating three-dimensional porous structure with pores running perpendicular to each other in the x,

Table 8.1 Molecular dimensions of various heavy compounds.

Molecule	Length, Å	Width, Å	Height, Å
C13 olefin	17.7	4.0	5.2
C13 paraffin	17.7	4.0	4.4
C13 aromatic ortho-C3,C4-benzene	11.1	4.0	9.8
Benzene	7.2	3.1	6.5
nC10 paraffin	15.7	4.0	4.4

y and z planes. The structure [6, 7] can be distilled down to its simplest repeating member or unit cell that is captured by the empirical formula: cation$_x$ [(AlO$_2$)$_y$(SiO$_2$)$_z$]*nH$_2$O.

Since the Sorbex process is a liquid-phase fixed-bed process, the selection of particle size is an important consideration for pressure drop and process hydraulics. The exact particle size is optimized for each particular Molex process to balance the liquid phase diffusion rates and adsorbent bed frictional pressure drop. The Sorbex process consists of a finite number of interconnected adsorbent beds. These beds are allocated between the following four Sorbex zones: zone 1 is identified as the adsorption zone, zone 2 is identified as the purification zone, zone 3 is identified as the desorption and zone 4 is identified as the buffer zone. The total number of beds and their allocation between the different Sorbex zones is dependent on the desired performance of the particular Molex process. Molex process performance is defined by two parameters; extract normal paraffin purity and degree of normal paraffin recovery from the corresponding feedstock. Details about the zone and the bed allocations for each Molex process are covered in subsequent discussions about each process.

The adsorbent consists of two components: (i) the active component (zeolite) that actually performs the desired separation and (ii) the binder used to form or "bind" the zeolite into the large mesh particles. The ratio between the quantity of zeolite and binder is set by process dynamics that dictate the strength requirements of the final adsorbent particle. Greater binder content results in a stronger particle. Binder composition can impart negative characteristics to the adsorbent particle and the quantity of binder should therefore be minimized. In practice, the particle composition maximizes the amount of active ingredient that performs the separation (zeolite) without compromising the strength and integrity of the adsorbent particle.

8.2.2
Desorbent Critical Characteristics

Like the criteria for adsorbent selection, there are criteria for desorbent selection. For the Molex process this is covered by the following five main performance characteristics: selectivity, stability, ease of separation (from feed components), availability and reactivity. A brief description of each characteristic follows.

8.2.2.1 Selectivity

Selectivity is a relative term that means a desorbent must have balanced strength or selectivity for the adsorbent relative to the desired feed normal paraffins. A desorbent that is more "strongly" held by the adsorbent than the feed normal paraffins will be difficult for the desired feed normal paraffins to displace from the adsorbent. This results in poor adsorbent utilization due to a loss in selective volume. At the same time, a desorbent cannot be too weakly attracted to the selective volume (relative to the feed normal paraffins) or excessive desorbent circulation is needed to displace the feed component from the adsorbent. In this situation, excessively high desorbent rates are needed to recover feed normal paraffins from the adsorbent. These high desorbent flush rates result in non-economical process conditions due to excessively high utility consumption in post-Sorbex product fractionation.

8.2.2.2 Compatibility

The second desorbent characteristic is that the desorbent material must be compatible with both the particular adsorbent and the feed mixture. Specifically, the desorbent must not reduce the capacity of the adsorbent or normal paraffin selectivity with respect to the raffinate components. Additionally, desorbent materials must not react with any feed component. Both the extract stream and the raffinate streams consist of a mixture of feed components with desorbent and any chemical reaction prevent product recovery.

8.2.2.3 Ease of Separation

The third desorbent characteristic is that the desorbent material must be easily separated from the two Sorbex process products: extract and raffinate. The adsorbent chamber's composition profile produces extract and raffinate streams comingled with desorbent. In order for the process to be economical, the separation of the feed components from the desorbent (achieved through fractionation) is set by the boiling point differences between the species. Depending on the selectivity possessed by the desorbent over that of the feed normals, the subsequent desorbent rates needed to flush feed normal paraffins from the adsorbent's selective volume and the resulting extract or raffinate streams from the Sorbex chambers could contain in some cases more than 50% desorbent. High concentration of desorbent demonstrates the importance of the desorbent characteristics when selecting a desorbent.

The *n*-paraffin separation process utilizes the conventional Sorbex flow scheme shown in Figure 8.1. There are three main sections of a Sorbex unit depicted: (i) adsorbent chambers, (ii) Sorbex rotary valve, (iii) product/desorbent fractionation columns. As indicated earlier, desorbent is typically recycled to the Sorbex adsorbent chambers via fractionation. Figure 8.1 depicts a post-Sorbex fractionation scheme using a "light" desorbent. This term refers to the boiling point of the

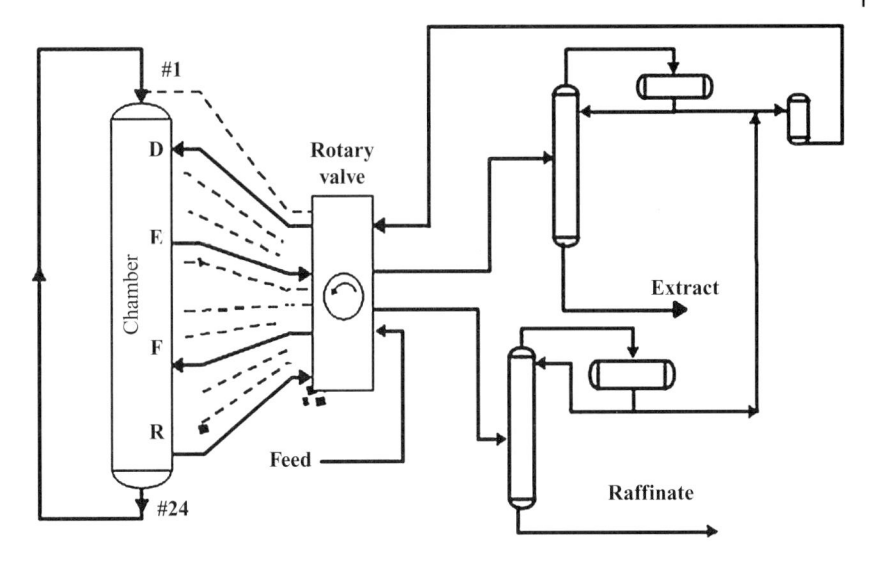

Figure 8.1 Molex process unit.

desorbent relative to the feed component where light desorbent is collected as a fractionation overhead product. In some n-paraffin separation processes such as the MaxEne process, UOP employs a "heavy" desorbent and, as one would expect, the feed components are collected from the fractionation column overhead while desorbent is collected as a bottoms product.

8.2.2.4 Availability
The fourth desorbent characteristic is that the desorbent material should be readily available (at a sustainable cost). It is important that the desorbent be readily available and not cost-prohibitive because desorbent is gradually lost during operations. Since the desorbent is separated from both the extract and raffinate components by fractionation, trace quantities (parts per million) of desorbent are present in the respective product streams due to fractionation tray efficiencies and fractionation control optimization of reflux to feed ratio. Ultimately, each n-paraffin separation facility must balance the operating expenses (utilities consumed during fractionation) to minimize desorbent loss against the replacement value of desorbent to maintain inventory.

8.2.2.5 Reactivity
The fifth and final desorbent characteristic is that the desorbent must not react with any feed components that would impart any negative characteristics on either the final extract and raffinate streams. This is important not only for the desired paraffin product purity but also for retaining the desorbent inventory. Therefore, a desorbent's reactivity must be quantified early in the desorbent selection process.

8.2.3
Simulated Moving Bed Operation: Sorbex Process

8.2.3.1 Adsorbent Allocation within the Molex Process

The adsorbent used in the Sorbex process is partitioned into discrete beds within the Sorbex chambers. These beds are then allocated among four main Sorbex zones. Table 8.2 lists these zones and their corresponding function.

Figure 8.2 depicts the four main zones and their immediate proximity to each other in the Molex process. As indicated earlier, the Sorbex process operates on a liquid–solid countercurrent contacting principle. Zone I is referred to as the

Table 8.2 Sorbex process zones and their corresponding function.

Zone	Identification	Function
I	Adsorption	Adsorption of normal paraffins from the feed with simultaneously displacement of desorbent normals from the adsorbent's selective volume
II	Purification	Enrichment of feed normals and displacement of non-normals
III	Desorption	Desorption of the normal paraffins from the adsorbent selective volume by Desorbent normals
IV	Buffer	Maintains reject or Raffinate non-normals hydrocarbons from entering Zone III and contaminating the Extract product (normal paraffins)

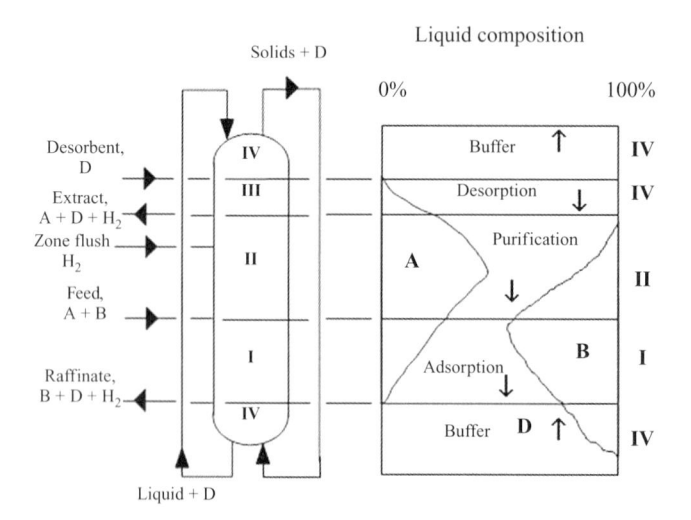

Figure 8.2 Adsorptive separation with moving bed.

"adsorption zone". This is usually the largest of the four zones. The adsorption zone performs the initial separation of normals from non-normal hydrocarbons. Since this is a countercurrent operation, the selective volume of the adsorbent enters the top of this zone containing only desorbent. As the adsorbent passes countercurrent against the feed, it ultimately reaches bottom of the zone which is the feed point. By this time, the adsorbent's selective volume is filled with normal paraffins. This zone may contain as many as eight beds and as few as three. Product purity and recovery requirements dictate how many beds are required. The purification zone (zone II) follows the adsorption zone. In this zone, the adsorbent's selective volume is loaded with normal paraffins but the interstitial volume contains feed non-normal components that enter the bottom of Zone II. Zone II is normally the second-largest zone in the Molex process. As the adsorbent moves countercurrent to the liquid, the interstitial voids are displaced by desorbent or non-normal flush desorbent material. The top of Zone II is the extract point and the bottom of Zone III is the desorption zone. In Zone III the adsorbent moves up the zone as desorbent normal paraffins displace the feed normal paraffins from the adsorbent's selective volume. This displacement continues until the adsorbent reaches the top of the desorption zone, which is the desorbent introduction point. Desorbent circulation rates are set to completely desorb all adsorbed normal paraffins from the adsorbent's selective volume prior to adsorbent exiting zone III. Adsorbent exiting zone III enters the buffer zone of zone IV. Zone IV acts as a buffer between the desorption and the adsorption zones (zone III, zone I, respectively). The buffer zone effectively isolates the high purity extract (consisting of normal paraffins) and the normal paraffin depleted raffinate stream. Adsorbent moves through zone IV until it reaches the top of zone IV (which is the raffinate withdrawal point) prior to entering the adsorption zone (zone I) again. This operation represents on "cycle" of the adsorbent inventory.

8.2.3.2 Critical Sorbex Zone Parameters

Common to all Molex processes are the following six Sorbex zone parameters: A, F_n, A/F_n, D_n, D_n/A and cycle time.

The first parameter A represents the selective pore rate (m^3/h). For a set volume of adsorbent contained in the Sorbex chambers, there is a known selective pore volume. This selective volume quantity is divided equally among the various adsorbent beds. Since Sorbex process simulates a moving bed process where adsorbent moves counter current to the process flow, the selective pore rate represents the quantity of selective volume that moves with every step or index of the rotary valve. One step of the rotary valve indexes the feed point from one bed to the next sequential bed position.

The second of the six Sorbex zone parameters is F_n, which represents the volumetric rate of feed normals introduced to the Molex process.

The third of the six Sorbex zone parameters is the ratio of A/F_n, which is a unitless value and represents the ratio of the selective pore rate (A) to the volumetric feed rate of n-paraffins (F_n). This means that the ratio A/F_n sets the amount of selective pores that circulate countercurrent to the rate of feed normals introduced to the Molex process. An A/F_n value of 1.0 means an equal volume of selective pores is contacted to an equal volume of feed normal paraffins. If a Molex unit circulates more selective pores per unit of feed n-paraffins introduced (where $A/F_n > 1.0$), greater normal paraffins recovery can be achieved. This phenomenon continues until there is no benefit derived from the circulation extra selective pore volume. This means that at optimum point, A/F_n increases does not increase paraffin recovery. This is due to inherent mass transfer limitation with the feed normal paraffins, Due to non-idealities introduced by the mechanical limitation of the simulated counter current moving bed, A/F_n ratios normally employed in commercial units are greater than 1.0.

The fourth of the six Sorbex zone parameters is the term D_n, which represents the volumetric rate (i.e., m^3/h) of the normals contained in the Molex desorbent introduced to the desorption zone.

The fifth of the six Molex zone parameter is the ratio D_n/A, which controls the performance of zone 3 (desorption zone). D_n/A represents the ratio of the rate of desorbent normals necessary to flush the selective pores in the desorption zone as they pass countercurrent to the desorbent.

The sixth and final Sorbex parameter is cycle time (theta), as represented in Eq. (8.2). This is defined as the time required for the feed to make a complete cycle through the total number of Sorbex beds starting at one feed location and stepping through all the beds and returning to the original starting location.

$$\text{theta} = (Va)/A = \text{volume}/(\text{volume}/\text{minute}) \tag{8.2}$$

8.2.4
Light Normal Paraffin Separation (Gasoline Range nC_{5-6})

The gasoline Molex process is the first of three processes since it separates the lowest molecular weight feed of the three Molex normal paraffin separation processes. Gasoline Molex was developed to optimize a Refiner's octane pool by extracting low octane value normal paraffins (specifically nC_{5-6}) from naphtha. In a typical refinery flow scheme, a gasoline Molex unit is integrated with a catalytic isomerization unit (Penex unit) which converts the Molex unit's extracted normal paraffins into desired iso-paraffins. These iso-paraffins are desirable because they possess higher octane value than their linear counterpart.

8.2.4.1 Industrial Use and Demand
Since its commercialization in 1989, the number of gasoline Molex units designed and commissioned has risen to 13, with a total extraction capacity of over 2080 KMTA (= × 10^3 t/year).

8.2.4.2 Unique Operating Parameters
In Sorbex terminology, a gasoline Molex unit is classified as a "low" purity and "low" recovery Sorbex unit since typical purity and recovery targets are greater than 85% n-paraffin purity while operating at greater than 90% feed n-paraffin recovery.

Table 8.3 Research octane number (RON) for C5–C6 hydrocarbons.

	RON calculated
iC4	100.2
nC4	95
iC5	93.5
nC5	61.7
CP	102.3
22DMB	94
23DMB	105
2MP	74.4
3MP	75.5
nC6	34
MCP	96
CH	84
BZ	120

These values are low in comparison to a detergent Molex unit, which exceeds 99% purity while recovering greater than 98% of the feed n-paraffins. The low purity and recovery target for a gasoline Molex unit requires a simplified design for a Sorbex unit and as such, the design of this type Sorbex unit has the fewest number of total beds of any of the Molex processes. The gasoline Molex unit possesses one-third of the standard bed allotment in Sorbex. Even with a reduced bed count, the gasoline Molex unit still contains the full complement of the four main Sorbex zones.

The naphtha processed by a gasoline Molex unit typically has a Enger distillation (D86) with an initial boiling point (IBP) of 26 °C and end point (EP) of 82 °C. Table 8.3 lists the typical components found in naphtha with their corresponding calculated research octane numbers (RON) and shows why nC_5 and nC_6 are undesirable in a gasoline pool due to their low RON value [8].

The gasoline Molex process extracts the majority of the low octane components (i.e., n-pentane, n-hexane) and after desorbent removal isomerizes them in the Penex process. The Penex process can easily isomerize nC_5 into isopentane, while nC_6 isomerization is somewhat more complex. This is due to fact that formation of 2- and 3-methylpentane and 2,3-dimethylbutane is limited by equilibrium. The net reaction mainly involves the conversion of normal hexane to 2,2-dimethylbutane. In addition to the nC_5 and nC_6 isomerization, any benzene contained in the extract and fed to the Penex unit is hydrogenated to cyclohexane. In a typical case, upgrading naphtha via a Molex/Penex process scheme can upgrade the octane rating by at least 14 octane numbers. This Penex isomerate is combined with fresh naphtha feed and then fed to the gasoline Molex unit for additional processing.

A refiner has many operation mode choices; if the required octane number of the pool can be met by recycle of the methylpentanes, the refiner can choose fractionation for capital reasons. Where the cost of utilities is high, the refiner might choose a Molex unit, which would separate both nC_{5-6} instead. In this case, the utility cost would be lower for separating both of these in a Molex unit than it would be for separating the methylpentanes by fractionation, and the refiner would ultimately achieve a higher octane value.

As in all Sorbex processes, the gasoline Molex process is both an isothermal and liquid-full process. The process employs sufficient operating temperature to overcome diffusion limitations with a corresponding operating pressure to maintain liquid-phase operation. The process also employs a "light" desorbent system. A light desorbent system recovers desorbent as an overhead product from the post-Sorbex product fractionation columns. Due to the low molecular weight of the desired feed components and the desorbent, this separation has few diffusion limitations. This enables the gasoline Molex process to operate at relatively quick Sorbex cycle times. A shorter cycle time is more cost effective since it requires less adsorbent per unit of normal paraffins separated.

8.2.5
Intermediate Normal Paraffin Separation (C_{6-10})

The second Molex process for normal paraffin extraction is the MaxEne process. Similar to the gasoline Molex process, the MaxEne process extracts normal paraffins from a naphtha mixture but instead of processing a narrow naphtha feed distillation range, it processes "full-range" naphtha. Typical full-range naphtha has a D86 Engler distillation that ranges from 95 to 175 °C. To take full advantage of the MaxEne process one would implement it at an integrated naphtha cracker and reforming facility. In this situation, normal paraffins are separated from full-range naphtha (prior to reforming) and employed as a feed to the naphtha cracker [9, 10]. Normal paraffins have been shown to crack with greater selectivity to the desired ethylene product than either a branched or cyclic compound. It is also well known in the industry that a cracker's ethylene yield can be increased by at least 30% when a highly normal paraffinic feed is substituted for full-range naphtha. At the same time, the normal paraffin-depleted naphtha or MaxEne raffinate is a superior catalytic reforming feed, since the aromatics or gasoline yields increase significantly when processing a normal paraffin-lean feed. Also, while processing this type of feed, it significantly decreases a reformer's byproducts such as coke and LPG. Thus the MaxEne process simultaneously benefits both the ethylene producer and the refinery operator.

8.2.5.1 Industrial Use and Demand
At this time (2009), the first commercial MaxEne unit is under design and will be capable of extracting 350 KMTA n-paraffins.

8.2.5.2 Unique Operating Parameters
The UOP MaxEne process for the separation of n-paraffins from branched and cyclic hydrocarbons found in full-range naphtha is based on the well established,

commercially proven Sorbex technology. As with all Sorbex processes, the MaxEne separation is effected in both the liquid phase and isothermal conditions. The process employs sufficient operating temperature to overcome diffusion limitations, with a corresponding operating pressure to maintain liquid-phase operation.

There are many similarities and differences between the gasoline Molex process and MaxEne process. Both the gasoline Molex and MaxEne process separations employ specially formulated proprietary adsorbent. The MaxEne process is similar to the gasoline Molex process in that they are both a "low" purity and recovery design relative to most Sorbex processes. The MaxEne process achieves purity and recovery targets dictated by process economics. The purity target is slightly greater for the MaxEne process than that of the gasoline Molex process because the products are routed directly to the naphtha cracking and reforming process units. The slightly elevated purity target requires half the standard Sorbex bed allotment and these are allocated among the full complement of the four Sorbex zones.

Unlike the gasoline Molex process that employs a iso-butane and n-butane desorbent mixture, the MaxEne process employs a "heavy" desorbent system. A heavy desorbent system means that the bottom product from both the Sorbex extract and raffinate fractionation columns is desorbent while the feed components are recovered as overhead products. In the MaxEne process case, heavy normal paraffin such as n-dodecane is employed as the desorbent though desorbents as light as n-decane and as heavy as n-tetradecane are possible candidates too.

Due to the wider carbon range of the naphtha feed (resulting in slower diffusion rates than the gasoline Molex process) and greater performance specifications, a MaxEne process unit is designed with more beds and operates at a slower cycle time than a gasoline Molex process unit. Depending on the feed, the gasoline Molex unit has few diffusion issues while the MaxEne unit must operate at slightly slower cycle times time to accommodate the heavier naphtha feed. Greater cycle time is also needed in the MaxEne unit to accommodate the slower diffusion rates of the heavy n-dodecane desorbent.

8.2.6
Heavy Normal Paraffin Separation (C_{10-18})

The detergent Molex process is the third Molex process for separating normal paraffins from non-normal hydrocarbons [11–13]. Like the gasoline Molex and MaxEne processes, it extracts normal paraffins from a comparable boiling mixture of linear and non-linear compounds but in the case of a detergent Molex process, it processes kerosene instead of naphtha and as such it extracts higher molecular weight normal paraffins. Just like naphtha, kerosene is a complex mixture of co-boiling iso-paraffins, cyclic paraffins and aromatics. The process is identified as detergent Molex process because it produces a C_{10-13} normal paraffins product used as one of the basic building blocks in the common anionic surfactant: linear alkylated benzene sulfonate (LAS). The surfactant is called "linear" alkylated benzene (LAB) because of the linear paraffin portion of the molecule. The process of making the surfactant LAB starts with isolation of C_{10-13} normal paraffins from kerosene. Subsequent processing steps dehydrogenate these paraffins into a mixture of linear mono-olefins which are then alkylated to benzene to produce the surfactant intermediate LAB. The intermediate LAB is sulfonated and neutralized to produce the anionic surfactant: LAS. The LAS molecule is essentially a surfactant or "surface-active" agent made of two portions: one moiety is hydrophilic and the other is hydrophobic. This means the straight chain and benzene ring are hydrophobic while the sulfonic acid anion portion is hydrophilic. As a result of these characteristics, when employed in an aqueous washing environment, they surround and concentrate at the interfaces separating immiscible phases, resulting in a decrease in the interfacial tension [14]. The detergent Molex process produces C_{10-13} normal paraffins but it is capable of separating through up C_{18} normal paraffins from kerosene/and or gasoil. This technique has been commercially demonstrated up to C_{18} normal paraffins and with adequate process conditions, is capable of extracting up to C_{30} normal paraffins.

8.2.6.1 Industrial Use and Demand

In 2009 the estimated worldwide detergent range normal paraffin capacity was approximately 3.6×10^6 t/year. With more than 27 licensed Molex units, liquid-phase adsorption separation is the predominant technology for heavy normal paraffin production.

8.2.6.2 Unique Operating Parameters

In the C_{10-13} or C_{11-14} detergent range applications, the feedstock boiling range is typically a Engler D86 distillation that ranges from 150 to 250 °C. If higher molecular weight normal paraffins are required, the endpoint of the feed can be extended to 300 °C. Higher molecular weight *n*-paraffins are more difficult to extract from their corresponding feed than naphtha-based normal paraffins because they are strongly held by the proprietary adsorbent. Polarity plays an important part in the performance of the adsorbent. The adsorbent is very selective for any polar molecule such as water and is therefore an excellent desiccant. Prior to processing any feed in a Molex unit one must hydroprocess it to remove the majority of polar contaminants (such as sulfur and nitrogen heteroatomic species) that are naturally found in kerosene. When left untreated, the polar contaminants compete with the normal paraffins for selective pore capacity and in some cases effectively "poison" the adsorbent. Some contaminants are temporary poisons while others are reactive with the adsorbent and result in permanent capacity loss. In any event, in order to demonstrate long adsorbent lives, it is prudent to hydroprocess the feed to remove any strongly adsorbed polar species.

Since the detergent Molex process is both a high normal paraffin purity and recovery process, it is designed with the full allotment of Sorbex beds in addition to the four basic Sorbex zones. The detergent Molex process utilizes a split desorbent. This means it consists of a mixture of *n*-pentane and iso-octane. The zone flush primarily consists of isooctane and its purpose is to keep the desorbent (*n*-

pentane) in the desorption zone from passing into the purification zone. Movement of desorbent into the purification zone results in competition between the feed normals and desorbent normals within the selective volume. This competition ultimately results in a reduction in capacity and reduction of n-paraffin recovery.

As with the other two Molex processes, the detergent Molex process operates under isothermal conditions and sufficiently high process pressure to ensure liquid-phase operation The detergent Molex process operates at a greater temperature than a gasoline Molex process since a higher operating temperature is necessary to overcome diffusion limitation inherent with higher molecular weight normal paraffins found in kerosene. The process employs a "light" desorbent system that consists primarily of normal pentane. Iso-octane is added to the pentane to help balance the desorbent strength and minimize plant operating pressure needed to maintain liquid-phase operation at the target operating temperature.

8.3
Mono-Methyl Paraffins Separation (C_{10-16})

8.3.1
Industrial Use and Demand

In 1998, UOP announced the development of a new Sorbex process called the MMP Sorbex process [15–19] that was capable of simultaneously separating both C_{10-16} mono-branched paraffins and C_{10-16} normal paraffins from a corresponding kerosene stream or n-paraffin-depleted Molex raffinate stream. Previously, no commercial process existed to isolate significant quantities of mono-methyl paraffin derived from either kerosene or n-paraffin depleted kerosene. Mono-methyl paraffins are desirable because they are needed for a new type of anionic surfactant.

The surfactant found in the majority of detergents worldwide is the anionic surfactant LAS. In 2000, Proctor & Gamble (P&G) applied for both United States and European patents [20–23] that identify the new surfactant: modified linear alkylated sulfonate (MLAS). In their patents, P&G claimed that this surfactant demonstrates superior detergent performance to that of the conventional detergent, LAS. P&G also teach that the controlled branching afforded by the mono methyl paraffins and exceptionally high 2 + 3-phenyl content found in MLAS result in a surfactant with superior cold- and hard-water performance (less hard-water cation deactivation) than the conventional surfactant, LAS. The process steps to produce MLAB follow many of the same steps that convert linear paraffin into LAB (as discussed in the section on the detergent Molex process). The methyl-branched paraffins supply the controlled branching while a unique catalytic processing step produces the "high" 2-phenyl alkylate (>75%).

8.3.2
Unique Operating Parameters

Early in the development of the process, UOP determined that an adsorbent other than the proprietary Molex adsorbent was necessary for this separation because diffusion simulations using ideal compounds encountered in kerosene revealed that the molecular diameter of desired mono-methyl-branched paraffins were excluded from this adsorbent. The simulated diffusion model involves advancing a probe molecule through the zeolite's main channel and using molecular mechanics methods to calculate the energy of the probe molecule at regular intervals as it interacts with the atoms in the sieve wall. The energy difference between the low- and high-energy positions of the probe molecule in the channel is defined as the energy barrier which can be used to predict the diffusion ease of the probe molecule in the zeolite channel – the larger the barrier, the more restrictive the diffusion.

Earlier in Section 8.2.1.1 we reviewed the laboratory pulse test technique used to confirm the suitability of adsorbent identified in the molecular modeling work with various desorbents in an adsorptive separation process [24]. Screening various adsorbents using this technique confirmed the new proprietary adsorbent had the best performance. This is a novel topologic type of tetrahedral framework that employs ten-member oxygen rings that produce straight channels along the x-axis with an elliptical cross-section. These channels are interconnected by zigzag channels along the y-axis, defined also by ten-membered oxygen rings. Based on the adsorbent characteristics, one would expect para-substituted aromatics and certain cyclopentanes and cyclohexanes would enter the channels in addition to both n-paraffins and mono-methyl paraffins too.

The MMP Sorbex process has many similarities but also some differences when compared to the detergent Molex process. As with all of Sorbex processes, the MMP Sorbex process operates in the liquid phase, employing suitable conditions (pressure, temperature) to overcome any diffusion constraints to achieve target performance. Table 8.4 highlights and contrasts the different characteristics of the detergent Molex and MMP Sorbex processes. The process was successfully demonstrated in a continuous countercurrent moving bed separation pilot plant using commercial n-paraffin-depleted kerosene (Molex raffinate) feedstock. A typical gas

Table 8.4 Comparison between the Molex and MMP Sorbex processes.

Process	Detergent Molex	MMP Sorbex
Feed	Kerosene	Kerosene or Molex raffinate
Product	C10–C16 n-paraffins	C10–C16 MMP and n-paraffins
Product % NP + MMP purity	>98.5%	>92%
% NP + MMP recovery	>95%	>70%

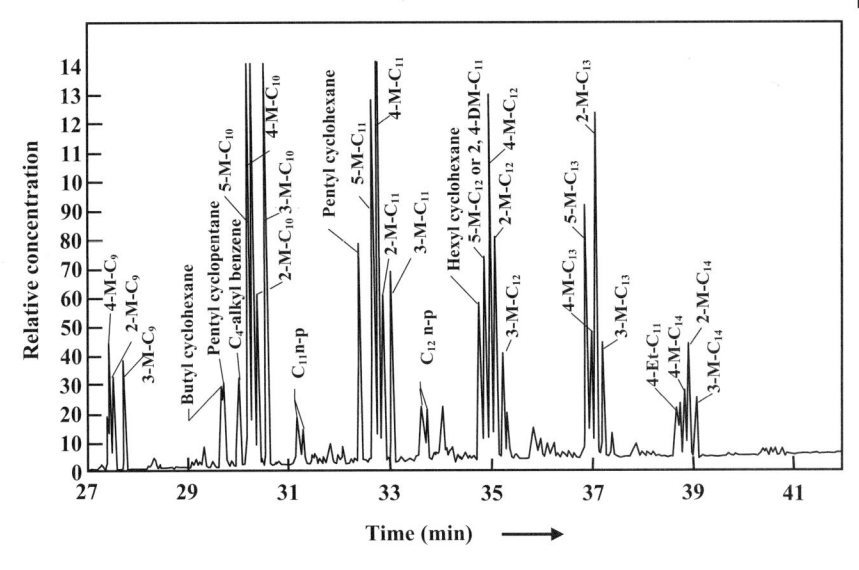

Figure 8.3 GCMS of MMP extract product.

chromatogram–mass spectrophotometry of the extracted mono-methyl paraffins is shown in Figure 8.3. As expected from the diffusion simulations, trace quantities of naphthenes and other small molecular compounds are present in the extracted product along with the desired product, normal and mono-methyl paraffins.

8.4
Olefin Separations

There are three liquid-phase adsorption Sorbex technology-based separation processes for the production of olefins. The first two are the UOP C_4 Olex™ and UOP Sorbutene™ processes and the third is the detergent Olex process(C_{10-16}) [25, 26]. The three olefin separation processes share many similarities. The first similarity between the three olefin separation processes is that each one utilizes a proprietary adsorbent whose empirical formula is represented by $Cation_x[(AlO_2)_y(SiO_2)_z]$ [27]. The cation type imparts the desired selectivity for the particular separation. This zeolite has a three-dimensional pore structure with pores running perpendicular to each other in the x, y and z planes [28]. The second similarity between the three olefin separation processes is the use of a mixed olefin/paraffin desorbent. The specifics of each desorbent composition are discussed in their corresponding sections. The third similarity is the fact that all three utilize the standard Sorbex bed allotment that enables them to achieve product purities in excess of 98%. The following sections review each process in detail.

8.4.1
C$_4$ Separations

8.4.1.1 Industrial Use and Demand

Mixed C$_4$ olefins (primarily iC$_4$) are isolated from a mixed C$_4$ olefin and paraffin stream. Two different liquid adsorption high-purity C$_4$ olefin processes exist: the C$_4$ Olex process for producing isobutylene (iC_4^{2-}) and the Sorbutene process for producing butene-1. Isobutylene has been used in alcohol synthesis and the production of methyl tert-butyl ether (MTBE) and isooctane, both of which improve octane of gasoline. Commercial 1-butene is used in the manufacture of both linear low-density polyethylene (LLDPE) and high-density polyethylene (HDPE)., polypropylene, polybutene, butylene oxide and the C$_4$ solvents secondary butyl alcohol (SBA) and methyl ethyl ketone (MEK). While the C$_4$ Olex process has been commercially demonstrated, the Sorbutene process has only been demonstrated on a pilot scale.

8.4.1.2 Unique Operating Parameters

The C$_4$ Olex process is designed with the full allotment of Sorbex beds in addition to the four basic Sorbex zones. The C$_4$ Olex process employs sufficient operating temperature to overcome diffusion limitations with a corresponding operating pressure to maintain liquid-phase operation. The C$_4$ Olex process utilizes a mixed paraffin/olefin "heavy" desorbent. In this case it is an olefin/paraffin mix consisting of n-hexene isomers and n-hexane. A rerun column is needed to remove heavy feed components such as C$_5$/C$_6$ because they would contaminate or dilute the hexene/hexane desorbent. Table 8.5 contains the typical feed and product distributions.

Table 8.5 C$_4$ Olex process typical feed and extract composition.

Component	C4 Olex	
	Feed, wt%	Extract, wt%
C3	0.18	0.01
C3=	0.00	0.01
iC4	50.23	1.91
nC4	1.93	0.07
iC4=	45.66	93.90
Tr2-butene	0.72	1.48
1-Butene	0.64	1.31
Cis2-butene	0.55	1.13
1,3-Butadiene	0.09	0.18
Neo-C5	0.00	0.00
nC5	0.00	0.00
Total	100.0	100.0

Table 8.6 Typical Sorbutene process feed and extract compositions.

Component	Sorbutene	
	Feed, wt%	Extract, wt%
C3	0.00	0.00
C3=	0.00	0.00
iC4	0.41	0.01
nC4	12.95	0.26
iC4=	0.50	0.01
Cis/Tr2-butene	35.64	0.72
1-Butene	50.50	99.00
1,3-Butadiene	0.00	0.00
Neo-C5	0.00	0.00
iC5	0.00	0.00
nC5	0.00	0.00
Total	100.0	100.0

The Sorbutene process [29] is the liquid adsorption process for separating butene-1 directly from a mixture of C_4 olefins and paraffins. The Sorbutene process replaces conventional multi-process unit schemes for isolating butene-1 which depend on chemical exploitation of any isobutylene co-product for economic viability. Sorbutene is capable of recovering butene-1 at purities exceeding 99% and at an extraction efficiency surpassing 90%. As with the C_4 Olex process, the Sorbutene process is designed with the full allotment of Sorbex beds in addition to the four basic Sorbex zones.. The Sorbutene process operates under the typical Sorbex conditions: an isothermal temperature profile and liquid-phase conditions. The operating temperature is sufficient to overcome any diffusion limitation with a corresponding operating pressure to maintain liquid-phase operation. Typical Sorbutene process feed and products are summarized in Table 8.6. The Sorbutene process desorbent consists of a blend of 75 vol% hexane-1 and 25 vol% cyclohexane. The olefin component performs the feed olefin desorption function while the cyclohexane portion occupies and flushes adsorbent voidage.

8.4.2
Detergent Range Olex C_{10-16}

8.4.2.1 Industrial Use and Demand

The C_{10-16} Olex process produces a range of high purity C_{10-16} internal linear olefin product. There are five licensed units with a combined capacity exceeding 340 KMTA. C_{10-13} olefin product is used as a feedstock for detergent LAB, C_{11-14} olefin finds use in C_{12-15} detergent alcohols and C_{15-16} olefin is processed into the lubricants formed from poly-internal olefins (PIO). The detergent Olex process

Table 8.7 Typical detergent Olex process feed and extract compositions.

	Detergent Olex	
	Feed, wt%	Extract, wt%
$nC10$	0.3	0.0
$nC11$	25.1	1.1
$nC12$	35.2	1.5
$nC13$	18.1	0.8
$nC14$	10.6	0.5
TNN	0.2	1.2
$nC10=$	0	0.3
$nC11=$	3.0	27.7
$nC12=$	4.2	38.4
$nC13=$	2.1	18.2
$nC14=$	1.3	10.4
	100.0	100.0
TNO purity		95.1

feedstock consists primarily of a mix of olefin and paraffins produced from the UOP Pacol™ process dehydrogenation [30] of the corresponding paraffins. Paraffin conversion in the Pacol process forms small quantities of byproducts such as diolefins and aromatics and when present in the feed to an Olex unit are in part co-extracted with the olefins. Typical detergent Olex process feed and extract compositions are outlined in Table 8.7.

8.4.2.2 Unique Operating Parameters

As with the C_4 Olex process, the detergent Olex process is designed with the full allotment of Sorbex beds in addition to the four basic Sorbex zones. Detergent Olex process employs sufficient operating temperature to overcome diffusion limitations with a corresponding operating pressure to maintain liquid-phase operation. The desorbent employed in detergent Olex process is dependent on the feed olefin carbon number distribution. The current desorbent mixture consists of *n*-hexene and *n*-hexane but historically, units have successfully operated with mixes of *n*-heptane and *n*-octene. As in any Sorbex process, there must be a balance between the adsorbent's selectivity for feed component over the desorbent and the desorbent must be easily separated from the feed components by fractionation. As the molecular weight (and carbon number) of the feedstock increases, the corresponding olefin behaves more like a comparable paraffin of the same carbon number and the adsorbent selectivity differences between the paraffin and olefin therefore decrease. This means that with increasing molecular weight, the paraffin and olefin separation becomes more difficult. A desorbent combination that works well for C_{10-13} paraffin/olefin separation will not perform as well in C_{15-16} paraffin/olefin service.

8.5
Carbohydrate Separation

8.5.1
Industrial Use and Demand

Carbohydrates are found in almost all plants and falls into three classifications: (i) monosaccharide or simple single sugars (such as glucose/dextrose and fructose), (ii) disaccharide or double sugars (such as sucrose or maltose) and (iii) polysaccharide or complex sugars (such as starch or glycogen). On a relative sweetness basis (with sucrose = 100), simple sugars have the following sweetness ranking: fructose 140, and glucose 70–80. The higher sweetness that fructose has over glucose merits its separation/isolation and the UOP Sarex™ process is a liquid-phase adsorption Sorbex process [31, 32] capable of separating fructose from mixed saccharides. The predominant source of fructose is derived from corn and corn syrup. Corn syrup is a glucose sweetener developed by treating cornstarch with enzymes. Corn syrup is not as sweet as sucrose, but is often used with or in place of sucrose to provide "mass" and "texture" in food. High-fructose corn syrup (HFCS) is made from corn syrup by converting glucose into fructose using an enzymatic process. The resulting product contains 42% fructose mixture (HFCS 42) that is much sweeter than regular corn syrup, allowing a reduction in the quantity used. The 42% fructose glucose mixture (HFCS 42) is then subjected to a liquid chromatography step or Sarex unit processing step where the fructose is enriched to approximately 90%. The 90% fructose (HFCS 90) is blended with 42% fructose to achieve a 55% fructose final product (HFCS 55). The soft drink industry uses HFCS 55 as its main nutritive sweetener. The average American consumed approximately 28.4 kg of HFCS in 2005, versus 26.7 kg of sucrose [33].

8.5.2
Unique Operating Parameters

The Sarex process employs the full allotment of Sorbex beds in addition to the four basic Sorbex zones. When a Sarex unit processes corn syrup that has an approximate composition (on a dry basis) of 42% fructose, 53% glucose and 5% other saccharides, it produces a fructose enriched corn syrup with >95% purity at >90% recovery [34, 35]. The Sarex process employs sufficient operating temperature to overcome diffusion limitations with a corresponding operating pressure to maintain liquid-phase operation.

8.6
Liquid Adsorption Acid Separations

Here, we review two organic acid separations; the first being citric acid from fermentation broth and the second separates saturated from unsaturated free fatty acids.

8.6.1
Citric Acid Separation

8.6.1.1 Industrial Use and Demand

Citric acid is a triprotic weak organic acid that finds uses as a natural antioxidant preservative and use as an acidifying agent in foods. Conventional citric acid production technique employs *Aspergillus niger* cultures fed on a sucrose- or glucose-containing medium. Citric acid is isolated from this fermentation broth by precipitation with calcium hydroxide to yield calcium citrate salt. A subsequent acid step regenerates the salt into citric acid. A less conventional but more direct method of citric acid production employs the liquid adsorption Sorbex process to isolate citric acid directly from the culture [36].

8.6.1.2 Unique Operating Parameters

Citric acid separation from fermentation broth employs the full allotment of Sorbex beds in addition to the four basic Sorbex zones. The process utilizes a resin instead of a zeolite based adsorbent. The resin is a nonionic cross-linked polystyrene polyvinyl benzene formulation. Operating temperatures for this process are sufficient to overcome diffusion limitations with a corresponding operating pressure to maintain liquid-phase operation. The desorbent consists of water blended with acetone. Subsequent processing steps remove the desorbent from the desired extract product: citric acid.

8.6.2
Free Fatty Acid Separation

8.6.2.1 Industrial Use and Demand

Fatty acids are aliphatic monocarboxylic acids, derived from or contained in esterified forms of animal or vegetable fat, oil or wax. Depending on their source, natural fatty acids commonly have a chain of 4–28 carbons that are usually unbranched, even numbered, and may or may not contain some degree of unsaturation. Recently there has been interest in omega-6 acids as a consumer diet supplement. Omega-6 acids have been identified as a essential dietary requirement. One omega-6 fatty acid of particular interest is linoleic acid, an 18-carbon chain with one of its two cis double bonds located six carbons from the omega end. There exists a pilot scale liquid adsorption Sorbex process that can separate linoleic acid from a mixture containing oleic acid (C_{18} monounsaturated omega-9 fatty acid) and other acids [37–40].

8.6.2.2 Unique Operating Parameters

Fatty acid separation uses standard allotment of Sorbex beds in addition to the four basic Sorbex zones. The process employs sufficient operating temperature to overcome diffusion limitations with a corresponding operating pressure to maintain liquid-phase operation. The desorbent consists of aqueous solution spiked with either acetone or methyl-ethyl ketone. As in all Sorbex processes, post-Sorbex process fractionation or evaporation separates the desorbent from the desired product.

8.7
Summary

As promised, this chapter outlines numerous liquid-phase non-aromatic adsorption processes that enable one to economically separate a commercially desirable component from a mixture when the separation is impossible (given the closeness of their relative volatilities) by conventional means such as distillation. We review process that can separate a wide range of normal paraffins from a mixture of their corresponding feedstocks. In addition to this, we also review how to separate mono branched paraffins and olefins from similar feedstocks. Finally we review liquid-phase adsorption processes to isolate desired carbohydrates, fatty acids and citric acid from their feed source; and for each separation we reveal insight on the corresponding operating conditions, process configuration and adsorbent necessary to achieve the separation.

References

1 Sohn, S.W. (2003) UOP molex process for production of normal paraffins, in *Meyer's Handbook of Petroleum Refining Processes*, 3rd edn (ed. R.A. Meyers), McGraw-Hill, pp. 10.75–10.77.

2 Gembicki, S.A. (2009) Adsorption, liquid separation, in *Kirk & Othmier Encyclopedia of Chemical Technology*, vol 1 (ed. S.A. Gembicki), John Wiley & Sons, pp. 672–673.

3 Jasra, R.V., et al. (1987) Sorption kinetics of higher n-paraffins on zeolite molecular sieve 5A. *Ind. Eng. Chem. Res.*, 26, 2546–2552.

4 Jones, E.K., et al. (1967) Biodegradable surfactants. U.S. Patent 3303233.

5 Stache, H.W. (1996) *Anionic Surfactants*, Marcel Dekker, pp. 6–9.

6 Breck, D.W. (1974) *Zeolite Molecular Sieves, Structure, Chemistry, Use*, John Wiley & Sons, New York, p. 133.

7 Xu, R., et al. (2007) *Chemistry of Zeolites and related Porous Material*, John Wiley & Sons, pp. 351–356.

8 Maxwell, I.E. (2001) *Introduction to Zeolite Science and Practice, HC Processing with Zeolites*, 2nd edn, Elsevier, Chapter 17, p. 757.

9 Rice, L.H., et al. (2006) Ethylene production by Steam Cracking normal paraffins. U.S. Patent Appl Pub US2006205988 A1.

10 Foley, T.D. et al. (2002) Ethylene production by Steam Cracking normal paraffins. PCT Int Appl WO 2002036716A1.

11 Raghuram, S. and Wilcher, S.A. (1992) The separation of n-paraffins from paraffin mixtures. *Sep. Sci. Technol.*, 27 (14), 1917–1954.

12 Broughton, D.B. (1968) Molex: case history of a process. *Chem. Eng. Prog.*, 64 (8), 60–65.

13 Bieser, H.J. (1977) Separating normal paraffins. U.S. Patent 4006197.

14 Broze, G. (1999) *Handbook of Detergents, Part A Properties*, vol. 7, Marcel Dekker, pp. 14–16, 40–41.

15 Kulprathipanja, S. and Johnson, J. (2002) Liquid separation, in *Handbook of Porous Solids* (eds F. Schuth, K. Sing, and J. Weitkamp), Wiley-VCH Verlag GmbH, Germany, pp. 2568–2613.

16 Kulprathipanja, S., et al. (2001) Process for monomethyl acyclic hydrocarbon

17 Sohn, S.W., et al. (2003) Mono-methyl Paraffin Adsorptive Separation. U.S. Patent 6670519.
18 Sohn, S.W., et al. (2002) Adsorptive separation process for recovery of two paraffin products. U.S. Patent 6407305.
19 Kulprathipanja, S., et al. (2001) Mono-methylparaffin adsorptive separation process. U.S. Patent 6222088.
20 Scheibel, J.J., et al. (2003) Process for making alkybenzenesulfonates surfactants and products thereof. U.S. Patent 6602840.
21 Kott, K.L., et al. (2003) Detergent compositions containing modified alkylaryl sulfonate surfactants. U.S. Patent 6589927B1.
22 Heltovics, G., et al. (2002) Detergent Composition with a Selected Surfactant System Containing a mid-chain branched surfactant. U.S. Patent 6380143.
23 Conner, D.S., et al. (2000) Process for preparing modified alkylaryl. U.S. Patent 6525233.
24 Hoering, T.C., et al. (1984) Shape-selective sorption of mono-methylalkanes by silicate, a zeolite form of silica. *J. Chromatogr.*, **316**, 333–341.
25 Tajbl, D.G., Kanofsky, J.S., and Braband, J.M. (1980) UOP's olex process: new applications. *Energy Process., Canada*, **72** (5), 61–63.
26 Sohn, S.W. (2003) UOP olex process for olefin recovery, in *Handbook of Petroleum Refining Process*, 3rd edn (ed. R.A. Meyers) McGraw-Hill, pp. 10.79–10.81.
27 Breck, D.W. (1974) *Zeolite Molecular Sieves, Structure, Chemistry, Use*, John Wiley & Sons, Inc., New York, p. 176.
28 Uppili, S., et al. (2000) Probing zeolites with organic molecules; Supercages of X and Y zeolites are superpolar. *Langumir*, **16**, 265–274.
29 Neuzil, R.W., et al. (1977) Desorbent for separation of butene-1 from a C4 hydrocarbon mixture using zeolite X. U.S. Patent 4119678.
30 Vora, B.V. (1985) Process for conversion of paraffin to olefin. U.S. Patent 4523045.
31 Broughton, D.B., Bieser, H.J., Berg, R.C., Connell, E.D., Korous, D.J., and Neuzil, R.W. (1977) High purity fructose via continuous adsorptive separation. *Sucr. Belg.*, **96** (5), 155–162.
32 Bieser, H.J. and De Rosset, A.J. (1977) Continuous countercurrent separation of saccharides with inorganic adsorbents. *Staerke*, **29** (11), 392–397.
33 Economic Research Service (2007) US per capita food availability – sugar and sweeteners, www. ERS.USDA.gov (accessed 15 February 2008).
34 Kulprathipanja, S. (1991) Process for separating Glucose and mannose with dealuminated Y zeolites. U.S. Patent 5000794.
35 Kulprathipanja, S., et al. (1984) Separation of sucrose from thick juice. U.S. Patent 4475954.
36 Kulprathipanja, S. (1988) Separation of citric acid from fermentation broth with neutral polymeric adsorbent. U.S. Patent 4720579.
37 Cleary, M.T., et al. (1985) Process for separating fatty acids from rosin acids. U.S. Patent 4522761.
38 Cleary, M.T., et al. (1985) Process for separating fatty acids. U.S. Patent 4524030.
39 Cleary, M.T., et al. (1985) Process for separating fatty acids. U.S. Patent 4524029.
40 Cleary, M.T., et al. (1985) Process for separating oleic from linoleic acid. US 4529551.

9
Industrial Gas Phase Adsorptive Separations
Stephen R. Dunne

9.1
Introduction

Adsorption as a gas phase separation process fills a space in the spectrum of separations processes that encompasses both purification and bulk separations. The market for gas phase adsorptive separations is of the order of several billion US dollars annually when all sorbent, equipment and related products are included.

Zeolite adsorbents play a dominant role in purifications owing to their ability to both adsorb large quantities of material and to achieve extremely low mole fractions of these adsorbed these compounds in product gas. Zeolites are the preferred adsorbent types for dehydration to low levels, purification and in several bulk separations. Zeolites also are employed in a significant portion of the PSA hydrogen purification market segment where they add value to bulk separations by achieving particularly high purity specifications.

The engineers and scientists of Union Carbide's Linde Division were indeed pioneers in separations technology using zeolite molecular sieves. The research and development laboratories at the Linde Tonawonda facility just outside of Buffalo (N.Y., USA) were in fact the birthplace of much of the fundamental science underlying design methodologies, process cycles and both adsorptive and catalytic processes that employ zeolite molecular sieves. Cyclic thermal swing processes were developed, design systems were defined and the processes were commercialized all within just a few years of the invention of the materials themselves. Fixed-bed adsorbers utilizing zeolite molecular sieve products were introduced commercially in 1957. The use of zeolite molecular sieves for such purposes has increased steadily over the ensuing decades. Before the end of the 1950s, the Linde laboratories had turned out commercial processes for drying, sweetening, CO_2 removal, air separation and hydrogen purification.

The business that grew out of the Linde Laboratories was the Union Carbide Molecular Sieve business which, since 1988, has been integral to UOP's Catalysts Adsorbents and Specialties business segment.

This chapter discusses adsorption fundamentals relating to the design and operation of large-scale industrial separations using zeolite molecular sieves.

Zeolites in Industrial Separation and Catalysis. Edited by Santi Kulprathipanja
Copyright © 2010 WILEY-VCH Verlag GmbH & Co. KGaA, Weinheim
ISBN: 978-3-527-32505-4

While the Union Carbide organization, which is now in UOP, was the leader in gas phase separations, liquid bulk separations were brought to a high degree of maturity by UOP's Don Broughton and his successors in UOP's liquid phase separations research and development groups.

In discussing gas phase separations, a few definitions will help in understanding the subject matter. Adsorbents, sometimes referred to here as sorbents, are solid chemical substances that possess micro-porous surfaces that can admit molecules to the interior surface of the structure. Zeolites in particular are solid, micro-porous, alumino-silicates with adsorption and or ion exchange capability. They affect separations by adsorbing molecules into their micro-structures.

Adsorption is an exothermic process wherein adsorbing molecules spontaneously condense on the surface of the sorbent. Adsorption always implies excess concentration on the surface of the sorbent over that concentration found in the gas phase surrounding the sorbent. That a molecule might enter the cage of a zeolite based sorbent does not constitute sorption. Only when that molecule lays down on the surface and concentrates there, giving up its heat of adsorption can we say that the molecule has been adsorbed. In order to liberate an adsorbed molecule from the surface of an adsorbent we must supply the same heat of adsorption that the molecule gave up during adsorption. The process of releasing sorbed molecules is often called de-sorption.

Sorbate or adsorbate is a term that has been used in the adsorption industry since its inception. A sorbate is simply a molecular species that has been adsorbed or can be adsorbed.

Loading is the term describing the level of adsorption. In UOP's standard parlance the loading is the mass concentration per unit mass of adsorbate free adsorbent. The most common units of measure for loading is wt% and by that we mean specifically kilograms of adsorbed matter per 100 kg of adsorbate-free adsorbent. Many in the scientific side of adsorption research prefer to use molar loadings, again on a mass basis, for example, so many moles of sorbate per unit mass of sorbent. If one is to think about adsorption in its proper sense it does help to use molar concentration per unit volume. As mentioned above a sorbate must concentrate on the sorbent surface. By using q, the molar concentration of sorbate per unit volume we clearly see the requirement. Speaking of water a loading of say 20 wt% on a product that has an intrinsic or particle density of about $1\,\mathrm{g\,cm^{-3}}$ will have a sorbed phase molar concentration given by:

$$q = (1\,\mathrm{g\,cm^{-3}})(20\text{ wt\%})(100 \times 18) \qquad (9.1)$$

or about $0.0111\,\mathrm{mol}$ water $\mathrm{cm^{-3}}$ finished bead volume. When one computes the concentration of saturated water vapor at say 298 K then one finds that water vapor has a characteristic concentration of about $1.27 \times 10^{-6}\,\mathrm{mol\,cm^{-3}}$ space. Thus the concentration of water on the surface has been increased by a factor of about 8800 over what it would be in the volume occupied by the bead. This concentration degree is by no means the maximum attainable. Consider for example that water might have a partial pressure at 298 K of only 1% of saturation and its loading owing to the sharp isotherm and long plateau can be almost equal to the saturation loading. This constitutes a concentration ratio that is almost 100× greater or ~8.8×10^6. Again this is not the limit, simply an illustration that zeolites can produce tremendous concentration ratios between the sorbed and gas phases. With this definition in mind it should be clear no real adsorption occurs unless that concentration ratio is substantially greater than 1.

9.2 Regeneration

Before discussion of the means of regeneration it is important to recognize certain features of the packed bed contactors that are generally employed in industrial separations. Packed-bed contactors are generally just that: adsorbent packed into a pressure vessel with bed supports and flow distributors. In general there is no provision for any indirect heat exchange with either the ambient or another process stream. The only thermal coupling to the ambient comes at the outside wall of the adsorber vessel and even then the coupling is weak. It is for this reason that packed-bed adsorbers can be termed adiabatic. They exchange heat only with streams that actually flow through the packed bed and thus heat transfer is termed direct. Heat can only be introduced through process streams or through the process of adsorption which having a negative heat of adsorption is exothermic. There are a hand-full of non-adiabatic packed bed adsorbers that have been mentioned in the literature but in general practice these are rare. This fact has serious implications regarding the efficacy of regeneration.

The method of regeneration is the primary classification for industrial adsorptive separations. In the field of gas phase separations we speak of pressure swing adsorption or thermal swing adsorption and thereby indicate the method of regeneration of the adsorbent. In thermal swing adsorption energy is added to the adsorbent in order to heat the bed and the gas surrounding the adsorbent. The resulting change in the adsorbent equilibrium promotes a change in the loading state. We provide the heat of adsorption through thermal energy transfer and thereby desorb the sorbates that reside on the adsorbent.

In pressure swing adsorption we rely upon a change in both the total pressure of the system and a change in the partial pressure of the adsorbed compounds to effect a change in the equilibrium. The molecules are desorbed as a result of the change in equilibrium. In general there is no heat being supplied to the adsorbent to desorb the molecules, the desorption does occur and therefore takes heat from the adsorbent and gas surrounding the adsorbent This heat absorption from the local environment is effectively adiabatic and therefore results in cooling the bed and the gas surrounding the desorbed portion of the bed. During adsorption the heat liberated by adsorption must go into heating the bed and the gas flowing through the bed. The net effect of heating of the bed during adsorption and cooling of the bed during regeneration is to diminish the working capacity of the adsorbent.

While inert and displacement purge regeneration is widely used in liquid phase separations, there are few industrially relevant inert purge systems employed in gas phase separations. It is sufficient to note that an inert purge regeneration can be done and it will generally be most effective at relatively high adsorption temperatures.

While adsorption equilibrium considerations do justify the possibility to regenerate an adsorbent bed completely and to uniform levels, this is rarely achieved in practice in either thermal swing adsorption and is almost never the case in pressure swing adsorption. Some residual sorbate is always left on the sorbent and in general, except for TSA with only one or two sorbates, the residual loadings are almost always found as a non-uniformly distributed profile across the length of the bed.

9.3
Adsorption Equilibrium

The description of the partial pressure exerted by a sorbate, or a mixture of sorbates, when they reside on the sorbent surface, at some given temperature is what we speak of as adsorption equilibrium. For a single adsorbate (adsorbing molecular species) we require three state variables to completely describe the equilibrium: the temperature, the sorbed phase concentration or loading and the partial pressure exerted by the sorbed phase are very convenient variables to use. As more adsorbable compounds are added to the problem we require additional information to adequately describe the problem. That information is the specification of the mole fractions of the adsorbable compounds in both the gas and sorbed states.

The reader will find adsorption equilibrium relationships presented in any of three typical forms. The form of equilibrium most frequently presented is the isotherm, the partial pressure as a function of the loading at a given constant sorbent temperature. An isobar implies a chart of the loading as a function of the temperature while the partial pressure of the sorbate is held constant.

An isostere is a line describing the equilibrium partial pressure, sometimes the dew point, as a function of the temperature of the sorbent. The most common form of isosteric plot is to display the log of the partial pressure as a function of the reciprocal absolute temperature with loading or sorbed phase concentration held fixed.

Equilibrium relationships are typically mathematical functions that serve to specify the pressure at a given loading and temperature or conversely they may serve to describe the loading state when given a temperature and partial pressure.

In addition to describing the partial pressure or loading states adsorption equilibrium must also include a description of the heat of adsorption. One may argue that the description of the partial pressure does, through the Clausius–Clapeyron equation, describe the heat of adsorption. It is generally wise however to break out the heat of adsorption as a separate but intimately linked parameter of our description of equilibrium. In the next section I lay out the basis and utility of a number of potentially useful adsorption equilibrium models.

9.3.1
Henry's Law: A Linear Isotherm

The simplest adsorption equation is Henry's Law, that is, the loading is directly proportional to the sorbate partial pressure. $X = KP$ This linear isotherm equation adequately describes some adsorbents and, in the limit of low coverage, it actually describes most sorbents. For adsorption that is truly described by Henry's linear relationship, the loadings are low, the adsorption is bound to be essentially isothermal and there are several published analytical solutions to describe both batch kinetics and column dynamic behavior for such systems.

9.3.2
Langmuir

The Langmuir equation for adsorption, derived from kinetic arguments, is another of the simpler equations for description of monolayer adsorption. In practice the Langmuir equation is often used to describe multi-layer adsorption with little error. It is interesting to note that the Langmuir equation and several other useful isotherm equations can be derived from the Gibbs isotherm equation together with some suitable assumption of an equation of state for the adsorbed phase. In the case of the Langmuir form that assumption is the equivalent of a simple modification of the ideal gas law written for the sorbed phase.

The Langmuir equation for loading as a function of pressure takes the form:

$$X_0 KP/(1+KP) \tag{9.2}$$

The inverse of this relationship describes the pressure in equilibrium with a given loading as:

$$P = X/[K(X_0 - X)] \tag{9.3}$$

The adsorption constant K is the equivalent Henry's Law constant and is found to be temperature- and heat-dependent. The most common way of describing the adsorption constant is through an expression using the heat and the sorbent temperature:

$$K = K_0 \exp(\Delta H/RT) \tag{9.4}$$

These equations as written imply a constant heat of adsorption irrespective of the loading state. Such an assumption is generally un-founded but none-the-less useful for obtaining an approximate loading relationship.

9.3.3
Potential Theory

Potential theory is first and most widely attributed to Polanyi. For reference to the original articles by Polanyi, Dubinin and Raduskevic, the reader is referred to Breck [1] and Ruthven [2]. Each lays out a useful explanation of the theory. Potential theory states, in essence, that the volume occupied by the sorbed phase in the sorbent structure will be a unique function of the Gibbs free energy change for the adsorption. One of the most widely used forms of potential theory is ascribed to Dubinin and Raduskevic (D-R). The D-R model fixes the power law dependency of the free energy change at 2. Thus D-R is the exponential of a constant multiplying the square of the Gibbs free energy. Specifically this model can be written as: $X = X_0 \exp(-\beta \Delta G^2)$ the potential theory can adequately describe the adsorption equilibrium for activated carbons and zeolites. One disadvantage of the potential theory models is that they do not reduce to a Henry Law form as loading approaches zero. Potential theory is particularly useful for extending the range of measured data. In potential theory the coefficients determined are expected to be independent of the sorbates. This property of the theory has been used successfully in *a priori* estimation of the adsorption of components that were not measured. The technique works especially well for estimation of homologous series such as the alkanes and alkenes. Potential theory also delivers one more tangible benefit: it describes for us the heat of adsorption as a function of the loading.

9.3.4
Universal Isotherm

The universal isotherm, an example of which is given by Carniglia and Ping [3], expresses the thermodynamic imperative that the equilibrium pressure must relate to the entropy and enthalpy changes that occur as the sorbate transitions from the gas phase to the sorbed state. The universal isotherm is written simply as:

$$\ln(P) = \Delta S(x)/R - \Delta H(x)/(RT) \tag{9.5}$$

This equation can be derived from potential theory. The entropy and enthalpy changes as functions of the loading state are the prime differentiators for various sorbent/sorbate pairs. These loading dependencies are indicated by the form of the functions $\Delta S(x)$ and $\Delta H(x)$. Here the dependencies are written strictly as functions of the loading (x) only. There may some modest temperature dependency as well. The heat and entropy changes with temperature tend to be small hence the universal form tends to be linear in reciprocal temperature over a wide range of temperatures.

It will be noted that the universal isotherm equation as written here has formal similarity to pressure explicit forms of Langmuir, Langmuir–Freundlich and LRC models. One key advantage of the universal form is that the heat of adsorption and the adsorption equilibrium are bound to be self-consistent.

The universal form is most often used to prepare isoteric charts. That is charts relating either equilibrium vapor pressure or dew point, in equilibrium, to the loading at a given temperature. The charts have lines of constant loading depicted as functions of reciprocal temperature. Such charts are particularly useful for estimation of the limiting dew point or partial pressure that can be achieved for a given separation problem.

9.3.5
Freundlich

The Freundlich isotherm by itself is not widely used to describe adsorption in zeolites. There are a handful of isotherms, for example, O_2 on Type A or X can often be fit to the Freundlich form owing to the small deviations from linearity that occur in the isotherm. This estimator has the mathematical form:

$$X = KP^{1/n} \tag{9.6}$$

The Freundlich form is often employed along with the Langmuir form and is referred to as the Langmuir–Freundlich equation. That model is the basis of a great many useful modifications including many empirical forms that serve to describe adsorption data quite accurately.

9.3.6
Langmuir–Freundlich

The Langmuir–Freundlich form is written as:

$$X = \frac{X_0 KP^{1/n}}{1 + KP^{1/n}} \tag{9.7}$$

It can often be inverted to allow either pressure explicit forms or loading explicit forms. The LRC model published by UOP is a form very similar to the Langmuir–Freundlich form.

9.3.7
Kelvin Equation and Capillary Condensation

The Kelvin equation is:

$$\text{Ln} P = \text{Ln} P_0 - 2\sigma V_m \text{Cos}\theta (rRT) \tag{9.8}$$

Though not a general adsorption equilibrium model the Kelvin equation does provide the relationship between the depression of the vapor pressure of a condensable sorbate and the radius (r) of the pores into which it is condensing. This equation is useful for characterization of pore size distribution by N_2 adsorption at or near its dew point. The same equation can also describe the onset of capillary condensation: the enhancement of sorption capacity in meso- and macro-pores of formed zeolite adsorbents.

Nearly any of the useful theories can be modified to allow for the computation of multi-component adsorption equilibrium. In addition to the LRC there are other specialized forms that may be employed.

The interested reader is referred to Ruthven [2], Yon and Turnock [4], Young and Crowell [5] and Yang [6] for a more comprehensive discussion of the various adsorption equilibrium theories that are available.

9.4
Mass Transfer in Formed Zeolite Particles

Mass transport on formed zeolites and in packed beds is described by diffusion equations. Whether one chooses to utilize a partial differential equation (pde) incorporating a diffusion coefficient or one chooses to use a simpler linear driving force approximation, one is always dealing with diffusion coefficients for the various mechanisms that promote or interfere with the transport of mass between the fluid and the adsorbent surface. In the following paragraphs I introduce the differential equations for the column behavior and the pde's describing local diffusion in particles or crystallites. I present a few approximating equations that will prove very useful in design of adsorption units and performance evaluation thereof. Finally in this section I provide an overview of the important mass transfer coefficients and diffusion mechanisms that impact mass transfer in formed zeolite particles.

Mass transfer in zeolite-based adsorbents can involve several mechanisms and can depend on a fairly wide variety of parameters that in turn depend on the crystallization, forming, bonding, heat treatments and ion exchange steps that went into the production of a formed zeolite particle. Mass transfer from the fluid phase into the zeolite structure can involve a gas phase or film transfer coefficient, diffusion of the adsorbing species in the macro-pore region of the particles and diffusion within the zeolite crystal structure. In addition to these widely recognized mass transfer mechanisms there are several transport resistances that can arise in the forming and firing steps as a result of damage to the crystallites or pore structure that create barriers to diffusion in either macro-pores or micro-pores. Some investigators have identified what have been called skin resistances that are associated with gradations in the porosity that appear under microscopy as skins or barrier layers at or near the outer surface of the beads or pellets.

Before an in-depth discussion of mass transfer models and coefficients we need to be explicitly clear that all mass transfer models are approximations that allow us to solve the partial differential equations (pde) describing an adsorption problem. There are a great many sources that derive and present the partial differential equations that describe adsorption of gases appropriate for column separations. *The Design Manual For Octane Improvement, Book I* [7] was among the earlier works to show them. The forms as presented by Ruthven [2] are shown here owing to the consistent and compact nomenclature that he has employed. There are a wider array of forms to choose from in the literature including [6, 7] and others. We cover the pde's for the gas and solid phase mass balances and discern what can be known about their solutions without actually providing the detailed solutions. The reader is cautioned that the description of the full problem of adiabatic adsorption will of course require gas- and solid-phase enthalpy balances for a complete description of the phenomena of interest. Again references [2, 6, 7] present these equations in a variety of useful forms.

$$\frac{\partial c}{\partial t} + \frac{\partial (vc)}{\partial z} + \frac{(1-\varepsilon)}{\varepsilon}\frac{\partial \overline{q}}{\partial t} - D_L \frac{\partial^2 c}{\partial z^2} = 0 \tag{9.9}$$

Equation (9.9) is the most general from of partial differential equation describing adsorption of some species from the gas phase flowing in a column. This equation is the so-called continuity or mass balance equation for a single component. Equation (9.9) allows for a significant change in sorbate concentration as the adsorption wave proceeds through the column in that it has a the spatial differential of the product of v and c. The pde includes a sink (or source) term written as: $\partial q / \partial t$ for the adsorbable component which is shown as a time derivative of q. This sink term in the pde is the mass transfer rate. It describes the time rate of change of the spatial average loading in the beads or pellets of adsorbent. In reference to Eq. (9.9) there is a second term involving mass transport: $D_L \partial^2 c / \partial z^2$. This term describes axial dispersion using an axial diffusivity of D_L. It should noted that axial dispersion is not a mass transport mechanism that seriously impacts the uptake rate of the sorbent except insofar as the axial dispersion may cause somewhat higher concentrations of the adsorbate to be transported downstream.

The solution for (Eq. 9.9) requires two boundary conditions on c, one on v; an initial condition on c and similarly one initial condition on q. Finally we must prescribe the sink/source term for the adsorption. This can be done in the most general case by writing another pde to describe adsorption, which is the transport of the adsorbing species into the crystal structure of the formed adsorbent. This model must be sufficiently broad to allow us to calculate the uptake at any location in the packed bed and at any time during the process. In many cases it will be found expedient and quite satisfactory to prescribe the uptake term as some kind of linear driving force model (LDF).

As a point of reference there are a great many adsorption problems where the full form (Eq. 9.9) may not be necessary. For small concentration changes across a mass transfer front and for sufficiently high velocity that axial dispersion is not of serious consequence we can write a much simpler first order pde as shown below:

$$\frac{\partial c}{\partial t} + v\frac{\partial c}{\partial z} + \frac{(1-\varepsilon)}{\varepsilon}\frac{\partial \overline{q}}{\partial t} = 0 \tag{9.10}$$

Equation (9.10) is written for problems where the adsorbable component represents a minor constituent of the flow. Purification type separations can generally be described by this pde.

9.4.3
Linear Driving Force Approximation and Resistance Modeling

In this section we deal with the following two approximations for adsorption uptake rate and how to go about calculation of the relationship between the mass transfer coefficients (k, kK) that appear and the solutions to diffusion problems.

$$\frac{\partial \overline{q}}{\partial t} = k(q^* - q) \qquad (9.16)$$

$$\frac{\partial \overline{q}}{\partial t} = kK(c - c^*) \qquad (9.17)$$

The introduction of linear driving force models also introduces the concept of the mass transfer coefficient. Having laid down the background we can now speak of mass transfer coefficients as they relate to the linear driving force approximations that employ them. The coefficients, k in Eq. (9.16) and kK in Eq. (9.17), are the mass transfer coefficients that describe the rates of adsorption. It is vitally important to remember that it is inappropriate to speak of "the mass transfer coefficient" without also specifying the linear driving force approximation that is being employed. A second point is that the mathematical form, units of measure and order of magnitude of the mass transfer coefficient depend on the form of the gas phase mass balance equation. Equations (9.9) and (9.10) are excellent forms to work with but be advised that there are many equivalent expressions and the comparison of the mass transfer coefficients that are introduced must be made very carefully in view of both the uptake rate model and the pde into which it is substituted.

Working from substitution of Eq. (9.17) into Eq. (9.10) we can with a little mathematical manipulation obtain an expression for the length of the mass transfer zone without resorting to the solution of the pde.

That result is presented without derivation as:

$$L_{\mathrm{MTZ}} = -\frac{v\varepsilon}{(1-\varepsilon)kK} \int_{c_\mathrm{f}}^{c_\mathrm{p}} \frac{1}{(c-c^*)} dc \cong \frac{v\varepsilon}{(1-\varepsilon)kK} \ln\left(\frac{c_\mathrm{f}}{c_\mathrm{p}}\right) \qquad (9.18)$$

Equation (9.16) can be used to obtain estimates of the length of the mass transfer zone when we know the mass transfer coefficient and the shape of the adsorption isotherm. Note that the first, general form of Eq. (9.17) does imply that the shape of the isotherm impacts the length of the mass transfer zone. A little explanation helps here. The end point of the gas phase driving force model, c^* is the concentration in the gas phase that will be in equilibrium with the local concentration in the sorbed phase at the sorbed phase temperature. In a very general sense $c^*(T,q)$, when written as a function of the loading state q, is the equilibrium line and the chord of the isotherm is the operating line on a McCabe–Thiele diagram that is familiar to most in the field of distillation. The shape of the isotherm then determines whether the wave that forms will be neutral as when the isotherm is linear, self sharpening as for a Langmuir-type isotherm, or Brunauer type I, (a positive first derivative and continuously negative second derivative) or an ever expanding

9.4 Mass Transfer in Formed Zeolite Particles

wave, which occurs in a Brunauer type III (a continuous positive first derivative and continuously positive second derivative). Both Ruthven [2] and Yang [6] display sketches of these isotherm types. Brunauer isotherm types II, IV and V (all positive first derivatives but including inflection points) can all display some artificial broadening of mass transfer zones as a result of inflections in the isotherm shapes. The Brunauer type I is the characteristic shape that arises from uniform microporous sorbents such as zeolite molecular sieves. It must be admitted though that there are indeed some deviations from pure Brunauer type I behavior in zeolites. From this we derive the concept of the favorable versus an unfavorable isotherm for adsorption. The computation of mass transfer coefficients can be accomplished through the construction of a multiple mass transfer resistance model. Resistance modeling utilizes the analogy between electrical current flow and transport of molecular species. In electrical current flow voltage difference represents the driving force and current flow represents the transport. In mass transport the driving force is typically concentration difference and the flux of the species into the sorbent is resisted by various mechanisms.

Equation (9.15) was written for macro-pore diffusion. Recognize that the diffusion of sorbates in the zeolite crystals has a similar or even identical form. The substitution of an appropriate diffusion model can be made at either the macro-pore, the micro-pore or at both scales. The analytical solutions that can be derived can become so complex that they yield little understanding of the underlying phenomena. In a seminal work that sought to bridge the gap between tractability and clarity, the work of Haynes and Sarma [10] stands out. They took the approach of formulating the equations of continuity for the column, the macro-pores of the sorbent and the specific sorption sites in the sorbent. Each formulation was a pde with its appropriate initial and boundary conditions. They used the method of moments to derive the contributions of the three distinct mass transfer mechanisms to the overall mass transfer coefficient. The method of moments employs the solutions to all relevant pde's by use of a Laplace transform. While the solutions in Laplace domain are actually easy to obtain, those same solutions cannot be readily inverted to obtain a complete description of the system. The moments of the solutions in the Laplace domain can however be derived with relative ease. The moments of the solutions thus obtained are then related to the individual mass transport diffusion mechanisms, dispersion mechanisms and the capacity of the adsorbent. The equation that results from this process is the model widely referred to as the three resistance model. It is written specifically for a gas phase driving force. Haynes and Sarma included axial diffusion, hence they were solving the equivalent of Eq. (9.10) with an axial diffusion term. Their results cast in the consistent nomenclature of Ruthven: first for the actual coefficient responsible for sorption kinetics as:

$$\frac{1}{kK} = \frac{R_p}{3k_f} + \frac{R_p^2}{15\varepsilon_p D_p} + \frac{R_c^2}{15KD_c} \tag{9.19}$$

The expression for the effective gas phase coefficient that would account for axial dispersion and hence give a proper mass transfer zone length is:

$$\frac{1}{kK} = \frac{(1-\varepsilon)}{\varepsilon}\frac{D_l}{v^2} + \frac{R_p}{3k_f} + \frac{R_p^2}{15\varepsilon_p D_p} + \frac{R_c^2}{15KD_c} \qquad (9.20)$$

The distinction here is that the kK calculated from Eq. (9.19) would be used in a linear driving force model for the actual uptake rate expression and an axial dispersion coefficient would be substituted into the pde. If however one simply desires to match the adsorption response or breakthrough curves then the kK' calculated according to Eq. (9.20) would provide very satisfactory results for estimation of the length of the mass transfer zone.

9.4.4
Diffusion Mechanisms in Formed Zeolites

With the forgoing as background we are now prepared to discuss the various diffusion mechanisms that occur in formed zeolites.

9.4.4.1 Fluid Film Diffusion
The transport of an adsorbable species from the bulk fluid flowing around an individual bead is a problem of molecular diffusion. With the fluid in motion the rate of transport to the surface of a bead or pellet of adsorbent material is generally treated as a linear driving force. For gas phase separations there are a variety of correlations available to describe the mass transport to the surface in terms of the particle Reynolds number, the Schmidt number, the size of the adsorbent particle and of course the binary diffusivity of the species of interest.

9.4.4.2 Macro-Pore Diffusion
Diffusion in the macro-pores of a formed particle is generally speaking a very important mechanism. If we speak in terms of resistances to mass transfer macro-pore resistance is often the largest of the resistances to mass transfer. For transport in the macro-pores we must introduce two parameters that influence the transport.

Macro-pore diffusion will generally exhibit itself as one of two mechanisms: (i) Maxwell diffusion, related only to porosity and molecular diffusion, or (ii) Knudsen diffusion, a diffusion mechanism wherein the diffusion coefficient is related to the size of the pores in which diffusion is taking place. Very often both mechanisms will be found to operate in such a manner that the inverses of the diffusion coefficients are additive. When we calculate diffusion coefficients we need to introduce two added parameters to allow for a proper description of the phenomena. The first is the internal porosity, generally labeled as ε_p. The internal porosity of a bead is related directly to its intrinsic or what is often called piece density. It is a measure of the free void space inside the bead. The second parameter generally introduced to aid in describing diffusion in the pores is called tortuosity. Tortuosity (τ) may be thought of as a factor that effectively increases the path length for diffusion. In traversing macro-pores a molecule may have to circumvent many partially saturated crystallites before it alights on the surface of a crystallite with a readily available sorption site. Tortuosity provides us the mathematical factor to describe that. Tortuosity is generally to be thought of a geometric property. Some authors have suggested a link between tortuosity and internal void space. Ruthven reports that for straight cylindrical pores with random orientation the tortuosity can be shown to be $\tau = 3$. In products where macro-pores may be plugged or not well connected, tortuosity factors can become quite large. Experimentally derived values are found between about 2 and 6 for commercial products. The forgoing discussion can be summarized mathematically with just a few relationships:

D_M is the molecular diffusivity of the species of interest. It can be estimated by any number of equations among the most general is that of Chapman and Enskog. Reid, Sherwood and Prausnitz [11] provide a wide variety of models for calculation of molecular diffusion. D_K is the Knudsen diffusion coefficient. It has been given in several articles as: $9700r(T/MW)^{1/2}$. Once we have both diffusion coefficients we can obtain an expression for the macro-pore diffusion coefficient: $1/D = 1/D_K + 1/D_M$. We next obtain the pore diffusivity by inclusion of the tortuosity: $D_p = D/\tau$; and finally the local molar flux J in the macro-pores is described by the familiar relationship: $J = -\varepsilon_p D_p \partial c/\partial z$. Thus flux in the macro-pores of the adsorbent product is related to the term $\varepsilon_p D/\tau$. This last quantity may be thought of as the effective macro-pore diffusivity. The resistance to mass transfer that develops due to macro-pore diffusion has a length dependence of R_p^2.

9.4.4.3 Intra-Crystalline Diffusion

Intra-crystalline diffusion is an activated process and its temperature dependence can generally be correlated by an equations such as: $D_c = D_{c0}\exp[-E/(RT)]$

The activation energy for diffusion is generally taken to be less than the heat of adsorption. It is through the choice of a zeolite with a pore structure that restricts diffusion that we obtain what is termed the molecular sieve effect. This effect is best achieved through the selection of small pore sieves where the aperture that admits molecules to the cage is close to the molecular diameter. In such a case the activation energy for diffusion will be large and diffusion correspondingly small.

The most widely recognized molecular sieving effect occurs with K^+ exchange. With K^+ exchange (monovalent) the cation site that in its native Na^+ form sits in a pore window partially occluding it.

The larger ionic radius of K^+ further occludes the pore openings into the type A and we effectively slow down or completely block access to the structure. Both CO_2 and N_2 demonstrate essentially this effect as exchange level is increased towards about 50% of the ions being K^+. Once we have that level the product is essentially type 3A.

Exchange with Ca^{2+} tends on the other hand to open pore windows. This effect occurs because the divalent Ca^{2+} actually must displace two Na^+ cations. This results in site sharing and actually pulls a cation from a site that at least partially occludes a pore opening, this opening the window. It has been known for several decades that ion exchange by Ca^{2+} in place of Na^+ in the type A structure opens up the pore apertures and that molecules that can diffuse slowly in NaA experience a dramatic drop in activation energy and corresponding increase in diffusivity as one-third of the pore apertures become opened due to the presence of the divalent Ca^{2+}.

Even with its large pore structure, type X structures can exhibit molecular sieving. There the nominal 9.5 Å pores admit considerably larger molecules. Curiously while Ca^{2+} exchange in A opens the pore window, the opposite is true in X. The degree of pore closure achievable with Ca^{2+} exchange in X is about the same magnitude as the change in the A structure: a change of the order of ~1 Å

When type X is utilized, in any of its ion exchange forms, for dehydration or possibly for sweetening (sulfur removal), there is little likelihood that the intra-crystalline diffusion will be the dominant resistance to mass transfer. Large aromatic sulfurs would of course be an exception. When type X is used for adsorption of hydrocarbons or aromatics then it is possible that the micro-pore diffusion might dominate. When type A is used there is always a distinct possibility that intra-crystalline diffusion will be slow and may dominate the mass transfer, even for relatively small molecules. This is especially true when the chosen structure is a K^+ A or type 3A. Selection of other small pore structures, for separations or purification applications can also create situations where the dominant resistance is found in the crystallites.

9.5
Industrial TSA Separations (Purification)

Thermal swing adsorption (TSA) has been practiced with zeolite molecular sieves for more than 50 years. In the following sections I discuss design methods and provide some reference data so that the design process can be understood.

The requirements for an effective thermal swing design include:

1) The basic adsorption process design. Sub-tasks within that include the adsorbent selection, made in view of all of the requirements imposed on the dehydration process. The adsorption step time, regeneration and cooling step times all need to be settled and these in view of mechanical details. The overall vessel configuration, for example, the vessel ID and length, which quantities are typically sized based on pressure drop. Finally we need to make some estimate of the expected service lifetime for the adsorbent product.

2) Mechanical design of the adsorber then takes up the remainder of the engineering effort to produce a workable adsorption process design. Once a vessel is sized to provide the required inventory of adsorbent, we need to provide the mechanical details, which include flow distribution devices, bed supports and the required vessel wall thickness to withstand the working pressure and added stresses encountered during regeneration and repeated de-pressurization and re-pressurization.

3) We need to make reliable estimates of the utilities requirements to run the process. Chief among these for a TSA process is the selection of the regeneration gas and the heat source that will supply the heat.

In the next few sections I provide a description of processes design and the regeneration requirements.

I provide the greatest level of detail on the subject of dehydration. Sweetening and CO_2 removal can then be handled as special cases where additional design considerations are present.

9.5.1
Dehydration

Dehydration is by far the largest industrial separation of interest here. Removal of water was the first commercial application of molecular sieves. Dehydration and related fixed-bed adsorptive separations in the process industries account for more than half of the commercial molecular sieve business volume.

Dehydration is applied to natural gas, air for feed to air separation units, cracked gas (e.g., ethylene), various other hydrocarbon streams just to name a few. Zeolite molecular sieves play a major role in dehydration because of their ability to reduce moisture content to very low levels attaining dew points of $-100\,°C$ or lower.

The process design requires certain basic data concerning the molecular sieves to be employed. The designer needs adsorption equilibrium information, including isotherms and isosteres, the heat of adsorption as a function of the loading and some basic data or correlations that predict the mass transfer parameters. In addition, vapor pressure relationships prove handy for specifying dew points or for converting dew points into partial pressures and then to mole fractions. These data historically were in graphical form, today designers can have access to a wide array of the required data in databases, through estimating equations or pre-programmed computer tools or spreadsheets.

It is worth noting than adsorption process design is a mature science and there is seldom a need to employ detailed numerical models to solve the pdes described in Section 9.4. There are some specialized circumstances that may fall outside the norms for which many of the today's design tools have been formulated. Only in these circumstances is a more rigorous process simulation required for obtaining a design.

The process I describe utilizes charts for isotherms and isosteres. The design sequence is intended to be done by hand and this rough sketch of the design procedure is meant for instruction rather than to produce a finished design. The first step in the process design is to assemble all of the requirements and the design basis. The design basis must start with the feed flow rate for adsorption. Further we must have the level of water in the feed, the maximum tolerable level of water in the product and often, in fact almost always, the allowable gas utilization for regeneration. Additional data that are required for the design also include the feed temperature, the pressure of operation that is to be assumed during adsorption and subsequently in regeneration. With all design basis data in hand we can begin.

Before starting any design calculations it is important to lay out the entire process cycle including the number of beds, the adsorption step, the de-pressurization step (if such is practiced), the hot gas purge for regeneration, the cool gas purge and re-pressurization if required. This exercise provides the designer with the appropriate range of pressures and temperature and at least starting points for the times required for each process step.

Our next step is to estimate the loading expected to be in equilibrium with the feed. That is accomplished by reference to the isotherms of Figure 9.1. We call this loading X_f. Next, surprisingly, we assume an average residual loading, X_r that exists in the bed after regeneration. The maximum average residual loading that we can tolerate can be quickly established by use of the isostere charts and the vapor pressure or dew point relationships. By reference to an isostere chart and with an assumption of a 4 wt% residual loading we can quickly establish the minimum product dew point that can be achieved. At 4 wt% from Figure 9.2 and assuming a 38 °C adsorption temperature we see that the minimum product dew point achievable after cooling is −67.7 °C. This dew point when at a pressure of 68.96 bar (6896 kPa) is equivalent to about 50 ppb (v/v). Generally this is adequate for many industrial dehydration designs. If this dew point is not adequate for our process purpose, then we try a lower residual loading. Bear in mind that the lower the residual loading the greater the expense in both heat and gas utilization for regeneration. The choice is not free.

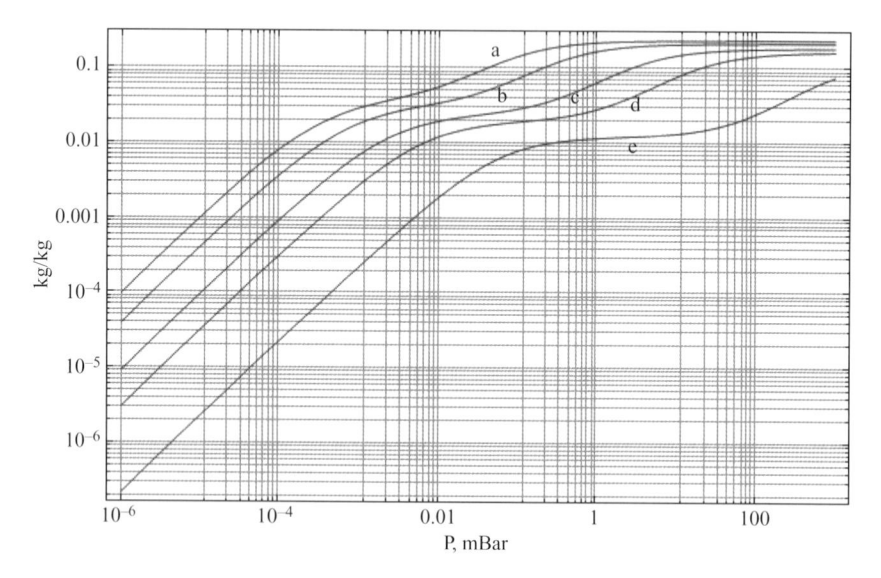

Figure 9.1 Adsorption isotherms for water on a type 4A molecular sieve for a = 25, b = 50, c = 100, d = 150, and e = 350 °C.

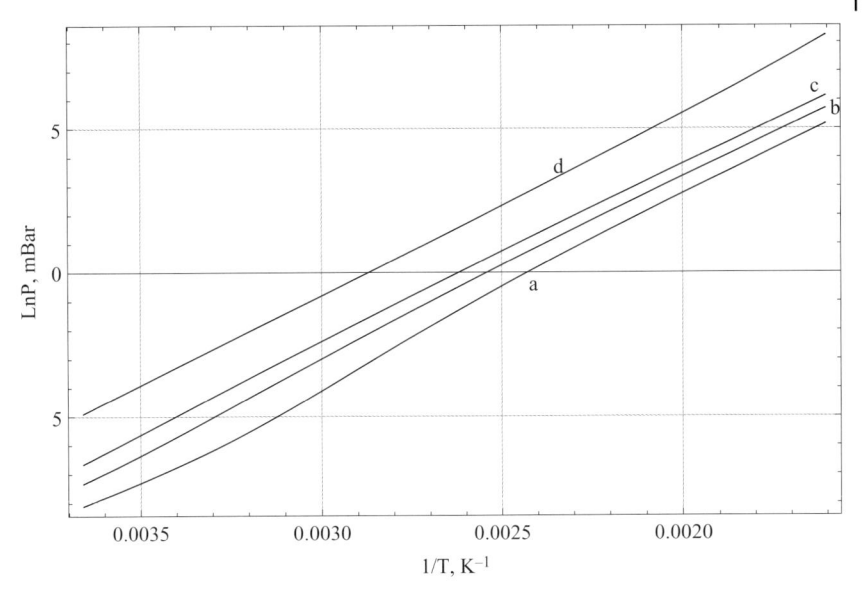

Figure 9.2 Abridged isosteres for water on a type 4A molecular sieve at a = 0.03, b = 0.04, c = 0.05, and d = 0.1 kg/kg.

Given our determination of X_r, our assumed X_r together with the feed specification, we now have the amount of molecular sieve that will be consumed, that is brought from X_r to X_f, per unit of time on stream during the adsorption step. This together with an assumed or specified adsorption time now defines the equilibrium section of the adsorber.

The designer now needs to make some estimates of mass transfer. These properties are generally well known for commercially available adsorbents, so the job is not difficult. We need to re-introduce the adsorber cross-section area and the gas velocity in order to make the required estimates of the external film contribution to the overall mass transfer. For spherical beads or pellets we can generally employ Eq. (7.12) or (7.15) of Ruthven's text to obtain the Sherwood number. That correlation is the mass transfer analog to the Nusselt number formulation in heat transfer:

$$\text{Sh} = 2k_f \frac{R_p}{D_m} = 2 + 1.1 Sc^{1/3} \text{Re}^{0.6} \tag{9.21}$$

Here the Reynolds number, Re, is formulated using the particle diameter and the superficial velocity.

From that relationship and together with D_m, k_f can be obtained. It should be obvious to the reader that, apart from the trivial constant = 2 at vanishing Reynolds number, this evaluation is an iterative procedure. One selects a bed diameter, hence a bed cross-sectional area, and that determines the fluid velocity. Evaluation of the Sh correlation is then possible. The user will generally find that above a given particle Reynolds number (Re) the film coefficient, k_f, will asymptote to a constant value. In general this process will converge to a fixed value for the fluid film coefficient very rapidly.

The rest of the terms in Eq. (9.18) are readily obtained from the literature and then the resulting value of the overall mass transfer coefficient kK can then be substituted into Eq. (9.18). We now have the required added length of the mass transfer zone and the bed sizing is complete.

The first pass at the adsorber sizing is now complete and a check on the overall pressure drop expected can be obtained from the Ergun equation. An adjustment in the bed diameter may be required and then the bed size can be fixed using one more pass through the sizing process outlined above.

At this point we have a specification of an adsorber bed under fresh conditions with a specified diameter, an inventory of molecular sieve that accounts for both equilibrium capacity (water load- and step time-dependent) and a mass transfer zone (velocity- and particle size-dependent) to assure that the design meets product specification. It is at this point that we must consider cyclic durability of the adsorbent product. Products used for dehydration will experience some loss of capacity for water as a result of steam de-alumination, particle breakup due to the cyclic nature of the process, in some cases organics in the feed stream may accumulate and foul the sieve. Depending on the service there may be other specialized concerns peculiar to the dehydration application that will be to be considered. There is no single universal aging or capacity curve. Over the years UOP has created reliable estimates of the aging process and has developed factors by which we will increase the adsorbent inventory to accommodate the changes in capacity as the adsorbent ages. These factors now provide a specification of the amount of sieve to be loaded and the extension of the bed length to accommodate that increased inventory so that at the expected design life the adsorber design still delivers the expected dehydration performance.

The designer now generally does some design work on the vessel, estimating the required wall thickness for both the cylindrical shell and the heads of the vessel. It is also at this point that some allowance in the tangent length of the shell must be made for bed supports and screens. Finally flow distribution nozzles should be specified. From this design work the designer is able to calculate the weight of the shell, heads and the bed supports. Each of these quantities is required in the estimation of the regeneration heating and cooling gas requirements.

Regeneration gas requirements are readily obtained once the adsorber is sized and the pressure vessel has been designed. The requirements for gas, a flow rate and duration for heating arise from a heat balance.

The heat requirement is defined as the sum of the following quantities:

1) The latent heat required to liberate the adsorbed water. This is obtained from the heat of adsorption for the water.

2) The sensible heat required to heat the dry adsorbent mass from the adsorption temperature to the regeneration temperature. This is generally a function of temperature and can be readily evaluated.

3) The sensible heat required to heat the adsorbed water on the molecular sieve again over the same temperature range. The properties of the adsorbed phase may safely be assumed to be those of liquid water and the calculations of the enthalpy change can be made from available data.

4) Depending on whether the bed is internally insulated or whether it is externally insulated we will have an additional sensible term that involves either the shell mass or the mass of the internal insulation.

5) Both heads must be heated irrespective of the choice of insulation.

6) All bed support and internals must be heated.

7) Any heat loss from the bed to the environment must also be accounted for.

These terms are not difficult to handle and once the sum of all heat loads is known then the quantity of hot purge gas required to supply that heat is obtained from the heat load, the heat capacity of the gas used and the temperature swing of the gas.

The quantity of cool purge gas is handled the same way with the exception that term 1 is omitted entirely and term 3 is significantly reduced because essentially all the water (except for a small residual) has been taken off the molecular sieve by the regeneration process.

Before leaving dehydration it is useful to point out some common practices in dehydration designs.

The first is that some designs are tightly constrained in pressure drop. When that is the case the designer can take one of two routes. Either a compound bed can be designed in which we employ a relatively large particle for installation in the equilibrium section and then specify a smaller particle for use in the mass transfer zone. This approach can cut pressure drop and still deliver a good product specification. Under more extreme circumstances the designer can elect to place several beds in parallel flow for adsorption and adjust the cycle times to allow for the required regeneration and cooling steps to be done on one bed while the others are involved in adsorption.

Yet another option exercised in gas dehydration service where continuous operation is required is the use of a lead/trim bed installation. In this design concept, each of two beds is designed to handle the full flow and the beds are operated with the lead bed (first in line for the adsorption flow) handling the dehydration and the trim bed being a very large guard zone. The trim bed may in some case be offline altogether and it is brought online only when the lead bed is nearly exhausted and likely spent. This concept can be employed in dehydration but it is of significantly greater value in non-regenerable guard bed designs where cycle times for adsorption can be measured in months or years.

In addition to these specialized adsorption designs there are a number of variations on regeneration and cooling processes. The pulse regeneration technique has been practiced for several decades. In this variation we introduce a pulse of hot purge gas over a somewhat limited duration and then follow that with a cooling gas purge. In the pulse regeneration process the cool purge serves to push the regeneration wave and its enthalpy through the bed and simultaneously cool the bed.

The process sketched out in this section can be used to create an adsorber design. There are often times when one wants to examine the performance of an existing unit. When performance analysis is needed there are several alternatives. Depending on the specifics of the problem there may be an analytical solution for the adsorption problem and that may enable the creation of a satisfactory description of the process to use in understanding phenomena that are observed in operation.

When no analytical solution can describe the process satisfactorily it may be possible, working from Eq. (9.18) (which describes the length of the wave) and either Eq. (9.11) or (9.13) (the expression for the velocity of the adsorption wave), to assemble a simple wave mechanics solution that approximates the length and movement of the mass transfer front in the bed. As with analytical solutions this method can deliver useful results that may approximate the wave shape inside the bed and thus can be used to describe the shape and duration of the breakthrough curve that occurs as the wave intercepts and crosses the end of the bed. Such methods are generally only applicable for one or at most two adsorbable components.

When these simple models fail to deliver a satisfactory description of the adsorption problem, then the investigator is faced with the need to produce a detailed numerical solution to the adsorption problem. Today there are a variety of adsorption simulation tools that can be employed for the purpose of studying adsorptive separations.

This discussion lays out the basic concepts involved in design using water as the subject contaminant. In the next sections we examine a handful of commercially relevant adsorptive separations. We find that the design ideas are all readily transferable but that in each application there are additional special considerations that enter into the sizing of the adsorption system.

9.5.2
De-Sulfurization

De-sulfurization, sometimes referred to as sweetening, can involve the removal of a range of sulfur types including mercaptans but the term is generally applied to removal of H_2S. Speaking specifically about H_2S removal we generally choose a Type 5A molecular sieve for this service. In most design problems one does need to perform a dehydration design along with the sulfur removal design.

The general method of doing the sulfur design is very similar to the dehydration design but owing to the relatively weak adsorption of H_2S as compared to water we are forced to: (i) dehydrate first and then (ii) also limit adsorbable hydrocarbons so that the co-loading of H_2S can be driven to a sufficiently high value. With the special concern for co-adsorption of the hydrocarbons the H_2S loading is somewhat smaller than is achieved for water on type 4A.

With that caveat the design proceeds in essentially the same manner as was outlined for water removal by 4A. We always do the dehydration design first, calculating the bed consumption for handling the water, allowing for a mass transfer zone and then factoring up the bed requirements in view of cyclic stability.

The same process is then followed for the H_2S section and in a somewhat conservative mode we simply add the bed requirements together to establish the design. There are occasionally requirements to remove mercaptans and even then some rather large ones. Type 5A can be chosen for some service and in the rare cases we may choose to us a NaX for large sulfur species.

Before discussing CO_2 removal it is worthwhile to speak of the problem of COS formation when removing H_2S from gas where CO_2 is present. COS is formed by a first order reaction, taking place in the sorbed phase, wherein:

$$H_2S + CO_2 \rightarrow COS + H_2O$$

The degree to which molecular sieves promote this reaction varies widely and UOP can offer a range of technical solutions to either drive or severely inhibit this reaction.

9.5.3
CO_2 Removal

As with the sweetening application our most common need for CO_2 removal is from natural gas prior to liquefaction. In this application we are often faced with amounts of oxygen in the feed that may range up to several hundred ppm by volume. The process is often limited to adsorption at a total pressure of about 35 bar. In this application however the feed gas is most often pipeline natural gas which gas will have been pre-dried to pipeline standards or about seven pounds (1 lb = 0.45 kg) of water per MMSCF of gas. In some cases the gas source may be other than pipeline and the water load needs to be estimated based on a given mole fraction. The liquefaction process, which runs at $-260\,°F$ ($127\,°C$), demands very low levels of water in the product as well as trace levels of CO_2 so that the heat exchangers in the downstream process remain clean.

As with sweetening the design process for removal of CO_2 is very similar to that of dehydration. Special consideration is given to the presence of O_2 in the feed and in some cases it is found necessary to catalytically remove the oxygen.

We can also offer a design for the removal of CO_2 from an ethylene stream. In this application there is the clear need to minimize co-adsorption of the ethylene and to ensure there is essentially no ethylene in the regeneration gas so as to minimize the possibility of polymerization during the hot regeneration step. CO_2 from ethylene is one of those applications were we generally do de-pressurize the bed because the regeneration gas, which is typically de-methanizer overhead (CH_4, H_2, minor contaminants; DMO). It is observed that the depressurization step can flash co-adsorbed hydrocarbons and ethylene. Bed cooling during de-pressurization can be rather extreme and this then can increase the regeneration gas requirements.

9.5.4
VOC Removal

For the adsorption of volatile organic compounds (VOC), there are but a relative handful of fixed-bed VOC adsorbers. All that operate are regenerated by a thermal swing process. The product of choice will be taken from the very hydrophobic high-silica zeolites and other high-silica molecular sieves. MFI, in this case UOP's silicalite (characterized by a minimum Si/Al ratio of about 75) and de-aluminated versions of Y (with a more modest Si/Al ratio) are the most widely employed adsorbents for VOC adsorption. In actual practice our silicalite is often synthesized at a much higher ratio of Si/Al, giving the product a much greater degree of hydrophobicity.

A wide range of organic pollutants can be adsorbed by MFI, Y and mixtures of the two structure types. In the VOC application fixed beds are comparatively few with most VOC adsorption being done by rotary adsorbers such as are built and sold by Seibu Giken. A few words about rotary adsorption applications follow in Section 9.7.

9.5.5
Mercury Removal

The need to remove mercury from natural gas has been known for many years. UOP is active as the only supplier to date to offer truly regenerable mercury adsorbents. The removal of Hg by UOP's HgSIV is actually a weak reversible chemi-sorption process.

UOP has sold HgSIV for more than 20 years and the product has done yeoman service in preventing Hg contamination of aluminum heat exchangers often found downstream. Corrosion protection is no longer the only concern as regards mercury. Environmental pressures to minimize mercury emissions to the atmosphere are growing daily. While the United States has no current legislation governing mercury emissions, many individual gas producers in the industry are acting responsibly by mandating that mercury is not simply removed from the gas stream but is also sequestered from the environment. UOP is unique in its ability to offer a broad range of product and process solutions to allow gas producers to practice the total containment of Hg and to offer a "cradle to grave" strategy for dealing with mercury.

9.6
Industrial PSA

The age of pressure swing adsorption (PSA) began so far as air separation is concerned in the late 1950s. The first process patent was that of Skarstrom [12]. The scientists and engineers of the Linde Laboratories were very active in this field. It is worth mentioning that the discovery of synthetic zeolite molecular sieves was driven by the Linde Company's desire to advance the state of air separation. It had been shown that zeolites could separate O_2 from air. Union Carbide's Linde Division were among the pioneers in commercial air separation using pressure swing adsorption. Following PSA air separation by less than a decade, the Linde group also introduced the process for purification of hydrogen via PSA. As with the drying technology the legacy of art in the purification of H_2 developed by the UCC Linde Division now resides in UOP.

PSA refers to the process of regeneration. It means pressure swing adsorption. As mentioned in Section 9.2 we reduce total pressure and generally introduce a contaminant-free purge, thus also lowering partial pressure in order to shift the adsorption equilibrium and effect the removal of the adsorbed species from the sorption sites. PSA is generally practiced with light fixed gases that have rather modest slopes to the isotherms as compared to those of water. This is not to say that water cannot be removed by PSA but it is a more challenging and energetic separation by far than PSA air or H_2 purification.

The principles underlying PSA separations are the same as any adsorption process: The movements of adsorption waves with characteristic wave velocities and finite mass transfer zones determine when and how much of the adsorbed contaminant will break through into the product gas. In general one can understand air separation or H_2 purification by some of the design principles laid out for dehydration. The key difference is the loadings achieved are lower, the isotherms have lower slopes making them suitable for PSA, the heats of adsorption of fixed gases are substantially lower than for dehydration and, while both dehydration and air separation are adiabatic, the temperature changes that occur have a larger effect on the adsorption equilibrium than what we find in dehydration.

Another key point of differentiation is the fact that nearly all PSA separations are bulk separations and any investigator interested in a high fidelity description of the problem of adsorption must solve a mass balance equation such as Eq. (9.9), the bulk separation equation, together with the uptake rate model and a set of thermal balance equations of similar form. In addition to the more complicated pde and its attendant boundary and initial conditions the investigator must also solve some approximate form of a momentum balance on the fluid flow as a whole. In PSA air separation the adsorbent, despite its low loading and shallow slope, is actually adsorbing and removing from the gas phase 79mol% of the dry feed stream. In H_2 purification there are generally a few more contaminants to remove, the H_2 is essentially non-adsorbed, high purity and recovery of the product gas is quite important to the process economics.

9.6.1
PSA Air Separation

PSA air separation is the problem of separating the O_2 contained in air from the N_2. Along with the N_2 there is likely to be water, CO_2, Ar and several other trace elements present in air. Of these trace elements Ar has a measurable impact on the purity limit on delivered O_2.

The critical performance parameters by which the quality of PSA air separation is judged are three:

1) The purity of oxygen that is achieved. The limit for O_2 purity with most first generation zeolites including Li is just over 95% O_2. This limit is imposed because Ar in the feed air having an adsorption isotherm close to that of O_2 tends to stay with the O_2 product and thus limits the purity. Both O_2 and Ar are concentrated to a significant degree in the PSA process.

2) The operating economics are characterized primarily by the recovery of oxygen. Recovery is defined as the quantity of O_2 delivered by the process divided by the quantity of O_2 that was present in the feed air.

3) A bed size factor which is typically expressed as either the mass or volume of adsorbent required to produce 1 tonne day^{-1} of O_2. In either case this factor can be reduced to the relative throughput; it is a measure of how long a unit must operate to produce its own volume of purified product O_2.

The definition of product oxygen recovery defines the minimum required feed rate for a specified product purity and recovery. That relationship is that the feed flow must be the product oxygen purity (%) divided by the recovery (% of feed O_2) and divide also by the mole fraction of O_2 in feed air. For 20.9 mol% O_2 in air this is:

$$\text{Feed rate} = 4.784\,(\text{product rate})(\text{purity\%})/(\text{recovery\%}).$$

The capital cost of air separation machinery is linked to both the size of the beds (which dictates the cost of piping valves), of course to molecular sieve inventory and to the size of the compressor required to run the process. A low product recovery may have little impact on the bed size factor but it has an enormous effect on the amount of gas required and on the cost of compressing that gas. Thus the recovery and bed size factors have direct links to the cost of capital and operations of air separation machines.

Air separation by PSA on a large scale is today dominated by machines in which the pressure swing may be from near atmospheric to substantially sub-atmospheric pressure. The industry typically calls these machines vacuum swing adsorption (VSA) separators. A second sub-class in air separation is the machines that use a pressure swing the ranges from somewhat super-atmospheric to sub-atmospheric and these may be called trans-atmospheric PSA. The distinctions made here have implications as to equipment specifications and performance limitations in both bed size factor and O_2 recovery.

Even though the feed is free, the energy cost to compress it is a significant portion of the cost of O_2 production and therefore a high recovery of O_2 is desirable. The theoretical limit on O_2 recovery is most strongly linked to the selectivity of the product for adsorbing N_2 in reference to O_2. Studies over the past ten years have shown that the introduction of Li as the preferred cation type and the introduction of LSX or low-silica X have both contributed to increasing the theoretically achievable recovery limit. That limit can under VSA conditions be as high as 90%.

Practical concerns, specifically mass transfer, limit the recovery to values in the high 70% range. All this said, the vast majority of operating air separation units are the small capacity medical oxygen concentrators. These operate under either PSA or VSA or trans-atmospheric process cycles. The key objective for medical O_2 is small unit size and power consumption prior to the push for portability was a secondary consideration. There are many such PSA air separation units that operate at recoveries as low as 35%.

It may not be obvious but driving selectivity to a high value is best done by driving N_2 adsorption to some acceptably high value and then driving O_2 to a minimum. This dramatically changes the volume of gas that must pass in and out of the macro-pore structure of the adsorbent. In all PSA separations it is the macro-pore diffusion that is the dominant resistance to mass transfer.

The field of PSA air separation is one that has been under research and development for half a century now. The field may be considered "well plowed." Further there is great deal of intellectual property that is owned by a wide range of companies.

The basic PSA air separation process cycle can be as simple as a rising pressure adsorption step, in which the bed is pressurized and product is taken, and a combined blow down or depressurization combined with a product purge. Such cycles can make compact machines and involve few valves. As the cost of operation becomes a major consideration then the process can become more complex, adding more beds, more valves and allowing for bed to bed transfers of either product or feed gas. More complex cycles will separate the process into a larger number of steps including a pressurization, a constant pressure production step in which the product O_2 is taken, a depressurization and followed by a purge. With poly-bed PSA air separation it is possible to depressurize one bed and feed at least a portion of that gas into a second bed. This process can, when done at the feed, move air from one bed to another, saving some compression. If the depressurization is done at the product end and the process has been run to a point short of N_2 breakthrough, then product quality O_2 can be captured and passed directly to a bed in a partial re-pressurization with product. Such a process is a product end equalization. It saves on compression cost as well and increases product O_2 recovery.

The key advancements in the art have been captured both in IP and in actual practice. These include driving recovery to high values through a number of specific techniques intended to reduce compression energy.

Design practices are the intellectual property of the major gas producing companies and a wide variety of other interested parties. In depth description of the design practices cannot be done with impunity because of the IP that exists.

9.6.2
PSA H2 Purification

As with air separation the purification of hydrogen via a PSA process is also a well developed field. The market place is highly competitive and UOP competes with

Linde AG and several other major suppliers. Purification of hydrogen from steam methane reforming (SMR) operations constitutes a the biggest applications of PSA. Purification of hydrogen from refinery off gas (ROG) is another large segment and finally ethylene off gas (EOG) is yet a third and together these constitute the largest majority of applications of hydrogen purification by PSA.

The contaminants in SMR that must be managed include CO_2, CO, CH_4, N_2 and water.

For ROG we mainly deal with H_2 and hydrocarbons up to about a C6. In EOG we must clean up CH_4 and CO_2. The total list of all contaminants that must be handled also includes argon and some mercaptans.

Unlike PSA air separation the adsorbents used in hydrogen purification are not limited to zeolite molecular sieves. Carbons and silica gel are used in many PSA installations. Zeolites are used for obtaining certain critical specifications where the nature of the isotherms that they possess helps in recovery, achieving purity and minimizing bed size factors.

It has been mentioned that purity and recovery are very significant drivers for the business of hydrogen purification. Today PSA hydrogen purifiers can deliver purities of several nines: 99.9% purity while simultaneously recovering nearly 90% of the hydrogen found in the feed gas.

As with air separation the design methodologies are proprietary, but in all likelihood empirical, being driven by a large base of design experience that each competitor in the field possesses. At UOP we have an excellent data base to draw from for sizing of the PSA adsorbent beds. In many cases detailed numerical models aid us in interpolating within our data base for design of machinery for which we have few predecessors.

It may be said that bed sizing is actually one of the smaller challenges for the high-pressure PSA H_2 purifiers that are designed. The mechanical details of valves, valve positioners and not the least the pressure vessels (working at high pressure and cycling between extremes of pressure over a fairly short period of time) constitute a significant work load in the design process.

9.6.3
PSA Dehydration

Dehydration by the pressure swing process is largely the province of activated alumina or silica gel type adsorbents. Owing to the nature of their isotherms they demonstrate more favorable desorption characteristics for use in PSA.

The problem with zeolites in PSA cycles is that the isotherms are so very sharp in the low pressure region that it is difficult to achieve any significant equilibrium loading shift by simple manipulation of either total or partial pressure. Differential loadings for water in PSA with type A are limited to about 1wt%. Type X can achieve a higher delta in loading owing to the reduced number of cations. Type Y can increase delta loading performance still further for the same reason.

Simply shifting to higher Si/Al ratios with the adsorbent type can improve delta loadings but that shift particularly with type X or Y also opens up the zeolite pore and increases the potential for co-adsorption of organic molecules. Finally PSA dehydration is one of the few applications where the adsorption performance of the separator can be improved by increasing the adsorption temperature. At higher temperatures PSA separation of water by zeolites is dramatically improved. The drying of ethanol to produce motor fuel grade ethanol is now widely practiced using molecular sieves at elevated temperature using a VSA process.

Even in more conventional PSA dehydration zeolites including types A, X and Y have all been employed in pressure swing drying. Compound beds of alumina and zeolites X or Y have been employed for PSA dehydration and CO_2 removal for pre-purification of feed to air separation units.

Compound beds of alumina and zeolite X have been employed successfully in industrial dehydration by PSA. In both types of applications, the more favorable shape of the zeolite isotherms shorten the mass transfer zone and simultaneously allow for achievement of lower mole fractions of water or lower dew points for the product gas.

9.7
Industrial Dehydration (Bulk Removal)

9.7.1
Desiccant Wheels

The use of rotary desiccant wheels for dehydration has been practiced for several decades. Carl Munters [13] created the first rotary adsorbers by impregnating corrugated paper with LiCl and then ran the rotor using direct heat exchange to regenerate the sorbent. More recently Munters, NTI and others have developed improved adsorbents for rotary dehydrators.

The work of Collier at the University of Florida [14] produced the finding that a modified Brunauer type I isotherm, with a more modest degree of curvature to the isotherm, was the theoretical optimum for deep dehydration cycles that were expected to be used in open cycle desiccant cooling cycles. The adsorbent was dubbed a type IM (M for moderate). To understand this designation zeolite type X with its incredible steep isotherm is designated a type IE (E for extreme).

This work prompted a flurry of activity in the mid- to late 1980s to find the type IM isotherm. A number of inventions can be found in which alumina, or silica gel are blended with zeolites type X or Y to mimic the shape of the isotherm that Collier defined. Mol Sieve type DDZ-70® is in fact one of only a few true type IM isotherms. This product and Engelhard's type ETS-10 both have the required isotherm shape for water and deliver the benefits expected, to wit: excellent capacity for water, self-sharpening mass transfer zone and low energy investment required to regenerate. Mol Sieve type DDZ-70® is used commercially in rotors produced by NTI, Seibu Giken and Nichias for specialty drying purposes. ETS-10 was the basis for the advanced desiccant rotor commercialized by Engelhard in the early 1990s for use in open cycle desiccant cooling. These rotors typically rotate at around 20 revolutions h^{-1}.

The use of desiccant wheels for dehydration is mostly limited to the HVAC industry and there the performance of desiccant wheels is not pushed to its limits. Rotors are optimized for minimizing pressure drop while delivering dew point depressions that are not very large. In fact most commercial desiccant wheels might take an air stream at close to ambient temperature from a relative humidity of about 80% down to perhaps 20–30%. Significantly more dehydration does not increase comfort levels. It was recognized that such rotors are in fact quite excellent high efficiency contactors for both bulk dehydration and for achieving very low dew points. In a demonstration project, actually commissioned in 2003, UOP did design, build and is now operating a dual desiccant rotor system operating at ambient pressure that pre-dries air prior to compression. The system takes air in Baton Rouge La during mid summer and drives the dew point to less than $-40\,°C$ after compression. Under more modest feed humidity the device can depress the dew point of incoming air to less than $-80\,°C$ again after compression. The device uses the final inter-cooler for the three-stage compressor that compresses the product gas from the rotor, to provide the heat required by both rotors.

Desiccant rotors are widely employed today in the HVAC industry, some in open cycle desiccant cooling, but most in air drying without a specific tie-in to other processes. An interesting market segment is the use of rotors to dry oil-free compressed air at high pressure. Both Atlas Copco and Kobelco employ rotary adsorbers at high pressure to obtain dew point control for compressed air units. There are tens of thousands of such units installed and operating throughout the world. These units (compared to ambient pressure rotors) are much more compact, easy to operate and offer good value for dew point control. The interested reader can find information on the design on the websites for these two companies.

9.7.2
Enthalpy Control Wheels

Enthalpy control rotors are a sub-segment of the desiccant rotor market place. These devices are not regenerated by a specific hot gas purge. They are instead total enthalpy exchangers. They typically rotate rapidly (20–30 revolutions min^{-1}) and exchange the air being vented from a building with incoming air. The presence of a desiccant on the surface of the rotors (or embedded as the body of the rotor), enables these rotors to capture heat and moisture from one air stream and deliver it to a counter-currently flowing air stream. There are many thousands of such rotors in service today throughout the HCAC industry. Here are some applications of these types of rotors in aircraft ventilation systems and in some processes where bulk water control may be required but the energy budget is unable to withstand the cost of a true deep desiccant rotor.

9.8
Non-Regenerable Adsorption

The discussion of gas phase adsorptive separations is not complete without at least mentioning non-regenerable adsorbents and their applications as guard beds for contaminant trapping. Zeolite molecular sieves do not play a major role in this field. Most non-regenerable guard beds employ rather strong chemi-sorbents that react with incoming contaminants and thereby sequester them from the gas stream. These reactions are for most practical purposes irreversible at or near operating conditions.

Such sorbents generally do not employ zeolites but they are used in overall separation schemes together with regenerable zeolite to provide broad spectrum protection or to completely sequester some offensive species. De-sulfurization, mercury removal and select trace metals removal can all be achieved by use of non-regenerable guard beds. These applications generally depend for economics on low concentrations of the contaminants in the feed and relatively high capacities of the non-regenerable sorbents. Adsorption cycles are generally measured in months to years on stream between change-outs of the sorbents.

9.9
Summary

Industrial gas phase separations span a wide variety of applications, from ultra-purifications to bulk separations. In the preceding sections I touched upon some of the more widely practiced separations technologies.

The mechanics of how adsorption waves and thermal waves move through the adsorption beds is what determines the success of any given separation. Some of the principles underlying design practices have been outlined and unfortunately for current PSA separations only the briefest outline could be given.

Nomenclature

c Concentration of an adsorbable species in the gas phase, kmol/m^3
c_f Feed concentration, kmol/m^3
c_p Product concentration, kmol/m^3
D Diffusivity, m^2/s
J Molar flux inside of the macro-pores, kmol/(m^2/s)
k Mass transfer coefficient for a solid phase driving force uptake rate model, 1/s
kK Mass transfer coefficient for a gas phase driving force uptake rate model, 1/s
K Adsorption constant, Henry's Law coefficient, slope of the chord of the isotherm, dimensionless

v	Interstitial velocity, m/s
v_0	Superficial velocity, m/s
q	Molar volumetric loading, kmol/m³
$\bar{q}$	Spatial average molar volumetric loading, kmol/m³
r	Spatial coordinate, radial position in a particle or crystallite, m
R_p	Radius of the adsorbent particle, m
t	temporal coordinate, time, s
W	Adsorption wave speed, m/s
X	Adsorbent loading, kg of adsorbate/kg of adsorbate-free adsorbent
X_f	Adsorbent loading in equilibrium with adsorption feed, kg/kg; molar loading, kmol/kg in Eq. (9.14)
X_0	Saturation loading of the sorbent
X_p	Adsorbent loading in equilibrium with residual loading at adsorption temperature, kg/kg; molar loading, kmol/kg in Eq. (9.14)
Y_f	Mole fraction of adsorbable species in the feed gas during adsorption
Y_p	Mole fraction of adsorbable species in product gas
z	Spatial coordinate, distance measured from adsorption inlet, (m)

Greek:

ε	Void fraction, external to the adsorbent particles
ε_p	Void fraction internal to the adsorbent particles, subscript p for pore
ρ	Gas density, kmol/(kg/m³)
ρ_b	Bulk density of the adsorbent, kg/m³
ρ_p	Particle density of the adsorbent material, kg/m³, subscript p for particle
τ	Tortuosity factor, dimensionless

References

1 Breck, D.W. (1974) *Zeolite Molecular Sieves*, John Wiley & Sons, New York
2 Ruthven, D.M. (1984) *Principles of Adsorption and Adsorption Processes*, John Wiley & Sons, New York
3 Carniglia, S.C. and Ping, W.L. (1989) Alumina desiccant isosteres: thermodynamics of desiccant equilibria. *Ind. Eng. Chem. Res.*, **28**, 1025–1030.
4 Yon, C.M. and Turnock, P.H. (1971) Multicomponent adsorption equilibria on molecular sieves. *AIChE Symp. Ser.*, No. 117, **67**, 75–83.
5 Young, D.M. and Cromwell, A.D. (1962) *Physical Adsorption*, Butterworths, London.
6 Yang, R.T. (1987) *Gas Separation by Adsorption Processes*, Butterworths, Boston.
7 Linde Company (1959) *The Design Manual for Octane Improvement, Book I*, Union Carbide Corporation, Tonawonda, N.Y.
8 Cooper, R.S. (1965) Slow particle diffusion in ion exchange columns. *Chem. Eng. Fund.*, **4**, 308.
9 Cooper, R.S. and Liberman, D.A. (1970) Fixed-bed adsorption kinetics with pore diffusion control. *Ind. Chem. Eng. Fund.*, **9**, 620.
10 Haynes, H.W. and Sarma, P.N. (1973) A model for the application of gas

chromatography to measurements of diffusion in bi-disperse structured catalysts. *AIChE J.*, **19**, 1043.
11 Reid, R.C., Prausnitz, J.M., and Sherwood, T.K. (1977) *The Properties of Gases and Liquids*, McGraw-Hill, New York.
12 Skarstrom, C.W. (1960) Method and apparatus for fractionating gaseous mixtures by adsorption, US Patent 2,944,627.
13 Munters, C.G. (1961) Method and apparatus of conditioning air, US patent Number 2,993,563.
14 Collier, R.K. and Cale, T.S. (1986) Advanced desiccant materials assessment. GRI Report Number 86/0182.

10
Zeolite Membrane Separations
Jessica O'Brien-Abraham and Jerry Y.S. Lin

10.1
Introduction

In recent years there has been a strong focus on developing continuous membrane processes to replace traditional separation methods for applications that require significant cost and energy. Zeolites are potential candidates for use in such processes due to their microstructure, which consists of uniform molecule-sized pores connected by continuous diffusion pathways. The pore sizes of these materials are on the order of the molecule size of many industrially important compounds, which provide molecular sieving capability and unique adsorption properties [1]. Additionally, the operating conditions able to be sustained by inorganic membranes far exceed that of polymeric membranes, which are commonly used in commercial separation processes. Development of new separation technologies such as high temperature gas separations and catalytic membrane reactors requires materials that are able to withstand extreme operating environments and zeolites present themselves as capable candidates.

There are many different zeolite structures but only a few have been studied extensively for membrane applications. Table 10.1 lists some of these structures and their basic properties. One of the most critical selection criterion when choosing a zeolite for a particular application is the pore size exhibited by the material. Figure 10.1 compares the effective pore size of the different zeolitic materials with various molecule kinetic diameters. Because the pores of zeolites are not perfectly circular each zeolite type is represented by a shaded area that indicates the range of molecules that may still enter the pore network, even if they diffuse with difficulty. By far the most common membrane material studied is MFI-type zeolite (ZSM-5, Al-free silicalite-1) due to ease of preparation, control of microstructure and versatility of applications [7].

The present review of zeolite membrane technology covers synthesis and characterization methods as well as the theoretical aspects of transport and separation mechanisms. Special attention is focused on the performance of zeolite membranes in a variety of applications including liquid–liquid, gas/vapor and reactive

---

Zeolites in Industrial Separation and Catalysis. Edited by Santi Kulprathipanja
Copyright © 2010 WILEY-VCH Verlag GmbH & Co. KGaA, Weinheim
ISBN: 978-3-527-32505-4

Table 10.1 Summary of common zeolite structures used in membrane applications along with important properties and typical synthesis conditions.

Zeolite	Maximum pore size (Å)	Thermal stability (°C)	Typical conditions for *in situ* synthesis methods				Ref.
			Composition ($Na_2O:SiO_2:Al_2O_3:H_2O$:template)	Temperature (°C)	Time (h)	Calcined condition	
LTA-NaA	4.2	600	2:2:1:120:0	100	3.5	N/A	[2]
MFI-ZSM5	5.3 × 5.6	600	9:50:1:3213:3 TPABr	125–175	8–48	480°C, 8h	[3]
MFI-Silicalite			1:10:0:110:2.4 TPAOH	130–170	2–15		[4]
FAU-NaX	7.4	660	4.3:3:1:150:0	100	6	N/A	[5]
FAU-NaY		710	14:10:1:840:0	100–120	3–10	N/A	[6]

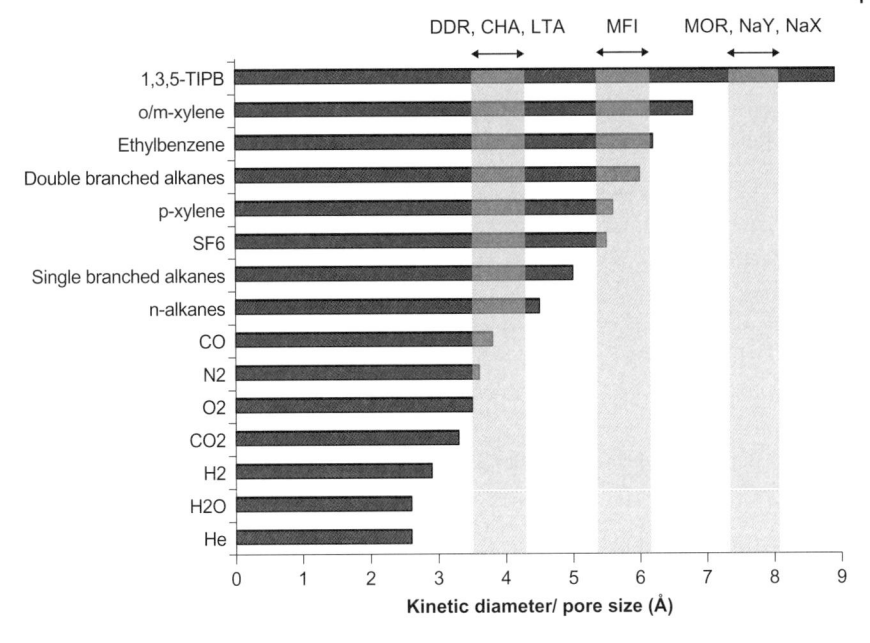

Figure 10.1 Kinetic diameter of common industrial molecules shown relative to the pore sizes of common zeolite structures; shaded areas represent the range of effective pore diameters for each group of zeolites.

systems. The existing challenges to facilitate commercialization are also discussed.

10.2
Synthesis and Properties of Zeolite Membranes

Zeolite membranes are generally synthesized as a thin, continuous film about 2–20 μm thick on either metallic or ceramic porous supports (e.g., alumina, zirconia, quartz, silicon, stainless steel) to enhance their mechanical strength. Typical supported membrane synthesis follows one of two common growth methods: (i) *in situ* crystallization or (ii) secondary growth. Figure 10.2 shows the general experimental procedure for both approaches.

10.2.1
In Situ Crystallization

In situ crystallization begins with a precursor synthesis sol consisting of Si, Al, Na, H_2O and a structure-directing agent (or template) [7]. The composition of the sol is dependent on the type of zeolite being synthesized and the final desired

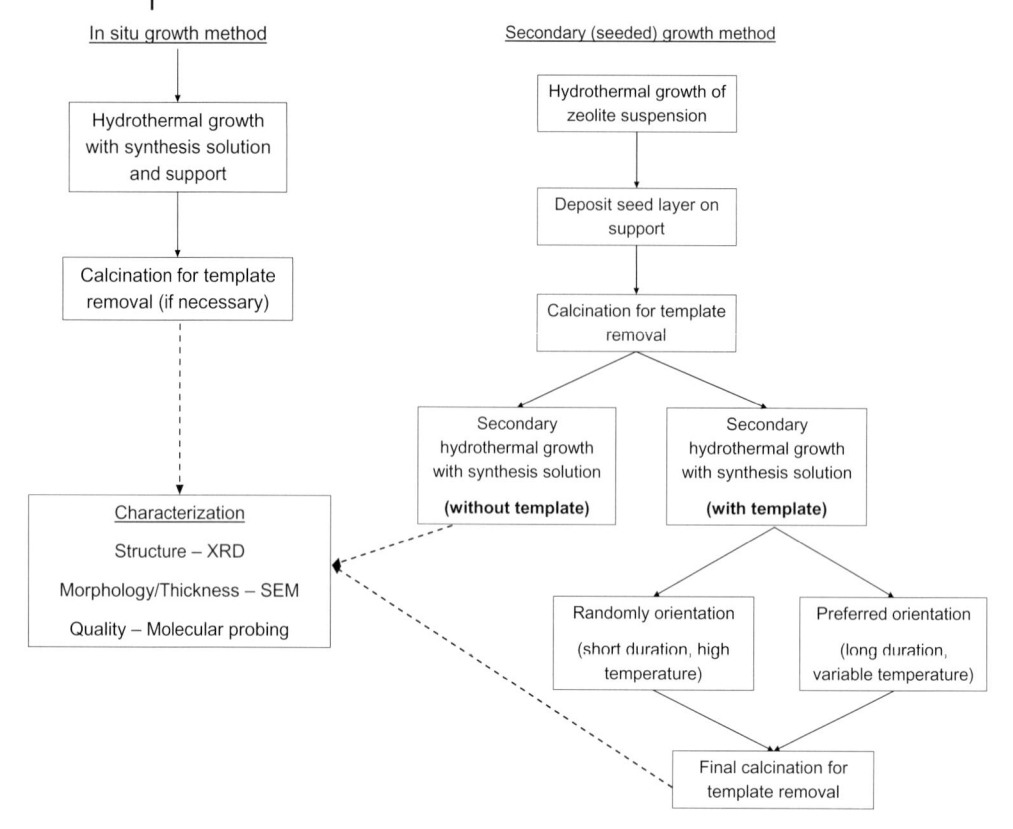

Figure 10.2 Schematic of the experimental procedures involved for two common zeolite synthesis methods including in situ crystallization and secondary (seeded) grow.

chemical make-up of the membrane. Both sol and support are added to an autoclave where hydrothermal synthesis is conducted at a prescribed temperature and duration. Table 10.1 gives typical synthesis conditions used for growth of the common zeolite structures used in membrane applications.

Polycrystalline membrane growth proceeds by initial formation of a gel layer on the surface of the support; crystallization takes place at the interface between the bulk liquid phase and the gel layer, resulting in deposition of zeolite nuclei and crystals formed [8]. Concurrently, the crystals deposited onto the support surface continue to grow, eventually resulting in a continuous membrane layer. Post-synthesis treatment is necessary when a template is used in synthesis to activate the zeolite and open the pores. Usually this is accomplished through calcination or burn out of the organic molecule.

In situ synthesis is extremely sensitive to a wide variety of factors, including the composition of the synthesis solution, support material, contact between the support and gel, as well as synthesis conditions (i.e., temperature, duration, sol aging, pH) [7]. Because the nucleation, deposition and growth are occurring at the same time, the membrane will be comprised of multiple layers of randomly oriented crystals, resulting in numerous intercrystalline gaps. In order to heal such defects multiple growths (resulting in thick membranes, 20–100 μm) or post-synthesis modification (e.g., coking, Si deposition) is necessary. These treatments can severely reduce the flux through the membrane due to increased mass transfer resistance.

10.2.2
Secondary (Seeded) Growth

The secondary growth method was first reported by Tsapatsis and co-workers [9] to address the challenge of producing thin, defect-free membranes with controlled microstructure (Figure 10.2). The initial step is to prepare a synthesis solution of nanocrystalline zeolite, which is then used for film deposition. The thin film, or seed layer, is composed of packed zeolite particles with interzeolitic porosity, which is sealed by a secondary hydrothermal growth. This approach serves to decouple the nucleation/deposition and crystal growth steps. By utilizing more dilute synthesis solutions and reduced synthesis times the nucleation in the bulk solution can be minimized, allowing for more control over the resulting structure [8]. This method has been used to synthesize not only MFI but zeolite L and zeolite A as well.

Tsapatsis and co-workers [10] identified conditions where the crystal growth proceeded in a preferentially oriented out-of-plane direction despite the presence of a randomly oriented seed layer. This growth mechanism is explained by the evolutionary Van-der-Drifts columnar growth where the crystal grains with the highest vertical velocity are most likely to dominate the structure [11]. Essentially the grains that grow fastest perpendicular to the surface will bury the slower growing grains and over time the membrane will adopt a preferred orientation.

The first reported synthesis of *c*-oriented MFI-type zeolite membranes required 24-h synthesis at 175 °C [10]. In these membranes the *c*-axis of the membrane crystallites is aligned normal to the substrate surface and the *b*- and *a*-axes are parallel to the surface. Figure 10.3a, e shows the SEM images and XRD spectra of a *c*-oriented membrane. Lovallo *et al.* [10] also reported the synthesis of *h*,0,*h*-oriented membranes where the *c*-axis is at a ~34° angle normal to the surface and can be synthesized at temperatures below 140 °C for extended growth times (shown in Figure 10.3b, f). More recently, Lai *et al.* [13] synthesized *b*-axis out-of-plane oriented MFI-type zeolite membranes by changing the structure-directing agent (SDA) in the hydrothermal growth solution from tetrapropylammonium (TPA) to a trimer-TPA. The *a*-axis out-of-plane can also be synthesized by using the trimer-TPA to synthesize the seeds used for deposition and conducting hydrothermal growth with TPA as the SDA, as also reported by Choi *et al.* [14].

Another recent modification to the secondary growth method as applied to MFI-type zeolite membranes is to perform the hydrothermal growth on the seeded

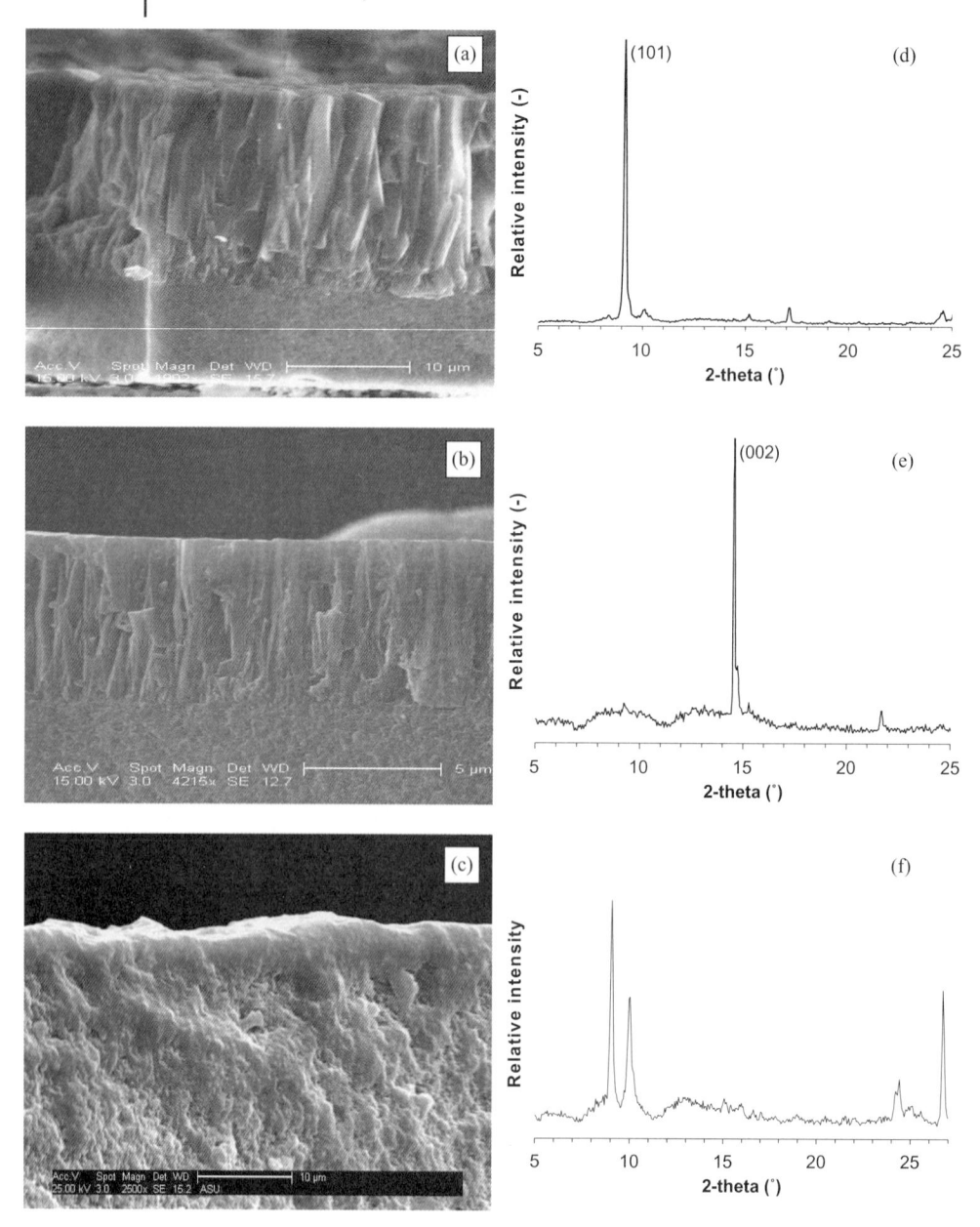

Figure 10.3 SEM cross-section image and XRD spectrum for c-oriented (a), h,0,h-oriented (b), and random (template-free) (c) MFI-type zeolite membranes along with associated XRD patterns (d, e, f) [12].

support without the use of a template. It is well known that during this template removal process, the mismatch in thermal behavior between the MFI-type layer and the α-Al_2O_3 support can cause mechanical stresses, which result in post-synthesis defect formation [15]. In template free synthesis, calcination is ultimately avoided. Figure 10.3c, g show an SEM micrograph and XRD spectra of a MFI-type zeolite membrane synthesized using this method [12]. These membranes have a growth rate of approximately $1\,\mu m\,h^{-1}$ and demonstrate a characteristic XRD spectrum indicating that the membrane continues to grow from a seed layer in the absence of a template.

The relationship between synthesis conditions and membrane properties has been the subject of significant study. Different growth conditions can result in significantly different morphologies as shown in Figure 10.3 [12]. This also has the effect of altering the defect formation, degree of intergrowth and stability. Ultimately, such differences can change the transport properties and separation behavior of the membrane.

10.2.3
Characterization of Zeolite Membranes

Despite improvement to synthesis methods used it is known that all membranes possess unavoidable defects due to intercrystalline porosity, which are formed during membrane growth [7, 8]. The goal of many characterization techniques is to determine the size and concentration of such defects and evaluate how their presence affects membrane performance.

There are numerous methods used to determine membrane quality, including permporosimetry, fluorescent confocal optical microscopy (FCOM), molecular probing and separation capability. Permporosimetry involves the measurement of a non-condensable gas (He) as a function of pressure of a condensable gas (hydrocarbon) [16]. An increase in the pressure of the hydrocarbon will result in capillary condensation in the larger defects, effectively blocking He from traveling down these pathways. The distribution of He flow through the defects can be estimated from the pressure–diffusion relationship. The latter characterization technique, FCOM, involves the use of a confocal microscope where the fluorophore chosen selectively adsorb into the defects or intercrystalline gaps in the membrane, which allows for imaging of these features [17].

By far the most widely used technique is molecular probing and separation capability, where molecules of specific sizes are chosen to probe the membrane for defects [7, 18]. Defects can be detected by comparison of two molecular permeances where one molecule should be too large to fit through the membrane pores and can only pass through defects. Generally, this feature is reported in terms of permselectivity or ideal selectivity, which is defined as the ratio of pure component flux. While use of this method is relatively straightforward, it is important to choose the right probe molecules to give desired information. The size/shape of the probe must be appropriate in order to evaluate either a range of defects or a critical defect size. Using Figure 10.1 probe molecules should be selected that fall decidedly outside of the shaded areas, one on either side, otherwise interactions between the probe molecule and zeolite surface may affect the assessment of quality. Similarly, molecules that adsorb strongly should also be avoided.

10.3
Transport Theory and Separation Capability of Zeolite Membranes

10.3.1
Permeation Through Zeolite Membranes

Transport through zeolite membranes occurs via three pathways: zeolitic micropores (~0.55 nm), intercrystalline microporous defect (<2 nm) and larger meso- and macroporous defects (>2 nm). Mass transport of single-component gas molecules through good quality zeolite membranes with no defects is controlled mainly by the diffusivity of the gas molecule through the zeolitic pores and the adsorption affinity between the gas and the zeolite framework [7, 19]. The relative importance depends on the membrane characteristics such as pore size, sorption strength and on permeating molecule properties such as size, shape and concentration. Additionally, operating conditions such as temperature and pressure can also play a large role [7].

For single-component gas permeation through a microporous membrane, the flux (J) can be described by Eq. (10.1), where ρ is the density of the membrane, Γ is the thermodynamic correction factor which describes the equilibrium relationship between the concentration in the membrane and partial pressure of the permeating gas (adsorption isotherm), q is the concentration of the permeating species in zeolite and x is the position in the permeating direction in the membrane. D_c is the diffusivity corrected for the interaction between the transporting species and the membrane and is described by Eq. (10.2), where E_d is the diffusion activation energy, R is the ideal gas constant and T is the absolute temperature.

$$J = -\rho D_c \Gamma \frac{dq}{dx} \tag{10.1}$$

$$D_c = D \exp\left(\frac{-E_d}{RT}\right) \tag{10.2}$$

At steady-state the permeance (F) can be described by Eq. (10.3):

$$F = \frac{\phi}{L(P_f - P_p)} \int_{q_f}^{q_p} D_c \Gamma \, dq \tag{10.3}$$

where ϕ is a constant describing the membrane porosity and tortuosity, L is the membrane thickness, P_f and P_p are the feed and permeate pressures, respectively, and q_p and q_f are the concentrations of the permeating species at the permeate ($x = L$) and feed ($z = 0$), respectively [7].

For high-quality MFI-type zeolite membranes either Knudsen or configurational diffusion is representative of the flow. Knudsen diffusion occurs in pores where the mean free path of the diffusing molecule is similar to the pore radius and the diffusivity can be described by the following equation [20]:

$$D = \frac{1}{3}\left[\frac{8kT}{\pi M}\right]^{1/2} d_p \quad (10.4)$$

where k is the Boltzmann constant (1.38×10^{-23} J K^{-1}), M and T are the molecular weight and temperature of the permeating species and d_p is the average pore size.

Configurational diffusion occurs when the diameter of the diffusing molecule is approaching the diameter of the pore size causing restrictive movement. This restriction is due to the fact that the diffusing species has to overcome the energy barrier posed by the walls during transport [21]. Essentially, the reduced pore size causes the molecule to interact with the potential field of zeolite channels. In this case, the diffusivity can be described by the following equation:

$$D = \frac{1}{z}\left[\frac{8kT}{\pi M}\right]^{1/2} \alpha \exp\left(\frac{-E_d}{RT}\right) \quad (10.5)$$

where z is the coordinate number and α is the distance between adjacent occupancy sites. In configurational diffusion, the molecules within the zeolite have restrictive movement because they have to overcome the energy barrier posed by the walls [21].

According to Xiao and Wei [21], the type of diffusion exhibited by a particular molecule within a zeolite framework is determined by the kinetic diameter of the molecule (d_m) as well as the diameter of the pore channel (d_p). They define a parameter λ as the ratio of d_m to d_c. The transition from Knudsen to configurational diffusion occurs in MFI-type zeolite at $0.6 < \lambda < 0.8$ at 300 K; at higher temperatures (500–700 K) the transition occurs at $\lambda \geq 0.8$. Small molecules such as H_2 ($d_m = 0.289$ nm) and He ($d_m = 0.260$ nm) can be expected to diffuse in a Knudsen-like manner, whereas larger molecules will exhibit configurational behavior in the MFI-type zeolite framework. The diffusion regime exhibited in a given separation is highly dependent on the diffusing molecules, zeolite pore structure, diffusion pathways and separation conditions. It is important to keep in mind that mass transfer through a real membrane will include transport through defects, which can exhibit configurational, Knudsen, or even viscous diffusion. The dominating mechanism depends on the size and quantity of the defects in the membrane and it is generally accepted that the flux through each can be considered additive [7].

The other component of molecular transport through zeolite membranes is the equilibrium sorption strength between the molecule and zeolite. In general, this information is reported in the form of adsorption isotherms, which are tabulated for many species and zeolite structures [22]. For example, Langmuir isotherm commonly describes the adsorption behavior of molecules transporting through MFI-type zeolite membranes and can be represented by the following equations:

$$\frac{q}{q_s} = \theta = \frac{KP}{1+KP} \quad (10.6)$$

$$\Gamma = \frac{1}{1-\theta} = \frac{\delta \ln P}{\delta \ln q} \tag{10.7}$$

where θ is the relative occupancy. The Langmuir adsorption constant can be represented by Eq. (10.8) below; Q_a is the heat of adsorption.

$$K = K_o \exp\left(\frac{Q_a}{RT}\right) \tag{10.8}$$

Equation (10.6) and can be directly substituted into Eq. (10.1) resulting in the expression for surface flux, J_s:

$$J_s = \rho q_s \frac{D_c}{L}\left(\frac{1+KP_f}{1+KP_p}\right) \tag{10.9}$$

In this case, it is assumed that D_c is independent of q. Equation (10.9) is valid at low temperatures where surface diffusion is dominant and transport occurs through molecules jumping from adsorption site to adsorption site.

One of the most important parameters when determining adsorption behavior is the operating temperature. For a linear isotherm, the permeance increases or decreases with increasing temperature, depending on the relative values of the heat of adsorption (Q_a) and the activation energy for diffusion (E_d) [7]. For a non-linear isotherm (e.g., Langmuir) the permeance increases with increasing temperature when $Q_a < E_d$. However, when $Q_a > E_d$, the permeance reaches a maximum and then decreases [7].

For diffusion of a small molecule with a weak adsorption affinity with zeolite pores, or at high temperatures, $K = 1/RT$, inserting Eq. (10.5) into Eq. (10.3) gives:

$$F = \frac{\phi \alpha}{L z}\left(\frac{8}{\pi MRT}\right)^{1/2} \exp\left(\frac{-E_d}{RT}\right) \tag{10.10}$$

Lin and co-workers prepared high-quality DDR-type [23] and MFI-type [24] zeolite membranes with defects and intercrystalline gaps sealed by a chemical vapor deposition method; and they measured pure gas permeantion of He, H_2, CO and CO_2 through these membranes at high temperatures (up to 500°C). The activation energy for diffusion for these four gases were obtained from the permeation data and are plotted against parameter λ (the ratio of d_m to d_p) in Figure 10.4. As shown the activation energy for diffusion increases substantially with λ, consistent with the prediction of Xiao and Wei [21]. With λ smaller than 0.5, the activation energy is close to zero and the diffusion is governed by the Knudsen mechanism.

10.3.2
Zeolite Membrane Separation Mechanisms

Separation through zeolite membranes proceeds through three different mechanisms: (i) molecular sieving, (ii) diffusion-controlled permeation, (iii) adsorption-controlled permeation [7, 8]. Figure 10.5 gives examples of common zeolite separations that fall within each category. The simplest of the three is molecular

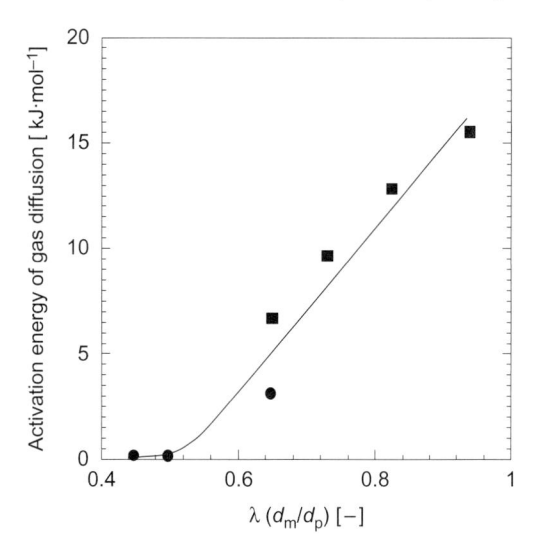

Figure 10.4 Calculated activation energy of gas diffusion for counter-diffusion CVD modified MFI-type (circle symbols) zeolite membranes and that of counter-diffusion CVD modified DDR-type (square symbols) zeolite membranes as a function of the ratio of kinetic diameter of the diffusion gas molecule to the zeolite pore diameter λ ($= d_m/d_p$) [24].

Separation Mechanism

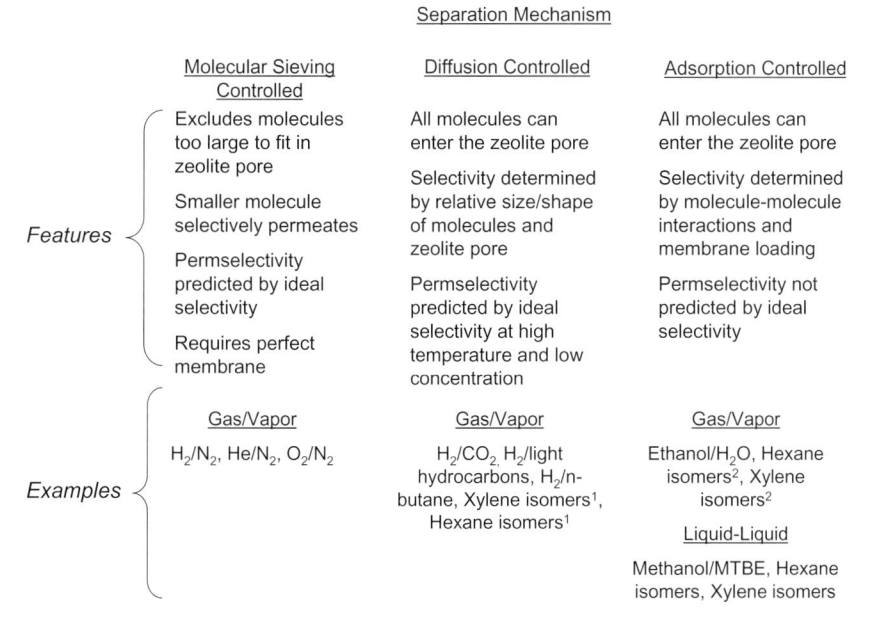

	Molecular Sieving Controlled	Diffusion Controlled	Adsorption Controlled
Features	Excludes molecules too large to fit in zeolite pore Smaller molecule selectively permeates Permselectivity predicted by ideal selectivity Requires perfect membrane	All molecules can enter the zeolite pore Selectivity determined by relative size/shape of molecules and zeolite pore Permselectivity predicted by ideal selectivity at high temperature and low concentration	All molecules can enter the zeolite pore Selectivity determined by molecule-molecule interactions and membrane loading Permselectivity not predicted by ideal selectivity
Examples	Gas/Vapor H_2/N_2, He/N_2, O_2/N_2	Gas/Vapor H_2/CO_2, H_2/light hydrocarbons, H_2/n-butane, Xylene isomers[1], Hexane isomers[1]	Gas/Vapor Ethanol/H_2O, Hexane isomers[2], Xylene isomers[2] Liquid-Liquid Methanol/MTBE, Hexane isomers, Xylene isomers

Figure 10.5 Summary of the features of separation mechanisms exhibited by zeolite membranes along with common examples. Superscripts: 1 high temperature and/or low partial pressure, 2 low temperature and/or high partial pressure.

sieving, which involves the exclusion of molecules too large to fit through the pore entrance [8]. For a mixture separation, only the molecule able to fit within the zeolite pore will preferentially adsorb. Expression of molecular sieving by a membrane requires that it be defect-free, which makes the mechanism the least observed [25].

In diffusion-controlled permeation, the components in a mixture differ largely in size but none is excluded from the pore network [8]. Separation is determined by the relative diffusivity (mobility) of a component within the zeolite. Expression of this mechanism requires low to medium loadings of the pore network where molecule–molecule interactions do not hinder diffusion. At high loadings, the mobility of all molecules is reduced and they become unable to pass one another in the pore network [8].

For the diffusion-controlled separation, the membrane selectivity is determined by the difference in diffusivity of various components. Equation (10.10) and Figure 10.4 show that the difference in permeability for various species depends on relative size of the diffusing molecule to the membrane pore size. Only for λ very close to 1, like diffusion of H_2 and CH_4 in SAPO-34 (CHA-type zeolite of eight-member ring opening), can the difference in the size of the diffusing molecules provide a reasonably good selectivity for the membrane. For example, Noble and co-workers reported molecule-size dependent H_2/CH_4 selectivity of about 20 for SAPO-34 zeolite membrane at temperatures around 100 °C. SAPO-34 has negligible adsorption affinity for H_2 and CH_4 and the molecule sizes of H_2 and CH_4, with H_2 being smaller than CH_4, are close to the pore size of SAPO-34 [26]. However, an inversed molecule size-dependent selectivity was observed for H_2/CO_2 for SAPO-34 zeolite membranes for which the smaller H_2 is less permeable than CO_2 [26, 27] due to strong adsorption of CO_2 on SAPO-34. For the membranes of DDR-type zeolite, also defined by eight-member rings with pore openings similar to SAPO-34, the molecule size-dependent permeability was also found for a number of small gases [28]. However there is an exception for H_2 and He for which the lighter, but larger H_2 permeates faster than He due to the presence of the defects in the DDR-type zeolite membrane [23].

The final separation mechanism is adsorption-controlled where separation is determined by both molecule–molecule interactions as well as molecule–zeolite interactions [8]. In general, this mechanism is observed at high loadings even when the permeating molecules do not necessarily adsorb strongly. Observed selectivity for a given component is related to the size/shape, adsorption strength, concentration and packing efficiency of the permeating molecules [19].

For multi-component systems it seems intuitive that single-component diffusion and adsorption data would enable one to predict which component would be selectively passed through a membrane. This is only the case where molecular sieving is observed; for all other separations where the molecules interact with one another and with the zeolite framework their behavior is determined by these interactions. Differences in membrane properties such as quality, microstructure, composition and modification can also play a large role in the observed separation characteristics. In many cases, these properties can be manipulated in order to tailor a membrane for a specific application or separation.

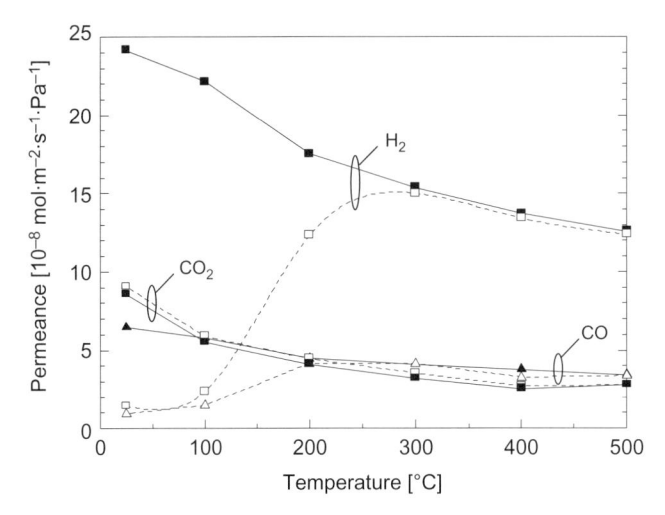

Figure 10.6 Temperature dependency of gas permeances for unmodified MFI-type zeolite membrane (closed symbols on solid line: gas permeances for single permeation, open symbols on broken line: those for ternary-component gas separation), feed gas composition (H_2:CO:CO_2 = 1:1:1, P_{up}: 300 kPa, P_{down}: 100 kPa) [24].

Figure 10.6 shows the temperature-dependency of gas permeances for a MFI-type zeolite membrane at 25–500 °C [24]. The permeances (H_2, CO, CO_2) obtained by ternary-component gas separation are quite similar to those for single-gas permeation above 300 °C. However, a clear difference between these permeances was observed below 300 °C. The permeances of H_2 and CO for ternary-component gas separation decrease drastically with decreasing temperature, lowering the H_2 permselectivity (H_2/CO). In contrast, the permeance of CO_2 for ternary-component gas separation is similar to that for single-gas permeation. It is expected that CO_2 preferential adsorption by MFI-type zeolitic pores should occur due to much stronger affinity between CO_2 molecules and the MFI-type zeolite pores. H_2 and CO molecules do not have such a strong affinity. Because CO_2 adsorption on MFI-type zeolitic pores decreases with increasing temperature, it is expected that H_2 separation performance for ternary-component gas separation should follow that of single-gas permeation at high temperatures, that is, there is no effect of the CO_2 adsorption on other gas permeation performance above 300 °C. This also indicates that small gas molecules can freely diffuse even in the presence of other molecules in the zeolite channels. At low temperature, the zeolitic pores are filled by adsorbed CO_2 molecules that block H_2 and CO molecules from permeating as freely through these pores in membrane.

10.3.3
Influence of Zeolite Framework Flexibility

It is known that the flexibility of the zeolite framework arises from the flexibility of the Si–O–Si joints linking rigid SiO_4 tetrahedra. This feature allows for the effective pore size of zeolite to vary continuously under the influence of the adsorbing species close in size to the pore diameter [29]. Such framework changes can affect both the inter- and intra-crystalline pathways through a zeolite membrane.

As an example, the loading of p-xylene into MFI-type zeolite induces structural changes that deform the pore openings of the framework in order to incorporate an even higher loading [30]. The saturation capacity of p-xylene in the MFI framework is reported to be approximately 7.8 molecules per unit cell where molecules are situated in both sinusoidal channels and channel intersections [30, 31]. This "high loading" condition causes displacement of framework atoms which increase the maximal sinusoidal pore dimension from 0.55 and 0.56 nm to 0.59 and 0.58 nm, respectively, while leaving the straight channel dimensions relatively unchanged [31, 32].

Recent studies have shown that these unit cell changes experienced for MFI-type zeolite saturated with p-xylene conditions cause an overall crystal expansion of 0.39%, with the a- and c-axes expanding 0.09% and 0.52% respectively. The b-axis experiences a contraction of 0.23% [29, 32]. These unit cell changes are enough to effectively "seal" 60–90% of intercrystalline defects <0.86 nm in randomly oriented MFI-type membranes and simultaneously expand the effective pore diameter [32]. However, when framework distortion is induced at high p-xylene loadings and both p-xylene and o-xylene must compete for transport through the zeolitic pores, the separation mechanism is no longer simple size/shape selectivity. At this point energetic and entropic effects become very important and ultimately determine membrane performance [25, 33].

Other MFI-type zeolite–sorbate systems are known to exhibit similar behavior. In a recent study, Yu et al. [34] reported that at saturated loadings of n-hexane a single MFI-type zeolite unit cell has an overall volume expansion of 2.3%, which can correlate to shrinkage in non-zeolitic pores up to 7 nm for a 1 µm crystal when isotropic expansion is assumed. It was demonstrated that, even in membranes with large number of defects, the crystallite swelling caused the membrane to achieve significant separation between n-hexane and trimethylbenzene, iso-octane and 2,2-dimethylbutane using pervaporation [34].

10.4
Zeolite Membranes in Separation and Reactive Processes

10.4.1
Liquid–Liquid Separation

One of the most common liquid–liquid separation methods investigated is pervaporation. This technique is a low-energy alternative for the separation of mixtures that are difficult and expensive to separate by traditional means such as azeotropes and isomers [35]. In pervaporation a liquid feed is vaporized as it travels through the membrane, producing a vapor permeate that is concentrated in the faster-transporting constituent [35]. In this operating mode, the membrane feed-side coverage is almost always higher than experienced during vapor permeation. Because of this the separation mechanism most commonly exhibited is that of adsorption-controlled diffusion.

Organic dehydration by hydrophilic zeolite membranes has been successfully demonstrated and has resulted in the first commercial application of zeolite membranes. Mitsui Engineering and Shipbuilding Co. [35, 36] developed an A-type zeolite membrane that separates water from ethanol (EtOH) at a permeation rate of $530\,L\,h^{-1}$ with selectivity well over 10 000. The zeolite A membrane (d_p ~0.4 nm) separates H_2O by two modes: size selectivity and selective adsorption. The pore size is smaller than most organics but is larger than H_2O and the low Si/Al content of the membrane provides high hydrophilicity. The latter enhances the adsorption of H_2O and essentially blocks the EtOH from being able to transport through the membrane. An added advantage is that non-zeolitic pores are also hydrophilic due to surface silanol groups which allows for higher fluxes without sacrificing selectivity [34].

Removal of organic from water uses hydrophobic membranes such as silicalite, Ge-ZSM5 and β-type [35]. In this case, preferential adsorption favors the larger organic and blocks the smaller H_2O molecule from passage. As expected, due to lower diffusivities, the transport fluxes are much lower for the organics than for H_2O but significant selectivity has been observed. An EtOH selectivity of 106 with a flux of $14\,mol\,m^{-2}\,h^{-1}$ was also reported for a silicalite membrane [37].

Work has also been conducted on separation of other organics from H_2O resulting in some commonalities. It has been observed that the specific functional groups of the organic being separated cause the adsorption strength and packing ability of the molecule to vary[35]. For molecules with the same functional groups the size of the molecule and degree of branching determines the extent to which separation from H_2O is possible. As an example, for alcohols the selectivity increases as the carbon number increases but decreases as the amount of branching increases. Larger molecules are able to better block H_2O from transporting but as the diffusivity of the alcohol decreases so does the observed selectivity [35].

Zeolite membranes have also been employed for organic–organic separations where selectivity is based on adsorption and diffusion differences of non-aqueous mixtures. NaX and NaY zeolite were used in the separation of methanol from MTBE and benzene ($800 < \alpha < 10\,000$) exploiting the more polar nature of methanol which is attracted to the electrostatic poles of the high Al content zeolites [38]. Other separations include: (i) separation of n-hexane from 2,2-DMB using ZSM5, (ii) benzene from p-xylene using MOR/FER and (iii) xylene isomers [34].

Another challenging and industrially important separation that utilizes pervaporation through zeolite membranes is acid removal from H_2O. In this case, the zeolite must have a high Si/Al ratio due to leaching of Al by the acid. Both Ge-ZSM5 and silicalite have demonstrated significant stability and separation capability for the removal of acetic acid from H_2O [35].

10.4.2
Gas/Vapor Separation

Binary separations of gas and vapors can be characterized into three categories according to the adsorption interaction between the permeating molecules and the zeolite framework: (i) weak–weak, (ii) weak–strong and (iii) strong–strong [7, 39]. For weak–weak separations neither component has strong adsorption properties; examples include H_2/N_2, N_2/O_2 and H_2/Cl. Separation is determined by diffusivity differences, either through molecular sieving or size/shape selectivity.

In weak–strong separations, one component does not strongly adsorb but has a high diffusivity; the other component is slow but strongly adsorbs. In this case, the separation mechanism exhibited is either diffusion-controlled permeation or adsorption-controlled permeation, depending on the operating conditions [39].

An example of this phenomenon is the separation of an eight-component mixture of hydrogen and light hydrocarbons (C_1–C_6~0.3–7.0%, H_2~84%) through an α-Al_2O_3 supported MFI-type zeolite membrane conducted by Lin and co-workers [40, 41]. Figure 10.7 shows permeance of each compound in the mixture through the membrane at two different temperatures. At low temperatures (<100°C), the enhanced adsorption of the hydrocarbons (strong) essentially blocked H_2 (weak) from permeating, resulting in excellent rejection of the smaller, faster diffusing molecule. However, as the temperature was increased (>500°C) the membrane transport favored H_2. This behavior is characteristic of a solution–diffusion model where the dominance of competitive adsorption and diffusion is dependent on the separation conditions [7]. This phenomenon is observed for numerous weak–strong separations including H_2/n-butane, N_2/CO_2, H_2/CO_2 [7].

Strong–strong mixture separations are most complex due to the interactions between the constituents and the zeolite framework as well as with each other [39].

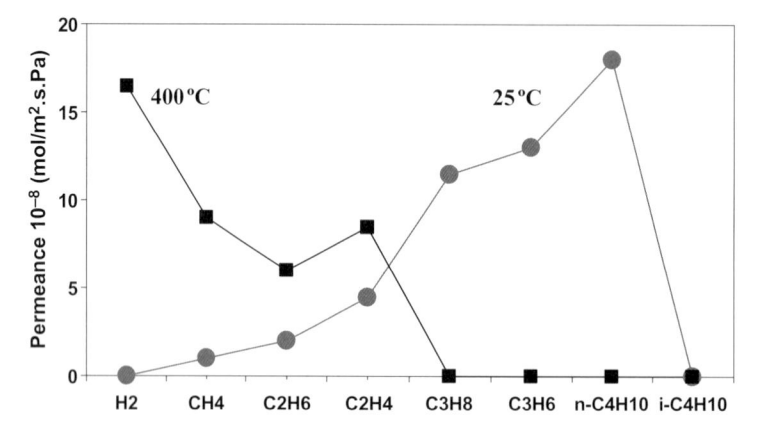

Figure 10.7 Permeance of various components with an eight component mixture feed (at 1 atm) for MFI-type zeolite membrane (data from ref. [41]).

Liu et al. [42] reported permeation of mixture of hexanes and octanes through silicalite membranes. It was found that the permeances of the mixture could not be predicted by the single-component data. In the separation of C_6 alkane isomers, the permeance of 2,2-DMB is significantly reduced in the presence of n-hexane resulting in a permselectivity much higher than the ideal separation factor [7].

Given the average pore size, a high quality MFI-type zeolite membrane is expected to be able to selectively pass p-xylene (kinetic diameter, $k_d = 0.59$ nm) while excluding the bulkier o-xylene and m-xylene isomers ($k_d = 0.68$ nm) [7]. While high selectivities ($\alpha = 20$–500) for p-xylene over o-xylene have been reported, they are only observed when operating at low xylene partial pressures (0.27–2.5 kPa, $\sim P_{actual}/P_{sat'd} < 1\%$) [43]. It should be pointed out that the p-xylene selectivity of these MFI-type zeolite membranes dropped significantly as the vapor pressure of the feed was increased [43]. The major reason cited for this behavior is that the shape/size selectivity of the MFI-type zeolite membrane is highly dependent on the p-xylene loading. At high loadings of p-xylene the MFI-type zeolite framework experiences distortion which changes the pore shape and ultimately allowing greater mobility of the larger isomers [33].

As the aforementioned example demonstrates, for strong–strong separations the selectivity can be dependent on the loading of the membrane. When the size of the adsorbed molecule is similar to the zeolite pore, at loadings near saturation the zeolite framework atoms will adjust to allow for entropically favorable packing. Under these conditions, constituents must compete with one another for adsorption sites and molecule–molecule interactions play a dominant role [33].

10.4.3
Reactive Separation Processes

Novel unit operations currently being developed are membrane reactors where both reaction and separation occur simultaneously. Through selective product removal a shift of the conversion beyond thermodynamic equilibrium is possible. The membrane itself can serve in different capacities including: (i) a permselective diffusion barrier, (ii) a non-reactive reactant distributor and (iii) as both a catalyst and permselective membrane [44].

For a packed-bed membrane reactor (PBMR) the membrane is permselective and removes the product as it is formed, forcing the reaction to the right. In this case, the membrane is not active and a conventional catalyst is used. Tavolaro et al. [45] demonstrated this concept in their work on CO_2 hydrogenation to methanol using a LTA zeolite membrane. The tubular membrane was packed with bimetallic Cu/ZnO where CO_2 and H_2 react to form EtOH and H_2O. These condensable products were removed by LTA membrane which increased the reaction yield when compared to a conventional packed bed reactor operating under the same conditions [45].

A catalytic membrane reactor (CMR) presents an alternate configuration where the membrane is both catalytically active and permselective. The reactant conversion in again enhanced through selective product removal but both reaction and separation occur in a single unit. Masuda *et al.* [46] used a ZSM-5 membrane reactor for methanol conversion reactions. As methanol permeated through the ZSM-5 it reacted at the acid sites, undergoing a series of reactions to form dimethylether, olefins and finally paraffins and aromatics. In order to favor recovery of olefins (the intermediate species), the pressure drop and number of acid sites in the zeolite was controlled to maintain the desired residence time and reaction rate, respectively. It was found that an olefin selectivity of 80–90% could be recovered from 60–98% methanol conversion [46].

Other membrane reactor configurations have been investigated where the membrane is both inert and non-permselective. In these cases the membrane acts as a reactant distributor, maintaining a stochiometric ratio between components and aiding in control of the reaction. Another type of reactor has permselective coatings on catalyst particles, which controls access of the reactants. This configuration has the advantage of significantly increased membrane surface area when compared to a conventional membrane reactor [44].

One of the main challenges for zeolite membrane reactors is that optimal reactor operation requires that the membrane flux be in balance with the reaction rate.. Whether acting in an inert or catalytic capacity, the extractive ability of the membrane needs to keep up with the production of the species being removed in order to fully participate in improving the reaction yield [44].

10.5
Summary

Current state-of-the-art research in zeolite membranes has shown the significant potential of these materials in numerous applications. However, there are a number of challenges that need to be addressed before zeolite membranes will be utilized commercially. These include the following [8, 39, 44]:

- Large area membranes that demonstrate high permeability and selectivity;
- Reproducibility of synthesis;
- Long-term stability without regeneration;
- High-temperature sealing;
- Scaling-up of membrane synthesis and modules;
- Cost of membranes and modules;
- Further understanding of transport and separation behavior through both zeolite pores and defects.

Novel manufacturing technologies are needed to address modern separation and reaction challenges. Zeolites possess molecular sieving, selective sorption properties and catalytic activity, as well as enhanced thermal and chemical stability which make them good candidates for these applications. Future focus needs to address the processing methods and theoretical understanding of these materials so their potential for commercial application can be realized.

Acknowledgment

The authors would like to thank the Department of Energy for their support (DE-PS36-03GO93007).

References

1 Auderbach, S.M., Carredo, K.A., and Dutta, P.K. (2003) *Handbook of Zeolite Science and Technology*, Marcel Dekker, New York.
2 Kita, H., Horii, K., Ohtochi, Y., Tanaka, K., and Okamoto, K. (1995) Synthesis of a zeolite NaA membrane for pervaporation of water/organic liquid mixture. *J. Mater. Sci. Lett.*, **14**, 206–208.
3 Li, S., Tuan, V.A., Falconer, J.L., and Noble, R.D. (2001) Separation of 1,3-propanediol from glycerol and glucose using a ZSM-5 zeolite membrane. *J. Membr. Sci.*, **191**, 53–59.
4 Arruebo, M., Coronas, J., Menendez, M., and Santamaria, J. (2001) Separation of hydrocarbons from natural gas using silicalite membranes. *Sep. Purif. Tech.*, **25**, 275–286.
5 Li, S., Tuan, V.A., Falconer, J.L., and Noble, R.D. (2001) Separation of 1,3-propanediol from aqueous solutions using pervaporation through an X-type zeolite membrane. *Ind. Eng. Chem. Res.*, **40**, 1952–1959.
6 Kita, H., Fuchida, K., Hority, T., Asamura, H., and Okamoto, K. (2001) Preparation of faujasite membranes and their permeation properties. *Sep. Purif. Tech.*, **25**, 261–268.
7 Lin, Y.S., Kumakiri, I., Nair, B.N., and Alsyouri, H. (2002) Microporous inorganic membranes. *Sep. Purif. Meth.*, **31**, 229–379.
8 Caro, J., Noack, M., Kolsch, P., and Membranes, Z. (2005) Zeolite membranes: from the laboratory scale to technical applications. *Adsorption*, **11**, 215–227.
9 Lovallo, M.C. and Tsapatsis, M. (1996) Preferentially oriented submicron silicalite membranes. *Separations*, **42**, 3020–3024.
10 Lovallo, M.C., Gouzinis, A., and Tsapatsis, M. (1998) Synthesis and characterization of oriented MFI membranes prepared by secondary growth. *AIChE J.*, **44**, 1903–1913.
11 Xomeritakis, G., Nair, S., and Tsapatsis, M. (2000) Transport properties of alumina-supported MFI membranes made by secondary (seeded) growth. *Micropor. Mesopor. Mater.*, **38**, 61–73.
12 O'Brien-Abraham, J., Kanezashi, M., and Lin, Y.S. (2007) A comparative study on permeation and mechanical properties of random and oriented MFI-type zeolite membranes, *Mesop. Microp. Mater.*, **105**, 140–148.
13 Lai, Z., Bonilla, G., Diaz, I., Nery, J.G., Sujaoti, K., Amat, M.A., Kokkoli, E., Terasaki, O., Thompson, R.W., Tsapatsis, M., and Vlachos, D.G. (2003) Microstructural optimization of a zeolite membrane for organic vapor separation. *Science*, **300**, 456–460.
14 Choi, J., Ghosh, S., King, L., and Tsapatsis, M. (2006) MFI zeolite membranes from a- and randomly oriented monolayers. *Adsorption*, **12**, 339–360.
15 Dong, J., Lin, Y.S., Hu, M.Z.C., Peascoe, R.A., and Payzant, E.A. (2000) Template-removal-associated microstructural development of porous-ceramic-supported MFI zeolite membranes. *Micropor. Mesopor. Mater.*, **34**, 241–253.
16 Tsuru, T., Takata, Y., Kondo, H., Hirano, F., Yoshioka, T., and Asaeda, M. (2003) Characterization of sol-gel derived membranes and zeolite membranes by nanopermporometry. *Separ. Purif. Technol.*, **32**, 23–27.
17 Bonilla, G., Tsapatsis, M., Vlachos, D.G., and Xomeritakis, G. (2001) Fluorescence confocal optical microscopy imaging of the grain boundary structure of zeolite

MFI membranes made by secondary (seeded) growth. *J. Membr. Sci.*, **182**, 103–109.

18 Xomeritakis, G., Lai, Z., and Tsapatsis, M. (2001) Separation of xylene isomer vapors with oriented MFI membranes made by seeded growth. *Ind. Eng Chem. Res.*, **40**, 544–552.

19 Dong, J., Lin, Y.S., Kanezashi, M., and Tang, Z. (2008) Microporou inorganic membranes for high temperature hydrogen purification. *J. Appl. Phys.*, **104**, 121301.

20 Jareman, F., Hedlund, J., Creaser, D., and Sterte, J. (2004) Modelling of single gas permeation in real MFI membranes. *J. Membr. Sci.*, **236**, 81–89.

21 Xiao, J. and Wei, J. (1992) Diffusion mechanism of hydrocarbons in zeolites. *Chem. Eng. Sci.*, **47**, 1123–1141.

22 Ruthven, D.M. (1984) *Principles of Adsorption and Adsorption Processes*, John Wiley & Sons, Inc., New York.

23 Kanezashi, M., O'Brien-Abraham, J., Lin, Y.S., and Suzuki, K. (2008) Gas permeation through DDR-type zeolite membranes at high temperatures. *AIChE J.*, **54**, 1478–1486.

24 Kanezashi, M. and Lin, Y.S. (2009) Gas permeation and diffusion characteristics of MFI-type zeolite membranes at high temperatures. *J. Chem. Phys. C*, **113**, 3767–3774.

25 Mohanty, S. and McCormick, A.V. (1999) Prospects for principles of size and shape selective separations using zeolites. *Chem. Eng. J.*, **74**, 1–14.

26 Hong, M., Li, S.G., Falconer, J.L., and Noble, R.D. (2008) Hydrogen purification using a SAPO-34 membrane. *J. Membr Sci.*, **307**, 277–283.

27 Carreon, M.A., Li, S.G., Falconer, J.L., and Noble, R.D. (2008) SAPO-34 seeds and membranes using multiple structure directing agents. *Adv. Mater.*, **20**, 729.

28 Tomita, T., Nakayama, K., and Sakai, H. (2004) Gas separation characteristics of DDR typ zeolite membranes. *Micropor. Mesopor. Mater.*, **68**, 71–75.

29 Nair, S. and Tsapatsis, M. (2000) The location of o- and m-xylene in silicalite by powder x-ray diffraction. *J. Phys. Chem*, **B104**, 8982–8988.

30 Mentzen, B.F. and Gelin, P. (1995) The silicalite/p-xylene system : part I – flexibility of the MFI framework and sorption mechanism observed during p-xylene pore-filling by X-ray powder diffraction at room temperature. *Mater. Res. Bull.*, **30**, 373–380.

31 van Koningsveld, H., Tuinstra, F., Van Bekkum, H., and Jansen, J.C. (1989) The location of p-xylene in a single crystal of zeolite H-ZSM-5 with a new, sorbate-induced, orthorhombic framework symmetry. *Acta Crystal.*, **B45**, 423–431.

32 van Koningsveld, H., Jansen, J.C., and Van Bekkum, H. (1990) The monoclinic framework structure of zeolite H-ZSM-5. Comparison with the orthorhombic framework of as-synthesized ZSM-5. *Zeolites*, **10**, 235–242.

33 O'Brien-Abraham, J., Kanezashi, M., and Lin, Y.S. (2008) Effects of adsorption-induced microstructural changes on separation of xylene isomers through MFI-type zeolite membranes. *J. Membr. Sci.*, **320**, 505–513.

34 Yu, M., Amundsen, T.J., Hong, M., Falconer, J.L., and Noble, R.D. (2007) Flexible nanostructure of MFI zeolite membranes. *J. Membr. Sci.*, **298**, 182–189.

35 Bowen, T.C., Noble, R.D., and Falconer, J.L. (2004) Fundamentals and applications of pervaporation through zeolite membranes. *J. Membr. Sci.*, **245**, 1–33.

36 Okamoto, K., Kita, H., Horii, K., Tanaka, K., Kondo, M., and NaA, Z. (2001) membrane: preparation, single-gas permeation, and pervaporation and vapor permeation of water/organic liquid mixtures. *Ind. Eng. Chem. Res.*, **40**, 163–175.

37 Lin, X., Chen, X., Kita, H., and Okamoto, K.I. (2003) Synthesis of silicalite tubular membranes by in situ crystallization. *AIChE J.*, **49**, 237–241.

38 Kita, H., Fuchida, K., Horita, T., Asamura, H., and Okamoto, K. (2001) Preparation of faujasite membranes and their permeatation properties. *Sep. Purif. Tech.*, **25**, 261–268.

39 Burggraaf, A.J. (1999) Single gas permeation of thin zeolite (MFI) membranes: theory and analysis of

experimental observations. *J. Membr. Sci.*, **155**, 45–65.

40 Dong, J., Liu, W., and Lin, Y.S. (2000) Multicomponent hydrogen/hydrocarbon separation by MFI- type zeolite membranes. *AIChE J.*, **46**, 1957–1966.

41 Pan, M. and Lin, Y.S. (2001) Template-free secondary growth synthesis of MFI-type zeolite membranes. *Micropor. Mesopor. Mater.*, **43**, 319–327.

42 Liu, Q., Noble, R.D., Falconer, J.L., and Funke, H.H. (1996) Organics/water separation by pervaporation with a zeolite membrane. *J. Membr. Sci.*, **117**, 163–174.

43 Xomeritakis, G. and Tsapatsis, M. (1999) Permeation of aromatic isomer vapors through oriented MFI-type membranes made by secondary growth. *Chem. Eng. Sci.*, **54**, 3521–3531.

44 McLeary, E.E., Jansen, J.C., and Kapteijn, F. (2006) Zeolite based films, membranes and membrane reactors: progress and prospects. *Micropor. Mesopor. Mater.*, **90**, 198–220.

45 Tavalaro, A. and Tavolaro, P. (2007) LTA zeolite composite membrane preparation, characterization and application in a zeolitic membrane reactor. *Catal. Commun.*, **8**, 789–794.

46 Masuda, T., Asanuma, T., Shouji, M., Mukai, S.R., Kawase, M., and Hashimoto, K. (2003) Methanol to olefins using ZSM-5 zeolite catalyst membrane reactor. *Chem. Eng. Sci.*, **58**, 649–656.

11
Mixed-Matrix Membranes

Chunqing Liu and Santi Kulprathipanja

11.1
Introduction

Membrane-based technologies have the advantages of both low capital cost and high energy efficiency compared to the much older and established techniques, such as cryogenic distillation, absorption and adsorption. Membranes, alone and in combinations with other methods, provide a comprehensive approach for solving energy, environmental resource recovery, medical and many other technical problems. Membrane-based separation processes are widely adopted today in petrochemical, semiconductor, food, pharmaceutical, biotechnology industries and a wide range of environmental applications. World-wide sales of membrane products and systems have more than doubled in a decade. The drive towards greater economic and environmental efficient separations will result in more aggressive future growth in membrane-based separations.

11.1.1
Scope of This Chapter

This chapter provides a brief introduction to polymer and inorganic zeolite membranes and a comprehensive introduction to zeolite/polymer mixed-matrix membranes. It covers the materials, separation mechanism, methods, structures, properties and anticipated potential applications of the zeolite/polymer mixed-matrix membranes.

11.1.2
Polymer Membranes

Polymer membranes are the most common commercial membranes for separations [1]. They have proven to operate successfully in many gas and liquid separations. For example, polymer membrane-based gas separation processes have undergone a major evolution since the introduction of the first polymer membrane-based industrial hydrogen separation process about two decades ago. The

UOP Separex™ membrane comprising cellulose acetate (CA) polymer has been extensively used for CO_2 removal from natural gas and currently holds the membrane market leadership for this application. The UOP Polysep™ membrane, a polymeric membrane, has been successfully applied to H_2 separation processes.

The membrane performance for separations is characterized by the flux of a feed component across the membrane. This flux can be expressed as a quantity called the permeability (P), which is a pressure- and thickness-normalized flux of a given component. The separation of a feed mixture is achieved by a membrane material that permits a faster permeation rate for one component (i.e., higher permeability) over that of another component. The efficiency of the membrane in enriching a component over another component in the permeate stream can be expressed as a quantity called selectivity or separation factor. Selectivity (α) can be defined as the ratio of the permeabilities of the feed components across the membrane (i.e., $\alpha_{A/B} = P_A/P_B$, where A and B are the two components). The permeability and selectivity of a membrane are material properties of the membrane material itself, and thus these properties are ideally constant with feed pressure, flow rate and other process conditions. However, permeability and selectivity are both temperature-dependent.

The polymer membrane separation is based on a solution–diffusion mechanism which involves molecule-scale interactions of the permeating component with the polymer [2]. The mechanism assumes that each feed component is sorbed by the membrane at one interface, transported by diffusion across the membrane through the voids between the polymer chains (or so-called free volumes) and desorbed at the other interface. According to this solution–diffusion mechanism, the permeation of feed components through a polymer membrane is controlled by two parameters: the solubility coefficient (S) and the diffusivity coefficient (D). The solubility coefficient equals to the ratio of sorption uptake normalized by some measure of uptake potential, such as partial pressure. The diffusivity coefficient is a measure of the mobility of a feed component passing through the voids between the polymer chains in a polymer membrane. Since the selectivity of the polymer membrane is the ratio of the permeabilities of the feed components across the membrane and the permeability is a product of solubility coefficient and diffusivity coefficient, the selectivity of the polymer membrane for a feed mixture of A and B ($\alpha_{A/B}$) is determined by the solubility selectivity (S_A/S_B) and diffusivity selectivity (D_A/D_B). Solubility selectivity is the ratio of the Henry's law sorption coefficients of A and B components. Diffusivity selectivity is the ratio of the diffusion coefficients of A and B components. The balance between the solubility selectivity and diffusivity selectivity determines whether a polymer membrane is selective for component A or B in a feed mixture of A and B. The feed components can have high permeability because of a high solubility coefficient, a high diffusivity coefficient, or both. Generally speaking, the diffusivity coefficient decreases and the solubility coefficient increases with increasing molecular size of the feed component. For a high-performance polymer membrane, both high permeability and high selectivity representing the membrane productivity and efficiency, respectively, are desirable.

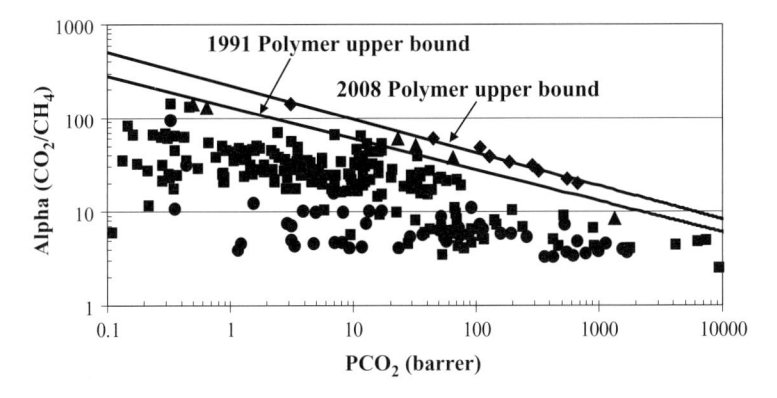

Figure 11.1 Polymer upper bound correlation for CO_2/CH_4 separation [3, 4].

Polymer membranes have a number of advantages for separations including low cost, ease of processability, good mechanical stability and reasonably good selectivity and permeability. In the past, both rubbery polymers and glassy polymers had been used for making commercial polymer membranes. These polymers include, but are not limited to, silicone rubber, cellulose acetate, polysulfone, polyethersulfone, polyamides, polyimides, polyetherimides, polypropylene, poly(vinyl chloride) and poly(vinyl fluoride). Polymer membranes made from these traditional polymers, however, are not without problems. Poor contaminant resistance, low chemical and thermal stability are among some of their limitations. In particular, there is a well-known trade-off between permeability and selectivity (or so-called polymer upper bound limit) of polymer membranes. By comparing the experimental data of hundreds of different polymers, Robeson demonstrated in 1991 that the selectivity and permeability of polymer membranes were inseparably linked to one another, in a relation where selectivity increases as permeability decreases and vice versa [3]. Figure 11.1 shows the polymer upper bound relationships between CO_2 permeability and CO_2/CH_4 selectivity for over 300 glassy and rubbery polymers [3]. Substantial research effort has been directed to overcoming the limits imposed by the polymer upper bound limit since 1991. There is not much success despite the intense research efforts to tailor polymer structure to improve separation properties. Robeson has revisited the polymer upper bound most recently and has reported that the polymer upper bound positions for the gas pairs studied in 1991 have only showed modest shifts since 1991 (Figure 11.1) [4].

11.1.3
Zeolite Membranes

In recent years, extensive work has been reported on the synthesis, characterization and applications of zeolite membranes [5]. Zeolite membranes are capable of overcoming some of the challenges facing polymer membranes. Under conditions where polymer membranes cannot be used zeolite membranes have the potential for separations with high efficiency and high productivity by taking advantages of their superior thermal and chemical stability, good erosion resistance and high plasticization resistance to condensable gases.

Membranes made from zeolite materials provide separation properties mainly based on molecular sieving and/or surface diffusion mechanism. Separation with large pore zeolite membranes is mainly based on surface diffusion when their pore sizes are much larger than the molecules to be separated. Separation with small pore zeolite membranes is mainly based on molecular sieving when the pore sizes are smaller or similar to one molecule but are larger than other molecules in a mixture to be separated.

A majority of the zeolite membranes reported to date are made from MFI, LTA, FAU, or MOR [6–14]. For example, the pore size of MFI zeolites is approximately 0.5–0.6 nm and is larger than CO_2, CH_4 and N_2. Lovallo and coworkers obtained a separation factor of about 10.0 for CO_2/CH_4 separation using a high-silica MFI membrane at 393 °K [9]. This separation can be explained by the competitive adsorption of CO_2 and CH_4. LTA zeolites have pores in the range of 0.3–0.5 nm and are able to distinguish small molecules such as H_2 and N_2. Guan and coworkers reported a H_2/N_2 ideal separation factor of 7.1 for a Na^+-type LTA zeolite membrane and improved the value to 7.5 by ion exchange with K^+ [12].

Some small-pore zeolite and molecular sieve membranes, such as zeolite T (0.41 nm pore diameter), DDR (0.36×0.44 nm) and SAPO-34 (0.38 nm), have been prepared recently [15–21]. These membranes possess pores that are similar in size to CH_4 but larger than CO_2 and have high CO_2/CH_4 selectivities due to a molecular sieving mechanism. For example, a DDR-type zeolite membrane shows much higher CO_2 permeability and CO_2/CH_4 selectivity compared to polymer membranes [15–17]. SAPO-34 molecular sieve membranes show improved selectivity for separation of certain gas mixtures, including mixtures of CO_2 and CH_4 [18–21].

The unique properties of zeolite materials combined with the continuous separation properties of membranes make zeolite membranes very attractive for a wide range of separation and catalysis applications. Zeolite membranes, however, have poor processability, poor mechanical stability and are much more expensive than the commercial polymer membranes with current state-of-the-art membrane manufacturing process. So far, the only large-scale commercial zeolite membrane is the A-type zeolite membrane and it has been used for dehydration of alcohols [22]. Further advancement in making thinner zeolite membranes and continuous improvement in membrane production techniques and reproducibility will make zeolite membranes more successful in commercial applications.

11.2
Compositions of Mixed-Matrix Membranes

In spite of their advantages, such as low cost, ease of processability, good mechanical stability and reasonably good selectivity and permeability, typical polymer membranes have poor contaminant resistance, low chemical and thermal stability and cannot overcome the polymer upper bound limit between permeability and selectivity. In contrast, zeolite membranes offer much higher permeability and selectivity than polymeric membranes, but are expensive and difficult for large-scale manufacture. Therefore, the membrane success in the growth of future membrane markets requires a breakthrough development of next generation cost-effective, high-performance membranes that can combine the advantages of both polymer and zeolite membranes. These new high-performance membranes will dramatically change the competitive position of membranes compared to the traditional separation technologies.

Based on the need for a membrane that is more efficient than polymer and zeolite membranes, a new type of membrane, the mixed-matrix membrane, has been developed recently. The mixed-matrix membrane is an organic–inorganic hybrid composite membrane containing dispersed inorganic particles such as zeolite particles in a continuous organic polymer matrix. Mixed-matrix membranes provide the advantages of both polymer membranes such as low cost and ease of processability and the inorganic membranes such as high selectivity and permeability. In addition, the fabrication of mixed-matrix membranes can be integrated into current polymer membrane manufacturing process.

11.2.1
Non-zeolite/Polymer Mixed-Matrix Membranes

Both zeolitic and non-zeolitic inorganic materials have been used as the dispersed phase for making mixed-matrix membranes.

Up to now, a variety of non-zeolite/polymer mixed-matrix membranes have been developed comprising either nonporous or porous non-zeolitic materials as the dispersed phase in the continuous polymer phase. For example, non-porous and porous silica nanoparticles, alumina, activated carbon, poly(ethylene glycol) impregnated activated carbon, carbon molecular sieves, TiO_2 nanoparticles, layered materials, metal–organic frameworks and mesoporous molecular sieves have been studied as the dispersed non-zeolitic materials in the mixed-matrix membranes in the literature [23–35]. This chapter does not focus on these non-zeolite/polymer mixed-matrix membranes. Instead we describe recent progress in molecular sieve/polymer mixed-matrix membranes, as much of the research conducted to date on mixed-matrix membranes has focused on the combination of a dispersed zeolite phase with an easily processed continuous polymer matrix. The molecular sieve/polymer mixed-matrix membranes covered in this chapter include zeolite/polymer and non-zeolitic molecular sieve/polymer mixed-matrix membranes, such as aluminophosphate molecular sieve (AlPO)/polymer and silicoaluminophosphate molecular sieve (SAPO)/polymer mixed-matrix membranes.

11.2.2
Zeolite/Polymer Mixed-Matrix Membranes

Zeolite/polymer mixed-matrix membranes are excellent candidates to address the issues of both polymer membranes and zeolite membranes. Kulprathipanja and coworkers at UOP first introduced the term "mixed-matrix membrane" and performed the pioneering study on zeolite/polymer mixed-matrix membranes in the mid-1980s [36, 37]. Zeolite/polymer mixed-matrix membranes combine the superior selectivity and permeability of inorganic zeolites with the low cost and ease of processability of polymer membranes. The dispersed zeolite phase in a zeolite/polymer mixed-matrix scenario can have a selectivity that is significantly higher than that of the continuous polymer matrix. Therefore, the addition of a small volume fraction of zeolite particles to the polymer matrix can significantly increase the overall separation efficiency. The superior properties of zeolite/polymer mixed-matrix membranes depend on the proper selection of the zeolite material used as the dispersed phase and the polymer material used as the continuous phase in the mixed-matrix membranes. Both rubbery polymers (e.g., silicone rubber) [38, 39] and glassy polymers (e.g., polyimides) [40–50] have been studied as the continuous polymer matrices in the zeolite/polymer mixed-matrix membranes. A variety of zeolites based on an aluminosilicate composition [40–42, 43, 45] and non-zeolitic molecular sieves such as aluminophosphate molecular sieve (AlPO) [48] and silicoaluminophosphate molecular sieve (SAPO) [39, 46, 49] has been used to prepare mixed-matrix membranes. Some of these mixed-matrix membranes have shown remarkably improved properties that have surpassed the polymer upper bound curve. Despite the intense research efforts, issues of material compatibility and adhesion at the zeolite/polymer interface of the mixed-matrix membranes are still not completely addressed.

11.3
Concept of Zeolite/Polymer Mixed-Matrix Membranes

Zeolite/polymer mixed-matrix membranes have the potential to achieve higher selectivity or/and greater permeability than their corresponding polymer membranes for separations, while maintaining the advantages of polymer membranes. Therefore, zeolite/polymer mixed-matrix membranes may open up new opportunities for membrane technology for separation and purification processes. The concept of zeolite/polymer mixed-matrix membranes for separations with improved selectivity or/and greater permeability is to combine the solution–diffusion mechanism of polymer membranes with the molecular sieving mechanism (small-pore zeolite membranes) or surface diffusion mechanism (large-pore zeolite membranes) of zeolite membranes. For the separation of a two-component mixture, enhanced selectivity can be accomplished by dispersing in the continuous polymer matrix zeolite particles with a micropore size smaller than one molecule but larger than the other molecule. The improved overall selectivity is due to the increased solubility selectivity or/and diffusivity selectivity.

It has been reported in the literature that the properties of mixed-matrix membranes can be predicted by using a Maxwell model [40]. The Maxwell model equation is as follows:

$$P_{MMM} = P_P \left(\frac{P_Z + 2P_P - 2\Phi_Z(P_P - P_Z)}{P_Z + 2P_P + \Phi_Z(P_P - P_Z)} \right) \quad (11.1)$$

In Eq. (11.1), P is permeability, Φ_Z is the volume fraction of the dispersed zeolite, the MMM subscript refers to the mixed-matrix membrane, the P subscript refers to the continuous polymer matrix and the Z subscript refers to the dispersed zeolite. The permeability of the mixed-matrix membrane (P_{MMM}) can be estimated by this Maxwell model when the permeabilities of the pure polymer (P_P) and the pure zeolite (P_Z), as well as the volume fraction of the zeolite (Φ_Z) are known. The selectivity of the mixed-matrix membrane for two molecules to be separated can be calculated from the Maxwell model predicted permeabilities of the mixed-matrix membrane for both molecules.

To improve the separation property of a selected polymer membrane using the zeolite/polymer mixed-matrix membrane approach, the Maxwell model can guide the selection of a proper zeolite from a set of zeolites with known permeabilities and selectivities. An example is given in Figure 11.2. Polymer A has CO_2 permeability of 30 Barrer and CO_2/CH_4 selectivity of 25 for CO_2/CH_4 separation. Zeolite B has CO_2 permeability of 50 Barrer and CO_2/CH_4 selectivity of 100. Zeolite C has CO_2 permeability of 1000 Barrer and CO_2/CH_4 selectivity of 90. The Maxwell model is used to select a proper zeolite from zeolites B and C for making a mixed-matrix membrane with much higher CO_2/CH_4 selectivity than that of the polymer A membrane. Using polymer A as the continuous polymer phase and 50 vol% of zeolite particles, the calculated CO_2 permeabilities and CO_2/CH_4 selectivities of A/B and A/C mixed-matrix membranes show that the A/B mixed-matrix membrane provides much higher CO_2/CH_4 selectivity than the A/C mixed-matrix membrane. The selectivity of the A/C mixed-matrix membrane is close to that of the polymer A membrane because the permeability of zeolite C is too high. Therefore zeolite B should be selected to make the desired mixed-matrix membrane. This

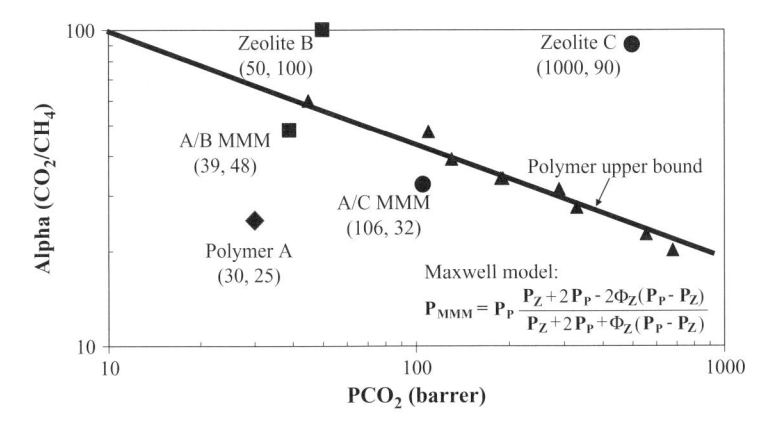

Figure 11.2 Selection of proper zeolite material for a mixed-matrix membrane (MMM) using the Maxwell model.

example suggests that the permeability of zeolite and polymer in the mixed-matrix membrane should match to take advantage of the high selectivity of zeolites.

The Maxwell model can also guide the selection of a proper polymer material for a selected zeolite at a given volume fraction for a target separation. For most cases, however, the Maxwell model cannot be applied to guide the selection of polymer or zeolite materials for making new mixed-matrix membranes due to the lack of permeability and selectivity information for most of the pure zeolite materials. In addition, although this Maxwell model is well-understood and accepted as a simple and effective tool for estimating mixed-matrix membrane properties, sometimes it needs to be modified to estimate the properties of some non-ideal mixed-matrix membranes.

11.4
Material Selection for Zeolite/Polymer Mixed-Matrix Membranes

The development of a successful zeolite/polymer mixed-matrix membrane with properties superior to the corresponding polymer membrane depends upon good performance match and good compatibility between zeolite and polymer materials, as well as small enough zeolite particle size for membrane manufacturing on a large scale.

11.4.1
Selection of Polymer and Zeolite Materials

11.4.1.1 Selection of Polymer Materials

Many organic polymers, including rubbery and glassy polymers, have been studied as the continuous phase in the zeolite/polymer mixed-matrix membranes in the literature. For example, Hennepa and coworkers used mixed-matrix membranes made from silicone rubber and silicalite. The good compatibility between the hydrophobic silicone rubber and the dispersed hydrophobic silicalite in this mixed-matrix membrane system resulted in significantly improved selectivity for ethanol/water separation using these membranes [38]. Most recently, Jha and coworkers have proved that the incorporation of SAPO-34 molecular sieve in a rubbery substituted polyphosphazene membrane increased the CO_2/H_2 separation factor from 8.5 for the pure polyphosphazene membrane to 12.0 at 22 °C [39].

Glassy polymers with much higher glass transition temperatures and more rigid polymer chains than rubbery polymers have been extensively used as the continuous polymer matrices in the zeolite/polymer mixed-matrix membranes. Typical glassy polymers in the mixed-matrix membranes include cellulose acetate, polysulfone, polyethersulfone, polyimides, polyetherimides, polyvinyl alcohol, Nafion®, poly(4-methyl-2-pentyne), etc.

It has been demonstrated by many studies that mixed-matrix membranes with a good match between the permeability of proper zeolite materials and these glassy polymers exhibit separation properties superior to the corresponding pure glassy polymer membranes. Some zeolite/glassy polymer mixed-matrix membranes have shown gas separation performance well above Robeson's polymer upper bound. For example, Mahajan and coworkers reported a series of mixed-matrix membranes prepared from glassy polymers such as Ultem® polyetherimide, Matrimid® polyimide and polyvinyl acetate [40–42]. With certain zeolite 4A loadings in these polymer matrices, the O_2/N_2 separation performance of these mixed-matrix membranes exceeded the polymer upper bound limit for O_2/N_2 separation. The O_2/N_2 selectivities of some of these mixed-matrix membranes were almost doubled compared to their corresponding polymer membranes. Some mixed-matrix membranes with property mismatch between zeolites and polymer matrix have shown no selectivity enhancement. For example, mixed-matrix membranes prepared from zeolite L and a polyimide (6FDA-6FpDA-DABA) synthesized from 4,4′-hexafluoroisopropylidenediphthalic anhydride (6FDA), 4,4′-hexafluoroisopropylidene dianiline (6FpDA) and 3,5-diaminobenzoic acid (DABA) did not show any selectivity improvement for O_2/N_2 separation due to the much higher O_2 permeability and much lower O_2/N_2 selectivity of large-pore zeolite L than the 6FDA-6FpDA-DABA polyimide matrix [51]. Mixed-matrix membranes prepared from 5A zeolite and silicone rubber rubbery polymer [52] or polyethersulfone (PES) glassy polymer [53] showed different separation properties. The 5A/PES mixed-matrix membranes showed significantly improved selectivity for O_2/N_2 separation. However, no significant selectivity enhancement was observed for the 5A/silicone rubber mixed-matrix membranes. These results suggest that the permeability of 5A zeolite matches well with the permeability of PES. However, silicone rubber has over 200 times higher gas permeability than PES and cannot match with 5A for permeability.

11.4.1.2 Selection of Zeolite Materials

With a good match between the zeolite and the polymer materials, the dispersed zeolites in the zeolite/polymer mixed-matrix membrane can provide the membrane improved selectivity by including selective holes or pores having a diameter that permits a particular molecule to pass through, but either does not permit another molecule to pass through, or permits it to pass through at a significantly slower rate. Zeolites in the mixed-matrix membrane need to have higher selectivity for the desired separation than the original polymer to enhance the selectivity.

Various zeolites have been studied as the dispersed phase in the mixed-matrix membranes. Zeolite performance in the zeolite/polymer mixed-matrix membrane is determined by several key characteristics including pore size, pore dimension, framework structure, chemical composition (e.g., Si/Al ratio and cations), crystal morphology and crystal (or particle) size. These characteristics of zeolites are summarized in Chapter 6.

Mixed-matrix membranes comprising small-pore zeolite or small-pore non-zeolitic molecular sieve materials will combine the solution–diffusion separation mechanism of the polymer material with the molecular sieving mechanism of the zeolites. The small-pore zeolite or non-zeolitic molecular sieve materials in the mixed-matrix membranes are capable of separating mixtures of molecular species based on the molecular size or kinetic diameter (molecular sieving mechanism). The highly selective separation is accomplished by allowing the smaller molecular species to enter the intracrystalline pore (or void space) while excluding larger species. Separation with the mixed-matrix membranes comprising large- or medium-pore zeolite or non-zeolitic molecular sieve materials is mainly based on solution–diffusion and competitive adsorption when the pores of the large- or medium-pore zeolite or non-zeolitic molecular sieve materials are much larger than all the molecules to be separated.

For example, Zeolite 4A is an attractive candidate dispersed phase for making mixed-matrix membranes for O_2/N_2 separation because zeolite 4A possesses an eight-membered ring pore with an effective pore size of 3.8 Å that falls between the lengths of O_2 and N_2 molecules with lengths of 3.75 and 4.07 Å, respectively. Mahajan and coworkers developed mixed-matrix membranes using small-pore zeolite 4A and these mixed-matrix membranes showed significantly improved O_2/N_2 selectivity compared to the corresponding polymer membranes [40, 42].

Small-pore zeolite Nu-6(2) has a NSI-type structure and two different types of eight-membered-ring channels with limiting dimensions of 2.4 and 3.2 Å [54]. Gorgojo and coworkers developed mixed-matrix membranes using Nu-6(2) as the dispersed zeolite phase and polysulfone Udel® as the continuous organic polymer phase [55]. These mixed-matrix membranes showed remarkably enhanced H_2/CH_4 selectivity compared to the bare polysulfone membrane. The H_2/CH_4 selectivity increased from 13 for the bare polysulfone membrane to 398 for the Nu-6(2)/polysulfone mixed-matrix membranes. This superior performance of the Nu-6(2)/polysulfone mixed-matrix membranes is attributed to the molecular sieving role played by the selected Nu-6(2) zeolite phase in the membranes.

The chemical compositions of the zeolites such as Si/Al ratio and the type of cation can significantly affect the performance of the zeolite/polymer mixed-matrix membranes. Miller and coworkers discovered that low silica-to-alumina molar ratio non-zeolitic small-pore molecular sieves could be properly dispersed within a continuous polymer phase to form a mixed-matrix membrane without defects. The resulting mixed-matrix membranes exhibited more than 10% increase in selectivity relative to the corresponding pure polymer membranes for CO_2/CH_4, O_2/N_2 and CO_2/N_2 separations [48]. Recently, Li and coworkers proposed a new ion exchange treatment approach to change the physical and chemical adsorption properties of the penetrants in the zeolites that are used as the dispersed phase in the mixed-matrix membranes [56]. It was demonstrated that mixed-matrix membranes prepared from the AgA or CuA zeolite and polyethersulfone showed increased CO_2/CH_4 selectivity compared to the neat polyethersulfone membrane. They proposed that the selectivity enhancement is due to the reversible reaction between CO_2 and the noble metal ions in zeolite A and the formation of a π-bonded complex.

Zeolites used for the preparation of mixed-matrix membranes not only should have suitable pore size to allow selective permeation of a particular molecular component, but also should have appropriate particle size in the nanometer range for the formation of commercially viable defect-free asymmetric mixed-matrix membranes. However, most of the studies on zeolite/polymer mixed-matrix membranes used zeolites with relatively large particle sizes in the micron range. Only a few mixed-matrix membranes comprising nano-sized zeolite particles are reported in the literature. For example, Moermans and coworkers reported the first incorporation of nano-sized silicalite-1 in membranes by dispersing colloidal silicalite-1 in polydimethylsiloxane polymer membrane [57]. Homogeneous zeolite/polymer mixed-matrix membranes were also fabricated by the incorporation of dispersible template-removed zeolite A nanocrystals into a polysulfone matrix [58].

11.4.1.3 Compatibility between Polymer and Zeolite Materials

Mixed-matrix membranes containing dispersed zeolites in a continuous polymer matrix may retain polymer processability and improved selectivity for separation applications due to the superior molecular sieving property of the zeolite materials. Most reported zeolite/polymer mixed-matrix membranes, however, have issues of aggregation of the zeolite particles in the polymer matrix and poor adhesion at the interface of zeolite particles and the polymer matrix. These issues resulted in mixed-matrix membranes with poor mechanical and processing properties and poor separation performance. Poor compatibility and poor adhesion between the polymer matrix and the zeolite particles in the mixed-matrix membranes resulted in voids and defects around the zeolite particles that are larger than the micropores of the zeolites. Mixed-matrix membranes with these voids and defects exhibited selectivity similar to or even lower than that of the continuous polymer matrix and could not match that predicted by Maxwell model [59, 60].

Research has shown that good compatibility and good adhesion between the polymer matrix and the zeolite particles in mixed-matrix membranes are of particular importance in forming successful mixed-matrix membranes with enhanced selectivity [41, 60, 61]. Despite all research efforts, issues of material compatibility and adhesion at the zeolite/polymer interface of the mixed-matrix membranes have not been completely addressed.

11.4.2
Modification of Zeolite and Polymer Materials

Most recently, significant research efforts have been focused on materials compatibility and adhesion at the zeolite/polymer interface of the mixed-matrix membranes in order to achieve enhanced separation property relative to their corresponding polymer membranes. Modification of the surface of the zeolite particles or modification of the polymer chains to improve the interfacial adhesion provide new opportunity for making successful zeolite/polymer mixed-matrix membranes with significantly improved separation performance.

Some small organic molecules (e.g., organosilanes), sizing agents (e.g. Ultem® polyetherimide), special surface treatment agents (e.g., Grignard reagent) and electrostabilizing additives have been used to modify the outside surface of the zeolite particles to improve the compatibility between the zeolite particles and the polymer matrix in the literature [50, 62–70].

Organosilane coupling agents have been used to functionalize the surface of the zeolites to improve the adhesion at the zeolite particle/polymer interface of the mixed-matrix membranes [62, 63, 65, 70]. The organosilane reacts with the hydroxyl groups on the surface of the zeolite particles via silylation reaction to form covalent bonds. Furthermore, the organosilanes with a second reactive functional group could be used to react with some functional groups on the polymer chains to form covalent bonds between the zeolite surface and the polymer, thus promoting adhesion between the zeolite and the bulk polymer phase in the membrane. Therefore, the defect-free mixed-matrix membranes prepared using this approach showed enhanced selectivity. In some cases, the silylation reaction also occurred on the inside surface of the micropores, resulting in their plugging, which is a concern for mixed-matrix membranes. The drawbacks of the organosilane surface functionalization method include the expense of the organosilane coupling agents and the complicated time-consuming zeolite purification and organosilane recovery steps after surface functionalization. The cost of making such defect-free organosilane functionalized zeolite/polymer mixed-matrix membranes on a commercially viable scale can be very expensive.

Electrostabilizing additives (e.g., an organic acid additive) were used to stabilize the mixed-matrix membrane spinning dope, providing mixed-matrix membranes with minimal macrovoids and defects [68]. The addition of electrostabilizing additives to the spinning dope formulation yielded mixed-matrix membranes with better and more consistent selectivity and improved mechanical strength over formulations without electrostabilizing additives. The electrostabilizing additives stabilized the zeolite particles in the concentrated spinning dopes. It is proposed that the use of electrostabilizing additives imparts an electrostatic charge to the zeolite particles in the spinning dope. Therefore, the spinning dope becomes stable to particle aggregation due to mutual electrostatic repulsion of the zeolite particles.

Husain and Koros modified the zeolite surface using a vigorous surface modification agent (Grignard reagent) to minimize zeolite–solvent/non-solvent interaction [69]. The surface of the zeolites was made hydrophobic by replacing the hydroxyl groups with methyl groups. Therefore, good interaction between the zeolites with hydrophobic surfaces and the hydrophobic polymers was achieved. Mixed-matrix hollow fibers incorporating Grignard treated HSSZ-13 zeolites in Ultem® polymer matrix showed a selectivity enhancement of 10% and 35% for O_2/N_2 and CO_2/CH_4 gas pairs, respectively. Due to the success of the Grignard-treated zeolites in the polymer matrix, the authors also envisioned that coupling of the polymer to the zeolite surface through formation of covalent bonds is not a prerequisite for successful mixed-matrix membrane formation.

To improve the adhesion and interaction in the zeolite/polymer interface, the surface of the zeolites can also be "sized" (or "primed") by coating the zeolite with an ultrathin layer of the matrix polymer or a different polymer. "Sizing" of the zeolite particles prior to dispersion in the polymer matrix reduced the stress at the zeolite/polymer interface. The mixed-matrix membranes comprising zeolites treated by this method showed enhanced selectivity for gas separations such as for O_2/N_2 separation compared to the neat polymer membranes [50].

Zeolite/polymer mixed-matrix membranes prepared from crosslinked polymers and surface-modified zeolite particles offered both outstanding separation properties and swelling resistance for some gas and vapor separations such as purification of natural gas. Hillock and coworkers reported that crosslinked mixed-matrix membranes prepared from modified SSZ-13 zeolite and 1,3-propane diol crosslinked polyimide (6FDA-DAM-DABA) synthesized from 2,2'-*bis*-(3,4-dicarboxyphenyl)hexafluoropropane dianhydride, *p*-dimethylaminobenzylamine- and 3,5-diaminobenzoic acid displayed high CO_2/CH_4 selectivities of up to 47 Barrer and CO_2 permeabilities of up to 89 Barrer under mixed gas testing conditions [71]. Additionally, these crosslinked mixed-matrix membranes were resistant to CO_2 plasticization up to 450 psia (3100 kPa).

11.5
Geometries of Zeolite/Polymer Mixed-Matrix Membranes

Zeolite/polymer mixed-matrix membranes can be fabricated into dense film, asymmetric flat sheet, or asymmetric hollow fiber. Similar to commercial polymer membranes, mixed-matrix membranes need to have an asymmetric membrane geometry with a thin selective skin layer on a porous support layer to be commercially viable. The skin layer should be made from a zeolite/polymer mixed-matrix material to provide the membrane high selectivity, but the non-selective porous support layer can be made from the zeolite/polymer mixed-matrix material, a pure polymer membrane material, or an inorganic membrane material.

11.5.1
Mixed-Matrix Dense Films

Reports on mixed-matrix membranes in the literature mainly focus on dense films. Mixed-matrix dense film has a symmetric structure and a thickness of more than 20 μm for most studies. Although dense films are not commercially attractive, they are used to measure the intrinsic separation properties including selectivity and permeability of the mixed-matrix membranes. Therefore, promising polymer and zeolite materials for making asymmetric mixed-matrix membranes for a particular separation can be identified through dense film study.

Typically, mixed-matrix dense film is prepared by a solution-casting method from a mixed-matrix casting dope comprising one or more organic solvents, zeolite particles or surface-treated zeolite particles and a polymer. To make defect-free mixed-matrix dense film with enhanced selectivity, sometimes the surface of the dispersed zeolite particles needs to be modified by organosilane coupling agent, polymer, or other agents, as discussed in Section 11.4.2. One method for making mixed-matrix dense film comprises: (i) dispersing the zeolite particles or surface-modified zeolite particles in an organic solvent or a mixture of two or more organic solvents by ultrasonic mixing and/or high-speed mechanical stirring or other method to form a stable zeolite slurry, (ii) dissolving a polymer that serves as a continuous polymer matrix in the zeolite slurry to form a stable zeolite/polymer casting dope, (iii) fabricating the mixed-matrix dense film by a solution-casting method, by pouring the stable zeolite/polymer casting dope into a glass ring on top of a clean glass plate or using a doctor knife to spread the stable zeolite/polymer casting dope on a flat and clean support, such as Teflon® or glass plate. The initial solvent evaporation after casting is generally done under controlled conditions and temperatures, followed by a drying step at higher temperatures to remove any remaining solvent. In some cases, a dense film post-treatment step such as annealing at high temperature can be added to further improve selectivity.

11.5.2
Asymmetric Mixed-Matrix Membranes

To maximize the flux, which represents the membrane productivity, through a membrane, it is essential to minimize the thickness of the membrane selective skin layer. The breakthrough development of Loeb–Sourirajan process [72] to make defect-free asymmetric cellulose acetate membranes with an ultrathin selective skin layer in the 1960s led to initial commercialization of reverse osmosis polymer membranes in the 1970s followed by intense research activity and further commercialization in the 1980s. Today, the commercial polymer membranes for gas separations have the geometry of an asymmetric flat sheet, a thin-film composite, or an asymmetric hollow fiber. These membranes have an ultrathin selective layer typically less than 100 nm on a highly porous non-selective support layer. The membrane selectivity is provided by the ultrathin selective layer and the membrane mechanical strength is provided by the highly porous nonselective support layer. The membranes with thinner selective layer will have higher productivity than those with thicker selective layer.

Extensive research has been done for mixed-matrix dense films, but only a few studies on asymmetric mixed-matrix membranes have been reported. Furthermore, most of the asymmetric mixed-matrix membranes have not shown dramatically improved selectivity (or so-called mixed-matrix membrane effect) compared to the corresponding asymmetric polymer membranes, although mixed-matrix dense films prepared from the same zeolite/polymer mixed-matrix materials have exhibited significantly enhanced selectivity. Another problem for making successful defect-free asymmetric mixed-matrix membranes that has been noticed by the researchers is the reproducibility, which is one of the biggest issues for the commercialization of asymmetric mixed-matrix membranes. Large zeolite particle size, poor adhesion (or compatibility) between the zeolite particles and the polymer matrix, the permeability and selectivity mismatch between the polymer matrix and the zeolites, as well as the formation of interface defects during the phase inversion process are some of the key issues that result in the formation of defective or defect-free but low performance asymmetric mixed-matrix membranes. The poor adhesion and formation of interface defects during the phase inversion process is possibly due to the elongation stress during hollow fiber spinning or due to the nucleation of non-solvent/polymer lean phase around the dispersed zeolite particles during the phase separation step.

The geometries for asymmetric mixed-matrix membranes include flat sheets, hollow fibers and thin-film composites. The flat sheet asymmetric mixed-matrix membranes are formed into spirally wound modules and the hollow fiber asymmetric mixed-matrix membranes are formed into hollow fiber modules. The thin-film composite mixed-matrix membranes can be fabricated into either spirally wound or hollow fiber modules. The thin-film composite geometry of mixed-matrix membranes enables selection of different membrane materials for the support layer and low-cost production of asymmetric mixed-matrix membranes utilizing a relatively high-cost zeolite/polymer separating layer on the support layer.

Most of the asymmetric mixed-matrix membranes reported to date were prepared from concentrated mixed-matrix dopes via a phase inversion technique [69, 70, 73–81]. During the phase inversion process, a one phase zeolite/polymer mixed-matrix dope is converted into a two-phase system consisting of a solid (zeolite/polymer-rich) phase which forms the membrane structure and a liquid (zeolite/polymer-poor) phase which forms the pores in the final membrane. Coating approach was also studied for a few mixed-matrix membranes with thin-film composite geometry [82, 83]. Table 11.1 summarizes some of the asymmetric zeolite/polymer mixed-matrix membranes reported in the literature. Typically the dopes (or formulations) for forming asymmetric mixed-matrix membranes are different from those for making mixed-matrix dense films because of the different membrane formation mechanism. The dope composition and the precipitation media determine the performance of the asymmetric mixed-matrix membranes formed via phase separation mechanism. To control the selective layer thickness and the morphology of the membranes, typically one or more non-solvents for the polymers are included in the formulations. The dopes for forming mixed-matrix membranes by the coating method are similar to those for making mixed-matrix dense films and have low concentration.

11.5.2.1 Flat Sheet Asymmetric Mixed-Matrix Membranes

A typical procedure for making asymmetric zeolite/polymer mixed-matrix membrane with flat sheet geometry via phase separation mechanism comprises: (i) setting the dope layer thickness with a doctor blade, (ii) pouring the zeolite/polymer mixed-matrix dope onto a support fabric (iii) letting a controlled amount of evaporation occur, (iv) immersing the partially dried nascent mixed-matrix membrane into a gellation (or precipitation) medium with controlled temperature, (v) waiting until all the solvents are replaced by the gellation fluid, (vi) immersing the membrane into a drying medium with controlled temperature, (vii) if a dry mixed-matrix membrane is required, the gellation fluid must be removed from the membrane without causing changes to the substructure of the membrane. In

Table 11.1 Asymmetric zeolite/polymer mixed-matrix membranes.

Membrane composition		Zeolite surface-treatment method	Geometry	Applications	Reference
Zeolite	Polymer				
Silicalite-1	CA	NA	Flat sheet	Gas separations (e.g., O_2/N_2)	[73]
SSZ-13	P84	Washing method	Flat sheet	Gas separations (e.g., CO_2/N_2, CO_2/CH_4, O_2/N_2)	[74]
SSZ-13	Sheath: Ultem® Core: Ultem® and Matrimid®	Silane coupling	Dual layer hollow fiber	Gas separations (e.g., O_2/N_2)	[75]
SSZ-13	Sheath: Ultem® Core: Ultem®	Grignard treatment	Dual layer hollow fiber	Gas separations (e.g., CO_2/CH_4, O_2/N_2)	[69]
Zeolite beta	Sheath: PSF Core: Matrimid®	NA	Dual layer hollow fiber	Gas separations (e.g., He/N_2, O_2/N_2)	[79]
Zeolite beta	Sheath: PES Core: P84	NA	Dual layer hollow fiber	Gas separations (e.g., CO_2/CH_4, O_2/N_2)	[80]
Zeolite beta	Sheath: PSF Core: Matrimid®	NA	Dual layer hollow fiber	Gas separations (e.g., CO_2/CH_4, O_2/N_2)	[81]
Silicalite-1	Selective layer: PDMS Support: PEI	NA	Thin-film composite flat sheet	Gas separations (e.g., O_2/N_2) Liquid separations (e.g., ethanol/water separation)	[82]
NaA	Selective layer: Polyamide Support: PSF	NA	Thin-film composite flat sheet	Liquid separations (e.g., water desalination)	[83]

some cases, a membrane post-treatment step such as coating a thin layer of high-flux material on top of the membrane to plug the minor defects is used to further improve membrane selectivity.

Kulprathipanja and coworkers reported the preparation of integrally skinned silicalite-1/cellulose acetate flat sheet asymmetric mixed-matrix membranes via phase inversion technique in 1992 [73]. The O_2/N_2 separation performance of these membranes was investigated. It was demonstrated that the separation factor of these silicalite-1/cellulose acetate flat sheet asymmetric mixed-matrix membranes for O_2/N_2 is higher than that of the neat cellulose acetate membrane.

Kulkarni and coworkers successfully made flat sheet asymmetric mixed-matrix membranes by incorporating zeolites (e.g., SSZ-13) into a polyimide polymer (P84) matrix [74]. They studied several approaches such as bonding of the zeolite to P84 polymer utilizing a silane, "sizing" of the zeolites using a polymer, or pretreating the zeolites by a washing method to improve the adhesion and eliminate the gaps at the zeolite/polymer interface. Some of these SSZ-13/P84 asymmetric mixed-matrix membranes comprising washed HSSZ-13 zeolites exhibited enhanced selectivities for several gas separations such as for CO_2/CH_4 and CO_2/N_2 separations. For example, the washed HSSZ-13/P84 asymmetric membranes showed 27–60% improvement in CO_2/N_2 selectivity over P84 asymmetric polymer membrane. It is worth noting that the CO_2/N_2 selectivity improvement for the HSSZ-13/P84 asymmetric membranes is similar to or better than that reported for HSSZ-13/P84 mixed-matrix dense films.

11.5.2.2 Hollow Fiber Asymmetric Mixed-Matrix Membranes

The procedure for making asymmetric zeolite/polymer mixed-matrix membrane with hollow fiber geometry via phase separation mechanism is similar to that used for forming asymmetric mixed-matrix membrane with flat sheet geometry, but no support fabric is needed. In addition, the dope for forming an integrally skinned asymmetric hollow fiber mixed-matrix membrane has a relatively higher viscosity than that for forming an integral skinned asymmetric flat sheet mixed-matrix membrane. An extrusion spinneret instead of a doctor blade is needed for spinning hollow fibers.

Most of the studies have been focused on the development of dual-layer hollow fiber mixed-matrix membranes via a one-step co-extrusion method. For example, Ekiner and Kulkarni reported the successful spinning of dual-layer asymmetric hollow fiber mixed-matrix membranes comprising a mixed-matrix sheath layer of surface-treated zeolites (e.g., SSZ-13) and Ultem® polymer and a core layer of Ultem®/Matrimid® blend polymers [75]. They demonstrated that this hollow fiber mixed-matrix membrane had good O_2 permeance and superior selectivity for O_2/N_2 compared to the homogeneous polymer membranes. Additionally, the selectivity gradually diminished as draw ratio increased but remained higher than any of the homogeneous polymer membranes. Husain and Koros prepared dual-layer asymmetric hollow fiber mixed membranes containing a sheath layer prepared by incorporating Grignard-treated HSSZ-13 zeolites into an Ultem® polymer matrix and a core layer of neat Ultem® polymer [69]. Good adhesion between the Grignard treated HSSZ-13 with hydrophobic surfaces and the hydrophobic polymers was achieved for these membranes. These membranes showed a selectivity enhancement of 10% and 35% for O_2/N_2 and CO_2/CH_4 gas pairs, respectively. Recently, Li, Jiang and coworkers conducted a series of studies on dual-layer asymmetric hollow fiber mixed-matrix membranes [78–81]. For example, Jiang and coworkers proposed a new approach to prepare dual-layer hollow fiber mixed-matrix membranes with a defect-free mixed-matrix skin layer [81]. They applied dual-layer co-extrusion technology to control particle distribution followed by annealing at a temperature above the glass transition temperature (T_g) of the skin layer for removing defects in the mixed-matrix skin layer after the formation of the hollow fibers. In addition to the annealing method, Jiang and coworkers investigated a p-xylenediamine/methanol soaking method to achieve an intimate zeolite/polymer interface. The mixed-matrix membranes formed using this method exhibited about 30% and 50% higher O_2/N_2 and CO_2/CH_4 selectivities, respectively, over the neat polymer membranes. However, the O_2 and CO_2 permeances are significantly lower than commercial polymer membranes for these gas separations.

11.5.2.3 Thin-Film Composite Mixed-Matrix Membranes

Thin-film composite mixed-matrix membranes can be made by a coating method or by a dual-layer co-extrusion technology, as discussed in Section 11.5.2.2.

Jia and coworkers prepared thin-film composite zeolite-filled silicone rubber membranes by a dip-coating method [82]. The membranes have a thin silicalite-1/silicone rubber mixed-matrix selective layer on top of a porous polyetherimide support.

Only a slight improvement in separation was obtained with the membrane having a mixed-matrix layer thickness of more than 18 mm.

Geong and coworkers reported a new concept for the formation of zeolite/polymer mixed-matrix reverse osmosis (RO) membranes by interfacial polymerization of mixed-matrix thin films *in situ* on porous polysulfone (PSF) supports [83]. The mixed-matrix films comprise NaA zeolite nanoparticles dispersed within 50–200 nm polyamide films. It was found that the surface of the mixed-matrix films was smoother, more hydrophilic and more negatively charged than the surface of the neat polyamide RO membranes. These NaA/polyamide mixed-matrix membranes were tested for a water desalination application. It was demonstrated that the pure water permeability of the mixed-matrix membranes at the highest nanoparticle loadings was nearly doubled over that of the polyamide membranes with equivalent solute rejections. The authors also proved that the micropores of the NaA zeolites played an active role in water permeation and solute rejection.

11.6
Applications of Zeolite/Polymer Mixed-Matrix Membranes

Zeolite/polymer mixed-matrix membranes have the potential to achieve higher selectivity with equal or greater permeability over the existing polymer membranes while maintaining their advantages of low cost and easy processability. Therefore, zeolite/polymer mixed-matrix membranes are promising for a wide range of applications such as for liquid, gas and vapor separations, fuel cell applications and catalysis. This section covers some highlighted applications for mixed-matrix membranes, including gas and liquid separations.

11.6.1
Gas Separation Applications

Gas separation processes with membranes have undergone a major evolution since the introduction of the first membrane-based industrial hydrogen separation process about two decades ago. The development of high selectivity mixed-matrix membranes will further advance the technology of membrane gas separation processes within the next decade.

Zeolite/polymer mixed-matrix membranes have been studied for a number of gas separations such as separation of N_2 from air [37, 73, 75, 81, 84, 85], H_2 and CO_2 removal from natural gas [51, 54, 69, 81, 86–88], CO_2 removal from N_2 [74], n-pentane/i-pentane separation [89] and separation of H_2 from CO_2 [65]. But a majority of the mixed-matrix membranes that have been evaluated for gas separations are mixed-matrix dense films.

Rubbery polymer, polydimethylsiloxane (PDMS), was used as the polymer matrix to prepare zeolite/PDMS mixed-matrix membranes [82, 89]. This type of mixed-matrix membrane, however, did not exhibit improved selectivity for n-pentane/i-pentane separation relative to the neat PDMS membrane.

Mixed-matrix membranes were developed using polyethersulfone matrix and several different zeolites such as hydrophilic zeolites 3A, 4A and 5A [90, 91]. Some of these membranes gave performance improvements at high zeolite loadings over neat polyethersulfone membranes. Mixed-matrix membranes, comprising zeolite 4A and poly [2,2'-bis-(3,4-dicarboxyphenyl)hexafluoropropane dianhydride-4,4'-hexafluoroisopropyl-idene dianiline/2,3,5,6-tetramethyl-1,4-phenylene diamine/3,5-diaminobenzoic acid)] (6FDA-6FpDA/4MPD/DABA) (2:2:1) or Matrimid®, have shown increased selectivity and decreased permeability over the neat polymers [40, 92]. It has been proposed that the permeability loss is attributed to polymer matrix rigidification at the zeolite/polymer interface.

Organosilane coupling agents have been widely used to improve the adhesion at the zeolite/polymer interface of the mixed-matrix membranes [62, 63, 65, 70]. The well-adhered interface resulted in defect-free mixed-matrix membranes with some improved performance for CO_2/CH_4 separation.

Mixed-matrix membranes prepared from small-pore zeolite Nu-6(2) and polysulfone showed significantly enhanced H_2/CH_4 selectivity over the neat polysulfone membrane [55]. The H_2/CH_4 selectivity increased from 13 for the neat polysulfone membrane to 398 for the Nu-6(2)/polysulfone mixed-matrix membranes.

11.6.2
Liquid Separation Applications

Zeolite/polymer mixed-matrix membranes have been investigated for liquid separations such as purification of p-xylene [76], separation of ethanol-water mixtures [93–96] and water desalination [83].

The new concept of using mixed-matrix membranes with commercially attractive thin-film composite geometry for desalination of water has been demonstrated by Jeong and coworkers [83].

Another potential application for zeolite/polymer mixed-matrix membranes is the separation of various liquid chemical mixtures via pervaporation. Pervaporation is a promising membrane-based technique for the separation of liquid chemical mixtures, especially in azeotropic or close-boiling solutions. Polydimethylsiloxane (PDMS), which is a hydrophobic polymer, has been widely used as the continuous polymer matrix for preparing hydrophobic mixed-matrix membranes. To achieve good compatibility and adhesion between the zeolite particles and the PDMS polymer, ZSM-5 was incorporated into the PDMS polymer matrix, the resulting ZSM-5/PDMS mixed-matrix membranes showed simultaneous enhancement in selectivity and flux for the separation of isopropyl alcohol from water. It was demonstrated that the separation performance of these membranes was affected by the concentration of the isopropyl alcohol in the feed [96].

Another type of mixed-matrix membranes for alcohol/water pervaporation applications was developed utilizing hydrophilic poly(vinyl alcohol) (PVA) and ZSM-5. The ZSM-5/PVA mixed-matrix membranes demonstrated increased selectivity and flux, compared to pure PVA, for the water/isopropyl alcohol separation [97]. This type of mixed-matrix membranes, however, may have membrane swelling issue due to the hydrophilic nature of the PVA polymer. Mixed-matrix membranes comprising modified poly(vinyl chloride) and NaA zeolite have shown both enhanced flux and selectivity for the ethanol/water separation at high NaA loadings [98].

11.7
Summary

Historical interest in zeolite/polymer mixed-matrix materials as the next generation of new membrane materials has been driven by the combination of low cost and ease of processing typical of polymers with the properties of high selectivity and high permeability typical of zeolite materials. The concept of mixed-matrix membranes has been developed over 20 years and research on mixed-matrix membranes has significantly advanced. Although intense research efforts have focused on modifying the surface of the zeolite particles and tailoring the polymer structures to improve separation properties, considerable effort is still required to bring mixed-matrix membranes to the point of commercial viability. The coming years should be a fruitful period for mixed-matrix membrane research, with advances in: (i) understanding the basic interactions between the polymer and the zeolite particles, (ii) discovering new methods to improve the adhesion and compatibility between the surface of the zeolites and the polymer, (iii) making less than 200 nm nano-sized zeolite particles without agglomeration, (iv) developing novel approaches to uniformly disperse nano-sized zeolite particles in the continuous polymer matrix, (v) understanding the intrinsic separation properties of zeolite materials and (vi) developing new modeling tools to guide the selection of zeolite and polymer materials with good performance matching.

References

1 Singh, R. (1998) Industrial membrane separation processes. *Chemtech*, **28** (4), 33–44.
2 Matsuura, T. (1994) *Synthetic Membranes and Membrane Separation Processes*, CRC Press, Boca Raton, FL, Chapter 2.
3 Robeson, L.M. (1991) Correlation of separation factor versus permeability for polymeric membranes. *J. Membr. Sci.*, **62**, 165–185.
4 Robeson, L.M. (2008) The upper bound revisited. *J. Membr. Sci.*, **320**, 390–400.
5 Caro, J. and Noack, M. (2008) Zeolite membranes-recent developments and progress. *Micropor. Mesopor. Mater.*, **115**, 215–233.
6 Chen, H., Li, Y., and Yang, W. (2007) Preparation of silicalite-1 membrane by solution-filling method and its alcohol extraction properties. *J. Membr. Sci.*, **296**, 122–130.
7 Gardner, T.Q., Martinek, J.G., Falconer, J.L., and Noble, R.D. (2007) Enhanced flux through double-sided zeolite membranes. *J. Membr. Sci.*, **304**, 112–117.
8 Sebastián, V., Kumakiri, I., Bredesen, R., and Menéndez, M. (2007) Zeolite membrane for CO2 removal: operating at high pressure. *J. Membr. Sci.*, **292**, 92–97.
9 Lovallo, M.C., Gouzinis, A., and Tsapatsis, M. (1998) Synthesis and characterization of oriented MFI membranes prepared by secondary growth. *AIChE J.*, **44**, 1903–1913.
10 Li, Y., Zhou, H., Zhu, G., Liu, J., and Yang, W. (2007) Hydrothermal stability of LTA zeolite membranes in pervaporation. *J. Membr. Sci.*, **297**, 10–15.
11 Bernal, M.P., Piera, E., Coronas, J., Menendez, M., and Santamaria, J. (2000) Mordenite and ZSM-5 hydrophilic tubular membranes for the separation of gas phase mixtures. *Catal. Today*, **56**, 221–227.
12 Guan, G., Kusakabe, K., and Morooka, S. (2001) Gas permeation properties of ion-exchanged LTA-type zeolite membranes. *Sep. Sci. Technol.*, **36**, 2233–2245.
13 Kyotani, T., Mizuno, T., Katakura, Y., Kakui, S., Shimotsuma, N., Saito, J., and Nakane, T. (2007) Characterization of tubular zeolite NaA membranes prepared from clear solutions by FTIR-ATR, GIXRD and FIB-TEM-SEM. *J. Membr. Sci.*, **296**, 162–170.
14 Weh, K., Noack, M., Sieber, I., and Caro, J. (2002) Permeation of single gases and gas mixtures through faujasite-type molecular sieve membranes. *Micropor. Mesopor. Mater.*, **54**, 27–36.
15 van den Bergh, J., Zhu, W., Gascon, J., Mouligin, J.A., and Kapteijn, F. (2008) Separation and permeation characteristics of a DD3R zeolite membrane. *J. Membr. Sci.*, **316**, 35–45.
16 Tomita, T., Nakayama, K., and Sakai, H. (2004) Gas separation characteristics of DDR type zeolite membrane. *Micropor. Mesopor. Mater.*, **68**, 71–75.
17 Nakayama, K., Suzuki, K., Yoshida, M., Yajima, K., and Tomita, T. (2004) Method for preparing DDR type zeolite film, DDR type zeolite film, and composite DDR type zeolite film, and method for preparation thereof. US Patent Application 2004/0173094 A1.
18 Li, S., Alvarado, G., Noble, R.D., and Falconer, J.L. (2005) Effects of impurities on CO2/CH4 separations through SAPO-34 membranes. *J. Membr. Sci.*, **251**, 59–66.
19 Li, S., Falconer, J.L., and Noble, R.D. (2008) SAPO-34 membranes for CO_2/CH_4 separations: effect of Si/Al ratio. *Micropor. Mesopor. Mater.*, **110**, 310–317.
20 Poshusta, J.C., Noble, R.D., and Falconer, J.L. (2001) Characterization of SAPO-34 membrane by water adsorption. *J. Membr. Sci.*, **186**, 25–40.
21 Li, S., Falconer, L., and Noble, R.D. (2006) Improved SAPO-34 membranes for CO_2/CH_4 separations. *Adv. Mater.*, **18**, 2601–2603.
22 Morigami, Y., Kondo, M., Abe, J., Kita, H., and Okamoto, K. (2001) The first large-scale pervaporation plant using tubular-type module with zeolite NaA membrane. *Sep. Purif. Technol.*, **25**, 251–260.

23 Rodgers, M.P., Shi, Z., and Holdcroft, S. (2008) Transport properties of composite membranes containing silicon dioxide and Nafion®. *J. Membr. Sci.*, **325** (1), 346–356.

24 Sadeghi, M., Khanbabaei, G., Saeedi Dehaghani, A.H., Sadeghi, M., Aravand, M.A., Akbarzade, M., and Khatti, S. (2008) Gas permeation properties of ethylene vinyl acetate–silica nanocomposite membranes. *J. Membr. Sci.*, **322** (2), 423–428.

25 Sitter, K.D., Andersson, A., D'Haen, J., Leysen, R., Mullens, S., Maurer, F.H.J., and Vankelecom, I.F.J. (2008) Silica filled poly(4-methyl-2-pentyne) nanocomposite membranes: similarities and differences with poly(1-trimethylsilyl-1-propyne)–silica systems. *J. Membr. Sci.*, **321** (2), 284–292.

26 Kulprathipanja, S. and Charoenphol, J. (2004) Mixed matrix membrane for separation of gases. US Patent 6,726,744.

27 Vu, D.Q., Koros, W.J., and Miller, S.J. (2003) Mixed matrix membranes using carbon molecular sieves. I. Preparation and experimental results. *J. Membr. Sci.*, **211** (2), 311–334.

28 Vu, D.Q., Koros, W.J., and Miller, S.J. (2003) Mixed matrix membranes using carbon molecular sieves. II. Modeling permeation behavior. *J. Membr. Sci.*, **211** (2), 335–348.

29 Hu, Q., Marand, E., Dhingra, S., Fritsch, D., Wen, J., and Wilkes, G. (1997) Poly(amide-imide)/TiO_2 nano-composite gas separation membranes: fabrication and characterization. *J. Membr. Sci.*, **135** (1), 65–79.

30 Yamasaki, A., Iwatsubo, T., Masuoka, T., and Mizoguchi, K. (1994) Pervaporation of ethanol/water through a poly(vinyl alcohol)/cyclodextrin (PVA/CD) membrane. *J. Membr. Sci.*, **89**, 111–117.

31 Liu, C. and Wilson, S.T. (2008) Mixed matrix membranes incorporating microporous polymers as fillers. US patent 7410525 B1.

32 Jeong, H.-K., Krych, W., Ramanan, H., Nair, S., Marand, E., and Tsapatsis, M. (2004) Fabrication of polymer/selective-flake nanocomposite membranes and their use in gas Separation. *Chem. Mater.*, **16** (20), 3838–3845.

33 Tsapatsis, M., Jeong, H.-K., and Nair, S. (2005) Layered silicate material and applications of layered materials with porous layers. US Patent 6,863,983.

34 Zhang, Y., Musselman, I.H., Ferraris, J.P., and Balkus, K.J., Jr. (2008) Gas permeability properties of Matrimid® membranes containing the metal-organic framework Cu–BPY–HFS. *J. Membr. Sci.*, **313** (1–2), 170–181.

35 Bello, M., Javaid Zaidi, S.M., and Rahman, S.U. (2008) Proton and methanol transport behavior of SPEEK/TPA/MCM-41 composite membranes for fuel cell application. *J. Membr. Sci.*, **322** (1), 218–224.

36 Kulprathipanja, S., Funk, E.W., Kulkarni, S.S., and Chang, Y.A. (1988) Separation of a monosaccharide with mixed matrix membranes. US Patent 4735193.

37 Kulprathipanja, S., Neuzil, R.W., and Li, N.N. (1988) Separation of fluids by means of mixed matrix membranes. US Patent 4740219.

38 Hennepe, H.J.C., Bargeman, D., Mulder, M.H.V., and Smolders, C.A. (1987) Zeolite-filled silicone rubber membranes: part 1. membrane pervaporation and pervaporation results. *J. Membr. Sci.*, **35** (1), 39–55.

39 Jha, P. and Way, J.D. (2008) Carbon dioxide selective mixed-matrix membranes formulation and characterization using rubbery substituted polyphosphazene. *J. Membr. Sci.*, **324** (1–2), 151–161.

40 Mahajan, R. and Koros, W.J. (2000) Factors controlling successful formation of mixed-matrix gas separation materials. *Ind. Eng. Chem. Res.*, **39**, 2692–2696.

41 Mahajan, R. and Koros, W.J. (2002) Mixed matrix membrane materials with glassy polymers. Part 1. *Polym. Eng. Sci.*, **42**, 1420–1431.

42 Mahajan, R. and Koros, W.J. (2002) Mixed matrix membrane materials with glassy polymers. Part 2. *Polym. Eng. Sci.*, **42**, 1432–1441.

43 Husain, S. and Koros, W.J. (2007) Mixed matrix hollow fiber membranes made with modified HSSZ-13 zeolite in polyetherimide polymer matrix for gas separation. *J. Membr. Sci.*, **288**, 195–207.

44 Chung, T.S., Jiang, L.Y., Li, Y., and Kulprathipanja, S. (2007) Mixed matrix membranes (MMMs) comprising organic polymers with dispersed inorganic fillers for gas separation. *Prog. Polym. Sci.*, **32** (4), 483–507.

45 Widjojo, N., Chung, T.S., and Kulprathipanja, S. (2008) The fabrication of hollow fiber membranes with double-layer mixed-matrix materials for gas separation. *J. Membr. Sci.*, **325** (1), 326–335.

46 Ciobanu, G., Carja, G., and Ciobanu, O. (2008) Structure of mixed matrix membranes made with SAPO-5 zeolite in polyurethane matrix. *Micropor. Mesopor. Mater.*, **115** (1–2), 61–66.

47 Kulkarni, S. and Hasse, D.J. (2007) Novel polyimide based mixed matrix composite membranes. US Patent Application 2007/0199445 A1.

48 Miller, S.J., Kuperman, A., and Vu, D.Q. (2006) Mixed matrix membranes with low silica-to-alumina ratio molecular sieves and methods for making and using these membranes. US Patent 7138006 B2.

49 Hasse, D.J., Kulkarni, S., Corbin, D.R., and Patel, A.N. (2003) Mixed matrix membranes incorporating chabazite type molecular sieves. US Patent 6626980 B2.

50 Miller, S.J., Kuperman, A., and Vu, D.Q. (2007) Mixed matrix membranes with small pore molecular sieves and methods for making and using these membranes. US Patent 7166146 B2.

51 Pechar, T.W., Kima, S., Vaughan, B., Maranda, E., Tsapatsis, M., Jeong, H.K., and Cornelius, C.J. (2006) Fabrication and characterization of polyimide–zeolite L mixed matrix membranes for gas separations. *J. Membr. Sci.*, **277** (1–2), 195–202.

52 Duval, J.M. (1995) Adsorbent filled polymeric membranes. PhD thesis. The Netherlands: University of Twente.

53 Li, Y., Chung, T.S., Cao, C., and Kulprathipanja, S. (2005) The effects of polymer chain rigidification, zeolite pore size and pore blockage on polyethersulfone (PES)-zeolite A mixed matrix membranes. *J. Membr. Sci.*, **260**, 45–55.

54 Whittam, T.V. (1983) US Patent 4,397,825.

55 Gorgojo, P., Uriel, S., Téllez, C., and Coronas, J. (2008) Development of mixed matrix membranes based on zeolite Nu-6(2) for gas separation. *Micropor. Mesopor. Mater.*, **115**, (1–2), 85–92.

56 Li, Y., Chung, T.S., and Kulprathipanja, S. (2007) Novel Ag^+-zeolite/polymer mixed matrix membranes with a high CO_2/CH_4 selectivity. *AIChE J.*, **53**, 610–616.

57 Moermans, B., Beuckelaer, W.D., Vankelecom, I.F.J., Ravishankar, R., Martens, J.A., and Jacobs, P.A. (2000) Incorporation of nano-sized zeolites in membranes. *Chem. Commun.*, **24**, 2467–2468.

58 Wang, H., Holmberg, B.A., and Yan, Y. (2002) Homogeneous polymer–zeolite nanocomposite membranes by incorporating dispersible template-removed zeolite nanocrystals. *J. Mater. Chem.*, **12** (12), 3640–3643.

59 Moore, T.T. and Koros, W.J. (2005) Non-ideal effects in organic-inorganic materials for gas separation membranes. *J. Mol. Struct.*, **739** (1–3), 87–98.

60 Mahajan, R., Burns, R., Schaefer, M., and Koros, W.J. (2002) Challenges in forming successful mixed matrix membranes with rigid polymeric materials. *J. Appl. Poly. Sci.*, **86**, 881–890.

61 Moore, T.T., Mahajan, R., Vu, D.Q., and Koros, W.J. (2004) Hybrid membrane materials comprising organic polymers with rigid dispersed phases. *AIChE J.*, **50** (2), 311–321.

62 Kulkarni, S.S., David, H.J., Corbin, D.R., and Patel, A.N. (2003) Gas separation membranes with organosilicone-treated molecular sieve. US Patent 6,508,860.

63 Marand, E., Pechar, T.W., and Tsapatsis, M. (2006) Mixed matrix membranes. US Patent 7,109,140 B2.

64 Pechar, T.W., Tspatsis, M., Marand, E., and Davis, R. (2002) Preparation and characterization of a glassy fluorinated polyimide zeolite mixed matrix membrane. *Desalination*, **146**, 3–9.

65 Guiver, M.D., Robertson, G.P., Dai, Y., Bilodeau, F., Kang, Y.S., Lee, K.J., and Jho, J.Y. (2002) Structural characterization and gas-transport properties of brominated matrimid polyimide. *J. Polym. Sci. Polym. Chem.*, **40**, 4193–4204.

66 Chung, T.S., Chan, S.S., Wang, R., Lu, Z., and He, C. (2003) Characterization of permeability and sorption in Matrimid/C60 mixed matrix membranes. *J. Membr. Sci.*, **211**, 91–99.

67 Vankelecom, I.F.J., Broeck, S.V.D., Mercks, E., Geerts, H., Grobet, P., and Yutterhoeven, J.B. (1996) Silylation to improve incorporation of zeolites in polyimides films. *J. Phys. Chem.*, **100**, 3753–3758.

68 Kulkarni, S.S., Hasse, D.J., and Kratzer, D.W. (2006) Novel method of making mixed matrix membranes using electrostatically stabilized suspensions. US Patent Application 2006/0117949.

69 Husain, S. and Koros, W.J. (2007) Mixed matrix hollow fiber membranes made with modified HSSZ-13 zeolite in polyetherimide polymer matrix for gas separation. *J. Membr. Sci.*, **288** (1–2), 195–207.

70 Ismail, A.F., Kusworo, T.D., and Mustafa, A. (2008) Enhanced gas permeation performance of polyethersulfone mixed matrix hollow fiber membranes using novel Dynasylan Ameo silane agent. *J. Membr. Sci.*, **319**, 306–312.

71 Hillock, A.M.W., Miller, S.J., and Koros, W.J. (2008) Crosslinked mixed matrix membranes for the purification of natural gas: effects of sieve surface modification. *J. Membr. Sci.*, **314**, 193–199.

72 Loeb, S. and Sourirajan, S. (1963) Sea water demineralization by means of an osmotic membrane. *Adv. Chem. Ser.*, **38**, 117–132.

73 Kulprathipanja, S., Neuzil, R.W., and Li, N.N. (1992) Separation of gases by means of mixed matrix membranes. US Patent 5127925.

74 Kulkarni, S.S. and Hasse, D.J. (2005) Novel polyimide based mixed matrix membranes. US Patent Application 2005/0268782 A1.

75 Ekiner, O.M. and Kulkarni, S.S. (2003) Process for making mixed matrix hollow fiber membranes for gas separation. US Patent 6,663,805.

76 Miller, S.J., Munson, C.L., Kulkarni, S.S., and Hasse, D.J. (2002) Purification of p-xylene using composite mixed matrix membranes. US Patent 6,500,233.

77 Koros, W.J., Wallace, D., Wind, J.D., Miller, S.J., Bickel, C.S., and Vu, D.Q. (2004) Crosslinked and crosslinkable hollow fiber mixed matrix membrane and method of making the same. US Patent 6,755,900.

78 Jiang, L.Y., Chung, T.S., Cao, C., Huang, Z., and Kulprathipanja, S. (2005) Fundamental understanding of nano-sized zeolite distribution in the formation of the mixed matrix single- and dual-layer asymmetric hollow fiber membranes. *J. Membr. Sci.*, **252** (1–2), 89–100.

79 Jiang, L.Y., Chung, T.S., and Kulprathipanja, S. (2006) An investigation to revitalize the separation performance of hollow fibers with a thin mixed matrix composite skin for gas separation. *J. Membr. Sci.*, **276**, 113–125.

80 Li, Y., Chung, T.S., and Huang, Z. (2006) Dual-layer polyethersulfone (PES)/BTDA-TDI/MDI co-polyimide (P84) hollow fiber membranes with a submicron PES–zeolite beta mixed matrix dense-selective layer for gas separation. *J. Membr. Sci.*, **277**, 28–37.

81 Jiang, L.Y., Chung, T.S., and Kulprathipanja, S. (2006) A novel approach to fabricate mixed matrix hollow fibers with superior intimate polymer/zeolite interface for gas separation. *AIChE J.*, **52**, 2898–2908.

82 Jia, M., Peinemann, K.V., and Behling, R.D. (1992) Preparation and characterization of thin-film zeolite–PDMS composite membranes. *J. Membr. Sci.*, **73**, 119–128.

83 Jeong, B.-H., Hoeka, E.M.V., Yan, Y., Subramani, A., Huang, X., Hurwitz, G., Ghosha, A.K., and Jawor, A. (2007) Interfacial polymerization of thin film nanocomposites: a new concept for reverse osmosis membranes. *J. Membr. Sci.*, **294**, 1–7.

84 Duval, J.M., Folkers, B., Mulder, M.H.V., Desgrandchamps, G., and Smolders, C.A. (1993) Adsorbent-filled membranes for gas separation. Part 1. Improvement of the gas separation properties of polymeric membranes by incorporation

of microporous adsorbents. *J. Membr. Sci.*, **80**, 189–198.

85 Tantekin-Ersolmaz, S.B., Atalay-Oral, C., Tatlier, M., Erdem-Senatalar, A., Schoeman, B., and Sterte, J. (2000) Effect of zeolite particle size on the performance of polymer-zeolite mixed matrix membranes. *J. Membr. Sci.*, **175** (2), 285–288.

86 Rojey, A.R., Deschamps, A., Grehier, A., and Robert, E. (1990) Process for separation of the constituents of a mixture in the gas phase using a composite membrane. US Patent 4925459.

87 Buttal, T., Bac, N., and Yilmaz, L. (1995) Effect of feed composition on the performance of polymer-zeolite mixed matrix gas separation membranes. *Sep. Sci. Technol.*, **30** (11), 2365–2384.

88 Duval, J.M., Kemperman, A.J.B., Folkers, B., Mulder, M.H.V., Desgrandchamps, G., and Smolders, C.A. (1994) Preparation of zeolite filled glassy polymer membranes. *J. Appl. Poly. Sci.*, **54** (4), 409–418.

89 Tantekin-Ersolmaz, S.B., Senorkyan, L., Kalaonra, N., Tatlier, M., and Erdem-Senatalar, A. (2001) n-Pentane/i-pentane separation by using zeolite-PDMS mixed matrix membranes. *J. Membr. Sci.*, **189** (1), 59–67.

90 Suer, M.G., Ba, N., and Yilmaz, L. (1994) Gas permeation characteristics of polymer-zeolite mixed matrix membranes. *J. Membr. Sci.*, **91** (1–2), 77–86.

91 Li, Y., Chung, T.-S., Cao, C., and Kulprathipanja, S. (2005) The effects of polymer chain rigidification, zeolite pore size and pore blockage on polyethersulfone (PES)-zeolite A mixed matrix membranes. *J. Membr. Sci.*, **260** (1–2), 45–55.

92 Yong, H.H., Park, H.C., Kang, Y.S., Won, J., and Kim, W.N. (2001) Zeolite-filled polyimide membrane containing 2,4,6-triaminopyrimidine. *J. Membr. Sci.*, **188** (2), 151–163.

93 Goldman, M., Fraenkel, D., and Levin, G. (1989) A zeolite/polymer membrane for separation of ethanol-water azeotrope. *J. Appl. Poly. Sci.*, **37** (7), 1791–1800.

94 Chen, X., Yang, H., Gu, Z., and Shao, Z. (2001) Preparation and characterization of HY zeolite-filled chitosan membranes for pervaporation separation. *J. Appl. Poly. Sci.*, **79** (6), 1144–1149.

95 Shah, D., Kissick, K., Ghorpade, A., Hannah, R., and Bhattacharyya, D. (2000) Pervaporation of alcohol–water and dimethylformamide–water mixtures using hydrophilic zeolite NaA membranes: mechanisms and experimental results. *J. Membr. Sci.*, **179** (1–2), 185–205.

96 Kittur, A.A., Kariduraganavar, M.Y., Kulkarni, S.S., and Aralaguppi, M.I. (2005) Preparation of zeolite-incorporated poly(dimethyl siloxane) membranes for the pervaporation separation of isopropyl alcohol/water mixtures. *J. Appl. Poly. Sci.*, **96** (4), 1377–1387.

97 Bartels-Caspers, C., Tusel-Langer, E., and Lichtenthaler, R.N. (1992) Sorption isotherms of alcohols in zeolite-filled silicone rubber and in PVA-composite membranes. *J. Membr. Sci.*, **70** (1), 75–83.

98 Goldman, M., Fraenkel, D., and Levin, G. (1989) A zeolite/polymer membrane for separation of ethanol-water azeotrope. *J. Appl. Poly. Sci.*, **37** (7), 1791–1800.

12
Overview and Recent Developments in Catalytic Applications of Zeolites

Christopher P. Nicholas

12.1
History of Catalytic Uses of Zeolites

12.1.1
R&D Uses Versus Industrial Application of Zeolite Catalysis

Zeolites have been used as catalysts since the late 1950s [1, 2], particularly in the oil refining and petrochemical industry where they are prized as shape-selective catalyst [3]. Faujasites were the first molecular sieves to be applied for catalytic purposes. Today, FAU is still the most important zeolite structure type used in catalytic transformations. Though most industrial catalysts are drawn from the FAU, MFI, MOR, FER or BEA structure types, R&D groups worldwide have attempted to synthesize new structure types to improve understanding of how the zeolite structure and physical properties such as crystal morphology, silica to alumina ratio and aluminum distribution affect catalytic transformations. Towards this goal, 191 zeotypes are currently known, many of which have been synthesized in the past 10 years [4–6]. With each new structure, understanding of what properties work best for a given reaction has been advanced. In addition, the community has learned, for example, that vastly different SiO_2/Al_2O_3 ratios in the same zeotype [7] often bring about changes in catalytic performance as great as synthesizing a new zeotype via developments such as alternate syntheses of known zeotypes.

Zeolites are bases for many of the catalysts in the petroleum refining industry, including for two of the most widely used processes, fluidized catalytic cracking and hydrocracking [8, 9]. In both these processes, the upgrading of heavy feeds to lighter, cleaner fuels is accomplished by using a zeolite with the proper shape-selective, coke and poison-resistant zeolite. Because of the ubiquitous nature of these processes in refining, zeolites today are made on a large scale. Indeed, some FAU and LTA zeolites are commodity materials.

Due to their prevalence in refining processes, fine chemical transformations, and continued R&D interest, I present a literature review of the catalytic uses of zeolites in industrially relevant reactions. The information is drawn from both the open literature and patent sources covering dates through June 2008 and is largely presented in table format separated by reaction type and zeolite structure type. Zeolite structure types are as catalogued by the International Zeolite Association on their website (http://www.iza-structure.org/databases/). Prior to each reaction type is a general discussion of the reaction and insights into the various zeolites used in each reaction.

12.2
Literature Review of Recent Developments in Catalytic Uses of Zeolites

12.2.1
Isomerization Reactions

Isomerization of olefins or paraffins is an acid-catalyzed reaction that can be carried out with any number of strong acids, including mineral acids, sulfated metal oxides, zeolites and precious metal-modified catalysts [10]. Often the catalyst contains both an acid function and a metal function. The two most prevalent catalysts are Pt/chlorided Al_2O_3 and Pt-loaded zeolites. The power of zeolites in this reaction type is due to their shape selectivity [11] and decreased sensitivity to water or other oxygenates versus $AlCl_3$. It is possible to control the selectivity of the reaction to the desired product by using a zeolite with the proper characteristics [12]. These reactions are covered in more detail in Chapter 14.

12.2.1.1 Butane Isomerization
Butane isomerization is usually carried out to have a source of isobutane which is often reacted with C3–C5 olefins to produce alkylate, a high octane blending gasoline [13]. An additional use for isobutane was to feed dehydrogenation units to make isobutene for methyl *tert*-butyl ether (MTBE) production, but since the phaseout of MTBE as an oxygenate additive for gasoline, this process has declined in importance. Zeolitic catalysts have not yet been used industrially for this transformation though they have been heavily studied (Table 12.1).

12.2.1.2 Pentane/Hexane Isomerization
Pentane and hexane isomerization are largely carried out to produce high octane gasoline blend components [13]. The most highly branched isomers of these alkanes have the highest octane values, so isomerization is a valued technology to upgrade otherwise difficult to use components of naphtha. Octane values of the various C5 and C6 isomers are presented in Table 12.2 below. The reaction is thermodynamically limited, with more highly branched isomers preferred at lower temperature, so industrially the isomerization is run at as low a temperature as possible. This is often achieved by using a precious metal-loaded zeolite and carrying out a cycle of dehydrogenation, isomerization, hydrogenation. An example of this technology is UOP's TIP™ process [19]. Data on *n*-heptane isomerization

Table 12.1 Butane isomerization catalysts.

Zeolite	SiO$_2$/Al$_2$O$_3$ ratio	Metal (wt%)	Feed	H$_2$/HC	W/HSV	Temperature	Pressure	Conversion	Product	Notes	Reference
AFI		Pt (2.0)	n-C4	4	1.5	450 °C	1 atm (101 kPa)	15%	42% to iC4	SAPO-5	[14]
Beta	22	Pt (2.0)	n-C4	1	1.6	350 °C	1 atm	28%	85% to iC4		[15]
MFI	70	Pt (2.0)	n-C4	1	1.6	350 °C	1 atm	12%	66% to iC4		[15]
MFI	15	Pd (0.82)	n-C4	8.3	1	350–430	1 atm	25%	73.3% to iC4		[16]
MOR	10	Pd (0.82)	n-C4	8.3	1	350–430	1 atm	25%	65.2% to iC4		[16]
MOR	45	Pd (0.82)	n-C4	8.3	1	350–430	1 atm	25%	71.3% to iC4		[16]
MOR	45	Pd (0.82)	n-C4	8.3	1	350–430	1 atm	25%	78.2% to iC4	Bound 35/65 zeolite/bentonite	[16]
MOR	19	Pt (0.47)	n-C4	5	90	550 °C		45%	10% iC4, 13% iC4=, 25% 1-C4=, 29% t-2-C4=; 18% c-2-C4=		[17]
MOR	10	—	n-C4	—		250 °C	1 atm		5% yield iC4	2% in He	[18]
MOR	15	—	n-C4	—		250 °C	1 atm		13% yield iC4	2% in He	[18]
MOR	20	—	n-C4	—		250 °C	1 atm		7% yield iC4	2% in He	[18]
MOR	20	Pt (2.0)	n-C4	4	1.5	350 °C	1 atm	32%	53% to iC4		[14]

Table 12.2 Research and motor octane numbers (RON, MON) of C5–C7 alkane isomers.

Compound	RON	MON	(R + M)/2
n-Pentane	62	63	62
i-Pentane (2-methylbutane)	92	90	91
Neopentane (2,2-dimethylpropane)	85	80	83
n-Hexane	25	26	25
Methylcyclopentane (MCP)	91	80	86
2-Methylpentane (2MC5)	73	73	73
3-Methylpentane (3MC5)	74	74	74
2,2-dimethylbutane (22DMC4)	92	93	93
2,3-Dimethylbutane (23DMC4)	101	94	98
n-Heptane	0	0	0
2-Methylhexane (2MC6)	42.4	46.4	44
2,3-Dimethylpentane	91.1	88.5	90
2,2,3-Trimethylbutane	112	101.3	106

is also included as this is a test reaction often used on new or unusual zeolites (Table 12.3).

12.2.1.3 C10+ Paraffin Isomerization

Paraffin isomerization of heavy alkane feeds is often used to alter the cloud or pour point of diesel or lube fractions. Catalysts for this reaction are almost always dual-function catalysts of Pt supported on a one-dimensional zeolite. Using a one-dimensional zeolite allows control of the isomerized product to contain few branches, usually methyl branches (Table 12.4).

12.2.1.4 Light Olefin Isomerization

The isomerization of light olefins is usually carried out to convert n-butenes to isobutylene [12] with the most frequently studied zeolite for this operation being FER [30]. Lyondell's IsomPlus process uses a FER catalyst to convert n-butenes to isobutylene or n-pentenes to isopentene [31]. Processes such as this were in larger demand to generate isobutene before the phaseout of MTBE as a gasoline additive. Since the phaseout, these processes often perform the reverse reaction to convert isobutene to n-butenes which are then used as a metathesis feed [32]. As double-bond isomerization is much easier than skeletal isomerization, most of the catalysts below are at equilibrium ratios of the n-olefins as the skeletal isomerization begins (Table 12.5).

12.2.2
Oligomerization Reactions

12.2.2.1 Light Olefin Oligomerization

The oligomerization of propylene and butenes to gasoline range olefins is a technology which began in the 1930s with the work of Ipatieff on phosphoric acid in

Table 12.3 Zeolitic pentane and hexane isomerization catalysts.

Zeolite	SiO$_2$/Al$_2$O$_3$ ratio	Metal (wt%)	Feed	H$_2$/HC	WHSV	Temperature	Pressure	Conversion	Product	Notes	Reference
AEL		Pt(0.4)	nC6	5	1	300 °C	1 atm	65	82.5% isom, 17.5% cracking		[20]
ATO		Pt(0.4)	nC6	5	1	325 °C	1 atm	63	85% isom, 15% cracking		[20]
BEA	25	Pt (0.5)	nC6	4–8	3.7	275 °C	1 atm	69	25.5 mol% 2MC5, 16.5%3MC5, 8% 22DMC4; 8% 23DMC4	11% cracking products	[21]
FAU/EMT	10	Pt(0.18)	nC5			350 °C	1 atm	45	35% yield isopentane	2:1 H$_2$/N$_2$ saturated in pentane at −3 °C	[22]
FAU	9.5	Pt(0.11)	nC5			350 °C	1 atm	28	25% yield isopentane	2:1 H$_2$/N$_2$ saturated in pentane at −3 °C	[22]
FAU	3.4	Pt (1.0)	nC7		Autoclave nC7/cat=10	260 °C	500 psig	77	71% isom, 6% cracking	X zeolite	[23]

Table 12.3 Continued

Zeolite	SiO$_2$/Al$_2$O$_3$ ratio	Metal (wt%)	Feed	H$_2$/HC	WHSV	Temperature	Pressure	Conversion	Product	Notes	Reference
FAU	5.8	Pt (0.5)	nC7		Autoclave nC7/cat=10	270 °C	500 psig	84	70% isom, 14% cracking	Y zeolite	[23]
FAU	60	Pt (0.5)	nC6	4–8	3.7	275 °C	1 atm	14.5	7 mol% 2MC5, 5% 3MC5, 0.5% 22DMC4; 1% 23DMC4	0.3% cracking products	[21]
FER	55	Pt (0.5)	nC6	4–8	3.7	275 °C	1 atm	4	3 mol% 2MC5, 1% 3MC5		[21]
MFI	80	Pt (0.5)	nC6	4–8	3.7	275 °C	1 atm	51	31.5 mol% 2MC5, 16% 3MC5, 0.5% 22DMC4; 1% 23DMC4	1.1% cracking products	[21]
MOR	20	Pt (0.5)	nC6	4–8	3.7	275 °C	1 atm	60.5	23.5 mol% 2MC5, 14.5% 3MC5, 7.5% 22DMC4; 7% 23DMC4	8.5% cracking products	[21]
MOR	90	Pt (0.5)	nC6	4–8	3.7	275 °C	1 atm	45	22 mol% 2MC5, 14% 3MC5, 1.5% 22DMC4; 5.5% 23DMC4	2% cracking products	[21]

Table 12.4 Isomerization of heavy paraffins.

Zeolite	SiO$_2$/Al$_2$O$_3$ ratio	Metal (wt%)	Feed	H$_2$/HC	LHSV	Temperature	Pressure	Yield	Product	Notes	Reference
BEA	23	Pt (1.0)	n-C10	20		220°C	30 bar (3 MPa)	57%	39% monobranch, 15% multibranch, 3% cracking		[24]
FAU	2.8	Pt (0.5)	n-C10	100	0.3	250°C	0.35 MPa	45%	Branched C10		[25]
FAU/TON	2.8/45	Pt (0.5)	n-C10	100	0.3	230°C	0.35 MPa	78%	Branched C10		[25]
MCM-22 (MWW)	34	Pt (1.0)	n-C10	20		220°C	30 bar	51%	24% monobranch, 18% multibranch, 9% cracking		[24]
MFI	50	Pt (1.0)	n-C10	20		220°C	30 bar	52%	30% monobranch, 3% multibranch, 19% cracking		[24]
MRE	200	Pt(0.6)	n-C10			240°C	1 atm (101 kPa)	81%	Iso-C10	ZSM-48	[26]
MRE	90	Pt(0.6)	n-C10			240°C	1 atm	84%	Iso-C10	ZSM-48	[26]
MRE	200	Pt(0.6)	600 N slack wax	2500	1	315°C	1000 psig (895 kPa)		370C+ pour point	ZSM-48 a) sccf/bbl	[26]
MTT		Pt (0.325)	500 N hydrocrackate; PP = 51°C	4000	1		2300 psig	97%	Viscosity index difference =2	SSZ-32X a) sccf/bbl	[27]
SFH	110	Pt (0.27)	n-C10	100	0.35 WH	220°C	1 atm	10%	10% cracking, 10% isom ~75% of isom is 2-Me C9	SSZ-53	[28]
SSZ-25 (MWW)	30	Pt (1.0)	Hydrocracked north slope VGO, pour point 100°F (63°C)			343°C	2150 psig		Pour point 10°F, Lube yield 69%		[29]
TON	45	Pt (0.5)	n-C10	100	0.3	220°C	0.35 MPa	68%	Branched C10		[25]

Table 12.5 Isomerization of light olefins.

Zeolite	SiO$_2$/Al$_2$O$_3$ ratio	Feed	SV	Temperature	Pressure	Conversion	Product	Notes	Reference
AEL		2-butenes	75 WHSV	530 °F	15 psia	29.4%	98.3% to 1-butene		[33]
ATN		1-pentene	1 LHSV	400 °F	12 psig	65.5%	67% to t-2-C5=; 27% to c-2-C5=; 2% to MC4=	N$_2$/HC=2.6 SAPO-39	[34]
CHA		1-pentene	1 LHSV	450 °F	12 psig	66%	58% to t-2-C5=; 38% to c-2-C5=; 4% to MC4=	N$_2$/HC=2.6 SAPO-34	[34]
CHA	0.32	2-butenes	60 WH	480 °F	40 psia	27%	96.7% to 1-butene	SAPO-34	[33]
CLI	11.5	1-butene	30000GH	400 °C	1 atm	20%	20% to iC4=	10.1 kPa butene in He	[35]
FER	17.5	1-butene	30000GH	400 °C	1 atm	42%	83% to iC4=	10.1 kPa butene in He	[35]
FER		2-butenes	1500 WHSV	480 °F	40 psia	34.5%	94.6% to 1-butene	Commercial catalyst partially Ca exchanged	[33]
FER	27	1-butene	20.6	350 °C	1 atm	14.6%	19.4% C$_3$H$_6$, 61.2% iC4=, 4.5% n-C$_4$H$_{10}$, 10.7% C5=		[36]
FER	17	1-butene		400 °C	1 atm	42%	81% to iC4=	10.1 kPa butene	[7a]

MFI	27	1-butene	30000GH	400°C	1 atm	88%	12% to iC4=	10.1kPa butene in He	[35]
MFS	49	1-butene	30000GH	400°C	1 atm	80%	20% to iC4=	10.1kPa butene in He	[35]
MTT	120	1-butene	5.34 WH	420°C	1 atm	80.7%	21.4% to iC4=	50% butene, 50% N_2	[37]
MTT	120	1-butene	21.4 WH	420°C	1 atm	65.7%	46.1% to iC4=	50% butene, 50% N_2	[37]
MTT	120	1-butene	85.4 WH	420°C	1 atm	45.2%	64.6% to iC4=	50% butene, 50% N_2	[37]
STI	14	1-butene		400°C	1 atm	50%	39% to iC4=	10.1kPa butene	[7a]
TON	67	1-butene	30000GH	400°C	1 atm	58%	40% to iC4=	10.1kPa butene in He; ZSM-22	[35]
TON	16	1-butene	30000GH	400°C	1 atm	32%	67% to iC4=	10.1kPa butene in He; SUZ-4	[35]
TON	28	1-butene	50 WHSV	400°C	1 atm	45%	32% iC4, 6% C4+, 6.5% cracking	0.42 atm butene	[38]
TON	28	1-butene	150 WHSV	400°C	1 atm	46%	60% to iC4=	K-form	[39]
TON	28	1-butene	150 WHSV	400°C	1 atm	45.5%	59% to iC4=	H-form	[39]

liquid [40] and solid [41] forms. The reaction is a simple acid-catalyzed dimerization of two olefins to form a longer olefin. More detail on the mechanism of oligomerization is provided in Chapter 15.

Following the discovery of zeolites, zeolites were used as solid acids to catalyze the oligomerization. The first large-scale use of the MFI structure type was to catalyze this reaction [42]. The process was named the Mobil Oil to Gasoline and Diesel (MOGD) and used ZSM-5 as the acid catalyst to oligomerize a propylene- and butene-containing feed into a liquid product containing gasoline and diesel. Since then, many other zeolites have been used. One of the new developments is Neste Oil's development of an isobutene dimerization process called the NExOCTANE process [43] (Table 12.6).

12.2.2.2 Heavier Olefin Oligomerization

The oligomerization of heavier olefins like the relatively non-valuable C5 and C6 olefins into diesel or lube products has been studied multiple times over the years, but has yet to be industrially implemented. Conditions and catalyst determine the product selectivity for this reaction, so both need to be optimized for the particular product molecular weight desired (Table 12.7).

12.2.3
Alkylation Reactions

12.2.3.1 Alkylation of Isobutane

The alkylation of isobutane with C3–C5 olefins to form isoalkane fractions called alkylate is a very important refining process to generate high octane value gasoline. Industrially, the reaction is currently carried out in HF- or H_2SO_4-catalyzed processes, but due to the toxicity of HF, the corrosiveness of H_2SO_4 and the general desire to move to a heterogeneously catalyzed process, many zeolites are studied for this reaction [49]. Industrial technology for this transformation is covered in Section 15.3. Conversion data is not presented below as all selectivity data is from the initial time on stream period of 100% olefin conversion.

Faujasites of the X and Y type are the most frequently studied zeolite structure type for this reaction. Because the key step in the reaction is a hydride transfer, zeolites with low SiO_2/Al_2O_3 ratios are favored. The other preferred characteristic is a large pore opening and hence a low diffusion barrier to product diffusion. These details and others were reviewed recently by Feller *et al.* [50] (Table 12.8).

12.2.3.2 Benzene Alkylation

The reaction of benzene with ethylene or propylene to form ethylbenzene or isopropylbenzene (cumene) is an industrially important transformation, with ethylbenzene as the key building block for polystyrene and cumene as the feedstock for phenol production [55]. Ethylbenzene was originally produced with a Lewis acid catalyst consisting of $AlCl_3$ or a Brønsted acidic solid phosphoric acid (SPA) catalyst [56]. Both catalyst systems suffered from equipment corrosion so, in the 1980s the Mobil–Badger vapor phase alkylation process was introduced, which

Table 12.6 Oligomerization of light olefins.

Zeolite	SiO$_2$/Al$_2$O$_3$ ratio	Feed	SV	Temperature	Pressure	Conversion	Selectivity	Notes	Reference
AEL		12% C3=, 88% C3P		200 °C	6.8 MPa	60%	54% dimer, 26% trimer, 13% tetramer, 7% cracking	28% silico-aluminate	[44]
AEL		16 wt% 2-C4=, 78% C3P, 6% C5P	0.56 WH	140 °C	6 MPa	73%	1.4% C6-7, 61% C8, 8% C9-11, 25% C12, 5% C13+		[45]
Beta	25	16 wt% 2-C4=, 78% C3P, 6% C5P	0.56 WH	120 °C	6 MPa	79%	1% C6-7, 51% C8, 14% C9-11, 29% C12, 5% C13+		[45]
FER	13	16 wt% 2-C4=, 78% C3P, 6% C5P	0.56 WH	140 °C	6 MPa	66%	1.5% C6-7, 69.5% C8, 7.5% C9-11, 20% C12, 1.5% C13+		[45]
MEL	90	16 wt% 2-C4=, 78% C3P, 6% C5P	0.56 WH	140 °C	6 MPa	90%	2% C6-7, 56.5% C8, 11% C9-11, 27% C12, 3.5% C13+		[45]
MOR	13	2% C3=, 2% C3P, 3% iC4P, 11% nC4P, 65% 1-C4=, 9% iC4=, 8% 2-C4=	3.5 WH	150 °C	51 bar	45%	76% dimer, 21% trimer, 3% tetramer		[46]
MRE	100	16 wt% 2-C4=, 78% C3P, 6% C5P	0.56 WH	110 °C	6 MPa	82%	1% C6-7, 60.5% C8, 10% C9-11, 25% C12, 4% C13+		[45]
MRE	100	12% C3=, 88% C3P		200 °C	6.8 MPa	80%	30% dimer, 31% trimer, 15% tetramer, 25% cracking	ZSM-48	[44]

Table 12.6 Continued

Zeolite	SiO$_2$/Al$_2$O$_3$ ratio	Feed	SV	Temperature	Pressure	Conversion	Selectivity	Notes	Reference
MFI	25	50% C$_2$H$_4$ 50% N$_2$	18.7 WH	400 °C	1 atm	90%	40% to gasoline (C5-12)		[47]
MFI	50	16 wt% 2-C4= 78% C3P, 6% C5P	0.56 WH	160 °C	6 MPa	72%	6% C6-7, 57% C8, 14% C9-11, 23.5% C12, 0% C13+		[45]
MFS	55	16 wt% 2-C4= 78% C3P, 6% C5P	0.56 WH	80 °C	6 MPa	89%	86% C8, 1% C9-11, 13% C12		[45]
MFS	55	7 wt% C3=, 7% 1-C4=, 76% iC4P	1.1 WH	150 °C	6 MPa	85%	1% C6, 4% C7, 24% C8, 23% C9, 29% C10, 8% C11, 7% C12, 4% C13+		[45]
MTT	89	12% C3= 88% C3P		200 °C	6.8 MPa	80%	60% dimer, 25% trimer, 15% tetramer		[44]
MWW	90	16 wt% 2-C4= 78% C3P, 6% C5P	0.56 WH	160 °C	6 MPa	95%	6% C6-7, 52% C8, 22% C9-11, 19.5% C12, 0% C13+	MCM-22	[45]
TON	84	12% C3= 88% C3P		200 °C	6.8 MPa	80%	71% dimer, 22.5% trimer, 6.5% tetramer		[44]
TON	60	16 wt% 2-C4= 78% C3P, 6% C5P	0.56 WH	150 °C	6 MPa	82%	1% C6-7, 57.5% C8, 4% C9-11, 30.5% C12, 7% C13+		[45]

Table 12.7 Oligomerization of C5 and C6 olefins.

Zeolite	SiO$_2$/Al$_2$O$_3$ ratio	Metal (wt%)	Feed	SV	Temperature	Pressure	Conversion	Selectivity	Notes	Reference
Beta	25		1-Pentene	1.8 WHSV	75 °C		26%	3% C9, 58% C10, 39% C12+	Bentonite bound	[48]
Beta	25	Ni (17.6)	1-Pentene	2.2 WHSV	75 °C		31%	2% C9, 81% C10, 17% C12+	Bentonite bound	[48]
MFI	50		1-Pentene	1.8 WHSV	75 °C		17%	2.5% C9, 85% C10, 12.5% C12+	Bentonite bound	[48]
MFI	90		1-Pentene	1.8 WHSV	75 °C		7.1%	1% C9, 79% C10, 20% C12+	Bentonite bound	[48]
MFI	90		1-Pentene	1.4 WHSV	125 °C		26%	1% C6-C8, 2% C9, 86% C10, 11% C12+	Bentonite bound	[48]
MFS	55		7 wt% 1-C4=, 10% 1-C5=, 10% C3P, 76% iC4P	0.35 WHSV	150 °C	6 MPa	88%	9% C8, 36% C9, 37% C10, 12% C12, 5% C13+		[45]
MOR	40		1-Pentene	1.8 WHSV	75 °C		55%	2% C9, 80% C10, 18% C12+	Bentonite bound	[48]

Table 12.8 Alkylation of isobutane with C4 olefins.

Zeolite	SiO$_2$/Al$_2$O$_3$ ratio	iC4/olefin ratio	Olefin	WHSV (olefin)	Temperature	Pressure	Selectivity	TMP/C8	Notes	Reference
Beta	26	5	1-Butene	0.11	75 °C	40 bar	70% C8	80%		[51]
Beta	27	107	2-Butene	0.33	140 °C	8 MPa	5% C3-C4, 35% C5-C7, 53% C8, 7% C9+	28%	supercritical iC4 50% DMH/C8	[52]
FAU	8	107	2-Butene	0.33	140 °C	8 MPa	10% C3-C4, 42% C5-C7, 48% C8, ~0% C9+	57%	USY, supercritical iC4 38% DMH/C8	[52]
FAU	6	5	1-Butene	0.11	75 °C	40 bar	82% C8	90%		[51]
FAU	2.6	20	2-Butene	0.5	60 °C	1.1 × 10^7 Pa	8.2% C5P, 6.8% C6P, 6.8% C7P, 76% C8P, 2.2% C9+	77%	USY	[53]
FAU	2.1	10	2-Butene	0.2	75 °C	20 bar	5% rC4, 21% C5-C7P, 74% C8P, ~0% C9+	91%	La X; 25.0 Å cell size	[54]
FAU	4.6	10	2-Butene	0.2	75 °C	20 bar	8% rC4, 19% C5-C7P, 70% C8P, ~3% C9+	67%	La Y; 24.8 Å cell size	[54]

utilizes MFI zeolite as the catalyst. Liquid-phase alkylation flowschemes such as UOP's EBOne process were introduced starting in the 1990s, so operation in either gas or liquid phase is possible.

Cumene was originally produced with SPA- [57], then FAU- or BEA-based catalysts, and most recently MWW. While most industrial processes use MWW-based catalysts [58], Dow and Kellog co-developed a dealuminated MOR based process called 3-DDM [59]. With each new process generation, conversion and selectivity to cumene has increased. These processes and the chemistry behind them are covered in Section 15.4. As the use of zeolites for alkylation reactions in industry increased, so did the study of the reaction and how the zeolite topology affects the mechanism and selectivity to products, so that now many zeotypes are tested for aromatic alkylation as a way of figuring out a new structure's reaction pattern. Therefore, many zeotypes have been used to catalyze aromatic alkylation (Tables 12.9–12.11).

12.2.4
Aromatics Reactions

Isomerization and transalkylation reactions to redistribute methyl groups on aromatic molecules are important processes in the production of benzene, toluene and xylenes (BTX). In particular, the production of *para*-xylene is preferred. The interconversion of C8 aromatics is covered in much greater depth in Section 14.3.

12.2.4.1 Transalkylation of Toluene to Xylene and Benzene
Processes such as UOP's Tatoray [2] process are used to increase xylene yield in an aromatics facility by converting two moles of toluene into one mole of xylenes and one mole of benzene. As the reaction is thought to be a bimolecular process, zeolites with side pockets or channel intersections are the most desired (Table 12.12).

12.2.4.2 Xylene and Ethylbenzene Isomerization
Processes such as UOP's Isomar process are used to carry out isomerization of C8 aromatic species so that *p*-xylene can be removed selectively from the mixture of xylenes. The reaction is equilibrium controlled, so a continuous isomerization process is used. As is seen below, both aluminophosphate and aluminosilicate zeotypes are capable of catalyzing the reaction (Table 12.13).

12.2.5
Chain-Breaking Reactions

12.2.5.1 Use of Zeolites in FCC Type Feeds
Thermal cracking was the first crude oil refining process to contribute more useful products from a barrel of oil than straight distillation [80]. Houdry's development of catalytic cracking shortly before World War II spurred then Standard Oil of New

Table 12.9 Ethylbenzene formation.

Zeolite	SiO_2/Al_2O_3 ratio	C_6H_6/olefin ratio	SV	Temperature	Pressure	C_2H_4 conversion	EB selectivity	DIEB selectivity	Other selectivity	Notes	Reference
Beta	27	8	1.2 WHSV	240 °C	3.5 MPa	87.2	88.9	10	1.1		[60]
Beta		7	4.6 WHSV	160 °C	3.1 MPa	100	91.1	7.9	1.0		[55]
FAU		7	4.6 WHSV	160 °C	3.1 MPa	100	82.0	8.4	9.5	Oligomers	[55]
MCM-22	30	8	1.2 WHSV	240 °C	3.5 MPa	85.7	88.3	11	0.7		[60]
MFI	36	8	1.2 WHSV	240 °C	3.5 MPa	31.3	42.8	3.5	51.3	Oligomers	[60]
SSZ-25	30	7	5.6 WHSV	205 °C	450 psig	100	92.4	6.6	1		[61]

Table 12.10 Cumene formation.

Zeolite	SiO$_2$/Al$_2$O$_3$ ratio	C$_6$H$_6$/olefin ratio	SV	Temperature	Pressure	Conversion	Cumene selectivity	DIPB selectivity	n-C3 C$_6$H$_6$ selectivity	Notes	Reference
Beta	27	3.5	18	125 °C	3.5 MPa	99%	87.4	12.4	1056 ppm		[62]
Beta	5	7.0	1 WHSV	150 °C	3.8 MPa	93%	93.3	6.6	190 ppm		[63]
Beta	27	7.2	2.6 WHSV	180 °C	3.5 MPa	97.3%	90.8	8.3	900 i/n ratio		[60]
ERB-1	7.5	7.0	WHSV	150 °C	3.8 MPa	95%	90.7	8.8	277 ppm		[63]
FAU	1.5	7.0	WHSV	150 °C	3.8 MPa	92%	77.6	21.5	140 ppm		[63]
ITQ-2											[64]
ITQ-21	25	3.5	18	125 °C	3.5 MPa	99%	88.1	11.9	351 ppm	Si/Ge=20	[62]
MCM-22	30	7.2	2.6 WHSV	180 °C	3.5 MPa	98%	90.6	8.8	830 i/n ratio		[60]
MOR	5	7.0		150 °C	3.8 MPa	92%	86.6	12.8	107 ppm		[63]
MTW	22	7.0	WHSV	150 °C	3.8 MPa	95%	94.3	4.6	406 ppm		[63]
SSZ-25	30	7.8	5.7 WHSV	163 °C	600 psig	100%	90.1	7.7	400 ppm		[61]

Table 12.11 Alkylation of biphenyl with propylene.

Zeolite	SiO$_2$/Al$_2$O$_3$ ratio	Biphenyl/olefin Ratio	SV	Temperature	Pressure	Conversion	i-Pr BP selectivity	Di-i-Pr BP selectivity	o-/m-/p-i-PrBP ratio	Notes	Reference
AFI		~0.001	N/A	250 °C	0.8 MPa	77%	36	61	~0/46/54	MgAPO-5 autoclave	[65]
AFI	172	~0.001	N/A	180 °C	0.8 MPa	88%	36	50	10/60/30	Al-SSZ-24 autoclave	[66]
AFI	768	~0.001	N/A	300 °C	0.8 MPa	88%	31	65		Al-SSZ-24 autoclave	[66]
ATS	73	~0.001	N/A	250 °C	0.8 MPa	97%	9	32		59% tri-iPr BP autoclave	[67]
Beta	30	4		250 °C	1 atm	4%	86	14.4	2.6/57/40.6		[68]
CON	39	4		250 °C	1 atm	4%	90	10	2.3/65/32.7	SSZ-26	[69]
FAU	10	4		250 °C	1 atm	4%	94.8	5.2	3/62.5/34.4	20% in heptane	[69]
IFR	280	~0.001	N/A	250 °C	0.8 MPa	74%	81	19		Autoclave	[67]
ITQ-2	100	4		250 °C	1 atm	4%	89.8	10.2	10.6/48/41.6	20% in heptane	[69]
MCM-22	100	4		250 °C	1 atm	4%	82	18	12.7/43/44	20% in heptane	[69]
MOR	25	~0.001	N/A	250 °C	0.8 MPa	55%	64	36		Autoclave	[70]
MOR	110	~0.001	N/A	250 °C	0.8 MPa	77%	32	65		Autoclave	[70]
MSE	70	~0.001	N/A	250 °C	0.8 MPa	77%	50	43	4,4'-DIBP 62% of DIBP	Autoclave	[71]
SSZ-25	59	4		250 °C	1 atm	4%	78	22	9.2/44/47		[61]
STF	45	~0.001	N/A	250 °C	0.8 MPa	40%	80	20		Autoclave	[67]

Table 12.12 Toluene transalkylation.

Zeolite	SiO$_2$/Al$_2$O$_3$ ratio	Feed	Metal (wt%)	SV	Temperature	Pressure	Conversion/selectivity	pX/X	Notes	Reference
Beta	25	50% Tol, 16% ethyl toluene, 28% trimethyl benzene, 6% A10+; H$_2$/Tol=8.5	Re (0.5)	4 WSHV	400°C	25 bar	58.9% conv.; 33.6% xylenes + C$_6$H$_6$, 16.1% lights, 5.6% heavy		80/20 with Al$_2$O$_3$	[72]
IMF	24	50% Tol, 16% ethyl toluene, 28% trimethyl benzene, 6% A10+; H$_2$/Tol=8.5	Re (0.25)	4 WSHV	400°C	25 bar	45.8% conv.; 33.1% xylenes + C$_6$H$_6$, 4.6% lights, 4.3% heavy		80/20 with Al$_2$O$_3$	[72]
ITH	100	Toluene; H$_2$/Tol=8.5		1	400°C 460°C	30 bar	47.9% conv.; 23.5% C$_6$H$_6$, 21.1% xylenes 53.1% conv., 25.3% C$_6$H$_6$, 22.2% xylenes			[73]
MFI	35	50% Tol, 16% ethyl toluene, 28% trimethyl benzene, 6% A10+; H$_2$/Tol=8.5	Re (0.5)	4 WSHV	400°C	25 bar	40.1% conv.; 23.8% xylenes + C$_6$H$_6$, 9.5% lights, 2% heavy		80/20 with Al$_2$O$_3$	[72]
MFI	35	Toluene; H$_2$/Tol=8.5		1	400°C	30 bar	23.5% conv.; 11.3% C$_6$H$_6$, 10.6% xylenes			[73]
MOR	20	Toluene; H$_2$/Tol=8.5		1	400°C	30 bar	39.1% conv.; 19.8% C$_6$H$_6$, 17.1% xylenes			[73]

Table 12.12 Continued

Zeolite	SiO$_2$/Al$_2$O$_3$ ratio	Feed	Metal (wt%)	SV	Temperature	Pressure	Conversion/selectivity	pX/X	Notes	Reference
MOR	20	50% Tol, 16% ethyl toluene, 28% trimethyl benzene, 6% A10+; H$_2$/Tol=8.5	Re (0.5)	4 WSHV	400 °C	25 bar	56.9% conv.; 38.1% xylenes + C$_6$H$_6$, 9.5% lights, 3.1% heavy		80/20 with Al$_2$O$_3$	[72]
MOR	96	20% Tol; 80% A9+ feed of 83% A9, 17% A10; H$_2$/HC=5.0	Re (0.29)		400 °C	30 barg	53% conv.; 5.8% C$_6$H$_6$, 33.3% xylenes, 12.5% light		79.7% MOR Presulfided w/DMDS	[74]
MOR	33.5	16.5 kPa tol in N$_2$		7.2 WH	300 °C	1 atm	25.8% conv.; 12.3% C$_6$H$_6$, 12.6% xylenes	24.1		[75]
MWW	33	16.5 kPa tol in N$_2$		7.2 WH	300 °C	1 atm	45% conv.; 21.8% C$_6$H$_6$, 23.4% xylenes	26.8	MCM-22	[75]
NES	69	20% Tol; 80% A9+ feed of 83% A9, 17% A10; H$_2$/HC=5.0	Re (0.31)		400 °C	30 barg	53% conv.; 5.5% C$_6$H$_6$, 34.4% xylenes, 11.8% light		69.8% NU-87 Presulfided w/DMDS	[74]
NES	67	50% Tol, 16% ethyl toluene, 28% trimethyl benzene, 6% A10+; H$_2$/Tol=8.5	Re (0.25)	4 WSHV	400 °C	25 bar	50.3% conv.; 37.8% xylenes + C$_6$H$_6$, 4.6% lights, 4.3% heavy		80/20 with Al$_2$O$_3$	[72]
TUN	40	16.5 kPa tol in N$_2$		7.2 WH	300 °C	1 atm	38.1% conv.; 18.4% C$_6$H$_6$, 17% xylenes	24.8		[75]

Table 12.13 C8 aromatic isomerization.

Zeolite	SiO$_2$/Al$_2$O$_3$ ratio	Feed	SV	Temperature	Pressure	Conversion/selectivity	pX/X	Reference
AEL		m-Xylene	3 WHSV	350 °C	1 atm	15% MX conv.; 0.2% Tol, 9.4% PX, 85% MX, 4.8% OX, 0.3% C9A		[76]
AFI		m-Xylene	3 WHSV	350 °C	1 atm	73% MX conv.; 1.2% C$_6$H$_6$, 19.9% Tol, 15.5% PX, 27% MX, 15.2% OX, 20.9% C9A	26.8	[76]
AFI		o-Xylene	pulse expt.	400 °C	1 atm	33% OX conv.; 0.5% C$_6$H$_6$, 2% Tol, 12.4% PX, 15% MX, 67% OX, 3% C9A	13.0	[77]
ATO		m-Xylene	3 WHSV	350 °C	1 atm	13% MX conv.; 0.5% Tol, 7.1% PX, 87% MX, 5.1% OX, 0.5% C9A		[76]
Beta	100	m-Xylene	3.4 WHSV	350 °C	1 atm	22.1% MX conv.; 14.1% Tol, 39.6% PX, 32.8% OX, 13.6% HA		[78]
ITH	100	m-Xylene	3.1 WHSV	350 °C	1 atm	22.8% MX conv.; 1.3% Tol, 53.1% PX, 44.6% OX, 1.1% HA		[78]
MFI	80	m-Xylene	3.6 WHSV	350 °C	1 atm	21.1% MX conv.; 1.4% Tol, 66.8% PX, 31.4% OX, 0.3% HA		[78]

Table 12.13 Continued

Zeolite	SiO$_2$/Al$_2$O$_3$ ratio	Feed	SV	Temperature	Pressure	Conversion/selectivity	pX/X	Reference
MFI	27	m-Xylene	4.1 WHSV	350 °C	1 atm	20.6% MX conv.; 1.0% Tol, 66% PX, 30.7% OX, 2.3% HA		[75]
MFI/MRE		60% EB, 27% MX, 13% OX; H$_2$/HC=4		400 °C	200 psig	60% EB conversion	29.0	[79]
MOR	33.5	m-Xylene	4.7 WHSV	350 °C	1 atm	18.8% MX conv.; 20% Tol, 35% PX, 30% OX, 15% HA		[75]
MCM-22 (MWW)	33	m-Xylene	2.4 WHSV	350 °C	1 atm	19% MX conv.; 4.7% Tol, 61% PX, 32.8% OX, 1.5% HA		[75]
SSZ-25 (MWW)	30	1.3% Tol, 9.8% EB, 9.6% PX, 54% MX, 23.1% OX, 1.8% HA	5 WHSV	400 °C	23 psig	1% non-aromatic, 2% C$_6$H$_6$, 3.7% Tol, 6.4% EB, 19.4% PX, 43% MX, 20% OX, 4.5% HA	23.5	[29]
NES	35	m-Xylene	21.4 WHSV	350 °C	1 atm	26.9% MX conv.; 14.9% Tol, 39.7% PX, 31.3% OX, 14% HA		[78]
TUN	40	m-Xylene	3 WHSV	350 °C	1 atm	22% MX conv.; 15% Tol, 58.6% PX, 16.4% OX, 10% HA		[75]

Jersey to develop fluidized catalytic cracking (FCC) as a replacement for thermal cracking [81]. The FCC has become a mainstay of the oil refining industry, particularly as the main gasoline producing unit. Many variants of FCC exist [82], but all FCC units exist to convert heavy feeds to lighter products. The products can be varied to favor kerosene or diesel fractions, gasoline, or today, even the production of propylene. The industrial application of FCC chemistry is covered in Section 16.4. All catalysts for these processes are based on FAU zeolites with a combination of binders and other solid acids such as silica–alumina or clays. Many catalysts today contain MFI additives to favor the production of propylene. Though FAU- and MFI-based catalysts are the only industrially used zeolites for FCC, many new zeolites are tested in FCC-type conditions to assess their viability as FCC additives (to complement/replace the MFI component) or as FCC catalyst bases (to replace the FAU component), so I have referenced tests of many other zeolites to allow comparison against the standard zeotypes (Table 12.14).

12.2.5.2 Olefin Cracking

Olefin cracking has been developed as a process to produce propylene in a highly selective manner from butenes and pentenes. Zeolites used in processes such as UOP's Olefin Cracking Process are often MFI-based in order to avoid coke build-up during the reaction, leading to longer times between catalyst regeneration (Table 12.15).

12.2.6
Dehydroaromatization

The dehydroaromatization of light alkane feeds (methane to butanes) into aromatics has come into prominence as a method of converting the unreactive light paraffins into useful chemical precursors. In many of the world's markets, light alkanes are very undesired off-gasses which can not be used other than as fuel. To accomplish this difficult transformation, catalysts typically are bifunctional, containing a dehydrogenating component such as Pt, Ga, Zn or Mo with an acidic zeolite. MFI is typically the preferred zeolite due to its coking resistance, but other zeolites have been used, as detailed in the tables below. The mechanism of the dehydroaromatization is covered in greater detail in Chapter 15, as are representative processes such as UOP's Cyclar process [2, 91].

12.2.6.1 Light Alkanes to Aromatics
Table 12.16

12.2.6.2 Methane to Aromatics
After commercial success with the propane/butane to aromatics processes, attempts were made to extend these catalysts to convert methane to aromatics. In 1993, Wang et al. [97] discovered that Mo/MFI is capable of catalyzing the difficult reaction. Almost all successful catalysts for this transformation are based on this bifunctional basis. Both strong acid sites and Mo are necessary to convert CH_4

Table 12.14 Fluidized catalytic cracking catalysts.

Zeolite	SiO$_2$/Al$_2$O$_3$ ratio	Catalyst/oil ratio	Feed	Temperature	Pressure	Conversion	Light selectivity	Gasoline selectivity	Diesel selectivity	Notes	Reference
BEA		0.35	Arabian light VGO	500°C	1 atm	53.9	28.4 (6% C$_3$H$_6$)	28	9		[83]
FAU		0.35	Arabian light VGO	500°C	1 atm	68	14.7 (3.5% C$_3$H$_6$)	39	14		[83]
FAU		30.0	Arabian light VGO	600°C	98 kPa	80	31.4 (10.7% C$_3$H$_6$)	45	16	3% coke	[84]
FAU		1.25	Arabian light VGO	520°C	1 atm	81	31 (5.5% C$_3$H$_6$)	45	11.5	3.5% coke	[85]
FAU	40	1.13	VGO; 79.2°C Aniline Point		1 atm	75	15.6 (3.4% C$_3$H$_6$)	39.2	13.1	2% coke	[86]
FAU	5.4	6.6	VGO; 139°F Aniline Point	524°C	1 atm	70	20.6	49.5	29.9	40% zeolite	[87]
FAU/Beta	5.4/35	7	VGO; 139°F Aniline Point	524°C	1 atm	70	22.8	47.2	29.9	75% FAU catalyst 25% Beta catalyst 40% zeolite	[87]
FAU/FER	40/120	1.49	VGO; 79.2°C Aniline Point		1 atm	75	16.5 (4% C$_3$H$_6$)	38.5	13.2	1.3% coke 20% FER	[86]

Catalyst	Ratio	Feed	Temp	Pressure	Col6	Col7	Col8	Notes	Ref
FAU/ITH	1.25	Arabian light VGO	520 °C	1 atm	90	40 (8% C_3H_6)	42	12	4.5% coke 10% ITH	[85]
FAU/ITH	40/160	VGO; 79.2 °C Aniline Point		1 atm	75	17.5 (5.2% C_3H_6)	37.9	13.1	1.5% coke 20% ITH	[86]
FAU/MFI		Arabian light VGO	600 °C	98 kPa	80	41.7 (18.4% C_3H_6)	34	16.4	3.5% coke 20% MFI	[84]
FAU/MFI	40/87	VGO; 79.2 °C Aniline Point		1 atm	75	22 (5.6% C_3H_6)	34.4	12	1.5% coke 20% MFI	[86]
FAU/MFI	1.25	Arabian light VGO	520 °C	1 atm	90	40 (8% C_3H_6)	42	12	4.5% coke 10% MFI	[85]
FAU/MSE	5.4/18	VGO; 139 °F Aniline Point	524 °C	1 atm	70	22.4	47.6	30.1	75% FAU catalyst 25% MSE catalyst 40% zeolite	[87]
ITQ-21	25	Arabian light VGO	500 °C	1 atm	72.5	24 (6% C_3H_6)	30	12	Si/Ge=10	[83]
MSE	18 (see Note a)	VGO; 139 °F Aniline Point	524 °C	1 atm	37.7	14.7	23	62.3	a) Before steaming 4h at 815 °C, 100% steam; 40% zeolite	[87]
SSZ-25 (MWW)	3.0	29.09 API, 0.3 Ramsbottom C	496 °C	1 atm	55.2		22	19	20 wt% in kaolin	[29]

Table 12.15 Olefin cracking catalysts.

Zeolite	SiO$_2$/Al$_2$O$_3$ ratio	Feedstock	Cat/Oil SV	Temperature	Pressure	Conversion	CH$_4$ + C$_2$H$_6$ selectivity	C2 + C3 selectivity	C4+ selectivity	Notes	Reference
Beta	27	1-Butene	3.5 WHSV	620 °C	0.1 MPa	97	22%	8% C$_2$H$_4$ 4% C$_3$H$_6$	44% C3P+C4P <1% C5+ 20% aromatics	Rapid coking	[88]
CHA		1-Butene	3.5 WHSV	620 °C	0.1 MPa	97	9%	28% C$_2$H$_4$ 45% C$_3$H$_6$	12% C3P+C4P 8% C5+ <1% aromatics	SAPO-34 rapid coking	[88]
FAU	6.8	1-Butene	3.5 WHSV	620 °C	0.1 MPa	97	33%	5% C$_2$H$_4$ 3% C$_3$H$_6$	52% C3P+C4P <1% C5+ 8% aromatics	Rapid coking	[88]
FER	30	1-Butene	3.5 WHSV	620 °C	0.1 MPa	97	18%	25% C$_2$H$_4$ 31% C$_3$H$_6$	13% C3P+C4P 9% C5+ 5% aromatics	ZSM-35 rapid coking	[88]
ITH	160	1-Hexene	0.9	500 °C	1 atm	54	0.2%	2.43% C$_2$H$_4$ 20.9% C3	0.7% C4P 9.8% C4O 18.4% C5+		[86]
MFI	87	1-Hexene	0.5	500 °C	1 atm	54	0.2%	2.7% C$_2$H$_4$ 11.9% C3	2.3% C4P 9.1% C4O 25.8% C5+		[86]
MFI	350	1-Butene	3.5 WHSV	620 °C	0.1 MPa	78	4.1%	21.4% C$_2$H$_4$ 46.2% C$_3$H$_6$ 4.1% C$_3$H$_8$	5.8% butanes 9.5% C5+ 9% aromatics	1 HOS data	[89]

Zeolite	SiO2/Al2O3	Feed	WHSV/LHSV	Temp	Pressure	Conversion	Selectivity	Products	Other	Ref
MRE	200	11.3% butanes, 54.4% 1-butene, 29.6% 2-butene, 4.7% i-butene	15 LHSV	550 °C	0.1 MPa	71.6	1.3%	13.3% C$_2$H$_4$ 43.2% C$_3$H$_6$ 4.3% C$_3$H$_8$	19.8% C5+	Some C4P and i-butene made	[90]
MRE	272	11.3% butanes, 54.4% 1-butene, 29.6% 2-butene, 4.7% i-butene	15 LHSV	550 °C	0.1 MPa	48.7	0.9%	8.0% C$_2$H$_4$ 45.2% C$_3$H$_6$ 1.4% C$_3$H$_8$	27.8% C5+	Some C4P and i-butene made	[90]
MRE	331	11.3% butanes, 54.4% 1-butene, 29.6% 2-butene, 4.7% i-butene	15 LHSV	550 °C	0.1 MPa	29.8	1.1%	5.9% C$_2$H$_4$ 41.9% C$_3$H$_6$ 0.7% C$_3$H$_8$	34.2% C5+	Some C4P and i-butene made	[90]
MTT	106	1-Butene	3.5 WHSV	620 °C	0.1 MPa	97	9%	28% C$_2$H$_4$ 29% C$_3$H$_6$	12% C3P+C4P 7% C5+ 14% aromatics	ZSM-23 rapid coking	[88]
MWW	30	1-Butene	3.5 WHSV	620 °C	0.1 MPa	97	16%	16% C$_2$H$_4$ 11% C$_3$H$_6$ 28% C$_3$H$_8$	12% butanes 2.2% C5+ 16% aromatics	MCM-22 initial data	[89]
MWW	30	1-Butene	3.5 WHSV	620 °C	0.1 MPa	76	4.1%	19% C$_2$H$_4$ 44% C$_3$H$_6$ 3.7% C$_3$H$_8$	10% butanes 7.3% C5+ 11% aromatics	MCM-22 1+ HOS data	[89]
TON	60	1-Butene	3.5 WHSV	620 °C	0.1 MPa	97	5%	24% C$_2$H$_4$ 38% C$_3$H$_6$	12% C3P+C4P 8% C5+ 13% aromatics	ZSM-22 rapid coking	[88]

Table 12.16 Propane or butane aromatization catalysts.

Zeolite	SiO₂/Al₂O₃ ratio	Metal (wt%)	Feed	SV	Temp	Pressure	Conversion	Aromatics Selectivity	Benzene Selectivity	Notes	Reference
FAU	6	Pt(0.66)	Propane	1200GH	525°C	1 atm	46.7%	67%		10:1 He:propane	[92]
FAU	6	Pt(0.66)	Propane	1200GH	550°C	1 atm	31.2%	37%		10:1 He:propane	[92]
MFI		Ga(10)	Propane	1.2WH	530°C	1 atm		9wt% yield		Ga₂O₃ + ZSM-5	[93]
MFI		Ga(5)	Propane	1.2WH	530°C	1 atm		6wt% yield		Ga₂O₃ + ZSM-5	[93]
MFI		Ga(2)	Propane	1.2WH	530°C	1 atm		0.5wt% yield		Ga₂O₃ + ZSM-5	[93]
MFI		Ga(2)	Propane	1.2WH	530°C	1 atm		11wt% yield		Ga₂O₃ + ZSM-5; H₂ pretreat	[93]
MFI	80	Si/Ga=50	Propane + ethene	6200GH	450°C	1 atm	27.5%	91%	7%, 22.6%Tol, 17.8% C8A	1:1 C2/C3 feed	[94]
MFI	80	Si/Ga=50	Propane + Hexane	6200GH	450°C	1 atm	23.1%	77%	4.4%, 15%Tol, 20.3% C8A	0.5 C6/C3 feed	[94]
MFI	56	None	n-Butane	14 WH	500°C	1 atm	44.7%	7%; mostly cracking	15% 42% Tol	34.2 kPa butane	[95]
MFI	56	Zn (1.2)	n-Butane	14 WH	500°C	1 atm	38.3%	46%	28%	34.2 kPa butane	[95]
MFI	56	Ga (1.9)	n-Butane	14 WH	500°C	1 atm	58%	28%	19%	34.2 kPa butane	[95]
MFI	80	None	n-Butane	0.25	530°C	1 atm	58%	10%			[96]
MFI	80	Ga (1.5)	n-Butane	0.21	530°C	1 atm	57%	20%			[96]
SSZ-25 (MWW)	30	Ga	n-Butane		500°C	1 atm	45%	>90%		2% feed in He	[29]
TON	75	None	n-Butane	0.23 WH	530°C	1 atm	52%	5%		Theta-1	[96]
TON	75	Ga (1.5)	n-Butane	0.21 WH	530°C	1 atm	46%	18%		Theta-1	[96]

into aromatics, while MFI seems to be necessary to avoid rapid coke buildup [98]. Bao's recent discovery of a Mo loaded MWW zeolite is starting to push the field forward [99] (Table 12.17).

12.2.7
Methanol to Olefins

The conversion of methanol into higher olefins (methanol to olefins, MTO) such as ethylene and propylene is an area that has been studied in great detail [103]. Chapter 15 goes into greater depth on details such as the formation of the first C–C bond and the derivation of the "carbon pool" mechanism. The technology was developed in the wake of the 1970s oil crisis and operated by Mobil as the methanol to gasoline or MTG process in New Zealand for several years before economic changes forced a shutdown of the plant. This technology used MFI as the catalyst, but it was not until the development of SAPO-34, an aluminophophate with the CHA structure, that MTO became highly selective to light olefins [104]. Most MFI based catalysts, such as that in Lurgi's MTP process, in addition to producing ethylene and propylene, also produce a gasoline co-product [105].

Today, MTO technology is becoming increasingly important as China seeks to convert their vast coal supplies into methanol for production of olefin building blocks. The two most important catalysts for the MTO reaction have historically been MFI and CHA, but many zeolites have been tested, particularly those containing cages and small pore openings. Table 12.18 records a wide range of zeolites and their performances.

12.2.8
Hydrotreating and Hydrocracking

Hydrotreating and hydrocracking are very important technologies which utilize hydrogen and a catalyst to convert heavy, sulfur- and nitrogen-containing, bottom of the barrel feedstocks to lighter valuable products such as diesel or gasoline free of sulfur or nitrogen content [113]. The first hydrocracking type process was developed in Germany during World War I to convert coal to gasoline to ensure a domestic supply of liquid fuels [114]. Today, hydrotreating processes such as the UOP Unionfining Process [115] or the Chevron RDS/VRDS process [116] are typically focused on removal of sulfur or nitrogen from a feedstock and may not contain a zeolitic component. Hydrocracking processes such as the UOP Unicracking [117, 118] or Chevron Isocracking Process [119] remove sulfur or nitrogen from a feedstock and crack feed components to convert heavy feeds into lighter products as well as to make higher cetane value diesel fuels. These processes are covered in greater detail in Chapter 16.

The complexity of the chemical pathways in these systems is due to the molecular complexity of the heavy feedstock rich in aromatic structures, heavy paraffins and S- and N-containing molecules as well as the multi-component catalyst used.

Table 12.17 Methane to aromatics catalysts.

Zeolite	SiO$_2$/Al$_2$O$_3$ ratio	Metal (wt%)	Feed	SV	Temp	Pressure	Conversion	Aromatics selectivity	Benzene selectivity	Notes	Reference
MCM-22 (MWW)	30	Mo (6)	CH$_4$	1500GH	700 °C	1 atm	10%		80%	9.5% N$_2$ in CH$_4$ feed	[99a]
MFI		Mo	CH$_4$		700 °C		9%	65%			[100]
MFI	30	Mo(3)	CH$_4$	840GH	700 °C	1 atm	12%		40%		[101]
MFI	35	Mo(2)	CH$_4$	810GH	720 °C	1 atm	14%		70%	90/10 CH$_4$/Ar feed	[102]

Table 12.18 Methanol to olefins catalysts.

Zeolite	SiO$_2$/Al$_2$O$_3$ ratio	SV	Temperature	Pressure	Conversion	Selectivity	Notes	Reference
AEI	1		420 °C	1 atm	100%	47% C$_2$H$_4$, 40% C$_3$H$_6$, 11% C$_4$H$_8$	60%N$_2$, 40% MeOH; SAPO-18	[106]
CHA	34		400 °C	1 atm	100%	54.7% C$_2$H$_4$, 35.1% C$_3$H$_6$, 2.8% C$_4$H$_8$	EtOH feed	[107]
CHA		2.85 WHSV	450 °C	101 kPa Fluidized bed	65%	2.5 wt% CH4, 36% C$_2$H$_4$, 32% C$_3$H$_6$, 15% C$_3$H$_8$, 11% C$_4$H$_8$, 2% C$_4$H$_{10}$, 4% C5+	SAPO-34 steamed 800 °C 8h DME feed	[108]
CHA		0.7 WHSV	350 °C	1 atm	100%	30% C$_2$H$_4$, 45% C$_3$H$_6$, 15% C$_4$H$_8$, 10% other	SAPO-34	[109]
ERI	11	0.7 WHSV	350 °C	1 atm		30% C$_2$H$_4$, 45% C$_3$H$_6$, 15% C$_4$H$_8$, 10% other	UZM-12	[109]
LTA	10	0.7 WHSV	350 °C	1 atm		25% C$_2$H$_4$, 40% C$_3$H$_6$, 10% C$_4$H$_8$, 25% other	UZM-9	[109]
MEL	72	7–12 WHSV	375 °C	1 atm	~100%	2.1 wt% C1+C2, 23.5% C3, 16.5% C4, 26.3% C5-8, 1% Tol, 4.8% Xy, 13.3%A9, 9.5% A10, 3% heavy		[110]

Table 12.18 Continued

Zeolite	SiO$_2$/Al$_2$O$_3$ ratio	SV	Temperature	Pressure	Conversion	Selectivity	Notes	Reference
MFI	65	7–12 WHSV	375 °C	1 atm	~100%	14.7 wt% C1+C2, 21% C3, 18.7% C4, 15% C5-8, 0.2% B, 2.7% Tol, 19.5% Xy, 6%A9, 1.5% A10, 0.5% heavy	Bound 50% SiO$_2$	[110]
MFI	30	19.6 WHSV	500 °C	1 atm	90%	46.1 wt% C3=, 15.4 wt% C2=, 13.7 wt%C4=, 11.5 wt% heavy	4 wt% P 2.3 w% DME	[111]
MFI	30	33 WHSV	500 °C	1 atm	97%	52.5 wt% C3=, 24.1 wt% C2=, 14.4 wt%C4=, 2.4 wt% heavy	1.1% CH4, 0.7% DME	[111]
MFI	280	12.4 WHSV	500 °C	1 atm	90%	50.4 wt% C3=, 8.4 wt% C2=, 16.6 wt%C4=, 14.5 wt% heavy		[111]
SSZ-25 (MWW)	30	0.625 LHSV	371 °C	1 atm	~100%	3.2 wt% C3=, 2.2 wt% C3P, 10.3 wt%C4, 76 wt% heavy	in He stream	[29]
STI	~100	2.8 WHSV	399 °C		100%	70% C2-C4, EB most prominent aromatic	SSZ-75	[112]
UFI	13	0.7 WHSV	350 °C	1 atm		35% C$_2$H$_4$, 50% C$_3$H$_6$, 10% C$_4$H$_8$, 5% other	UZM-5	[109]

The catalyst sytem usually contains a metal sulfide phase based on Ni, Mo or Co and an acidic component such as a zeolite or silica–alumina, along with binders such as clays or alumina. Reactions occurring in a hydrotreating or hydrocracking system include olefin or aromatic saturation, desulfurization or denitrogenation, and chain scission reactions such as ring-breaking, paraffin cracking or dealkylation of aromatics [120]. Because the conditions of the reaction and the condition/position of the metal sulfide phase are as important as the zeolite used, only a few references are given below to illustrate the effect of the zeolite (Table 12.19).

12.2.9
Reactions Using Heteroatom Substituted Zeolites

Following the discovery of TS-1 [125], a titanium-substituted MFI, the use of zeolitic materials for oxidation increased significantly. The presence of the Ti atom in the framework of a zeolite structure provides a site-isolated Ti center, a situation not possible with other Ti-containing materials while also allowing shape-selective oxidations. The combination of the two effects gives highly active and selective oxidation reactions [126].

12.2.9.1 Epoxidation

The epoxidation of propylene to propylene oxide is a high-volume process, using about 10% of the propylene produced in the world via one of two processes [127]. The oldest technology is called the chlorohydrin process and uses propylene, chlorine and water as its feedstocks. Due to the environmental costs of chlorine and the development of the more-efficient direct epoxidation over TiO_2/SiO_2 catalysts, new plants all use the hydroperoxide route. The disadvantage here is the co-production of stoichiometric amounts of styrene or t-butyl alcohol, which means that the process economics are dependent on finding markets not only for the product of interest, but also for the co-product. The hydroperoxide route has been practiced commercially since 1979 to co-produce propylene oxide and styrene [128], so when TS-1 was developed, epoxidation was looked at extensively [129].

Using TS-1 as the catalyst and H_2O_2 as the oxidant allows the epoxidation process to give high selectivity to propylene oxide with good conversion without the production of a co-product as indicated in Table 12.20. However, the price differential of H_2O_2 and propylene oxide has not been high enough to allow commercial operation, so other oxidants have been investigated [130].

12.2.9.2 Other Oxidations

In addition to the epoxidation of olefins, zeolitic materials have been studied for other fine chemical transformations. Table 12.21 indexes the zeolites used for oxidative dehydrogenation of propane, direct hydroxylation of benzene to phenol and ε-caprolactam synthesis. A recent review summarizes other reactions for which there is not enough space in the table [138, 139].

Table 12.19 Zeolites used in hydrotreating or hydrocracking processes.

Zeolite	SiO$_2$/Al$_2$O$_3$ ratio	Metal (wt%)	Feed	H$_2$/HC	SV	Temperature	Pressure	Yield information	Notes	Reference
AEL	1.95% Si	Pt (0.5)	Hydrogenated sunflower oil	1500	2.5 LHSV	300 °C	55 bar	96% yield, 23% i-paraffins, 76% n-paraffins	SAPO-11 90 cetane	[121]
Beta	29	Pt (0.5)	Heavy VGO	1500	1 LHSV	350 °C	5.5 MPa	21% conversion, 18% at 150–328 °C BP	Large crystal	[122]
Beta	33	Pt (0.5)	Heavy VGO	1500	1 LHSV	350 °C	5.5 MPa	33% conversion, 8% C5 at 150 °C BP, 19% at 150–328 °C BP	Small crystal	[122]
FAU	5.5	NiO(1.7) MoO$_3$ (6.7)	Arabian heavy atmospheric residue		Autoclave	410 °C	9.8 MPa	74.9% liquid product, 52% HDS activity, 26.8% HDN activity	Contains mesopores	[123]
FAU	8	NiO(1.7) MoO$_3$ (6.7)	Arabian heavy atmospheric residue		Autoclave	410 °C	9.8 MPa	72.3% liquid product, 33% HDS activity, 12.7% HDN activity		[123]
SFO		Pd (0.5)	nC16	200	1.5 WHSV	315 °C 353 °C	1200 psig	30% at 315 °C, 90% at 353 °C, indicative of large pores	~2% Co in SSZ-51	[124]

Table 12.20 Epoxidation catalysts.

Zeolite	Si/Ti ratio	Feed	Oxidant	SV	Temperature	Pressure	Conversion	Selectivity	Notes	Reference
Beta	40	2-hexene	H_2O_2	Autoclave	60 °C	1 atm	15.9	96% epoxide, 4% diol	Ti-Beta; CH_3CN solvent	[131]
Beta	35	Cyclohexene	H_2O_2	Autoclave	60 °C	1 atm	16.5%	78.4% epoxide, 21.6% others	Ti-Beta; CH_3CN solvent, 1:1 ene:H_2O_2	[132]
CZP		Limonene	H_2O_2	Autoclave	60 °C	1 atm	20.5	24% epoxide, 20% diol	Chiral zinc phosphate	[133]
MEL	95	2-Hexene	H_2O_2	Autoclave	60 °C	1 atm	13.6	96% epoxide, 4% diol	TS-2; CH_3CN solvent	[131]
MFI	42	2-Hexene	H_2O_2	Autoclave	60 °C	1 atm	29.1	96% epoxide, 4% diol	TS-1; CH_3CN solvent	[131]
MFI	Not given	1-Pentene	H_2O_2	Autoclave	25 °C		94%	91% epoxide	60 min rxn MeOH soln	[134]
MFI	Not given	1-Hexene	H_2O_2	Autoclave	25 °C		88%	90% epoxide	70 min rxn MeOH soln	[134]
MFI	Not given	Cyclohexene	H_2O_2	Autoclave	25 °C		9%	Not determined	90 min rxn MeOH soln	[134]

Table 12.20 Continued

Zeolite	Si/Ti ratio	Feed	Oxidant	SV	Temperature	Pressure	Conversion	Selectivity	Notes	Reference
MFI	Not given	1-Octene	H_2O_2	Autoclave	45 °C		81%	91% epoxide	45 min rxn MeOH soln	[134]
MFI	40	Propylene	$H_2 + O_2$ See note	13.6 WH	43 °C	50 bar		2.25% yield epoxide 0.15% yield methylformate 0.02% yield acetone	Pd/Pt substituted 33% N_2, 23% CH_3OH, 18.7% C_3H_6, 13.2% H_2O, 7.2% O_2, 4.6% H_2	[135]
MFI	45.7	Propylene	H_2O_2	Autoclave	60 °C	0.4 MPa	56%	83.5% epoxide, 16.5% diol		[136]
MFI	45.7	Propylene	Urea/H_2O_2 (5:1)	Autoclave	60 °C	0.4 MPa	90.4%	97.2% epoxide, 2.8% diol		[136]
MFI	100	Propylene	$H_2 + O_2$ See note		90 °C	1 atm	0.295	94.1% epoxide	5 wt% Ag, 1 wt% Na on TS-1; 49% He, 30% C_3H_6, 10% O_2, 11% H_2	[137]
MWW	46	2-Hexene	H_2O_2	Autoclave	60 °C	1 atm	50.8%	99% epoxide, 1% diol	Ti-MWW; CH_3CN solvent	[131]
MWW	240	Cyclohexene	H_2O_2	Autoclave	60 °C	1 atm	21.2%	90.8% epoxide, 9.2% others	Ti-YNU-1; CH_3CN solvent, 1:1 ene:H_2O_2	[132]

Table 12.21 Other oxidation reactions catalyzed by zeolites.

Zeolite	Feed	Oxidant	SV	Temperature	Pressure	Conversion	Product	Selectivity	Notes	Reference
AFI	Cyclohexanone NH$_3$	Air	Autoclave	80 °C	35 bar	68.3%	ε-Caprolactam	77.9% lactam, 5.7% oxime, 16.4% other	Mn and Mg AlPO$_4$-5	[140]
AFI	Cyclohexanone NH$_3$	Air	Autoclave	80 °C	35 bar	71.9%	ε-Caprolactam	72% lactam, 11.5% oxime, 16.3% other	Fe and Mg AlPO$_4$-5	[140]
AFI	Cyclohexanone NH$_3$	Air	Autoclave	80 °C	35 bar	76.5%	ε-Caprolactam	3.3% lactam, 86.6% oxime, 10% other	Mn and Si AlPO$_4$-5	[140]
AFI	Propane	O$_2$	0.9 WH	520 °C	1 atm	21%	Propylene	64% C$_3$H$_6$, 24% CO, 12%CO$_2$	VAPO-5; 0.38% V	[141]
MFI	Benzene	H$_2$O$_2$	Autoclave	60 °C		52.3%	Phenol	89.5%	1.5 C$_6$H$_6$/H$_2$O$_2$ Triphase cond.	[142]
MFI	Propane	H$_2$ + O$_2$	4500GH	170 °C	0.3 MPa	6.5%	Acetone	90% acetone, 5% 2-propanol, 5% CO$_2$	0.1 wt% Au C$_3$H$_8$:H$_2$:O$_2$:Ar = 0.5:3:1:5.5	[143]
MFI	Benzene	N$_2$O	30000 GHSV	425 °C	1 atm	16%	Phenol	>99%	1 vol% C$_6$H$_6$, 4% N$_2$O in He 800 °C calcination Si/Al=25, Fe/Al=0.33	[144]

392 12 Overview and Recent Developments in Catalytic Applications of Zeolites

Table 12.21 Continued

Zeolite	Feed	Oxidant	SV	Temperature	Pressure	Conversion	Product	Selectivity	Notes	Reference
MFI	Benzene	N_2O	30000 GHSV	425 °C	1 atm	7%	Phenol	42%	1 vol% C_6H_6, 4% N_2O in He 550 °C calcination	[144]
MFI	Benzene	Ascorbic acid	Autoclave	80 °C	0.4 MPa		Phenol	1.5% yield	5.6 C_6H_6/ascorbic acid	[145]
MFI	Benzene	Ascorbic acid	Autoclave	80 °C	0.4 MPa		Phenol	3% yield	5.6 C_6H_6/ascorbic acid; 0.5 wt% La	[145]
MFI	Benzene	N_2O	17.2 WH	482 °C		4.2%	Phenol	96% C_6H_6 to phenol 90% N_2O to phenol	55% C_6H_6, 3% N_2O, 42% N_2 Si/Al=6, 250 ppm Fe	[146]
MFI	Propane	O_2	1200 GH	500 °C	1 atm	65%	Propylene	53% C_3H_6, 15% CO, 12% CO_2, 19% arom, 1% cracking	VS-1; Si/V=230; 18.7% C_3H_8, 10.7% O_2, 70.6% He	[147]
MWW	Propane	O_2		500 °C	1 atm	10%	Propylene	27% C_3H_6, 36% CO, 25% CO_2, 9% C_2H_4, 3% CH_4	V-MCM-22 0.57% V_2O_5; 2.5% C_3H_8, 5% O_2, 92.5% He	[148]

12.3
Future Trends in Catalysis by Zeolites

As documented in the preceding pages, the past few years have seen a large increase in the number of known zeolite types. In addition, the application of theoretical approaches have developed large libraries of possible zeolite structures [149]. The future for zeolites will be driven not only by the continued discovery of new structure types but also by a few trends. Foremost among them will be the development of large-pore zeolites such as ITQ-33 and SSZ-53 [28, 150]. Another approach to provide large pores is to incorporate mesoporosity with the zeolite microporosity [151]. Both approaches serve to provide better diffusion for the catalyst so that heavy molecules are better able to reach the active sites and diffuse away. This serves to facilitate both conversion and selectivity. Incumbent in the approach of designing large-pore materials is to develop thermally stable materials that are robust to regeneration. Since the discovery of VPI-5, the first material with a pore larger than a 12-ring, the lack of thermal stability has held back the industrial use of large pore zeolites.

A second area that will be important in the future is the continued development of MOFs and ZIFs [152]. Much as the discovery of $AlPO_4$-based materials revolutionized the catalytic use of zeolites when only aluminosilicates were known, MOFs and ZIFs have the potential to revolutionize low temperature processes such as oxidations and organic reactions [153]. Newly discovered materials along these same lines are covalent organic frameworks, the so-called COFs [154]. These materials have similar channels to those known for MOFs and ZIFs but tend to have higher thermal stability.

To create new materials, whether they be MOFs, ZIFs, COFs, new zeotypes or simply new routes to a known structure, the development of alternate synthesis methodologies such as ionothermal synthesis will be important [155]. Synthesizing the same structure type material using two different synthesis methodologies alters properties such as morphology and Al distribution, so new synthesis routes are often as effective as developing whole new zeolite structure types. In these and other ways, the catalytic uses of zeolites will continue to progress toward higher conversion and greater selectivity for each reaction in which they are used.

References

1 Rabo, J.A., Pickert, P.E., Stamires, D.N., and Boyle, J.E. (1960) *Proceedings of the Second International Congress on Catalysis, Editions Technol., Paris*, p. 2055.

2 Sherman, J.D. (1999) Synthetic zeolites and other microporous oxide molecular sieves. *Proc. Natl. Acad. Sci. U. S. A.*, **96**, 3471–3478.

3 Marcilly, C.R. (2000) Where and how shape selectivity of molecular sieves operates in refining and petrochemistry catalytic processes. *Top. Catal.*, **13**, 357–366.

4 (a) Earl, D.J., Burton, A.W., Rea, T., Ong, K., Deem, M.W., Hwang, S.-J., and Zones, S.I. (2008) Synthesis and monte-carlo structure determination of

S.S., and Yurchak, S. (1992) U.S. Patent 5,166,455.
32 Weidert, D.J. (2000) Olefin transformation – a means to ease your MTBE woes. Talk given at AIChE Spring 2000 Meeting, Session T9021.
33 Brown, S.H., Vaughn, S.N., Santiesteban, J.G., and Strohmaier, K.G. (2003) Method for isomerizing a mixed olefin feedstock to 1-olefin. US Patent Application 20030233018.
34 Miller, S.J. (2001) Process for olefin isomerization. US Patent 6281404.
35 Lee, S.-H., Shin, C.-H., and Hong, S.B. (2004) Investigations into the origin of the remarkable catalytic performance of aged H-ferrierite for the skeletal isomerization of 1-butene. *J. Catal.*, **223**, 200.
36 Guisnet, M., Andy, P., Gnep, N.S., Benazzi, E., and Travers, C. (1996) Skeletal isomerization of n-butenes. *J. Catal.*, **158**, 551–560.
37 Xu, W.-Q., Yin, Y.-G., Suib, S.L., and O'Young, C.-L. (1994) Selective conversion of n-butene to isobutylene at extremely high space velocities on ZSM-23 zeolites. *J. Catal.*, **150**, 34–45.
38 Byggningsbacka, R., Kumar, N., and Lindfors, L.-E. (1999) Kinetic model for skeletal isomerization of n-butene over ZSM-22. *Ind. Eng. Chem. Res.*, **38**, 2896–2901.
39 Byggningsbacka, R., Lindfors, L.-E., and Kumar, N. (1997) Catalytic activity of ZSM-22 zeolites in the skeletal isomerization reaction of 1-butene. *Ind. Eng. Chem. Res.*, **36**, 2990–2995.
40 (a) Ipatieff, V.N. and Pines, H. (1935) Polymerization of ethylene under high pressures in the presence of phosphoric acid. *Ind. Eng. Chem.*, **27**, 1364–1369. (b) Ipatieff, V.N. and Pines, H. (1936) Propylene polymerization: under high pressure and temperature with and without phosphoric acid. *Ind. Eng. Chem.*, **28**, 684–686.
41 (a) Ipatieff, V.N., Corson, B.B., and Egloff, G. (1935) Polymerization, a new source of gasoline. *Ind. Eng. Chem.*, **27**, 1077–1081. (b) Ipatieff, V.N. and Corson, B.B. (1936) Gasoline from ethylene by catalytic polymerization. *Ind. Eng. Chem.*, **28**, 860–863.

42 Quann, R.J., Green, L.A., Tabak, S.A., and Krambeck, F.J. (1988) Chemistry of Olefin Oligomerization over ZSM-5. *Ind. Eng. Chem. Res.*, **27**, 565–570.
43 Birkhoff, R. and Nurminen, M. (2004) NExOCTANE™ Process for isooctane production, in *Handbook of Petroleum Refining Processes*, 3nd edn (ed. R.A. Myers), McGraw-Hill, pp. 1.3–1.9.
44 Martens, J.A., Verrelst, W.H., Mathys, G.M., Brown, S.H., and Jacobs, P.A. (2005) Tailored catalytic propene trimerization over acidic zeolites with tubular pores. *Angew. Chem. Int. Ed.*, **44**, 5687–5690.
45 Martens, J.A., Ravishankar, R., Mishin, I.E., and Jacobs, P.A. (2000) Tailored olefin oligomerization with H-ZSM-57 zeolite. *Angew. Chem. Int. Ed.*, **39**, 4376–4379.
46 Kojima, M., Rautenbach, M.W., and O'Connor, C.T. (1988) Butene oligomerization over Ion-exchanged mordenite. *Ind. Eng. Chem. Res.*, **27**, 248–252.
47 Yamamura, M., Chaki, K., Wakatsuki, T., Okado, H., and Fujimoto, K. (1994) Synthesis of ZSM-5 zeolite with small crystal size and its catalytic performance in ethylene oligomerization. *Zeolites*, **14**, 643–650.
48 Schmidt, R., Welch, M.B., and Randolph, B.B. (2008) Oligomerization of C5 olefins in light catalytic naphtha. *Energy Fuels*, **22**, 1148–1155.
49 Weitkamp, J. and Traa, Y. (1999) Isobutane/butene alkylation on solid catalysts. Where do we stand? *Catal. Today*, **49**, 193–199.
50 Feller, A. and Lercher, J.A. (2004) Chemistry and technology of isobutane/alkene alkylation catalyzed by liquid and solid acids. *Adv. Catal.*, **48**, 229–295.
51 Sarsani, V.R. and Subramaniam, B. (2009) Isobutane/butene alkylation on microporous and mesoporous solid acid catalysts: probing the pore transport effects with liquid and near critical reaction media. *Green Chem.*, **11**, 102–108.
52 Mota Salinas, A.L., Sapaly, G., Ben Taarit, Y., Vedrine, J.C., and Essayem, N. (2008) Continuous supercritical iC4/C4= alkylation over H-Beta and H-USY Influence of the zeolite structure. *Appl. Catal. A*, **336**, 61–71.

53 Petkovic, L.M. and Ginosar, D.M. (2004) The effect of supercritical isobutane regeneration on the nature of hydrocarbons deposited on a USY zeolite catalyst utilized for isobutane/butene alkylation. *Appl. Catal. A*, **275**, 235–245.

54 Sievers, C., Liebert, J.S., Stratmann, M.M., Olindo, R., and Lercher, J.A. (2008) Comparison of zeolites LaX and LaY as catalysts for isobutane/2-butene alkylation. *Appl. Catal. A*, **336**, 89–100.

55 Perego, C. and Ingallina, P. (2002) Recent advances in the industrial alkylation of aromatics: new catalysts and new processes. *Catal. Today*, **73**, 3–22.

56 Ipatieff, V.N. and Schmerling, L. (1946) Ethylation of benzene in the presence of solid phosphoric acid. *Ind. Eng. Chem.*, **38**, 400–402.

57 Ipatieff, V.N., Pines, H., and Komarewsky, V.I. (1936) Phosphoric acid as the catalyst for alkylation of aromatic hydrocarbons. *Ind. Eng. Chem.*, **28**, 222–223.

58 (a) Bentham, M.F. (1997) UOP Q-max process for cumene production, in *Handbook of Petroleum Refining Processes*, 2nd edn (ed. R.A. Myers), McGraw-Hill, pp. 1.67–1.69.
(b) Hwang, S.Y., and Chen, S.S. (2004) *Cumene in the Kirk-Othmer Encyclopedia of Chemical Technology*, 5th edn, vol. 8, John Wiley & Sons, pp. 147–157.

59 Wallace, J.W., and Gimpel, H.E. (1997) The Dow-Kellog cumene process, in *Handbook of Petroleum Refining Processes*, 2nd edn (ed. R.A. Myers), McGraw-Hill, pp. 1.15–1.20.

60 Corma, A., Martinez-Soria, V., and Schnoeveld, E. (2000) Alkylation of benzene with short-chain olefins over MCM-22 zeolite: catalytic behaviour and kinetic mechanism. *J. Catal.*, **192**, 163–173.

61 Holtermann, D.L. and Innes, R.A. (1992) Alkylation using zeolite SSZ-25. US Patent 5149894.

62 Corma, A., Diaz-Cabanas, M.J., Martinez, C., and Rey, F. (2008) Zeolite ITQ-21 as catalyst for the alkylation of benzene with propylene. *Stud. Surf .Sci. Catal.*, vol. 174, Elsevier, Amsterdam, pp. 1987–1990.

63 Perego, C., Amarilli, S., Millini, R., Bellussi, G., Girotti, G., and Terzoni, G. (1996) Experimental and computational study of beta, ZSM-12, Y, mordenite and ERB-1 in cumene synthesis. *Micropor. Mater.*, **6**, 395–404.

64 Onida, B., Geobaldo, F., Borello, L., and Garrone, E. (2001) Acidity of ITQ-2 zeolite as studied by FT-IT spectroscopy of adsorbed molecules in comparison with that of MCM-22. *Stud. Surf. Sci. Catal.*, vol. 135, Elsevier, Amsterdam, pp. 1984–1990.

65 Maekawa, H., Saha, S.K., Mulla, S.A.R., Waghmode, S.B., Komura, K., Kubota, Y., and Sugi, Y. (2007) Shape-selective alkylation of biphenyl over metalloaluminophosphates with AFI topology. *J. Mol. Catal. A*, **263**, 238–246.

66 Ito, A., Maekawa, H., Kawagoe, H., Komura, K., Kubota, Y., and Sugi, Y. (2007) Shape-selective alkylation of biphenyl over H-[Al]-SSZ-24 zeolites with AFI topology. *Bull. Chem. Soc. Jpn.*, **80**, 215–223.

67 Maekawa, H., Shibata, T., Niimi, A., Asaoka, C., Yamasaki, K., Naiki, H., Komura, K., Kubota, Y., Sugi, Y., Lee, J.-Y., Kim, J.-H., and Seo, G. (2008) The isopropylation of biphenyl over one-dimensional zeolites with corrugated channels. *J. Mol. Catal. A*, **279**, 27–36.

68 Aguilar, J., Corma, A., Melo, F.V., and Sastre, E. (2000) Alkylation of biphenyl with propylene using acid catalysts. *Catal. Today*, **55**, 225–232.

69 Aguilar, J., Pergher, S.B.C., Detoni, C., Corma, A., Melo, F.V., and Sastre, E. (2008) Alkylation of biphenyl with propylene using MCM-22 and ITQ-2 zeolites. *Catal. Today*, **133–135**, 667–672.

70 Sugi, Y., Tawada, S., Sugimura, T., Kubota, Y., Hanaoka, T., Matsuzaki, T., Nakajima, K., and Kunimori, K. (1999) Shape-selective isopropylation of biphenyl over H-mordenites Relationships of bulk products and encapsulated products in the pores. *Appl. Catal. A*, **189**, 251–261.

71 Shibata, T., Suzuki, S., Kawagoe, H., Komura, K., Kubota, Y., Sugi, Y., Kim, J.-H., and Seo, G. (2008) Synthetic investigation on MCM-68 zeolite with MSE topology and its application for shape-selective alkylation of biphenyl. *Micropor. Mesopor. Mater.*, **116**, 216–226.

72 Serra, J.M., Corma, A., and Guillon, E. (2008) Catalyst comprising a 10MR zeolite and a 12MR zeolite, and its use in transalkylation of aromatic hydrocarbons. US Patent 7419931.

73 Corma, A., Alfaro, J.M.S., Segui, V.F., and Sanchez, R.C. (2008) Method and Catalyst for the transalkylation/dealkylation of organic compounds. US Patent Application 2008/0021253.

74 Merlen, E., Alario, F., Ferrer, N., and Martin, O. (2005) Catalyst comprising at least one zeolite with structure type NES and rhenium, and its use for transalkylation of aromatic hydrocarbons. US Patent 6864400.

75 Hong, S.B., Nam, I.-S., Min, H.-K., Shin, C.-H., Warrender, S.J., Wright, P.A., Cox, P.A., Gramm, F., Baerlocher, C., McCusker, L.B., Liu, Z., Ohsuna, T., and Terasaka, O. (2007) TNU-9: a novel mediuim-pore zeolite with 24 topologically distinct tetrahedral sites. Stud. Surf. Sci. Catal., vol. 170A, Elsevier, Amsterdam, pp. 151–159.

76 Sinha, A.K., Sainkar, S., and Sivasanker, S. (1999) An improved method for the synthesis of the silicoaluminophosphate molecular sieves, SAPO-5, SAPO-11 and SAPO-31. Micropor. Mesopor. Mater., **31**, 321–331.

77 Akolekar, D.B. (1994) Thermal Stability, acidity, catalytic properties, and deactivation behaviour of SAPO-5 catalysts: effect of silicon content, acid treatment and Na exchange. J. Catal., **149**, 1–10.

78 Llopis, F.J., Sastre, G., and Corma, A. (2006) Isomerization and disproportionation of m-xylene in a zeolite with 9- and 10-membered ring pores: molecular dynamics and catalytic studies. J. Catal., **242**, 195–206.

79 Buchanan, J.S., Feng, X., Mohr, G.D., and Stern, D.L. (2003) Isomerization of ethylbenzene and xylenes. US Patent 6660896.

80 Yergin, D. (1992) *The Prize: The Epic Quest for Oil, Money and Power*, Simon and Schuster, pp. 80–97.

81 (a) Letzsch, W. (2006) Fluid catalytic cracking, in *Handbook of Petroleum Processing* (eds D.S.J. Jones and P.R. Pujado), Springer, pp. 239–286.

(b) Yergin, D. (1992) *The Prize: The Epic Quest for Oil, Money and Power*, Simon and Schuster, pp. 350–370.

82 Wojciechowski, B.W., and Corma, A. (1986) *Catalytic Cracking: Catalysts, Chemistry and Kinetics*, Marcel Dekker, New York.

83 Corma, A., Diaz-Cabanas, M.J., Martínez-Triguero, J., Rey, F., and Rius, J. (2002) A large-cavity zeolite with wide pore windows and potential as an oil refining catalyst. Nature, **418**, 514–517.

84 Fujiyama, Y., Al-Tayyar, M.H., Dean, C.F., Aitani, A., and Redhwi, H.H. (2007) Development of high-severity FCC process: an overview. Stud. Surf. Sci. Catal., vol. 166, Elsevier, Amsterdam, pp. 1–12.

85 Castañeda, R., Corma, A., Fornes, V., Martínez-Triguero, J., and Valencia, S. (2006) Direct synthesis of a 9×10 member ring zeolite (Al-ITQ-13): a highly shape-selective catalyst for catalytic cracking. J. Catal., **238**, 79–87.

86 Corma, A. (2004) Catalytic cracking process. US Patent 6709572.

87 Chester, A., Green, L.A., Dhingra, S.S., Mason, T., and Timken, H.K.C. (2007) Catalytic cracking processing using an MCM-68 catalyst. US Patent 7198711.

88 Zhu, X., Liu, S., Song, Y., and Xu, L. (2005) Catalytic cracking of C4 alkenes to propene and ethene: influences of zeolite pore structures and Si/Al_2 ratios. Appl. Catal. A, **288**, 134–142.

89 Zhu, X., Liu, S., Song, Y., Xie, S., and Xu, L. (2005) Catalytic cracking of 1-butene to propene and ethene on MCM-22 zeolite. Appl. Catal. A, **290**, 191–199.

90 Zhao, G., Teng, J., Zhang, Y., Xie, Z., Yue, Y., Chen, Q., and Tang, Y. (2006) Synthesis of ZSM-48 zeolites and their catalytic performance in C4-olefin cracking reactions. Appl. Catal. A, **299**, 167–174.

91 (a) Ono, Y. (1992) Transformation of lower alkanes into aromatic hydrocarbons over ZSM-5 zeolites. Catal. Rev. Sci. Eng., **34**, 179.

(b) Giannetto, G., Monoque, R., and Galliasso, R. (1994) Transformation of LPG into aromatic hydrocarbons and hydrogen over zeolite catalysts. Catal. Rev. Sci. Eng., **36**, 271.

92 Katranas, T.K., Triantafyllidis, K.S., Vlessidis, A.G., and Evmiridis, N.P. (2007) Propane reactions over faujasite structure zeolites type-X and USY: effect of zeolite silica over alumina ratio, strength of acidity and kind of exchanged metal ion. *Catal. Lett.*, **118**, 79–85.

93 Kanazirev, V., Price, G.L., and Dooley, K.M. (1990) Enhancement in propane aromatization with Ga_2O_3/HZSM-5 catalysts. *J. Chem. Soc. Chem. Commun.*, 712–713.

94 Choudhary, V.R., Kinage, A.K., and Choudhary, T.V. (1996) Simultaneous aromatization of propane and higher alkanes or alkenes over H-GaAlMFI zeolite. *Chem. Commun.*, 2545–2546.

95 Ono, Y. and Kanae, K. (1991) Transformation of butane over ZSM-5 zeolites. *J. Chem. Soc. Faraday Trans.*, **87**, 669–675.

96 Lukyanov, D.B. and Vazhnova, T. (2007) Aromatization activity of gallium containing MFI and TON zeolite catalysts in n-butane conversion: effects of gallium and reaction conditions. *Appl. Catal. A*, **316**, 61–67.

97 Wang, L., Tao, L., Xie, M., Xu, G., Huang, J., and Xu, Y. (1993) Dehydrogenation and aromatization of methane under non-oxidizing conditions. *Catal. Lett.*, **21**, 35.

98 Tessonnier, J.-P., Louis, B., Rigolet, S., Ledoux, M.J., and Pham-Huu, C. (2008) Methane dehydroaromatization on Mo/ZSM-5: about the hidden role of Bronsted acid sites. *Appl. Catal. A*, **336**, 79–88.

99 (a) Shu, Y., Ma, D., Bao, X., and Xu, Y. (2000) Methane dehydro-aromatization over Mo/MCM-22 catalysts: a highly selective catalyst for the formation of benzene. *Catal. Lett.*, **70**, 67–73.
(b) Sobalik, Z., Tvaruzkova, Z., Wichterlova, B., Fila, V., and Spatenka, S. (2003) Acidic and catalytic properties of Mo/MCM-22 in methane aromatization: an FTIR study. *Appl. Catal. A*, **253**, 271–282.
(c) Ma, D., Shu, Y., Han, X., Liu, X., Xu, Y., and Bao, X. (2001) Mo/HMCM-22 for methane dehydroaromatization: a multinuclear MAS NMR study. *J. Phys. Chem. B*, **105**, 1786–1793.
(d) Ma, D., Zhu, Q., Wu, Z., Zhou, D., Shu, Y., Xin, Q., Xu, Y., and Bao, X. (2005) The synergic effect between Mo species and acid sites in Mo/MCM-22 catalysts for methane aromatization. *Phys. Chem. Chem. Phys.*, **7**, 3102–3109.

100 Wang, D., Lunsford, J.H., and Rosynek, M.P. (1997) *J. Catal.*, **169**, 347–358.

101 Liu, B., Yang, Y., and Sayari, A. (2001) Non-oxidative dehydroaromatization of methane over Ga-promoted Mo/HZSM-5-based catalysts. *Appl. Catal. A*, **214**, 95–102.

102 Matus, E.V., Ismagilov, I.Z., Sukhova, O.B., Zaikovskii, V.I., Tsikova, L.T., Ismagilov, Z.R., and Moulijn, J.A. (2007) Study of methane dehydroaromatization on impregnated Mo/ZSM-5 catalysts and characterization of nanostructured molybdenum phases and carbonaceous deposits. *Ind. Eng. Chem. Res.*, **46**, 4063–4074.

103 Haw, J.F., Song, W., Marcus, D.M., and Nicholas, J.B. (2003) The mechanism of methanol to hydrocarbon catalysis. *Acc. Chem. Res.*, **36**, 317–326.

104 (a) Wilson, S.T. and Barger, P. (1999) The characteristics of SAPO-34 which influence the conversion of methanol to light olefins. *Micropor. Mesopor. Mater.*, **29**, 117–126.
(b) Chang, C.D. (1983) *Hydrocarbons from Methanol*, Marcel Dekker.
(c) Chen, J.Q., Bozzano, A., Glover, B., Fuglerud, T., and Kvisle, S. (2005) Recent advancements in ethylene and propylene production using the UOP/Hydro MTO process. *Catal. Today*, **106**, 103–107.

105 Koempel, H. and Liebner, W. (2007) Lurgi's Methanol to Propylene (MTP®) report on a successful commericialization. *Stud. Surf. Sci. Catal.*, vol. 167, Elsevier, Amsterdam, pp. 261–267.

106 Wendelbo, R., Akporiaye, D., Andersen, A., Dahl, I.M., and Mostad, H.B. (1996) Synthesis, characterization and catalytic testing of SAPO-18, MgAPO-18, and ZnAPO-18 in the MTO reaction. *Appl. Catal. A*, **142**, L197–L207.

107 Itakura, M., Inoue, T., Takahashi, A., Fujitani, T., Oumi, Y., and Sano, T. (2008) Synthesis of high-silica CHA

zeolite from FAU zeolite in the presence of benzyltrimethylammonium hydroxide. *Chem. Lett.*, **37**, 908–909.

108 Chen, Y., Zhou, H., Zhu, J., Zhang, Q., Wang, Y., Wang, D., and Wei, F. (2008) Direct synthesis of a fluidizable SAPO-34 catalyst for a fluidized dimethyl ether to olefins process. *Catal. Lett.*, **124**, 297–303.

109 Park, J.W., Lee, J.Y., Kim, K.S., Hong, S.B., and Seo, G. (2008) Effects of cage shape and size of 8-membered ring molecular sieves on their deactivation in methanol-to-olefin (MTO) reactions. *Appl. Catal. A*, **339**, 36–44.

110 Derouane, E.G., Dejaifve, P., Gabelica, Z., and Vedrine, J.C. (1981) Molecular shape selectivity of ZSM-5, modified ZSM-5 and ZSM-11 type zeolites. *Faraday Discuss. Chem. Soc*, **72**, 283.

111 da Silva Barros, Z., Zotin, F.M.Z., and Henriques, C.A. (2007) Conversion of natural gas to higher valued products: light olefins production from methanol over ZSM-5 zeolites. *Stud. Surf. Sci. Catal.*, vol. 167, Elsevier, Amsterdam, pp. 255–260.

112 Zones, S.I., Burton, A.W., Maesen, T.L.M., Smit, B., and Beerdsen, E. (2007) Hydrocarbon conversion using molecular sieve SSZ-75. US Patent Application 20070284284.

113 Furimsky, E. (2007) Catalysts for upgrading heavy petroleum feeds. *Stud. Surf. Sci. Catal.*, vol. 169, Elsevier, Amsterdam, pp. 1–290.

114 (a) Gruia, A. (2006) Distillate hydrocracking, in *Handbook of Petroleum Processing* (eds D.S.J. Jones and P.R. Pujado), Springer, pp. 287–320.(b) Yergin, D. (1992) *The Prize: The Epic Quest for Oil, Money and Power*, Simon and Schuster, pp. 311–332.

115 Thompson, G.J. (1997) UOP RCD unionfining process, in *Handbook of Petroleum Refining Processes*, 2nd edn (ed. R.A. Myers), McGraw-Hill, pp. 8-39–8-48.

116 Brossard, D.N. (1997) Chevron RDS/VRDS hydrotreating–transportation fuels from the bottom of the barrel, in *Handbook of Petroleum Refining Processes*, 2nd edn (ed. R.A. Myers), McGraw-Hill, pp. 8.3–8.26

117 Reno, M. (1997) UOP unicracking process for hydrocracking, in *Handbook of Petroleum Refining Processes*, 2nd edn (ed. R.A. Myers), McGraw-Hill, pp. 7.41–7.49.

118 Genis, O. (1997) UOP catalytic dewaxing process, in *Handbook of Petroleum Refining Processes*, 2nd edn (ed. R.A. Myers), McGraw-Hill, pp. 8.49–8.53

119 Bridge, A.G. (1997) Chevron isocracking–hydrocracking for superior fuels and lubes production, in *Handbook of Petroleum Refining Processes*, 2nd edn (ed. R.A. Myers), McGraw-Hill, pp. 7.21–7.40.

120 Kennedy, J.E. (1997) UOP unionfining technology, in *Handbook of Petroleum Refining Processes*, 2nd edn (ed. R.A. Myers), McGraw-Hill, pp. 8.29–8.38.

121 Hancsok, J., Krar, M., Magyar, S., Boda, L., Hollo, A., and Kallo, D. (2007) Investigation of the production of high quality biogasoil from pre-hydrogenated vegetable oils over Pt/SAPO-11/Al2O3. *Stud. Surf. Sci. Catal.*, vol. 170B, Elsevier, Amsterdam, pp. 1605–1610.

122 Landau, M.V., Vradman, L., Valtchev, V., Lezervant, J., Liubich, E., and Talianker, M. (2003) Hydrocracking of heavy vacuum gas oil with a Pt/H-beta/Al2O3 catalyst: effect of zeolite crystal size in the nanoscale range. *Ind. Eng. Chem. Res.*, **42**, 2773–2782.

123 Sato, K., Nishimura, Y., Honna, K., Matsubayashi, N., and Shimada, H. (2001) Role of HY mesopores in hydrocracking of heavy oils. *J. Catal.*, **200**, 288–297.

124 Zones, S.I. and Harris, T.V. (2006) Hydrocarbon conversion using molecular sieve SSZ-51. US Patent 7115198.

125 (a) Taramasso, M., Perego, G., and Notari, B. (1983) Preparation of porous crystalline synthetic material composed of silicon and titanium oxides. US Patent 4410501.
(b) Taramasso, M., Manara, G., Fattore, V., and Notari, B. (1987) Silica-based synthetic material containing titanium in the crystal lattice and process for its preparation. US Patent 4666692.
(c) Bellussi, G. and Fattore, V. (1991) *Stud. Surf. Sci. Catal.*, vol. 69, Elsevier, Amsterdam, pp. 79.

(d) Notari, B. (1987) Stud. Surf. Sci. Catal., vol. 37, Elsevier, Amsterdam, pp. 413.
(e) Notari, B. (1991) Titanium silicalite: a new selective oxidation catalyst. Stud. Surf. Sci. Catal., vol. 60, Elsevier, Amsterdam, pp. 343.

126 Bellussi, G. and Rigutto, M.S. (2001) Metal ions associated to molecular sieve frameworks as catalytic sites for selective oxidation reactions. Stud. Surf. Sci. Catal., vol. 137, Elsevier, Amsterdam, pp. 911–955.

127 Nijuis, T.A., Makkee, M., Moulijn, J.A., and Weckhuysen, B.M. (2006) The production of propene oxide: catalytic processes and recent developments. Ind. Eng. Chem. Res., 45, 3447–3459.

128 Buijink, J.K.F., van Vlaanderen, J.J.M., Crocker, M., and Niele, F.G.M. (2004) Propylene epoxidation over titanium-on-silica catalyst – the heart of the SMPO process. Catal. Today, 93–95, 199–204.

129 Lane, B.S. and Burgess, K. (2003) Metal-catalyzed epoxidations of alkenes with hydrogen peroxide. Chem. Rev., 103, 2457–2474.

130 Parmon, V.N., Panov, G.I., Uriarte, A., and Noskov, A.S. (2005) Nitrous oxide in oxidation chemistry and catalysis: application and production. Catal. Today, 100, 115–131.

131 Wu, P. and Tatsumi, T. (2002) Unique trans-selectivity of Ti-MWW in epoxidation of cis/trans-alkenes with hydrogen peroxide. J. Phys. Chem. B., 106, 748–753.

132 Fan, W., Wu, P., Namba, S., and Tatsumi, T. (2004) A titanosilicate that is structurally analogous to an MWW-type lamellar precursor. Angew. Chem. Int. Ed., 43, 236–240.

133 Santa, A.M., Vergara, J.C., Palacio, L.A., and Echavarria, A. (2008) Limonene epoxidation by molecular sieves zincophosphates and zincochromates. Catal. Today, 133–135, 80–86.

134 Clerici, M.G. and Ingallina, P. (1993) Epoxidation of lower olefins with hydrogen peroxide and titanium silicate. J. Catal., 140, 71–83.

135 Jenzer, G., Mallat, T., Maciejewski, M., Eigenmann, F., and Baiker, A. (2001) Continuous epoxidation of propylene with oxygen and hydrogen on a Pd-Pt/TS-1 catalyst. Appl. Catal. A, 208, 125–133.

136 Cheng, W., Wang, X., Li, G., Guo, X., and Zhang, S. (2008) Highly efficient epoxidation of propylene to propylene oxide over TS-1 using urea + hydrogen peroxide as oxidizing agent. J. Catal., 255, 343–346.

137 Bowman, R.G., Kuperman, A., Clark, H.W., Hartwell, G.E., and Meima, G.R. (2001) Process for the direct oxidation of olefins to olefin oxides. US Patent 6323351.

138 Clerici, M.G. (2000) Zeolites for fine chemical production. Topic. Catal., 13, 373–386.

139 Raja, R., Thomas, J.M., Greenhill-Hooper, M., Ley, S.V., and Almeida Paz, F.A. (2008) Facile, one-step production of niacin (Vitamin B3) and other nitrogen-containing pharmaceutical chemicals with a single-site heterogeneous catalyst. Chem. Eur. J., 14, 2430–2438.

140 Thomas, J.M., and Raja, R. (2005) Design of a "green" one-step catalytic production of e-caprolactam (precursor of nylong-6). Proc. Natl. Acad. Sci. U. S. A., 102, 13732–13736.

141 Blasco, T., Concepcíon, P., López Nieto, J.M., and Pérez-Pariente, J. (1995) Preparation, characterization, and catalytic properties of VAPO-5 for the oxydehydrogenation of propane. J. Catal., 152, 1–17.

142 Kumar, R., Mukherjee, P., and Bhaumik, A. (1999) Enhancement in the reaction rates in the hydroxylation of aromatics over TS-1/H_2O_2 under solvent free triphase conditions. Catal. Today, 49, 185–191.

143 Bravo-Suárez, J.J., Bando, K.K., Akita, T., Fijitani, T., Fuhrer, T.J., and Oyama, S.T. (2008) Propane reacts with O_2 and H_2 on gold supported TS-1 to form oxygenates with high selectivity. Chem. Commun., 3272–3274.

144 Sun, K., Xia, H., Feng, Z., van Santen, R., Hensen, E., and Li, C. (2008) Active sites in Fe/ZSM-5 for nitrous oxide decomposition and benzene hydroxylation with nitrous oxide. J. Catal., 254, 383–396.

145 Ichihashi, Y., Taniguchi, T., Amano, H., Atsumi, T., Nishiyama, S., and Tsuruya,

S. (2008) Liquid phase oxidation of benzene to phenol by molecular oxygen over La catalysts supported on HZSM-5. *Top. Catal.*, **47**, 98–100.

146 Notté, P.P. (2000) The AlphOx™ process or the one-step hydroxylation of benzene into phenol by nitrous oxide. Understanding and tuning the ZSM-5 catalyst activities. *Top. Catal.*, **13**, 387–394.

147 Centi, G. and Trifiro, F. (1996) Catalytic behavior of V-containing zeolites in the transformation of propane in the presence of oxygen. *Appl. Catal. A*, **143**, 3–16.

148 Teixeira-Neto, A.A., Marchese, L., Landi, G., Lisi, L., and Pastore, H.O. (2008) [V,Al]-MCM-22 catalyst in the oxidative dehydrogenation of propane. *Catal. Today*, **133–135**, 1–6.

149 (a) Treacy, M.M.J., Rivin, I., Balkovsky, E., Randall, K.H., and Foster, M.D. (2004) Enumeration of period tetrahedral frameworks. II. Polynodal graphs. *Micropor. Mesopor. Mater.*, **74**, 121.
(b) Zwijnenberg, M.A., Cora, F., and Bell, R.G. (2007) Dramatic differences between the energy landscapes of SiO_2 and SiS_2 zeotype materials. *J. Am. Chem. Soc.*, **129**, 12588–12589.
(c) Mellot-Draznieks, C., Serre, C., Surblé, S., Audebrand, N., and Férey, G. (2005) Very large swelling in hybrid frameworks: a combined computational and powder diffraction study. *J. Am. Chem. Soc.*, **127**, 16273–16278.

150 (a) Moliner, M., Díaz-Cabañas, M.J., Fornes, V., Martinez, C., and Corma, A. (2008) Synthesis methodology, stability, acidity, and catalytic behavior of the 18 × 10 member ring pores ITQ-33 zeolite. *J. Catal.*, **254**, 101–109.
(b) Corma, A., Díaz-Cabañas, M.J., Rey, F., Nicolopolous, S., and Boulahya, K. (2004) ITQ-15: the first ultralarge pore zeolite with a bi-directional pore system formed by intersecting 14- and 12-ring channels, and its catalytic implications. *Chem. Commun.*, 1356–1357.
(c) Paillaud, J.-L., Harbuzaru, B., Patarin, J., and Bats, N. (2004) Extra-large-pore zeolites with two-dimensional channels formed by 14 and 12 rings. *Science*, **304**, 990–992.

151 (a) Wang, H., and Pinnavaia, T.J. (2006) MFI zeolite with small and uniform intracrystal mesopores. *Angew. Chem. Int. Ed.*, **45**, 7603–7606.
(b) Ogura, M. (2008) Towards realization of a Micro- and Mesoporous composite catalyst. *Catal. Surv. Asia*, **12**, 16–27.

152 Park, K.S., Ni, Z., Cote, A.P., Choi, J.Y., Huang, R., Uribe-Romo, F.J., Chae, H.K., O'Keefe, M., and Yaghi, O. (2006) Exceptional chemical and thermal stability of zeolitic imidazolate frameworks. *Proc. Natl. Acad. Sci. U. S. A.*, **103**, 10186–10191.

153 Llabrés i Xamena, F.X., Casanova, O., Galiasso Tailleur, R., Garcia, H., and Corma, A. (2008) Metal organic frameworks (MOFs) as catalysts: a combination of Cu^{2+} and Co^{2+} MOFs as an efficient catalyst for tetralin oxidation. *J. Catal.*, **255**, 220–227.

154 (a) Mastalerz, M. (2008) The next generation of shape-persistent zeolite analogues: covalent organic frameworks. *Angew. Chem. Int. Ed.*, **47**, 445–447.
(b) Cote, A.P., Benin, A.I., Ockwig, N.W., O'Keefe, M., Matzger, A.J., and Yaghi, O.M. (2005) Porous, crystalline, covalent organic frameworks. *Science*, **310**, 1166–1170.

155 (a) Parnham, E.R. and Morris, R.E. (2007) Ionothermal synthesis of zeolites, metal-organic frameworks, and inorganic-organic hybrids. *Acc. Chem. Res.*, **40**, 1005–1013.
(b) Parnham, E.R. and Morris, R.E. (2006) The ionothermal synthesis of cobalt aluminophosphate zeolite frameworks. *J. Am. Chem. Soc.*, **128**, 2204–2205.
(c) Xing, H., Li, J., Yan, W., Chen, P., Jin, Z., Yu, J., Dai, S., and Xu, R. (2008) Cotemplating ionothermal synthesis of a new open-framework aluminophosphate with unique Al/P ratio of 6/7. *Chem. Mater.*, **20**, 4179–4181.
(d) Kuhn, P., Antonietti, M., and Thomas, A. (2008) Porous, covalent triazine-based frameworks prepared by ionothermal synthesis. *Angew. Chem. Int. Ed.*, **47**, 3450–3453.

13
Unique Aspects of Mechanisms and Requirements for Zeolite Catalysis in Refining and Petrochemicals
Hayim Abrevaya

13.1
Introduction

Before discussing the mechanisms and zeolitic requirements for some specific reactions, including shape selectivity effects, this chapter discusses the governing principles of performance parameters like adsorption, diffusion, acidity and how these parameters fundamentally influence zeolite catalysis. The section on adsorption starts by describing Langmuir isotherms and Langmuir–Hinshelwood rate expressions for some simple reactions. This is followed by description of N_2 adsorption measurements used for characterizing microporosity in zeolites, as well as mesopore surface area and pore volume distribution. Then adsorption of various hydrocarbon probe molecules is discussed for characterizing zeolite topology. In addition to pure hydrocarbon work, adsorption of mixtures is discussed with the ultimate objective of understanding shape selectivity effects. The section on diffusion focuses on different techniques for measuring intracrystalline diffusivity with illustrations of how diffusion impacts catalytic performance. On the topic of acidity, first, generation of Bronsted and Lewis acid sites are described, followed by a discussion of the significance of acid strength and acid site density for catalysis. Then there is a section on the formation and characterization of carbocations in zeolites, including carbenium ions and carbonium ions. Formation of surface alkoxides is also discussed. This is followed by brief descriptions of the elementary steps of hydrocarbon conversion over zeolites: protonation, deprotonation, hydride shift, beta scission, alkylation, hydride transfer, protonated cyclopropane ring closure and ring opening. Various tools are then provided for shape selectivity analysis. These include the list of footprints for characterizing the size of molecules, the list of constraint indices, modified constraint indices and spaciousness indices for further characterization of zeolite topology. Discussion of various shape selectivity effects are broken down into three groups. Selectoforming is used for illustrating reactant shape selectivity effect. Transition state shape selectivity is illustrated by *meta*-xylene disproportionation, methylcyclohexane ring opening, alkane hydroisomerization and dimethylether carbonylation. Product shape selectivity is illustrated by alkane hydroisomerization, alkene oligomerization, alkylation of naphthalene, hydrocracking, *meta*-xylene isomerization and methanol-to-olefins. Finally, the mechanism of some selected reactions are analyzed in some detail. These include alkene double bond isomerization, alkene skeletal isomerization, alkene oligomerization, isobutane–butylene alkylation, aromatic–alkene and aromatic–alcohol alkylation, xylene dispropotionation, ethylbenzene isomerization to xylenes, bimolecular alkane cracking, monomolecular alkane cracking and methanol-to-olefins. The intent here has been not to have an extensive review of the field of zeolite catalysis, but to select a sufficiently large subset of published literature through which key points can be made about reaction mechanisms and zeolitic requirements.

13.2
Adsorption

Adsorption is the opening act for zeolite catalysis. The interaction between an approaching organic molecule and a zeolite involves van der Waals, polar and chemical components. Van der Waals interaction between the organic and the framework oxygens is sensitive to the pore geometry. The interaction is maximized when the size of the sorbate molecule approaches the pore size. Polar interaction is sensitive to the polarity of the sorbate, as well as to the charges on the zeolite. The latter arises from the presence of Bronsted acid sites, nonframework cations and silanols. External silanols enable preservation of the tetrahedral coordination at the crystal exterior. Internal silanols often are indication of defect sites and may be generated as a result of framework dealumination. Chemical interaction between a Bronsted acid site and the incoming organic is essentially an acid–base reaction.

13.2.1
Langmuir Isotherm and Reaction Kinetics

At a given temperature adsorption isotherms measure the number of adsorbed molecules as a function of pressure for the fluid that is in contact with the zeolite. The simplest form is the Langmuir isotherm which treats the zeolite as a collection of equivalent adsorption sites in the absence of adsorbate–adsorbate interactions:

$$C_A = C_t K_A P_A / (1 + K_A P_A)$$

Here C_A is the number of adsorbed molecules A per unit volume, C_t is the maximum loading of A, K_A is the adsorption equilibrium constant and P_A is the partial pressure of A. During adsorption there could be several energetically and topologically inequivalent adsorption sites and, in that case, total adsorption may simply become the sum of multiple Langmuir isotherms with contributions from different adsorption sites. The latter and several other isotherms that model the inequivalency of adsorption sites in different ways are discussed in the literature [1].

The derivative of the equilibrium constant with respect to temperature at constant surface coverage is the isosteric heat of adsorption:

$$Q_{St} = \partial K/\partial T$$

In the limit of low pressure the front end of the adsorption isotherm is approximated by the Henry regime which states that the number of adsorbed molecules per unit volume is proportional to the pressure and to the Henry coefficient, K_H:

$$K_H = C_t K_A$$
$$C_A = K_H P_A$$

Based on Langmuir–Hinshelwood kinetics the rate expression for a first order reaction (A → R) that is surface reaction-controlled becomes equal to the following expression [2]:

$$r_A = \frac{k_{sr} K_A C_t (P_A - P_R/K_P)}{1 + K_A P_A + K_R P_R}$$

Here k_{sr} is the surface reaction rate constant, K_R is the adsorption equilibrium constant for product R, P_R is the partial pressure of R and K_P is the reaction equilibrium constant. At low loading the reaction rate simply becomes proportional to the product of the intrinsic rate constant and the Henry coefficient.

For the same A → R reaction, when kinetics are controlled by adsorption of reactant A rather than by surface reaction, the Langmuir–Hinshelwood rate expression becomes equal to the following [2]:

$$r_A = \frac{k_A C_t (P_A - P_R/K_P)}{1 + \dfrac{K_A}{K_P} P_R + K_R P_R}$$

Here k_A is the rate constant for the adsorption of A. Conversely, desorption control of kinetics for A → R leads to the following rate expression [2]:

$$r_A = \frac{k_R K_P K_R C_t (P_A - P_R/K_P)}{1 + K_A P_A + K_P K_R P_A}$$

Here k_R is the desorption rate constant for R. Another rate expression that is often applicable is when kinetics are controlled by impact of gas phase A on adsorbed B for the A + B → R reaction. For this case:

$$r_A = \frac{k_A K_B C_t (P_A P_B - P_R/K_P)}{1 + K_B P_B + K_R K_R}$$

A does not adsorb. These and other rate expressions for some simple reactions have been first derived by Yang and Hougen and later summarized by Froment and Bischoff [2, 3].

13.2.2
Nitrogen Adsorption

Microporous substances with pore widths smaller than twenty Angströms exhibit type I isotherms (IUPAC classification) in the absence of mesopores, as illustrated in Figure 13.1 [4]. The steep uptake of N_2 at very low relative pressures is due to the capillary condensation in micropores. Following the filling of micropores, adsorption isotherms become nearly flat because further N_2 uptake can now only occur at the external surface area, which is typically much lower relative to the total surface area.

A typical N_2 adsorption measurement versus relative pressure over a solid that has both micropores and mesopores first involves essentially a mono-layer coverage of the surface up to point B shown in isotherm IV (IUPAC classification) in Figure 13.1. Up to and near point B the isotherm is similar to a Langmuir isotherm for which equilibrium is established between molecules adsorbing from the gas phase onto the bare surface and molecules desorbing from the adsorbed layer. The volume of adsorbed N_2 that covers a monolayer volume, hence the surface area of N_2 can then be determined from the slope of the linearized Langmuir plot when P/V is plotted against P:

$$\frac{P}{V} = \frac{P}{V_m} + \frac{1}{bV_m}$$

Here P is the equilibrium pressure during the adsorption measurements, V the adsorbed volume of N_2, V_m the monolayer volume and b is a constant.

Following mono-layer uptake, further increase in pressure results in multi-layer adsorption of N_2. For this part of the isotherm, condensation–evaporation equilibrium is assumed to take place, instead of adsorption–desorption equilibrium for each individual layer other than the first layer. This dynamic equilibria for the first and higher layers and some simplifying assumptions form the basis for the BET treatment of the multi-layer adsorption isotherm. A lengthy derivation leads to the BET relation between adsorbed volume of N_2 and relative pressure. Here relative pressure is defined as the ratio of the equilibrium pressure to the

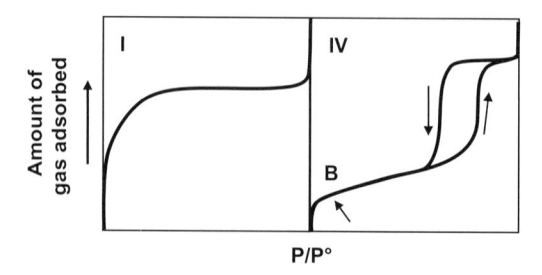

Figure 13.1 Type I and IV adsorption isotherms [4].

saturation pressure, P/P^0. Since adsorption measurements are done at the boiling point of liquid N_2, 77°K, the saturation pressure is 101 kPa. This derivation also produces a constant related to the heat of adsorption of N_2 on the surface, c. A linearized form of the BET plot, typically in P/P^0 range of 0.05 to 0.25, can then be used to calculate V_m, hence the BET surface area from the slope and intercept:

$$\frac{p}{V(p^0-p)} = \frac{1}{V_m c} + \frac{c-1}{V_m c}\frac{p}{p^0}$$

Making a change in the independent variable from P/P^0 to t, the thickness of the adsorbate layer provides a method for calculating the mesopore plus external surface area. This t-plot surface area is obtained from the slope, s_t, of the amount of adsorbed N_2, n (moles adsorbed/g), plotted against the thickness of the adsorbed layer, t (nm). The latter is equal to the thickness of a single layer multiplied by the number of layers, n/n_m where n_m is the amount of adsorbed N_2 for monolayer coverage. The thickness of a single layer is 0.354 nm based on hexagonal close packing of N_2:

$$t = (0.1399/(0.034 - \log_{10} P/P^0))^{1/2}$$
$$n = n_m * t / 0.354$$
$$s_t = n_m / 0.354$$

at 77°K [4]. In the t-plot method the slope is $n_m/0.354$. Multiplying the slope by 0.354 and the area occupied by 1 mole of N_2, hence gives the t-plot area. Here the area (m²/g) occupied by 1 mole of N_2 is $16.2*10^{-20} * L$, where L is the Avogadro's number. Micropore surface area is given by the difference between BET and t-plot surface areas. The micropore volume (V_{MP} in cm³/g) is calculated from the intercept of the t-plot, i_t:

$$A_t = s_t * 0.354 * 16.2 * 10^{-20} * L \quad (t\text{-plot surface area in m}^2/\text{g})$$
$$A_{BET} = V_m * 16.2 * 10^{-20} * L / 22410 \quad (\text{BET surface area in m}^2/\text{g})$$
$$V_{MP} = i_t * 28.01 / 0.8081 \quad (\text{micropore volume in cm}^3/\text{g})$$

As progressively higher pressures are used during N_2 adsorption, capillary condensation will occur in pores that are increasingly larger. The Kelvin equation illustrates that the equilibrium vapor pressure is lowered over a concave meniscus of liquid, which is why N_2 is able to condense in catalyst pores at pressures lower than the saturation pressure:

$$\ln \frac{P}{P^0} = \frac{-2\gamma V_L}{RT}\frac{1}{r_m}$$

The smaller the radius of curvature, the lower the vapor pressure. Here r_m is the radius of curvature, which is equal to the pore radius minus the thickness of the adsorbed N_2 layer, γ is the surface tension and V_L is the molar volume. This capillary condensation as a function of pressure helps establish the pore size distribution when the volume of adsorbed N_2 is plotted against P/P^0. A sharp increase in the N_2 uptake is then observed at the pressure corresponding to the filling of mesopores. This type of isotherm is known as type IV, as illustrated in Figure 13.1.

As relative pressure approaches one, condensation of N_2 becomes complete and the entire pore volume, including the larger pores, becomes filled with liquid N_2. At that point, if the pressure is reduced by a small increment a small amount of N_2 evaporates from the meniscus formed at the ends of the largest pores. The radii of pores emptied by the condensate is given by the Kelvin equation. With catalysts exhibiting type IV isotherms it is typical to desorb a larger quantity of vapor than that corresponding to adsorption at a given relative pressure, the so-called hysteresis observation. Usually desorption is used to calculate the pore size distribution, rather than adsorption, because during desorption the shape of the meniscus is believed to be close to hemispherical, that is, the contact angle is zero. Only in that case can the curvature of the meniscus be used to calculate the radius of the pore.

13.2.3
Hydrocarbon Adsorption

13.2.3.1 Pure Component Adsorption and Specificity with Respect to Zeolite Topology

Pure component adsorption over three-dimensional medium size ten-membered ring (10-MR) MFI framework type has been the most widely studied system. Over MFI the Henry coefficient for linear alkanes increases with carbon number, so does the isosteric heat of adsorption. These increases should continue until the size of the alkane becomes too big for the molecule to fit without creating an energetically unfavorable configuration. The maximum loading is similar for linear alkanes and approximately corresponds to the free volume of MFI. The multi-step adsorption isotherms observed with some linear alkanes over MFI has been interpreted as arising from independent contributions of the straight and zigzag channels. A similar phenomenon was also observed with FER framework type. Over FER, while linear alkanes with fewer than five carbons adsorb both in 8-MR and 10-MR channels, linear alkanes with more than five carbons prefer to adsorb in the 10-MR channels [5]. Configurational bias Monte Carlo (CBMC) simulations for sorption of n-alkanes with carbon numbers $n = 1$–4 over MFI were in good agreement with experimental data, as illustrated in Figure 13.2 for propane and n-butane [6].

CBMC simulations indicated that over MFI branched alkanes preferentially adsorb at the intersections of the straight and zigzag channels and these have a maximum capacity of four molecules per unit cell [6]. Only at sufficiently high pressures adsorption for mono-branched alkanes starts occurring in between the intersections, that is, in the straight and zigzag channels and that may cause an inflection point in the adsorption isotherm. This inflection point has been observed both experimentally and by CBMC simulations for i-butane around 1 to 10 kPa, as illustrated in Figure 13.3 and by other mono-branched alkanes up to a carbon number of eight [6–8]. CBMC simulations over MFI at carbon number six are illustrated in Figure 13.4. n-Hexane and 3-methylpentane adsorption isotherms are similar to n-butane and i-butane, respectively. Figure 13.5 gives schematic

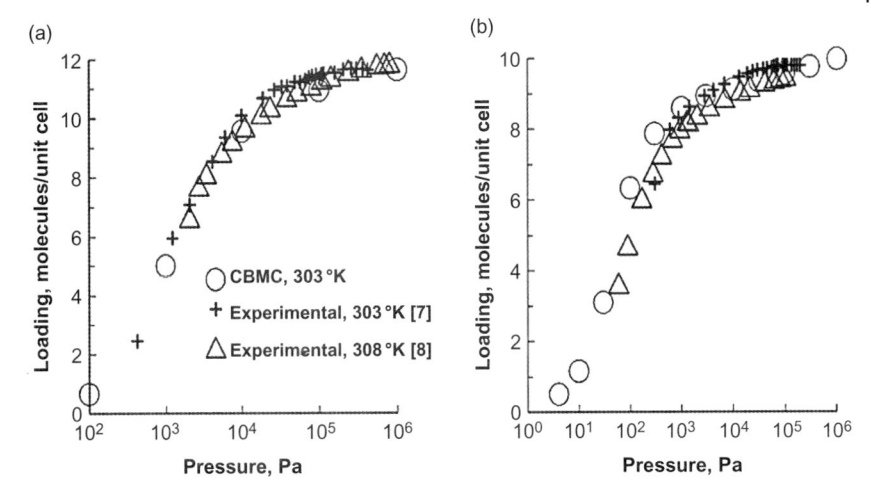

Figure 13.2 Comparison of experimental and CBMC simulations for adsorption of (a) propane and (b) n-butane over MFI [6].

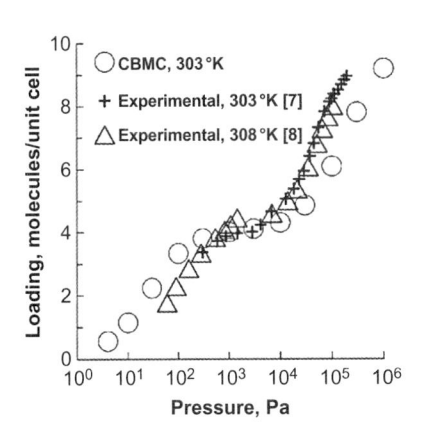

Figure 13.3 Comparison of experimental and CBMC simulations for adsorption of i-butane over MFI [6].

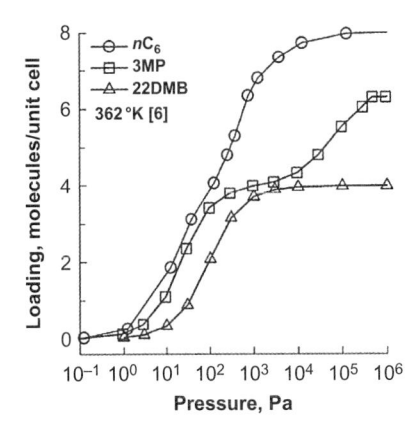

Figure 13.4 CBMC simulations for adsorption of linear and branched hexanes over MFI [6].

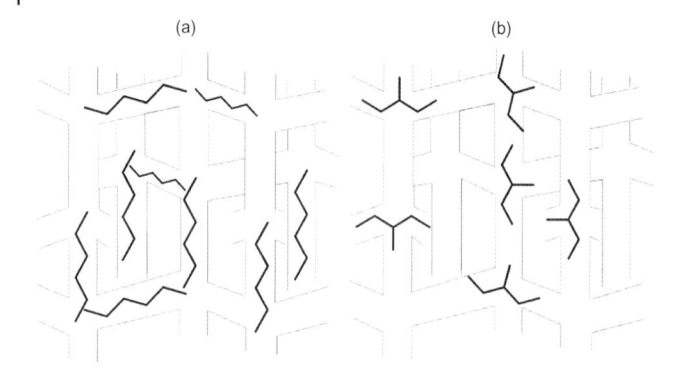

Figure 13.5 Sitting of (a) *n*-hexane and (b) 3-methylpentane in MFI [6].

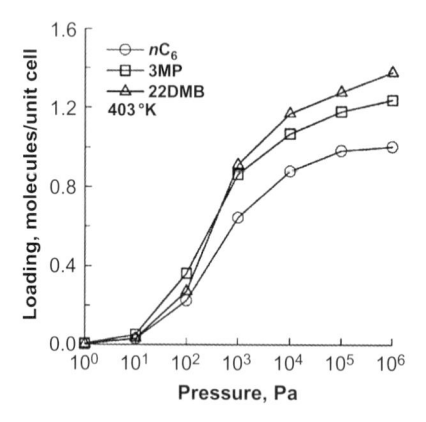

Figure 13.6 CBMC simulations for linear and branched hexane adsorption over AFI [6].

drawings for the siting of *n*-hexane and 3-methylpentane in MFI. 2,2-Dimethylbutane also prefers to sit at the intersections. However, since it is bulkier than 3-methylpentane it cannot be pushed to the channel interiors, even at high pressures. Accordingly, the maximum loading remains at four molecules per unit cell [6]. Over MFI the maximum loading for alkanes follows the order linear > mono-branched > dibranched [9]. A similar ranking was observed over one-dimensional small 10-MR TON and MTT framework types [10, 11]. Interestingly, the same inflection point at four molecules per unit cell was also observed experimentally for *para*-xylene adsorption over MFI at around 30 kPa and 313°K [1]. In contrast, over three-dimensional 12-MR AFI framework type the reverse order is predicted by CBMC simulations, as illustrated in Figure 13.6, presumably because bulky dibranched alkanes can be more effectively packed in the large channels than mono-branched alkanes, which in turn pack more efficiently than linear alkanes [6, 9].

Adsorption capacities for a large number of gases and hydrocarbons on various zeolites has been reported by Breck [12]. More recent references exist that discuss the specificity of adsorption with various zeolites, including more recently discovered zeolites [13]. The ability to adsorb n-alkanes but not i-alkanes is typically a marker for 8-MR zeolites. The ability to adsorb benzene and cyclohexane, but not larger, typically classifies MTT, MFI, MEL and MRE framework types as medium-pore. The relatively lower benzene and cyclohexane adsorption capacities observed with MTT and MRE are attributed to their one-dimensional nature. MFI can adsorb mono-branched alkanes but not 2,2-dimethylbutane, can adsorb 2-methylnaphthalene but not 1-methylnaphthalene. Adsorption of bulky 1,3,5-trimethylbenzene (TMB) is typically used to distinguish 12-MR zeolites from 10- and 8-MR zeolites. The lower 1,3,5-TMB adsorption with MTW versus MOR is attributed to the smaller pore opening. The lower hydrocarbon adsorption capacity, in general, with 12-MR MTW and MOR relative to FAU is also attributed to the one-dimensional nature of MTW and MOR. The pore size of extra-large zeolites was characterized by adsorption of 1,3,5 triisopropylbenzene. The latter has little capacity with 12-MRs, but high capacity with 14-MR zeolites like DON and VFI framework types.

13.2.3.2 Energetics of Adsorption

In situ infrared (IR) spectroscopy and calorimetry were used by the Lercher group during alkane adsorption measurements by gravimetry for various zeolites [5, 14]. IR spectra of the OH-streching vibrations of Bronsted acid sites in Ferrierite during adsorption of n-butane at 333°K are illustrated in Figure 13.7. The data showed that, as the partial pressure of n-butane increased, the 3600 cm^{-1}

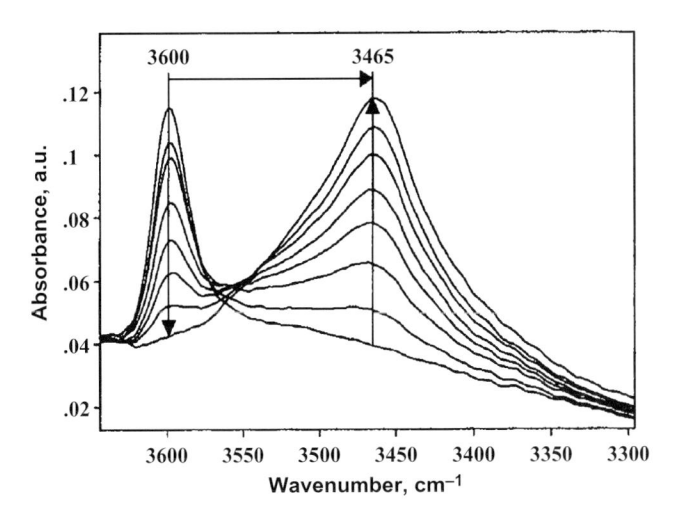

Figure 13.7 Infrared spectra of the OH-stretching vibrations of Bronsted acid sites over FER during adsorption of n-butane at 333°K [5].

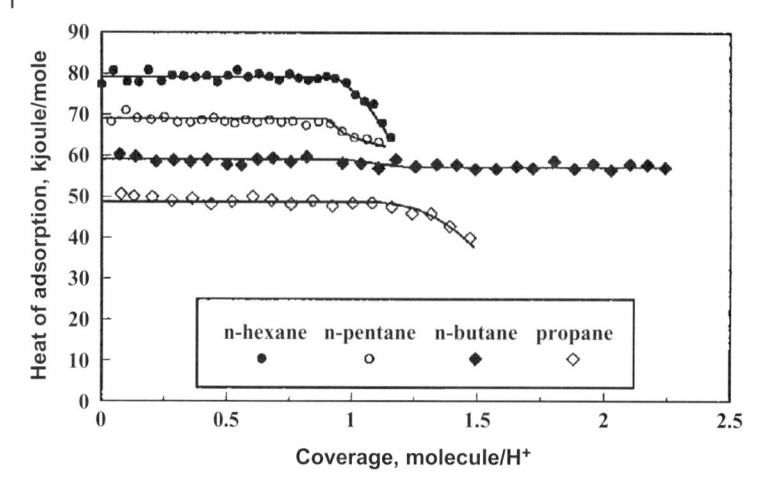

Figure 13.8 Differential heats of adsorption on n-alkanes over FER as a function of coverage at 333°K [5].

band (representative of the O–H stretching frequency in Bronsted acid sites) broadened and shifted to lower wavenumbers (by 135 cm^{-1}) due to the perturbation of the O–H bond upon interaction with n-butane: the larger the perturbation (the stronger the bond with the Bronsted acid site), the larger the shift in frequency [5].

The differential heat of adsorption, Q_d, for n-butane over FER framework type was 59 kJ/mol until a coverage of about one n-butane per acid site was reached, beyond which there was 1–2 kJ/mol lower heat, as illustrated in Figure 13.8. The differential heat of adsorption is related to the isosteric heat of adsorption according to the following relation.

$$Q_{st} = Q_d + RT$$

For propane, n-pentane and n-hexane the differential heats of adsorption over FER dropped more rapidly, right after ~1 molecule was adsorbed per Bronsted acid site. Similar results were obtained with TON. In contrast, with MOR and FAU the drop in the differential heats of adsorption for n-alkanes occurred at lower coverages, indicating that only a certain fraction of the Bronsted acid sites were accessible to the adsorbing alkane probe molecules. With MFI the drop did not occur until ~2 molecules of n-alkane were adsorbed per Bronsted acid site, suggesting perhaps a higher stoichiometry of about two n-alkanes per Bronsted acid site. In the cases of i-butane and i-pentane the drop occurred around one alkane per Bronsted acid site. Finally, n-butane adsorption isotherms measured over TON framework type catalysts having three different Al contents (Si/Al$_2$ = 90, 104, 128) showed Henry coefficients to increase with increase in the Al content [5]. Based

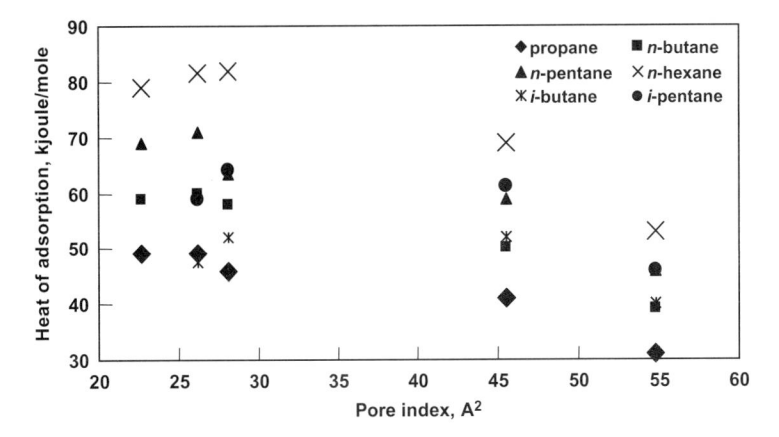

Figure 13.9 Effect of pore size on heat of adsorption for alkanes [5, 14].

on the experimental evidence presented in this series of work it was then concluded that, at temperatures below 373°K, alkanes adsorbed on Bronsted acid sites via hydrogen bonding of the dipole induced in the alkane with the hydroxyl group.

From the series of work described above from the Lercher group the differential heats of adsorption for various alkanes are summarized in Figure 13.9, which plots the heat of adsorption as a function of the pore index (PI), the product of the two dimensions for the pore opening in the largest zeolite channel. The zeolite framework type with the smallest pore displayed in Figure 13.9 is FER, followed by TON, MFI, MOR and FAU [5, 14]. The first key observation is, for any given zeolite, the differential heat of adsorption increases with increase in the chain length for the n-alkane. The second key observation is, for any n-alkane, beyond a pore index of 26–28 Å^2, there is a decrease in the differential heat of adsorption. The third observation is, for small pores, the linear alkane generates a higher differential heat of adsorption relative to branched alkane (though this difference seems to disappear with larger pores). These results, in general, indicate that the heat of adsorption is very much influenced by the fit between the alkane and the zeolite pore size and shape, and there is an optimum interaction occurring at different pore sizes with different alkanes.

13.2.3.3 Adsorption of Mixtures

Often the ratio of Henry coefficients, related to adsorption at zero loading, is used for predicting the selectivity of adsorption for mixtures. The ratio of Henry coefficients for linear and mono-branched alkanes with carbon number $n = 5–8$ are summarized for various zeolites in Figure 13.10 [15]. The Henry coefficient ratios were ~1 for FAU, ~2 for BEA, MOR and MFI, 6–9 for TON and 10–14 for MTT. Interestingly, CBMC simulations suggest that the ratio of Henry coefficients, actu-

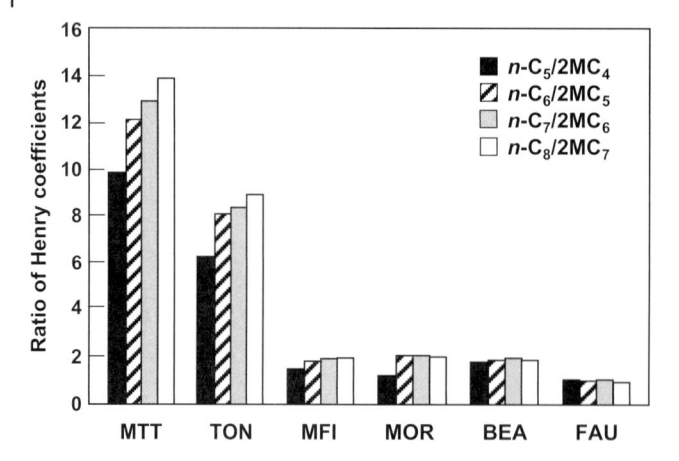

Figure 13.10 Ratio of Henry coefficients for linear and branched alkanes over various framework types [15].

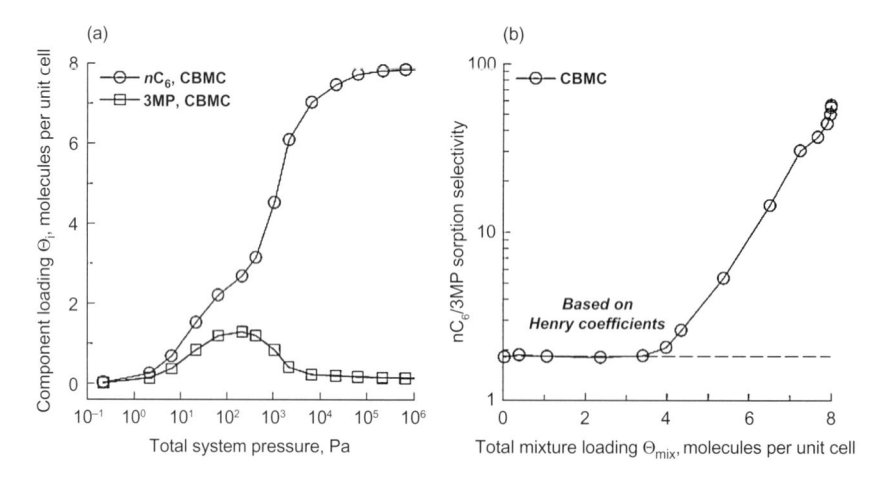

Figure 13.11 CBMC simulations of adsorption isotherm (a) and adsorption selectivity (b) for a 50:50 mixture of n-hexane and 3-methylpentane over MFI at 362°K [6].

ally, is not a good measure of the adsorption selectivities at high loadings, as illustrated in Figure 13.11 for a 50:50 mixture of n-hexane and 3-methylpentane. Here the selectivity for the linear alkane significantly increases above the value predicted by the Henry coefficients, simply because at high pressures not only the vacant sites in MFI are more easily filled with the smaller molecule, but also n-hexane preferentially displaces 3-methylpentane [6]. The variation of linear:monobranched alkane adsorption selectivity ratio as a function of pressure for various framework types are illustrated in Figure 13.12.

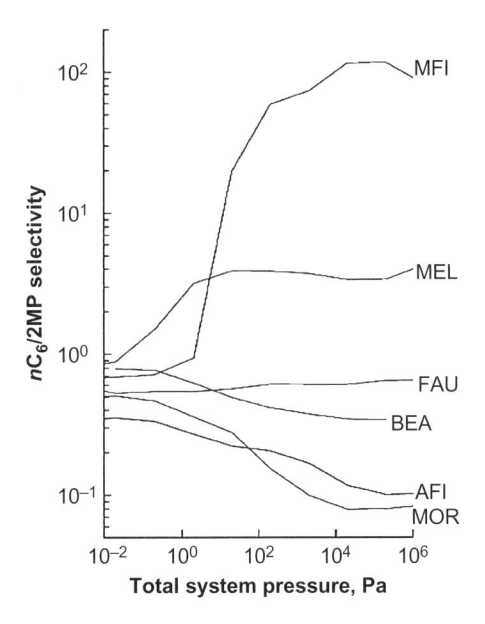

Figure 13.12 CBMC simulations of adsorption selectivity in a 50:50 mixture of n-hexane and 2-methylpentane at 300°K over various framework types [6].

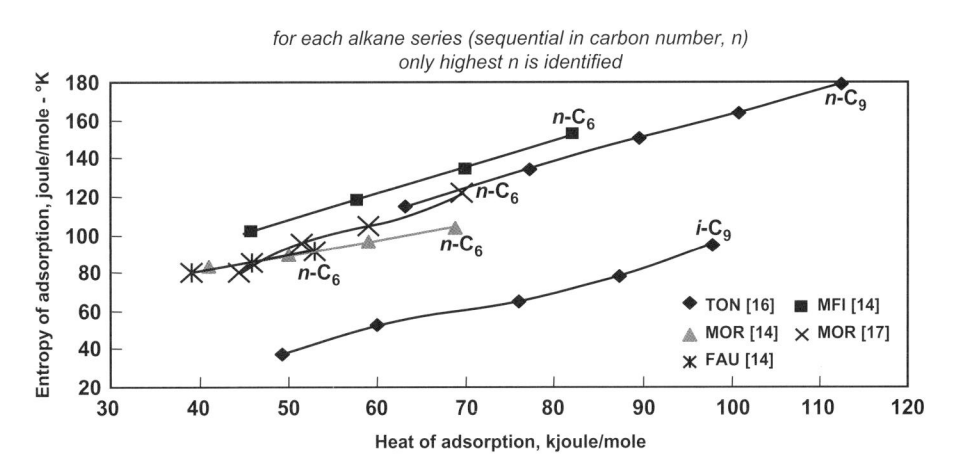

Figure 13.13 Compensation relation between differential heat and differential entropy of adsorption.

13.2.3.4 Compensation between Enthalpy and Entropy

For various alkanes over TON, MFI, MOR and FAU framework types the entropy of adsorption was calculated based on the measured values for the adsorption equilibrium constant and for the heat of adsorption. Figure 13.13 indicates that each time a linear relation is found between the entropy and the heat of adsorption [14, 16, 17]. While the increase in the heat of adsorption with carbon number typically makes the equilibrium constant larger for higher alkanes, the corresponding increase in the entropy of adsorption with carbon number limits this increase in the adsorption equilibrium constant. The differences in adsorption equilibrium constant between lighter and higher alkanes become even smaller at higher temperatures since, while adsorption equilibrium constant for alkanes decreases with increase in temperature, the higher alkanes decrease even more due to their higher heats of adsorption. Accordingly, a temperature may exist at which the large differences measured between the adsorption equilibrium constants at relatively low temperatures may essentially vanish, making this compensation relation quite significant for catalysis.

13.3
Diffusion

Understanding diffusion of molecules adsorbed in the pores of zeolites is important partly because intracrystalline diffusion, also known as configurational diffusion, may become rate-limiting when the chemical reaction is sufficiently fast. Intracrystalline diffusion is an activated process for which the activation energy originates from steric hindrance. For a given organic molecule, slight variations in molecular sieve pore size, that is, 0.3–0.4 nm can cause as much as ten orders of magnitude variations in diffusivity [18]. Such high sensitivity of diffusivity to zeolite pore size is critical in determining the extent of reactant and product shape selectivity effects, as discussed in the following sections.

Micropore mass transfer resistance of zeolite crystals is quantified in units of time by r_c^2/D_c, where r_c is the crystal radius and D_c is the intracrystalline diffusivity. In addition to micropore resistance, zeolitic catalysts may offer another type of resistance to mass transfer, that is resistance related to transport through the surface barrier at the outer layer of the zeolite crystal. Finally, there is at least one additional resistance due to mass transfer, this time in mesopores and macropores: R_p^2/D_p. Here R_p is the radius of the catalyst pellet and D_p is the effective mesopore and macropore diffusivity in the catalyst pellet [18].

Tendency of gas molecules A to spread and migrate in the presence of background gas molecules B, in such a way as to eliminate a macroscopic concentration gradient of A in B, is called transport diffusion. Transport diffusion and spreading of A are mainly controlled by random collisions between A and B and the rate is determined by measuring the flux of A, as given by Fick's first law of diffusion. In the absence of a macroscopic concentration gradient of A in B, individual A molecules still undergo, a thermal energy-induced sequence of rapid and random movements called Brownian molecular motion. Depending on the system, the distance covered may range from atomic dimensions to macroscopic dimensions. Brownian motion may be followed by tagging (isotopic labeling) some of the A molecules and following their trajectories. Here, while the concentration gradient of A in B is zero, the concentration gradient of tagged A molecules is not necessarily zero and is equal to and opposite to the concentration gradient of untagged A molecules. Self-diffusion is controlled by random collisions between tagged A and untagged A or B and the rate is determined by applying Fick's first law to tagged A molecules [19]. Magnetization associated with individual nuclei rotating about the direction of an externally applied magnetic field, as detected in a pulse field gradient (PFG) NMR experiment and the number of neutrons scattered by nuclei, as detected by quasi-elastic neutron scattering (QENS), measure the mean square value of the molecular displacement during the individual motions of molecules, hence self-diffusivity [20]. In both cases, though these methods are microscopic, the movements of molecules are followed over macroscopic distances. Although transport diffusion and self-diffusion occur by the same mechanism, the coefficients of transport diffusion and self-diffusion are the same only when the mobility of untagged molecules is unaffected by the presence of tagged molecules and this only happens in dilute solutions.

A derivation for self-diffusivity involves treating diffusion as a random walk resembling the motion of a "drunken sailor coming out of a bar" and taking a sequence of steps, each at arbitrary angles [21]. The problem of calculating the mean square displacement of the sailor from the bar, as a function of time, was solved by Einstein, who showed that the mean square displacement is proportional to the self-diffusivity and to the observation time. Self-diffusivity was proportional to the square of the step length divided by the time interval between steps. Diffusion in small pore zeolites can be treated as a hopping motion, through random collisions, from one adsorption site to another, that is, from a cage in zeolite, which represents the potential energy minima to a neighboring cage, with the transition state occurring in between the neighboring cages where the potential energy reaches a maximum. Einstein's expression for self-diffusivity, based on the random walk model, can be applied to diffusion in zeolites. Here, the step length becomes equal to the characteristic zeolite dimension, that is, the distance between the centers of adjacent cages. An Arrhenius-type behavior is then observed with an activation barrier representing the difference between the local potential energy maximum and minimum. At low loading the presence of molecules inside the cages does not influence the hopping rate other than setting it to zero in the unlikely event that the jump is occurring to an occupied site. At high loadings intermolecular interactions can alter the activation barrier in either direction. Actually, random walk is an oversimplification of intracrystalline diffusion in zeolites and individual jumps may be correlated, that is, depending on the geometrical configuration of the diffusion path.

Over most zeolites the diffusivity of n-alkanes decreases monotonically by several orders of magnitude with increasing carbon number due to the increasing activation energy barrier [9]. Exceptions to this rule are discussed in the literature [9]. The PFG NMR and QENS-derived self-diffusivities for n-alkanes over NaZSM-5 (Si/Al_2 = 64) at 300 K show a monotonic decrease with carbon number, as illustrated in Figure 13.14 [19]. The same ZSM-5 sample was used for measuring diffusivity of methane by both techniques, but different samples were used for longer alkanes. PFG NMR is expected to be sensitive to the presence of defects,

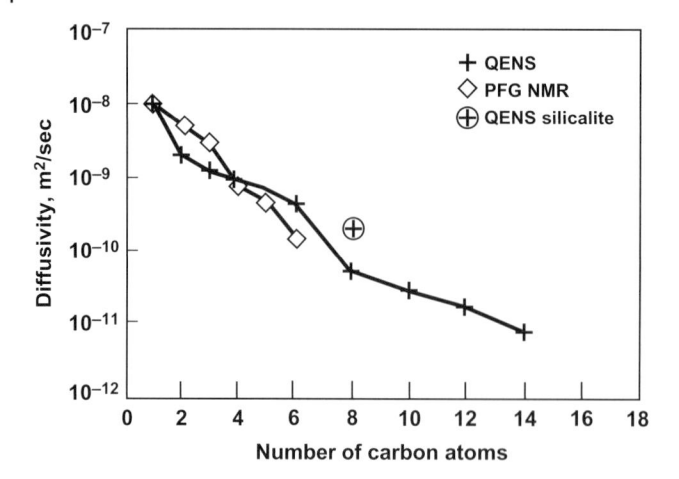

Figure 13.14 Self-diffusivity measurements for alkanes in MFI by NMR and neutron scanning [19].

but QENS is not expected to be due to the shorter diffusion path probed during the measurement. QENS yields information on molecular mobility at the 1 nm scale. Nevertheless, the results obtained with these two techniques were within a factor of three to four. In addition, at $n = 8$, a QENS diffusivity at 300 K, is shown for Silicalite to be four times larger than NaZSM-5 [19].

For the diffusion of n-alkanes in zeolites with cages connected by narrow windows like with an ERI framework type, a surprising increase in diffusion rate was observed between $n = 8$ and $n = 12$. It has been argued that, while a C_8 molecule fits well within a cage, C_{12} adopts a configuration with one end of the molecule sticking out of the cage into the window. The part of the hydrocarbon chain in the window then reduces the activation energy for hopping between neighboring cages and gives the n-alkane a higher mobility [22]. A similar periodicity in diffusivity was reported in LTA framework type [19]. There have also been reports that this phenomenon is an experimental artifact [1]. Meanwhile the debate continues.

Single file diffusion may occur in one-dimensional zeolite channels that are so narrow not to allow molecules to pass each other and this can cause a drastic reduction in the mobility. In the extreme case of a completely full single file system with N molecules, the molecule at one end of the channel needs to desorb first in order to create a vacancy that has to jump $N-1$ times in one direction to allow the file to move just by a single lattice spacing in the opposite direction. Relating this to Einstein's expression for self-diffusivity, many jumps are necessary in order to create enough clearance between molecules of one step length. The self-diffusivity here becomes inversely proportional to the file length (channel length or crystal size) and fractional site occupancy (coverage). As opposed to the drunken sailor, in single file diffusion with high occupancy, a displaced molecule is more likely to return to its original position than to proceed further, since the latter would cause a further concentration of the molecules ahead. Accordingly, the mean square displacement no longer varies linearly with observation time, but is now proportional to the square root of the observation time. The significance of single file diffusion in catalysis has been reported in the literature [9].

While microscopic techniques like PFG NMR and QENS measure diffusion paths that are no longer than dimensions of individual crystallites, macroscopic measurements like zero length column (ZLC) and Fourier Transform infrared (FTIR) cover beds of zeolite crystals [18, 23]. In the case of the popular ZLC technique, desorption rate is measured from a small sample (thin layer, placed between two porous sinter discs) of previously equilibrated adsorbent subjected to a step change in the partial pressure of the sorbate. The slope of the semi-log plot of sorbate concentration versus time under an inert carrier stream then gives D/R^2. Provided micropore resistance dominates all other mass transfer resistances, D becomes equal to intracrystalline diffusivity while R is the crystal radius. It has been reported that the presence of other mass transfer resistances have been the most common cause of the discrepancies among intracrystalline diffusivities measured by various techniques [18].

Figure 13.15 illustrates the diffusivity by ZLC for various branched and linear C_6 alkanes and $n-C_5$, as compiled by Ruthven and Post [18]. The results indicate that diffusivities vary by four orders of magnitude between different isomers. Linear alkanes show lower activation energy than branched alkanes. The absence

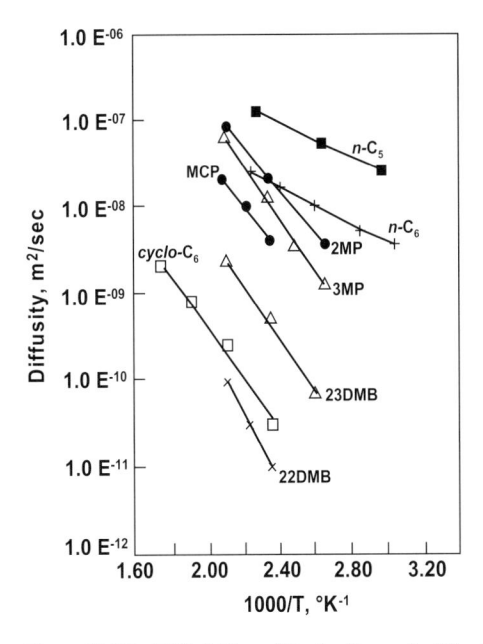

Figure 13.15 Diffusivities of C_5–C_6 alkanes in Silicate by ZLC [18].

of large differences in activation energy between mono-branched and dibranched alkanes indicate that the very low diffusivity of dibranched alkanes is an entropic effect.

13.4
Acidity

13.4.1
Bronsted Acidity

In a calcined zeolite Bronsted acidity in the form a proton, the bridging hydroxyl, is generated to balance the charge when tetrahedral silicon is isomorphously substituted with aluminum or another trivalent ion like gallium and boron. While with FAU framework type there is one unique T-site surrounded by four different oxygen atoms (O-site) leading to four possible locations for the proton, MOR has four, FER has five and MFI has 24 topologically inequivalent T-sites [8]. In the case of MFI there are 96 distinct Bronsted acid sites [9]. The proton occupancy of these sites is expected to correlate with the deprotonation energy, the lower the energy the more abundant the acid site associated with the particular O-site.

The NMR signature for tetrahedral Al is a chemical shift at 50–60 ppm in ^{27}Al MAS experiments. For a given zeolite, the resolution of the aluminum atoms at different T sites by ^{27}Al MAS NMR spectroscopy is not possible due to the similarity of the chemical shifts. Extra-framework Al gives a chemical shift around 0 ppm. ^{29}Si MAS NMR provides information on the coordination environment of Si in relation to the number of neighboring Al's. The bridging hydroxyls also give a signal at 4–5 ppm in ^{1}H MAS NMR experiments. The O-sites pointing to large cavities in FAU have chemical shits at 3.8 and 3.9 ppm. The chemical shifts for O-sites pointing to smaller cavities in FAU are 4.3 and 5.0 ppm [24].

The IR signature for the bridging hydroxyl is an absorption band around 3600 cm^{-1}. FAU is one of the less common cases where two distinct IR bands can be clearly observed: 3640 cm^{-1} for the O-sites pointing into the large cages and 3540 cm^{-1} for the O-sites pointing into the small cages. The bridging hydroxyl band can be perturbed by weakly basic probe molecules through hydrogen bonding interactions, leading to an increase in the IR intensity and a bathochromic shift proportional to the strength of the perturbation [24]. For a given zeolite, the more basic the probe molecule, the higher is the shift in the frequency. For a given probe molecule, the stronger the Bronsted acidity, the higher is the shift in the frequency. The IR band around 3750 cm^{-1} is assigned to the terminal silanols.

The positively charged structure directing agent during zeolite synthesis is actually the counterion for the negatively charged framework. However, during synthesis some of the positive charge may also be balanced by silanol defects, that is O_3SiO^- groups. Defects are typically minimized by carrying the zeolite synthesis in fluoride media which provides the F^- ions as an alternate means for balancing the excess positive charge.

13.4.2
Significance of Acid Strength

While average deprotonation energy is a good measure of the intrinsic Bronsted acid strength of a zeolite, it is the extrinsic acidity, also impacted by the chemical interaction between the protonated basic probe molecule and the deprotonated zeolite, that really counts for catalysis.

In the case of alkenes, 1-pentene reactions were studied over a catalyst with FAU framework (Si/Al_2 = 5, ultrastable Y zeolite in H-form; USHY) in order to establish the relation between acid strength and selectivity [25]. Both fresh and selectively poisoned catalysts were used for the reactivity studies and later characterized by ammonia temperature programmed desorption (TPD). It was determined that for alkene reactions, cracking and hydride transfer required the strongest acidity. Skeletal isomerization required moderate acidity, whereas double-bond isomerization required weak acidity. Also an apparent correlation was established between the molecular weight of the hard coke and the strength of the acid sites that led to coking.

Among commonly used reactions, alkane cracking is generally considered to be the reaction that requires the strongest acidity. The question has been consistently raised as to whether relative alkane cracking activities can be used to rank zeolite acid strength. In the case of monomolecular cracking, there is general agreement that protonation is the rate-controlling step. Interestingly, when apparent activation energies for monomolecular cracking of propane were corrected by their corresponding heats of adsorption it was found that the resulting intrinsic activation energies were, in fact, quite similar over MFI, MOR, FAU and BEA framework types, all in the range of 190–198 kJ/mol [26]. Similar activation energies for protonating propane over various zeolites were then interpreted as implying that these zeolites have a similar Bronsted acid strength. The rate variations observed among these zeolites for propane cracking have been previously attributed to variations in the adsorption equilibrium constants, which are quite large at low temperatures, where adsorption equilibrium constants are commonly measured [26]. More recently, it was shown that the compensation between the heat of adsorption and the entropy of adsorption, along with the exponential temperature dependency of the adsorption equilibrium constant, may cause unexpected variations in the adsorption equilibrium constants [27]. In the case of propane cracking over MFI and MOR, by making the reasonable assumptions of constant heat of adsorption and constant entropy of adsorption with respect to temperature, the adsorption equilibrium constant over MFI was estimated to be lower by a factor of 4.5 relative to MOR at the reaction temperature of 823°K, as illustrated in Table 13.1. Based on the observed cracking rates (that were quite similar) and the apparent activation energies published in the literature plus the estimated adsorption equilibrium constants (at the reaction temperature) it then became possible to estimate the intrinsic rate constant over MFI to be, in fact, greater by a factor of 4.7 relative to MOR. Since MFI and MOR exhibited very similar intrinsic activation energies for cracking propane, the observed difference in cracking rate constants was then

Table 13.1 Kinetics of propane cracking over H-MFI and H-MOR[27].

	TOF mol mol H⁺-sec-bar	ΔH_{ads} kjoule mol	ΔS_{ads} joule mole-°K	K (at 823°K)	E_{ads} kjoule mol	k_{int} mol mol H⁺-sec
MFI	0.024	−45.3	−102	0.00355	147	6.6
MOR	0.022	−41.3	−84.6	0.0159	145	1.4

attributed to a difference in the pre-exponential terms (for the rate constant), that is, to a higher activation entropy for MFI. This difference could then be interpreted to imply stronger Bronsted acidity for MFI relative to MOR framework type not because of a lower activation energy, but because of a higher activation entropy. These results are summarized in Table 13.1.

Acid strength is critical not only because it determines which pathways are going to be enabled, but also because it determines the temperature that is required to reach a certain level of activity for a given reaction: the weaker the acidity, the higher the temperature. The selectivity possessed by the catalyst at the higher temperature then determines the practical feasibility of the reaction. For example, in general, due to the weaker acidity observed for crystalline silicon aluminum phosphates relative to zeolites, Pt/SAPO-11 requires approximately 100°C higher operating temperature than Pt/ZSM-22 and Pt/ZSM-23 for hydroprocessing n-alkanes [28, 29]. In this case it helps that SAPOs, in general, do not have as high a cracking activity as zeolites, since temperature can then be effectively raised in order to increase activity without suffering from selectivity loss due to cracking during hydroprocessing of n-alkanes [28, 29].

Since aluminum is less electronegative than silicon, high aluminum content in the zeolite leads to weaker T–O bonds and, by bond-order conservation, to stronger O–H bonds, hence to weaker acidity. Acidity is also impacted by the isomorphous substitution of Ga (or B) into the tetrahedral framework. In the case of MFI framework type, Ga substitution for Al lowers the strength of the Bronsted acid sites. Through IR measurements following low-temperature CO adsorption, T. Mezza from UOP measured the bathochromic shift in H-GaMFI and in H-AlMFI to be 292 cm^{-1} and 326 cm^{-1}, respectively. The lower shift in the case of Ga is consistent with the weaker Bronsted acidity relative to MFI with Al. This weaker acidity is readily observed by the lowered propane aromatization activity for the H form of Ga-MFI relative to Al-MFI [30]. Interestingly, the lowered activity is accompanied by increased aromatization selectivity in favor of light alkenes and light alkanes. The Al version of the zeolite is likely to possess sufficiently strong Bronsted acidity for activating propane through a carbonium ion intermediate, followed by cracking to methane + ethylene or hydrogen + propylene. The Ga version of the zeolite, in contrast, is postulated to activate propane through hydride abstraction over Lewis acid sites composed of coordinatively unsaturated Ga^{+3}, framework or extra-framework. It appears that these Lewis acid sites benefit from being next to Bronsted acid sites for effectively enabling the formation of molecular hydrogen, the key marker for a selective aromatization catalyst. The propyl carbenium ion resulting from the hydride ion abstraction step can then combine with another propylene molecule to make a larger carbenium ion, followed by dehydrocyclization to make an aromatic. The cyclization reaction is also shown to benefit from a short distance between Ga^{+3} and Bronsted acid sites [30].

13.4.3
Significance of Acid Site Density

Total concentration of acid sites and the proximity of acid sites become critical selectivity parameters for the zeolite, particularly when there are competing pathways. An illustrative example is monomolecular versus bimolecular cracking reaction pathways. It is generally expected that the higher is the concentration of Al, the more favored become the bimolecular pathways. Cracking of 1-butene at 500 °C over HZSM-5 provides such competing pathways. Light alkenes can be obtained by monomolecular cracking of 1-butene or of 1-butene dimer. Light alkenes, in turn, can make aromatics through a bimolecular pathway. Measurement of the zeolite exchange capacity for Co(II), along with UV-Vis characterization of exchanged Co(II) were used for determining the concentration of neighboring Al atoms in various HZSM-5 having different Si/Al_2 ratios [31]. It was observed that catalysts with higher concentration of neighboring Al atoms consistently made more aromatics relative to catalysts having the same amount of Al, but which was more isolated. IR characterization of adsorbed H_2 and CO, as well as the corresponding shifts of the OH vibrations did not show any significant differences in acid strength for these samples. Accordingly, the distribution of Al in the framework, more specifically the concentration of neighboring Al pairs was suggested to be the most likely factor in determining the relative selectivity towards aromatics versus light alkenes. More specifically, enhancement of bimolecular aromatization pathway was attributed to the presence of neighboring Al atoms. The significance of acid site density is further illustrated in Section 13.8.4 by comparing monomolecular cracking of alkanes with bimolecular cracking for which the rate-controlling step is likely to involve a hydride transfer from an alkane physisorbed on an acid site to a neighboring acid site having a carbenium ion.

13.4.4
Lewis Acidity

Lewis acid sites may be formed following dehydroxylation of zeolite surface in H-form. At sufficiently high temperatures two Bronsted acid sites can drive off a water molecule and leave behind a coordinatively unsaturated Al^{+3} site, as illustrated in Figure 13.16 [32]. Here not only the resulting tri-coordinated Al but also the tri-coordinated positively charged Si can act as a Lewis acid. Furthermore dehydroxylation may be followed by framework dealumination, leading to cationic extra-framework species like AlO^+, $Al(OH)_2^+$ that can act as Lewis acids [33–37].

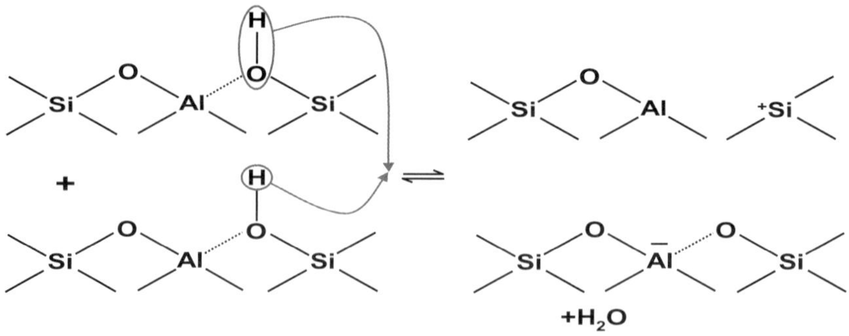

Figure 13.16 Dehydroxylation of a zeolite surface.

Lewis acid sites have empty orbitals able to accept electron density from the occupied orbitals of a Lewis base, in parallel with back-donation from the catalyst to the empty anti-bonding orbitals of the base [33]. This interaction leads to the formation of an activated acid–base adduct. In the case of alkanes activation may proceed by hydride abstraction [38]. Y and Beta are good examples of zeolites with Lewis acidity, often quite significant for catalysis [39, 40].

13.4.5
External acidity

Since shape selectivity arises from the molecular sieving action provided by zeolite micropores the zeolite external surface may be considered to lead, in general, to non-shape selective reactions. Nevertheless, for certain reactions it has been postulated that the pore mouths may be sufficiently structured to provide some form of shape selectivity [41]. In the case of 2,2,-dimethylhexane adsorption at the pore mouth for TON framework type, for example, it was computed that the heat of adsorption is in fact quite sensitive to the configuration of the adsorbing molecule. The heat of adsorption was larger when the longer hydrocarbon chain part penetrated the pore mouth relative to the configuration which had the methyl group penetrating the pore mouth, as illustrated in Figure 13.17. This example demonstrates that adsorption on the external surface can be quite specific, due to the zeolite structure and geometry of the pore mouth and suggests that some level of shape selectivity may indeed take place at the external surface [42].

Hydroprocessing of long chain n-alkanes over bifunctional zeolitic catalysts having TON and MTT framework types has been the subject of intense debate concerning the significance of pore mouth catalysis [16, 43–45]. One group of researchers claimed that partial penetration of the reactant molecules does occur into several pore openings with two separate hydrocarbon fragments simultaneously (key-lock adsorption), the key to shape selective catalysis [46]. However, other research groups concluded that adsorption and shape selective catalysis inside the zeolite channels dominate and the contribution from the exterior surface is negligible [47]. These latter conclusions were reached by showing that Henry coeffi-

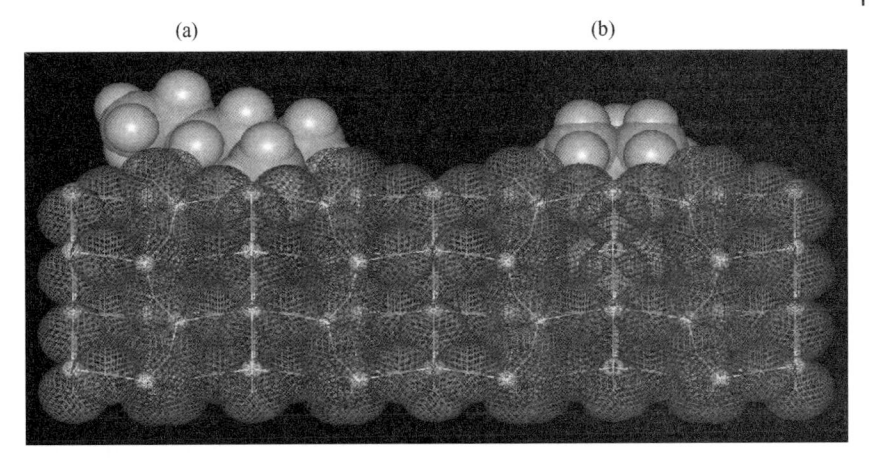

Figure 13.17 Optimized adsorption structure of 2,2-dimethylhexane over TON framework type. (a) methyl group penetrating the pore mouth (18 kJ/mol). (b) longer hydrocarbon chain penetrating the pore mouth (65 kJ/mol) [42].

cients for n-alkanes (reactants) and branched alkanes (products) were more than an order of magnitude higher at the interior than at the exterior surface of these two zeolites.

In any case, it is generally accepted that external acidity provides more opportunity for non-shape selective reactions. Zeolite selectivation by chemical modification of the external surface through tetraethyl *ortho*-silicate has been commonly employed [48]. Such procedures typically lead to enhanced para-selectivity in alkylation of alkylbenzenes with alcohols. Similarly, neutralizing the external acidity of H-ZSM-5 was found to improve *para*-selectivity during methylation of methylnaphthalenes [49]. ZSM-5 for which the external acidity was neutralized through 2,6-di-tertbutyl pyridine showed $C_{20}{}^+$ oligomers with a low degree of methyl branching from processing of propylene or 1-decene. In the absence of the surface modification a significant amount of the reaction is not taking place inside the zeolite channels, leading to high degree of methyl branching [50]. In the case of ZSM-22 the external surface was neutralized by collidine (2,4,6-trimethyl pyridine) or by lanthanum oxide or by yttrium oxide in order to obtain dimers, trimers and tetramers with significantly reduced branching from oligomerization of 2-butene or propylene [51]. Several such procedures for neutralizing the external acidity in zeolites are discussed in the literature [47].

13.5
Carbocations

Protonation of alkenes, cycloalkanes and alkanes lead to the formation of carbenium ions, carbonium ions and protonated cyclopropane rings (also called cycloproponium ions) as either reaction intermediates or transition states during hydrocarbon conversion. The relative stabilities of these carbocations are discussed by Martens and Jacobs [46].

13.5.1 Carbenium Ions and Alkoxides

Following physical adsorption, the chemistry between an organic molecule and a Bronsted acid site is initiated through the formation of a pi complex, as illustrated in Figure 13.18a for the case of *i*-butene. A slight increase occurs for the bond distance between C_a (interacting with the proton connected to the acidic oxygen, O_a) and C_b (interacting with the basic oxygen, O_b), in parallel with an increase in O_a–H bond length, in accordance with bond-order conservation. It is important to note that steric constraints imposed by the zeolite channels may result in different C_b–H and C_a–H distances, leading an asymmetric pi complex, as was shown in the case of 2-pentene adsorption in TON framework type [52].

Transfer of the electron pair between C_a and C_b of *i*-butene to make a bond with the zeolitic proton can lead to the transition state for the tertiary or for the primary carbenium ion, depending on whether C_b is tertiary or primary, as illustrated in Figure 13.18 and Figure 13.19, respectively. Here the transition state for the primary is higher in energy than the transition state for the tertiary [53]. The end result is either a trivalent carbenium ion interacting with the negatively charged zeolite framework, as illustrated in Figure 13.18e or an alkoxide, as illustrated in Figure 13.18d. The latter involves a covalent interaction between C_b and O_b.

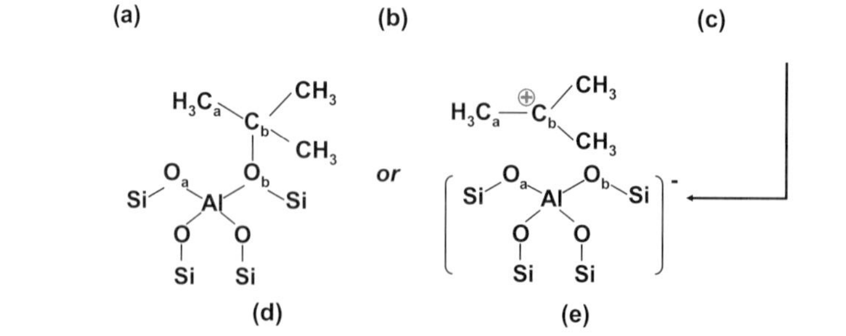

Figure 13.18 Protonation mechanism for *i*-butylene over a Bronsted acid site–tertiary carbenium ion or an alkoxide.

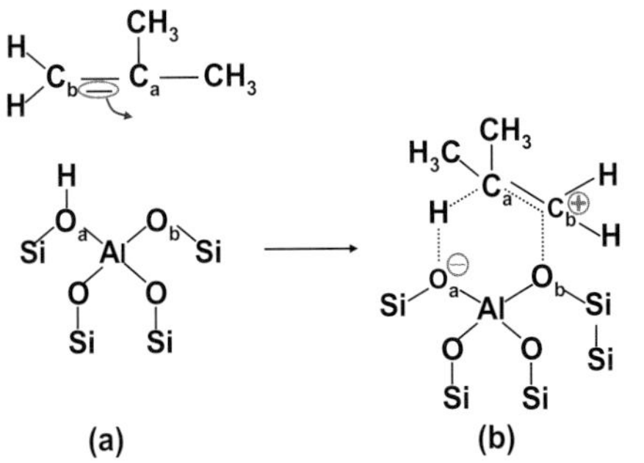

Figure 13.19 Protonation mechanism for i-butylene over a Bronsted acid site – primary carbenium ion.

Carbenium ions are more likely to form when charge can be effectively delocalized and steric constraints prevent covalent bond formation with framework oxygens. The extent of charge separation between the organic fragment and the zeolite framework is an indication of whether a carbenium ion or a neutral alkoxide is formed, both of which can act as reaction intermediates or transition states in the subsequent elementary steps. Interestingly, while in gas phase the primary carbenium ion is much less stable than the tertiary, interaction with the zeolite can change this rule. For example, in the case of i-butylene, TON interaction was found to preferentially stabilize the primary isobutyl cation relative to the tertiary butyl cation, as well as the primary iso-butoxide relative to the tertiary butoxide [53]. This reversal does not occur in ranking of the activation energies though, with primary transition states (TS) typically requiring higher activation energy relative to the secondary, which are higher than for the tertiary TSs [54].

If there is sufficient electron delocalization it may become possible to have stable carbenium ions that can be characterized by IR. In the case of aromatics good illustrations exist of experimental evidence for protonation of tetra- and hexamethylbenzenes over Bronsted acid sites in zeolites. 1,3,5-Trimethylbenzene did not undergo significant protonation, presumably due to its limited proton affinity [55, 56]. Figure 13.20 shows a series of IR spectra of the H-form of Beta zeolite (Si/Al$_2$ = 24) with decreasing coverage of tetramethylbenzene (spectrum 1 – initially) obtained by evacuation first at 300°K (spectra 2–10) and then at 400°K (spectrum 11). Spectra 12 and 13 belong to zeolite-only and tetramethylbenzene-only cases, respectively. In spectrum 12 the band at 3782 cm^{-1} is assigned to O–H stretching vibration of Al–OH species, the band at 3746 cm^{-1} to free silanols and the bands at 3667 and 3610 cm^{-1} to defective and regular bridging hydroxyls.

Spectrum 1 illustrates how the interaction with tetramethylbenzene (by hydrogen-bonding) shifts the silanol band in Beta to 3563 cm^{-1} and the bridging hydroxyls to even lower frequencies where overlapping absorption from hydrogen-bonded silanols make them difficult to detect. In spectrum 1 the bands between 3100 and

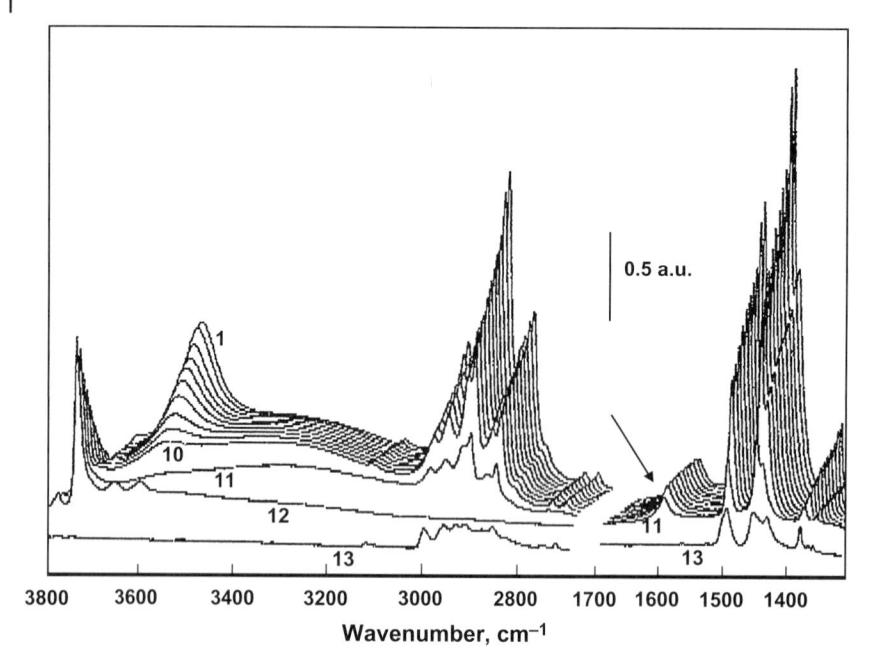

Figure 13.20 IR spectra of tetramethylbenzene over H-form of Beta zeolite at 300°K and 400°K. Spectrum 1: initial. Spectra 2–10: obtained by evacuation at 300°K. Spectrum 11: evacuation at 400°K. Spectrum 12: zeolite-only. Spectrum 13: tetramethylbenzene-only. The arrow points to the 1604 cm^{-1} band assigned to the tetramethylbenzenium ion [55].

2700 cm^{-1} represent the C–H stretching region of tetramethylbenzene, while the low frequency region at 1700–1350 cm^{-1} shows the stretching vibrations of the aromatic ring, as well as the C–H bendings of the methyl groups. In the low frequency region the most remarkable difference between spectrum 1 with adsorbed tetramethylbenzene and spectrum 13 with free tetramethylbenzene is the presence of the band at 1604 cm^{-1} assigned to tetramethylbenzenium ion. The fact that this band does not disappear upon evacuation is a good evidence of the strong interaction between the probe molecule and the Bronsted acid site [55]. In the case of hexamethylbenzenium over *Beta* this latter band position was quite similar. As expected, when another *Beta* zeolite, this time with much lower Bronsted acidity was used (Si/Al$_2$ = 420) the band at ~1600 cm^{-1} disappeared, further supporting that this band is indeed attributed to the benzenium ion. It is important to note that pyridine, which is commonly used as a basic probe molecule, shows a characteristic band around 1540 cm^{-1} when protonated by zeolitic Bronsted acid sites [32]. In another work IR spectra of 1-methylcyclopentene first adsorbed at 150°K and then gradually heated to 298°K were measured over Y zeolite, mordenite, ZSM-5 and *Beta*. In each case the band at 1513 cm^{-1}, characteristic of the protonated species, was observed [32].

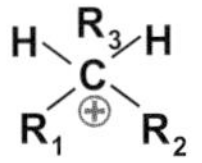

Figure 13.21 Pentacoordinated carbonium ion.

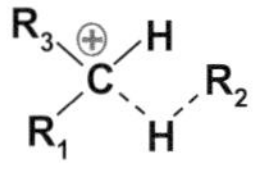

Figure 13.22 Carbonium ion as a two-electron three-center bridge.

13.5.2
Carbonium Ions

Protonation of adsorbed alkanes leads to the formation of penta-coordinated carbonium ions, as shown in Figure 13.21. Here R_1, R_2 and R_3 are either alkyl groups or hydrogen. The carbonium ion can also be illustrated as a two-electron three-center C–H-R bridge, as shown in Figure 13.22 [53]. A H-alkonium or a C-alkonium ion is obtained depending on whether a C–H bond or a C–C bond is protonated, respectively. The five bonds of the charged C-atom contain only eight electrons, hence the basic rule of organic chemistry is not violated. With carbonium ions, also, a Coulombic interaction is expected to occur between the carbocation and the negatively charged zeolite framework.

13.6
Elementary Steps of Hydrocarbon Conversion over Zeolites

This section makes only brief remarks about some of the elementary steps of hydrocarbon conversion. More detailed analyses of these steps are done within the context of various reaction mechanisms.

Mechanism for protonation of alkenes was previously discussed in Section 13.5.1. In general, protonation of alkenes is an exothermic process. Protonation of alkanes was discussed in Section 13.5.2. There will be further discussion on this step in Section 13.8.4 within the context of alkane cracking mechanisms. The formation of a penta-coordinated carbonium ion from alkane protonation is typically an endothermic process, the reverse being true for deprotonation.

Hydride shift is a fast elementary step that enables the positive charge to move around the carbocation ion to ultimately achieve thermodynamically the most favorable configuration. 1,2 hydride shift in the secondary butyl carbenium ion is illustrated in Figure 13.23. The use of this common step during alkene skeletal isomerization is further discussed in Section 13.8.1.

Methyl shift in carbenium ions is expected to require three elementary steps (ring closure, ring opening, hydride shift), enabling the positive charge to move

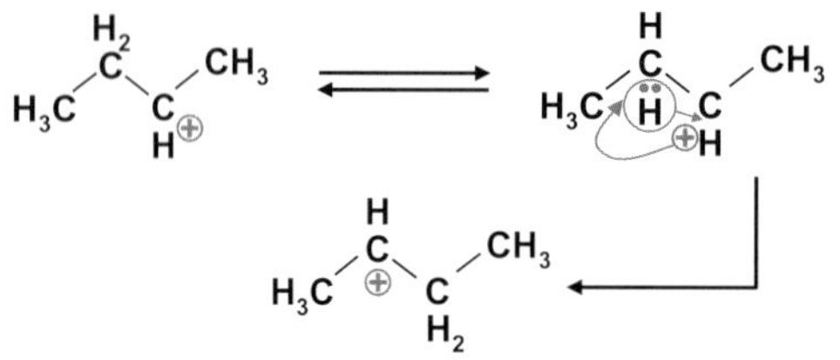

Figure 13.23 Hydride shift.

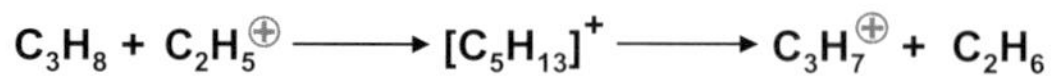

$$C_3H_8 + C_2H_5^\oplus \longrightarrow [C_5H_{13}]^+ \longrightarrow C_3H_7^\oplus + C_2H_6$$

Figure 13.24 Hydride transfer.

around, to ultimately achieve thermodynamically the most favorable configuration. This is shown in Section 13.8.1.

Beta scission of a carbenium ion is an elementary step that is initiated by the weakening of the bond *beta* to the positive charge, leading to a smaller carbenium ion and an alkene. This elementary step is further discussed in Sections 13.8.1, 13.8.3.1 and 13.8.4 within the context of alkene skeletal isomerization, isobutane-2-butene alkylation and alkane cracking, respectively.

Intermolecular hydride transfer (HT) is an elementary reaction between an alkane and a relatively small carbenium ion to give a relatively larger, typically more stable carbenium ion and a smaller alkane, hence responsible for chain propagation in many acid-catalyzed transformations. This is illustrated in Figure 13.24. The transition state is believed to be a penta-coordinated carbonium ion for which the formation is endothermic and largely responsible for the high activation energy that is typical of HT [57]. Hydride transfer enables alkylation of isobutane with 2-butene, as described in Section 13.8.3.1 and bimolecular cracking of alkanes, as described in Section 13.8.4.1.

Protonated cyclopropane ring closure and ring opening steps are discussed in Section 13.8.1 within the context of alkene skeletal siomerization. Carbonium ion decomposition is further discussed in Section 13.8.4.2 within the context of monomolecular cracking of alkanes.

13.7
Shape Selectivity

13.7.1
Tools

Adsorption of various molecules that are similar in size to micropores have been used to characterize pore size and shape in zeolites, as summarized in Section 13.2.3.1, with the ultimate objective of understanding shape selectivity effects. Also

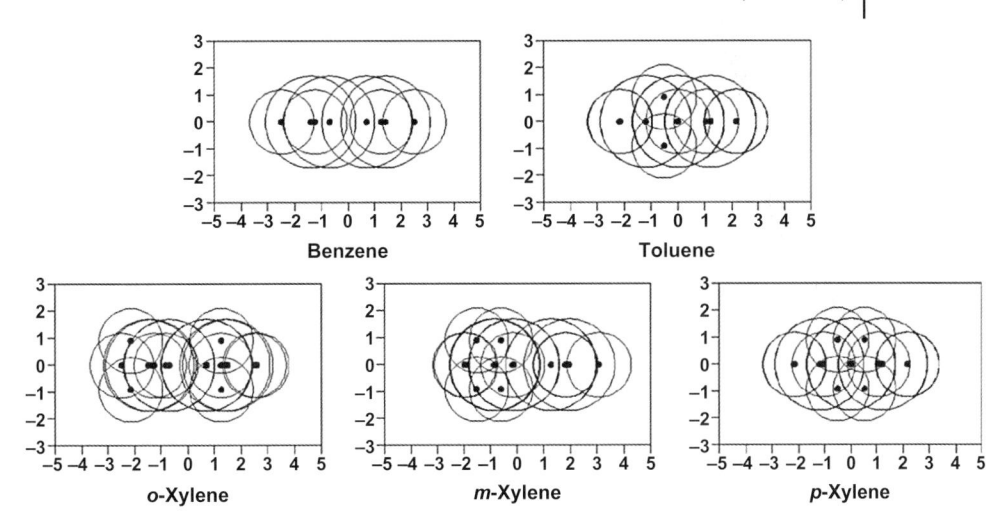

Figure 13.25 Footprints for various aromatics (Å) [58].

the kinetic diameters and footprints for various potential reactant and product molecules have been summarized (see below). Finally, three types of test reactions have been applied to a large number of zeolites for better characterization of pore size and shape (see below).

13.7.1.1 Footprints and Kinetic Diameters

Shape selectivity arises from the molecular sieving action provided by the zeolite micropores which inhibit the formation of large transition states and excludes molecules beyond a certain critical size from diffusing inside, diffusing outside and adsorbing onto the zeolite channels and cavities. The kinetic diameter for various molecules were previously compiled by Breck [12]. While the kinetic diameter concept has been quite useful for describing molecular sieve effects, it is nevertheless an oversimplification due to the spherical representation of the molecules. More recently the concept of molecular footprint was introduced [58]. The footprint is a two-dimensional projection of the smallest area of the molecule, in preparation for penetration into a zeolitic channel. Footprints for various aromatics are displayed in Figure 13.25. Zeolites have a variety of channel shapes, hence the molecule properly rotates in order to provide the projection that is most suitable for the channel. The footprint dimensions for elliptical projection of various molecules are summarized in Table 13.2 [58]. For comparison, kinetic diameters published by Breck are also compiled here. According to the analysis presented in that work, if the major diameter of the zeolite channel is greater than the major diameter of the footprint and the minor diameter of the zeolite channel is greater than the minor diameter of the footprint then there is free passage. In case the fit is not good then there could be a certain level of strain for the molecule to distort itself to be able to fit into the channel. These strain energies have been calculated for various molecule–zeolite combinations and reported [58].

Table 13.2 Footprint for various molecules.

	Footprint (Å) [58]	Kinetic diameter (Å) [12]
Hydrogen	2.40 × 2.40	2.2
Oxygen	2.80 × 2.80	3.5
Nitrogen	3.00 × 3.00	3.6
Water	3.24 × 2.88	2.7
Carbon monoxide	3.40 × 3.40	3.8
Carbon dioxide	3.40 × 3.40	3.3
Acetylene	3.40 × 3.40	3.3
Ammonia	3.98 × 3.32	2.6
Hydrogen sulfide		3.6
Methane	4.22 × 4.22	3.8
Ethylene	4.24 × 3.40	3.9
Ethane	4.48 × 4.48	
SF_6	5.22 × 4.22	5.5
Cyclopropane		4.2
Propane	5.38 × 5.02	4.3
n-Butane	5.40 × 5.04	4.3
n-Pentane	5.40 × 5.04	
Propylene	5.52 × 4.22	4.5
1-Butene		4.5
i-Butane	6.30 × 5.34	5.0
i-Pentane	6.36 × 5.36	
Neopentane	6.70 × 5.70	6.2
Toluene	6.72 × 4.28	
p-Xylene	6.72 × 4.28	
Benzene	6.76 × 3.76	5.9
Cyclohexane		6.0
Ethylbenzene	7.10 × 4.58	
m-Xylene	7.78 × 4.60	
o-Xylene	7.92 × 4.82	

13.7.1.2 Constraint Index

The Constraint index (CI) is the ratio of the first order rate constant for cracking n-hexane to 3-methylpentane (present in a 1:1 mixture with 90% He as carrier gas) over the mono-functional form of the zeolite [13]. For this test the reaction temperature is between 290 °C and 510 °C, the liquid hourly space velocity (LHSV) is between 0.1 and 1.0 h^{-1} and the conversion is between 10 and 60%. As long as the catalyst pores are sufficiently large, branched alkanes crack at higher rates. However, reactant shape selectivity effects in smaller pores lower the cracking rate for branched alkanes. CI allows classification of zeolites into large-pore (CI < 1), medium-pore (1 < CI < 10) and small pore (12 < CI), as illustrated in Table 13.3. This is consistent with corresponding classification of zeolites into 8-, 10- and 12-membered ring zeolites, with the only exception of MTW which behaves more like a large 10-membered ring due to the puckered nature of its pores. Another noteworthy comment is related to the meaning of CI for MWW. The latter pos-

Table 13.3 Constraint index for various zeolites (CI based on the patent literature, except for numbers in parentheses that are from the open literature)[13].

Zeolite	CI
Eronite (ERI)	38
ZSM-23 (MTT)	9.1
ZSM-22 (TON)	7.3 (7.4)
ZSM-5 (MFI)	6.0–8.3 (4.6)
ZSM-11 (MEL)	5.0–8.7
ZSM-50 (EUO)	2.1
MCM-22 (MWW)	1.5
ZSM-12 (MTW)	2.3
Mordenite (MOR)	0.5 (1.0)
Beta (BEA)	0.6–2.0
X or Y (FAU)	0.4 (0.2)

Table 13.4 Constraint index revisited with new framework types[59].

	Framework	MR	CI
SSZ-13	CHA	8	>100
ZSM-23	MTT	10	7.0–10.6
ZSM-22	TON	10	5.0–11.0
SSZ-20	TON	10	6.9
ZSM-5	MFI	10	4.9–7.7
SSZ-28	DDR	8	3.7–4.0
ZSM-50	EUO	10	2.3
ZSM-12	MTW	12	2.1
SSZ-23	STT	9 × 7	3.2
SSZ-36	ITE/RTH	8	1.1
SSZ-31	STO	12	0.9
SSZ-25	MWW	10	0.8
SSZ-35	STF	10	0.6
LZY-82	FAU	12	0.4
CIT-5	CFI	14	0.4
SSZ-24	AFI	12	0.3
UTD-1	DON	14	0.3

sesses large supercavities (0.71 nm diameter, 1.8 nm height) that are accessible via 10-MR openings (0.40*0.55 nm), as well as large hemicavities (0.71 nm diameter, 0.70 nm depth). Accordingly, CI = 5 is a weighted average value characterizing both 10-MR pores and large cavities.

CI measurements were later revisited by remeasuring the familiar zeolites, but this time including some of the new structures that were invented. These results are in Table 13.4. While most of these results were within expected ranges, there were still some anomalies. For example, CIs for SSZ-28 and SSZ-36 were lower than expected. Furthermore, it was not possible to distinguish between 12-MR zeolites and some of the new 14-MR zeolites [59].

13.7.1.3 Modified Constraint Index

Modified constraint index (CI*) was developed to characterize bifunctional zeolitic catalysts with medium size pores [13]. CI* is the selectivity ratio of 2-methylnonane to 5-methylnonane at low conversion from hydroprocessing of n-decane. 2-methylnonane becomes the preferred product with decrease in pore size. CI* values range from 14 (TON) to 3 (MEL) over 10-MR zeolites and hence provide an opportunity to distinguish within the 10-MR family. The range for 12-MR zeolites, 1 to 2, however, is much narrower, as illustrated in Figure 13.26. There is more to be gained from pore characterization data provided by n-decane conversion, however. As the zeolite pore size decreases, the selectivities to bulkier conversion products (4-propylheptane, 3- and 4-ethyloctane) decrease. Furthermore, some cracking occurs during n-decane conversion and the ratio of dibranched to mono-branched alkanes or the selectivity to i-pentane is another indication of the constraint provided by the zeolite topology on the formation of large transition states or on the diffusion of the bulky products [60].

13.7.1.4 Spaciousness Index

Spaciousness index (SI) also was developed for characterizing bifunctional catalysts, this time with large pores. SI is the selectivity ratio of i-butane to n-butane from the cracking of butylcyclohexane [13]. With a much larger feed probe molecule it now becomes possible to distinguish among various 12-MR zeolites. SI values range from 3 (MTW) to 20 (FAU), as illustrated in Figure 13.27. The

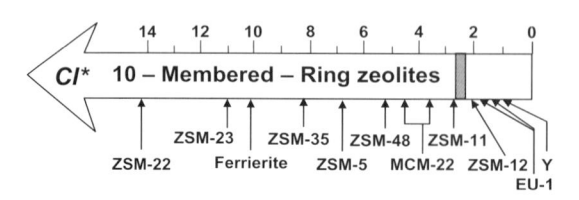

Figure 13.26 Modified constraint index for various zeolites [13].

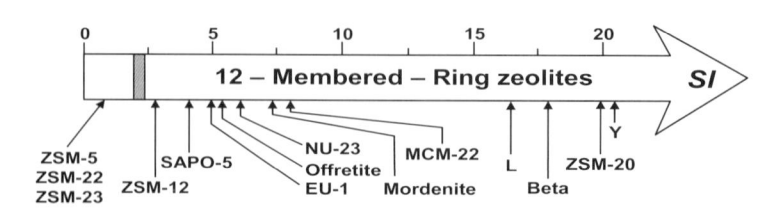

Figure 13.27 Spaciousness index for various zeolites [13].

spaciousness index has proven to be useful for characterizing internal cage or void space in zeolite channels.

13.7.2
Reactant Shape Selectivity

A good example for reactant shape selectivity includes the use of catalysts with ERI framework type for selective cracking of linear alkanes, while excluding branched alkanes with relatively large kinetic diameters from the active sites within the narrow 8-MR zeolite channels [61, 62]. Here molecular sieving occurs both because of the low Henry coefficient for branched alkanes and because of the intracrystalline diffusion limitations that develop from slow diffusivities for branched alkane feed molecules.

13.7.3
Transition State Shape Selectivity

It is often difficult to distinguish restricted transition state shape selectivity from product shape selectivity due to the lack of clear experimental evidence that the pore geometry and local spatial environment are actually influencing the reaction rate [63]. The following test reactions are more likely be impacted by transition state selectivity effects.

13.7.3.1 *Meta*-Xylene Disproportionation

An example of restricted transition state shape selectivity involves *meta*-xylene disproportionation to trimethylbenzenes (TMBs) and toluene. The tests were carried out at 317–318 °C and 0.5 kPa pressure of *meta*-xylene in the presence of He carrier gas. The flow rates were adjusted in order to achieve 10% *meta*-xylene conversion each time. A lower selectivity to xylene isomerization products relative to disproportionation is indicative of sufficiently large reaction cavities for accommodating large transition states, as illustrated in Figure 13.28 [64]. The isomerization:disproportionation ratio, in general, decreases with increase in pore size. However whether the zeolite is uni- or multi-dimensional also is a significant factor. Further characterization of reaction cavity restriction becomes possible by the breakdown of the TMBs: 1,2,4-TMB (smallest product), 1,2,3-TMB (intermediate size product) and 1,3,5-TMB (largest product). BEA, MOR, FAU framework types led essentially to equilibrium distribution of TMBs, whereas with MTW, MFI, EUO, OFF only 1,2,4-TMB was observed [63]. It is suggested that, of the two seemingly operative mechanisms, the larger transition state diphenylmethane-mediated pathway is more likely to be valid for the larger 12-membered ring zeolites than the smaller transition state methoxide-mediated pathway.

It is interesting to note that, in contrast with experiments, computational work on transition state selectivities unexpectedly showed the intermediate-size 1,2,3-TMB to be kinetically preferred over the smallest-size 1,2,4-TMB. This result was

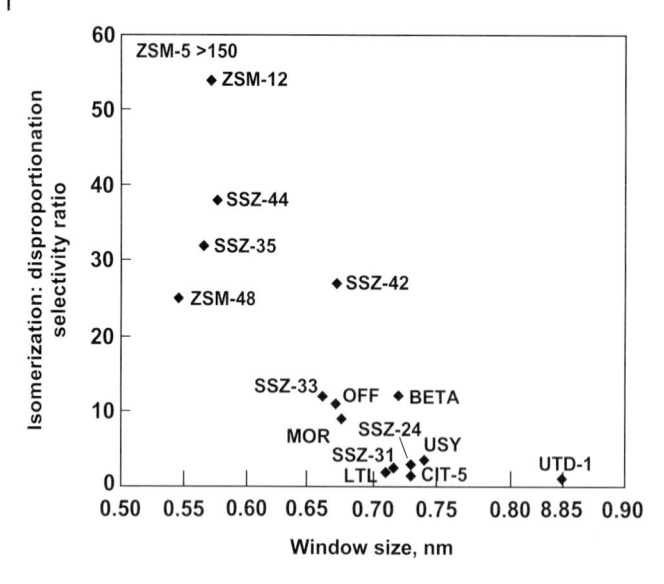

Figure 13.28 *Meta*-xylene disproportionation: isomerization selectivity ratio over various zeolites at 317–318 °C and 10% conversion [64].

presented as evidence to support the argument that product shape selectivity (favoring 124) and restricted transition state selectivity (favoring 123) effects often may be occurring simultaneously, making it difficult to decouple them. Interestingly, there have also been reports for the formation of larger 1,2,3- and 1,3,5-TMB in EUO framework type which is a ten-membered ring [13]. The presence of spacious side pockets running perpendicular to the 10-MR channels was provided as the rationale for the occurrence of these bulky compounds.

13.7.3.2 Methylcyclohexane Ring Contraction

Over bifunctional zeolitic catalysts methylcyclohexane can be activated through dehydrogenation to methylcyclohexene, followed by formation of the carbenium ion. The critical reaction that follows is ring contraction to dimethylcyclopentenes (DMCP) and ethylcyclopentene, with some minor amount of cracking. The distribution of DMCP isomers gives a good indication of restricted transition state selectivity effects [64]. 1,1-DMCP has the largest kinetic diameter, followed by the *trans* and *cis* isomers of 1,2-DMCP, which are larger than the 1,3-DMCPs (*c*, *t*). Over 12-MR FAU and MOR framework types, the 1,2-DMCP:1,3-DMCP ratio at 593K and 14±7% conversion was similar to that observed with amorphous silica–alumina, that is 2.2–2.3. However, this ratio was much smaller with 10-MR zeolites: MTT = 0.51, AEL = 0.34, TON = 0.16, MFI = 0.02 [65].

13.7.3.3 Alkane Hydroisomerization

Dewaxing is the final example of a reaction illustrated here with possibly multiple restricted transition state shape selectivity effects. Bifunctional zeolitic catalysts

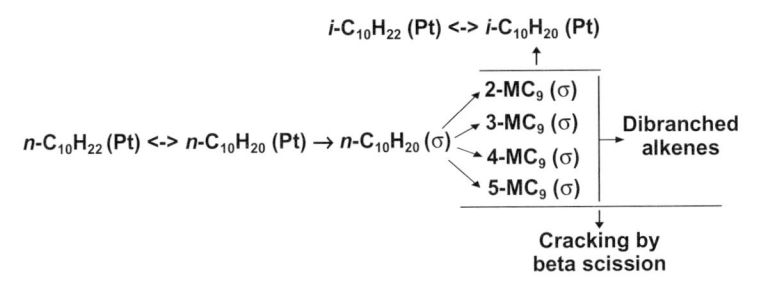

```
                    i-C_{10}H_{22} (Pt) <-> i-C_{10}H_{20} (Pt)
                                    ↑
                            ┌─────────────────┐
                            │ → 2-MC_9 (σ)    │
                            │ → 3-MC_9 (σ)    │   Dibranched
n-C_{10}H_{22} (Pt) <-> n-C_{10}H_{20} (Pt) → n-C_{10}H_{20} (σ) ─┤                  ├──→  alkenes
                            │ → 4-MC_9 (σ)    │
                            │ → 5-MC_9 (σ)    │
                            └─────────────────┘
                                    ↓
                              Cracking by
                              beta scission
```

Figure 13.29 Reaction mechanism for *n*-decane hydroisomerization.

with MTT, TON and AEL framework types are used for regulating skeletal isomerization of wax range *n*-alkanes [28, 29, 43, 66–68]. While creating branching in alkanes lowers the pour point, the total amount of branching should not be too much in order to maintain high viscosity index [68]. Also multibranched alkene intermediates are more prone for cracking. Accordingly, carefully regulating the extent of skeletal isomerization is critical to achieve the optimal product properties. In relatively small pores restricting the formation of bulky transition states that lead to excessive branching is the first type of shape selectivity effect possibly governing here. Cracking, in turn, can also be limited by inhibiting secondary to tertiary *Beta* scission pathways that typically require large transition states and this is the second type of restricted transition state shape selectivity that may play a role here.

The reaction mechanism is illustrated in Figure 13.29 for the case of *n*-decane, the most frequently used model feed compound for alkane hydroisomerization. The first step is the dehydrogenation of *n*-decane to make *n*-decene over the metal function (here designated as Pt), which is then protonated over the acid function to generate a carbenium ion. Here the requirement for the proximity of the acid sites to the metal sites is discussed in the literature [46]. The key advantage of bifunctional catalysis is the ability to generate carbenium ions at relatively lower temperatures than would otherwise be required for the generation of carbonium ions from alkanes. It is important to note that such bifunctional catalysis typically occurs at mild temperatures at which dehydrogenation of alkanes is not thermodynamically very favorable. Nevertheless, the tendency of the alkene to immediately get protonated over the Bronsted acid sites ensures further dehydrogenation despite unfavorable thermodynamic equilibrium. The *n*-decyl carbenium ion can undergo skeletal isomerization to produce various monomethylnonyl carbenium ions. These monobranched species can undergo further skeletal isomerization and cracking. The carbenium ions eventually get deprotonated and hydrogenated over the metal function to produce the branched alkane products. The quick hydrogenation step ensures a low alkene concentration in the reaction mixture which, in turn, helps minimize the undesirable consecutive alkene reactions. It is generally accepted that dehydrogenation–hydrogenation steps over metal function are equilibrated and kinetics are actually controlled by acid-catalyzed skeletal isomerization.

This mechanism is supported by reports of comparable reaction rates obtained during skeletal isomerization with the corresponding alkenes over metal-free zeolites [52].

13.7.3.4 Dimethylether Carbonylation

Transition state selectivity effect, however, does not have to be limited to cases of exclusion by size. In the case of dimethylether (DME) carbonylation to methyl acetate for which either reactants, products or the transition state are not large the high reaction rate observed with certain zeolites was attributed to the stabilization of the specific transition state by the local structure prevailing in the 8-MR channels. More specifically, while several zeolites were shown to decompose DME stoichiometrically over Bronsted acid sites to methoxys, based on infrared spectroscopy, only the bridging hydroxyls in the eight-membered ring channels were shown to be effective for reacting CO with the methoxys. Such reaction specificity for the bridging hydroxyls within the 8-MR channels was not observed in other reactions, for example, during isotopic exchange with CD_4, presumably because of the less ionic character of the transition state (stabilized by interaction with framework oxygens) involved [69].

13.7.4 Product Shape Selectivity

13.7.4.1 Alkane Hydroisomerization

While restricted transition state selectivity effects are likely to play significant role during alkane hydroisomerization, selective formation of mono-branched alkanes, in favor of the larger kinetic diameter multi-branched alkanes during dewaxing may also be explained by product shape selectivity effects. Figure 13.30 shows CBMC simulations over various framework types of the Gibbs free energy of adsorption for various decane isomers. These simulation results indicate that FAU does not distinguish among various decane isomers. Based on these results we expect that it is unlikely to have product shape selectivity over FAU during hydroprocessing of *n*-decane. However, with 10-MR framework types, like MFI and MEL and especially with TON, the situation is quite different, showing significant inhibition for adsorption of branched alkanes, the higher the branching the higher the inhibition [70]. These simulation results then suggest low Henry coefficients for branched alkanes over 10-MR zeolites. Low adsorbed concentration for mono-branched alkanes implies minimal consecutive reactions to multi-branched alkanes. Product shape selectivity effects can also be observed through diffusional effects. In relatively small-pore zeolites the selectivity to multi-branched products may be lower due to the low diffusivity possessed by bulky products.

Octadecane hydroprocessing behavior of Pt-containing bifunctional catalysts with TON and MTT framework types was compared, as illustrated in Figure 13.31 [28]. While the two zeolitic catalysts showed similar activities, the selectivity vs conversion performances were different. At any given conversion, the selectivity to dibranched isomers was lower and the selectivity to mono-branched isomers

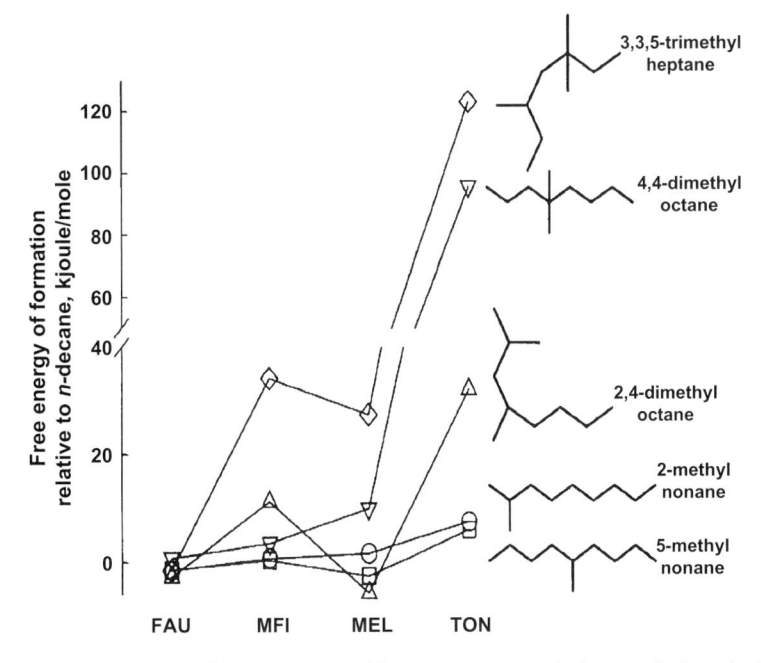

Figure 13.30 The Gibbs free energy of formation (CBMC calculations) for branched decane isomers relative to linear decane over various framework types [70].

was higher over MTT. The question was raised whether differences in Henry coefficients and/or diffusivities for branched alkanes between TON and MTT can explain the selectivity results. Based both on simulation and experimental work, it was previously reported that the mono-branched alkanes have lower Henry coefficients in MTT relative to TON, which would be consistent with less consecutive reactions to multi-branched alkanes over MTT [15, 66]. However, a more recent examination of one of the earlier experimental work reported the Henry coefficients for mono-branched C_5–C_8 alkanes to be similar, perhaps slightly higher with MTT versus TON [47]. Molecular mechanics calculations done by M. Gatter at UOP using Accelrys Insight II solids diffusion modeling package showed the energy minima to be higher by 42–55 kJ/mol for 2,5-, 2,6- and 2,7-dimethyloctane in MTT relative to TON. This lower heat of adsorption for dibranched alkanes in MTT relative to TON suggests a more restricted pore geometry for MTT. There have also been reports of lower diffusivities simulated for the dibranched alkanes in MTT relative to TON [71]. As illustrated in Figure 13.32, the pore for MTT is less elliptical and slightly smaller, which would be consistent with slower simulated diffusivities reported for dibranched alkanes in MTT channels. At the time this chapter was written the issue had not yet been resolved as to what extent restricted transition state shape selectivity and product shape selectivity effects are responsible for the higher mono-branched alkane selectivity over MTT. In the case of product shape selectivity, it is not yet clear whether inhibition of mono-branched

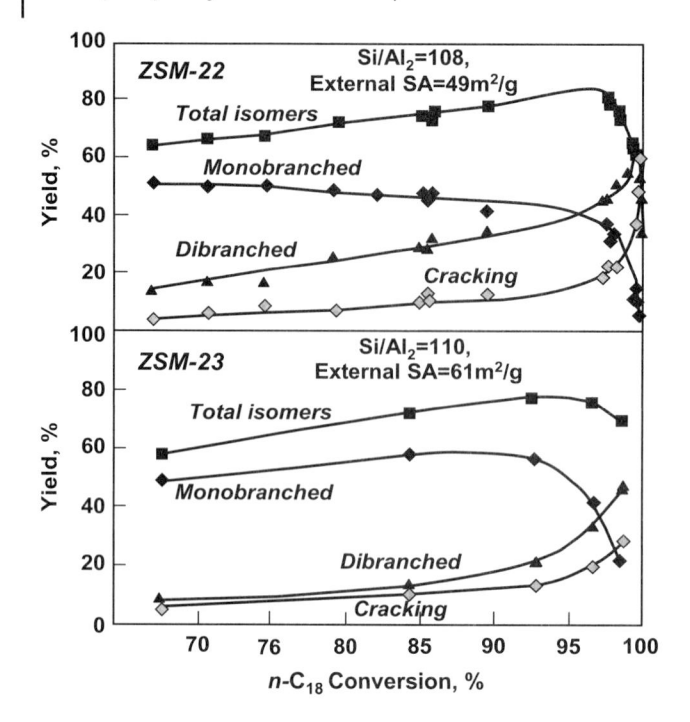

Figure 13.31 Hydroprocessing octadecane (2% feed diluted with heptane) over bifunctional catalysts (contains Pt) with TON and MTT framework types at 230 °C, 450 kPa, H_2:HCBN = 13 and varying flow rates [28].

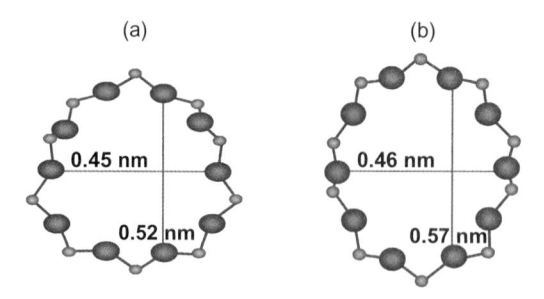

Figure 13.32 Views of 10-MR channels in (a) MTT and (b) TON.

alkanes adsorption and/or inhibition of dibranched alkane adsorption and/or slow diffusion of dibranched alkanes are responsible for observed selectivity differences. Interestingly, crystal size effects have actually been reported with MTT during hydroprocessing of alkanes, with smaller crystals leading to less cracking, which is consistent with the presence of product shape selectivity suggested here [72].

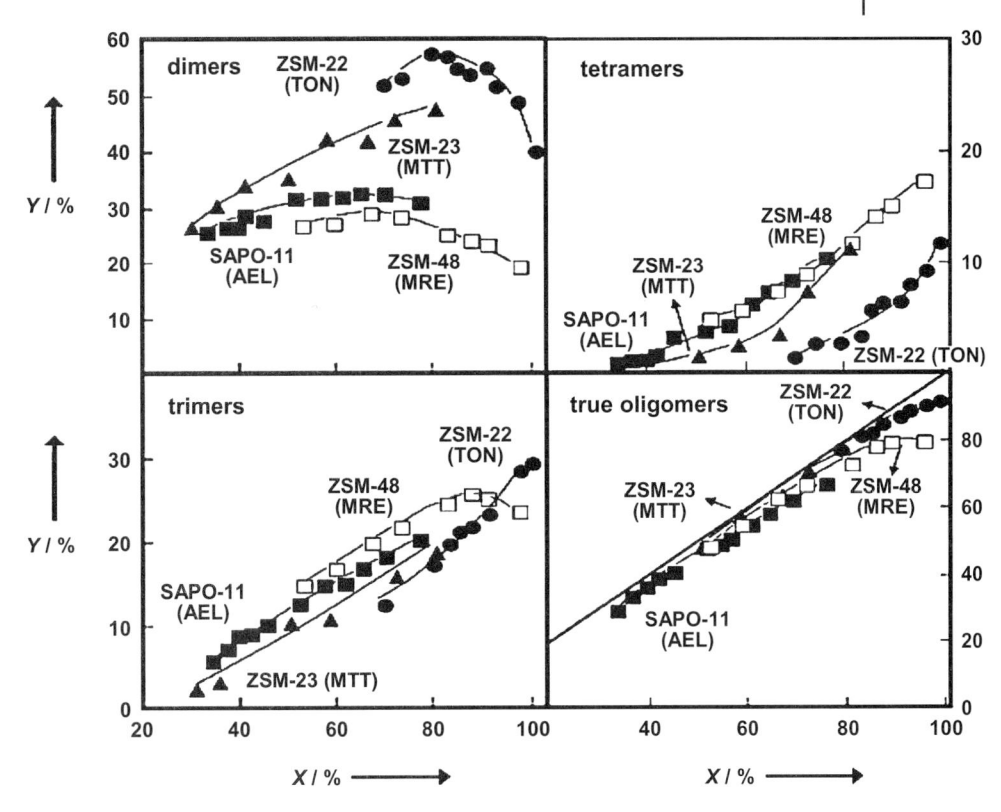

Figure 13.33 Yield (Y) of propene dimers, trimers, tetramers and total true oligomers (di- to hexamers, without products with intermediate carbon numbers obtained by cracking and recombination of fragments) againet propene conversion (X) with ZSM-22 (TON), ZSM-23 (MTT), ZSM-48 (MRE) and SAPO-11 (AEL), test conditions: 200–240 °C, 6–8 µPa with 12% propylene in propane feed [73].

13.7.4.2 Oligomerization of Propylene

The distributions of oligomers from processing of propylene over TON, MTT, AEL and MRE framework types were compared as a function of conversion in Figure 13.33 [73]. For any given framework type selectivities are a strong function of conversion, as illustrated in Figure 13.34, which is why selectivities for different zeolites need to be compared at the same conversion level. All four of these 10-MR zeolites showed enhanced selectivity for dimerization, in contrast to forming trimers and tetramers. TON ($Si/Al_2 = 84$, channel dimensions 0.46 * 0.57 nm, pore index 26.2 Å^2) showed highest dimer selectivity and lowest trimer + tetramer. MTT ($Si/Al_2 = 88$, 0.45 * 0.52 nm, pore index = 23.4 Å^2) showed the second highest dimer selectivity, followed by AEL ($Si_{0.21}Al_{0.43}P_{0.36}$, 0.64 * 0.39 nm, pore index = 25.0 Å^2) and MRE ($Si/Al_2 = 100$, 0.53 * 0.56 nm, pore index = 29.7 Å^2). A different ordering of framework types was obtained when the degree of branching was compared. The degree of branching, as characterized in the trimer fraction, was lowest over MTT, followed by TON, MRE and AEL, as illustrated in Figure 13.35. These results

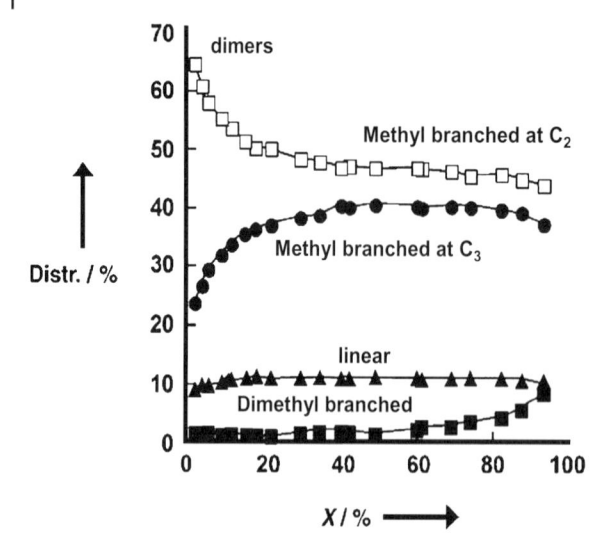

Figure 13.34 Skeletal distribution of the dimer fraction obtained at increasing propene conversion with ZSM-22 (TON). Test conditions: 200–240 °C, 6.8 µPa with 12% propylene in propane feed [73].

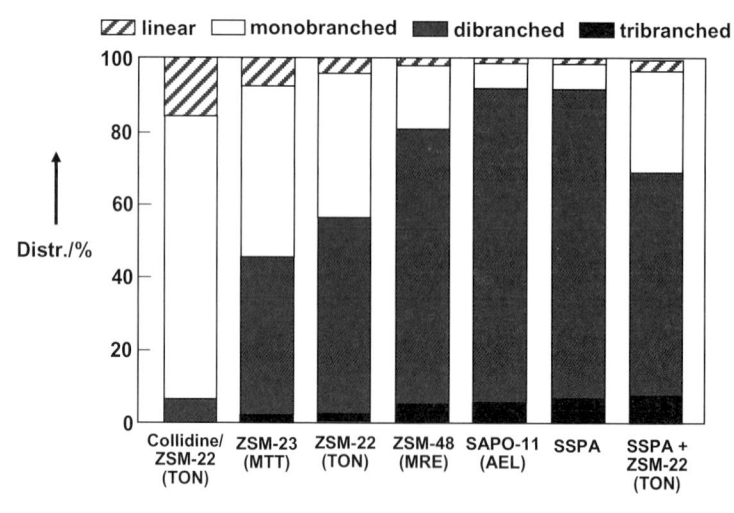

Figure 13.35 Skeletal distribution of trimer fractions at 30% propene conversion, obtained on collidine-treated ZSM-22 (TON), ZSM-23 (MTT), ZSM-22 (TON), ZSM-48 (MRE), SAPO-11 (AEL), SSPA and on successive layers of SSPA and ZSM-22 (TON). Test conditions: 200–240 °C, 6.8 µPa with 12% propylene in propane feed [73].

indicate shape selectivity effects are often a complex function of zeolite topological properties. In this case not only product shape selectivity effects, but also restricted transition state shape selectivity effects are likely to govern at the same time. Interestingly, neutralizing the external acidity of TON framework type with collidine significantly lowered the degree of branching by forcing the reaction to the zeolite channels. These four zeolites all showed lower branching than silica-supported phosphoric acid (SSPA).

13.7.4.3 Alkylation of Naphthalene

Some medium- and large-pore zeolites were compared for selective synthesis of 2,6-dimethylnaphthalene (DMN) in favor of 2,7-DMN through alkylation of naphthalene with methanol or through isomerization of other DMNs. Both diffusion simulations and actual alkylation experiments in the presence of 1,2,4-trimethylbenzene (TMB) as a solvent were performed [74]. 10-MR framework types like MFI and EUO were found to be unsuitable since reactants could not diffuse in without a high energy barrier. Among 12-MRs MTW was found to provide the highest activation energy difference between 2,6-DMN diffusion and 2,7-DMN diffusion relative to BEA, MOR, MAZ, FAU, LTL and OFF. MTW did not allow the diffusion of 1,5-DMN and 1,6-DMN, both with very large kinetic diameters. MTW then provides a typical case of product shape selectivity during alkylation of naphthalene, whereby undesired isomers eventually undergo isomerization to the least bulky 2,6-DMN before elution. The same authors also report experimentally MTW does indeed provide the highest 2,6-DMN:2,7-DMN ratio in the range of 2.0–2.7, much higher than the thermodynamic equilibrium value that is close to 1.0 obtained over other zeolites. Interestingly, in this work the alkylation is proposed to occur through a bimolecular mechanism that involves an electrophilic attack of the TMB-derived cation on naphthalene. Due to the bulky nature of the transition state involved, then, restricted transition state selectivity may also be governing along with product shape selectivity effects to explain the observed product distribution.

13.7.4.4 Hydrocracking of *n*-Hexadecane

Selective formation of dimethylbutanes relative to n-hexane during hydrocracking of *n*-hexadecane over AFI and MAZ framework types in preference to larger and smaller pore size zeolites has been attributed to the bell-shaped nature for the Gibbs free energy of adsorption versus pore size [75]. Figure 13.36 summarizes the experimental and computational dimethylbutane (DMB): *n*-hexane ratios at 577 K and 3000 kPa for various framework types normalized with respect to the results obtained over AFI. The left bars are computational ratios for adsorption of 2,2-DMB:n-C$_6$ and the middle bars for 2,3-DMB:n-C$_6$. The right bars are the experimental DMB:n-C$_6$ ratios obtained during hydrocracking of *n*-hexadecane. For each framework type zeolite topology information, including dimensionality and pore channel dimensions are also given in Figure 13.36. Interestingly, while simulations at low pressures did reveal, as expected, the repulsion of dibranched paraffins in small pore framework types like MTW and BEA, no selectivity differences relative to *n*-hexane were observed for larger pores. However, simulations at the high pressures showed preference for the adsorption of dimethylbranched butanes relative to *n*-hexane with AFI and MAZ framework types, consistent with the experimental data. More specifically, it was stated that intermediate pore size AFI and MAZ framework types minimize the drop in entropy of 2,2-DMB and

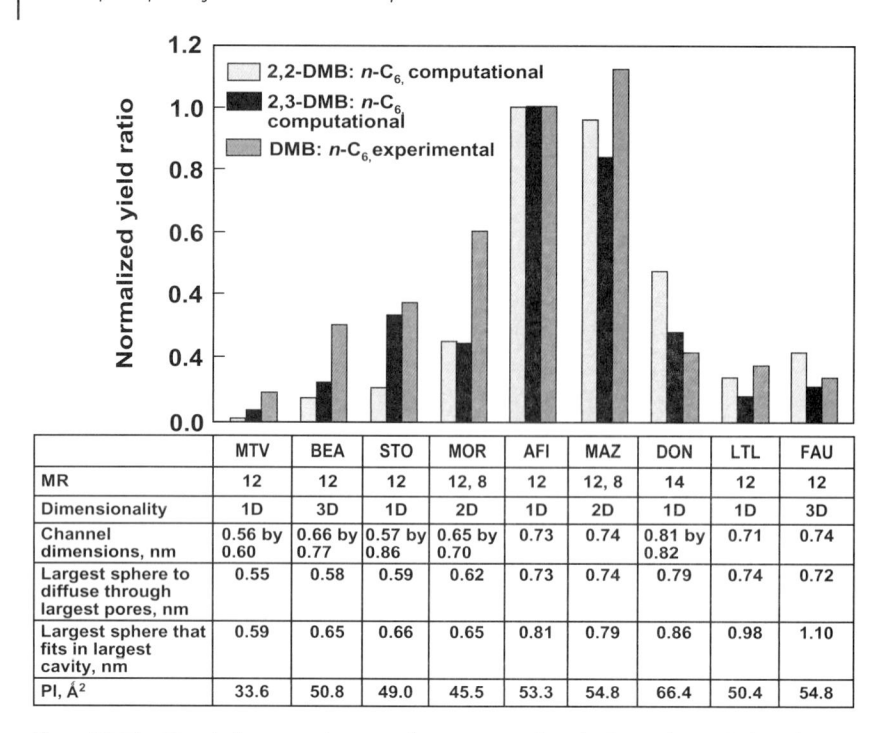

	MTV	BEA	STO	MOR	AFI	MAZ	DON	LTL	FAU
MR	12	12	12	12, 8	12	12, 8	14	12	12
Dimensionality	1D	3D	1D	2D	1D	2D	1D	1D	3D
Channel dimensions, nm	0.56 by 0.60	0.66 by 0.77	0.57 by 0.86	0.65 by 0.70	0.73	0.74	0.81 by 0.82	0.71	0.74
Largest sphere to diffuse through largest pores, nm	0.55	0.58	0.59	0.62	0.73	0.74	0.79	0.74	0.72
Largest sphere that fits in largest cavity, nm	0.59	0.65	0.66	0.65	0.81	0.79	0.86	0.98	1.10
PI, Å^2	33.6	50.8	49.0	45.5	53.3	54.8	66.4	50.4	54.8

Figure 13.36 Dimethylbutane: n-hexane selectivity ratios from hydrocracking of n-hexadecane over various framework types [75].

2,3-DMB upon adsorption at high adsorbate loading, leading to the highest selectivity for adsorbing DMBs relative to n-hexane, hence providing an alternate explanation for the experimental results previously referred to as representing inverse shape selectivity in earlier literature [13]. Furthermore, it was stated that, it is the compatibility between an adsorbate and an adsorbent that determines the extent to which the pores are filled with a given molecule and that a branched isomer may pack more efficiently than a linear isomer in the presence of significant adsorbate–adsorbate interactions at high pressures, leading to a higher Henry coefficient for the branched species [75]. In case the kinetics for hydrocracking of n-hexadecane are controlled by desorption of products the reaction rate would indeed increase with increase in Henry coefficient of the product (see Section 13.2.1).

While the above analysis suggested product shape selectivity may be governing here, transition state shape selectivity effects could not be ruled out. Further analysis of this problem was done at UOP with the objective of comparing the dimensions of the zeolite cavities where reaction may be taking place. For each framework type the size of the largest sphere that fits in the largest cavity, as well as the diameter of the largest sphere that can diffuse through the largest pores were calculated. These largest sphere calculations were done by R. Broach by using the cavities display mode of the software Atoms V6.2 (Shape Software). For each

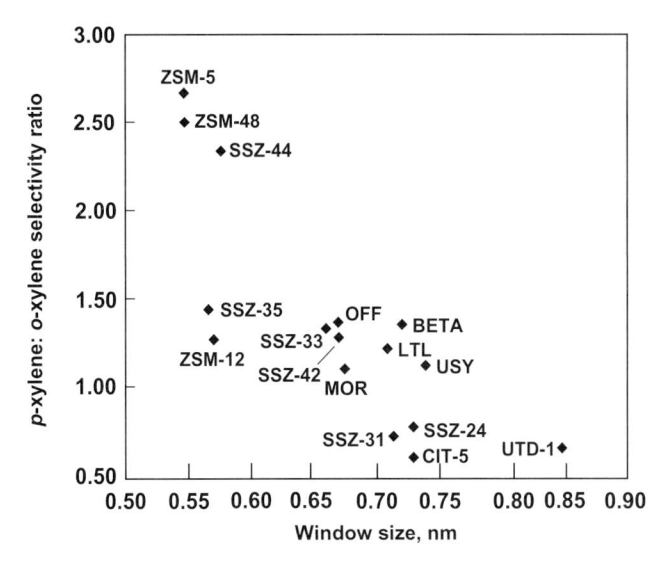

Figure 13.37 Para-xylene: ortho-xylene selectivity ratio from conversion of meta-xylene over various zeolites at 317–318 °C and 10% conversion [64].

framework type the all-silica version available from the IZC Structure Commission web site was used. Atomic radii were set to 0.026 nm for Si and 0.135 nm for O. A grid spacing of 0.01 nm was used for the calculations. To estimate the largest sphere that would fit in the zeolite cavities, the particle radius was varied until the largest sphere that still displayed cavity volume was found. To estimate the largest sphere that would diffuse, the radius was reduced until channels appeared. The largest sphere calculations have been inserted to the table at the bottom of Figure 13.36. Also the pore index for each zeolite was included. The largest sphere that fits in the largest cavity had a diameter of 0.80 ± 0.01 nm for AFI and MAZ framework types for which the experimentally measured selectivity of dimethylbutanes: n-hexane was highest. For larger and smaller cavity sizes the selectivity was lower. Since transition state shape selectivity effects could not be ruled out during hydrocracking of n-hexadecane, it is conceivable that 0.8 nm zeolite cavity may be providing a more favorable interaction for the transition state in the case of the reaction pathway that leads to dimethylbutanes in favor of n-hexane.

13.7.4.5 Meta-Xylene Isomerization

Meta-xylene isomerization to ortho- and para-xylene over 10- and 12-MR zeolites is another illustration of product shape selectivity effects [13]. The two products are essentially equally favorable from the standpoint of thermodynamics. With decreasing pore size, however, kinetics come into play and the selectivity to para-xylene increases, as illustrated in Figure 13.37 for results obtained at 317–318 °C, 0.5 kPa meta-xylene pressure (in the presence of He carrier gas) and 10% conversion [64]. While the para: ortho ratio is typically 1.0–1.5 with multi-dimensional

12-MRs, with 10-MRs the *para*: *ortho* ratio is typically >2, the smaller the pore size, the higher the *para*-xylene selectivity. More recent molecular dynamics simulations verify that the diffusivity ratio for para: ortho is much higher for ZSM-5 (7.4) than for *Beta* (2.3), consistent with the higher *para*-xylene selectivity obtained over ZSM-5 [76].

It is important to note here that product shape selectivity effects are crystal size-dependent. Accordingly, comparisons in Figure 13.37 become meaningful only to the extent that crystal morphology is not an independent variable. For example, the higher *para*:*ortho* ratio observed with SSZ-44 relative to SSZ-35 having essentially identical pore size is attributed to the three times larger crystal size for SSZ-44 [64]. Interestingly, the *para*:*ortho* ratio is reported to be <1 with extra-large one-dimensional zeolites like SSZ-31, SSZ-24, CIT-5 and UTD-1, presumably due an *ortho*-xylene selective bimolecular isomerization pathway. Finally, it is important to mention that pore size as the only independent variable even with same crystal size often becomes an over-simplification of the zeolite characteristics from the standpoint of shape selectivity effects. Whether the zeolite is uni- or multi-dimensional and other topological features often become significant [64].

13.7.4.6 Methanol to Olefins

Methanol to light alkenes in favor of methanol to aromatics is another example of a reaction that is governed by product shape selectivity effects [13]. While aromatics are formed within the cavities of CHA and ERI framework types, as the reaction mechanism would require, aromatics would be trapped within the 8-MR channel systems. With medium-pore size MFI aromatics are indeed observed as products, but the carbon number is limited to 10 (durene). Bulkier aromatics are observed with MOR and BEA.

13.7.4.7 Other Examples of Product Shape Selectivity

Other examples of systems that are likely to be governed by product shape selectivity effects include toluene disproportionation to *para*-xylene + benzene in favor of other xylenes + benzene [61]. Toluene alkylation by methanol to give *para*-xylene in favor of other xylenes is yet another such example [76].

13.7.5
Crystal Size Effects

If intracrystalline diffusional effects are observed then reactant and product shape selectivity effects may be occurring. For example cracking rates of gem-branched dimethyl alkanes were found to decrease with increase in crystal size, implying reactant shape selectivity effects may occur during cracking of alkanes with different degrees of branching [18]. In the case of product shape selectivity, good examples of the impact of crystal size are in toluene disproportionation and in toluene alkylation by methanol over ZSM-5. For both cases, *para*-xylene selectivity is increased with increase in crystal size, resulting from the diffusional limitations imposed upon the production of *ortho*-xylene and *meta*-xylene by the restricted pore geometry of MFI framework type [18]. However, transition state shape selectivity is not impacted by crystal size. In the case of cracking of alkanes over HZSM-5 the higher reaction rate observed with n-hexane relative to 3-methylpentane was first attributed to reactant shape selectivity effects. However, subsequent diagnostic experiments with ZSM-5 having two different crystal sizes (0.05 μm, 2.7 μm) did not show any difference in reaction rate, strongly suggesting that these shape selectivity effects were, in fact not reactant-based, but restricted transition state-related. It was further suggested that, in the rate-controlling hydride transfer step between a small carbenium ion and a feed molecule, more space is required for the transition state when the feed is branched [41]. On the other hand, *para*-xylene selectivity during toluene disproportionation over HZSM-5 was found to increase with crystal size, strongly suggesting that the effect was product shape selectivity, as suspected [41].

Zeolite crystal size can be a critical performance parameter in case of reactions with intracrystalline diffusion limitations. Minimizing diffusion limitations is possible through use of nano-zeolites. However, it should be noted that, due to the high ratio of external to internal surface area nano-zeolites may enhance reactions that are catalyzed in the pore mouths relative to reactions for which the transition states are within the zeolite channels. A 1.0 μm spherical zeolite crystal has an external surface area of approximately $3\,m^2/g$, no more than about 1% of the BET surface area typically measured for zeolites. However, if the crystal diameter were to be reduced to 0.1 μm, then the external surface area becomes closer to about 10% of the BET surface area [41]. For example, the increased 1,2-DMCP:1,3-DMCP ratio observed with decreased crystallite size over bifunctional SAPO-11 catalyst during methylcyclohexane ring contraction was attributed to the increased role of the external surface in promoting non-shape selective reactions [65].

13.8
Reaction Mechanisms

For the sake of simplicity, carbenium ions, carbonium ions or protonated cyclopropane rings were used as reaction intermediates, omitting the anionic zeolite framework in the illustration of the reaction mechanisms for the reactions discussed here. Furthermore, it is conceivable that many such reaction paths involve alkoxide intermediates, instead of carbenium and carbonium ions.

13.8.1
Alkene Skeletal Isomerization

13.8.1.1 Unimolecular Mechanism
The unimolecular mechanism involves formation of a protonated cyclopropane ring first, which avoids the formation of a primary carbenium ion until after skeletal rearrangement has taken place. Such reaction intermediates were first

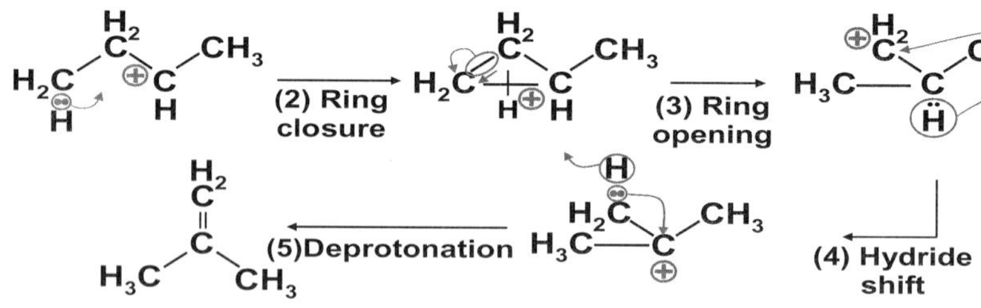

Figure 13.38 Mechanism for unimolecular skeletal isomerization of n-butylene.

proposed for skeletal isomerization in liquid superacid medium [77]. The most likely elementary steps are: (i) protonation of the internal alkene to give a secondary linear carbenium ion intermediate, (ii) formation of the protonated cyclopropane ring, (iii) ring opening to give a primary carbenium ion – the most likely candidate for the rate-determining step, (iv) hydride shift to give a tertiary branched carbenium ion, (v) decomposition of the primary carbenium ion to give the iso-alkene adsorbed on a Bronsted acid site, in essence the deprotonation step. For the sake of simplicity these elementary steps are illustrated in Figure 13.38 for the skeletal isomerization of n-butene through carbenium ion intermediates only without showing the active site each time. A similar mechanism was proposed for the skeletal isomerization of 2-pentene over TON [52]. Skeletal isomerization through formation of larger than three-rings for the formation of ethyl and larger n-alkyl branching is discussed in the literature [46].

13.8.1.2 Bimolecular Mechanism

The bimolecular mechanism, (also called dimerization–isomerization) on the other hand, is likely to involve the following elementary steps for n-butylene conversion: (i) protonation of the internal alkene to give a secondary linear carbenium ion, (ii) alkylation by an internal alkene to give a secondary 3,4 dimethylhexyl carbenium ion, (iii) methyl shift to give a secondary 2,4 dimethylhexyl carbenium ion, (iv) hydride shift to give a tertiary 2,4 dimethylhexyl carbenium ion, (v) beta scission to give isobutylene and a secondary linear carbenium ion. This mechanism is illustrated in Figure 13.39.

13.8.2
Alkene Oligomerization

The mechanism for alkene oligomerization is illustrated here by dimerization of butylene as an example due to the technological significance of making high octane C_8 oligomer species. Given an adsorbed i-butyl tertiary carbenium ion., three different C_8 dimers can be produced depending on the nature of the incoming butylene molecule, as shown in Figure 13.40 and Figure 13.41. The oligomerization product is the tertiary 2,2,4- trimethylpentyl carbenium ion when i-butyl carbenium ion reacts with i-butene. 2,2,4-Trimethyl 2-pentene is then formed by

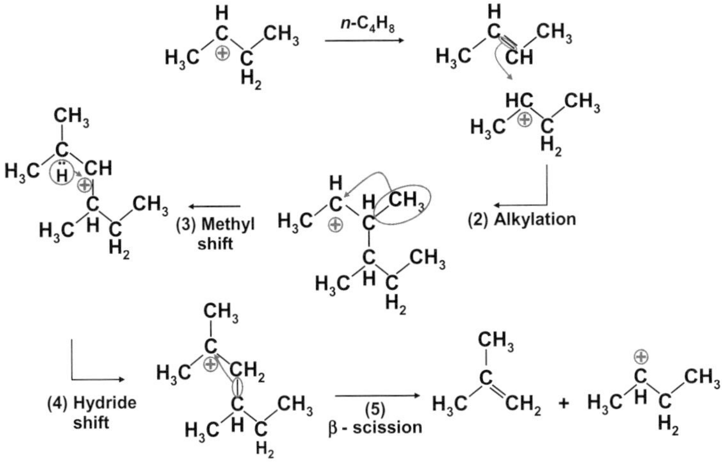

Figure 13.39 Mechanism for bimolecular skeletal isomerization of *n*-butylene.

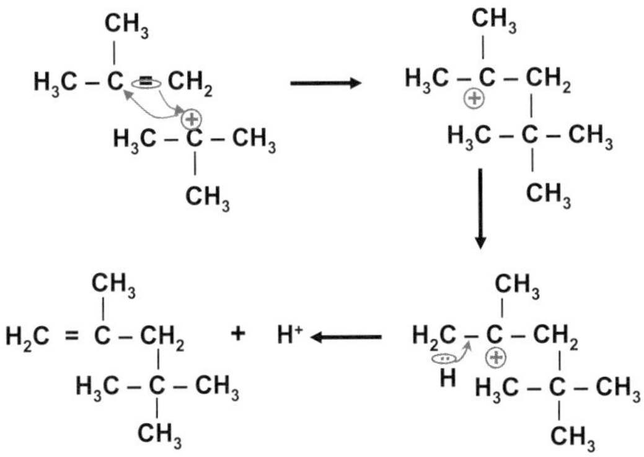

Figure 13.40 Mechanism for oligomerization of *i*-butylene with *i*-butylene.

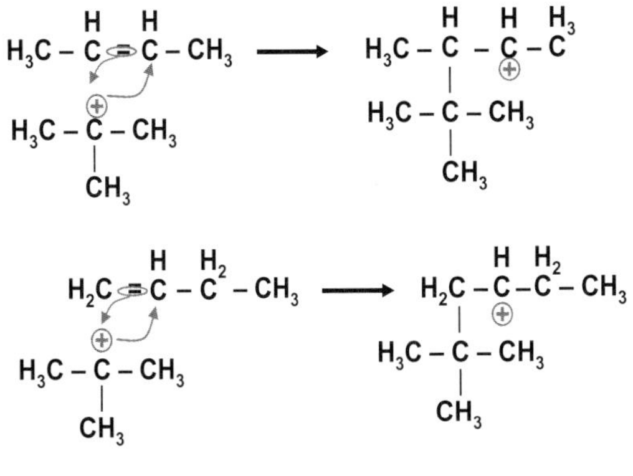

Figure 13.41 Mechanism for oligomerization of *i*-butylene with 1-butylene or with 2-butylene.

a simple deprotonation step. But, the product is the secondary 2,2,3-trimethylpentyl carbenium ion when the reaction occurs between *i*-butyl carbenium ion and 2-butene. When the reaction occurs between *i*-butyl carbenium ion and 1-butene the product is the secondary 2,2-dimethylhexyl carbenium ion. While the research and motor octane numbers, RON/MON, are high for the corresponding alkanes of 2,2,4-trimethylpentene and 2,2,3-trimethylpentene, 100/100 and 110/100, respectively, they are low in the case of 2,2-dimethylhexene (72.5/77.4). In case of reactions with adsorbed secondary butyl carbenium ion, the products are 3,4-dimethylhexyl carbenium ion when the second reactant is 2-butene, 3-methylheptyl carbenium ion when the second reactant is 1-butene. RON/MON for the corresponding alkanes are quite low with the latter: 35/27.

13.8.3
Alkylation

13.8.3.1 Isobutane Alkylation by 2-Butylene

The isobutane alkylation mechanism that includes both 1-butene and 2-butene reactants, showing the desired as well as some of the undesired reaction pathways is illustrated in Figure 13.42 [78, 79]. A simplified version of the mechanism limited to *i*-butane + 2-butene and outlining only the desired elementary steps (over a fresh catalyst) is illustrated in Table 13.5. This is an example of a mechanism which involves carbenium ions as chain carriers. The reaction is initiated by protonation of 2-butene to form a secondary carbenium ion (1), which then abstracts a hydride ion from *i*-butane and forms the more stable tertiary butyl carbenium ion plus *n*-butane (2). This tertiary carbenium ion is the key intermediate for the subsequent alkylation with 2-butene to give the secondary 2,2,3-trimethylpentyl (TMP) carbenium ion (3). If the tertiary *i*-butyl cation were to react with 1-butene, then the secondary 2,4-dimethyhexyl (DMH) carbenium ion could eventually be obtained. 2,2,3-TMP$^+$ can undergo methyl shift to give 2,3,3,-TMP$^+$ (4) or 2,2,4-TMP$^+$ (5), which, in turn, can give 2,3,4-TMP$^+$ by another methyl shift (6). The initial hydride transfer step (2), hence provides the opportunity for *i*-butane to take part in the C_8 formation process. However, at that point in the sequence of events there is still no net consumption of butanes since one mole of *n*-butane was made for each mole of *i*-butane that was activated (steps 1–6). In the next step, however, the C_8 carbenium ion abstracts a hydride ion from another *i*-butane to form a trimethylpentane, along with the tertiary *i*-butyl carbenium ion (intermolecular hydride transfer; 7). If the C_8 carbenium ion were 2,4-dimethylhexyl, then 2,4 dimethylhexane would be formed. At that point, the sum of steps 1–7 indicates that two moles of *i*-butane have been consumed along with two moles of 2-butene plus one proton to give one mole of TMP, one mole of *n*-butane and one mole of tertiary *i*-butyl carbenium ion, the key intermediate for sustaining the alkylation reaction. Once the tertiary *i*-butyl carbenium ion is available, unless it is deprotonated to make *i*-butylene, it can be alkylated with another 2-butylene molecule to make another C_8 species. At that point in the mechanism *i*-butane can be consumed without making *n*-butane. This can be understood from the sum

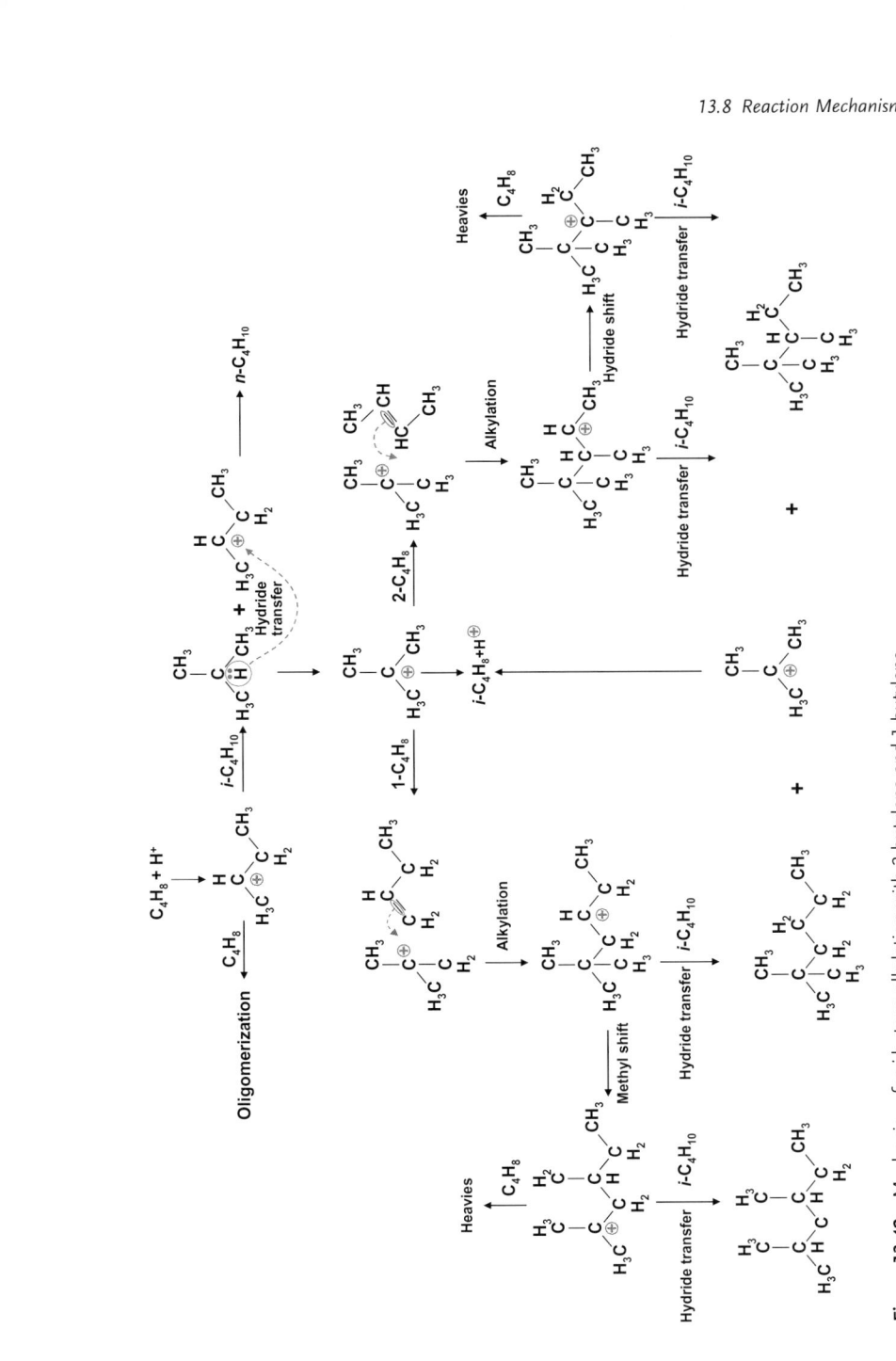

Figure 13.42 Mechanism for *i*-butane alkylation with 2-butylene and 1-butylene.

Table 13.5 Steps for i-butane alkylation with 2-butylene to make TMP.

Protonation	1.	$2\text{-}C_4H_8 + H^\oplus \rightleftharpoons n\text{-}C_4H_9^\oplus$
Hydride transfer	2.	$n\text{-}C_4H_9^\oplus + i\text{-}C_4H_{10} \rightleftharpoons n\text{-}C_4H_{10} + i\text{-}C_4H_9^\oplus$
Alkylation	3.	$i\text{-}C_4H_9^\oplus + 2\text{-}C_4H_8 \rightleftharpoons 2,2,3\text{-}TMP^\oplus$
Methyl shift	4.	$2,2,3\text{-}TMP \rightleftharpoons 2,3,3\text{-}TMP^\oplus$
Methyl shift	5.	$2,2,3\text{-}TMP \rightleftharpoons 2,2,4\text{-}TMP^\oplus$
Methyl shift	6.	$2,2,4\text{-}TMP^\oplus \rightleftharpoons 2,3,4\text{-}TMP^\oplus$
	1-6	$2(2\text{-}C_4H_8) + (i\text{-}C_4H_{10}) + H^\oplus \rightleftharpoons n\text{-}C_4H_{10} + TMP^\oplus$
Hydride transfer	7.	$TMP^\oplus + i\text{-}C_4H_{10} \rightleftharpoons TMP + i\text{-}C_4H_9^\oplus$
	1-7	$2(2\text{-}C_4H_8) + 2(i\text{-}C_4H_{10}) + H^\oplus \rightleftharpoons n\text{-}C_4H_{10} + TMP + i\text{-}C_4H_9^\oplus$
Desired alkylation	3-7	$2\text{-}C_4H_8 + i\text{-}C_4H_{10} \rightleftharpoons TMP$

of steps 3–7 which now indicates that one mole of i-butane is consumed for every mole of 2-butylene, which is the desired reaction stoichiometry, the so-called true alkylation. When i-butyl carbenium ion is akylated with another i-butylene molecule 2,2,4-trimethylpentene carbenium ion is formed, the so-called undesired self-alkylation step.

It is important to note that, while the primary product C_8 carbenium ions that are formed (after reaction with 2-butene or 1-butene) are secondary, they can undergo hydride shift or methyl shift and form a tertiary carbenium ion in each case. In that case the driving force is diminished for either of the two tertiary C_8 carbenium ions to abstract a hydride ion from i-butane since this now becomes a transition from a large tertiary carbenium ion to a smaller tertiary carbenium ion. Nevertheless, this hydride transfer can still occur due to the high ratio of i-butane to tertiary C_8 carbenium ion that exists in the reaction medium. At the same time the tertiary C_8 carbenium ion may get alkylated with another butylene molecule to make the more stable C_{12} carbenium ion, which would then lead to heavies.

In case the catalyst's protonation and alkylation activities remain high, but its hydride transfer activity is lowered due to aging, then steps 2 and 7 would be slow. Accordingly, there will be more opportunity for the secondary butyl cation made in the first step now to react with 2-butene to make 3,4-dimethylhexyl carbenium ion. This is now becoming more and more like an oligomerization reaction between two alkenes. 3,4-DMH$^+$ may have hard time abstracting a hydride from i-butane to make the alkane and instead may deprotonate to make an alkene or continue the alkylation to make heavies. Reduced hydride transfer activity also implies that primary 2,2,3-TMP$^+$ has less opportunity to desorb as alkane after hydride transfer with i-butane and hence undergoes further consecutive reactions to 2,3,3-TMP$^+$, 2,2,4-TMP$^+$ and 2,3,4-TMP$^+$ [79]. In summary, the key implications of a loss of hydride transfer function are difficulty in activating i-butane and desorbing TMP$^+$ as an alkane, resulting in reduced consumption of i-butane, resulting in more alkenes, more DMH, less TMP, less 2,2,3-TMP, more 2,3,4-TMP and more heavies.

Beta zeolite has been extensively studied for i-butane/2-butylene alkylation. Regular post-synthesis treatments like calcination of the organic template, followed by ammonia exchange and a second calcination result in some framework dealumination in Beta. This extra-framework aluminum (EFAL) was shown to benefit the hydride transfer activity of Si/Al_2 = 26 Beta, as determined by a lower TMP/DMH ratio and a higher alkene content upon removal of the EFAL through chemical treatments. This behavior suggests a synergy exists between EFAL and the Bronsted acid sites for enabling hydride transfer, the key requirement for the alkylation reaction [80]. Lowering the Bronsted acid concentration in H-Beta by Na^+ exchange shortened the catalyst lifetime and lowered the hydride transfer activity, hence increased the likelihood of TMP^+s to undergo consecutive reactions, including to heavy products [81]. It is interesting to note that hydride transfer is generally slow over Beta and a diagnostic alkylation test performed in the presence of adamantane showed the hydride transfer rate to increase significantly, enabling faster desorption of the primary products, as verified by the increased selectivity to 2,2,3-TMP in favor of the consecutive reaction product 2,3,4-TMP [81].

BEA (Si/Al_2 = 30), FAU (Si/Al_2 = 8.6) and EMT (Si/Al_2 = 8.6) framework types were compared for i-butane/2-butylene alkylation. During the lifetime of the catalyst the butylene turnover number (TON) was approximately the same for each of the three zeolites and the acid sites were equivalent from the standpoint of stability in each case. With EMT the lowered selectivity to consecutive reaction products 2,2,4-TMP + 2,3,4-TMP relative to 2,2,3-TMP + 2,3,3-TMP and the lowered selectivity to heavies relative to BEA was interpreted as higher hydride transfer activity. The latter was attributed to the presence of large cages present in EMT, but not in BEA, for enabling bulky transition states for hydride transfer [82].

A series of La–H–X zeolites were prepared by various levels of La^{+3} exchange of Na–X (Si/Al_2 = 2.4) and evaluated for i-butane/2-butylene alkylation. Upon calcination Bronsted acid sites bound to La^{+3} (3520 cm^{-1}), as well as regular bridging hydroxyls (3640 cm^{-1}, 3600 cm^{-1}) are generated based on IR. With increased level of La^{+3} exchange the fraction of strong Bronsted acid sites, as measured by pyridine adsorption, increased and that correlated with a significant increase in the catalyst lifetime [83]. The best catalyst in the series yielded a TON of approximately 100 during its lifetime, arbitrarily defined as the regime with 100% butylene conversion [84]. For this catalyst about 80% of the alkylate was C_8 isomers with little DMH production. Interestingly, while the primary product 2,2,3-TMP was very low in concentration, the most abundant product was 2,3,3-TMP that requires a single methyl shift. The 2,2,4-TMP and 2,3,4-TMPs, obtained by multiple methyl shifts were lower in concentration. These results suggest that this catalyst has a modest level of hydride transfer activity.

13.8.3.2 Aromatic–Alkene and Aromatic–Alcohol

Liquid phase alkylation of benzene with propylene to make cumene can be used to illustrate the general reaction mechanism for the alkylation of aromatics with

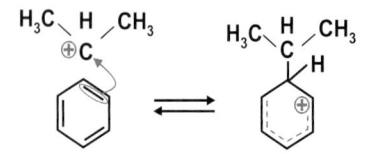

Figure 13.43 Mechanism for alkylation of benzene with propylene.

alkenes. In this case, the first step is the protonation of propylene to make the secondary propyl carbenium ion. In the next step, presumably the kinetically slow step, the carbenium ion exerts an electrophilic attack on the pi electrons of the aromatic ring, as illustrated in Figure 13.43. Alkylation leads to a benzenium type cation.

Cumene is generated following a deprotonation step [85]. One of the undesired reactions is multiple alkylation with propylene. Oligomerization of propylene is also undesired. *Beta* zeolite is a typical catalyst for this reaction. A series of *Beta* zeolites synthesized with Si/Al_2 ratios ranging from 20 to 350 were evaluated for the alkylation of benzene with propylene at 423°K and 3 MPa in the presence of benzene:alkene molar ratio of 7.0. The benzene:alkene molar ratio was kept high in order to minimize the undesired reactions. The selectivity to the mono-alkylate product was 92–93% in every case with the balance being the dialkylated product. The activity decreased with increase in Si/Al_2 but the selectivity was independent of the Si/Al2 ratio [86].

In the case of alkylation of benzene with 1-octene measurements were done over USY-zeolites with Si/Al_2 = 12, 26 and 60 at 343–373°K and 2 MPa. For this work the Si/Al_2 ratios were varied by dealumination. The range for benzene: octene molar ratios was 1–10. Here, a secondary carbenium ion is first formed upon protonation of 1-octene. A series of hydride shift–deprotonation steps then move the double bond around. While double bond isomerization is faster than alkylation, internal equilibrium is not achieved among octene isomers. The largest sphere than can fit inside the cavity for FAU framework type should be sufficiently large to accommodate the bulky transition state for the alkylation reaction [87]. The absence of 1-phenyloctane among the reaction products verifies that the primary carbenium ion is not favored during protonation of 1-octene. Among the alkene isomers 2-octene is obtained in highest yield, followed by 3-octene and 4-octene. Consistent with the alkene isomer distribution, 2-phenyloctane is the most abundant alkylbenzene, followed by 3- and 4-phenyloctanes. Selectivity is a function of 1-octene conversion. The extent of double bond isomerization and the alkylation of internal alkenes become increasingly favored at higher conversions. The essential absence of skeletal isomerization, cracking and hydride transfer reactions are attributed to the low reaction temperature that is typically used for liquid phase alkylation. While activity for this series of USY-zeolites increased with increasing Si/Al_2 ratio, the selectivities were similar.

The mechanism for aromatic–alcohol alkylation is analogous to aromatic–alkene reaction. In this case the first step is likely to be the protonation of the alcohol.

When methanol is used as the alkylating agent a methoxonium ion is formed. This is followed by an electrophilic attack of the carbocation on the aromatic ring. In the case of toluene alkylation with methanol an opportunity exists for *para* selectivity. *Para*-xylene : *ortho*-xylene ratio was 3.1 over MFI and 0.6 over BEA framework types.

13.8.4
Alkane Cracking

13.8.4.1 Classic Cracking Mechanism, Bimolecular

Alkane activation during bimolecular, so-called classic cracking, is believed to be initiated via intermolecular hydride transfer from alkanes to smaller carbenium ions that are formed by protonation of light alkenes present in the feed in trace quantities or made through thermal reactions. The driving force for hydride transfer is the formation of the more stable carbenium ions generated from feed alkanes. In this mechanism carbenium ions are the chain carriers, as was the case in isobutane-butylene alkylation. The larger carbenium ions then undergo *beta* scission, with or without prior skeletal rearrangements, to give each time, an alkene and a smaller carbenium ion. The possible *beta* scission pathways for the carbenium ions include, starting from the most favorable going to the least favorable, tertiary-to-tertiary, secondary-to-tertiary, tertiary-to-secondary and secondary-to-secondary transitions. These pathways are illustrated in Figure 13.44. The cracking pattern is determined not only by the stability of the carbenium ions that

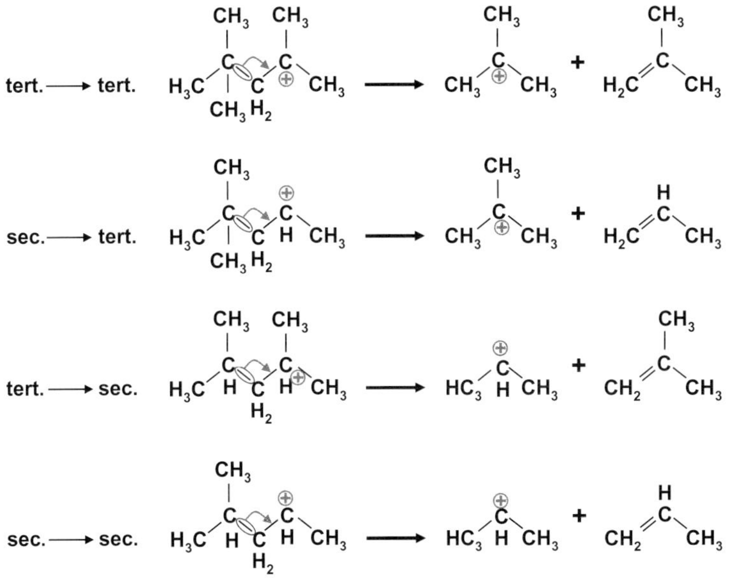

Figure 13.44 Beta scission pathways for carbenium ions.

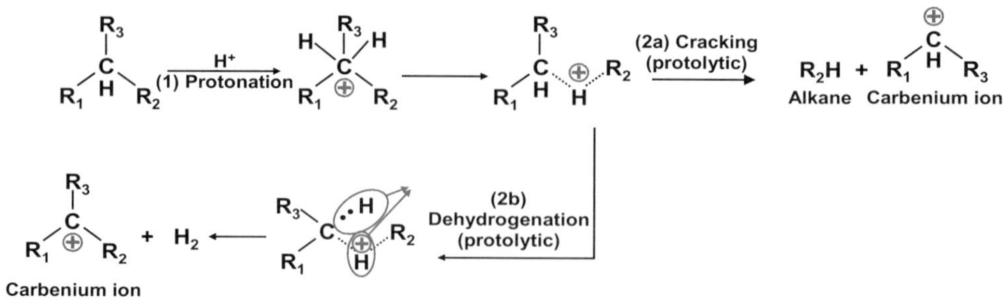

Figure 13.45 Mechanism for alkane monomolecular cracking.

are formed but also by the zeolite topology which may inhibit certain pathways that require large transition states.

This bimolecular mechanism also applies to cycloalkanes which can be activated by intermolecular hydride transfer to small carbenium ions to form cyclohexyl cations prior to cracking. Alternately, the cyclohexyl cations can deprotonate and form cyclohexene. With two similar intermolecular hydride transfers an aromatic can also form [46].

13.8.4.2 Monomolecular Cracking Mechanism

During the *Eighth International Congress on Catalysis* in Berlin Haag and Dessau proposed an alternate mechanism to classical cracking for alkanes that was initiated by direct protonation over Bronsted acid sites to make a carbonium ion, did not involve assistance from another molecule and became increasingly dominant at high temperatures, low pressures and low conversions [88]. More specifically, the following elementary steps have been outlined, as shown in Figure 13.45: (1) protonation of an alkane to form a penta-coordinated carbonium ion, typically illustrated as a two-electron three-center C–H–C bridge, (2a) decomposition of the carbonium ion to form a carbenium ion and an alkane, the so-called protolytic cracking or (2b) decomposition of the carbonium ion to form a carbenium ion and H_2, the so-called protolytic dehydrogenation pathway. Subsequently the carbenium ion deprotonates to give an alkene. One of the earlier kinetic analysis on this latter reaction came from the Lercher group using C_3–C_6 n-alkanes over H-MFI with $Si/Al_2 = 70$ at 460–550 °C [89]. That work concluded that protolytic dehydrogenation kinetics were controlled by the desorption of the alkene, partly based on increase in the intrinsic activation energies measured for the reaction as the carbon number of the alkane increased.

In later work by Haag and Dessau product selectivity data were provided for n-butane cracking at 426–523 °C over HZSM-5 with $Si/Al_2 = 70$ [90]. The selectivity results at 496 °C and 1–10 kPa for n-butane were extrapolated to zero percent conversion in Table 13.6 to be able to identify the primary products and to assess the decomposition pattern of the n-butyl carbonium ion. Similar selectivities to methane and propylene implied, as expected, that the decomposition of the car-

Table 13.6 n-Butane cracking over MFI at 496 °C [90].

Temperature (°C)	496	496
n-Butane partial pressure (kPa)	1–10	10–100
n-Butane conversion (%)	~0	17
Selectivity (mol%)		
Hydrogen	15	6
Methane	20	9
Ethane	17	13
Ethylene	15	14
Propane	0	40
Propylene	16	10
Butylene	17	8
Total	100	100

bonium ion to methane + propyl carbenium ion was one of the reaction pathways. The selectivities to ethane and ethylene were nearly the same, implying that the decomposition to ethane and ethyl carbenium ion also occurred. Furthermore, the selectivities to hydrogen and butylenes were nearly the same, implying that decomposition to hydrogen and n-butyl carbenium ion occurred too. Finally, the similarity of the selectivities to methane, ethane and hydrogen suggested that all three decomposition pathways (for the carbonium ion) were essentially equally favorable. From the temperature dependencies of the cracking patterns also similar apparent activation energies were calculated for the different decomposition products: 134–149 kJ/mol. The authors, accordingly, concluded that decomposition of the n-butyl carbonium ion was statistical in nature and without any significant energy barrier. The point was then made that, had there been an actual energy barrier these different pathways would have had quite different activation energies, resulting in different selectivities, as expected from the Polanyi relationship between energy barrier and the stability of the expected carbenium ion that was quite different for each pathway. It remains to be seen whether future work will show decomposition of n-butyl and higher carbonium ions during cracking of alkanes to be statistical in nature at higher temperatures and with different framework types. More specifically, whether the zeolite framework type can introduce shape selectivity during the decomposition of the carbonium ion is likely to be explored.

As part of the same study selectivity data were provided at 10–100 kPa partial pressures of n-butane at 0–17% conversion over HZSM-5 [90]. With increase in pressure and conversion secondary reactions started to occur. These results are also summarized in Table 13.6. The lowered selectivity to hydrogen, methane and ethane was attributed to increasingly less favorable conditions for monomolecular cracking. The dramatic increase in selectivity to propane which was absent at zero conversion, along with decrease in propylene was considered as signature for bimolecular cracking. More specifically, it was suggested that hydride transfer

Table 13.7 Biomolecular reactions during n-butane cracking over MFI at 496 °C [90].

3 n-C_4H_{10} + 3 C_3H_6	→	3 n-C_4H_8 + 3 C_3H_8
2 n-C_4H_8	→	C_3H_6 + C_5H_{10}
n-C_4H_8 + C_5H_{10}	→	3 C_3H_6
....		....
3 n-C_4H_{10}	→	3 C_3H_8 + C_3H_6 (overall stoichiometry)

occurred from n-butane to primary product propylene in order to generate n-butene which then dimerized and cracked to give propylene and pentene. The latter reacted with n-butene to make nonene which cracked to make more propylene. The overall stoichiometry that was then suggested involved n-butane converting to a mixture of 3 propane : 1 propylene. These secondary reactions are summarized in Table 13.7.

While it was possible to identify the primary cracking products from n-butane monomolecular cracking over HZSM-5 by extrapolating the selectivities to zero conversion, with n-hexane feed, even at zero conversion, it was not possible to determine the primary cracking products [91]. Over HZSM-5 with Si/Al_2 = 50 at 480–540 °C, the high selectivity to hydrogen at zero conversion suggests a large amount of hexene was probably formed, which then may have immediately cracked to two moles of propene or to one mole of ethylene plus a mole of butylene. The latter may then have recracked, making it difficult to model the primary cracking pattern in the absence of very low conversion data. Monomolecular cracking is also applicable to cycloalkanes which can be activated by direct protonation to form a carbenium ion, followed by cracking. This is also called endocyclic protolytic cracking [46].

13.8.4.3 Kinetics of Cracking

Having assumed that impact of a gas phase alkane molecule with a Bronsted acid site (for protonation) controls the cracking kinetics for monomolecular cracking, the simplified rate expression (r_1) was then indicated to involve a rate constant, a gas phase concentration term for feed alkanes and a concentration term for Bronsted acid sites [92]. But, assuming hydride transfer between a gas phase alkane and a surface carbenium ion controls the kinetics for bimolecular cracking, the corresponding rate expression (r_2) then involves a rate constant, a gas phase concentration term for feed alkanes and a concentration term for adsorbed carbenium ions. Utilizing Langmuir–Hinshelwood formalism and assuming adsorbed alkane concentration is negligible, rate expressions in terms of measurable independent variables were then developed, as illustrated in Table 13.8. At high temperatures, low hydrocarbon pressure and low alkane conversions the concentration of adsorbed carbenium ions become low, making it increasingly unfavorable for the bimolecular hydride transfer step to occur. Accordingly, under these conditions monomolecular cracking becomes more dominant. The monomolecular cracking rate expression derived in Table 13.8 assumed that rate was controlled by impact of a gas phase alkane molecule with a Bronsted acid site. If monomolecular crack-

Table 13.8 Monomolecular and bimolecular kinetics of alkane cracking [92].

$r_2 = k_2 \,[RH]\,[R^+]$	RH, alkane
$= k_2 \,[RH] \dfrac{K_o[O]}{1+K_o[O]}$	R^+, adsorbed carbenium ion
	O, alkene
$r_1 = k_1 \,[RH]\,[H^+]$	k_2, hydride transfer rate constant
$= k_1 \,[RH] \left[1 - \dfrac{K_o[O]}{1+K_o[O]}\right]$	K_o, adsorption equilibrium constant for alkenes
$= k_1 \,[RH] \left[\dfrac{1}{1+K_o[O]}\right]$	H^+, proton
	k_1, protonation rate constant
$r = r_1 + r_2 = \dfrac{k_1[RH] + k_2[RH]K_o[O]}{1+K_o[O]}$	

When K_o and [O] are low $r \approx r_1 = k_1\,[RH]$.

Table 13.9 Monomolecular kinetics for alkane cracking when surface reaction controls the rate.

$r_1 = k_1\,K_{RH}\,[RH]\,[H^+]/(1+K_o\,[O]+K_{RH}\,[RH])$

$= k_1\,K_{RH}\,[RH]\left(1 - \dfrac{K_o[O]}{1+K_o[O]+K_{RH}[RH]} - \dfrac{K_{RH}[RH]}{1+K_o[O]+K_{RH}[RH]}\right)/(1+K_o[O]+K_{RH}[RH])$

When $K_{RH} \ll K_o$

$r_1 \approx k_1\,K_{RH}\,[RH]\left(1 - \dfrac{K_o[O]}{1+K_o[O]}\right)/(1+K_o[O])$

$r_1 \approx k_1\,K_{RH}\,[RH]\left(\dfrac{1}{1+K_o[O]}\right)^2$

When K_o is small

$r_1 \approx k_1\,K_{RH}\,[RH]$

$E_{app} = E_1 + \Delta H_{RH}$

ing kinetics were instead controlled by surface reaction of the adsorbed alkane with a Bronsted acid site and the adsorption equilibrium constant for the alkane were not too low, then the rate expression (r_1) would have a surface reaction rate constant multiplying the adsorption equilibrium constant for the alkane, the gas phase concentration term for feed alkanes, the concentration term for Bronsted acid sites, and an adsorption group in the denominator as illustrated in Table 13.9. According to this rate expression alkane cracking benefits from a high adsorption equilibrium constant for that specific alkane and accordingly, differences in rates are often due to differences in alkane surface coverages [17]. Furthermore, according to this analysis, the true activation energies for monomolecular cracking exceed the apparent activation energies by the corresponding enthalpy of adsorption. Interestingly, while the apparent activation energies for cracking various n-alkanes

13 Unique Aspects of Mechanisms and Requirements

Table 13.10 Activation energies for cracking alkanes over H-MFI.

Alkanes	Framework type	$E_{apparent}$ (kJ/mol)	$E_{intrinsic}$ (kJ/mol)	Reference
n-C_4	MFI	142	205	[93]
n-C_6	MFI	126	205	[93]
n-C_8	MFI	92	197	[93]
n-C_9	MFI	84	197	[93]
n-C_{10}	MFI	67	193	[93]
C_3	MFI	155	198	[17]
n-C_4	MFI	135	197	[17]
n-C_5	MFI	120	197	[17]
n-C_6	MFI	105	197	[17]
C_3	MOR	149	190	[26]
C_3	FAU	165	196	[26]
C_3	BEA	156	198	[26]

Table 13.11 Kinetics of cracking n-alkanes over H-MFI at 773 °K [27].

Carbon Number	ΔH_{ads} kJ/mol	ΔS_{ads} J/mol-°K	K (at 373°K)	K (at 773°K)	TOF mol/mol H⁺-s-bar	$A/10^{13}$ s⁻¹
3	−45	−102	10	55	0.0013	0.57
4	−58	−119	78	51	0.0058	2.3
5	−69	−135	494	46	0.022	6.2
6	−83	−152	4410	46	0.060	27

(n = 3–10) over MFI were found to differ by as much 88 kJ/mol, after correction with the corresponding enthalpies of adsorption, the intrinsic activation energies all fell in the range of 193–205 kJ/mol, as illustrated in Table 13.10 [17]. Size-independent activation energy for protonating alkanes suggests that the transition state is essentially the same for various n-alkanes. This result has been interpreted as the absence of Bronsted acid strength effects for cracking alkanes of different sizes. The increase in turnover rate for cracking with increase in carbon number was then attributed to the increase in the concentration of adsorbed alkanes within the zeolite channels (physisorbed). However, recently the same data were reanalyzed, this time by calculating the adsorption equilibrium constants at reaction temperature through the use of the enthalpy and entropy of adsorption. While the adsorption equilibrium constants at 373°K varied by as much as a factor of 441 between propane and n-hexane, the difference was found to be no more than a factor of 0.84 at 773°K, as illustrated in Table 13.11. Since the intrinsic activation energies were comparable for cracking these alkanes the factor of 46 increase in turnover rate from propane to n-hexane was attributed to the increase in pre-exponential factor for the intrinsic rate constant [27]. More specifically, it was stated that the stronger adsorption of larger alkanes also leads to more constrained adsorption, which, in turn, results in more negative adsorption entropies.

13.8.4.4 Effect of Pore Size and Acid Site Density on Cracking

To what extent the monomolecular pathway is governing the overall cracking process depends not only on the reaction conditions, but also on the catalyst properties. For example, bimolecular hydride transfer was reported to be sterically hindered in small-pore zeolites [94]. More recent work compared the cracking selectivities for n-octane at 500 °C for DDR (8-MR), MFI (10-MR) and BEA (12-MR) framework types [95]. In order to make the comparisons more meaningful catalysts with similar Si/Al_2 ratios were selected. Also, the space velocities were adjusted in order to obtain similar conversion levels. The results displayed in Table 13.12 indicate that, as the pore dimension became smaller, the product distribution shifted to lighter hydrocarbons. Very little methane and ethane were observed over BEA, as would be expected under classical cracking. Methane and ethane are some of the decomposition products of the carbonium ion resulting from protonation of n-octane and increased from BEA to MFI and then to DDR. Isobutane, a typical product from classical cracking, as would be expected from significant skeletal isomerization prior to *beta* scission, was highest for BEA. The cracking mechanism ratio (CMR), the ratio for the yields of methane + ethane divided by

Table 13.12 Effect of framework type on cracking selectivities for n-octane at 500 °C [95].

Framework type	DDR	MFI	BEA
Pore dimensions (nm)	0.36 × 0.44	0.51 × 0.55	0.66 × 0.67
		0.53 × 0.56	0.56 × 0.56
MR	8	10	12
Si/Al_2	46	42	48
n-C8 conversion at 500 °C (wt%)	13.4	11.9	10.2
Selectivities (mol%)			
Methane	4.0	1.8	0.4
Ethane	4.2	4.7	1.5
Ethylene	27.8	12.5	2.4
Propane	3.9	7.7	8.4
Propylene	37.7	28.0	14.1
i-Butane	0.2	0.3	10.8
n-Butane	2.8	7.6	12.1
n-Butylene	12.2	12.6	16.2
i-Butylene	3.8	9.2	12.4
Pentanes and pentenes	3.1	14.8	21.3
Hexanes and hexenes	0.3	0.8	0.5
Total	100	100	100
CMR, yield ratio $\dfrac{\text{Methane} + \text{Ethane}}{i\text{-Butane}}$	205	62	0.4

13 Unique Aspects of Mechanisms and Requirements

i-butane, dramatically increased from BEA to MFI and then to DDR. The selectivities to ethylene and propylene became increasingly high with decrease in pore dimension, as would be expected from the increased inhibition of hydride transfer from *n*-octane to light alkenes. An increase in the propylene:propane ratio from BEA to MFI and then to DDR suggests that propylene has been a key recipient of hydride transfer.

As part of the same study, the effect of the Al content on the product distribution was determined by varying the Si/Al_2 ratio from 54, to 46 and then to 36 with DDR framework type [95]. The results indicated that the CMR progressively became smaller with increase in Al content. That is, given a framework type with small pores, bimolecular cracking became more dominant relative to monomolecular cracking as the Al content increased. These results suggest that, during bimolecular cracking, the initial activation of *n*-octane by hydride transfer to an adsorbed carbenium ion is likely to be mediated by a Bronsted acid site in the vicinity of another Bronsted acid site that generated the carbenium ion.

13.8.5
Aromatic Transformation

13.8.5.1 Transalkylation and Disproportionation

As an example for aromatic transformation the mechanism for *meta*-xylene disproportionation to toluene + trimethylbenzene is illustrated in Figure 13.46. In the first step the zeolite extracts a hydride from *meta*-xylene to form a carbenium ion at one of the methyl groups, presumably the rate-controlling step. This mechanism is likely to involve a Lewis acid site. The carbenium ion then adds to a second

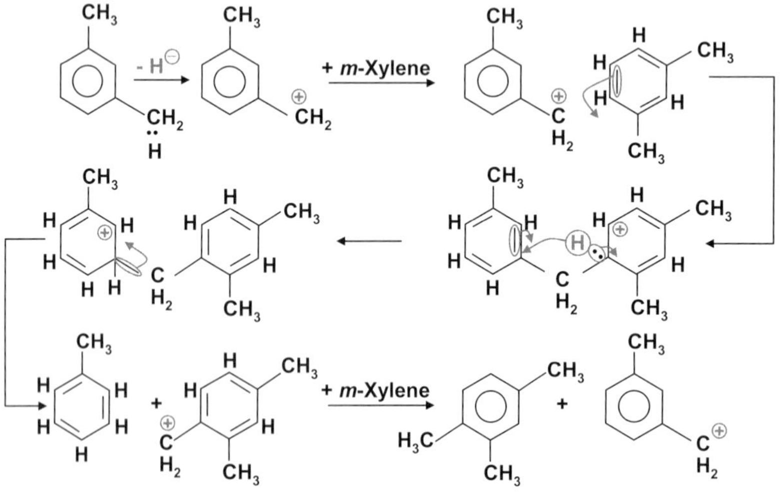

Figure 13.46 Mechanism for disproportionation of *meta*-xylene to toluene and trimethylbenzene via diphenlymethane intermediate.

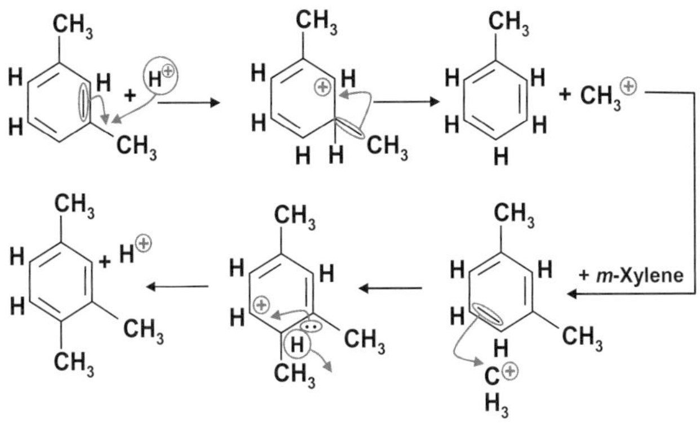

Figure 13.47 Mechanism for disproportionation of *meta*-xylene to toluene and trimethylbenzene via methoxy intermediate.

meta-xylene by transferring the positive charge and forms a methyl-substituted diphenylmethane carbenium ion species. In the next step a proton at one of the methyl rings in the carbenium ion is transferred back to the first *meta*-xylene ring. Then dissociation of the two rings occurs, leading to toluene + trimethylbenzene carbenium ion. The catalytic cycle is completed when the trimethylbenzene carbenium ion extracts a hydride ion from a third *meta*-xylene [63].

Another activation mechanism that involves protonation of the aromatic ring followed by evolution of H_2 to generate the carbenium ion has also been reported [96]. Alternately, *meta*-xylene is activated by protonation to generate a dialkylbenzenium ion. This is followed by transfer of the positive charge to one of the methyl groups. As a result toluene is formed along with a methoxide, presumably the rate-controlling step. This methoxide can add to another *meta*-xylene and then deprotonate to generate a trimethylbenzene, as illustrated in Figure 13.47 [63].

13.8.5.2 Ethylbenzene Conversion to Xylenes

Ethylbenzene conversion to xylenes requires a bifunctional zeolitic catalyst. The first step is hydrogenation of ethylbenzene over a metal function to ethylcyclohexene, which then forms a carbenium ion (ethylcyclohexyl cation) by protonation. The latter can undergo a series of ring contraction and ring expansion reactions to form xylenes through methylethylcyclopentene (MECP) and trimethylcyclopentene (TMCP) intermediates. Ring contractions and ring expansions are enabled through protonated cyclopropane intermediates, as previously illustrated in Figure 13.38. Some of these ring contraction and expansion routes through MECP intermediates have been illustrated in Figure 13.48 as examples of reaction pathways to dimethylhexenes. Dehydrogenation over the metal function then convert these to xylenes. It is noteworthy to state that some of the reaction intermediates outlined in these steps (not all steps are elementary), especially those that are leading to o-xylene and m-xylene (1,1-MECP, 1,2-MECP), would be kinetically hindered in small-pore zeolites. Such shape selectivity effects have been discussed for the

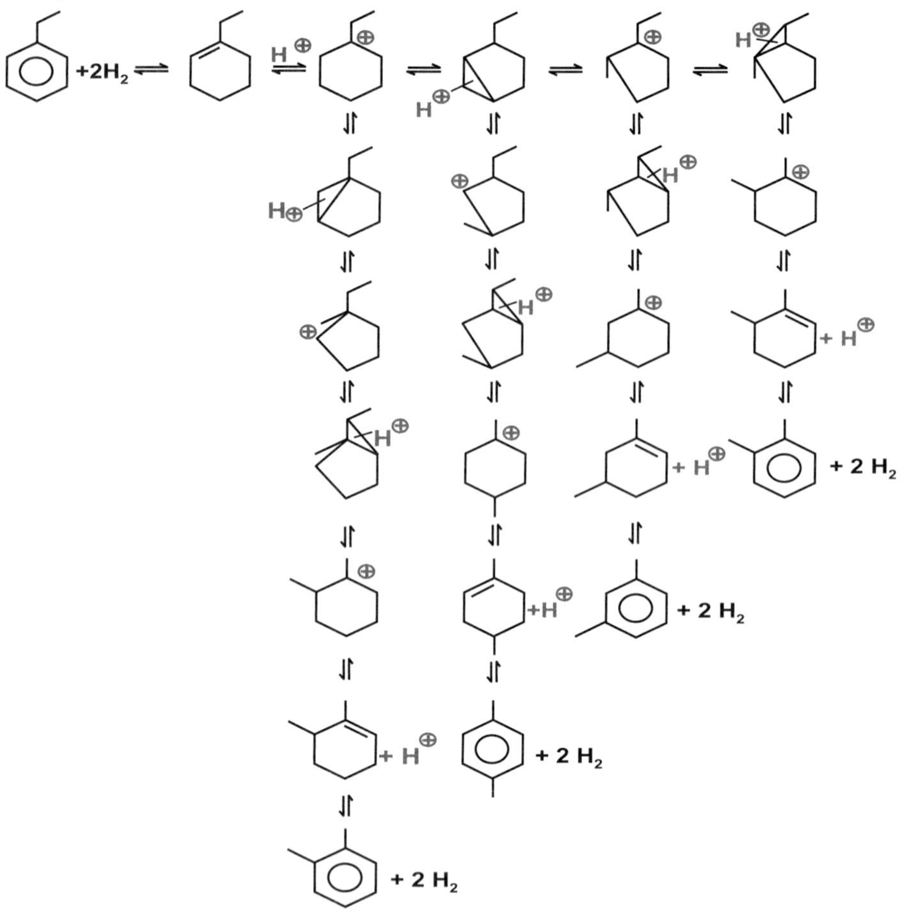

Figure 13.48 Bifunctional mechanism for isomerization of ethylbenzene to xylenes via methylethylcyclopentene intermediates.

analogous case of dimethylcyclopentenes [65]. Accordingly, by selection of the proper framework type with restricted pore dimensions, it may become possible to minimize the paths to *ortho*-xylene and to some extent to *meta*-xylene. Finally, it is important to note that such ring expansion and ring contraction steps illustrated here can also enable xylene isomerization.

13.8.6
Methanol to Olefins

Light alkene selectivities from MTO over SAPO-34 at 400–450 °C (ethylene > propylene >> butylenes > pentenes) are quite different than those predicted from thermodynamic equilibrium (butylenes > propylene > pentenes > ethylene). Over

ZSM-5 propylene and higher alkene selectivities are typically higher and ethylene selectivity is typically lower compared to SAPO-34 [97]. Over *Beta* light alkene breakdown is closer to thermodynamic equilibrium compared with ZSM-5 and SAPO-34 [97]. While the SAPO-34 pore opening is not large enough to let aromatics and branched alkanes out, di- and trimethylbenzenes and even some tetramethylbenzenes are observed over ZSM-5 as products in the effluent and hexamethylbenzenes are observed over *Beta* as products in the effluent [98, 99]. The conditions that lead to the formation of the first C–C bond during MTO, the mechanism for making ethylene and propylene from methanol, the critical catalyst parameters that are responsible for the wide variation in light alkene selectivities observed among different framework types and between fresh versus aged catalysts are some of the most intriguing questions of the catalysis field today [100–105].

Methoxy groups can easily form from methanol over SAPO-34. However, initially, that is, in the absence of a hydrocarbon pool, no evidence is found in such experiments for the breaking of C–H bonds in methoxy groups and for the formation of C–C bonds between neighboring methoxys [106]. The first C–C bond formation in MTO is very slow. However, once this bond is formed or a compound with a C–C bond is acquired through trace impurities in the feed or through residual organic template in the molecular sieve, then a faster reaction path is enabled that eventually leads to the formation of the organic reaction center (ORC), which acts as a co-catalyst, shown for SAPO-34, for ZSM-5 and for Beta MTO catalysts [107]. This is also referred to as "hydrocarbon pool" mechanism.

Computational work was done in search of the critical reaction intermediate that leads to the formation of the first C–C bond in MTO [108]. For these calculations a methanol dimer was used as the reactant due to its ease of protonation. Protonated methanol dimer converted into protonated dimethylether with a low activation energy barrier of 70 kJ/mol. This oxonium ion can then deprotonate to give dimethylether and water, as commonly observed. However, subsequent reaction paths that led to the formation of the first C–C bond from dimethylether were found to have very high activation energy barriers. Accordingly, molecular dynamics were used to search for the C–C bond-containing compound that would require the lowest activation energy barrier. Interestingly, that search led to the identification of ethanol which required the lowest activation energy of 180 kJ/mol. This reaction path for making ethanol from methanol is illustrated in Figure 13.49.

Whether it is indeed ethanol or another small hydrocarbon with a C–C bond, a few simple transformations can certainly turn this molecule into the ORC. ORC

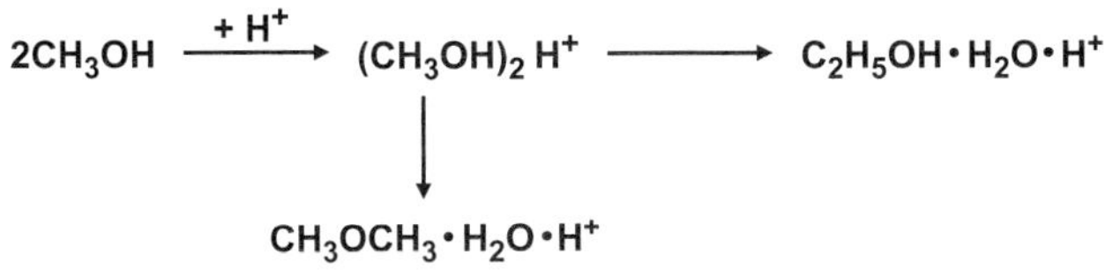

$$2CH_3OH \xrightarrow{+ H^+} (CH_3OH)_2 H^+ \longrightarrow C_2H_5OH \cdot H_2O \cdot H^+$$
$$\downarrow$$
$$CH_3OCH_3 \cdot H_2O \cdot H^+$$

Figure 13.49 A possible reaction pathway for formation of the first C–C bond during MTO.

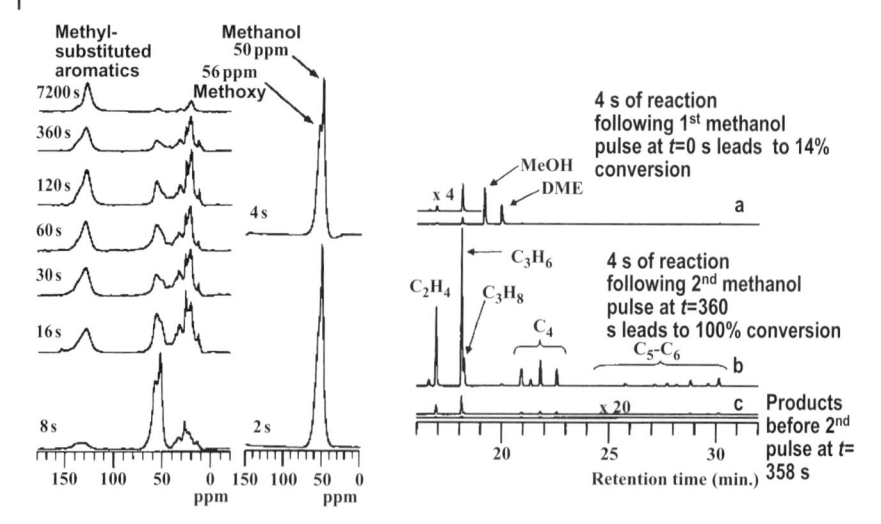

Figure 13.50 ^{13}C CP/MAS NMR spectra of products retained in SAPO-34 and GC analyses of volatile products formed after various reaction times following a methanol pulse (0.053 g/g cat) at 673°K [109].

for MTO is believed to be a methylated benzene species, as evidenced by NMR analyses of trapped hydrocarbons in used catalysts [109–111]. This is illustrated in Figure 13.50. ^{13}C CP/MAS NMR spectra of products retained in SAPO-34 during 2 s of reaction following a pulse of methanol at 673°K showed only methanol, as evidenced by the NMR signal at 50 ppm. A subsequent independent experiment (with a fresh load of catalyst), this time with 4 s of reaction, showed a shoulder peak now appearing at 56 ppm that belonged to methoxy species in the retained products. A series of similar independent experiments (with a fresh load of catalyst each time) were then implemented with various other reaction times. After 8 s the methoxy peak grew at the expense of methanol and two bands now appeared at ~130 ppm and ~25 ppm that belonged to methylated benzenes. After 16 s methanol peak was not clearly visible any more. The methylated benzenes grew at the expense of the methoxy following reaction times of 16, 30, 60, 120, 360 and 7200 s. On the other hand, GC analysis of volatile products formed after 4 s of reaction only showed 14% conversion of methanol, mostly to DME with some small amount of propylene. Interestingly, over the same load of catalyst, following a waiting period of 360 second, a second pulse of methanol showed 100% methanol conversion, mostly to ethylene and propylene, along with small amounts of propane and C_4–C_6 hydrocarbons. There were no volatile products observed just prior to the second pulse. These results indicate that methanol conversion to light alkenes required the formation of the ORC. The NMR analyses of retained products (in the used catalyst) suggest that ORC was a methylated benzene species.

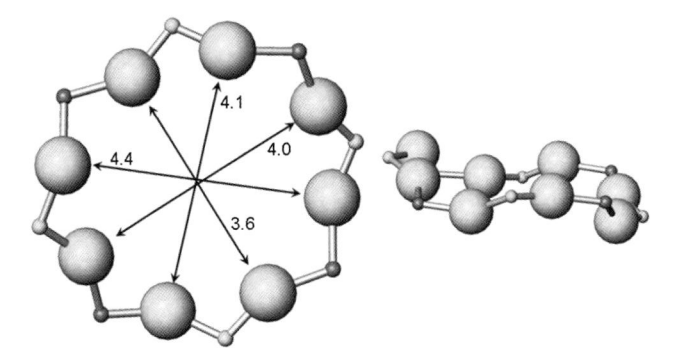

Figure 13.51 The O–O distances across the eight-ring channel in SAPO-34.

If methanol feeding stopped and the catalyst was flushed with a carrier gas methylated benzene species disappeared within a couple of minutes, some faster than others (as observed by NMR of retained products in the used catalyst). When isotopic composition of methanol was changed to ^{13}C, incorporation of labeled carbon into both volatile products and into retained products (in the used catalyst) occurred, indicating that these methylated benzenes are not just pore-filling but participating in the reaction. Over SAPO-34 incorporation of labeled carbon occurred fastest with hexamethylbenzene, followed by pentamethylbenzene > tetramethylbenzenes > trimethylbenzenes > xylenes. Interestingly, the order was exactly opposite over ZSM-5 [112]. These indicate that over SAPO-34 hexamethylbenzene was the dominant component of ORC. But, methyl substitution on the benzene ring was less with ORC over ZSM-5 [98, 111]. The number of methyl substitution of ORC is believed to be partly responsible for the relative ethylene and propylene selectivities during MTO [113]. While the differences in the nature of ORC for different framework types may partly explain the observed selectivity differences, transition state and product shape selectivity effects are also likely to be critical factors in determining the light alkene breakdown.

In the case of SAPO-34, while the pore channel is not large enough to process hexamethylbenzene, the cage is sufficiently large to enable the in situ formation of ORC during the kinetic induction period [114]. The pore dimensions in SAPO-34 were calculated by S. Wilson from UOP based on the atomic coordinates given in the literature [114]. These are illustrated in Figure 13.51 and Figure 13.52. The O–O distances across the eight-ring vary from 3.6 to 4.4 Å. Edge-on view of the eight-ring shows it is not planar. The height of the cage is based on O–O distance from six-ring on the floor of the cage to the corresponding six-ring on the roof of cage (dotted line). This distance is 10.9–11.1 Å, depending on the oxygen pairs chosen. After substracting 2× the oxygen radius, the height of the cage that is available for catalysis becomes 8.2–8.4 Å. The width of the cage is calculated in a similar manner. The oxygens chosen for the measurement are connected by the dotted line. The cage width available for catalysis based on the O–O distance of 10.3 Å is then 7.6 Å. The dimensions of the cage are then 8.2–8.4 Å by 7.6 Å.

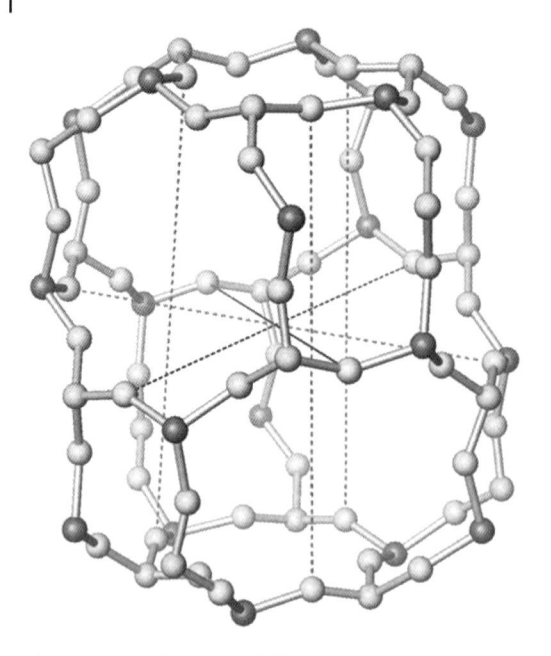

Figure 13.52 The cage in SAPO-34.

Figure 13.53 Paring mechanism for making ethylene from hexamethylbenzene during MTO.

In the case of hexamethylbenzene, following protonation, hexamethylbenzenium ion is formed and can undergo a series of ring contraction, ethylene elimination and ring expansion reactions that eventually lead to ethylene + tetramethylbenzene, as illustrated in Figure 13.53. Alkylation of tetramethylbenzene with methanol then regenerates the original ORC. It is important to note that during MTO the ORC can also be alkylated with primary reaction products, that is ethylene and propylene [115].

However, methylation of hexamethylbenzenium ion can also occur, leading to the formation of heptamethylbenzenium ion. The latter can split off, this time propylene, by a series of ring contraction and ring expansion reactions [116, 117].

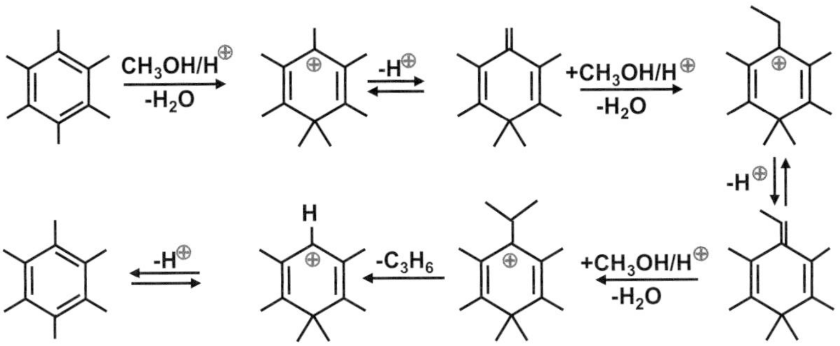

Figure 13.54 Mechanism for making propylene during MTO from hexamethylbenzene via heptamethylbenzenium ion intermediate.

With ethyl- or isopropylmethyl benzenes expansion to a seven-membered ring intermediate, followed by contraction has also been suggested as a possible route for enabling the paring mechanism [118].

An alternate mechanism exists for making light alkenes from hexamethylbenzene. Alkylation of hexamethylbenzene with methanol forms the heptamethylbenzenium ion. Upon deprotonation an exocyclic double bond is formed which can then be methylated to form a higher alkyl chain. The alkyl chain is then split off to give an alkene, as illustrated in Figure 13.54 [119]. This specific mechanism is favored over *Beta* and is unlikely to be operative over ZSM-5 [120, 121]. The formation of ethylene and propylene by feeding hexamethylbenzene or other methylated benzenes over *Beta* zeolite has actually been shown [122]. Benzenes with less methyl substitution were less effective in forming ethylene and propylene over *Beta* [121].

In the case of SAPO-34, the ORC evolves during MTO and methylbenzenes have been observed to gradually convert to naphthalenes [123, 124]. The evolution of the ORC is significant also because it may be responsible for the enhanced ethylene: propylene selectivity observed with aging of SAPO-34 [125]. It is conceivable that ethylene and propylene can also alkylate methylbenzenes, hence contribute to the evolution of ORC and selectivities during MTO [115]. In fact, experimental evidence exists that shows co-feeding ethylene lowers ethylene selectivity in favor of propylene selectivity over SAPO-34 during MTO and co-feeding propylene lowers propylene selectivity in favor of ethylene selectivity [97].

Finally, an additional reaction pathway exists and this does not seem to be operative with SAPO-34 and *Beta* under regular processing conditions. This path seems to be operative with ZSM-5 and that may involve successive methylations of propene, followed by cracking to yield higher alkenes [111]. A similar mechanism that involves successive methylations of ethylene followed by cracking to yield higher alkenes over ZSM-5 does not seem to be as important [125]. It is conceivable that this mechanism may be partly operative during the MTO experiments over SAPO-34 described above that used co-fed ethylene or co-fed propylene [126].

13.9
Key Remaining Questions

The identification of a methylated benzene species to be the organic reaction center (ORC) makes the mechanism for the MTO uniquely fascinating. How the first C–C bond is formed during MTO and how this enables the synthesis of the ORC are critical questions to be further addressed. Potential alkylation of the ORC with primary reaction products ethylene and propylene certainly adds to the complexity of the MTO chemistry. More mechanistic work is expected on the reactivity of the ORC with primary reaction products, possibly providing further insight on how MTO selectivity is impacted by the conversion level and process conditions. Work is expected to continue on decoupling the contributions of intrinsic rate constant and adsorption equilibrium constant to the overall kinetic rate constant for various reactions. Such effort would require methods for better estimation of adsorption constants at high temperatures. For example, it will be important to compare the rate constants for cracking alkanes over various framework types with the ultimate objective of comparing the acid strengths for these particular zeolites. To what extent activation entropy and activation energy play a role in determining Bronsted acid strength differences will be also quite insightful to know. It remains to be seen whether the decomposition of carbonium ions formed during monomolecular cracking of higher alkanes over various framework types at a range of process conditions remains to be statistical in nature, as was the case during cracking of n-butane over MFI at modestly high temperatures. The potential of regulating the selectivity to specific light alkenes from monomolecular cracking of alkanes through selection of zeolite topology is quite intriguing. To what extent transition state selectivity and product shape selectivity play a role in selective cracking of hexadecane to dimethylbutanes over AFI and MAZ framework types and in selective hydroprocessing of n-alkanes to mono-branched alkanes over MTT versus TON are likely to be further investigated, with the ultimate objective of better defining these shape selectivity effects. Last but not least is characterization of the active site in isobutane-2-butylene alkylation catalyst that plays a key role in the hydride ion abstraction step from i-butane.

References

1 Rees, L.V.C. and Shen, D. (2001) Adsorption of gases in zeolite molecular sieves, in *Introduction to Zeolite Science and Practice*, 2nd edn (eds H. Van Bekkum, E.M. Flanigen, P.A. Jacobs, and J.C. Jensen), Stud. Surf. Sci. Catal., vol. 137, Elsevier, Amsterdam, pp. 579–631.

2 Froment, F.G. and Bischoff, B.K. (1990) *Chemical Reactor Analysis and Design*, John Wiley & Sons, Inc., New York.

3 Yang, K.H. and Hougen, O.A. (1950) Determination of mechanism of catalyzed gaseous reactions. *Chem. Eng. Prog.*, **46**, 146–157.

4 Gregg, J.S. and Sing, S.W.K. (1982) *Adsorption, Surface Area and Porosity*, Academic Press, San Diego.

5 Eder, F. and Lercher, J.A. (1997) On the role of the pore size and tortuosity for sorption of alkanes in molecular sieves. *J. Phys. Chem. B*, **101**, 1273–1278.

6 Krishna, R., Smit, B., and Calero, S. (2002) Entropy effects during sorption of alkanes in zeolites. *Chem. Soc. Rev.*, **31**, 185–194.

7 Zhu, W., Kapteijn, F., and Moulijn, J.A. (2000) Adsorption of light alkanes on silicalite-1: reconciliation of experimental data and molecular simulations. *Phy. Chem. Chem. Phys.*, **2**, 1989–1995.

8 Sun, M.S., Shah, D.B., Xu, H.H., and Talu, O. (1998) Adsorption equilibria of C_1 to C_4 alkanes, CO and SF on silicalite. *J. Phys. Chem. B*, **102**, 1466–1473.

9 Smit, B. and Maesen, L.M. (2008) Molecular simulations of zeolites: adsorption, diffusion, and shape selectivity. *Chem. Rev.*, **108** (10), 4125–4184.

10 Ocakoglu, R.A., Denayer, F.M., Marin, G.B., Martens, J.A., and Baron, G.V. (2003) Tracer chromatographic study of pore and pore mouth adsorption of linear and monobranched alkanes on ZSM-22 zeolite. *J. Phys. Chem. B*, **107**, 398–406.

11 Maesen, T.L.M., Krishna, R., van Baten, J., Smit, B., Calero, S., and Sanchez, J.M.C. (2008) Shape-selective n-alkane hydroconversion at exterior zeolite surfaces. *J. Catal.*, **256**, 95–107.

12 Breck, D.W. (1974) *Zeolite Molecular Sieves*, John Wiley & Sons, Inc., New York.

13 Traa, Y., Sealy, S., and Weitkamp, J. (2007) Characterization of the pore size of molecular sieves using molecular probes. *Mol. Sieves*, **5**, 103–154.

14 Eder, F., Stockenhuber, M., and Lercher, J.A. (1997) Bronsted acid site and pore controlled sitting of alkane sorption in acidic molecular sieves. *J. Phys. Chem. B*, **101**, 5414–5419.

15 Denayer, J.F., Ocakoglu, A.R., Huybrechts, W., Martens, J.A., Thybaut, J.W., Marin, G.B., and Baron, G.V. (2003) Pore mouth versus intracrystalline adsorption of isoalkanes on ZSM-22 and ZSM-23 zeolites under vapour and liquid phase conditions *Chem. Com. R. Soc. Chem.*, **15**, 1880–1881.

16 Ocakoglu, R.A., Denayer, J.F.M., Marin, G.B., Martens, J.A., and Baron, G.V. (2003) Tracer chromatographic study of pore and pore mouth adsorption of linear and monobranched alkanes on ZSM-22 zeolite. *J. Phys. Chem.*, **107**, 398–406.

17 Van Bokhoven, J.A., Williams, B.A., Ji, W., Koningsberger, D.C., Kung, H.H., and Miller, J.T. (2004) Observation of a compensation relation for monomolecular alkane cracking by zeolites: the dominant role of reactant sorption. *J. Catal.*, **224**, 50–59.

18 Ruthven, D.M. and Post, M.F.M. (2001) Diffusion in zeolite molecular sieves, in *Introduction to Zeolite Science and Practice*, 2nd edn (eds H. Van Bekkum, E.M. Flanigen, P.A. Jacobs, and J.C. Jensen), Stud. Surf. Sci. Catal., vol. 137, Elsevier, Amsterdam, pp. 525–577.

19 Jobic, H. and Méthivier, A. (2005) Intracrystalline diffusion in zeolites studied by neutron scattering techniques. *Oil Gas Sci. Technol.*, **60** (5): 815–830.

20 Kärger, J. (2002) The random walk of understanding diffusion. *Ind. Eng. Chem. Res.*, **41** (14), 3335–3340.

21 Feynman, R.P., Leighton, R.B., and Sands, M. (1989) Brownian motion, in *The Feynman Lectures in Physics*, Addison–Wesley, Redwood City, pp. 41-1–41.10.

22 Dubbeldam, D. and Snurr, R.Q. (2007) Recent developments in the molecular modeling of diffusion in nanoporous materials. *Mol. Simul*, **33**, 4–5, 15–30.

23 Chimie, C.R. (2005) Infrared spectroscopic investigation of diffusion, co-diffusion and counter-diffusion of hydrocarbon molecules in zeolites. *Elsevier, Comptes Rendus Chimie*, **8**, 303–319.

24 Makarova, M., Ojo, A., Karim, K., Hunger, M., and Dwyer, J. (1994) FTIR study of weak hydrogen bonding of bronsted hydroxyls in zeolites and aluminophoshates. *J. Phys. Chem.*, **98** (14), 3619–3623.

25 Wang, B. and Manos, G. (2008) Role of strong zeolitic acid sites on hydrocarbon reactions. *Ind. Eng. Chem. Res.*, **47** (9), 2948–2955.

26 Van Bokhoven, J.A. and Xu, B. (2007) Towards predicting catalytic performances of zeolites, in *From Zeolites to*

Porous MOF Materials – the 40th Anniversary of International Zeolite Conference (eds R. Xu, Z. Gao, J. Chen and W. Yan), Elsevier, Amsterdam, pp. 1167–1173.

27 Bhan, A., Gounder, R., Macht, J., and Iglesia, E. (2008) Entropy considerations in monomolecular cracking of alkanes on acidic zeolites. *J. Catal.*, **253**, 221–224.

28 Huybrechts, W., Vanbutsele, G., Houthoofd, K.J., Bertinchamps, F., Laxmi Narasimhan, C.S., Gaigneau, E.M., Thybaut, J.W., Marin, G.B., Denayer, J.F.M., Baron, G.V., Jacobs, P.A., and Martens, J.A. (2005) Skeletal isomerization of octadecane on bifunctional ZSM-23 zeolite catalyst. *Catal. Lett.*, **100**, 235–242.

29 Blasco, T., Chica, A., Corma, A., Murphy, W.J., Agúndez-Rodríguez, J., and Pérez-Pariente, J. (2006) Changing the Si distribution in SAPO-11 by synthesis with surfactants improves the hydroisomerization /dewaxing properties. *J. Catal.*, **242**, 153–161.

30 Bayense, C.R., van der Pol, A.J.H.P., van Hooff, J.H.C. (1991) Aromatization of propane over MFI-gallosilicates. *Appl. Catal.*, **72**, 81–98.

31 Sazama, P., D de ek, J., Gábová, V., Wichterlová, B., Spoto, G., and Bordiga, S. (2008) Effect of aluminum distribution in the framework of ZSM-5 on hydrocarbon transformation. Cracking of l-butene. *J. Catal.*, **254**, 180–189.

32 Normura, J. and Shima, H. (2008) Adsorption of hydrocarbons and formations of carbocations over zeolites studied by IR spectroscopy. *J. Jpn. Pet. Inst.*, **51** (5), 274–286.

33 Boronat, M., Concepcion, P., Corma, A., Navarro, M.T., Renz, M., and Valencia, S. (2009) Reactivity in the confined spaces of zeolites: the interplay between spectroscopy and theory to develop structure-activity relationships for catalysis. *Phys. Chem. Chem. Phys.*, **11**, 2876–2884.

34 Elanany, M., Koyama, M., Kubo, M., Broclawik, E., and Miyamoto, A. (2005) Periodic density functional investigation of lewis acid sites in zeolites: relative strength order as revealed from NH_3 adsorption. *Appl. Surf. Sci.*, **246**, 96–101.

35 Chen, T., Men, A., Sun, P., Zhou, J., Yuan, Z., Guo, A., Wang, J., Ding, D., and Li, H. (1996) Lewis acid sites on dehydroxylated zeolite HZSM-5 studied by NMR and EPR. *Catal. Today*, **30**, 189–192.

36 Sonnemans, M.H.W., den Heijer, C., and Crocker, M. (1993) Studies on the acidity of mordenite on ZSM-5. 2. Loss of Bronsted acidity by dehydroxylation and dealumination. *J. Phys. Chem.*, **97**, 440–445.

37 Gonzales, N.O., Bell, A.T., and Chakraborty, A.K. (1997) Density functional theory calculations of the effects of local composition and defect structure on the proton affinity of HZSM-5. *J. Phys. Chem. B*, **101**, 10058–10064.

38 Hafner, J., Benco, L., and Bucko, T. (2006) Acid-based catalysis in zeolites investigated by density-functional methods. *Top. Catal.*, **37** (1), 41–54.

39 Mota, C.J.A., Bhering, D.L., and Rosenbach, N. (2004) A DFT study of the acidity of ultrastable Y zeolite: where is the Bronsted/Lewis acid synergism. *Angew. Chem. Int. Ed.*, **43**, 3050–3053.

40 Busco, C., Barbaglia, A., Broyer, M., Bolis, V., Foddanu, G.M., and Ugliengo, P. (2004) Characterization of Lewis and Bronsted acidic sites in H-MFI and H-BEA zeolites: a thermodynamic and ab initio study. *Thermochim Acta*, **418**, 3–9.

41 Weitkamp, J., Ernst, S., and Puppe, L. (1999) Shape-selective catalysis in zeolites, in *Catalysis and Zeolites*, vol. 131 (eds J. Weitkamp, and L. Puppe), Springer, Heidelberg, pp. 327–376.

42 Pieterse, J.A.Z., Veefkind-Reyes, S., Seshan, K., and Lercher, J.A. (2000) Sorption and ordering of dibranched alkanes on medium-pore zeolites ferrierite and TON. *J. Phys. Chem. B*, **104** (24), 5715–5723.

43 Martens, J.A., Vanbutsele, G., Jacobs, P.A., Denayer, J., Ocakoglu, R., Baron, G., Arroyo, J.A.M., Thybaut, J., and Marin, G.B. (2001) Evidences for pore mouth and key-lock catalysis in

hydroisomerization of long n-alkanes over 10-ring tubular pore bifunction zeolites. *Elsevier J. Catal.*, **65**, 111–116.

44 Denayer, J.F., Baron, G.V., Vanbutsele, G., Jacobs, P.A., and Martens, J.A. (1999) Modeling of absorption and bifunctional conversion of n-alkanes on Pt/H-ZSM-22 zeolite catalyst. *Chem. Eng. Sci.*, **54**, 3553–3561.

45 Laxmi Narasimhan, C.S., Thybaut, J.W., Marin, G.B., Jacobs, P.A., Martens, J.A., Denayer, J.F., and Baron, G.V. (2003) Kinetic modeling of pore mouth catalysis in the hydroconversion of n-octane on Pt-H-ZSM-22. *J. Catal.*, **220**, 399–413.

46 Martens, J.A. and Jacobs, P.A. (2001) Introduction to acid catalysis with zeolites in hydrocarbon reactions, in *Introduction to Zeolite Science and Practice*, 2nd edn (eds H. Van Bekkum, E.M. Flanigen, P.A. Jacobs, and J.C. Jensen), Stud. Surf. Sci. Catal., vol. 137, Elsevier, Amsterdam, pp. 525–577.

47 Maesen, T.L.M., Krishna, R., van Baten, J., Smit, B., Calero, S., and Sanchez, J.M.C. (2008) Shape-selective n-alkane hydroconversion at exterior zeolite surfaces. *J. Catal.*, **256**, 95–107.

48 Halgeri, A. and Das, J. (2002) Recent advances in selectivation of zeolites for para-distributed aromatics. *Catal. Today*, **73**, 65–73.

49 Tasi, G., Pálinko, I., Molnár, A., and Hannus, I. (2003) Molecular shape selective catalysis. *J. Mol. Struct. (Theochem)*, **666–667**, 69–77.

50 Hsia Chen, C.S. and Bridger, R. (1996) Shape-selective oligomerization of alkenes to near-linear hydrocarbons by zeolite catalysis. *J. Catal.*, **161**, 687–693.

51 Cheng, J.C., Miseo, S., Soled, S., and Buchanan, J.S. (2009) United States Patent Application 20090093663.

52 Demuth, T., Rozanska, X., Benco, L., Hafner, J., van Santen, R.A., and Toulhoat, H. (2003) Catalytic isomerzation of 2 pentene in H-ZSM-22–a DFT investigation. *J. Catal.*, **214**, 68–77.

53 Boronat, M. and Corma, A. (2008) Are carbenium and carbonium ions reaction intermediates in zeolite-catalyzed reactions? *Sci. Dir. Appl. Catal. A*, **336**, 2–10.

54 Boronat, M., Virrruela, M., and Corma, A. (2004) Reaction intermediates in acid catalysis by zeolites: prediction of the relative tendency to form alkoxides or carbocations as a function of hydrocarbon nature and active site structure. *J. Am. Chem. Soc.*, **126** (10), 3300–3309.

55 Lamberti, C., Groppo, G., Spoto, S., Bordiga, S., and Zecchina, A. (2007) Infrared spectroscopy of transient surface species. *Adv. Catal.*, **51**, 1–74.

56 Bjørgen, M., Bonino, F., Kolboe, S., Lillerud, K.-P., Zecchina, A., and Bordiga, S. (2003) Spectroscopic evidence for a persistent benzenium cation in zeolite H-beta. *J. Am. Chem. Soc.*, **125**, 15863–15868.

57 Boronat, M., Virrruela, M., and Corma, A.A. (1998) Theoretical study of the mechanism of the hydride transfer reaction between alkanes and alkenes catalyzed by an acidic zeolite. *J. Phys. Chem.*, **102** (48), 9863–9868.

58 Gounaris, C., Floudas, C.A., and Wei, J. (2006) Rational design of shape selective separation and catalysis–I: concepts and analysis. *Chem. Eng. Sci.*, **61**, 7933–7948.

59 Zones, S.I. and Harris, T.V. (2000) The constraint index test revisited: anomalies based upon new zeolite structure types. *Micropor. Mesopor. Mater.*, **35–36**, 31–46.

60 Corma, A., Chica, A., Guil, J.M., Llopis, F.J., Mabilon, G., Perigon-Melon, A., and Valencia, S. (2000) Determination of the pore topology of zeolite IM-5 by means of catalytic test reactions and hydrocarbon absorption measurements. *J. Catal.*, **189**, 382–394.

61 Corma, A. (2003) State of the art and future challenges of zeolites as catalysts. *J. Catal.*, **216**, 298–312.

62 Chen, N.Y., Maziuk, J., Schwartz, A.B., and Weisz, P.B. (1968) Selectoforming–new process to improve octane and quality. *Oil Gas J.*, **66** (47), 154.

63 Clark, L., Sierka, M., and Sauer, J. (2004) Computational elucidation of the transition state shape selectivity phenomenon. *J. Am. Chem. Soc.*, **126** (3), 936–947.

64 Jones, C.W., Zones, S.I., and Davis, M.E. (1999) M-xylene reactions over

zeolites with unidimensional pore systems. *Appl. Catal. A*, **181**, 289–303.

65 McVicker, G., Feeley, O.C., Ziemiak, J.J., Vaughan, D.E.W., Strohmaier, K.C., Kliewer, W.R., and Leta, D.P. (2005) Methylcyclohexane ring-contraction: a sensitive solid acidity and shape selectivity probe reaction. *J. Phys. Chem. B*, **109** (6), 2222–2226.

66 Maesen, L.M., Schenk, M., Vlugt, T.J.H., de Jonge, J.P., and Smit, B. (1999) The shape selectivity of paraffin hydroconversion on TON-MTT-, and AEL-type type sieves. *J. Catal.*, **188**, 403–412.

67 Laxmi Narasimhan, C.S., Thybaut, J.W., Marin, G.B., Jacobs, P.A., Martens, J.A., Denayer, J.F., and Baron, G.V. (2003) Kinetic modeling of pore mouth catalysis in the hydroconversion of n-octane on Pt-H-ZSM-22. *J. Catal.*, **220**, 399–413.

68 Miller, S.J., O'Rear, D.J., and Rosenbaum, J. (2005) Processes for producing lubricant base oils with optimized branching. United States Patent Application Publication US 0077209 A1

69 Bhan, A. and Iglesia, E.A. (2008) Link between reactivity and local structure in acid catalysis on zeolites. *Acc. Chem. Res.*, **41** (4), 559–567.

70 Smit, B. and Krishna, R. (2003) Simulating absorption of alkanes in zeolites, in *Handbook of Zeolite Science and Technology* (ed. S.M. Auerbach, K.A. Carrado, and P.K. Dutta), CRC, Boca Raton, pp. 317–340.

71 Webb, E.B., III and Grest, G. (1998) Influence of intracrystalline diffusion in shape selective catalytic test reactions. *Catal. Lett.*, **56**, 95–104.

72 Zones, S.I., Zhang, G., Krishna, K.R., Biscardi, J.A., Marcantononoi, P., and Vittoratos, E. (2005) Preparing small crystal SSZ-32 and its use in a hydrocarbon conversion process. United States Patent Application Publication US 0092651 A1.

73 Martens, J.A., Verrelst, W.H., Mathys, G.M., Brown, S.H., and Jacobs, P.A. (2005) Tailored catalytic propene trimerization over acidic zeolites with tubular pores. *Angew. Chem. Int. Ed.*, **44**, 5687–5690.

74 Millini, R., Frigerio, F., Bellussi, G., Pazzuconi, G., Perego, C., Pollesel, P., and Romano, U. (2003) A priori selection of shape-selective zeolite catalysts for the synthesis of 2,6-dimethylnaphthalene. *J. Catal.*, **217**, 298–309.

75 Schenk, M., Calerno, S., Maesen, T.L.M., Vlugt, T.J.H., van Benthem, L.L., Verbeek, M.G., Schnell, B., and Smit, B. (2003) Shape selectivity through entropy. *J. Catal.*, **214**, 88–89.

76 Llopis, F., Sastre, G., and Corma, A. (2004) Xylene isomerization and aromatic alkylation in zeolites NU-87, SSZ-33, β, and ZSM-5: molecular dynamics and catalytic studies. *J. Catal.*, **227**, 227–241.

77 Brouwer, D.M. (1968) HF-SbF$_5$ catalyzed isomerization of n-butane-1-13C. *Recueil*, 87.

78 Stocker, M. (2005) Gas phase catalysis by zeolites. *Micropor. Mesopor. Mater.*, **82**, 257–292.

79 Corma, A. and Martínez, A. (1993) Chemistry, catalysis, and processes for isoparaffin-olefin alkylation: actual situation and future trends. *Catal. Rev., Sci. Eng.*, **35** (4), 483–570.

80 Corma, A., Martínez, A., Arroyo, P.A., Monteiro, J.L.F., and Sousa-Aguiar, E.F. (1996) Isobutane/2-butene alkylation on zeolite beta: influence of post-synthesis treatments. *Appl. Catal. A*, **142**, 139–150.

81 Nivarthy, G.S., Seshan, K., and Lercher, J.A. (1998) The influence of acidity on zeolite H-BEA catalyzed isobutene/N-butene alkylation. *Micropor. Mesopor. Mater.*, **22**, 379–388.

82 Nivarthy, G.S., Feller, A., Seshan, K., and Lercher, J.A. (2000) The role of hydride transfer in zeolite catalyzed isobutane/butene alkylation, in *Proceedings of the 12th ICC* (eds A. Corma, F.V. Melo, S. Mendioroz, and J.L.G. Fierro), Stud. Surf. Sci. Catal., vol. 130, Elsevier, Amsterdam, pp. 2561–2566.

83 Guzman, A., Zuazo, I., Feller, A., Olindo, R., Sievers, C., and Lercher, J.A. (2005) On the formation of the acid sites in lanthanum exchanged X

zeolites used for isobutane/Cis-2-butene alkylation. *Micropor. Mesopor. Mater.*, **83**, 309–318.

84 Sievers, C., Zuazo, I., Guzman, A., Olindo, R., Syska, H., and Lercher, J.A. (2007) Stages of aging and deactivation of zeolite LaX in isobutane/2-butene alkylation. *J. Catal.*, **246**, 315–324.

85 Craciun, I., Reyniers, M.-F., and Marin, G.B. (2007) Effects of acid properties of Y zeolites on the liquid phase alkylation of benzene with 1-octene: a reaction analysis. *J. Mol. Sci. A Chem.*, **277**, 1–14.

86 Wang, H. and Xin, W. (2001) Surface acidity of H-beta and its catalytic activity for alkylation of benzene with propylene. *Catal. Lett.*, **76**, 225–229.

87 Olsen, D.H. and Dempsey, E. (1969) The crystal structure of the zeolite hydrogen faujasite. *J. Catal.*, **13**, 221–231.

88 Haag, W.O. and Dessau, R.M. (1984) *Proceedings of the 8th International Congress on Catalysis*, vol. 2, Verlag Chemie, Weinheim, p. 305.

89 Narbeshuber, T.F., Brait, A., Seshan, K., and Lercher, A.J. (1997) Dehydrogenation of light alkanes over zeolites. *J. Catal.*, **172**, 127–136.

90 Krannila, H., Haag, W.O., and Gates, B.C. (1992) Monomolecular and bimolecular mechanisms of paraffin cracking: n-butane cracking catalyzed by HZSM-5. *J. Catal.*, **135**, 115–124.

91 Babitz, B.A., Williams, J.T., Snurr, R.Q., Haag, W.O., and Kung, H.H. (1999) Monomolecular cracking of n-hexane on Y, MOR, and ZSM-5 zeolites. *Appl. Catal. A*, **179**, 71–86.

92 Haag, W.O., Dessau, R.M., and Lago, R.M. (1991) Kinetics and mechanism of paraffin cracking with zeolite catalysts. Stud. Surf. Sci. Catal., vol. 60, Elsevier, Amsterdam, pp. 255–265.

93 Haag, W.O. (1994) Catalysis by zeolites – science and technology. Stud. Surf. Sci. Catal., vol. 84, Elsevier, Amsterdam, pp. 1375–1394.

94 Wiellers, A.F.H., Vaarkamp, M., and Post, M.F.M. (1991) Relation between properties and performance of zeolites in paraffin cracking. *J. Catal.*, **127**, 51–66.

95 Altwasser, S., Welker, C., Traa, Y., and Weitkamp, J. (2005) Catalytic cracking of n-octane on small-pore zeolites. *Micropor. Mesopor. Mater.*, **83**, 345–356.

96 Rozanska, X., Saintigny, X., van Santen, R.A., and Hutschka, F.A. (2001) DFT study of isomerization and transalkylation reactions of aromatic species catalyzed by acidic zeolites. *J. Catal.*, **202**, 141–155.

97 Wu, X. and Anthony, R.G. (2001) Effect of feed composition on methanol conversion to light olefins over SAPO-34. *Appl. Catal. A*, **218**, 241–250.

98 Svelle, S., Joensen, F., Nerlov, J., Olsbye, U., Lillerud, K.-P., Kolboe, S., and Bjørgen, M. (2006) Conversion of methanol into hydrocarbons over the zeolite H-ZSM-5: ethene formation is mechanistically separated from the formation of higher alkenes. *J. Am. Chem. Soc.*, **128**, 14770–14771.

99 Mikkelsen, Ø. and Kolboe, S. (1999) The conversion of methanol to hydrocarbons over zeolite H-beta. *Micropor. Mesopor. Mater.*, **29**, 173–184.

100 Dahl, I. and Kolboe, S. (1993) On the reaction mechanism for propene formation in the MTO reaction over SAPO-34. *Catal. Lett.*, **20**, 329–336.

101 Dahl, I.M. and Kolboe, S. (1996) On the reaction mechanism for hydrocarbon formation from methanol over SAPO-34 2. Isotopic labeling studies of the co-reaction of propene and methanol. *J. Catal.*, **161**, 304–309.

102 Olsbye, U., Bjørgen, M., Svelle, S., Lillerud, K.-P., and Kolboe, S. (2005) Mechanistic insight into the methanol-to-hydrocarbons reaction. *Catal. Today*, **106**, 108–111.

103 Svelle, S., Rønning, O., Olsbye, U., and Kolboe, S. (2005) Kinetic studies of zeolite-catalyzed methylation reactions. Part 2. Co-reaction of [12C] propelyene or [12C] N-butene and [13C]. methanol. *J. Catal.*, **234**, 385–400.

104 Svelle, S., Kolboe, S., Swang, O., and Olsbye, U. (2005) Methylbenzenes by dimethyl ether or methanol on acidic zeolites. *J. Phys. Chem. B*, **109**, 12874–12878.

105 Svelle, S., Aravinthan, S., Bjørgen, M., Lillerud, K.-P., Kolboe, S., Dahl, I.M.,

and Olsbye, U. (2006) The methyl halide to hydrocarbon reaction over H-SAPO-34. *J. Catal.*, **241**, 243–254.

106 Marcus, D.M., McLachlan, K.A., Wildman, M.A., Ehresmann, P.W., and Haw, J.F. (2006) Experimental evidence from H/D exchange studies for the failure of direct C-C coupling mechanisms in the methanol-to-olefin process catalyzed by HSAPO-34. *Angew. Chem. Int. Ed.*, **45**, 3133–3136.

107 Song, W., Marcus, D.M., Fu, H., Ehresmann, J.O., and Haw, J.F. (2002) An oft-studied reaction that may have never been: direct catalytic conversion of methanol or dimethyl ether to hydrocarbons on the solid acids HZSM-5 or HSAPO-34. *J. Am. Chem. Soc.*, **124**, 3844–3845.

108 Nachtigall, P. and Sauer, J. (2007) Applications of quantum chemical methods in zeolite science. *Stud. Surf. Catal.*, vol. 168, Elsevier, Amsterdam, pp. 659–700.

109 Song, W., Haw, J.F., Nicholas, J.B., and Heneghan, C. (2000) Methybenzenes are the organic reaction centers for methanol-to-olefin catalysis on HSAPO-34. *J. Am. Chem. Soc.*, **122**, 10726–10727.

110 Bjørgen, M., Arstad, B., and Kolboe, S. (2001) Methanol-to-hydrocarbons reaction over SAPO-34. Molecules confined in the catalyst cavities at short time on stream. *Catal. Lett.*, **71**, 209–212.

111 Bjørgen, M., Svelle, S., Joensen, F., Nerlov, J., Kolboe, S., Bonino, F., Palumbo, L., Bordiga, S., and Olsbye, O. (2007) Conversion of methanol to hydrocarbons over zeolite H-ZSM-5: on the origin of the olefinic species. *J. Catal.*, **249**, 195–207.

112 Dahl, I.M. and Kolboe, S. (1996) On the reaction mechanism for hydrocarbon formation from methanol over SAPO-34 2. Isotopic labeling studies of the co-reaction of propene and methanol. *J. Catal.*, **161**, 304–309.

113 Song, W., Fu, H., and Haw, J.F. (2001) Supramolecular origins of product selectivity for methanol-to-olefin catalysis on HSAPO-34. *J. Am. Chem. Soc.*, **123**, 4749–4754.

114 Woodcock, D.A. and Lightfoot, P. (1999) Negative thermal expansion in the siliceous zeolites chabazite and ITQ-4: a neutron power diffraction study. *Chem. Mater.*, **11**, 2508–2514.

115 Arstad, B., Kolboe, S., and Swang, O. (2004) Theoretical investigation of arene alkylation by ethene and propene over acidic zeolites. *J. Phys. Chem. B*, **108**, 2300–2308.

116 Arstad, B., Kolboe, S., and Swang, O. (2005) Theoretical study of the heptamethylbenzenium ion intramolecular isomerizations and C_2, C_3, C_4 alkene elimination. *J. Phys. Org. Chem.*, **109**, 8914–8922.

117 Svelle, S., Bjørgen, M., Kolboe, S., Kuck, D., Letzel, M., Olsbye, U., Sekiguchi, O., and Uggerud, E. (2006) Intermediates in the methanol-to-hydrocarbons (MTH) reaction: a gas phase study of the unimolecular reactivity of multiply methylated benzenium cations. *Catal. Lett.*, **109**, 25–35.

118 Arstad, B., Kolboe, S., and Swang, O. (2006) Theoretical study of carbon atom scrambling in benzenium ions with ethyl or isopropyl groups. *J. Phys. Org. Chem.*, **19**, 81–92.

119 Arstad, B., Nicholas, J.B., and Haw, J.F. (2004) Theoretical study of the methylbenzane side-chain hydrocarbon pool mechanism in methanol to olefin catalysis. *J. Am. Chem. Soc.*, **126**, 2991–3001.

120 Bjørgen, M., Olsbye, U., Petersen, D., and Kolboe, S. (2004) The methanol-to-hydrocarbons reaction: insight into the reaction mechanism from [12C] and [13C] methanol co-reactions over zeolite H-beta. *J. Catal.*, **221**, 1–10.

121 Sassi, A., Wildman, M.A., Ahn, H.J., Prasad, P., Nicholas, J.B., and Haw, J.F. (2002) Methylbenzene chemistry on zeolite h-beta: multiple insights into methanol-to-olefin catalysis. *J. Phys. Chem. B*, **106**, 2294–2303.

122 Bjørgen, M., Olsbye, U., and Kolboe, S. (2003) Coke precursor formation and zeolite deactivation: mechanistic insights from hexamethylbenzene conversion. *J. Catal.*, **215**, 30–44.

123 Fu, H., Song, W., and Haw, J.F. (2001) Polycyclic aromatics formation in

HSAPO-34 during methanol-to-olefin catalysis: *ex situ* characterization after cryogenic grinding. *Catal. Lett.*, **76**, 89–94.

124 Song, W., Fu, H., and Haw, J.F. (2001) Selective synthesis of methylnaphthalenes in HSAPO-34 cages and their function as reaction centers in methanol-to-olefin catalysis. *J. Am. Chem. Soc.*, **105**, 12839–12843.

125 Haw, J.F. and Marcus, D. (2005) Well-defined (supra) molecular structures in zeolite methanol-to-olefin catalysis. *Top. Catal.*, **34**, 41–48.

126 Svelle, S., Rønning, O., and Kolboe, S. (2004) Kinetic studies of zeolite-catalyzed methylation reactions 1. Coreaction of [12C] ethane and [13C] methanol. *J. Catal.*, **224**, 115–123.

Figure References

Many figures in this chapter originate from other sources: The following figures were generously provided by scientists and companies. I gratefully acknowledge them:

Figure 13.1 was published in *Adsorption, Surface Area and Porosity*, Gregg, J.S. and Sing, S.W.K., p. 4, Copyright Elsevier (Academic Press) (1982)

Figures 13.2, 13.3, 13.4, 13.5, 13.6, 13.11 and 13.12 are reproduced with permission from the Royal Society of Chemistry from *Chemical Society Reviews*, 31, Krishna, R., Smit, B., and Calero, S., Entropy effects during sorption of alkanes in zeolites, p. 185–194, Copyright The Royal Society of Chemistry (2002) http://www.rsc.org/Publishing/Journals/CS/article.asp?doi=b101267n

Figures 13.7 and 13.8 are reprinted with permission from the *Journal of Physical Chemistry B*, 101 (8), Eder, F., Lercher, J.A., On the role of the pore size and tortuosity for sorption of alkanes in molecular sieves, p. 1273–1278, Copyright 1997 American Chemical Society

Figure 13.10 is reproduced with permission of the Royal Society of Chemistry from *Chemical Communications*, 15, Denayer, J.F., Ocakoglu, A.R., Huybrechts, W., Martens, J.A., Thybaut, J.W., Marin, G.B., and Baron, G.V., Pore mouth versus intracrystalline adsorption of isoalkanes on ZSM-22 and ZSM-23 zeolites under vapour and liquid phase conditions, p. 1880–1881, Copyright The Royal Society of Chemistry (2003) http://www.rsc.org/Publishing/Journals/CC/article.asp?doi=b304320g

Figure 13.14 was published as Figure 11 from the article Intracrystalline Diffusion in Zeolites Studied by Neutron Scattering Techniques by H. Jobic and A. Methivier, *Oil and Gas Science and Technology – Rev. IFP*, Vol. 60 (2005), No. 5, pp. 815–830 DOI: 10.2516/ogst:2005058

Figure 13.15 was published in *Studies in Surface Science and Catalysis*, 137, Ruthven, D.M., Post, M.F.M., Diffusion in zeolite molecular sieves, p. 525–577, Copyright Elsevier (2001)

Figure 13.17 is reprinted with permission from the *Journal of Physical Chemistry B*, 104 (24), Pieterse, J.A.Z., Veefkind-Reyes, S., Seshan, K., and Lercher, J.A., On the role of pore size and tortuosity for sorption of alkanes in molecular sieves, p. 5715–5723, Copyright 2000 American Chemical Society

Figure 13.20 was published in *Advances in Catalysis*, 51, Lamberti, C., Groppo, G., Spoto, S., Bordiga, S., and Zecchina, A., Infrared spectroscopy of transient surface species, p. 1–74, Copyright Elsevier (2007)

Figure 13.25 was published in *Chemical Engineering Science*, 61, Gounaris, C., Floudas, C.A., Wei, J., Rational design of shape selective separation and catalysis – I: concepts and analysis, p. 7933–7948, Copyright Elsevier (2006)

Figures 13.26 and 13.27 are reprinted with kind permission from Springer Science+Business Media: *Molecular Sieves*, 5, Traa, Y., Sealy, S., Weitkamp, J., Characterization of the pore size of molecular sieves using molecular probes, p. 103–154, Copyright 2007

Figure 13.28 and 13.37 were published in *Applied Catalysis A: General*, 181, Jones, C.W., Zones, S.I., Davis, M.E., Metaxylene reactions over zeolites with unidimensional pore systems, p. 289–303, Copyright Elsevier (1999)

Figure 13.30 was published in *Handbook of Zeolite Science and Technology*, Smit, B., Krishna, R., Simulating absorption of alkanes in zeolites, p. 317–340, Copyright Taylor & Francis (2003)

Figure 13.31 is reprinted with kind permission from Springer Science+Business Media: *Catalysis Letters*, 100, Huybrechts, W., Vanbutsele, G., Houthoofd, K. J., Bertinchamps, F., Laxmi Narasimhan, C.S., Gaigneau, E. M., Thybaut, J. W., Marin, G.B., Denayer, J. F. M., Baron, G.V., Jacobs, P.A., Martens, J. A., Skeletal isomerization of octadecane on bifunctional ZSM-23 catalyst, p. 235–242, Copyright 2005

Figures 13.33, 13.34 and 13.35 were reproduced with permission: *Angew. Chem. Int. Ed.*, 44, Martens, J. A., Verrelst, W. H., Mathys, G. M., Brown, S. H., Jacobs, P. A., Tailored catalytic propene trimerization over acidic zeolites with tubular pores p. 5687–5690, Copyright Wiley-VCH Verlag GmbH & Co. KGaA (2005)

Figure 13.36 was published in *Journal of Catalysis*, 214, Schenk, M., Calerno, S., Maesen, T. L.M., Vlugt, T. J.H., van Benthem, L.L., Verbeek, M. G., Schnell, B., Smit, B., shape selectivity through entropy, p. 88–89, Copyright Elsevier (2003)

Figure 13.50 is reprinted with permission from the *Journal of the American Chemical Society*, 122, Song, W., Haw, J. F., Nicholas, J.B., Heneghan, C., Methybenzenes Are the Organic Reaction Centers for Methanol-to-Olefin Catalysis on HSAPO-34, p. 10726–10727, Copyright 2000 American Chemical Society

14
Industrial Isomerization
John E. Bauer, Feng Xu, Paula L. Bogdan, and Gregory J. Gajda

14.1
Introduction

For hydrocarbons of more than three carbons, multiple isomers are possible. Among those isomers, the natural or equilibrium distributions rarely match the commercial demand. Isomerization technology provides the means to convert the less valuable isomers into more valued ones. Specific isomerization reaction mechanisms involve species of relatively similar size, so zeolites, with their precise morphologies, can be made into exceptional catalysts with high selectivity. The ability to adjust zeolite chemistry through innovative synthesis or post-synthesis treatments further enhances their versatility in isomerization applications.

14.2
Metal–Zeolite Catalyzed Light Paraffin Isomerization

The legislative requirements of lead additive elimination, methyl *tert*-butyl ether (MTBE) phase-out and reduction of benzene and other aromatics make isomerate a more attractive blending component for reformulated gasoline. There are three types of industrial metal–acid catalysts for isomerizing light naphtha to a high octane gasoline blend. A few excellent reviews have been published in this area [1–5]. A metal–zeolite catalyst, comprising a large-pore zeolite such as a mordenite and a Pt function, was first commercialized by Shell in 1970 [1]. Since the 1990s, UOP LLC (a Honeywell company) and Cosmos oil have pioneered a new-generation industrial catalyst based on sulfated zirconia [6], whose activity and robustness is between Pt/chlorided alumina and zeolite-based catalysts. This section only discusses metal–zeolite bifunctional catalysts in light paraffin isomerization with an emphasis on C6 paraffin isomerization.

Zeolites in Industrial Separation and Catalysis. Edited by Santi Kulprathipanja
Copyright © 2010 WILEY-VCH Verlag GmbH & Co. KGaA, Weinheim
ISBN: 978-3-527-32505-4

14 Industrial Isomerization

Table 14.1 Physical properties of selected molecules in isomerization feed and product.

Components	Boiling point (°C)	RONC	Density (g/cm³)	RVP (psi; 10 psi = 68.95 kPa)
Isopentane	27.8	92	0.625	18.9
n-Pentane	36.1	62	0.631	14.4
2,2-Dimethylbutane	49.7	92	0.664	9.1
2,3-Dimethylbutane	58.0	104	0.664	6.9
2-Methylpentane	60.3	73	0.667	6.3
3-Methylpentane	66.3	75	0.652	5.7
n-Hexane	68.7	25	0.657	4.6
Methylcyclopentane	71.8	91	0.754	4.2
Benzene	80.1	100+	0.882	3.0
Cyclohexane	80.7	83	0.783	6.0
n-Heptane	98.4	0	0.687	1.5

14.2.1
General Considerations

Light naphtha consists of primarily C_5s and C_6s, plus a smaller fraction of C_7s and C_4s. The feed is highly paraffinic, but also contains benzenes and naphthenes. It has a blend research octane number in the 60s to low 70s. Isomerization to boost octane level is based on the fact that branched paraffins have significantly higher anti-knock ratings, as measured by research octane number-clear (RONC), shown in Table 14.1. The branched isomers also have lower densities and lower boiling points and therefore higher Reid vapor pressures (RVP) than their straight-chain counterparts. Thermodynamically, branched isomers are favored at low temperature. This is illustrated by the equilibrium hexane isomers distribution and RONC as a function of temperature in Figure 14.1. Industrial zeolitic catalysts are very active, able to achieve equilibrium conversion per pass at relatively mild temperature. Full conversion of n-hexane to mono- and di-branched isomers via physical separation and recycle can increase octane rating further by ~10RONC at 250 °C.

14.2.2
Bifunctional Paraffin Isomerization Mechanism

Industrial metal–zeolite catalysts undergo a bifunctional, monomolecular mechanism [1–5, 7]. Carbenium ions are the critical reaction intermediates to complete chain reactions. In the zeolite channels, carbenium ions likely exist as an absorbed alkoxyl species, rather than as free-moving charged ions [8]. Figure 14.2 illustrates the accepted reaction mechanism, using hexanes as an example.

Five key steps for the transformations are indicated by the roman letters. In step I, normal paraffins adsorb on Pt sites and undergo dehydrogenation to form small

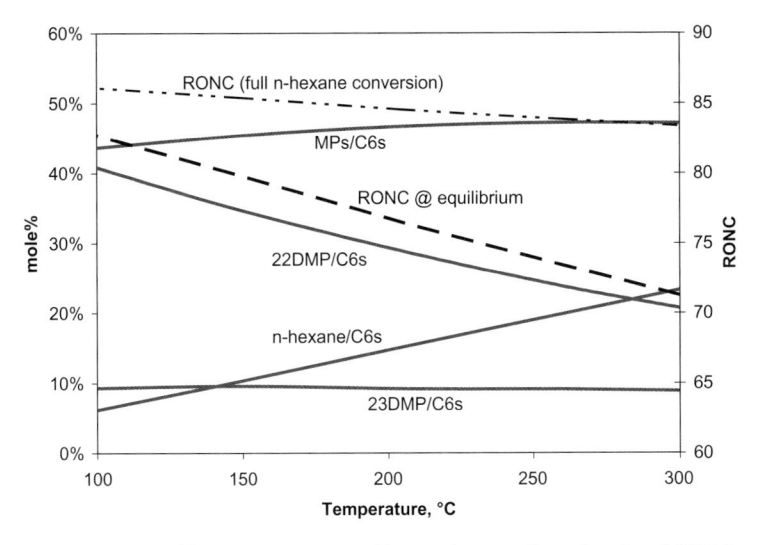

Figure 14.1 Equilibrium composition of hexane isomers (gas phase) and RONC.

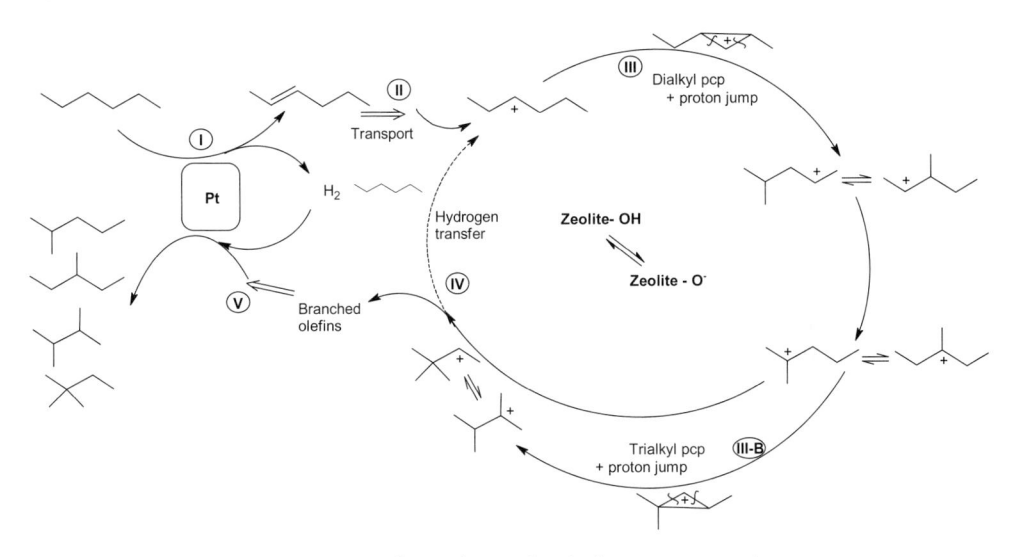

Figure 14.2 A bifunctional mechanism of Pt/zeolite catalyzed n-hexane isomerization.

amount of olefins and hydrogen. In step II, olefins transport to zeolite acid sites and react with acidic bridging hydroxyl protons to form secondary carbenium ions, which is the chain initiation step for carbenium ion transformation cycle. Step III is the skeletal isomerization of carbeniums to form mono-branched carbenium ions. This requires the formation of dialkyl protonated cyclopropane (PCP) intermediate, a proton jump step and a bond cleavage step. In step III-B, similar to in step III, further increase of branching involves trialkyl PCP intermediates to di-branched carbenium species. The formation of 2,2-dimethylbutane incurs the formation of a secondary carbenium ion, which is considered the most difficult step. In step IV, carbenium ions undergo hydrogen transfer to desorb branched olefin species; at the same time, the acid sites are regenerated, thereby closing the carbenium ion catalytic cycle. Finally in step V, olefins transfer to Pt sites and hydrogenate to branched paraffins and desorb from zeolite channels. The apparent activation energy for skeletal isomerization is around 130 kJ/mol and is considered the rate-determining step under most circumstances [9, 10]. Strong Bronsted acidity is critical for carbenium ion transformation chemistry. Cracking via β-scission (or via PCP intermediates [3, 4]) generates a carbenium ion and a smaller olefin. The propensity of cracking is $C_5 < C_6 \ll C_7$ due to the formation of more stable tertiary carbenium species from C_7. β-Scission for C_6 and C_5 are energetically difficult, explaining the high yield of C_5–C_6 isomerization. Product selectivity is determined by the life-time of and competing reactions from carbenium ions.

A strong metal function is to maintain a controlled concentration of olefins for sustaining a sufficient chain initiation reaction and avoiding chain termination. Isomerization chain termination occurs when a carbenium ion undergoes hydride transfer with an olefin rather than with a paraffin. As a result, allylic carbenium ions and even more unsaturated species are species leading to fast deactivation via further oligomerization or aromatization. A strong metal function is therefore essential for good catalyst activity and stability. A small amount of benzene in the feed is completely hydrogenated to cyclohexane under process conditions. Isomerization of cyclohexane to methylcyclopentane is a facile reaction and equilibrium between the two usually quickly establishes.

14.2.3
Zeolitic Paraffin Isomerization Catalysis

Acid site type, strength, density, distribution, acid/metal site ratio, as well as transport properties all influence catalyst activity, selectivity and stability. High activity is expected from a strong Bronsted acid, plus a high density of sites and a fast turnover frequency on such sites. Feed molecules need easy access to the active sites. Branched products quickly desorb from catalyst surface once isomerization completes. Kinetic diameters of di-branched C_6 isomers have larger kinetic diameters than n-hexane (e.g., 2,2-dimethylbutane of 0.63 nm vs n-hexane of 0.43 nm). Among large-pore (12MR) zeolites, strongly acidic mordenite, mazzite, ZSM-12 and beta zeolites show higher activities over faujasite or L zeolite [4]. Mazzite is reported to be more active and to produce higher RONC products than mordenite [11, 12]. Several groups have reported beta zeolite is more selective and less cracking than mordenite for n-heptane isomerization [10, 13]. Medium-pore 10MR zeolites such as MFI [14], ZSM-22, ZSM-23 and EU-1 [15] have diffusion limitation for large isomers. This property of MFI, coupled with its high acidity and channel structure, favors cracking reactions over isomerization.

An optimum isomerization activity versus Si/Al_2 ratio was reported for mordenite [5, 13, 16, 17]. At too high an acid density, the acid strength decreases and olefin oligomerization reaction is favored. On the other side, too few acid sites impair activity. Reported optimum Si/Al_2 ratios are around 20, which is close to the value where all the acid sites are isolated. The optimum ratio does vary with the dealumination procedure used. Acid leaching and steaming are found to generate mesoporosity in the zeolites, leading to faster diffusion of multi-branched isomers [5, 18, 19].

Introduction of Pt significantly enhances zeolite isomerization catalyst stability and alters the reaction pathways. The Pt/acid ratio not only changes the isomerization/cracking ratio, but also changes the ratio of mono/di-branched isomers in Pt/Y [14]. High Pt dispersion and close proximity to acid sites correlate with high n-hexane conversion as well as high isomerization selectivity [20, 21].

Like in any catalytic process, process variables crucially impact reaction kinetics, conversion efficiency and catalyst stability. Increasing temperature favors cracking, thus decreasing the isomerate yield. It is preferred to have a high-activity catalyst and operate at the lowest possible temperature to achieve the highest RONC. Hydrogen shifts the equilibrium concentrations of olefins and carbenium ions. High H_2 partial pressure suppresses bimolecular reactions [4]. These together underline the importance of maintaining high selectivity and stability. At high total pressure, high boiling point components exert an inhibition effect on the hydroisomerization of more volatile paraffins. They may fill zeolite pores, thus limiting the transport of olefins to metal to acid site [10]. In industrial practice, pressure and space velocity are optimized to achieve optimum activity, yield, life and ease of system integration.

14.2.4
Industrial Zeolitic Isomerization Catalysts and Processes

The earliest industrial zeolitic isomerization process was the Hysomer process, formerly offered for license by Shell. Currently UOP offers a zeolite- and Pt-containing catalyst: HS-10 in the fixed-bed UOP TIP™ process [3]. A similar catalyst Hysopar® was introduced by Sud-Chemie [22] in the CKS Isom process (Cepsa–Kellogg–Sud Chemie). Recently there were reports of IMP-02 and CI-50 commercial catalysts from China [23] and Russia [24].

Industrial zeolitic isomerization process operates at 240–370 °C, total pressure of 0.8–3.5 MPa, with H_2 cofeed. Coked catalyst can be regenerated by simple carbon burn. Industrial catalysts have accomplished a service life longer than ten years. Physical separation and recycle are needed to achieve close to complete isomerization. Selective zeolite adsorption to separate iso/normal paraffins is advantageous to remove low amount of n-paraffins after single pass conversion. Chapter 8 discusses the UOP IsoSiv™ process using 5A zeolite to selectively absorb smaller molecular n-paraffins. Desorption is achieved by purging with hydrogen, which is readily available in a refinery. Combining aspects of the Hysomer process and the IsoSiv process is the TIP process, a flow scheme of which is shown in Figure 14.3. The TIP process enhances the product RONC by approximately 20. This is significant octane gain compared with once-through (~82 RONC) zeolitic isomerization.

Membrane reactors [25] and catalytic distillation [26] are two options for even closer integration of isomerization and separation. The membrane concept is

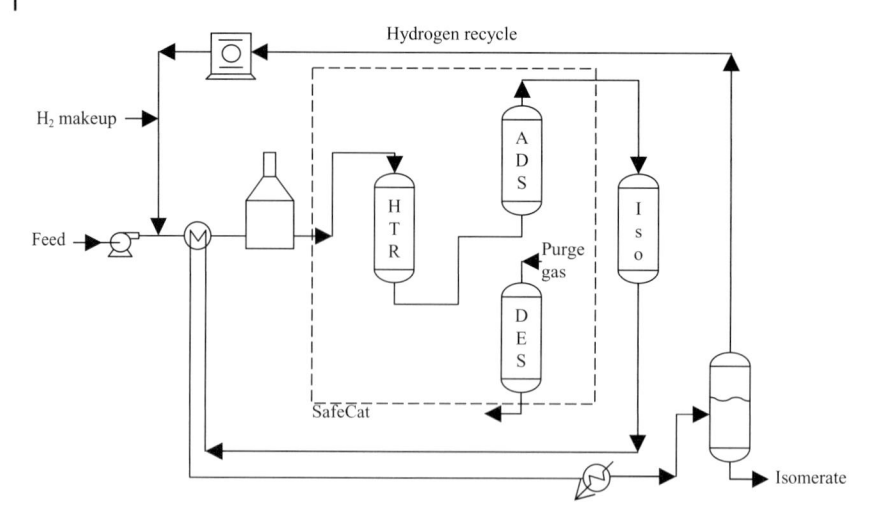

Figure 14.3 UOP SafeCat isomerization scheme. HTR = Hydrotreating Reactor, ADS = Adsorption, DES = Desorption, ISO = Isomerization Reactor. Reprinted from [103], with permission from Elsevier.

demonstrated by using a silicalite-1 membrane and a platinum on chlorided alumina catalyst [4]. Zeolites with the CFI and ATS structure combined with catalytic distillation concept have been documented [27]. In principle, above-equilibrium conversion in one reactor may be achievable with these concepts.

14.2.5
Summary

The reformulated gasoline pool faces ever changing blending challenges. The blending of more ethanol and a lower benzene requirement [28] are two drivers for refiners to reconfigure an isomerization unit. Future directions for development are higher activity, ability to process heavier feedstock with a high isomerate yield and more contaminant tolerance. Sulfated zirconia catalysts have recently surpassed zeolitic catalysts as the more preferred commercial paraffin isomerization catalysts [29–31].

14.3
Olefin Isomerization

14.3.1
General Considerations

Olefin isomerization consists of three possible reactions (see Figure 14.4): cis–trans, double bond and skeletal. As noted by Damon, et al., [32, 33], cis–

Figure 14.4 Olefin isomerization reactions.

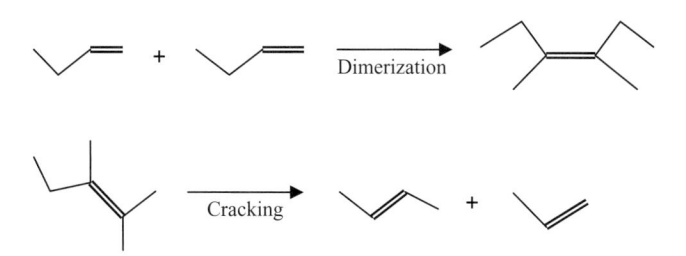

Figure 14.5 Byproduct reactions.

trans isomerization requires acid sites with $-H_R > -0.82$, as does double bond isomerization from the '1' to the '2' position. Further positional isomerization requires acid sites with $-H_R > 4.04$ and skeletal isomerization requires sites with $-H_R > 6.63$. The major side reactions (see Figure 14.5) are dimerization or higher oligomerization and cracking. As a result, olefin isomerization reactions are generally carried out at the lowest practical operating pressures and moderate temperatures.

14.3.2
Cis–Trans and Double Bond Isomerization

Other than a few fine chemicals reactions, *cis–trans* isomerization is seldom practiced using molecular sieve catalysts.

Double bond isomerization using molecular sieves (5A) was reported in the patent literature by Fleck and Wight of Union Oil Company [34] only a few years after synthetic zeolites became commercially available. More recently [35] ferrierite has also been claimed. The major initial uses were to convert α-olefins (1-olefins) into mixtures of internal olefins for further conversion, usually by oligomerization into various products – lube oil base stocks predominating. Inevitably, patents were issued noting the ability to convert internal olefins into mixtures containing greater concentrations of 1-olefins (e.g., [36]), but few practical processes have resulted.

Table 14.2 Research octane numbers (RON) for selected compounds.

Compound	RON	Compound	RON
n-Hexane	25	Trans-3-hexene	94
2,3-Dimethylbutane	104	2,3-Dimethyl-2-butene	97
		MTBE	118

Source: "Knocking Characteristics of Pure Hydrocarbons", ASTM Special Technical Publication number 225.

14.3.3
Skeletal Isomerization (Butenes, Pentenes, Hexenes)

Olefins, unlike paraffins, do not show significant gains in octane number with skeletal isomerization (see Table 14.2). As a result, olefin isomerization is not a useful octane boosting strategy. However, tertiary olefins (olefins with three alkyl substituents on the double bond), do react fairly readily with olefins to form ethers, which do have good octane numbers – for example, methyl *tert*-butyl ether (MTBE).

Several review articles [37–40] have been written discussing the use of molecular sieves for the isomerization of light olefins, especially butene. The major driving force was the requirement in the 1990 amendments to the Clean Air Act in the United States that required the addition of oxygenates to gasoline in amounts up to 2.7 wt% oxygen in the final gasoline product [37]. The primary additive chosen to meet these requirements was MTBE, with lesser amounts of the ethyl ether (ETBE) or *tert*-amyl methyl ether (TAME) as supplements. One major possible route to meet these requirements was the isomerization of linear butenes (1-butene, *cis*-2-butene and *trans*-2-butene) into isobutene (2-methyl-1-propene).

The thermodynamics are indicated in Figure 14.6, using data from [41]. There is a considerable drop in the equilibrium isobutene concentration, from almost 60 mol% at 400 K to only about 30% at 1000 K. This places a premium on an active catalyst to obtain high conversions at temperatures that yield reasonably high equilibrium conversions.

Early reports of reaction over borosilicate [42], silicalite [43], Theta-1 [44], ZSM-22 [45] and ZSM-23 [46] gave relatively low selectivities for isobutene. With the use of the aluminophosphates SAPO-11 [47] and MgAPSO-31 [48] improved selectivities were reported, reaching a final plateau with ferrierite [49].

Commercial processes for the isomerization of butene and/or higher olefins were announced by, among others, UOP (UOP Butesom™ and Pentesom™ processes[50]), MW Kellogg (ISOFIN™ process[40]), Texaco (ISOTEX™ process [51]) and Lyondell [52]. The process scheme is quite simple and the Butesom process scheme is shown in Figure 14.7.

All these processes faced competition from the on-demand production of isobutene through a combined process of isomerization of *n*-butane to isobutane

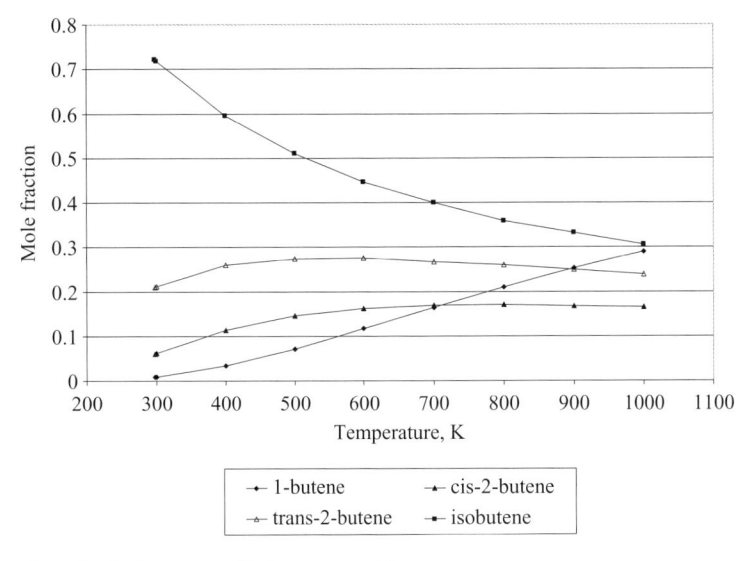

Figure 14.6 Butene equilibrium composition.

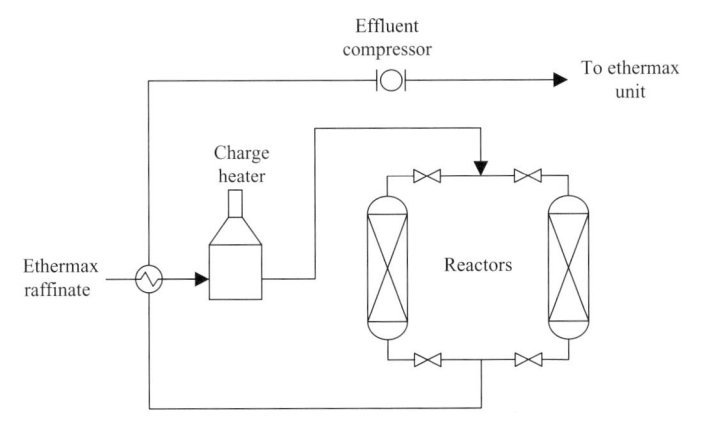

Figure 14.7 Butesom process.

followed by dehydrogenation of isobutane to isobutene (e.g., UOP Butamer™ and Oleflex™ processes). The scale of demand for MTBE and the availability of butanes from the FCC unit in refineries led to the butane isomerization/dehydrogenation route providing essentially all of the MTBE demand. The remainder was met by reacting the isobutene already present in the FCC butene product stream. With the removal of the oxygenate requirements from gasoline, the demand for MTBE and, hence, isobutene has declined to a level that can be supported by the amount of isobutene readily available from FCC off gas and steam crackers.

14.3.4
Skeletal Isomerization (Longer-Chain Olefins)

Skeletal isomerization has been claimed to reduce the degree of olefin branching to improve the viscosity of synthetic lubricating oil feedstocks using ZSM-type zeolites, beta or rare earth-exchange Y [53], 13X [54] and SAPO-11 [55], among others. Conversely, isomerization of long-chain *n*-alkenes to *iso*-alkenes using zeolites (beta [56], mordenite [57], SAPO-11 [58], among others) as a method of dewaxing higher petroleum fractions has also been claimed.

14.3.5
Olefin Isomerization Summary

Olefin isomerization reactions range from some of the most facile using acid catalysts to moderately difficult and, as components of more complex reaction schemes such as catalytic cracking, may be among the most common reactions in hydrocarbon processing. As stand-alone reactions, they are primarily used to shift the equilibrium between terminal and internal olefins or the degree of branching of the olefin. While olefin isomerization was considered for the production of MTBE, today stand-alone olefin isomerization processes are only considered for a few special situations within a petrochemical complex.

14.4
C8 Aromatics Isomerization

Over the past several decades, there has been a continuing growth in the worldwide demand for plastics, films and fibers, particularly polyesters. The raw materials that make up these polymers are based primarily on the C8 family of aromatics (C8A) – ethylbenzene (EB), *para*-xylene (PX), *meta*-xylene (MX), and *ortho*-xylene (OX). Polyester (PET plastic), derived from PX, in particular has experienced rapid growth and is projected to see continue rapid growth as many developing countries desire to have the lifestyle flexibility that such readily available, versatile plastics support. While the markets for MX- and OX-derived plastics are smaller (plasticizers and specialty polyesters, respectively), all C8A markets continue to increase with population growth [59].

The desired C8A isomer is separated out from the full blend. Depending on the isomer, this separation may be done by fractionation, by crystallization, or by selective absorption (see Chapters 5–7). Whatever the method, the separation of that species leaves a mixture depleted in the desired isomer. C8A isomerization technology converts that mixture back to an equilibrium or near-equilibrium mixture, which is then fed back into the separation unit within the separation–isomerization loop. In this way, the full range of C8 aromatics can be converted to extinction to the desired end-product. Although EB could also be separated from the C8A mixture, this is usually too costly, so EB is typically generated by the ethylation of benzene, rather than separation and isomerization from the other C8A.

There are two general types of C8A isomerization, defined by the method of converting EB. There is the EB dealkylation type, which converts EB into benzene and ethane, and the EB isomerization type, which converts EB into xylenes. The differences between these two categories are addressed in Section 14.4.1.3.

14.4.1
The Chemistry of C8 Aromatics Isomerization

14.4.1.1 Feed Composition and Characteristics

The feed to a C8A isomerization unit is typically the effluent from a series of fractionation columns, so that it has a high concentration of C8A, with only small amounts of lighter materials, such as toluene, or heavier materials, such as trimethylbenzenes or methyl ethylbenzenes. This feed is depleted in one or more of the C8A isomers, depending on which type of separation units are feeding it. For example, given that the equilibrium composition of C8A between 350–400 °C is approximately 8% EB, 22% PX, 48% MX, 22% OX (which is fed to the separation section), the resulting product going to the isomerization unit may be 10% EB, 1% PX, 61%MX, and 28% OX (if the separated product were PX) [59]. In addition to single product systems, separate adsorptive separation of PX or MX could be combined with fractionation of OX, so a wide range of compositions is possible.

C8 naphthenes (C8N) may be present in the recycled portion of the feed, for cases where it is desirable to convert EB to Xylenes. C8N are not typically present in the fresh feed to the separation/isomerization loop, but equilibrium reactions between C8A and C8N at the isomerization unit operating conditions of temperature and pressure drive the creation of C8N from C8A, and these C8N are then recirculated. EB dealkylation catalysts do not require C8N intermediates, and in fact the strong acid sites on that type catalyst easily crack C8N, leading to undesirable losses. For that reason, EB dealkylation catalysts optimally produce few C8N. For EB isomerization catalysts, however, C8N are key intermediates in the conversion of EB to xylenes, and are present in the range of ~5% to 15% (usually to the lower end, but it depends on the particular catalyst system and plant conditions). C8N boil slightly lighter than C8A, so recovering the C8N with fractionation also brings along some of the toluene. Hence, EB isomerization systems usually have a higher toluene content in the recycled feed than do EB dealkylation systems.

The C8A isomerization system is typically a clean system, with low levels of impurities (having been removed during upstream hydrotreating, fractionation, or other treatments). Adsorptive recovery systems in particular require low impurities, since adsorption is more sensitive to contaminants than crystallization or fractionation, especially when the contaminants are polar. Basic contaminants may adsorb strongly to the catalyst's acid sites, causing loss of activity and premature deactivation. Olefins could alkylate with the aromatics, forming heavies that need to be fractionated out of the system and cause a yield loss. The olefins could also polymerize on the catalyst surface, blocking pores and reducing activity.

Therefore, common olefin removal steps such as clay treating may be required. Heavy aromatics that enter in the feed could react to form polynuclear aromatics, which could also react and coat the catalyst.

14.4.1.2 Reaction Product Composition and Characteristics

The ideal product from the C8A isomerization reaction is pure xylenes and depleted EB. However, as discussed in Section 14.4.1.3, there are numerous side reactions that deviate from this desired goal. The amount of deviation is typically defined as xylene loss, C8 aromatic loss, or C8 ring loss, each of which represents particular yield losses across the reactor. They are defined by the general equation:

$$Q \text{ loss} = (Q \text{ in feed} - Q \text{ in product})/(Q \text{ in feed}) * 100\% \tag{14.1}$$

where "Q" = mol% xylenes, mol% C8 aromatics, or moles of (C8 aromatics + C8 naphthenes), respectively.

These losses require additional feed to make a given amount of product. Feed costs are 80–95% of the final product cost, so yield losses significantly reduce profitability [59].

The most common by-product losses are due to transalkylation, dealkylation, saturation and cracking. Transalkylation results in toluene, trimethylbenzenes, methylethyl benzenes, benzene and C10As. These are the "best" by-products to have, because they are the easiest to react back into C8A in a transalkylation unit (if the aromatics complex is so equipped) without any loss of carbon atoms [59–61]. Dealkylation results in benzene, toluene, methane and ethane. The benzene and toluene are aromatics and represent valuable by-products, but the C1–C6 non-aromatics represent carbons that are lost from the complex as less valuable LPG and fuel gas.

Saturation of the aromatic ring generates C8 naphthenes. In EB isomerization systems, C8N are necessary intermediates in the EB to xylene pathway. During startup of EB isomerization systems, C8N production are typically high, leading to a large exotherm and significant hydrogen consumption. However, at steady state, C8N production is low, being only that which is needed to replace C8N that are lost to cracking or fractionation in other parts of the complex.

Cracked by-products are C7-paraffins. Due to the difficulty in converting such paraffins back into aromatics, cracked by-products are the least desirable by-products, and catalyst and process design go to great length to avoid their formation. Since it is easier to crack C8N than C8A, the level of C8N must be carefully balanced between their benefit to assist EB isomerization versus their contribution to cracking. In EB dealkylation systems, C8N are rapidly cracked by the stronger acid sites on such catalysts.

Historically, the earliest C8 aromatic isomerization catalysts tended to use amorphous supports with a halogen such as chloride or fluoride. Due to water sensitivity and corrosion issues, these were replaced by large-pore zeolites such as mordenite. The larger pore size was more favorable toward bimolecular transalkylation, whereas the chlorided alumina support tended to promote cracking. In both cases the net gain in xylenes across the catalyst was comparatively low. Newer catalysts have moved toward smaller pore materials, reducing transalkylation and thus reducing overall losses (see Section 14.4.2). Other by-products present in smaller quantities tend to be heavier aromatics and some double-bonded compounds. Species with double bonds can adsorb and/or dimerize across an adsorbent or other catalysts in the complex. Dimerized products are often removed by clay treating, wherein the double-bond species react with other aromatics to create heavy compounds that are removed from the complex by fractionation. The EB dealkylation systems generate significantly less olefinic by-products such that most commercial operations do not bother to clay treat the resulting heavy C8 aromatic recycle

14.4.1.3 Isomerization Reactions

Xylene Isomerization There are several mechanisms by which the three xylene isomers can be interconverted. The one that is of the greatest interest with respect to industrial applications is the so-called monomolecular or direct xylene isomerization route. This reaction is most commonly catalyzed by Bronsted acid sites in zeolitic catalysts. It is believed to occur as a result of individual protonation and methyl shift steps.

$$OX \leftrightarrow MX \leftrightarrow PX \tag{14.2}$$

With increasing reaction severity, the concentrations of the individual isomers approach their equilibrium values. The monomolecular route is the most effective for achieving high yields of PX, which is typically the most desirable for petrochemical applications. The schematic above shows the stepwise interconversion of OX to MX and MX to PX, which is consistent with a 1,2-methyl shift route. However, the results of kinetics studies provide some indications in favor of a reaction step that directly converts OX to PX [62]. It is not clear what the form of the reaction intermediate for this transformation is. Some *in situ* time-resolved spectroscopic methods have been used to look at how modification of zeolites like MFI affects the monomolecular mechanism by constraining the diffusion of MX [63].

A second method for interconverting xylene isomers occurs by way of a bimolecular mechanism. This can be catalyzed by materials such as Y-zeolite. The reaction proceeds through a bridged, diphenylmethane-like intermediate, so that the space requirements are more demanding than the monomolecular route (Figure 14.8).

The distribution of isomers formed as a result of this reaction tends to be higher in OX at the expense of PX, so catalysis through this route is less desirable from an industrial perspective. Comparisons of the monomolecular versus bimolecular reaction have been made, providing insight into the properties of zeolitic catalysts that favor one route over the other [64, 65]. Mechanistic aspects in MOR and TON structure zeolites have been evaluated using *ab initio* calculations, which suggest that the initiation step involves a defect site rather than an acidic proton [66]. It is

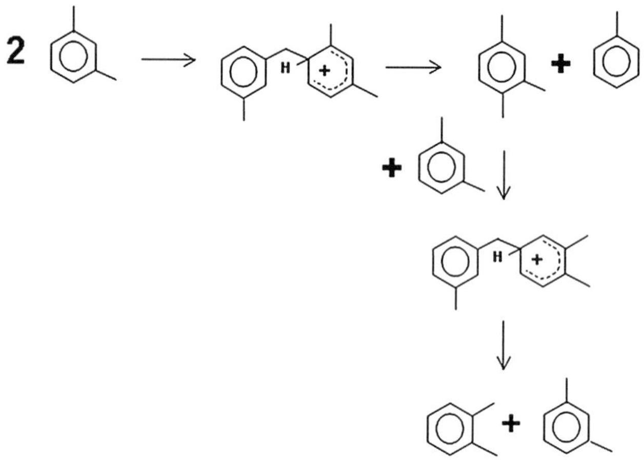

Figure 14.8 Xylene disproportionation.

interesting that, even on medium-pore zeolites like MFI, there is evidence that the bimolecular pathway contributes to xylene isomerization [67]. The MWW structure, for example, has three independent pore systems that have different tendencies for intramolecular and bimolecular xylene isomerization [68, 69].

A third method of performing xylene isomerization is through the same bifunctional catalyst mechanism used to convert EB to xylenes. This is covered in more detail in the next section. This is not generally a preferred method for xylene isomerization because it would require significant heat and hydrogen management. To achieve near-equilibrium isomer distributions would mean saturation of a large fraction of the feed stream to naphthenes with hydrogen. This is highly exothermic and control of the catalyst bed temperature would be difficult. A subsequent step is needed to remove hydrogen and supply heat to transform the product back to aromatics. Though isomerization of xylenes through naphthenic intermediates occurs to some extent when EB isomerization is effected, this mechanism is not typically employed for streams that contain predominantly xylene. There have been some studies of xylene hydroisomerization in the literature aimed at understanding the role of metal and zeolite pre-treatment in improving *para*-selectivity [70].

Ethylbenzene Isomerization Isomerization of EB requires both metal and acid function. Hydrogenation results in an intermediate naphthene. The acid function is required to isomerize the naphthene to a methyl-ethyl-substituted five-membered ring species that can further convert to a dimethyl-substituted six-membered ring naphthene. This can be dehydrogenated by the metal function to a xylene isomer, OX in the example shown in Figure 14.9.

The dotted lines indicate the formation of a new bond, while the double line indicates bond breaking. This mechanism can provide isomerization for all of the

Figure 14.9 Ethylbenzene isomerization.

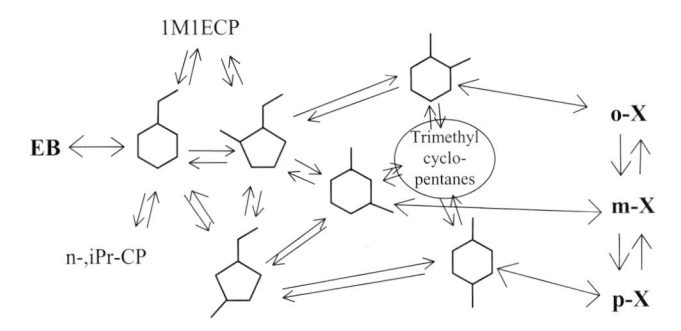

Figure 14.10 Naphthene interconversion.

C$_8$ aromatics. The schematic in Figure 14.10 shows how the saturated species can interconvert, including formation of mono-, di- and tri-substituted alkylcyclopentanes. The trimethylcyclopentanes are intermediates that can be invoked for isomerization among the different xylenes.

The process for isomerization of EB requires that some fraction of the feed be maintained in a saturated state, as described in Sections 14.4.1.1 and 14.4.1.2. The ability to isomerize the EB is affected by the naphthene concentration and constrained by the equilibrium ratio of EB/xylenes at the reaction temperature. Use of a pore-restricted molecular sieve can be used to eliminate more sterically demanding species from the reaction network and can effectively remove these from consideration. In this manner, one can achieve higher levels of a desired species, such as PX from EB, than could ordinarily be obtained by isomerization over a larger-pore zeolite.

In some applications, it is desirable to convert EB to benzene. Dealkylation reactions are described in the next section.

Side Reactions One of the major side reactions that occurs during isomerization of C$_8$ aromatics is transalkylation. This reaction produces species such as toluene, trimethylbenzene, methylethylbenzene, dimethylethylbenzene, benzene and diethylbenzene. The types and specific isomers of transalkylated products formed depend on the acidity and spatial constraints of the zeolitic catalyst used. These reactions can be controlled through modification of catalyst properties, especially pore size and external acidity, though these reactions are still among the major contributors to xylene losses.

Dealkylation of the ethyl group of EB is a side reaction that is undesirable if isomerization to xylene is desired. This is an acid-catalyzed reaction that is especially facile over pentasil zeolites. In the presence of metal and hydrogen, ethane is formed. This keeps the olefin from re-alkylating onto another aromatic ring. In circumstances where benzene co-production is of value, the catalyst type is chosen to facilitate this reaction.

The presence of metal may catalyze demethylation and can occur to some extent in catalysts where the metal function is under-passivated, as by incomplete sulfiding. This would convert valuable xylenes to toluene. The demethylation reaction is usually a small contributor to xylene loss. Metal also catalyzes aromatics saturation reactions. While this is a major and necessary function to facilitate EB isomerization, any aromatics saturation is undesirable for the process in which xylene isomerization and EB dealkylation are combined. Naphthenes can also be ring-opened and cracked, leading to light gas by-products. The zeolitic portion of the catalyst participates in the naphthene cracking reactions. Cracked by-products can be more prevalent over smaller pore zeolite catalysts.

14.4.2
C8 Aromatics Isomerization Catalysts

14.4.2.1 General Aspects

The catalysts used for isomerization of C_8 aromatics contain an acidic function to perform xylene isomerization and naphthene isomerization for EB conversion to xylenes. Relatively high metal activity is needed to maintain the naphthene/aromatic equilibrium that allows isomerization of EB. For conversion of EB by dealkylation, an acidic function is required along with metal activity capable of capturing and hydrogenating the ethylene by-product before it can re-alkylate another aromatic ring.

The acid function is typically supplied by a molecular sieve or zeolite. The size of the sieve used needs to be sufficient to allow diffusion of the xylene isomers, though large pore materials may be prone to higher yields of transalkylated by-products. Often, the zeolite is modified by treatment with cations (alkali [71] or alkaline earth), silicon [72] or phosphorus [73] reagents or pre-coking [74] in an attempt to either cut down on bimolecular side reactions or induce *para*-selectivity. Besides MFI zeolite, IM-5 is another zeolite reported to show good EB dealkylation activity [75]. The types used for EB isomerization need to be able to accommodate the naphthenic intermediates, which can be more sterically demanding than xylenes. The zeolites of interest for C_8 aromatics range in size from mordenite and omega to small structures like EUO and TON. One literature report suggests that EB isomerization is favored by the combination of large zeolite channels with mesoporosity and a relatively low density of Bronsted acid sites [76]. For EB isomerization, platinum is routinely used to provide the aromatics hydrogenation activity. Since just olefin, rather than aromatics, saturation is desired for xylene isomerization with EB dealkylation, a wider range of metals can be used. This includes bimetallic catalysts containing platinum, palladium, nickel, molybdenum and rhenium. Reference [77] provides a good contrast of the performance characteristics in xylene-containing feed of mordenite and MFI zeolite catalyst types, which isomerize and dealkylate EB, respectively.

Full catalyst formulations consist of zeolite, metal and a binder, which provides a matrix to contain the metal and zeolite, as well as allowing the composite to be shaped and have strength for handling. The catalyst particle shape, size and porosity can impact the diffusion properties. These can be important in facile reactions such as xylene isomerization, where diffusion of reactants and products may become rate-limiting. The binder properties and chemistry are also key features, as the binder may supply sites for metal clusters and affect coke formation during the process. The binders often used for these catalysts include alumina, silica and mixtures of other refractory oxides.

Catalyst stability with time on stream is an important characteristic. Acidic catalysts can be deactivated by basic poisons such as nitrogen. Carbonaceous species can build up on both metal and acid sites. These are the two prevalent mechanisms for catalyst deactivation. Other ways that a catalyst can be damaged, such as a temperature excursion, may be more likely to occur during the initial start up or during coke burning regenerations. Regeneration is discussed in the next section.

Coke formation during xylene isomerization has been studied using *in situ* infrared spectroscopy [78]. A study done on EB isomerization with a bifunctional catalyst containing EUO zeolite indicated that poor initial selectivity of the catalyst improves after a period of fast deactivation, during which micropores are blocked [79].

14.4.2.2 Regeneration

The build-up of carbonaceous residue on the catalyst can reduce its ability to effect the desired reactions. Regeneration of bifunctional catalysts for C_8 aromatics isomerization from the fouling that occurs with time on stream usually involves a step to burn the carbon off the catalyst. This is done with carefully controlled conditions of inlet temperature and oxygen concentration to avoid causing hydrothermal damage to the molecular sieve component. The oxidation step used to remove carbon from the catalyst may also impact the metal component present. Non-noble metal components may be rejuvenated by an oxidation treatment similar to that used to burn off carbon residue. However, metals like platinum may be subject to sintering during the carbon burn step. Noble metals may require a redispersion treatment, which could include the introduction of a gaseous chlorine-containing species. On alumina-bound catalysts, platinum dispersion can be readily restored through such a treatment. After removal of carbon and optional rejuvenation of the catalyst's metal function, the regenerated catalyst is typically returned to its reduced state under hydrogen pressure at elevated temperatures. Some catalysts may also require another post-treatment, such as sulfiding, to return to the desired state for a second cycle of feed processing.

14.4.3
C8 Aromatics Isomerization Processes

14.4.3.1 Process Variables

The key process variables for C8A isomerization are the temperature, pressure, weight- or liquid-hourly space velocity (WHSV or LHSV) and hydrogen partial pressure. Hydrogen partial pressure is usually not an explicitly measured variable. Rather, it is controlled through the hydrogen to hydrocarbon (H_2/HC) ratio or the gas to oil ratio and the hydrogen purity.

Temperature affects the reactions as described by the Arrhenius relationship. All reaction rates increase as temperature increases, but equilibrium considerations and reversibility of the reaction are also important. For example, as temperature increases, C8A isomerization approaches equilibrium, whereas cracking can continue to increase. For that reason, it is desirable to keep the commercial operating temperature as low as practical to achieve the desired activity. Excessively high temperatures (>400–450 °C) may also lead to sintering of the metal on the catalyst. This is more of a risk during regeneration than during typical operation. Both EB dealkylation and EB isomerization systems operate in the range 350–425 °C. Note that these bounds are usually due to equipment or operation constraints (maintaining vapor phase, furnace maximum output, etc.), not limitations of the catalyst.

In gas phase, pressure affects the concentration of the reactants and thus the reaction rates. Higher pressure favors reactions that reduce the total moles in the system, making dealkylation less favorable but saturation more favorable. Specifically, saturation increases the amount of C8 naphthenes available, which are key intermediates for EB isomerization reactions. Therefore, in EB isomerization systems it is common to adjust pressure as temperature increases, to maintain a constant level of C8Ns in the system – typically 80–200 psig (550–1380 kPa). EB dealkylation systems are not constrained by naphthene levels, so they typically operate at higher pressures to increase activity and allow high WHSV. Isomerization and transalkylation do not change the number of moles, so any effect of pressure on those two reactions is purely a concentration–activity effect.

In a liquid-phase system, pressure has no direct effect on activity – the pressure must simply be sufficient to maintain a liquid phase at the appropriate temperature. However, if there is dissolved H_2 in the liquid-phase system, then the pressure affects the solubility and can affect the activity or stability of the system.

In vapor-phase systems, the hydrogen purity of the recycle gas affects the H_2 partial pressure within the reactor. The H_2 partial pressure affects the concentration of C8Ns and can also affect stability, since in general increased hydrogen reduces the rate of deactivation due to saturation of coke precursors. If there is significant production of methane and ethane, these are present in the recycle hydrogen. As catalyst selectivity has improved, fewer light ends are produced and hydrogen purity has increased.

14.4.3.2 Commercial Catalysts

There are two distinct types of catalysts used for C_8 aromatics isomerization. These differ in the manner that EB is converted. The catalysts that are used to isomerize EB tend to have higher metal activity and larger-pore molecular sieve components capable of accommodating naphthenic species without cracking them. EB isomerization type catalysts tend to require more regular regenerations than EB dealkylation catalysts. Depending upon the operating severity and the separation unit's inability to handle higher levels of EB in the feed, EB isomerization type catalysts may require regeneration as often as every couple of years. Milder operations with high separation capacity have stretched out the need for regeneration more similar to the EB dealkylation catalyst types, which are typically at 4+ years. The earliest catalysts used in this service were similar in nature to naphtha reforming catalysts and were operated at severities approaching those used in reforming. As zeolitic materials were developed for industrial applications, these were found to be useful for EB and xylene isomerization. Starting in the 1970s, mordenite zeolite-based catalysts came into use. Over the next few decades, new catalysts were developed based on structures with smaller pore sizes. These would include modifications of mordenite [80] and other zeolite framework types. The patent literature suggests that Axens' Oparis catalysts may be based on an EUO structure or mixtures of EUO and NES-type zeolites [81, 82]. Offerings in Chinese markets may be based on modified mordenite zeolites. There have been several generations of catalysts mentioned in patents by UOP. Structures covered range from mordenite in early patents to non-zeolite molecular sieves and Ga-in-framework zeolites (MFI, MTW). Recent catalyst performance is described in [83].

The catalysts for xylene isomerization with EB dealkylation are dominated by MFI zeolite. The de-ethylation reaction is particularly facile over this zeolite. There have been several generations of catalyst technology developed by Mobil, now ExxonMobil [84]. The features in their patents include selectivation and two-catalyst systems in which the catalysts have been optimized separately for de-ethylation of EB and xylene isomerization [85–87]. The crystallite size used for de-ethylation is significantly larger than in the second catalyst used for xylene isomerization. Advanced MHAI is one example. The Isolene process is offered by Toray and their catalyst also appears to be MFI zeolite-based, though some patents claim the use of mordenite [88, 89]. The metal function favored in their patents appears to be rhenium [90]. Bimetallic platinum catalysts have also been claimed on a variety of ZSM-type zeolites [91]. There are also EB dealkylation catalysts for the UOP Isomar™ process [92]. The zeolite claimed in UOP patents is MFI in combination with aluminophosphate binder [93].

14.4.3.3 Modeling/Optimization of Commercial Units

Due to the growing demand for C8 aromatics and the competitive nature of the business, there is great incentive to optimize production through better prediction of the isomerization processes. Roughly 90% of the flow through the isomerization reactor consists of the four C8A isomers. This small number of components makes it simpler to build a solid database of activity, selectivity and stability across a given catalyst type, as opposed to broader boiling range processes which often require lumping of components. The number of components is more complex for EB isomerization systems, since saturation of the four C8A isomers and subsequent ring rearrangement yield over 20 different C8N with five- and six-ring backbones. For that reason there are more studies available for xylene isomerization than EB isomerization [94–98].

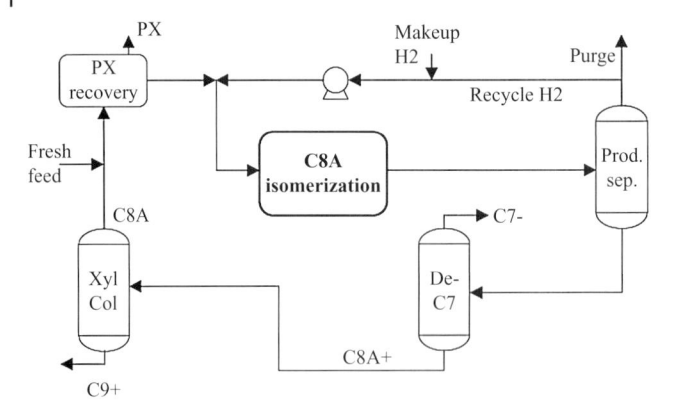

Figure 14.11 C8 aromatics process flow diagram.

14.4.3.4 Process Flow Schemes

A typical commercial C8A process flow diagram is shown in Figure 14.11. The C8A isomerization reactor is usually combined with a recovery unit for one or more xylene isomers. PX recovery by adsorption or crystallization is the most common. For the sake of this discussion, PX recovery is assumed, although MX recovery by adsorption is possible, as are OX or EB recovery by fractionation.

The PX depleted feed is combined with recycled hydrogen and make-up hydrogen. The make-up hydrogen replaces the small amount of hydrogen consumed in the isomerization reactor by saturation, dealkylation or cracking reactions, so the amount needed depends on the selectivity of the catalyst. This combined mixture is heated and then sent to the isomerization reactor. The effluent from the isomerization reactor is sent to a product separator, which recycles most of the hydrogen back to the reactor. A small amount of gas may be purged to prevent build-up of light gases in the recycle. Improved selectivity catalysts can reduce or eliminate the purge.

After the separator, the liquid product is sent to a deheptanizer to remove toluene, benzene and other lighter products. If this is an EB isomerization-style process, the deheptanizer operation may be constrained by the need to send the C8N to the bottoms, which also results in more toluene in the bottoms than would be present in an EB dealkylation system (which does not require C8N recirculation). The elevated toluene is not generally detrimental to catalyst performance, primarily acting as a diluent, although in some cases it may actually be beneficial, by pushing the toluene + C9A transalkylation equilibrium back toward C8A.

Following the deheptanizer, the recycled C8A+ fraction is combined with fresh C8A feed, then sent to a xylene splitter. If OX is a desired product, the xylene splitter sends much of the OX into the bottoms, which is sent to a secondary column to separate OX from C9A+. If OX is not a product, then only the C9A+ are sent into the xylene column bottoms. In either case, the xylene column overhead is a high-purity C8A stream that is sent to the PX recovery unit.

The PX depleted feed then repeats the cycle into the C8A isomerization unit. In this way, the C8A feed is eventually converted into the desired product, PX.

14.4.4
Future Developments

While higher-activity catalysts are valuable, future improvements will likely focus more on selectivity improvements. Higher activity means less catalyst needs to be purchased, which is an infrequent, rather than continuous, investment. Selectivity improvements, in contrast, represent a reduction in the feed needed for a given amount of product, so that is a reduction in the feed needed for every day of operation. Since feedstock is often the largest contribution to product cost, selectivity improvements can be quite valuable. Unlike activity improvements, which may shrink the size of the isomerization reactor, better selectivity can reduce the size of all units in the system due to reduced feed and reduced recycle. In an existing unit, the effective throughput would increase.

As noted in Section 14.4.2.1, catalytic selectivity is influenced by many things – the particular zeolite used, the binder, the metals used, the conditions and treatments during catalyst finishing and the operating conditions. The first zeolites used for C8A isomerization were larger-pore materials like MOR, which more easily allowed bimolecular reactions and gave significant transalkylation losses. Smaller-pore zeolites such as MFI and EUO have since given reduced transalkylation. Increasingly, catalysts use still smaller pore sizes, or treatments such as coking or silanation to reduce the pore sizes [99]. However, efforts to tune the pore sizes must be balanced with the eventual diffusion problems this could cause, most notably with OX, which has the largest kinetic radius of the C8A.

Precious metals (Section 14.4.2.1) are generally more expensive than the other catalyst components, so reduced metal content is an ongoing effort. This is more of an issue with EB isomerization catalysts since they have higher metal content than EB dealkylation catalysts.

Since the chemistry of xylene isomerization and EB isomerization follow different pathways, there have also been attempts to optimize the two reactions independently, either by a simple mixture of two catalysts, two catalysts separated within a single reactor, or two entirely different beds operating at different conditions [93, 99]. These changes are usually targeted to improving the selectivity or approach to equilibrium of the traditional processes. Some go beyond this, attempting to use a true molecular "sieving" approach to allow PX concentrations beyond what is considered equilibrium in the normal sense (greater than 24% at typical industrial conditions) [100, 101]. This may also be accomplished through the use of membranes [102]. This could be of great benefit to a PX production, since it will improve the efficiency of the PX separation unit.

Greater cycle lengths are also on the horizon. Where one-year cycle lengths between regenerations were typical for the early generations of EB isomerization catalysts, customers now expect two-year life as a minimum, and the demand for even longer cycle lengths will continue to drive catalyst innovations. EB dealkylation catalysts have been reported to have cycle lengths of 10+ years.

14.4.5
C8 Aromatics Isomerization Summary

C8 aromatics are the fundamental building blocks for many commercially important materials. Demand for *para*-xylene, which is used to create polyesters, has risen significantly and will continue to grow for many years. Zeolite-based catalysts are an essential part of the isomerization technology required to support this growth. Both types of C8A isomerization catalysts – EB isomerization type and EB dealkylation type – have demonstrated increased activity and selectivity since they were introduced over 20 years ago. Improvements in the understanding of zeolite synthesis and application will continue to be critical, both for improvements in the existing technologies and to enable innovative technologies in the future.

References

1 Maxwell, I.E. (1987) Zeolite catalysis in hydroprocessing technology. *Catal. Today*, **1**, 385–413.
2 Rice, L.H., Kuchar, P.J., and Gosling, C.D. (1999) Tutorial: Upgrading light naphtha and refinery light ends. *2nd International Conference on Refinery Processing, AIChE Spring National Meeting*, pp. 153–164.
3 Cusher, N.A. (1997) UOP TIP and once-through zeolitic isomerization processes, in *Handbook of Petroleum Refining Processes*, 2nd edn (ed. R.A. Meyers), McGraw-Hill, pp. 9.29–9.39.
4 Sie, S.T. (2008) Isomerization, in *Handbook of Heterogeneous Catalysis*, vol. 6, 2nd edn (eds G. Ertl, H. Knözinger, F. Schüth, and J. Weitkamp), Wiley-VCH Verlag GmbH, pp. 2809–2830.
5 (a) Corma, A. and Martínez, A. (2005) Zeolites in refining and petrochemistry, in *Studies in Surface Science and Catalysis*, vol. 157 (eds J. Čejka and H. van Bekkum), Elsevier B. V., pp. 337–366. (b) Jiménez, C., Romero, F.J., and Gómez, J.P. (2001) Isomerization of paraffins. *Rec. Res. Devel. Pure Appl. Chem.*, **5**, 1–21.
6 Gosling, C., Schimizu, T., Imai, T., Rosin, R., and Bullen, P. (1997) Revamp opportunities for isomerization units. *Pet. Tech. Q.*, **2** (4), 55–59.
7 Weitkamp, J. (1980) New evidence for a protonated cyclopropane mechanism in catalytic isomerization of n-alkane, in *Studies in Surface Science and Catalysis*, vol. 17 (eds T. Seiyama and K. Tanabe) Elsevier, pp. 1404–1405.
8 Kazansky, V.B. (1994) The catalytic site from a chemical point of view, in *Studies in Surface Science and Catalysis*, vol. 85 (eds J.C. Jansen, M. Stocker, H.G. Karge and J. Weitkamp), Elsevier, pp. 251–271.
9 van de Runstraat, A., van Grondelle, J., and van Santen, R.A. (1997) Microkinetics modeling of the hydroisomerization of n-hexane. *Ind. Eng. Chem. Res.*, **36**, 3116–3125.
10 Holló, A., Hancsók, J., and Kalló, D. (2002) Kinetics of hydroisomerization of C_5-C_7 alkanes and their mixtures over platinum containing mordenite. *Appl. Catal. A.*, **229**, 93–102.
11 Allain, J.F., Magnoux, P., Schulz, P., and Guisnet, M. (1997) Hydroisomerization of n-hexane over platinum mazzite and platinum mordenite catalysts: kinetics and mechanism. *Appl. Catal. A.*, **152**, 221–235.

12 Calero, S., Schenk, M., Dubbeldam, D., Maesen, T.L.M., and Smit, B. (2004) The selectivity of n-hexane hydrocarversion on MOR-, MAZ-, and FAU-type zeolites. *J. Catal.*, **228**, 121–129.

13 Chao, K.-J., Wu, H.-C., and Leu, L.-J. (1996) Hydroisomerization of light normal paraffins over series of platinum-loaded mordenite and beta catalysts. *Appl. Catal. A.*, **143**, 223–243.

14 Guisnet, M., Alvarez, F., Giannetto, G., and Perot, G. (1987) Hydroisomerization and hydrocracking of n-heptane on PtH zeolites. Effect of the porosity and of the distribution of metallic and acid sites. *Catal. Today*, **1**, 415–433.

15 Raybaud, P., Patrigeon, A., and Toulhoat, H. (2001) The origin of the C_7-hydroconversion selectivities on Y, β, ZSM-22, ZSM-23, and EU-1 zeolites. *J. Catal.*, **197**, 98–112.

16 Koradia, P., Kiovsky, J.R., and Asim, M.Y. (1980) Optimization of SiO_2/Al_2O_3, mole ratio of mordenite for n-pentane isomerization. *J. Catal.*, **66**, 290–293.

17 Guisnet, M. and Fouche, V. (1991) Isomerization of n-hexane on platinum dealuminated mordenite catalysts II. Kinetic study. *Appl. Catal.*, **71**, 295–306.

18 van Donk, S., Broersma, A., Gijzeman, O.L.J., Tromp, M., van Bokhoven, M.T., Bitter, J.H., and de Jong, K.P. (2001) Combined diffusion, adsorption, and reaction studies of n-hexane hydroisomerization over Pt/H-mordenite in an oscillating microbalance. *J. Catal.*, **204**, 272–280.

19 Chica, A. and Corma, A. (1999) Hydroisomerization of pentane, hexane, and heptanes for improving the octane number of gasoline. *J. Catal.*, **187**, 167–176.

20 Jao, R.-M., Leu, L.-J., and Chang, J.-R. (1996) Effects of catalyst preparation and pretreatment on light naphtha isomerization over mordenite-supported Pt catalysts. *Appl. Catal. A.*, **135**, 301–316.

21 Blomsma, E., Martens, J.A., and Jacobs, P.A. (1997) Isomerization and hydrocracking of heptane over bimetallic bifunctional PtPd/H-beta and PtPd/USY zeolite catalysts. *J. Catal.*, **165**, 241–248.

22 Weyda, H. and Kohler, E. (2003) Modern refining concepts – an update on naphtha-isomerization to modern gasoline manufacture. *Catal. Today*, **81** (1), 51–55.

23 Wu, Z. (2007) Industrial application of catalyst for C_5/C_6 isomerization. *Jingxi Shiyou Huagong Jinzhan*, **8** (6), 42–46.

24 Kuznetsov, P.N., Kuznetsova, L.I., Tverdokhlebov, V.P., and Sannikov, A.L. (2005) C_4-C_6 n-alkane isomerization catalysts. *Khimicheska Tekhnologiya*, **2**, 7–14.

25 Nijmeijer, A. and Den Otter, G.J. (2007) Process for the separation and isomerization of n-paraffins in the presence of a zeolite membrane. WO 2007/135042.

26 Boyer, C.C., Loescher, M.E., Xu, J., and Dautzenberg, F.M. (2006) Normal heptane isomerization. US 2006/0270885.

27 (a) Maesen, T. and Harris, T. (2006) Process for producing high RON gasoline using CFI zeolite. US 7037422. (b) Maesen, T. and Harris, T. (2006). Process for producing high RON gasoline using ATS zeolite. US 7029572.

28 Jensen, S.D. and Tamm, D.C. (2006) Impact of ethanol mandate on U.S. gasoline supply, NPRA March 2006, AM-06-38, Salt Lake City, UT.

29 Xu, F., Bauer, L.J., Gillespie, R.D., Bricker, M.L., and Bradley, S.A. (2008) Modified Pt/Ru for ring opening and process using the catalyst. US 7345214.

30 Rosin, R.R., Stine, M.A., Anderson, G., and Hunter, M.J. (2004) New solutions for light paraffin isomerization. NPRA March 2004, AM-04-46, San Antonio, TX.

31 Watanabe, K., Chiyoda, N., and Kawakami, T. (2008) Development of new isomerization process for petrochemical by-products. 18th Saudi Arabia-Japan Joint Symposium, Dhahran, Saudi Arabia, November 16–17, 2008.

32 Damon, J.-P., Bonnier, J.-M., and Delmon, B. (1976) *J. Colloid Interface Sci.*, **55**, 381.

33 Damon, J.-P., Delmon, B., and Bonnier, J.-M. (1977) *J. Chem. Soc. Faraday Trans. I*, **73**, 372.

34 Fleck, R. and Wight, C.G. (1956) US Patent 2,988,578.
35 Hamilton, D.M., Jr. (1988) US patent 4,727,203.
36 Vora, B.V. (2000) US patent 6,156,947.
37 Butler, A.C., and Nicolaides, C.P. (1993) *Catal. Today*, **18**, 443–471.
38 Corma, A. (1994) Roles of the Zeolite Catalysts in the New Refing Strategies, in Studies in Surface Science and Catalysis, vol. 83 (eds T. Hattori and T. Yashima), Elsevier, pp. 461–472.
39 O'Conner, C.T., van Steen, E., and Dry, M.E. (1996) New Catalytic Applications of Zeolites for Petrochemicals, in Studies in Surface Science and Catalysis, vol. 102 (eds H. Chon, S.I. Woo, and S.-E. Park), Elsevier, pp. 323–362.
40 Maxwell, I.E. and Stork, W.H.J. (2001) Hydrocarbon Processing with Zeolites, in Studies in Surface Science and Catalysis, vol. 137 (eds R. van Bekkum, E.M. Flanagan, P.A. Jacobs, and J.C. Janson), Elsevier, pp. 747–819.
41 Kilpatrick, J.E., Prosen, E.J., Pitzer, K.S., and Rossini, F.D. (1946) *J. Res. Nat. Bur. Stand.*, **36**, 559.
42 Sikkenga, D.L. (1985) US Patent 4,550,091.
43 de Clippeleir, G.E.M.J., Leeuw, S.P., Cahen, R.M., and Debras, G.L.G. (1985) German patent 3,512,057.
44 Barri, S.A.I., Walker, D.W., and Tahir, R. (1987) European patent 247,802.
45 Rahmin, I., Huss, A., Lissy, D.N., Klocke, D.J., and Johnson, I.D. (1992) US patent 5,157,194.
46 Haag, W.O., Harandi, M.N., and Owen, H. (1992) US patent 5,132,467.
47 Gajda, G.J. (1991) US patents 5,057,635 and 5,132,484 (1992).
48 Gaffney, A.M. and Jones, A. (1992) US patent. 5,107,050
49 Grandvallet, P., de Jong, K.P., Mooiweer, H.H., Kortbeek, A.G.T.G., and Kraushaar-Czarnetzki, B. (1992) European patent 501,577.
50 Davis, S. (1996) UOP Olefin Isomerization, in *Handbook of Petroleum Refining Processes* (ed. R.A. Meyers), McGraw-Hill, New York, pp. 13.13–13.17.
51 Sawicki, R.A., Pellet, R.J., Casey, D.G., Kessler, R.V., Huang, H.-M., O'Young, C.-L., and Kuhlmann, E.J. (1998) *Intl. J. Hydrocarbon Eng.*, **3** (3), 44, 2246–2248, 50.
52 Wise, J.B. and Powers, D. (1994) *ACS Symp. Ser.*, **552** (Environmental Catalysis), 273–285.
53 Dessau, R.M. and Olson, D.H. (1987) US patent 4,650,917.
54 Mattox, W.J. (1965) US patent 3,214,487.
55 Gajda, G.J. and Barger, P.T. (1995) US patent 5,463,161.
56 Weitkamp, J., Kumar, R., and Ernst, S. (1989) *Chem. Ing. Teck.*, **61** (9), 731–733.
57 Akiyama, N. and Mori, M. (1990) WO patent 9,003,354.
58 Miller, S.J. (1990) WO patent 9,009,362.
59 Nexant/ChemSystems (2006) Nexant/ChemSystems PERP Report Xylenes, 05/06-8, August.
60 Ali, M.A., Al-Saleh, M.A., Aitani, A., Ali, S.A., Al-Nawad, K., Okomoto, T., and Ishikawa, K. (2005) Transalkylation of heavy aromatics for enhanced xylene production. Proceedings of 15th Saudi-Japan Joint Symposium, Dhahran, Saudi Arabia, November 27–28.
61 Jeanneret, J.J. (1996) *Handbook of Petroleum Refining Processes* (ed. R.A. Meyers), McGraw-Hill, New York., pp. 2.55–2.62.
62 Al-Khattaf, S., Tukur, N.M., and Al-Amer, A. (2005) *Ind. Eng. Chem. Res.*, **44**, 7957.
63 Zheng, S., Jentys, A., and Lercher, J.A. (2006) *J. Catal.*, **241**, 304.
64 Morin, S., Gnep, N.S., and Guisnet, M. (1996) *J. Catal.*, **159**, 296.
65 Corma, A. and Sastre, E. (1991) *J. Catal.*, **129**, 177.
66 Demuth, T., Raybaud, P., Lacombe, S., and Toulhoat, H. (2004) *J. Catal.*, **222**, 323.
67 Bauer, F., Bilz, E., and Freyer, A. (2005) *Appl. Catal. A. Gen.*, **289**, 2.
68 Laforge, S., Martin, D., and Guisnet, M. (2004) *Appl. Catal. A. Gen.*, **268**, 33.
69 Laforge, S., Martin, D., and Guisnet, M. (2004) *Micropor. Mesopor. Mater.*, **67**, 235.
70 Aboul-Gheit, A.K., Abdel-Hamid, S.M., and El-Desouki, D.S. (2001) *Appl. Catal. A. Gen.*, **209**, 179.

71 Guisnet, M., Moreau, F., and Gnep, N.S. (2002) *Appl. Catal. A. Gen.*, **230**, 253.
72 Rao, B.S., Shaikh, R.A., and Ramaswamy, A.V. (2000) *ACS Symp. Ser.*, **738**, 225–235.
73 Angelescu, E., Constantinescu, F., Gârgel-Ropot, M., and Georghe, G. (1996) *Prog. Catal.*, **5**, 25.
74 Bauer, F., Chen, W.-H., Ernst, H., Huang, S.-J., Freyer, A., and Liu, S.-B. (2004) *Micropor. Mesopor. Mater.*, **72**, 81.
75 Serra, J.M., Guillon, E., and Corma, A. (2004) *J. Catal.*, **227**, 459.
76 Fernandez, L.D., Monteiro, J.L.F., Sousa-Aguiar, E.F., Martinez, A., and Corma, A. (1998) *J. Catal.*, **177**, 363.
77 Silva, J.M., Ribeiro, M.F., Ramoa Ribeiro, F., Benazzi, E., and Guisnet, M. (1995) *Appl. Catal.*, **125**, 15.
78 Thibault-Starzyk, F., Vimont, A., and Gilson, J.-P. (2001) *Catal. Today*, **70**, 227–241.
79 Moreau, F., Moreau, P., Gnep, N.S., Magnoux, P., Lacombe, S., and Guisnet, M. (2006) *Micropor. Mesopor. Mater.*, **90**, 327.
80 Benazzi, E., De Tavernier, S., Beccat, P., Joly, J.F., Nédez, C., Choplin, A., and Basset, J.M. (1994) *ChemTech*, **24**, 13–18.
81 Rouleau, L., Kolenda, F., Merlen, E., and Alario, F. (2004) US Patent 6,723,301, assigned to Institute Francais du Petrole.
82 Guillon, E., Sanchez, E., and Lacombe, S. (2007) US Patent Application 2007/0167660 from Institute Francais du Petrole, 7/19/2007.
83 Carimati, A., Lim, C., Fung, J.L.P., and Marr, G. (2007) *PTQ Catal.*, **12** (2), 59–62.
84 Mohr, G. (2002) XyMax[sm]: ExxonMobil state-of-the-art xylenes isomerization technology, Pre-Print Archive – American Institute of Chemical Engineers, [Spring National Meeting], New Orleans, LA, United States.
85 Olsen, D.H. and Haag, W.O. (1979) US Patent 4,159,282, assigned to Mobil Oil Corporation.
86 Beck, J.S., Borghard, W.G., Chester, A.W., Kennedy, C.L., and Stern, D.L. (2007) US Patent 7,238,636, assigned to ExxonMobil Chemical Patents Inc.
87 Stern, D.L. (2007) US Patent 7,247,762, assigned to ExxonMobil Chemical Patents Inc.
88 Iwayama, K., Ebitani, A., Inoue, T., and Kani, A. (1984) US Patent 4,467,129, assigned to Toray Industries, Inc.
89 Iwayama, K. and Inoue, T. (1983) US Patent 4,409,413, assigned to Toray Industries, Inc.
90 Tada, K., Minomiya, E., and Watanabe, M. (1991) US Patent 5,004,855, assigned to Toray Industries, Inc.
91 Onodera, T., Sakai, T., Yamasaki, Y., and Sumitami, K. (1982) US Patent 4,331,822, assigned to Teijin Petrochemical Industries, Ltd.
92 Ebner, T., O'Neil, K., and Silady, P. (2002) UOP's new xylene isomerization catalysts (I-300™ series). Pre-Print Archive – American Institute of Chemical Engineers, [Spring National Meeting], New Orleans, LA, United States, 2002.
93 Sharma, S.B. and Imrie, A.J. (2000) U.S. patent 6,576,581, assigned to UOP LLC.
94 Guisnet, M., Gnep, N.S., and Morin, S. (2000) *Micropor. Mesopor. Mater.*, **35–36**, 47–59.
95 Al-Khattaf, S., Tukur, N.M., and Al-Amer, A. (2005) *Ind. Eng. Chem. Res.*, **44** (21), 7957–7968.
96 Norman, G.H., Shigemura, D.S., and Hopper, J.R. (1976) *Ind. Eng. Chem. Prod. Res. Dev.*, **15** (1), 41–45.
97 Polinski, L.M. and Baird, M.J. (1985) *Ind. Eng. Chem. Prod. Res. Dev.*, **24** (4), 540–544.
98 Deshayes, A.L., Miro, E.E., and Horowitz, G.I. (2006) *Chem. Eng. J.*, **122**, 149–157.
99 Abichandani, J.S., Beck, J.S., Olson, D.H., Reischman, P.T., Stern, D.L., and Venkat, C.R. (1997) U.S. Patent 5,689,027, assigned to Mobil Oil Corporation.
100 Zhou, L., Maher, G.F., Johnson, J.A., and Bauer, J.E. (2008) U.S. Patent 7,368,620, assigned to UOP LLC.

101 Buchanan, J.S., Feng, X., Mohr, G.D., and Stern, D.L. (2003) U.S. patent 6,660,896, assigned to ExxonMobil Chemical Patents Inc.

102 Miller, S.J., Munson, C.L., Kulkarni, S.S., and Hasse, D.J. (2002) US Patent 6,500,233 B1, assigned to Chevron, U.S.A, Inc., and Medal, L.P.

15
Processes on Industrial C–C Bond Formation

Deng-Yang Y. Jan and Paul T. Barger

15.1
Introduction

The acid properties of zeolites have made them valuable catalysts for industrial processes involving the formation of new carbon–carbon bonds. This chapter describes some of the key technologies involving carbon–carbon bond formation in which zeolites are either used as the commercial catalyst of choice or being actively researched as a potential improvement over the existing catalyst. Zeolite catalysts are well established in industrial olefin oligomerization and aromatic alkylation and transalkylation processes. The current technologies in these areas are reviewed. More recently, zeolites or other molecular sieves, such as silicoaluminophosphates (SAPOs), have found commercial application for paraffin cyclization to aromatics (in conjunction with metal-function to aid dehydrogenation) and the conversion of methanol to hydrocarbons ranging from light olefins to aromatics. The molecular sieve catalysts employed in the emerging technologies and their key characteristics are summarized. The replacement of liquid acid catalysts for the alkylation of paraffins with olefins with a solid catalyst, such as a zeolite, has been a topic of significant interest over the past 15 years. Although no zeolite-based process has been developed to the full commercial scale to date, the recent advances in this area are discussed.

15.2
Olefin Oligomerization

15.2.1
$C_2/C_3/C_4$ Olefin Oligomerization

15.2.1.1 Process Chemistry: Feeds, Products and Reactions
The light olefins are derived from FCC (C_3–C_4) olefin or high temperature F–T (C_3–C_6) olefin processes. The feeds contain various levels of contaminants such as dienes, sulfur-, oxygen- and nitrogen-containing compounds, which significantly

Zeolites in Industrial Separation and Catalysis. Edited by Santi Kulprathipanja
Copyright © 2010 WILEY-VCH Verlag GmbH & Co. KGaA, Weinheim
ISBN: 978-3-527-32505-4

shorten catalyst life, if not treated. Water is frequently introduced into the feed streams at low levels and is claimed to improve product selectivity and catalyst life. The oligomerization is aimed to produce higher alkenes for fuel (gasoline and diesel) and as intermediates for producing plasticizers, surfactants and freeze-point depressants for lube. The degrees of oligomerization and branching are controlled to meet product specifications. This is done by controlling the operating pressures, temperatures, catalysts and the reactivity of olefin feed components. Optionally the product is hydrogenated to the corresponding paraffin, part of which is recycled to the oligomerization reactor to control the heat of reaction and to improve the catalyst stability by minimizing coke formation on the catalyst surface. The UOP InAlk™ process recycles part of hydrogenated products to maintain the process operating in liquid phase conditions to improve catalyst stability and selectivity [1] possibly by mitigating the hydride transfer reaction as shown by Johan Martens, et al. [2]. The oligomerization of C_2 olefin from FCC off-gas is not practiced commercially, since it is typically diluted with methane and ethane and is saddled with relatively high levels of contaminants.

15.2.1.2 Catalysts

Commercial oligomerization processes use solid phosphoric acid (SPA), as shown in the Figure 15.1, and acid resins mainly due to their efficiency in selectively converting C_3–C_4 olefin to C_8–C_9 olefin and iso-C_4 olefin to C_8 olefin, respectively, and the low catalyst cost. A low level of H_2O or oxygenates is typically injected into the feed stream to increase selectivity of primary dimerization possibly via either a more facile desorption of products or by lowering the surface acid site density. However, the catalysts are not regenerated and need to be changed out. Nickel-containing catalysts have also been used in both homogenous and heterogeneous catalyst systems by IFP.

The use of zeolite–base catalysts have only emerged relatively recently. Mobil olefin to gasoline/distillate (MOGD) developed by Mobil and conversion of olefin to distillate (COD) of PetroSA use ZSM-5 zeolite to convert olefins to gasoline and

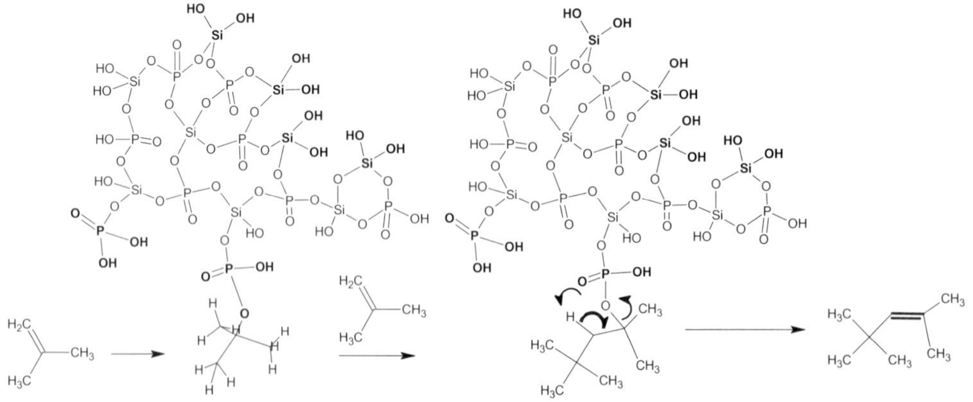

Figure 15.1 Olefin oligomerization using SPA catalyst.

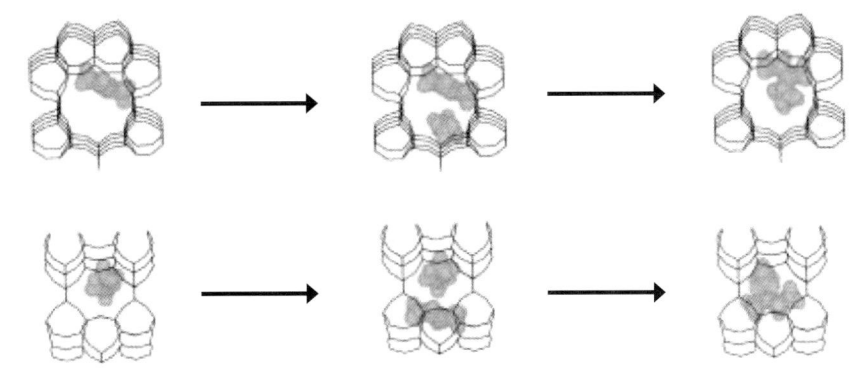

Figure 15.2 Crosswise arrangement of trans-but-2-ene molecules in ZSM-57 pores giving rise to the formation trans-3,4-dimethylhex-2-ene views along ten-ring (above) and eight-ring (below) channels.

diesel fuels. The recently developed COD process by PetroSA purportedly uses COD-900 catalyst based on a proprietary zeolite. IFP Polynaphtha and Selectopol processes also uses a zeolite-based catalyst designated as IP501. ExxonMobil olefins to gasoline (EMOGAS) appears to be based on 10-MR zeolite and claims to give good quality gasoline and catalyst stability.

15.2.1.3 Physicochemical Characterization of Active Sites

The high C_8–C_9 olefin selectivity of SPA and acid resins is probably due to the ready desorption of relatively hydrophobic C_8–C_9 olefin product from the hydrophilic surface. Olefin oligomerization by zeolites utilizes the microporous pores to effectively control the degrees of branching, but is not particularly effective in controlling the degrees of oligomerization in most zeolite. ZSM-57, however, has been shown to give high C_8–C_9 olefin selectivity (Figure 15.2)[3]. It was proposed that selective C_8 formation from 2-C_4 olefin is due to the favorable shape selectivity of the reaction intermediate (crosswise) derived from intersection of 8-MR and 10-MR channels. Specifically, the side pocket (lobe) at the intersection of 8-R with 10-R provides a site for an adsorbed olefin to interact with another olefin to proceed with the selective dimerization reaction [3]. Furthermore, this space at the intersection is limited and does not favor the interaction of C_4 olefin with the primary C_8 olefin product. Consequently, ZSM-57 showed high activity and C_8 olefin selectivity.

15.3 Paraffin/Olefin Alkylation

15.3.1 Motor Fuel Alkylation

The alkylation of light isoparaffins, such as isobutane, with light olefins, such as propene and butenes, is a key process in modern gasoline production. The C_7–C_8

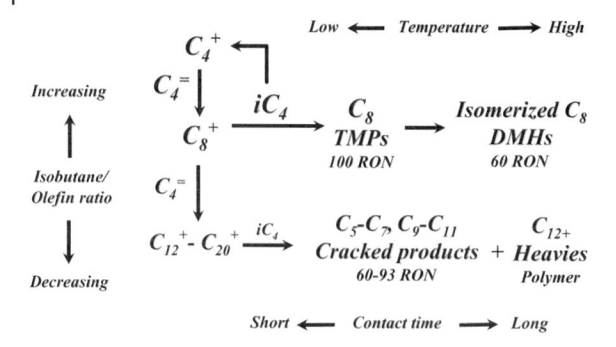

Figure 15.3 Paraffin alkylation reaction pathways.

alkylate formed is an ideal gasoline blending component that is clean-burning with high octane and low vapor pressure. The process also allows the conversion of light refinery by-products back into the gasoline boiling range.

15.3.1.1 Process Chemistry: Feeds, Products and Reactions

Paraffin alkylation involves the acid-catalyzed addition of an olefin to a branched paraffin to give a highly branched, paraffinic product. The representative reaction is that of isobutane with 2-butene to give 2,2,4-trimethylpentane:

$$\text{(isobutane)} + \text{(2-butene)} \longrightarrow \text{(2,2,4-trimethylpentane)} \tag{15.1}$$

The alkylation proceeds via carbenium ion intermediates initially formed by olefin protonation, in a catalytic cycle involving dimerization with a free olefin followed by hydride transfer from the isoparaffin to form a heavier paraffin and the original carbenium ion as shown in Figure 15.3. In addition to these steps, isomerization of the olefin feed, that is, 1-butene to 2-butene, and the alkylate product, in particular conversion of trimethylpentanes (TMPs) to dimethylhexanes (DMHs), can occur and impact the isomer distribution of the final alkylate product. Skeletal isomerization can have profound effect on the octane number of the final alkylate. Maintaining a high concentration of TMPs with research octane numbers (RONs) of 100–109 is necessary to obtain an alkylate which is valuable for gasoline blending. Secondary oligomerization which builds heavier C_{9+} carbenium ions is another important side reaction. These species and the products arising from them, including dienes and aromatics, can strongly interact with the acidic sites and cause catalyst deactivation. One of the key challenges for a motor fuel alkylation catalyst and process is balancing the rates of the oligomerization and hydride transfer steps to minimize the formation of heavies. In practice, it has not been possible to avoid some secondary oligomerization from occurring and an efficient procedure for removing the heavy by-products from the catalyst is required to maintain commercially viable activity stability.

15.3.1.2 Catalysts

At the current time, all industrial motor fuel alkylation units employ the liquid-phase acid catalysts HF or H_2SO_4 that were developed in the 1930s and 1940s. These acids have high activity at low temperatures (35 °C for HF, 0 °C for H_2SO_4) and give an alkylate product with 93–96 RON, depending on the feedstock used. Both processes use two-phase, liquid-filled reactors that are highly engineered (i.e., use of feed injectors and reactor internals) to ensure thorough mixing of the acid and hydrocarbon phases. In addition to the mineral acid, the acid phase contains 5–15% of organic diluent (OD), consisting of heavy olefin oligomer by-products, which attenuates acid strength and increases hydrocarbon solubility in the acid phase. The OD also acts as an available source of hydride (H^-) in the reaction media increasing the rate of hydride transfer and inhibiting secondary oligomerization. While OD builds up with time in the acid phases, techniques have been developed to maintain acid purity. In the case of HF, the low-boiling acid can be distilled away from the OD and recycled back to the process. In sulfuric acid alkylation, a drag stream of the impure acid phase must be removed from the process and sent to a sulfuric acid plant where is burned and reconstituted as fresh H_2SO_4. The relatively low cost of these two acids means that small acid losses across these recovery steps have minimal economic impacts on the overall processes. Technical details of these processes have been published [4–6].

The safety hazards associated with the large-scale use of HF and H_2SO_4 in pressurized refinery units are the major drawback of the use of these acids for motor fuel alkylation. In the case of HF, the potential for aerosol cloud formation in an event of a reactor or storage vessel rupture has been recognized [7]. The need to transport large quantities of concentrated sulfuric acid between the alkylation and sulfuric acid units, typically by truck, affords the opportunity for accidental acid release. Significant steps have been taken over the past 20 years to increase the safety of liquid acid alkylation processes, including the use of minimal acid designs, rapid acid drain to storage and water deluge systems for HF alkylation [4] and the construction of on-site sulfuric acid plants for H_2SO_4. However, it has been a long-standing goal of the refining industry to identify an alternative solid acid catalyst to replace HF and H_2SO_4 liquid acids in motor fuel alkylation [8].

A commercially viable alkylation process employing a solid acid catalyst must match the alkylate yield, product quality (particularly octane number) and economics of the established HF and H_2SO_4 processes, in addition to eliminating the hazard of liquid acid handling. Alkylate yield, expressed as grams of C_{5+} alkylate produced per gram of olefin converted, needs to be in excess of 2.0. The stoichiometric limit for the reaction of butene with butane is 2.04. Yields less than 2.0 indicate the presence of some olefin oligomerization, while greater values typically occur is there is excessive cracking of the desired C_8 product to C_{5-7} paraffins. The value of the alkylate product for gasoline blending is directly related to its octane number with RONs in the 94–96 range or higher preferred.

The ability of zeolites to catalyze paraffin–olefin alkylation has been known since 1968. Garwood and Venuto of Mobil described the use of rare-earth hydrogen X faujasite for the alkylation of isobutane with ethylene to give branched C_5–C_8

paraffins in short batch experiments, which included evidence of catalyst coking by the olefin feed causing rapid catalyst deactivation [9]. The advantage of large-pore sized zeolites for isobutane–butene alkylation was reported by Chu and Chester in the comparison of the performances of several faujasite-based catalysts with one based on H-ZSM-5 [10]. The medium-pore zeolite was found to be inactive at temperatures up to 97 °C. This paper also quantified the rapid aging of rare-earth exchanged Y, including its impact on the product distribution, and suggested that the coke deposits were formed via conjunct olefin polymerization. More recently, the use of rare-earth exchanged X and Y zeolites has been re-visited [11].

Large-pore zeolites other than faujasite have been reported to catalyze the alkylation of isobutane with butenes. EMT (hexagonal faujasite) has been reported to give slightly higher alkylate and TMP yields than the comparable FAU zeolite [12], with further improvement on dealumination or La-exchange [13, 14]. However, many of these tests gave alkylate yields of less than 1.6, suggesting that significant olefin oligomerization was occurring at the conditions being compared. The ten-ring zeolite MCM-22 (MWW structure) can be expanded and pillared to give MCM-36 which has greater mesoporosity to better accessibility of the framework acid sites. The pillared material showed superior performance in paraffin alkylation with better activity and stability [15, 16]. Beta zeolite has also been shown to be an effective catalyst for the alkylation isobutane with a number of light olefins in a well-mixed slurry-phase reactor [17]. Shell workers were able to demonstrate catalyst lifetimes of up to 500 h in this type of reactor, but could not identify a regeneration procedure other than a carbon burn to restore fresh catalyst activity after eventually deactivation [18].

The need for an effective method to regenerate a zeolitic alkylation catalyst by the removing the heavy, probably poly-unsaturated, hydrocarbon deposits was recognized in the 1970s. Yang at Union Carbide found that the alkylation of zeolite-based catalyst could be maintained for up to 85 h by adding a small amount of group VIII metal to the catalyst and periodically contacting it with hydrogen either in the gas phase or dissolved in isobutane [19, 20]. More recently, Akzo claimed that it is advantageous to limit the extent of catalyst deactivation occurring in an individual process cycle by frequently subjecting the used catalyst bed into a liquid-phase hydrogen–isobutane regeneration step well before catalyst activity drops below a specific olefin conversion level [21]. Frequent extractions of deactivated USY catalysts with supercritical isobutane have been shown to extend catalyst longevity, but does not completely restore catalyst activity to fresh performance [22]. Lercher and co-workers have reviewed the reaction pathways involved in the deactivation of zeolite catalysts during paraffin alkylation [23, 24].

A variety of reactor designs have been proposed to meet the requirement for frequent regeneration of zeolitic paraffin alkylation catalysts by moving the catalyst bed between process and regeneration zones. Fixed-bed swing reactors have been the most commonly employed, due to their operating simplicity. Alkylation processes by Akzo/ABB Lummus [25] and Excelus [26] employ this type of reactor design with swing times up to a few hours. A key disadvantage of this design is the changing product composition as the fixed catalyst bed deactivates, which typically require the uses of several process reactors operating in parallel. In addition, the exposure of fresh feed to deactivated catalyst in the inlet of a partially deactivated bed introduces the potential for undesirable olefin oligomerization before the feed reaches the catalyst zone active for paraffin alkylation.

Circulating catalyst reactors can help overcome these problems by contacting regenerated catalyst with fresh feed in a co-current flow through the alkylation zone of the reactor. At least some of the liquid is then separated from the solid to recover the alkylate product while the reminder is transported with the catalyst through a regeneration zone. Shell proposed a slurry reactor process for isobutane alkylation over low Si/Al Beta zeolite [18], while Mobil has patented a circulating slurry reactor using small (20–2000 µm) zeolite catalyst particles [27]. A solid–liquid riser reactor design has been described by UOP in which the catalyst is transported, in essentially plug flow, up a tubular reactor by the liquid flow [28]. This design also incorporates two simultaneous catalyst regeneration operations; a mild liquid-phase wash with hydrogen–isobutane on the entire recirculating catalyst bed and a vapor-phase H_2 stripping operation on a slip stream of catalyst removed from the recirculating reactor/regenerator vessel. Lurgi has proposed carrying out slurry-phase alkylation on the trays of a distillation column in order to control and recover the exothermic heat of reaction [29]. To date, none of these reactor designs have been commercialized for motor fuel alkylation.

15.3.1.3 Physicochemical Characterization of Active Sites

The nature of the acid sites of the zeolite has the most significant impact on the alkylation performance of these alkylation catalysts. The acid site density, acid strength and Brønsted versus Lewis acid character have been shown to impact the product selectivity and lifetimes of catalysts based on Beta and faujasite zeolites. Increasing the concentration of Brønsted acid sites by lowering the extent of exchange of Beta zeolite with Na^+ cations gives a higher alkylation turn-over number (mole alkylate produced per mole Brønsted acid site prior to deactivation) by favoring the desired hydride transfer versus secondary oligomerization [30]. It was also suggested that weak adsorption of olefins on the Na^+ cations might increase their local concentration in the vicinity of the Brønsted sites, leading to heavy olefin formation. A relationship between increased acid site strength and reduced catalyst deactivation has been shown with FAU catalysts by the correlation of alkylation lifetime with the percentage of strong Brønsted acid sites (those capable of adsorbing pyridine at 450 °C, as measured by IR) out of the total Brønsted sites (those that adsorb pyridine at 100 °C) [31]. A similar correlation is observed between alkylation lifetime and the fraction of Brønsted acid sites among the total Brønsted and Lewis sites measured by pyridine IR at 450 °C. One approach for controlling the Brønsted/Lewis acid site ratio is to lower the temperature of activation so that a trace of residual water remains on the catalyst. Lercher et al. have shown by in situ IR and ^1H MAS NMR spectroscopy that activation of La-X zeolite at 150–180 °C leaves some adsorbed water molecules on zeolite, which promotes the rate of hydride transfer over oligomerization and affords a longer alkylation lifetime than the same material activated at 280 °C [32].

15.4
Benzene Alkylation

15.4.1
Ethylbenzene (Ethylene Alkylation), Cumene and Detergent Linear Alkylbenzene

15.4.1.1 Process Chemistry: Feeds, Products and Reactions

The reactions of benzene with C_2, C_3 and C_{9-14} olefins to form ethylbenzene (EB), cumene and linear alkylbenzenes (LAB), respectively, are processes to produce intermediates to styrene, phenol (and acetone) and sulfonated LAB (detergent). These processes are typically operated under liquid or mixed phase conditions to achieve desirable product quality and catalyst life. The liquid phase operation renders high product purity, that is, lower xylenes in EB, lower n-propylbenzene and EB in cumene and high linearity in detergent alkylation. The catalysts undergo regeneration periodically, ranging from operating cycle lengths of days for detergent alkylation to several years for EB and cumene. In commercial EB and cumene processes, the olefin feed is divided and injected into individual reactors for better control of the heat of reaction to imporve product quality and catalyst life. Due to more demanding product specifications in cumene, part of product effluent is recycled as well to control the heat of reaction to improve the product purity and catalyst life.

Sources of benzene are mainly from reforming, or steam cracking of naphtha, or gas oil. Guard beds are typically used to remove olefinic and organic nitrogen compounds, both of which will contribute to significantly shorter catalyst life. The olefinic impurities are generated in the cracking process (benzene co-boilers), while organic nitrogen compounds may be introduced during the separation of aromatics from non-aromatics via a liquid–liquid extraction process. Ethylene is mainly derived from stream crackers and can be high purity (polymer grade) or lower concentration streams. Propylene is mainly from steam crackers or FCCs with the latter being in the range of 50–75 wt% C_3 olefin. Refinery C_3 olefin needs to be treated to remove sulfur and nitrogen compounds, before entering the alkylation unit. C_9–C_{14} olefin is typically derived from a kerosene fraction through a series of purification, separation and conversion processes such as the UOP Detal™ process.

15.4.1.2 Catalysts

Mobil/Badger vapor phase EB technology was commissioned in 1980 and is based on ZSM-5. The vapor phase process gives higher xylene impurity make, resulting in higher operating costs in downstream styrene plant, and relatively shorter catalyst life due to carbonaceous deposits. The Lummus/UOP EB*One*™ process for liquid phase EB alkylation was first commissioned in 1990, with a catalyst first based on zeolite Y (UOC-4120™) and then later based on zeolite beta (UOP EBZ-500™). The liquid phase process gives high on-stream efficiency, often greater than 99%, resulting in significant cost saving. The EBMax process was commissioned by Mobil/Badger in 1995 and is a liquid-phase alkylation process based on zeolite MCM-22 (MWW). CDTECH EB™ is based on catalytic distillation and

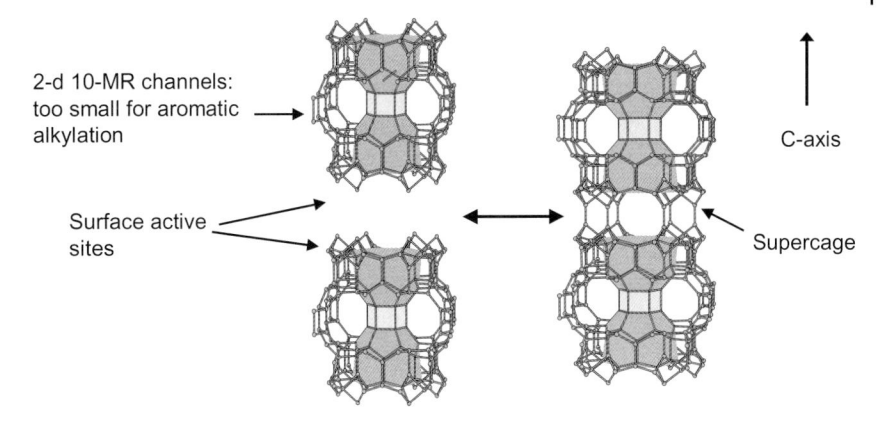

Figure 15.4 Origin of external surface sites in MWW.

operates in mixed phase. The zeolitic catalyst is packaged into specially engineered bales to perform chemical reaction and distillation process. Similar technologies and catalysts are offered in cumene processes by UOP (the UOP Q-Max™ process) and Mobil/Badger. The Polimeri Europa cumene process, formerly by Enichem, is based on a proprietary beta zeolite catalyst (PBE-1). A more detailed description of advancement in industrial alkylation of aromatics can be found in reviews by Degnan [33], Perego [34], Woodle [35] and Schmidt [36]. It is believed that detergent alkylation process is not currently based on a zeolite catalyst.

15.4.1.3 Physicochemical Characterization of Active Sites

Chronologically, EB alkylation has evolved from aluminum chloride, to ZSM-5, then to faujasite, followed by BEA and MWW (PSH-3, SSZ-25, MCM-22/-49/-56, ITQ-2). In contrast, cumene alkylation evolved from SPA, to faujasite, followed by BEA and MWW. Active sites of zeolites ZSM-5, faujasite and BEA are characterized as microporous, while those of MWW are external surface acid sites in the 0.7×0.7 nm cups derived from a half moiety of the super-cage of the MWW structure (Figure 15.4). Molecular modeling has shown a high diffusion barrier for cumene (Perego) [37] and later low diffusivity for benzene (Sastre) [38] in the pores of MCM-22. The presence of the surface sites has been verified structurally by DIFFaX simulation [39] and catalytically by the reaction of benzene (Corma) [40] and toluene (Guisnet) [41] with propylene, with and without selective poisoning.

Taking the literature data as a whole, it appears that: (i) the large pores of zeolites with a three-dimensional topology, preferably without cages, and (ii) external surface acid sites are most desirable for benzene alkylation reactions. It is interesting to note that, while the literature data is generally in good agreement, there appear to be some fine differences. The differences might be due to: (i) testing methodology (fixed bed versus stirred reactor), (ii) diffusion characteristics of the testing such as the size of pellets used in the test (partition between alkylation and oligomerization) [42], (iii) porosity of the catalysts (activity and stability), (iv) composition, (v) morphology of zeolite and/or (vi) thermal and hydrothermal damage incurred during the zeolite/catalyst activation. For example, Ercan and co-authors showed that EB and cumene alkylation are significantly limited by intra-particulate (pore) diffusion. This was based on the testing results of a catalyst of varying sizes in a liquid-phase differential fixed bed loop reactor. Fajula [43] and co-authors showed that the optimal beta zeolite for catalytic activity has Si/Al_2 ratio of about 22 and average crystal diameters of 0.3–1.0 μm (by SEM). Lobo and co-workers also showed that, while MCM-56 has a significantly lower stacking structurally than MCM-22, the external surface sites do not appear to be readily available for toluene disproportion due to the stacking of these low stacking and curled layers.

Although zeolite catalysts have not been used in industrial detergent alkylation processes, there are a good number of reviews in the literature describing their use in detergent alkylation [34, 44–48]. In kinetic studies of benzene alkylation with terminal linear olefins, the formation of primary products, that is, 2-olefins and 2-phenylalkanes, is observed at low olefin conversions and the rate of double bond isomerization is faster than alkylation. It is generally shown that large-pore zeolites with three-dimensional topology are effective in producing linear alkylbenzene (LAB) with a reasonably high turnover frequency. Smaller 12-MR pore zeolites with three-dimensional topology, such as zeolite BEA, and one-dimensional topology, such as MTW, and 10-MR zeolite, such as MFI, are not effective in providing stable production of LAB. One-dimensional 12-MR zeolites, such as mordenite, appeared to be effective in carrying out the alkylation reaction, but give higher than thermodynamic levels of 2-phenylalkane in the alkylate. The interpretation of the kinetic results and reactivity patterns resides in: (i) acid strength controlling the activation energy of the protonation step and, consequently, the concentration of surface carbenium ions and (ii) the barriers to the products diffusion from the microporous interior through pore mouths of various sizes and shapes. It is interesting to note most studies have been conducted under reaction conditions where poly-alkylated benzene was produced in lower concentration. Guisnet and co-workers showed that trans-alkylation appears to be another pathway to produce mono-alkylated benzene in toluene alkylation with 1-heptene using a relatively low molar ratio of toluene to olefin at 90 °C.

15.4.2
Para-Xylene (Methylation of Toluene)

15.4.2.1 Process Chemistry: Feeds, Products and Reactions

No commercial processes are in operation currently for the production of *para*-xylene by the methylation of toluene with methanol. As methanol becomes more favorable feedstock over time, toluene methylation might become competitive for *para*-xylene production. This process has the potential advantages of increasing *para*-xylene production throughput and lowering the operating and capital cost via the adoption of single-stage crystallization for *para*-xylene purification. However, there are still several technical hurdles that need to be resolved before commercial viability is realized. The reaction typically requires operating at toluene to methanol molar ratios greater than one and at a high degree of dilution under atmospheric pressure to improve methanol utilization [49]. In conjunction with the high degree of dilution, the process typically operates at temperatures greater than 400 °C to avoid methanol conversion to hydrocarbons via MTO/MTG pathways instead of toluene alkylation.

15.4.2.2 Catalysts

The catalysts are predominantly modified ZSM-5 zeolite. In general, the modifications are intended to restrict pore mouth size to promote the shape selective production of *para*-xylene within the microporous structure. The same modifications also serve to remove external acid sites and eliminate the consecutive isomerization of *para*-xylene. Methods used to modify the zeolite pore openings have included silation [50], incorporation of metal oxides such as MgO, ZnO and P_2O_5 [51, 52], steaming and the combination of steaming and chemical modification [53].

15.4.2.3 Physicochemical Characterization of Active Sites

The adsorption of methanol on the catalyst is required to have methylation taking place on aromatic ring instead on the methyl functional group [54]. The rate of ring methylation appears to increase with the acidity of the active sites as shown by Yashima in 1970 [55]. Yashima also showed that above equilibrium *para*-xylene selectivity can be obtained at 225 °C as the primary kinetic product. Under similar reaction conditions Dumitriu and co-worker showed greater *para*-xylene selectivity is observed at approximately 20% toluene conversions [56]. The selective *para*-xylene formation is evidently governed by the kinetics of the ring alkylation of toluene and the more constraining 10-MR pore mouth of MCM-22. Epitaxial growth of silicalite onto ZSM-5 zeolite showed high para-xylene selectivity, while maintaining very good activity and activity stability in comparison with other means of selectivation [57]. It seems plausible that using a silicate layer to impart shape selective formation of *para*-xylene might have worked around the coke formation at the pore mouth.

15.4.3
Styrene and Ethylbenzene from Methylation of Toluene

15.4.3.1 Process Chemistry: Feeds, Products and Reactions

No commercial process is offered at this time for side chain alkylation of toluene with methanol for styrene and ethylbenzene production. In the literature the reaction is typically carried out at toluene to methanol molar ratios from 1.0:7.5 to 5:1 from 350 to 450 °C at atmospheric pressures. In some cases inert gas is introduced to assist vaporizing the liquid feed. In other cases H_2 is co-fed to improve activity, selectivity and stability. Exelus recently claimed 80% yields in their ExSyM process at full methanol conversion using a 9:4 toluene:methanol feed ratio at 400–425 °C and 1 atm (101 kPa) in a bench-scale operation. This performance appears to be significantly better than what has been shown in the literature of about 25–55% yields at 50–100% methanol conversions, but at more diluted toluene to methanol ratios.

15.4.3.2 Catalysts

The catalyst consists of basic and acid sites in a microporous structure provided by zeolite and microporous materials [58–62]. Basic sites are provided by framework oxygen and/or occluded CsO. Acid sites are provided by the Cs cation and, possibly, additives such as boric and phosphoric acids. The addition of Cu and Ag increased the activity [63, 64]. Incorporation of Li, Ce, Cr and Ag also has been shown to increase the styrene to ethylbenzene product ratio [65]. The reactivity of catalysts is sensitive to the presence of occluded CsO, which is in turn influenced by the preparative technique as shown by Lacroix and co-authors [64] and pointed out by Lercher [61].

15.4.3.3 Physicochemical Characterization of Active Sites

Basic sites carry out the dehydrogenation of methanol to formaldehyde and the activation of the methyl function group of the toluene molecule. The acid sites provide adsorption sites for the aromatic ring and also adsorption sites to stabilize formaldehyde. The requirement of basic and acid sites in close proximity is critical for the aldol condensation-type reaction mechanism based on adsorption of toluene and formaldehyde molecules by infrared spectroscopy. This supports a surface reaction proceeding in a concerted manner, as illustrated in Figure 15.5 by Lercher [61]. It also explains why poisoning of the catalyst with an acidic or basic reagent can shut down the reaction and greater than 40% exchange of CsY is required to trigger the onset of side chain alkylation, as observed by Borgna and coauthors [66, 67].

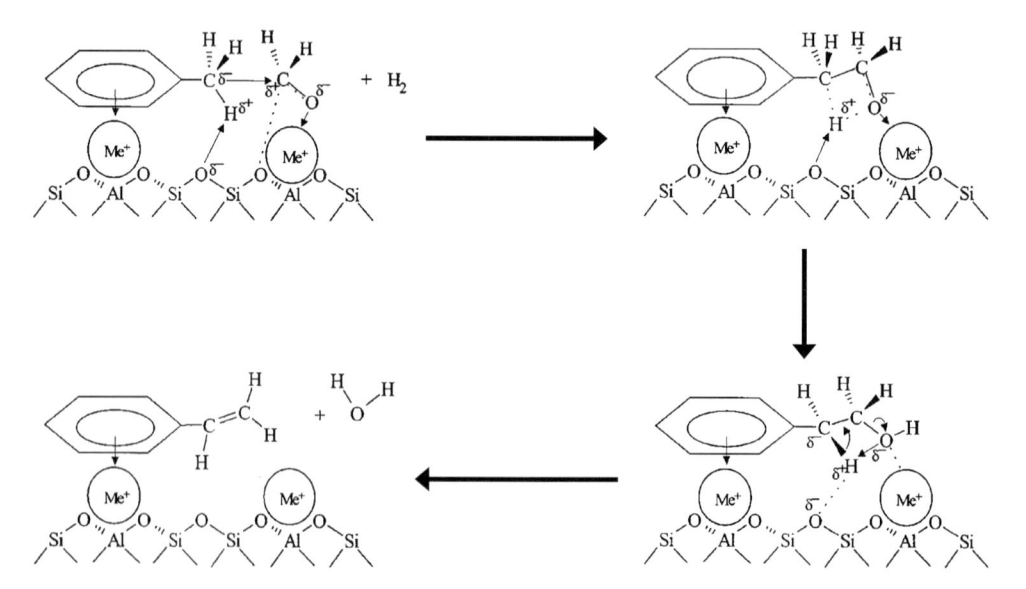

Figure 15.5 Mechanism for toluene side-chain alkylation by formaldehyde [61].

15.5
Alkylbenzene Disproportionation and *Trans*-Alkylation

15.5.1
Process Chemistry: Feeds, Products and Reactions

Trans-alkylation and disproportionation are primary conversion pathways to obtain xylene in an aromatic refinery complex. The typical feeds are from reformate and hydro-treated pyrolysis gas (pygas). The reactants are a mixture of toluene and aromatics containing nine, ten and higher carbons (A_9, A_{10}, A_{11+}). The A_9 consists of trimethylbenzenes, methyethylbenzenes and propylbenzenes. The *trans*-alkylation process typically operates in vapor phase at around 400 psig (2760 kPa) and about 2–3 h^{-1} WHSV using H_2:hydrocarbon ratios of two to four. The start of run (SOR) temperature is around 350 °C and the temperature is adjusted to maintain conversion, xylene and benzene production. Commercially, the products consist of mainly of C_8 aromatics and benzene. The methyl to phenyl ratio of the feed blend is an important variable, since the product distribution is controlled primarily by thermodynamics and the catalyst stability is controlled by the amount of C_{11+} in the feed. In addition to disproportionation and *trans*-alkylation, hydro-dealkylation reactions of the heavy aromatics to remove ethyl and propyl side chains are important reaction pathways for xylene formation. In the same process, benzene co-boilers, consisting of C_6 and C_7 naphthenes such as cyclohexane, methylcyclopentane, methylcyclohexane and dimethylcyclopentane, are cracked to light paraffins to improve benzene purity in downstream recovery processing. Some technologies appear to operate with *trans*-alkylation of heavy aromatics and cracking of non-aromatics or naphthenes as the predominant reactions [68]. Critical reviews of industrial vapor phase *trans*-alkylation processes can be found in recent publications by Tsai [69–71], Oh [72–74] and Corma [75].

15.5.2
Catalysts

The suitable catalysts for alkylbenzene *trans*-alkylation are large-pore zeolites such as mordenite, ZSM-12 or beta. In some cases MFI is used in addition to mordenite. In addition to the zeolite acid function, a weak hydrogenation function is often included to assist the saturation of olefins generated from de-alkylation of ethyl and propyl side chains of heavy aromatics. The weak metal function can consist of low levels of a noble metal such as Pt or Pd, attenuated noble metals such as Pt-Sn and Pt-Pb or varying levels of other metals such as rhenium. The zeolite functions primarily to de-alkylate ethyl and propyl side chains off the heavy aromatics, *trans*-alkylate the heavy aromatics to lower aromatics and disproportionate toluene to xylene and benzene. Strong zeolite acidity appears to play a role in cracking the non-aromatic benzene co-boilers to increase its purity. A study by Tseng-Chang Tsai and co-workers using a cascade/stack arrangement of bi-functional followed by an acid function only catalyst showed that selective naphthene cracking is carried out by the back-end catalyst to improve benzene purity [71]. This approach is effective at temperatures less than about 350 °C, above which the aromatic hydrogenation is greatly inhibited.

The catalyst used in the TransPlus process jointly developed by ExxonMobil and CPC uses a proprietary zeolite [76]. From the literature, the SK Energy ATA-11 catalyst appears to be based on either mordenite or beta zeolite using attenuated noble metals. The UOP TAC9™ process was developed by UOP and Toray and is licensed by UOP (outside Japan and Korea). The UOP Tatoray™ process is based on a bi-functional catalyst with long catalyst life, minimal ring loss and hydrogen consumption in comparison with noble metal catalyst systems.

15.5.3
Physicochemical Characterization of Active Sites

The optimal pore size for processing heavy aromatics through *trans*-alkylation and hydro-dealkylation reactions is not clear and can depend greatly on amounts of heavy aromatics in the feed and the type of weak hydrogenation function. It is believed that a large-pore zeolite without cages or large intersecting channels is desirable. The cages or intersecting large pores might tend to promote formation of poly-aromatic rings, leading to fast catalyst deactivation. This is probably why the most cited materials are mordenite and ZSM-12. ZSM-12 was shown to be more effective in converting 135-trimethylbenzene (TMB) than toluene in a study using a model feed blend by Tsai and co-workers. In the same study mordenite showed relatively even activity in converting TMB and toluene, but poorer stability. The interpretation of these results is not straightforward, because the morphological differences might be more important than the structure differences between ZSM-12 and mordenite. In a recent paper, MCM-56 impregnated with NiO was shown to have good activity and selectivity in producing xylene along with benzene and toluene from commercial heavy aromatics [77]. Qunbing Shen and coauthors attributed this efficiency to a moderate interaction between the metal and acid sites. Although the operating temperature and conversion appeared to be significantly higher than typical, the results appeared to imply that the surface sites residing on MCM-22 and MCM-56 can effectively carry out the conversions of heavy aromatics.

15.6
Paraffin/Olefin to Aromatics

15.6.1
C_3/C_4 Paraffin to Aromatics and C_3/C_4 Paraffin/Olefin to Aromatics

15.6.1.1 Process Chemistry: Feeds, Products and Reactions
C_{3-5} feeds or LPG typically come either from the field or from naphtha crackers and FCC units. Hydrogenation to minimize dienes is required to limit severe

coke formation. In certain cases, the feed can be extended to light- or full-range gasoline with varying amounts of paraffins and olefins. The reaction consists of dehydrogenation or cracking of paraffin to olefin, coupling of olefin to form a higher carbon chain, dehydrocyclization to naphthenes and, finally, dehydrogenation to aromatics. The process yields 45–65% aromatics consisting up mostly benzene, toluene, xylene and a relatively small amount of heavy aromatics with the balance being mostly lighter paraffins and large amounts of hydrogen. These processes represent special cases for aromatic production and are typified by (UOP-BP Cyclar™) [78–80] and (Asahi ALPHA-ARO™) [81, 82]. The concentration of aromatics is high in liquid product and therefore separation by simple distillation is usually sufficient to recover aromatics without adopting solvent extraction. The light paraffin by-products from the cracking of the feed and saturated, higher carbon number intermediates are excellent feeds for a steam cracker. The methane-rich byproduct stream can be used for fuel gas for the process, and the co-product hydrogen can be used for other petrochemical processing downstream. The process operates from about 2 to 5 bars (200–500 kPa) above atmospheric pressure at temperatures ranging from about 500 to 600 °C. The reaction is endothermic when a paraffin-rich feed is used. With a feed having a balanced paraffin to olefin ratio, the reaction can be maintained thermally neutral, alleviating the need for inter-stage re-heating. When a paraffin-rich gasoline-boiling hydrocarbon from FCC was processed at atmospheric pressure and 550–600 °C, Philips Petroleum Company reported the co-production of aromatics and light olefin [83, 84].

15.6.1.2 Catalysts

The catalysts in general consist of metals or metal oxides supported on ZSM-5 containing catalysts. The UOP-BP Cyclar catalyst is based on a modified zeolite, which promotes dehydrogenation of the paraffinic reactant and dehydrocyclization of intermediates to aromatics. Asahi's patent applications describe zinc supported on zinc aluminate-bound ZSM-5 zeolite with specific acidity characteristics. The catalyst is claimed to impart good aromatic yields, low coke formation and good regenerability. Philips Petroleum Company has patents claiming the use of zinc containing spinel bound ZSM-5 catalyst for converting paraffin rich gasoline blend stock to light olefin and aromatics. The benefits of improved activity and lower coke formation were realized after treating the catalyst under a reducing atmosphere. In a separate application similar claims were made for converting similar feedstock to light olefin and aromatics at ratios from 0.28 to 0.66. The catalysts have the characteristics consisting of acid leached ZSM-5, zinc and a second metal components selected from Si, Cr, Mo, B, Ti and P.

15.6.1.3 Physicochemical Characterization of Active Sites

Metals and/or metal oxides are required to dehydrogenate paraffinic reactants to olefins, which are dimerized or oligomerized via acid sites of the ZSM-5 to increase carbon number and then dehydrocyclized, first to napthenes and then aromatics. The dehydrocyclization to aromatics most likely takes place on metal or metal oxide within the microporous structure of zeolite. It appears that a stronger interaction of hydrocarbon with metal or metal oxide than H_2 is preferred to limit the hydrogenation and hydrogenolysis reactivity. Furthermore, the low coke formation at the relatively severe operating temperatures is most likely due to the use of basic binders such as spinels and other zinc containing mixed metal oxides.

15.6.2
C_6/C_7 Paraffin to Aromatics (Zeolitic Reforming)

15.6.2.1 Process Chemistry: Feeds, Products and Reactions

The feed is either light naphtha containing C_{6-7} paraffins or aromatics extraction raffinates. These feeds are converted to benzene and toluene via a dehydrocyclization reaction. Paraffins greater than C_7 are generally converted to aromatics using conventional catalytic reforming catalysts. The zeolitic reforming process operates in a semi-regenerative mode under conditions similar to semi-regenerative reforming. The feed sulfur content is reduced to extremely low levels to ensure optimal catalytic efficiency. The catalyst is also very sensitive to water. The reaction is highly endothermic and inter-stage re-heating is required. Commercial operation is represented by the UOP RZ Platforming™ process and Chevron Phillips Chemical's (CP Chem) AROMAX process. Along with Asahi Chemical's ALPHA process and the Cyclar process, they can be categorized as nonconventional routes to aromatic production.

15.6.2.2 Catalysts

The catalyst is made up of platinum dispersed on a basic zeolite, K-L or Ba-K-L [85]. The conversion of linear paraffins from C_6 to C_9 is much faster over Pt-KL or Pt-KBaL catalyst than over conventional reforming catalysts. However, the advantage diminishes as carbon number increases, so these technologies are primarily of interest for benzene production. It is therefore more efficient to complement conventional reforming with the zeolitic reforming process when a broader range of aromatic products is desired. Relatively large crystal size has been claimed to be beneficial for example in CP Chem's AROMAX process. Residual acidity on the catalyst has been shown to be detrimental [86].

15.6.2.3 Physicochemical Characterization of Active Sites

The zeolite provides the environment for shape selective chemistry and is also a high surface area support on which to disperse platinum in a relatively confined environment. The small platinum crystals within the zeolite channels and the orientation effect of the channel window are responsible for the high efficiency of the Pt-KL catalyst to convert linear paraffin to aromatics. Zeolite KL also provides an electron rich environment to enhance stronger platinum–substrate interaction via stronger platinum–support interaction. A review on the subject can be found in the article written by Meriasdeau and Naccache [85].

15.7
Methanol to Olefins and Aromatics

The discovery that zeolite catalysts could effectively convert light oxygenates, in particular methanol, to higher hydrocarbons introduced a new method for the catalytic formation of carbon–carbon bonds. It has been long known that passing methanol over a strong acid catalyst at severe conditions results in an ill-defined mixture of heavier products. But, Chang and Silvestri's 1977 paper on the conversion of methanol over H-ZSM-5 demonstrated that a well-defined mixture of C_1–C_{10} hydrocarbons, highly concentrated in aromatics, could be obtained [87]. This product was very suitable for gasoline blending at a time when conventional oil supplies were undergoing significant disruption. Mobil's methanol-to-gasoline (MTG) process was eventually commercialized in New Zealand. Subsequently, it was found that small-pore zeolites, in particular SAPO-34, favor the formation of light, unbranched olefins which are a preferred feedstock for commercial polymers [88]. In the recent years the construction of several methanol-to-olefins (MTO) units have been announced. Several reviews of these technologies have been published [89–91].

15.7.1
Methanol to C_2–C_4 Olefins

The monetization of remote natural gas has been a key economic driver for catalysis research over the past 20 years. Significant reserves of natural gas exist in remote locations, distant from available gas pipelines, which cannot be readily brought to market. The conversion of these resources to higher-valued, transportable products, such as methanol or polyolefins can allow the economical utilization of these "stranded" assets. Other low-valued natural gas streams, such as associated gas from oil production, could also provide feedstocks to such a technology. The conversion of remote gas, typically valued at US$ 0.50–1.50 per MMBTU, into polyolefins, valued at more than US$ 1000/t, via methanol has sparked the development of several MTO technologies.

MTO processes fall into two categories based on the molecular sieve catalyst employed: small and medium pore-size sieves. Small-pore sieves, such as SAPO-34, give a mixture of ethylene and propylene, the composition of which can be varied by processing conditions. Small amounts of higher, mostly linear, olefins, are also produced which can be selectively cracked back to propylene to increase the yield of the desired light olefins. Over 90% selectivity to ethylene and propylene can be obtained at complete methanol conversion [92]. These catalysts deactivate rapidly by coke deposition, which is trapped within the intracrystalline cavities of the molecular sieve. Therefore, the process design must accommodate frequent catalyst regeneration by carbon burn. The UOP/HYDRO MTO process uses a fluidized bed reactor and regenerator design to maintain steady-state production, provide temperature control of the exothermic reactions and allow facile movement of the catalyst between the reactor and regenerator zones [93].

With the selection of appropriate operating conditions, medium-pore MFI zeolite catalysts can give propylene as the major product. Mobil workers in the mid-1980s scaled-up a MTO process from laboratory to pilot plant scale [94]. A combination of low zeolite acidity, obtained with high zeolite Si/Al ratios, and the use of high temperatures shifted the product slate from aromatics to light olefins [95]. At near complete methanol conversion, propylene selectivity ranged over 22–39% with 15–24% C_{6+}, about equally split between paraffins/olefins and aromatics. At the time, propylene demand was insufficient to justify commercialization. However, over the past ten years, propylene demand has increased greatly with the increased production of polymers. Lurgi have re-activated the development of this technology and are offering a methanol conversion process that produces propylene in up to 71% yield with 25% going to C_{4+} by-products that can be used for gasoline blending [96]. Per pass propylene selectivity is about 50%, so in order to maximize propylene production several olefin product streams are recycled to the main synthesis reactor for inter-conversion. The Lurgi MTP process uses multiple fixed-bed reactors, swinging about every 500 h into a mild carbon burn regeneration to maintain activity. Since temperature control of the exothermic MTO reaction is more difficult in this type of reactor, methanol to dimethylether pre-conversion is employed to isolate about one half of the exothermic heat of reaction where it can be used for feed pre-heating [97].

15.7.2
Methanol to Aromatics

The initial application of zeolite-catalyzed methanol conversion to hydrocarbons in the 1970s and early 1980s was for the production of highly aromatic gasoline. At conditions favoring high methanol conversion, the MFI zeolites discovered by Mobil give primarily C_6–C_{10} mono-ring aromatics suitable for gasoline blending and light C_3–C_{5+} paraffins. The shape selectivity of the ten-ring MFI zeolite prevents the formation of heavier and multi-ring aromatics hydrocarbons that would fall above the gasoline boiling range. Early high-space velocity experiments established that methanol conversion over MFI proceeded through at least three stages: (i) dehydration to dimethylether, (ii) further dehydration to light olefins and finally (iii) hydrogen transfer to yield light paraffins and aromatics. The final product composition is highly dependent upon the process conditions (temperature, pressure) and catalyst composition (Si/Al ratio) [98]. Representative product distributions as a function of reaction severity are summarized in Table 15.1 [99].

Fixed-bed and fluidized-bed reactor configurations were developed by Mobil in the early 1980s. The fixed bed design was commercialized in a 570 000 t/year gasoline production plant in New Zealand in 1985 [100]. Five parallel reactors were used, four staggered in the process to maintain a uniform product consistency and one in regeneration at any given time [101]. Pre-reaction of some of the methanol to DME and light gas recycle was required to control the exothermic heat of reaction. The unit ran in the gas to gasoline (GTG) configuration for a few years until

Table 15.1 MTG production distributions as a function of space velocity (643 K, 101.3 kPa).

	LHSV (h⁻¹)		
	1080	108	1
Product distribution (wt%)			
Water	8.9	33.0	56.0
Methanol	67.4	21.4	0.0
Dimethylether	23.5	31.0	0.0
Hydrocarbons	0.2	14.6	44.0
Hydrocarbon distribution (wt%)			
Methane	1.5	1.1	1.1
Ethane	0.0	0.1	0.6
Ethylene	18.1	12.4	0.5
Propane	2.0	2.5	16.2
Propylene	48.2	26.7	1.0
Butanes	13.9	7.8	24.3
Butenes	11.9	15.8	1.3
C5+ aliphatics	4.4	27.0	14.0
Aromatics	0.0	6.6	41.1

the economic constraints of lower oil prices led to the shut-down of the MTG portion.

15.7.3
Catalysts

Molecular sieve catalysts that have been used for the conversion of methanol to hydrocarbons fall into two general classifications, medium-pore and small-pore zeolites. Most of the initial research was done using ZSM-5 (MFI), a medium-pore zeolite with a three-dimensional pore system consisting of straight (5.6 × 5.3 Å) and sinusoidal channels (5.5 × 5.1 Å). In the early 1980s researchers at Union Carbide discovered that small-pore silicoaluminophosphate (SAPO) molecular sieves were effective for converting methanol to ethylene and propylene. The best performances were obtained with SAPO-34 and SAPO-17 catalysts [102]. SAPO-34 has the CHA structure with a three-dimensional pore system consisting of large cavities (about 9.4 Å in diameter) separated by small windows (3.8 × 3.8 Å).

The structures of SAPO-34 and ZSM-5 along with small organic molecules that are key to the MTO process are shown in Figure 15.6. The ionic radii of the oxygen atoms of one of the pore mouths of each structure are highlighted. While linear hydrocarbons, such as ethylene and propylene, are able to pass through the pores of both structures, branched molecules, such as i-butene and benzene, cannot pass through the eight-ring pores of SAPO-34.

The impact of the structural differences of ZSM-5 and SAPO-34 on representative MTO product compositions is shown in Figure 15.7. The medium-pore sieve

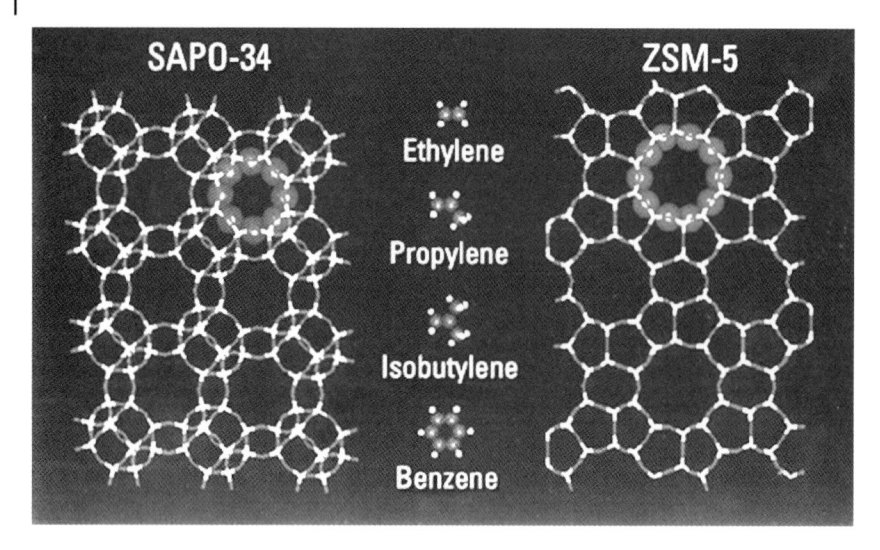

Figure 15.6 Framework structures of SAPO-34 (CHA) and ZSM-5 (MFI).

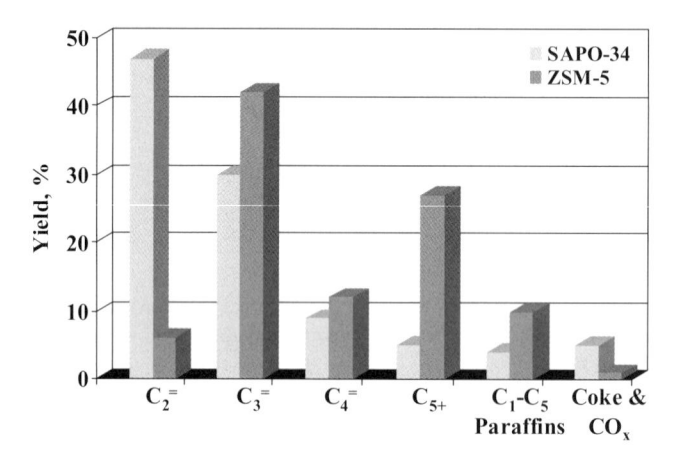

Figure 15.7 MTO product compositions for ZSM-5 and SAPO-34.

gives propylene as the major light olefin and a significant amount of C_{5+} hydrocarbons, much of which is aromatic and accounts for the high light paraffin production [94]. In contrast, the small pore-size SAPO-34 gives predominantly ethylene and propylene with less heavy hydrocarbons and paraffins. Coke formation is higher with SAPO-34, since any aromatics formed within the pore structure are trapped, whereas some aromatics are able to diffuse through and desorb from the ZSM-5 structure.

15.7.3.1 Physicochemical Characterization of Active Sites
The acid strength of protons in the crystalline molecular sieve structure plays a key role in of MTO catalysis. The acid sites of silicoalumina-based zeolites

Table 15.2 Comparison of SAPO-34 and SSZ-13 (chabazite) catalysts for MTO.

Material (T atom %)	SAPO-34 (10% Si)	SSZ-13 (18% Al)	SSZ-13 (10% Al)	SSZ-13 (3.3% Al)
Selectivity (% at 2 h)				
C_2–C_4 olefins	96	69	75	87
CH_4	1.4			
C_2H_6	0.3			
C_3H_8	0.9			
Stability (h at >50% conversion)	>40	6	13	7
Coking (% carbon on used catalyst)	19 after 54 h	16.6 after 18 h	19.3 after 18 h	15.0 after 18 h

tend to have higher acid strengths than those of SAPO-based sieves which accelerate the rates of side reactions that lead to aromatics and coke formation. The zeolitic acid sites arise from the substitution of Al^{+3} for Si^{+4} in the mostly siliceous crystalline lattice. An acidic proton, as a bridging hydroxyl between Si and Al framework atoms, must also be added to the framework in order to balance the charge. In the case of SAPO molecular sieves the basic structure is $AlPO_4$. Substitution of Si^{+4} and H^+ for a portion of the framework P^{+5} introduces acid sites to the material. The greater electron affinity of aluminum bonded to –O–Si(OR)$_3$ groups in the zeolite, compared to –O–P(OR)$_3$ for the silicoalumino-phosphate, more effectively stabilizes the conjugate anion of the bridging hydroxyl. The impact of this acidity difference on MTO performance has been shown by the comparison of the catalytic performances of SAPO-34 and SSZ-13, which is a synthetic aluminosilicate that also has the chabazite structure [103]. Table 15.2 shows that SAPO-34 has significantly better stability, due to a lower rate of coke formation, than SSZ-13 samples with comparable tetrahedral-atom substitutions and acid site densities. SSZ-13 also shows greater production of light paraffins which is consistent with accelerated hydride transfer for the catalyst with higher acidic strength.

In addition to shape selectivity and acid-site strength, other catalyst characteristics that influence the catalytic performance of SAPO-34 have also been identified. Variation in the SAPO-34 gel composition and synthesis conditions have been were used to prepare samples with different median particle sizes and Si contents (Tables 15.3 and 15.4) [104]. In these samples the median particle size was varied from 1.4 to 0.6 µm, and the Si mole fraction in the product was varied from 0.14 down to 0.016. A comparison of samples B and E (which have similar particle size distributions) shows that reducing Si content decreases propane formation and increases catalyst life. A comparison of samples B and C (which have similar Si contents) illustrates an increase in catalyst life with a reduction in particle size.

Table 15.3 Preparative conditions for SAPO-34 in Table 15.4. All samples were stirred during crystallization. DPA = di-n-propylamine.

Catalyst	Synthesis formulation (molar basis)	Temperature (°K)	Time (h)
A	2.0 DPA · 0.5 TEAOH · 0.6 SiO_2 · Al_2O_3 · P_2O_5 · 50 H_2O	448	17
B	1.0 TEAOH · 0.4 SiO_2 · Al_2O_3 · P_2O_5 · 35 H_2O	423	72
C	1.0 TEAOH · 0.4 SiO_2 · Al_2O_3 · P_2O_5 · 35 H_2O	423	72
D	1.0 TEAOH · 0.1 SiO_2 · Al_2O_3 · P_2O_5 · 35 H_2O	423	72
E	1.0 TEAOH · 0.05 SiO_2 · Al_2O_3 · P_2O_5 · 35 H_2O	448	72

Table 15.4 Influence of SAPO-34 particle properties and composition on initial catalyst performance.

Catalyst[a]	Particle size (μm)		Si[d]	Initial performance[e]		Catalyst life[f] (h)
	50%[b]	90%[c]		$C_2^{2-} + C_3^{2-}$	C_3	
A	1.2		0.14	70.6	7.3	15
B	0.7	1.3	0.09	73.2	3.5	25
C	1.4	2.5	0.09	75.3	2.9	14
D	0.9	2.0	0.03	73.9	1.1	36
E	0.6	1.2	0.016	72.9	0.7	33

a) Preparative conditions in Table 15.3.
b) 50% of particles are smaller than listed values.
c) 90% of particles are smaller than listed values.
d) Si framework mole fraction, expressed as Si/(Si + Al + P).
e) Feed consists of $MeOH/H_2/H_2O$ in 1/5.3/4.6 molar ratio; 673°K, 136 kPa and 1 h^{-1} MeOH WHSV. Initial performance measured at 45 min on-stream.
f) Time to first appearance of MeOH or DME in reactor effluent.

The data show that improved performance, in terms of reduced propane by-product formation and increased catalyst life, is obtained by using a catalyst having an average particle size of less than 1.0 μm or less than 0.05 mole fraction Si or both. Superior performance is obtained with catalysts that combine these two properties.

Because of the variety of Si locations (isolated Si and Si islands) in SAPO molecular sieves, frequently no correlation exists between Si content and the number of acid sites [105]. However, for SAPO with low Si content, Si sites are usually isolated and there is one acid site per Si. In general, the framework charge and, thus, the maximum number of acid sites in a SAPO should be related to the value of framework (Al-P) [106]. This relationship is true of zeolites too, because (Al-P) is equal to Al if there is no framework P. Based on this relationship, the

Table 15.5 Observed acid-site density based on NH_3-TPD, compared to calculated values based on SAPO-34 stoichiometry.

Catalyst	Si	Observed Meq NH_3 (cm^3 micropore volume)	Calculated Meq NH_3 (cm^3 micropore volume)[a]	Acid sites per CHA cage[a]
A	0.14	5.1	7.0	1.38
B	0.09	3.2	3.2	0.64
C	0.09	3.2	3.9	0.77
D	0.03	1.4	2.3	0.47
E	0.016	0.9	1.6	0.31

a) Calculated based on mole fraction (Al-P).

number of acid sites were calculated for catalysts A to E and compared with the observed acid site density measured by NH_3-TPD (Table 15.5).

15.7.4
Reaction Mechanism of Methanol to Hydrocarbons

The mechanism of higher hydrocarbon formation from methanol over acidic molecular sieves has been a subject of interest ever since the first discovery of MTG chemistry in the 1970s. In the past few years the Haw and Kolboe/Olsbye groups have demonstrated the importance of a hydrocarbon pool of adsorbed molecules, in particular methylated aromatics, in the conversion methanol to higher olefins [107, 108]. Isotopic labeling studies have shown exchange between side chain and ring carbons of adsorbed aromatics at typical MTO conditions [109]. This is consistent with the involvement of aromatics in methanol conversion, probably via methylation, ring contraction and expansion and dealkylation. Direct methylation of ^{12}C-ethylene with ^{13}C-methanol has also been observed at low conversion [110]. Thus, it appears that the formation of higher hydrocarbons from methanol involves a complex series of steps including aromatic methylation, rearrangement and dealkylation as well as olefin methylation and cracking [108]. These pathways appear to occur on a wide variety of molecular sieves ranging from large-pore beta zeolite to medium-pore ZSM-5 and small-pore SAPO-34. The key differences in the performance of these materials are relative steric constraints on product diffusion out of the molecular sieve structure. In the case of SAPO-34, this constraint favors the formation of the small olefins, ethylene and propylene.

Chen and co-workers have studied the role of coke deposition in the conversion of methanol to olefins over SAPO-34 [111]. They found that the coke formed from oxygenates promoted olefin formation while the coke formed from olefins had only a deactivating effect. The yield of olefins during the MTO reaction was found to go through a maximum as a function of both time and amount of coke. Coke was found to reduce the DME diffusivity, which enhances the formation of olefins, particularly ethylene. The ethylene to propylene ratio increased with intracrystalline coke content, regardless of the nature of the coke.

15.8
Summary

Zeolite catalysts play a vital role in modern industrial catalysis. The varied acidity and microporosity properties of this class of inorganic oxides allow them to be applied to a wide variety of commercially important industrial processes. The acid sites of zeolites and other acidic molecular sieves are easier to manipulate than those of other solid acid catalysts by controlling material properties, such as the framework Si/Al ratio or level of cation exchange. The uniform pore size of the crystalline framework provides a consistent environment that improves the selectivity of the acid-catalyzed transformations that form C–C bonds. The zeolite structure can also inhibit the formation of heavy coke molecules (such as medium-pore MFI in the Cyclar process or MTG process) or the desorption of undesired large by-products (such as small-pore SAPO-34 in MTO). While faujasite, mordenite, beta and MFI remain the most widely used zeolite structures for industrial applications, the past decade has seen new structures, such as SAPO-34 and MWW, provide improved performance in specific applications. It is clear that the continued search for more active, selective and stable catalysts for industrially important chemical reactions will include the synthesis and application of new zeolite materials.

References

1 Luebke, C., Meister, J., Duttlinger, M., and Krupa, S. (2001) Continue producing premium blendstocks. Paper AM-01-24, NPRA, March 18–21, at New Orleans, LA.
2 Pater, J.P.G., Jacobs, P.A., and Martens, J.A. (1998) 1-Hexene oligomerization in liquid, vapor and supercritical phases over beidellite and ultrastable Y zeolite catalysts. *J. Catal.*, **179**, 477.
3 Martens, J.A., Raman Ravishankar, I.E.M., and Jacobs, P.A. (2000) Tailored alkene oligomerization with H-SM-57 zeolite. *Angew. Chem. Int. Ed.*, **39** (23), 4376.
4 Detrick, K.A., Himes, J.F., Meister, J.M., and Nowak, F.-M. (2004) UOP HF alkylation technology, in *Handbook of Petroleum Refining Processes* (ed. R.A. Meyers), McGraw-Hill, New York, pp. 1.33–1.56.
5 Graves, D.C. (2004) STRATCO Effluent refrigerated H2SO4 alkylation process, in *Handbook of Petroleum Refining Processes* (ed. R.A. Meyers), McGraw-Hill, New York, pp. 1.11–1.24.
6 Corma, A. and Martinez, A. (1993) Chemistry, catalysts, and processes for isoparaffin-olefin alkylation: actual situation and future trends. *Catal. Rev. Sci. Eng.*, **35**, 483–570.
7 Schatz, K.W. and Koopman, R.P. (1989) Effectiveness of Water Spray Mitigation Systems for Accidental Releases of Hydrogen Fluoride, summary report and volumes I-X, NTIS, Springfield, VA.
8 Feller, A. and Lercher, J.A. (2004) Chemistry and technology of isobutane/alkene alkylation catalyzed by liquid and solid acids. *Adv. Catal.*, **48**, 229–295.
9 Garwood, W.E. and Venuto, P.B. (1968) Paraffin-olefin alkylation over a crystalline aluminosilicate. *J. Catal.*, **11**, 175.
10 Chu, Y.F. and Chester, A.W. (1986) Reactions of isobutane with butene over zeolite catalysts. *Zeolites*, **6**, 195–200.

11 Sievers, C., Liebert, J.S., Stratmann, M.M., Olindo, R., and Lercher, J.A. (2008) Comparison of zeolites LaX and LaY as catalysts for isobutane/2-butene alkylation. *Appl. Catal. A*, **336**, 89–100.

12 Stöcker, M., Mostad, H., and Rørvik, T. (1994) Isobutane/2-butene alkylation on faujasite- type zeolites. *Catal. Lett.*, **28**, 203–209.

13 Stöcker, M., Mostad, H., Karlsson, A., Junggreen, H., and Hustad, B. (1996) Isobutane/ 2-butene alkylation on dealuminated H EMT and H FAU. *Catal. Lett.*, **40**, 51–58.

14 Rørvik, T., Mostad, H.B., Karlsson, A., and Ellestad, O.H. (1997) Isobutane/ 2-butene alkylation on fresh and regenerated La-EMT-51 compared with H-EMT. The catalysts selectivity changes at high butene conversion in a slurry reactor. *Appl. Catal. A*, **156**, 267–283.

15 Hellring, S.D., Huss, A., and Thomson, R.T. (1995) Hydrogen transfer and isoparaffin- olefin alkylation process. U.S. Patent 5, 461,182.

16 He, Y.J., Nivarthy, G.S., Eder, F., Seshan, K., and Lercher, J.A. (1998) Synthesis, characterization and catalytic activity of the pillared molecular sieve MCM-36. *Micropor. Mesopor. Mater.*, **25**, 207–224.

17 Nivarthy, G.S., Feller, A., Seshan, K., and Lercher, J.A. (2000) Alkylation of isobutane with light olefins catalyzed by zeolite beta. *Micropor. Mesopor. Mater.*, **35–36**, 75–87.

18 de Jong, K.P., Mesters, C.M.A.M., Peferoen, D.G.R., van Brugge, P.T.M., and de Groot, C. (1996) Paraffin alkylation using zeolite catalysts in a slurry reactor: chemical engineering principles to extend catalyst lifetime. *Chem. Eng. Sci.*, **51**, 2053–2060.

19 Yang, C.-L. (1974) Hydrocarbon alkylation process using catalyst regeneration. U.S. Patent 3, 851,004.

20 Yang, C.-L. (1975) Isoparaffin alkylation process with periodic catalyst regeneration. U.S. Patent 3, 893,942.

21 Van Broekhovem, E.H., Cabre, F.R.M., Bogaard, P., Klaver, G., and Vonhof, M. (1999) Process for alkylating hydrocarbons. U.S. Patent 5, 986,158.

22 Thompson, D.N., Ginosar, D.M., Burch, K.C., and Zalewski, D.J. (2005) Extended catalyst longevity via supercritical isobutane regeneration of a partially deactivated USY alkylation catalyst. *Ind. Eng. Chem. Res.*, **44**, 4534–4542.

23 Feller, A., Barth, J.-O., Guzman, A., Zuazo, I., and Lercher, J.A. (2003) Deactivation pathways in zeolite-catalyzed isobutane/butene alkylation. *J. Catal.*, **220**, 192.

24 Sievers, C., Zuazo, I., Guzman, A., Olindo, R., Syska, H., and Lercher, J.A. (2007) Stages of aging and deactivation of zeolite LaX in isobutane/butene alkylation. *J. Catal.*, **246**, 315–324.

25 Dautzenberg, F. and Angevine, P.J. (2004) Encouraging innovation in catalysis. *Catal. Today*, **93–95**, 3–16.

26 Mukherjee, M., Nehlsen, J., Sundaresan, S., Suciu, G.D., and Dixon, J. (2006) Scale-up strategy applied to solid-acid alkylation process. *Oil Gas J.*, 48–54.

27 Huang, T.J. and Shinnar, R. (1994) Isoparaffin-olefin alkylation process. U.S. Patent 5, 292,981.

28 Zhang, S.Y., Gosling, C.D., Sechrist, P.A., and Funk, G.A. (1996) Fluidized solid bed motor fuel alkylation process. U.S. Patent 5, 489,732.

29 Eberhardt, J., Boll, W., Buchold, H., and Dropsch, H. (2002) Method for catalytically reacting isoparaffins with olefins to from alkylates. WO Patent 02/094747A1.

30 Nivarthy, G., Seshan, K., and Lercher, J.A. (1998) The influence of acidity on zeolite H-BEA catalyzed isobutane /n-butene alkylation. *Micropor. Mesopor. Mater.*, **22**, 379–388.

31 Feller, A., Zuazo, I., Guzman, A., Barth, J.-O., and Lercher, J.A. (2003) Common mechanistic aspects of liquid and solid acid catalyzed alkylation of isobutane with n-butene. *J. Catal.*, **216**, 313–323.

32 Guzman, A., Zuazo, I., Feller, A., Olindo, R., Sievers, C., and Lercher, J.A. (1998) Influence of the activation temperature on the physical properties and catalytic activity of La-X zeolites for isobutane/n-butene alkylation. *Micropor. Mesopor. Mater.*, **22**, 379–388.

33. Degnan, T., Jr., Smith S.M., and Venkat, C.R. (2001) Alkylation of aromatics with ethylene and propylene: recent development in commercial processes. *Appl. Catal. A*, **221**, 283–294.
34. Perego, C. and Ingallina, P. (2002) Recent advance in the industrial alkylation of aromatics: new catalysts and new processes. *Catal. Today*, **73** (1/2), 3–22.
35. Woodle, G.B. (2005) Ethylbenzene Encyclopedia of Chemical Processing by Sunggyu Lee, CRC Press.
36. Schmidt, R.J. (2005) Industrial catalytic processes-phenol production. *Appl. Catal. A*, **280** (1), 89–93.
37. Perego, C., Amarilli, S., Millinni, R., Bellussi, G., Girotti, G., and Terzoni, G. (1996) Experimental and computational study of beta, ZSM-12, Y, mordenite and ERB-1 in cumene synthesis. *Micropor. Mater.*, **6**, 395.
38. Sastre, G., Catlow, C.R.A., and Corma, A. (1999) Diffusion of benzene and propylene in MCM-22 zeolite: a molecular dynamics study. *J. Phys. Chem. B*, **103** (25), 5187.
39. Juttu, G.G. and Lobo, R.F. (2000) Characterization and catalytic properties of MCM-56 and MCM-22 zeolites. *Micropor. Mesopor. Mater.*, **40**, 9–23.
40. Corma, A., Martinez-Soria, V., and Schnoeveld, E. (2000) Alkylation of benzene with short-chain olefins over MCM-22 zeolite: catalytic behavior and kinetic mechanism. *J. Catal.*, **192**, 163.
41. Rigoreau, J., Laforge, S., Gnep, N.S., and Guisnet, M. (2005) Alkylation of toluene with propene over H-MCM-22 zeolite. location of the main and secondary reactions. *J. Catal.*, **236**, 45–54.
42. Erecan, C., Dautzenberg, F., Yeh, C.Y., and Barner, H.E. (1998) Mass transfer effects in liquid phase alkylation of benzene with zeolite catalysts. *Ind. Eng. Chem. Res.*, **37**, 1724–1728.
43. Vaudry, F., Di Renzo, F., Fajula, F., and Schulz, P. (1998) Origin of the optimum in catalytic activity of zeolite beta. *JCS Faraday Trans.*, **94**, 617–621.
44. Cracium, I., Reyniers, M.-F., and Marin, G.B. (2007) Effects of acid properties of Y zeolites on the liquid phase alkylation of benzene with 1-octene: a reaction path analysis. *J. Mol. Catal. A*, **277**, 1–14.
45. Boveri, M., Marquez-Alvarez, C., Laborde, M.A., and Sastre, E. (2006) Steam and acid dealumination of mordenite characterization and influence on the catalytic performance in linear alkylbenzene synthesis. *Catal. Today*, **114**, 217–225.
46. Cao, Y., Kessas, R., Naccache, C., and Taarit, Y.B. (1999) Alkylation of benzene with dodecene. The activity and selectivity of zeolite type catalysts as a function of the porous structure. *Appl. Catal.*, **184**, 231–238.
47. Magnoux, Z.D. and Guisnet, M. (1999) Liquid phase alkylation of toluene with 1-heptene over a HFAU zeolite: evidence for transalkylation between toluene and non-desorbed products. *Appl. Catal. A*, **182**, 407–411.
48. Meriaudeau, P., Taarit, Y.B., Thangaraj, A., Almeida, J.L.G., and Naccache, C. (1997) Zeolite based catalysts for linear alkylbenzene production: dehydrogenation of long chain alkanes and benzene alkylation. *Catal. Today*, **38**, 243–247.
49. Das, J., Char, P.R., Halgeri, A.A., Mewada, R.K., and Subrahmanyam, N. (2003) Kinetics of toluene methylation over silica modified H-ZSM-5 zeolites. *Indian J. Chem. Technol.*, **9**, 334–240.
50. Reddy, K.M., Sun, L., and Song, C. (1998) Effect of physico-chemical properties of ZSM-5 on shape selective methylation of toluene. 215 National Meeting, American Chemical Society at Dallas, Texas, March 1998.
51. Xie, Y., Zhao, B., Long, X., and Tang, Y. (2000) Dispersion of oxides on H-ZSM-5 and threshold effect on shape selective methylation of toluene. *ACS Symp. Ser.*, **738** (Shape Selective Catalysis), 188–200.
52. Ghosh, A.K. and Harvey, P. (2005) Toluene methylation process. WO Patent 2005/033071A2.
53. Brown, S.H., Mathias, M.F., Ware, R.A., and Olson, D.H. (2002) Selective para-xylene production by toluene methylation. U.S. Patent 6, 423,879.
54. Rep, M., Palomares, A.E., Eder-Mirth, G., Van Ommen, J.G., and Lercher, J.A.

(1999) On selectivity aspects of alkylation of toluene with methanol over zeolites, DGMK tagungsbericht, 9903. Proceedings of DGMK-Conference The Future Role of Aromatics in Refining and Petrochemistry, 1999, pp. 279–286.
55 Yashima, T., Ahmad, H., Yamazaki, K., Katsuta, M., and Hara, N. (1970) Alkylation on synthetic zeolites, I. alkylation of toluene with methanol. *J. Catal.*, **16**, 273–280.
56 Dumitriu, E., Fechete, I., Caullet, P., Kessler, H., Hulea, V., Chelaru, C., Hulea, T., and Bourdon, A. (2002) Conversion of aromtic hydrocarbons over MCM-22 and MCM-36 catalysts, Stud. Surf. Sci. Catal., vol. 142, Elsevier, Amsterdam, pp. 951–958.
57 Vu, D.V., Miyamoto, M., Nishiyama, N., Ichikawa, S., Egashira, Y., and Ueyama, K. (2008) Catalytic activities and structures of silicalite-1/H-ZSM-5 zeolite composites. *Micropor. Mesopor. Mater.*, **115**, 106.
58 Yashina, T., Sato, K., Hayasaka, T., and Hara, N. (1972) Alkylation on synthetic zeolites: III. Alkylation of toluene with methanol and formaldehyde on alkali cation exchange zeolite. *J. Catal.*, **26**, 303.
59 Itoh, H., Miyamoto, A., and Murakami, Y. (1980) Mechanism of the side-chain alkylation of toluene with methanol. *J. Catal.*, **64**, 284–294.
60 Wieland, W.S., Davis, R.J., and Garces, J.M. (1996) Solid base catalysts for side-chain alkylation of toluene with methanol. *Catal. Today*, **28**, 443–450.
61 Palomares, A.E., Eder-Mirth, G., Rep, M., and Lercher, J.A. (1998) Alkylation of toluene over basis catalysts – key requirements for side chain alkylation. *J. Catal.*, **180**, 56–65.
62 Hathaway, P.E. and Davis, M.E. (1989) Base catalysis by alkali modified zeolite. *J. Catal.*, **119**, 497–507.
63 Unland, M.L. and Barker, G.E. (1978) Zeolite catalyst. U.S. Patent 4, 115,424.
64 Lacroix, C., Deluzarche, A., and Kiennemann, A. (1984) Promotion role of some metals (Cu, Ag) in the side chain alkylation of toluene with methanol. *Zeolites*, **4**, 109–111.

65 Liu, H.-C. (1985) Modified zeolite catalyst composition and process for alkylating toluene with methanol to form styrene. U.S. Patent 4, 499,318.
66 Borgna, A., Speulveda, J., Magni, S.I., and Apesteguia, A.R. (2004) Active sites in alkylation of toluene with methanol: a study by selective acid-base poisoning. *Appl. Catal. A*, **276**, 207–215.
67 Borgna, A., Magni, S., Speulveda, J., Padro, C.L., and Apesteguia, A.R. (2005) side chain alkylation of toluene with methanol on Cs-exchanged Na-Y zeolite: effect of Cs loading. *Catal. Lett.*, **102**, 15–21.
68 Choi, S., Oh, S.H., Kim, Y.S., Seong, K.H., Lim, B.S., and Lee, J.H. (2006) APUTM technology for the production of BTX and LPG from pyrolysis gasoline using metal promoted zeolite catalyst. *Catal. Surv. Asia*, **10** (2), 110–116.
69 Tsai, T.-C., Liu, S.-B., and Wnag, I. (1999) Disproportionation and trans-alkylation of alkylbenzenes over zeolite catalysts. *Appl. Catal. A*, **181**, 355–398.
70 Tsai, T.-C., Chen, W.-H., Liu, S.-B., Tsai, T.S., and Wnag, I. (2002) Metal zeolite for trans-alkylation of toluene and heavy aromatics. *Catal. Today*, **73**, 39–47.
71 Tsai, T.-C., Chao, P.-H., and Zeng, W.L. (2007) Selective naphthene cracking over cascade dual catalyst in heavy aromatics trans-alkylation. *Zeolite to Porous MOF Materials International Zeolite of Conference in Beijing*, pp. 1611–1616.
72 Oh, S.-H., Lee, S.I., Seong, K.H., Kim, Y.S., Lee, J.H., Woltermann, J., Cormier, W.E., and Chu, Y.F. (2002) Heavy aromatics upgrading using noble metal promoted zeolite catalyst. *Stud. Sur. Sci. Catal.*, **142**, 887.
73 Oh, S.-H., Lee, S.I., Seong, K.H., Kim, Y.S., Lee, J.H., Woltermann, J., Cormier, W.E., and Chu, Y.F. (2003) Heavy aromatics upgrading using noble metal promoted zeolite catalyst. *Stud. Sur. Sci. Catal.*, **145**, 487.
74 Oh, S.-H., Lee, S.I., Seong, K.H., and Park, S.H. (2008) Catalyst for the disproporation/transalkylation of

75 Serra, J.M., Guillon, E., and Corma, A. (2004) A rational design of alkyl-aromatics dealkylation-transalkylation catalysts using C8 and C9 alkyl-aromatics as reactants. *J. Catal.*, **227**, 459–469.
76 Stachelczyj, D.A. (2000) TransPlusTM: a flexible approach for upgrading heavy aromatics. Paper AM-00-60, *NPRA, March 26–28, at San Antonio, TX*.
77 Shen, Q., Zhu, X., and Dong, J. (2009) Hydrodealkylation of C9+ aromatics to BTX over zeolite-supported nickel oxide and molybdenum oxide catalysts. *Catal. Lett.*, **129**, 170–180.
78 Imai, T., Kocal, J.A., and Gosling, C.D. (1991) The UOP-BP cyclar process. *Catal. Sci. Technol.*, **1**, 399–402.
79 Dave, D. and Hall, A.H.P. (1981) Production of Aromatics from ethane and/or ethylene. EP 0050021B1 to BP.
80 Johnson, J.A. and Hilder, G.K. (1984) Cyclar process for aromatization of LPG. *NPRA* Annual Meeting at San Antonio, TX, pp. 84–45.
81 Kawase, M., Nomura, K., Nagamori, Y., and Kinishita, J. (2001) High silica content zeolite based catalyst. U.S. Patent 6, 207,605.
82 Kiyama, K., Tsunoda, T., and Kawase, M. (1999) Method for producing aromatic hydrocarbons. U.S. Patent 5, 877,368.
83 Yao, J., Kimble, J.B., and Drake, C.A. (2000) A method of making such improved zeolite material and the use thereof the conversion of non-aromatic hydrocarbon to aromatics and light olefins. U.S. Patent 6, 048,815.
84 Drake, C.A. (1998) Improved catalyst composition useful for conversion of non-aromatic hydrocarbons to aromatics and light olefins. WO Patent 98/51409.
85 Meriasdeau, P. and Naccache, C. (1997) Dehydrocyclization of alkanes over zeolite-supported metal catalysts: monofunctional or bifunctional route. *Cat. Rev. Sci Eng.*, **39**, 5–48.
86 Bernard, J.R. (1980) Hydrocarbons aromatization on platinum alkaline zeolites, in *Proceedings of the Fifth International Conference on Zeolites* (ed. L.V. Rees), Heyden, London, p. 686.
87 Chang, C.D. and Silvestri, A.J. (1977) The conversion of methanol and other O-compounds to hydrocarbons over zeolite catalysts. *J. Catal.*, **47**, 249.
88 Lewis, J.M.O. (1988) Methanol to olefins process using silicoalumino-phosphate molecular sieve catalysts, in *Catalysis 1987* (ed. J.W. Ward), Elsevier, Amsterdam, p. 199.
89 Stocker, M. (1999) Methanol-to-hydrocarbons catalytic materials and their behavior. *Micropor. Mesopor. Mater.*, **29**, 3.
90 Froment, G.F., Dehertog, W.J.H., and Marchi, A.J. (1992) Zeolite catalysis in the conversion of methanol into olefins. *Catalysis*, **9**, 1.
91 Kvisle, S., Fuglerud, T., Kolboe, S., Olsbye, U., Lillerud, K.P., and Vora, B.V. (2008) Methanol to hydrocarbons, in *Handbook of Heterogeneous Catalysis* (eds G. Ertl, H. Knozinger, and J. Weitkamp), Wiley-VCH Verlag GmbH, Weinheim, pp. 2950–2965.
92 Chen, J.Q., Bozzano, A., Glover, B., Fuglerud, T., and Kvisle, S. (2005) Recent advancements in ethylene and propylene production using the UOP/hydro MTO process. *Catal. Today*, **106**, 103–107.
93 Vora, B.V., Marker, T.L., Barger, P.T., Nilsen, H.R., Kvisle, S., and Fuglerud, T. (1997) Economic route for natural gas conversion to ethylene and propylene, Stud. Surf. Sci. Catal., vol. 107, Elsevier, Amsterdam, pp. 87.
94 Socha, R.F., Chang, C.D., Gould, R.M., Kane, S.E., and Avidan, A.A. (1987) Fluid-bed studies of olefin production from methanol, in *Industrial Chemical via C_1 Processes* (ed. D.R. Fahey), ACS, Washington D.C, p. 34.
95 Chang, C.D., Chu, C.T.-W., and Socha, R.F. (1984) Methanol conversion to olefins over ZSM-5. *J. Catal.*, **86**, 289–296.
96 Rothaemel, M. and Holtmann, H.-D. (2001) C1 to propylene: a long awaited economical route from natural gas to chemical base feedstock, in *Proceedings AIChE Spring National Meeting*,

American Institute of Chemical Engineers, New York, p. 37.

97 Liebner, W. (2005) Gas to propylene, GTP/MTP technology. Presented at Propylene Trade & Derivatives Markets, Singapore, October 24–25.

98 Chang, C.D. (1997) Methanol to hydrocarbons, in *Handbook of Heterogeneous Catalysis* (eds G. Ertl, H. Kozinger, and J. Weitkamp), Wiley-VCH Verlag GmbH, Weinheim, pp. 1894–1907.

99 Chang, C.D. (1992) The New Zealand gas-to-gasoline plant: an engineering tour de force. *Catal. Today*, **13**, 103.

100 Maiden, C.J. (1988) The New Zealand Gas-to-gasoline project, Stud. Surf. Sci. Catal., vol. 36, Elsevier, Amsterdam, p. 1.

101 Krohn, D.E. and Melconian, M.G. (1988) The first fixed-bed methanol-to-gasoline (MTG) plant: design and scale-up considerations, Stud. Surf. Sci. Catal., vol. 36, Elsevier, Amsterdam, p. 679.

102 Kaiser, S.W. (1985) Methanol conversion to light olefins over silicoaluminophosphate molecular sieves. *Arab. J. Sci. Eng.*, **10**, 361–366.

103 Yuen, L-T., Zones, S.I., Harris, T.V., Gallegos, E.J., and Auroux, A. (1994) Product selectivity in methanol to hydrocarbons conversion for isostructural compositions of AEI and CHA molecular sieves. *Micropor. Mater.*, **2**, 105.

104 Wilson, S.T. and Barger, P.T. (1999) Characteristics of SAPO-34 that influence the conversion of methanol to light olefins. *Micropor. Mesopor. Mater.*, **29**, 117–126.

105 Barthomeuf, D. (1994) Acidity and Basicity of Solids. *NATO ASI Ser. Ser. C*, **444**, 375.

106 Vomscheid, R., Briend, M., Peltre, M.J., Man, P.P., and Barthomeuf, D. (1994) The role of the template in directing the Si distribution in SAPO zeolites. *J. Phys. Chem.*, **98**, 9614.

107 Haw, J.F., Song, W., Marcus, D.M., and Nicholas, J.B. (2003) The mechanism of methanol to hydrocarbons catalysis. *Acc. Chem. Res.*, **36**, 317.

108 Olsbye, U., Bjorgen, M., Svelle, S., Lillerud, K.-P., and Kolboe, S. (2005) Mechanistic insight into the methanol-to-hydrocarbons reaction. *Catal. Today*, **106**, 108.

109 Bjorgen, M., Olsbye, U., Petersen, D., and Kolboe, S. (2004) The methanol-to-hydrocarbon reaction: insight into the reaction mechanism from [^{12}C]benzene and [^{13}C]methanol coreactions over zeolite H-beta. *J. Catal.*, **221**, 1.

110 Svelle, S., Rooning, P.O., and Kolboe, S. (2004) Kinetic studies of zeolite-catalyzed methylation reactions: 1. coreaction of [^{12}C]ethene and [^{13}C] methanol. *J. Catal.*, **224**, 115.

111 Chen, D., Rebo, H.P., Moljord, K., and Holmen, A. (1997) The role of coke deposition in the conversion of methanol to olefins over SAPO-34, Stud. Surf. Sci. Catal., vol. 111, Elsevier, Amsterdam, p. 159.

16
Bond Breaking and Rearrangement
Suheil F. Abdo

16.1
Introduction

Bond breaking and rearrangement of hydrocarbon molecules are the two essential chemical transformations underlying major refining processes, including fluidized catalytic cracking (FCC), hydrocracking (HC) and isomerization. Over the years these processes have increasingly relied on zeolitic catalysts due to their strong acidity and molecular sieving properties. This chapter reviews key structural and catalytic properties of the most important zeolites employed in these processes, starting with a brief historical perspective to describe the key developments which led to their application in these processes and to the explosive growth in the use of zeolites in refining in general. Many of the historical and fundamental aspects of zeolite technology have been covered in earlier chapters, however, some aspects specific to these applications are covered in this chapter in order to provide a coherent and more complete description of the field.

Control of the multitude of pathways which feed molecules can take is the primary objective of all catalyst and process developments. The work covered in this chapter focuses primarily on describing the approaches in material and catalysis development which have led to major advances in zeolite application in hydrocarbon conversion. The breaking and formation of carbon–carbon and carbon–hydrogen bonds constitute the majority of the chemical transformations involved here with the less prevalent, but very important, breaking of carbon bonds with sulfur, nitrogen and oxygen taking place in parallel.

Many books, reviews and treatises have been published on related subjects [1–7]. Thus the objective of this chapter is the delineation of the key features of the catalytic surface and the process conditions which enable better control of the reaction pathways for more efficient and environmentally friendly processes and minimal utilization of precious natural resources. As it stands today, hundreds of known framework types have been synthesized and scaled-up [8], but only a handful have found significant application in the hydrocarbon processing industries. They are zeolite Y and its many variants, ZSM-5, Mordenite and zeolite Beta. Other very important crystalline materials (including aluminophosphates (ALPOs),

Zeolites in Industrial Separation and Catalysis. Edited by Santi Kulprathipanja
Copyright © 2010 WILEY-VCH Verlag GmbH & Co. KGaA, Weinheim
ISBN: 978-3-527-32505-4

silicoaluminophosphates (SAPOs), and many other crystalline materials [9–14] have found some commercial application but their use is much less common than the group above. Thus, while many scientifically interesting and fundamentally elegant studies of the catalytic applications of zeolites have been published, a comprehensive review of this voluminous literature is beyond the scope of this chapter. Instead, we focus on the established knowledge and principles which led to successful application and resulted in technologically significant contributions. A key objective here is to identify some unifying themes of the role of zeolites in bond making and breaking to improve our understanding of how they function and how their properties are modified and controlled to make them suitable for new applications.

16.2
Critical Zeolite Properties

Zeolite science and technology have progressed since the early discovery phase to the point where we have a much deeper understanding of the key properties required to effectively function in various applications. However, there remains much room for discovery of the specific characteristics needed for effective use in different process applications. Since the role of zeolites in catalytic processes is to act as solid acids and provide geometrically well defined environments for reactants, reactive intermediates, and products, it comes as no surprise that the nature of acidity and the local site geometry have received the most attention in zeolite catalysis literature. Determination of the intrinsic strength and strength distribution of acid sites and their spatial distribution are central to the understanding of their precise functional attributes. Given today's more precise performance requirements, sophisticated tools such as infrared, nuclear magnetic resonance, gravimetric methods and thermal analysis are critical tools applied along with the older traditional methods for acidity measurement to build a more comprehensive picture of the catalyst. In addition to assessing the acidity, the crystal morphology, pore geometry, micro- and meso-porosity and composition of the framework and extra-framework species must all be controlled and accounted for in any comprehensive description of a catalytic zeolite. For example, volumes of scientific and marketing literature have been generated in the FCC technology area, demonstrating significant advantages from subtle variations in one or more of these attributes [15–20].

16.2.1
Framework Types and Compositions

As mentioned above, the main types of zeolites in industrial catalytic application today are Zeolite Y (FAU) and its variants, ZSM-5 and Silicalite (MFI), Mordenite (MOR) and Beta (BEA). After the first commercial synthesis of X and Y zeolites in the mid- to late 1940s by workers at the Union Carbide Corporation [21, 22], their tremendous catalytic potential was unlocked a few years later [23]. While the intrinsic power of their acidity was recognized early on, it was the development of methods to reliably unlock this acidity and provide stability under process conditions which permitted full realization of this potential. It was hypothesized early on that the enhanced acidity of zeolites compared to amorphous silica–aluminas is due to the presence of aluminum cations in framework positions substituting for Si in the extended SiO_2 network [1, 23, 24]. As synthesized, the excess negative charge associated with the Al^{3+} cations in lattice positions is neutralized by the Na^+ cations present in the synthesis media. Early attempts to unlock the acidity involved ion exchange with multivalent cations such as Ca^{2+} and Mg^{2+} and with ammonium cations which after thermal treatment converted to the hydrogen form HY [21–23]. However, it was the discovery of two major methods of stabilization that eventually led to wide-scale application in the FCC process. Workers at the Mobil Oil company [25] discovered the activity-enhancing and stabilizing effect of rare-earth exchange which not only rendered an acidic form of Y zeolites but imparted hydrothermal stability to these structures permitting their use in the FCC process where they must withstand the harsh environment of the regenerator where temperatures of around 750 °C and high water partial pressure prevail. This development had such a tremendous impact on FCC technology and the United States gasoline production capacity that the contribution of these scientists was acknowledged by their eventual induction into the Inventors Hall of Fame. Meanwhile, workers at W.R. Grace Company discovered the so-called steam stabilization effect which resulted in the wide-spread use of the highly acidic and very stable USY form [26, 27]. Much debate took place in the scientific literature following these discoveries regarding the origin of the enhanced acidity and stability of the USYs [28, 29], but it is well established by now that the dealumination of the framework leading to more spatial separation between Al^{3+} cations and fewer Al next nearest neighbors enhances the intrinsic acid strength per site [30]. The evolution of Al occupancy of framework positions is well illustrated by ^{29}Si NMR measurements (Chapter 4) of Y zeolites which have undergone stabilization treatments of varying severity. As the steaming severity is increased, the number of Si linked to multiple Al species (Q^4, Q^3) decreases in favor of those linked to a lower number of aluminums (Q^2, Q^1). Aluminum-27 NMR provides complimentary data showing Al migration from tetrahedral framework positions to the octahedral extraframework. As the framework Si/Al_2 increases with dealumination, the higher charge on Si^{4+} causes a decreases in unit cell size and in the angle between framework Si or Al cations (T–O–T angle) The added strain is thought to lower the dissociation energy of the bridging hydroxyl protons at the Si–O–Al linkages, thereby increasing acid strength [31]. This evolution can also be monitored by IR spectroscopy in the hydroxyl region as well as the framework vibrations which shift to a higher energy with dealumination due to this increased strain on the T–O–T angle. Today's improved understanding of the mechanism of acidity enhancement in Y zeolites has led to a great deal of focus by researchers in the field on the acidity assessment and control in Y zeolites, and various schemes to control framework composition by secondary synthesis methods have been developed.

Zeolites with the MFI framework are the next most versatile class in use. Two most widely studied and utilized variants of MFI are ZSM-5, discovered by workers at Mobil [32], and Silicalite discovered by the Union Carbide group [33]. The pore geometry of this structure type consists of two systems of intersecting straight and zig-zag channel with elliptical pore openings if about 5.5×5.6 Å. Naturally, entry and exit of reactants, intermediates and products are more hindered in this narrow pore system than in the interconnected cavities of Y zeolites with their 12-membered ring (7.3 Å), but it is thought that the channel intersections in MFI provide a much larger reaction space and behave as a pseudo-supercage accommodating larger molecules than can be expected based on pore mouth dimensions alone. ZSM-5 is most commonly manufactured with a Si/Al_2 ratio in the range of 23–38 whereas Silicalite is essentially a pure crystalline silicate with Si/Al_2 ratios typically above about 250. This vast difference in framework composition leads to significant differences in reactivity and selectivity pattern in hydrocarbon processing. The lower-ratio MFIs have much higher total acidity leading to higher activity and, at times, higher overcracking. Due to their geometric similarities both of these structures have been employed in paraffin conversion processes such as paraffin cracking and isomerization [34]. ZSM-5 is also widely used as an FCC process additive to improve octane quality of the gasoline product by cracking and isomerization of the low-octane normal paraffins [35, 36]. A performance advantage resulting from the restrictive pore dimensions of this structure is the low deactivation rate and enhanced stability of ZSM-5 catalysts attributed to their ability to limit growth of coke domains inside the channel system.

Other zeolites have been either employed or proposed for bond breaking and rearrangement applications. Key among these are Mordenite, zeolite Beta and some silico-alumino phosphates, especially SAPO-11. Despite its high acidity and large pore size, Mordenite has seen limited use in hydrocracking and essentially none in FCC technology due to its tendency to undergo rapid coking and deactivation. This phenomenon is often attributed to its uni-dimensional pore structure consisting of two sets of non-intersecting 12-membered ring and eight-membered ring channels. It is accepted in the literature that the eight-membered ring channels are too small to participate in catalytic reactions, whereas the large 12-membered ring channels are blocked off by rapid deposition of coke, thus preventing access of reactant molecules to the interiors of the channels [37]. This phenomenon appears especially problematic with heavier feeds containing larger molecules which would be more prone to slow diffusion through the straight channel pores than the smaller molecules present in feeds for aromatic alkylation and transalkylation processes where this zeolite is widely employed.

Zeolite Beta with its three-dimensional channel system appears more suitable for conversion of gas oil feeds containing large molecules in FCC and hydrocracking processes and has indeed received much attention in the scientific and technical literature [38–40]. It exhibits very strong acidity which exceeds that of stabilized Y zeolites. While the three dimensional pore system of this zeolite with its 12-member pore mouth makes it suitable for use in processing heavier feedstocks, its more tortuous channel system appears to impart a tendency to selectively crack normal paraffins as compared to Y-zeolite. A close look as the channel geometry reveals the presence of a set of straight channels along with another set of obstructed channels presenting a more tortuous path for the reactants or intermediates to diffuse through. The combination of this pore geometry along with the strong intrinsic acidity result in a tendency for selective cracking of paraffins and excessive production of light products. This latter tendency has limited its application in gas-oil conversion processes where products boiling in the naphtha and distillate ranges are more valuable than the light ends.

Silicoaluminophosphates (SAPOs), along with their crystalline aluminum phosphate counterparts (ALPOs), first discovered by Union Carbide workers in the early 1970s [41, 42], derive their acidity through the substitution of framework phosphorous by silicon thereby creating the charge imbalance which, when compensated for by protons, creates acidic centers. SAPOs in general have seen limited use in bond-breaking applications primarily due to weaker acidity, framework stability, or technoeconomic reasons. Of the rich variety of structures available, SM-3 (a variant of SAPO-11) has been commercially applied in the Isodewaxing™ process marketed by the Chevron Company [43]. The particular combination of pore geometry and moderate acidity make this structure suitable for this application where the objective is to form mono- and dimethyl branches on long straight-chain paraffins. The weaker acidity of this structure leads to minimal cracking of the long-chain paraffins while retaining enough isomerization to improve the cold-flow properties.

16.2.2
Stabilization Methods

As mentioned above, Y zeolites did not realize their full potential in catalysis until the methods of rare-earth exchange and steam stabilization were introduced. The rare-earth method was of such tremendous economic impact that its inventors, Plank and Rosinski, entered the inventor hall of fame. The underlying mechanism of stabilization by rare-earth exchange is thought to come from their ability to enter the sodalite cages and, due to their multivalent oxidation states, they are thought to coordinate to more than one framework aluminum and hinder their ability to leave the framework. Other methods have also been introduced including one or combinations of steam stabilization, acid washing, chelation with EDTA and other chelates 44], dealumination by treatment with $SiCl_4$ [45] and treatment with ammonium hexafluorosilicate (AFS) [46, 47]. The latter two are distinguished from the others in that they attempt to heal the framework upon aluminum removal by insertion of silicon atoms in the hydroxyl nests created as shown in the scheme below:

$$\begin{array}{c} & \text{Si} & & & \\ \text{Si} & \text{OH} & & & \text{Si} \\ \text{Si-O-Al-O-Si} \rightarrow \text{Si-OH} & \quad \text{HO-Si} + (NH_4)_6\,SiF_6 \rightarrow \text{Si-O-Si-O-Si} \\ \text{Si} & \text{OH} & & & \text{Si} \\ & \text{Si} & & & \end{array}$$

The AFS approach has led to the commercial development of a new family of Y zeolites by the Union Carbide Corporation called LZ-210. It allowed for the synthesis of Y products with framework Si/Al_2 ratios ranging as high as 18 before the onset of significant loss of crystallinity. These products were shown to exhibit significant advantage in fundamental model compound testing [48] and eventually found limited application in FCC technology due to catalyst cost constraints and much greater use in hydrocracking where they are mostly used today [3].

By far the bulk of Y zeolites in commercial application today are steam-stabilized with occasional further treatments to remove residual alkali metal cations or extra-framework species generated by the steam treatment. Variables including temperature, steam partial pressure and treatment time are manipulated to control the final product characteristics. In a typical approach, the as-synthesized sodium form of the zeolite is ion exchanged with an ammonium salt to remove some of the Na cations followed by steam treatment and a second ion exchange step to remove the residual Na down to a suitable level. Such treatment typically yields a zeolite with a sodium level below 0.2 wt% calculated as Na_2O, a cubic unit cell size around 24.50 ± 0.05 Å, a chemical overall Si/Al_2 of around 5.0–5.5 and a framework ratio of 10–12. All of the steps in this modification scheme can be altered to yield variants of Y zeolite with a wide range of performance characteristics illustrating the high versatility of this structure. For example, in production of transportation fuels in FCC or hydrocracking, the Y zeolite component of the catalyst is typically modified for selective naphtha production. On the other hand, it can be subjected to more severe steaming conditions and, in some cases, acid washing treatments to produce distillate selective catalysts. In a landmark paper published in the early 1980s, Pines showed that adjustment of steaming conditions to produce very low unit cell size Y reflecting a very high framework Si/Al_2 results in FCC catalysts which produce high octane gasoline [30]. During the same time period, Y zeolites were being modified by severe hyrothermal treatments to produce a highly dealuminated Y called ultrahydrophobic Y (UHPY) which, when formulated into hydrocracking catalysts, resulted in distillate selective [49]. A large amount of work in this field carried out both in industrial and academic laboratories has contributed to our current level of understanding of the fundamental processes responsible for these performance characteristics. It is now well established that calcination in the presence of steam generates both framework Bronsted acid sites as well as Lewis sites associated with the extra-framework aluminous species. Thus, tailoring Y acidity to suit a particular application involves precise control of the level of one or a combination of steps involving steam calcination, ammonium ion exchange, acid washing, chelation, or silicon reinsertion into the framework. Even the order in which these steps are carried out may be varied to create the required performance attributes.

Control of strength and location of acid sites and tailoring these properties for a particular application are the key objectives of zeolite modification schemes and the main focus of zeolite technologists in the field. Overall strength and strength distribution of acid sites are determined by framework Al content for Bronstead acidity, and framework versus extra-framework distribution for Lewis acidity. Different combinations and order of ion exchange steps, heat treatment with or without steam, acid treatments, and Si reinsertion treatments are applied to

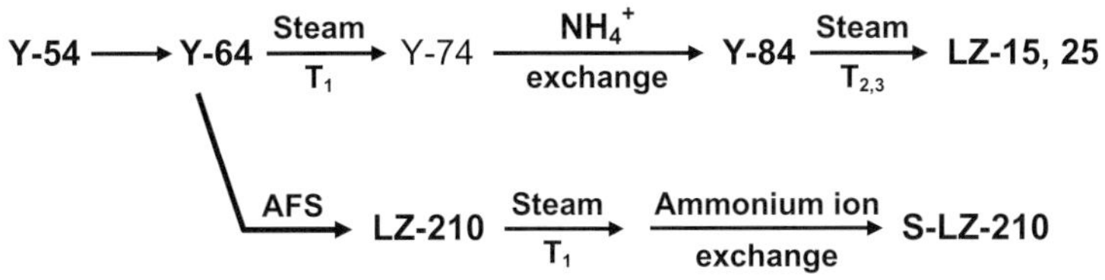

16.2 Critical Zeolite Properties | 541

```
Y-54 ⟶ Y-64  ──Steam/T₁──▶  Y-74  ──NH₄⁺ exchange──▶  Y-84  ──Steam/T₂,₃──▶  LZ-15, 25
              ╲
               ──AFS──▶  LZ-210  ──Steam/T₁──▶  ──Ammonium ion exchange──▶  S-LZ-210
```

AFS: Ammonium fluorosilicate, $(NH_4)_2SiF_6$

Scheme 16.1 Y Zeolite modification schemes.

produce active and stable forms of Y suitable for a particular application, as can be seen in Scheme 16.1.

This scheme illustrates the many steps required for stabilization and alternate modification approaches. Careful control of temperature, time, and steam content of the calcination atmosphere must be exercised to achieve the required degree of dealumination without suffering the consequence of structural collapse. But the acidity potential of zeolites must be unlocked by removal of alkali cations introduced in the primary synthesis step at various points during the modification process. Ammonium ion exchange steps reduce the sodium to a very low level compared to the as-synthesized zeolite, and the required sodium level varies depending upon the intended application. The two heat-treatment steps may be carried out separately or combined depending on the manufacturer, equipment design, or intended use and cost constraints. The nomenclature USY, shorthand for ultrastable Y, has been adopted since the first introduction of steam stabilization [26, 27]. UHPY stands for ultrahydrophobic Y and it was employed by the Carbide group [49] to describe a severely steamed variant where the framework is highly dealuminated with low water adsorption capacity. These two classes are best distinguished by unit cell size ranges they fall into. We use a range of 24.30–24.60 Å for US-Y and a range of ≤24.30 Å for UHP-Y. The USY and UHPY in this scheme are often lumped together in the literature as USY. We make the distinction here because these two variants differ enormously in their acid distribution and strength as well as in their performance characteristics and the value of much work found in the literature is greatly diminished by the use of the simplistic description USY to describe the catalyst used in a given communication. In fact much of the literature references to USY should fall into the UHP-Y category. We maintain this distinction here for better clarity of discussion. Ideally, of course, the stabilized Y zeolite should best be identified by the unit cell constant and, when appropriate, the chemical and framework Si/Al_2 ratio.

In addition to the thermal and ion exchange treatments described above, another rich area in the study of zeolite stabilization and acidity modification involves a broad class of post stabilization chemical treatments to remove extra-framework aluminum, and silicon, species partially or completely after steaming. In actual commercial practice, such treatments primarily involve use of mineral acids such as HCl or HNO$_3$, but occasional use of chelates such EDTA, oxalates, or other polydentate ligands has been disclosed [44]. Initially the sequence of steaming and ion exchange, leave the chemical composition of the zeolite nearly the same as that of the starting material. However, as the framework is depleted of Al cations the framework Si/Al$_2$ ratio increases, and to a significant discrepancy between the chemical (overall) ratio and that of the framework. The framework composition in typically determined through the use of various spectroscopic techniques and correlations based on X-ray diffraction measurements of the lattice parameter, infrared frequencies of key bands characteristic of framework Si(Al)–O vibrations, or NMR measurements, *vide supra* (Chapter 4). Typical ratios for the USY types range over about 10–30, whereas the UHPYs possess framework ratios of ≥30. A considerable uncertainty is associated with the latter due to the insensitivity of the above mentioned techniques to framework compositions when highly depleted of Al. These are best assessed by wet chemical analysis or atomic absorbtion after exhaustive treatments to remove extra-framework species. Such acid treatment are often critical in controlling catalytic behavior, but are used only when necessary due to complexity and costs involved.

Removal of the aluminum from the framework leaves behind a cation vacancy which must be "healed" to maintain framework crystallinity. In steam stabilization, this "healing" process is thought to begin with the formation of additional Si–OH groups to create what is referred to as hydroxyl nests. Evidence for the formation of such nests can be seen in the appearance of a new Si–OH band in the infrared spectra of dealuminated zeolites at about 3736 cm^{-1} which has been attributed to internal Si–OH group as distinguished form the external Si–OH group whose IR band falls at about 3744 cm^{-1}. Extra-framework Si species derived from some amorphous domains formed during calcination as a result of some framework collapse may migrate to these nests and occupy the Al vacancies which helps maintain crystallinity. The approach of Skeels and Breck to dealuminate and heal the framework by reinsertion of silicon [46, 47] involved treating a sodium or ammonium form of Y zeolite with a solution of ammonium hexafluorosilicate (AFS) which immediately inserts Si in place of Al as it is extracted from the framework resulting in a highly crystalline dealuminated zeolite. As illustrated in Scheme 16.1, this can then be further processed if necessary to further modify the framework and adjust its catalytic properties. This approach is the only one in large commercial practice other than steam stabilization and rare-earth exchange.

16.2.3
Property–Function Relationship

One of the most studied aspects of catalysis science is the relationship between structure and function. Some general themes are well established by now, but specific connections between catalyst characteristics and performance attributes remain elusive in most cases. The crystalline geometry of zeolites makes them relatively more amenable to study by a variety of powerful modern characterization tools, but there remain many key unanswered questions in the catalytic application

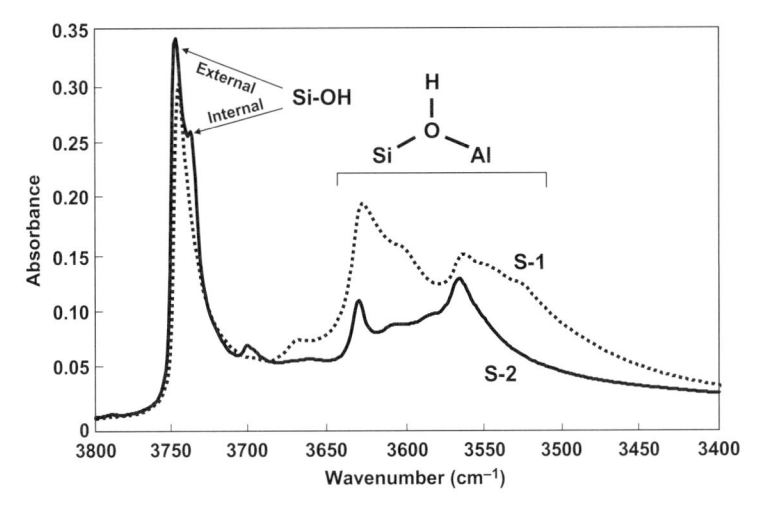

Figure 16.1 Hydroxyl IR spectra of stabilized Ys: S-1, steamed at 750 °C; S-2, steamed and acid-washed.

of zeolites. This is largely due to the fact that converting the as-synthesized materials into their catalytically active forms involves modification, secondary synthesis, and formulation steps rendering them significantly different from their parent material, especially at the local active site level.

To illustrate, the hydroxyl infrared spectrum of a NaHY zeolite derived by the ammonium exchange and thermal treatment of as synthesized NaY zeolite (Scheme 16.1) reveals two bands centered around 3640 and 3550 cm^{-1}. These bands are attributed to O–H groups located in the supercage and sodalite cages, respectively [50]. Literature assignments of these bands assume that the 3550 cm^{-1} band is also more acidic based on the hypothesis that the lower vibrational energy reflects weaker O–H bond which should be more acidic. However, catalytic activity of this form is very low despite the presence of these acidic OH sites. It is not until this intermediate undergoes steam stabilization that it becomes catalytically active. Stabilization of Y zeolites discussed above yields as many as seven different type of Si–O–Al related hydroxyl groups present in different relative concentrations depending on the procedure employed. Figure 16.1 shows the hydroxyl infrared spectra of two steam-stabilized Y zeolite samples made using different sequence of steaming and ion exchange steps. The OH band envelop in these spectra represent acid sites of different acid strength and, possibly, spatial location within the framework [50]. Understanding the relationship between the structural attribute of the various sites represented in these spectra and their performance characteristics are critical to good catalyst design.

MOR and MFI infrared spectra are not nearly as complex as those of Y zeolites. Single bands around 3620 and 3605 cm^{-1}, respectively, are typical for the acid forms of these framework types. The lower energy of these bands compared to those of zeolite Y is usually cited in explaining their higher intrinsic acidity and consequent cracking activity consistent with the hypothesis that weaker oscillator strength correlates with more acidic hydroxyl.

16.2.3.1 Acid Strength Requirements for Product Control and Influence of Spatial Distribution on Selectivity

Product control in processes that depend on bond breaking and rearrangement depends, primarily, on the ability to control the depth of cracking reaction and the cracking-to-isomerization ratios. This ability in turn is critically dependent on good control of both the strength and spatial distribution of acid sites. Not only must the acid sites be accessible to reactants but also they need to be located in positions where the primary reaction products are able to exit the system before encountering other active sites and undergoing undesirable secondary reactions. This was illustrated early on by the elegant work of Weitkamp and coworkers who demonstrated the significant tendency of zeolitic catalysts to produce secondary cracking products compared to amorphous silica–alumina cracking catalysts with their more open pore structures [51–53]. The higher acid site density in zeolite coupled with the small pore dimensions increase the probability of re-cracking of intermediates or primary cracking products to secondary products due to multiple contact with acid sites as they move through the tortuous diffusion paths. Thus, zeolite catalysis has focused on the control of not only the strength of acid sites, but also on their spatial distribution and accessibility to reactants [54]. A good deal of scientific and technical literature in the field is concerned with framework site density, speciation between framework and extra-framework sites, and introduction of mesopores to provide better access to the interiors of zeolite crystals.

A good illustration of the importance of framework Al density on product distribution is found in the FCC technology where control of hydrogen transfer reactions is manipulated to control gasoline selectivity, octane quality and coke make [18, 55]. The bimolecular hydrogen transfer reactions discussed below require adjacent acid sites to facilitate interaction between two adsorbed molecules. These reactions result in higher gasoline make, but decrease olefin production which results in lower octane gasoline and higher stability due to the absence of reactive olefins.

Thus, a large focus of FCC catalyst research involves control of the density and location of acid sites in order to control product selectivity, product quality and coke make. In a landmark publication Pines [30] demonstrated how widely spaced framework Al sites may be utilized to produce high octane gasolines by decreasing hydrogen transfer reactions which in turn results in preservation of olefins.

An interesting consequence of hydrogen transfer chemistry has been reported by FCC catalyst technologists who find that high hydrogen transfer catalysts limit the impact of ZSM-5 additives in production of light olefins [59, 60]. With high-hydrogen transfer catalysts, donation of hydrogen by naphthenes to reactive olefins decreases their concentration and limits the opportunity for cracking the longer-chain olefins in the ZSM-5 to produce light C3 and C4 olefins. Many examples of the critical importance of spatial control of framework Al, which act as hydrogen transfer sites, are available in the FCC literature and can be found in several major reviews of the subject [4, 5, 19].

In the hydrocracking process, this phenomenon is exploited to shift catalyst selectivity from the naphtha to the distillate products. Here the wide separation of sites is exploited to minimize the potential for secondary cracking in initial products and intermediates. This, along with the introduction of escape routes for the primary product tends to preserve the higher molecular weight hydrocarbons, thereby producing more distillates [49, 61, 62].

For optimum results, modification of material properties must be coupled to process chemistry requirement and a good illustration of the importance of this concept the can be seen when comparing the specific acidity requirements of the FCC and hydrocracking processes. Both processes are based on acid catalysis, but hydrocracking requires a hydrogenation function provided by added metals since it is bifunctional. The acid function for both processes is provided by a combination of stabilized Y zeolites in their acid form and amorphous silica–alumina in various proportions as determined by the requirements of the specific application. Upon more careful examination of the acid function requirement for the two processes it turns out that there are significant differences in the level of acidity required. Specifically, the last step in the stabilization process for Y zeolites involves an ammonium ion exchange step to remove the residual sodium. FCC catalysts specifications require Na_2O levels below 1% [5] whereas hydrocracking catalysts require levels well below about 0.2–0.3 wt% in order to have the appropriate level of acidity to be effective. Process condition differences put different demands for different level of acidity required. FCC's short catalyst life requirements, high operating temperatures and short contact times evidently put much less demand on the zeolite compared to the hydrocracking process which typically operates at much lower temperatures, and needs high enough acid strength to provide activity to meet the 2–3 year cycle length requirements.

16.2.3.2 Pore Geometry and Framework Composition

Due to their crystalline nature and the wide variety of framework types available, zeolite catalysis literature is highly focused on the geometric effects imposed by zeolitic pores on reactants, products and intermediates. The so-called shape selectivity effect has been well documented and has served as a basis for development of many important processes as has been illustrated in other chapters in this book. However, pore geometries that the reactants, intermediates and products experience are not always the idealized ones deduced from the crystallographic attributes, and the geometric effects are not readily decoupled from compositional influences. For example, the discussion above illustrated the variety of reaction pathways and how catalyst activity and selectivity vary considerably by compositional modification of a single structure type, Y zeolite. In addition, a secondary pore system with sizes in the mesopore range develops as a result of the various modification schemes employed, Figure 16.2. Such a system not only influences diffusion path lengths for reactants and products with dimensions compatible with the primary

546 | *16 Bond Breaking and Rearrangement*

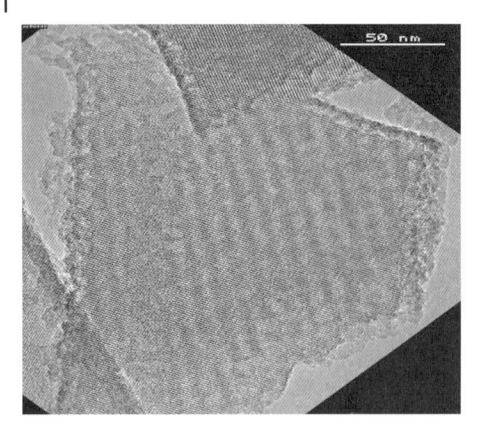

Figure 16.2 Transmission electron microscopy image of steam-stabilized Y zeolite showing mesopore network.

pores, but may allow much larger molecules to diffuse into the interior of the zeolite crystals and react at sites located at the periphery of the interior mesopores creating fragments small enough to diffuse through the primary geometric pores. Such possibilities must be given serious consideration when assessing performance characteristics of zeolitic catalysts.

16.3
Chemistry of Bond Scission Processes

Petroleum derived hydrocarbons which, by far, constitute the majority of feeds to processing units producing transportation fuels contain hydrocarbons of varying molecular weights and carbon chain lengths up to ~45–50 carbon atoms along with sulfur-, nitrogen- and oxygen-containing compounds. Bond scission and rearrangement reactions important in production of transportation fuels include heteroatom removal, carbon–carbon bond scission and rearrangement of carbon skeletons to produce higher quality products. The following sections address each of these classes of reactions separately in an attempt to better identify the fundamental principles involved.

16.3.1
Heteroatom Removal: Desulfurization Denitrogenation and Deoxygenation

Heteroatom removal from petroleum feeds is important for improving their processability and for eventual production of clean environmentally friendly fuels.

Zeolites are not typically employed in converting these classes of compounds, but their impact on zeolitic catalyst performance in FCC and hydrocracking, coupled with the occasional use of zeolite based catalysts to assist in their conver-

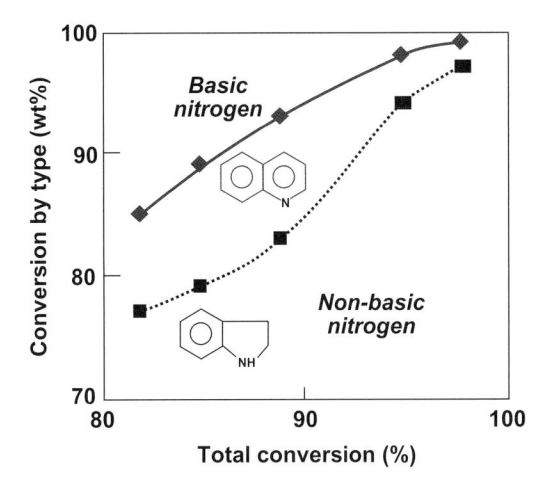

Figure 16.3 Relative reactivity of basic and non-basic nitrogen compounds in the hydrocracker environment.

sion, renders them worthy of some brief discussion in this context. Organo-sulfur compounds vary greatly in reactivity with mercaptans and sulfides being easiest to react while substituted dibenzothiophenes are the most difficult to convert.

Reactivities of different compounds vary greatly within each class and require different mechanistic pathways among classes. For example, nitrogen compound reactivities line up in the following order [63]:

> Anilines > Indoles > Quinolines > Carbazoles

This order is roughly proportional to their basicity and molecular weight. Likewise, sulfur compound reactivities roughly increase in the order:

> Mercaptans, Sulfides, Disulfides > Thiophenes >
> Benzothiophenes > Dibenzothiophenes

The relative reactivity of basic and non-basic compounds in the hydrocracker is illustrated in Figure 16.3 where the residual nitrogen compounds in the products from a hydrocracking test are plotted against the degree of conversion which is varied by increasing processing severity. The higher reactivity of the basic nitrogen compound seen in this figure is clearly due to their affinity for acid sites and their consequent conversion via an acid-catalyzed mechanism. This is also well illustrated by the high ratio of non-basic to basic nitrogen present in light cycle oils derived from FCC units where they have undergone cracking over the monofunctional acid cracking catalysts utilized in these units.

Optimum catalysts in use today for hydrodesulfurization typically consist of Mo or W metals promoted with Co and supported on weakly acidic alumina supports. Increasingly stringent regulatory requirements have pushed sulfur specifications for transportation fuels to less than 10 ppm for diesel fuels and 30 ppm for gasoline in most industrialized countries, making it extremely difficult to meet with conventional technologies. This has spurred a high level of interest in the development of catalysts and processes to help meet these specifications. Accepted mechanisms for the conversion of the very refractory substituted dibenzothiophenes involve hydrogenating one of the benzene ring which renders the C–S bond more labile and easier to break [64]. In the specific case of 4,6-dimethyldibenzothiophene it is accepted that the presence of methyl group in the 4 and 6 positions adjacent to the sulfur atom inhibits the removal of the sulfur due to steric hindrance. Addition of an acidic function to the catalyst to promote methyl group shift out of the 4 or 6 positions and carbon–sulfur bond scission is occasionally proposed. However, there are no credible reports of successful application of this promotion mechanism. Supports with acid function are also occasionally encountered when a modest degree of simultaneous boiling point conversion is desired in hydrodesulfurization units and is commonly referred to as mild hydrocracking. The acid function employed in such applications is typically non-zeolitic and most often is an amorphous silica–alumina. It is worth noting that the H_2S byproduct of the desulfurization chemistry exhibits only a modest influence on the zeolite cracking catalysts typically present downstream.

Nitrogen contaminants, unlike their sulfur counterparts, have a much stronger impact on zeolite catalyst activities either due to their basic nature or to the ammonia byproduct they produce. These compounds are typically grouped in two classes: basic and non-basic. The basic compounds are aliphatic or aromatic amines which act as poisons to the acid function of acidic cracking catalysts. Non-basic nitrogen compounds, in contrast, do not initially poison acid sites, but may still act as inhibitors by a simple adsorptive interaction with the metal function of hydrocracking catalysts. Both of these classes, however, produce ammonia upon C–N bond scission which in turn acts as an acid poison and inhibit cracking activity. As seen in the section below, this effect has major implications on the design and practice of the hydrocracking process as imposes requirements for larger catalyst volumes (large reactors), higher operating temperatures, or the implementation of separation and scrubbing sections in order to remove or dilute the ammonia.

Oxygen-containing compounds behave in a manner analogous to sulfur compounds and their conversion follows a parallel pathway. Water produced by hydrodeoxygenation reactions can exert a mild inhibitive role to zeolitic cracking reactions, but more importantly, may play a more destructive role of zeolite modification while in service.

Heterocyclic nitrogen compounds, amines and the ammonia generated by C–N bond scission in the presence of hydrogen exert a strong inhibitive influence on acid-catalyzed reactions and can significantly alter the dominant reaction pathways in a manner and to an extent that depends on the specific reactor environment. Similarly, sulfur compounds and hydrogen sulfide produced in processing exert a parallel, though more complicated, influence on the hydrogenation reactions. Hydrogen sulfide can act as an inhibitor or promoter depending on its partial pressure and the type of reaction involved [65]. The degree of influence it exerts under normal hydrocracking conditions also depends on the type of metal employed. Thus it exerts a significant deactivating influence on noble metal catalyst activity, but has a much lesser impact on base metal catalysts.

Due to the much larger influence of feed nitrogen compounds and their byproducts on bond scission and rearrangement chemistry, major design elements, process configurations and operating practices are incorporated in conversion processes specifically to deal with this influence. In the FCC process for example, hydrotreating of feeds to remove nitrogen and sulfur compounds from the feed prior to introduction into the FCC unit has become a well established practice [66]. The key benefits derived from this processing step stem from the removal of the nitrogen compounds leading to higher catalyst activity and from the hydrogenation of aromatics and other hydrogen deficient compounds to enhance gasoline yields. FCC feed pretreating severity is typically set to achieve a product nitrogen level of around 0.3 wt%, even though lower levels would benefit FCC catalyst cracking activity, for technoeconomic reasons related to costs associated with excessive hydrogen consumption and the consequent impact on the FCC unit operation [66].

This level of FCC feed pretreatment is optimum, achieving the desired impact of removal of acid poisons without incurring the economic penalty associated with parallel undesirable reactions of other feed components. However, the optimum lies at a very different point for hydrocracking process since deeper treatment of the feed prior to its contact with the hydrocracking catalyst may be highly beneficial as it leads to the desired impact not only on catalyst activity and stability, but also on product quality and selectivity. Thus, nitrogen management strategies are key to the design and operation of hydrocrackers and, especially, to catalyst selection. Specialized catalysts for nitrogen removal are typically installed in a lead reactor or lead beds followed by a hydrocracking catalyst installed in one or more downstream reactors (Figure 16.4). In some designs the lead reactor, or beds, may be loaded with catalysts capable of carrying out both hydrotreating and hydrocracking functions. In either case a key function of catalysts in the front of the unit is to destroy the nitrogen and sulfur compounds in the feed to very low levels in order to minimize their poisoning effect on the hydrocracking catalyst downstream.

Conditions in the first reactors are typically set so as reduce the level of nitrogen in the effluent to a level well below 100 ppm and often below 20 ppm. The impact of the depth of hydrotreating on hydrocracking catalyst activity and product selectivity is illustrated in Figure 16.5, which shows results from a test comparing two feeds treated to nitrogen levels of 40 and 10 ppm and then processed over the same hydrocracking catalyst. These data clearly show improved activity, that is, a lower temperature requirement for a constant conversion level, as well as a shift to a heavier product with the feed which has been preprocessed to a lower nitrogen level. These data are especially interesting because the test conditions include doping with ammonia and hydrogen sulfide precursors at high levels intended to simulate the gas phase composition prevailing in a commercial reactor environment. Thus, these results suggest that unconverted organic nitrogen compounds exert a poisoning influence at certain acid sites over and above that of ammonia,

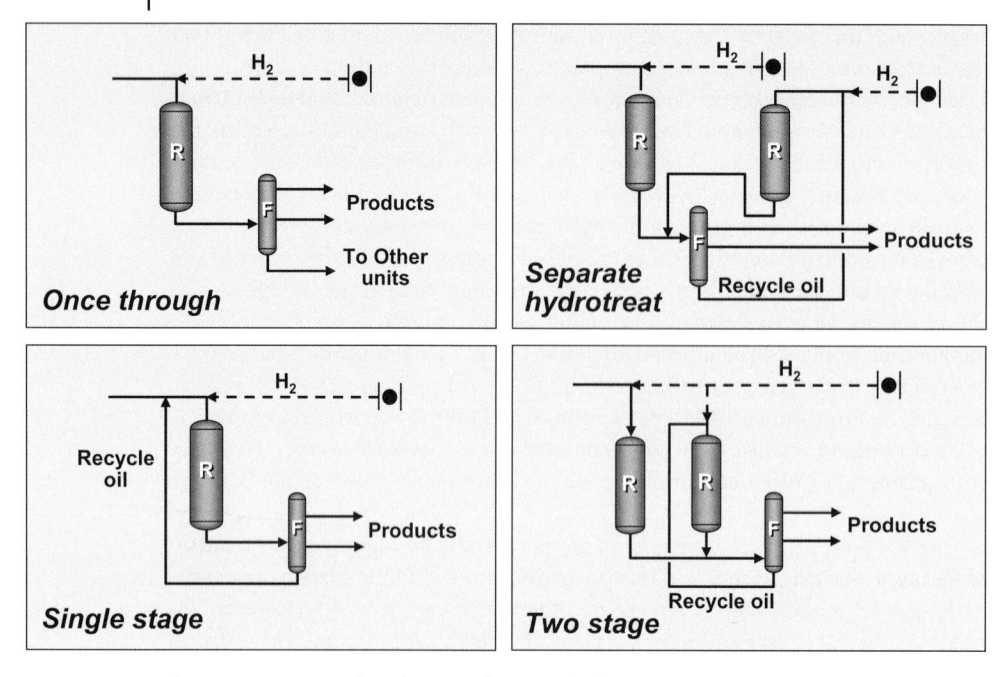

Figure 16.4 Process flow diagrams for major hydrocracker designs.

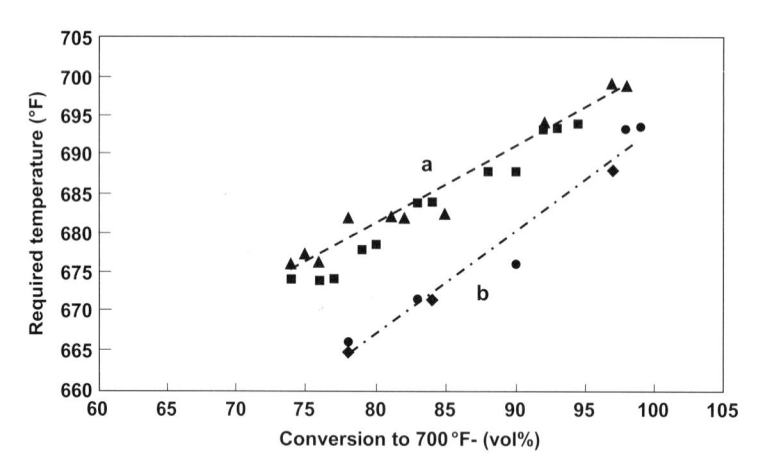

Figure 16.5 Impact of pretreating severity on the activity of a hydrocracking catalyst: temperature required for conversion of a feed hydrotreated to: (a) 40 ppm nitrogen, (b) 10 ppm nitrogen.

most likely because they adsorb more strongly at acid sites and likely accumulate at the catalyst surface under process conditions. For this reason control of these nitrogen compounds at very low levels is key to the maintenance of high activity and long catalyst life.

Ammonia produced by the HDN reactions also functions as an acid site moderator-depressing catalyst activity and influencing product selectivity in a similar manner to unconverted nitrogen compounds, with the main difference being that its impact is completely reversible whereas that is not always the case with organic nitrogen compounds. The deactivating influence of gas phase ammonia was dealt with early on by process designs where the gas and liquid effluents of the first reactor are separated and the gases are then scrubbed free of ammonia and hydrogen sulfide and returned to a second reactor along with the liquid product to react in a cleaner environment. The ammonia-free environment of this second reactor permits operation under much milder conditions due to the lack of the inhibition effect of the acid sites. This was the standard design for early hydrocracker which employed amorphous silica–aluminas with their weaker acidity [67]. Introduction of zeolites with their much stronger acidity permitted the elimination of the then costly intermediate separation step and the development of a new single stage process design, as discussed later in this chapter.

16.3.2
Boiling Point Reduction

16.3.2.1 Paraffin Cracking

It is well accepted that the fundamental reaction steps underlying all hydrocarbon chain cracking (bond scission) processes involve the formation of carbenium ion intermediates and their subsequent reactions involving carbenium ion rearrangement, beta bond scission and hydride transfer or hydrogenation [51–53, 68]. Mechanisms involving penta-coordinated carbonium ion formation on strong zeolite acid sites have also been postulated to take place in rare instances such as the cracking of short chain alkanes over ZSM-5 [69]. However, under the relatively mild conditions of the industrial cracking processes such as hydrocracking and FCC and due to the overwhelming importance of carbenium ion chemistry in these processes, we dedicate most of the discussion in this section to mechanistic pathways involving this chemistry. In the FCC process where monofunctional acid catalysis is predominant, carbenium ions are postulated to form by a sequence of steps involving the adsorption and subsequent protonation of feed olefins at Bronsted acid sites, or hydride abstraction from an alkane adsorbed at a Lewis site [5–7]. Once formed, the carbenium ion can undergo reactions of isomerization to a more stable carbenium ion, beta scission, or hydrogen transfer. Which of these pathways dominates is dependent on the strength and spatial distribution of the acid sites. Isomerization of the initially formed carbenium ions proceeds immediately to the more energetically favored secondary and tertiary carbenium ions. Cracking of the longer-lived carbenium ions takes place via scission of the bonds beta to the positively charged carbon, producing another tertiary ion and a shorter paraffinic fragment. As discussed above, hydrogen transfer reactions are another important class of reactions in FCC which determine gasoline selectivity and octane quality as well as coke make and catalyst life:

$$\text{Olefins} + \text{Napthenes} \rightarrow \text{Paraffins} + \text{Aromatics}$$

The transformation of naphthenes to aromatics starts with the formation of mono- and polyunsaturated cyclo-olefins and continues to form polyaromatics which eventually condense to form carbonaceous deposits referred to as coke.

A similar mechanism for the initiation of carbon–carbon bond scission and rearrangement is invoked for the hydrocracking process [51–53, 68]. However, some key differences between these two processes must be considered in order to explain the differences in their product compositions. First, FCC catalysis is primarily monofunctional acid catalysis essentially devoid of any hydrogenation function, while hydrocracking is a bifunctional catalytic process involving not only an acid function but a strong hydrogenation function as well, provided by noble metals such as Pt or Pd, or sulfided base-metals such as Mo or W promoted with Ni or Co. It also operates at high hydrogen pressure to enable the hydrogenation of aromatic components in the feed which in turn provides high product quality and the long-term stability needed for this fixed process.

The carbenium ion theory has by far been the predominant theory accepted by most workers in the field due to its ability to explain the major difference in product distribution between acid catalyzed cracking and thermal cracking. However, given the strong hydrogenation function and the overwhelming presence of hydrogen in the hydrocracking process, it is difficult to reconcile the postulated initiation step for its formation which invokes olefin formation and subsequent protonation over acid sites and or by hydride abstraction from paraffin. To be sure, this is the most plausible mechanism offered. However, an appealing alternative mechanism was advanced by Sie which plausibly explains the observed cracking chemistry and product distribution. It also resolves some of the problems of the Greensfelder theory, including the necessary formation of highly energetically unfavored primary carbenium ions immediately after the beta scission step [70–72].

In the carbenium ion mechanism a small amount of olefinic intermediates are formed by dehydrogenation (hydride abstraction) of paraffins at metal sites followed by immediate protonation at nearby acid sites. This starts the chain of carbenium ion rearrangement to the most energetically favored secondary and tertiary ions followed by isomerization and beta scission to produce smaller fragments in a manner similar to that described above for FCC. Chain termination eventually takes place by hydrogenation of the cations at a metal site followed by diffusion of the products outside the zeolite system. In order to promote the desired reaction pathways it is critically important to facilitate good communication between the metal and acid site, ensure proper balance of the hydrogenation and acid functions and provide the least tortuous escape path for the desired products. The term ideal hydrocracking was coined by Weitkamp and coworkers to describe the situation where the formation of the carbenium ion, subsequent rearrangements and formation of the cracked product from the catalyst surface are rate-determining [51]. In the absence of internal and external mass transfer limitations, these primary steps determine the reaction pathways and the product composition and, to the extent that active site geometry is involved, it can be manipulated by the choice of zeolite type or modification scheme to control

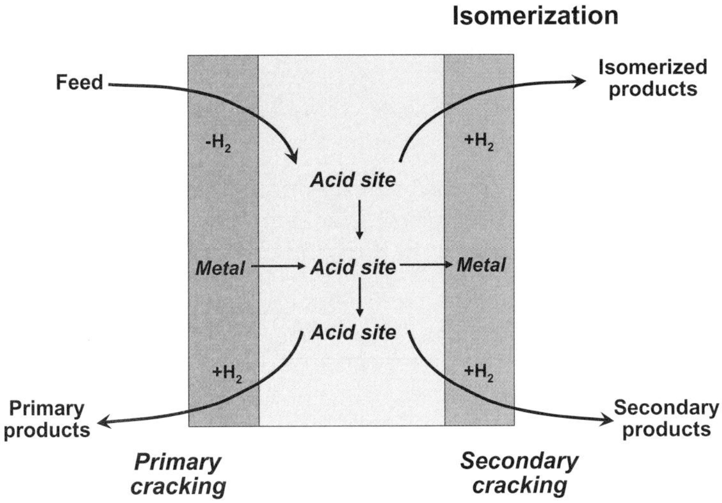

Isomerization

Figure 16.6 Schematic illustration of the elementary steps in bifunctional hydrocracking.

product selectivity. Thus, the proximity of the acid and metal sites and the relative strength of these two functions become critical in determining selectivity.

An illustration of these mechanistic events taking place in the reaction is provided in Figure 16.6. Control of the product distribution in this process is highly dependent on the proper choice of zeolite type, zeolite modification scheme, hydrogenation function and the placement of these components in a manner to facilitate optimum feed and product pathways. This model helps explain the observed product distribution differences between hydrocracking catalysts with acidic components having on more open pore structures such as amorphous silica–aluminas or narrower pores such as zeolites. It can even be invoked to explain distribution differences between zeolite catalysts with wider pores such as Y and medium-pore zeolites such as MFI. The combined effect of support porosity and acid strength on hydrocracking product distributions is illustrated by the data shown in Figure 16.7, which compares the carbon number distribution of products from pilot plant tests of hydrocracking catalysts with varying ratios of amorphous and Y zeolites with a constant hydrogenation (metal) function. These data show that the carbon number distribution of the products shifts higher when the proportion of amorphous component in the catalyst increases, suggesting a closer approach to "ideal cracking" as the probability of secondary cracking to form lighter products is lowered due to the combined effects of more open porosity and, possibly, weaker acid strength. Thus, to approach ideality where the cracking products are mainly products of the cracking of the feed molecules, the catalyst must possess large enough pores and adequate hydrogenation function to allow escape of the primary products. In actual practice, the critical pore dimensions required are not possible to determine *a priori* because of the complexity of real feed composition which usually contain naphthenic and aromatic compounds in

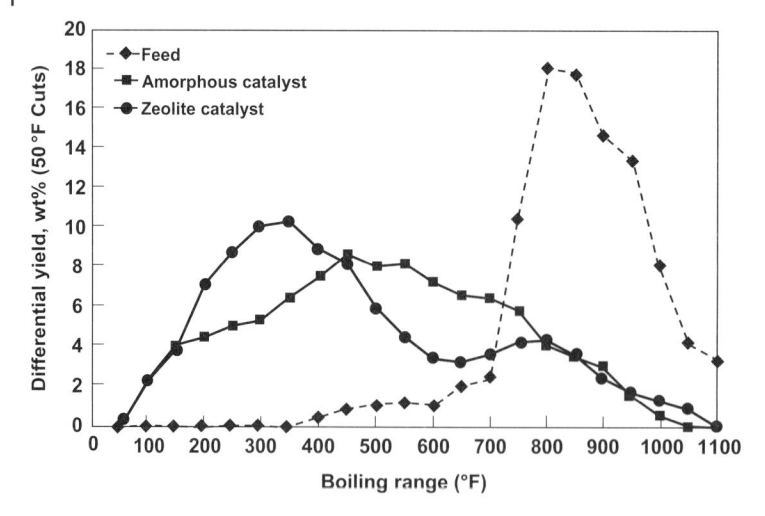

Figure 16.7 Influence of support type on product distribution in hydrocracking: hydrotreated light Arabian gasoil feed hydrocracked over amorphous and high zeolite catalysts; differential yields measured in 50 °F (10 °C) increments.

addition to paraffins of varying chain lengths. Thus, one must resort to studies such as this to map out the support characteristics as well as the metal:support ratios. In addition, paraffin cracking rates increase sharply with chain length [73–75], making it quite difficult to precisely control the product distributions without implementing process steps designed to deal with these issues.

16.3.2.2 Aromatic and Naphthene Ring Opening

Aromatic and saturated ring compounds (naphthenes) are major components in many feedstocks and their conversion through a combination of hydrogenation and ring opening is of great importance when processing gas oils and cycle oils. The mechanisms of ring opening of aromatic and naphthenes differ from paraffin cracking mechanisms. The conversion of the electron-rich aromatic rings, and even saturated cyclic compounds, would in a bifunctional mechanism depend much more strongly on the type and strength of the metal hydrogenation function. Thus, processing feedstocks which possess large proportions of ring compounds, such as light cycle oils, requires a different strategy for catalyst design than for highly paraffinic feeds because the demanding requirement for the hydrogenation function to be strong enough to saturate aromatic compounds is high.

Among the classes of feedstock processed in the hydrocracker the most highly aromatics feed are light cycle oils produced in the FCC unit. Once formed by cyclization and the hydrogen transfer mechanism discussed above, they accumulate in the product due to the absence of a metal function in the FCC catalyst and adequate hydrogen in the process environment. They are typically sold as low-value fuel oil, or hydrotreated to reduce sulfur content and improve their quality as diesel blend stocks. Another approach to upgrade their value even further

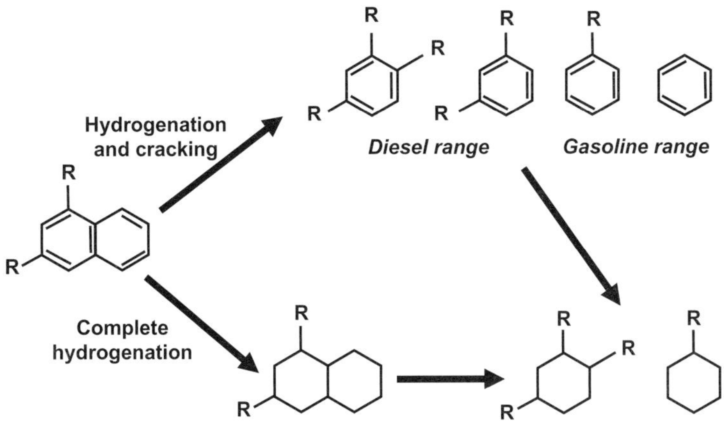

Scheme 16.2 Reaction scheme for cyclic compounds in light cycle oil.

involves selective opening of rings in a specialized hydrocracking process where they are converted to produce high octane gasoline or even more valuable aromatic complex feedstocks [59, 60, 76]. The chemical pathways, shown in Scheme 16.2, involve the dealkylation of substituents and the opening of one or two rings while preserving the last ring with short-chain substituents.

Much work has been published in the literature on the mechanisms of this ensemble of reaction, but the approaches followed to achieve the desired product objectives remain in the realm of proprietary knowledge [59, 60, 77]. Clearly, a successful approach involves selection of the proper type and level of hydrogenation metals which are strong enough to promote second or third ring hydrogenation but not overly strong so as to promote hydrogenation of the benzene ring. The zeolitic component in these catalysts must allow access of the reactant to the reaction sites and easy egress of the substituted single-ring aromatics from the system before they react and convert to a less-desirable product.

A comparison of the feed and product compositions achievable by this approach is shown in Figure 16.8, which shows the depletion of multi-ring aromatics from the feed in favor of a variety of single ring aromatics with short alkyl chains. A more challenging approach that leads to a higher-value product involves optimization of the catalyst and process conditions to maximize xylene and toluene production for aromatic complex feeds [60].

A particularly interesting observation regarding the reactivity of naphthenes was made early on by Sullivan and coworkers, who observed that these saturated cyclic compound are difficult to ring open in the hydrocracking environment [78, 79]. They observed that substituted naphthenes tend to dealkylate without ring opening and coined the name "paring reactions" to describe this phenomenon. This observation has been explained by the unfavorable orientation of the fixed *p*-orbitals of the bond beta to the naphthenic endocyclic carbenium ion which make beta scission within the ring very unfavorable.

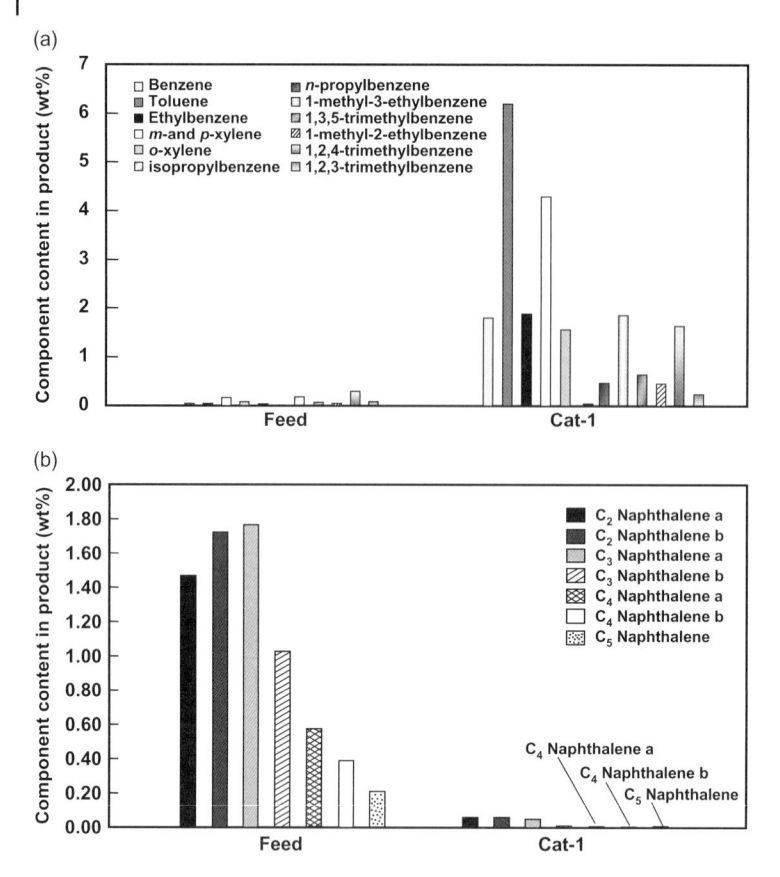

Figure 16.8 Comparison of feed and product composition from light cycle oil hydrocracking showing: (a) essentially complete conversion of feed two-ring compounds on four different catalysts, (b) the formation of different distributions of substituted single-rings depending on catalyst composition. Catalysts 1–4 differ in metal and acid function.

16.4
Fluidized Catalytic Cracking

FCC is one of the two major "bond breaking and rearrangement" processes in industrial practice. It produces about 40–45% of the world gasoline production [4, 5, 7]. Atmospheric and vacuum gasoils are processed to produce a product slate with gasoline as the major product, C3 and C4 olefins and paraffins and the heavier highly aromatic and less valuable light cycle oil (LCO) byproduct. This process was first commercialized in the 1940s as a result of a combination of innovative technological brought about by the collective effort of Houdry, Mobil Oil and the Standard Oil of New Jersey Corporation [5]. Since then there have been major innovations in process and catalyst design, including the introduction of zeolites and the riser reactor [4, 5]. With the increasing need to process cheap heavy feedstocks many new innovations in process and catalyst design have been introduced to meet these needs. The so-called Resid Cracking has required major changes in catalyst compositions as well as significant changes to the unit operation and hardware to handle the additional heat generated by additional coke make due to the use of the resid feeds [5]. Many other key innovations have come along over the past 20–30 years and have involved catalysts and additives intended to adjust product quality, such as the introduction of ZSM-5 additives to enhance octane, or to address emission problems from the unit, including CO combustion promoters and SO_x additives to control sulfur oxide emissions. The brief process and catalyst descriptions provided in the following two sections focus on the major features in use today with a special emphasis on the role of zeolites and zeolite modification. The reader is referred to other treatises for a detailed study of this process as it is outside the scope of this section.

16.4.1
Process Configuration and Catalysts

As practiced today, FCC is a fluidized-bed process with continuous catalyst regeneration which relies on short contact in a riser reactor between the feed and catalyst, fluidized with an inert gas, followed by disengagement and catalyst regeneration to burn off coke deposits and return the catalyst to near-fresh activity. A simplified diagram of a typical FCC unit is shown in Figure 16.9. The reaction chemistry described above is carried out in this process at temperatures in the range of 500–540 °C by contacting the fluidized catalyst in the form of particles in the range of 30–120 μm in diameter with the hot feed injected near the top of a riser reactor followed by rapid disengagement after a short contact time (on the

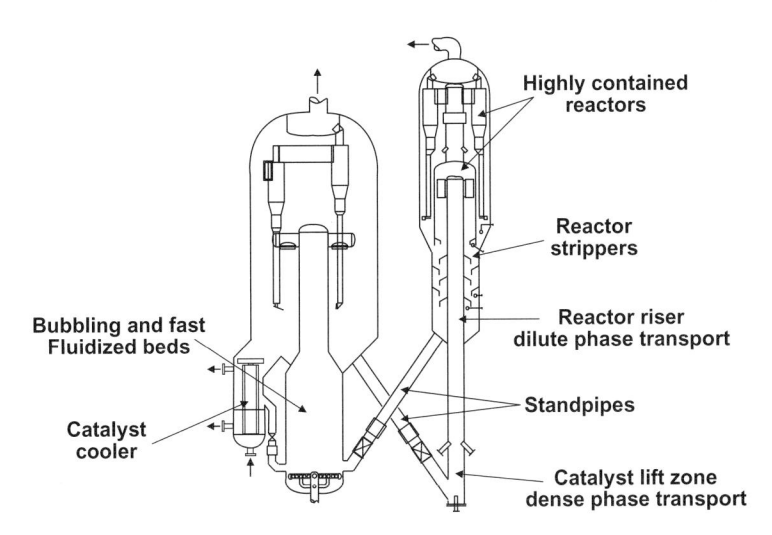

Figure 16.9 Diagram of a standard fluidized catalytic cracking process unit.

order of fractions of a second) and separation of the catalyst from the reaction products. From a catalytic perspective, critical process parameters include control of riser reactor temperatures and contact time along with regenerator temperatures and residence time. A great deal of technological development has been incorporated into the FCC process to optimize the ability to process increasingly heavier feedstocks and improve the yield and octane quality of the primary gasoline product. On the process side, short contact times, high riser temperatures, improved feed injection strategies, improved regenerator efficiency and better cyclone efficiency and fines recovery are some of the key developments incorporated into this process since its early introduction. Many variations in unit design introduced by major licensors such as UOP, Mobil, MW Kellog, Shell, etc. include changes from down-flow to up-flow designs, stacked designs, side by side designs and more recently multiple risers introduced by UOP to adjust product yield structure. Most of the recent developments have targeted an improved ability to handle residuum oil feeds and to produce lighter products for petrochemical feeds. Residuum oil feeds bring with them a special set of challenges such as higher coke make and metals on catalyst, which require special measures and operating practices to improve heat management in the unit.

Major FCC catalyst developments include: (i) the ability to fine-tune the rare-earth content to tailor the product properties and control the equilibrium unit cell size of the catalyst which in turn allow better control of hydrogen transfer reactions, (ii) improvements in particle strength to minimize fines production and catalyst replacement costs, (iii) the use of metal traps to capture Ni and V metal contaminants in heavy feeds and prevent them from damaging the zeolite and (iv) the use of CO combustion promoters and octane enhancing ZSM-5 additives. As discussed above, early FCC catalyst developments after the introduction of zeolites involved adjustment of zeolite levels and rare-earth content to control equilibrium catalyst activity, gasoline yields and octanes. Rare-earth exchange into Y zeolite promotes hydrogen transfer reactions by slowing down the progress of dealumination under operating conditions. Another hypothesis for their ability maintains high zeolite acidity invokes the high polarization of former protonic sites adjacent to the trivalent rare earth cations which in turn creates special high electric field gradients and high acidity [4, 5]. Enhanced hydrogen transfer enhances gasoline make and coke production, however the octane quality of the gasoline is lower than can be optimally achieved from the unit due to the loss of gasoline range olefins by this mechanism. Octane enhancement has been achieved through two significant catalyst developments in this technology area. First, it was realized the zeolite dealumination and the consequent decrease in hydrogen transfer, while diminishing catalyst activity, can act to preserve the olefins in the product thereby improving octane. Pines and coworkers discovered a strong relationship between equilibrium catalyst unit cell size and gasoline octane [30]. These low-rare-earth octane catalysts are also suitable for residual oil processing because their low hydrogen transfer characteristics also results in low coke make and better stability in these bottoms upgrading applications [5]. Aside from control of rare-earth content and unit cell size control, much work has focused on controlling the level and location of extra-framework aluminum species (EFAL) formed during steam stabilization, porosity [19] and matrix activity.

Many studies have attempted to unravel the nature and disposition of the EFAL and to determine their impact on performance. Acid washing treatment is frequently applied to remove this so-called "debris" after steaming and is reported to decrease dry gas (C_1, C_2) make and improve selectivity [4, 5]. Extractive approaches to dealumination have also been applied with the objective of decreasing the amounts initially formed. Of these, extraction with EDTA was disclosed early on by Kerr [44], but was not commonly practiced in the FCC catalyst technology area. However treatment with $(NH_4)_2SiF_6$ was commercially practiced after its discovery by Union Carbide workers [46, 47] and actually introduced into commercial FCC catalyst by the Katalistics International Company. The performance benefits of this approach have been reported in the scientific literature and in technological publcations [18, 54]. In addition to lowering zeolite aluminum content, these varied methods of dealumination result in different aluminum profiles through the zeolite crystal with significant consequences on performance. Generally, dealumination by steaming results in an aluminum gradient where the exterior of the zeolite crystal is aluminum-rich whereas chemical dealumination and silicon reinsertion treatments result in an aluminum-depleted exterior. In a study comparing steam-dealuminated to AFS-treated Y zeolites, a lower activity for cracking large gasoil molecules was reported for the AFS-treated material [80]. This behavior was attributed to the fact that these large molecules are unable to diffuse into the interior of the zeolite crystals and thus are only able to react near the Si-rich surface. N-heptane reaction rates were found to be similar for the same two zeolites in this study. In contrast, aluminum-rich exteriors produced by steaming are thought to result in less selective cracking and excessive coke and gas make, possibly due to the Lewis acidity of the extra-frame alumina [19]. An alternative explanation for the lower activity for gas oil cracking of the AFS-treated Y may be related to the absence of mesoporosity normally created in the steaming step and the consequent poor access to the crystal interior by the large molecules. Such observations led FCC catalyst technologists to propose subsequent treatments to produce more even aluminum profiles and improve activity throughout the crystal [18]. Many of the developments in FCC catalyst technology have been the result of focusing on framework and extra-framework Al density and distribution. This continues as a major focus today and it is likely that all manufacturer have incorporated such technologies in their production practices.

In parallel to efforts to optimize dealumination methods and aluminum distribution, major development approaches center on the adjustment of matrix characteristics and activity. A wide range of matrix compositions have been incorporated in FCC catalyst formulations over the years, including clays and acid-washed clays, silica–magnesias, amorphous silica–aluminas and aluminas. Today's catalyst matrices are largely composed of Kaolin clay fillers along with active silica–alumina and alumina matrices and are optimized not only for their catalytic properties, but also for their physical characteristics, including bonding strength density. Alumina was introduced into these matrices as residuum upgrading took hold due to its good performance in bottoms upgrading.

A major trend in FCC catalyst technologies which began in the mid- to late 1980s was the introduction of the additive approach to adjusting the unit performance. Here additives formulated for a specific function began to be introduced into the unit's catalyst inventory as separate particles rather than being incorporated into the main catalyst formulation. This allowed the unit to address special performance needs without significantly impacting the main functions or properties of the main catalyst and added a tremendous level of flexibility. Major additives still in use today are CO combustion promoters and ZSM-5 octane additives. Combustion promoters facilitated lower regenerator temperatures by addressing the problem of CO removal encountered when operating under these conditions. They typically consist of very small amounts of platinum in the range of a few parts per million supported on an alumina matrix. Octane additives are in much wider use today due to the increased need for higher octane quality from the FCC naphthas and olefins used in gasoline blending. As discussed above, octane enhancement by ZSM-5 results from the shape-selective cracking of feed paraffins with the production of light C3 and C3 olefins and iso-olefins which are used for blending or as alkylation unit feeds. This performance feature led to the establishment of a new trend in FCC process which is the use of this unit for light olefin productions, as discussed in the next section.

16.4.2
The Changing Role of the FCC: Transportation Fuel Production or Petrochemical Feed Production

Once the stable, highly shape-selective, paraffin cracking performance of ZSM-5 additives in FCC became well established in the production of high octane gasoline blend stock, and as the light olefin demand grew, refiners began exploring the potential of this unit for the production of propylene. This trend developed to the point where several technology licensors have offered this as revamps or new unit designs. However, recognition is developing that a propylene make in the range of 12–20% is the upper limit currently achievable. This appears limited by the nature of the feedstock processed and by the large impact on unit operation, heat balance and byproduct mix. Equilibrum constaints are also thought to limit propylene make.

16.5
Hydrocracking and Hydroisomerization

Hydrocracking and hydroisomerization are related bond breaking and rearrangement processes which rely on the use of dual function catalysts operating under high hydrogen pressure to achieve their objectives. In fact, they share the same fundamental mechanistic steps and differ mainly in the degree to which some reaction pathways progress (Fig. 16.6). Much of the discussion in the following sections applies to both types; and distinctions are pointed out where required. Even though the hydrocracking process has evolved greatly since its early introduction in Germany in the 1930s some essential elements remain at its foundation and are critical to its successful application. Much of the evolution relied on an improved understanding of reaction fundamentals and the development of new materials with combined acidic and geometric properties suitable for use in this application. Whereas FCC was the main gasoline machine in the refinery, hydrocracking was best suited for the production of high-quality distillate suitable for jet and diesel production, due to the molecular composition of the products with their high hydrogen content and relatively low aromatics. Increased demand for diesel and the expected continuation of the drop in the gasoline:diesel ratio around the world has strengthened the demand for hydrocracking installations. Its ability to produce clean fuels with properties meeting the most stringent regulatory requirements has accentuated the demand even further.

16.5.1
Process Configurations and Catalysts

Many hydrocracking process configurations and flow schemes are in operation today in installations around the world producing a variety of diesel, jet and even naphtha products. The latter are produced mainly in North American units which were designed to process light cycle oils, to remove sulfur and nitrogen contaminants and to send the highly hydrogenated ring-products to catalytic naphtha reformers for octane enhancement. Process flow schemes are traditionally divided into four classes: once-through, single-stage recycle, two-stage recycle and separate-hydrotreat designs, Figure 16.4.

Technology licensors recently made significant advances in process design and they now offer many variations and improvements of these basic scheme, however the discussion below only examines these four basic designs because their reaction environments encompass all conditions that impact the elementary catalytic steps and their implications on catalyst and materials requirements. From a catalytic perspective these four configurations may be grouped into two sets which differ significantly in the reaction environments the catalysts operate in as discussed in the section below.

16.5.2
Catalyst Requirements for the Hydrocracker

Single-stage and once-through designs may be grouped into one category while the two-stage designs are grouped into another. All of these four schemes include a first reactor which is intended primarily for hydrodenitrogenation (HDN) and hydrodesulfurization (HDS) of the feed to prepare it for introduction over the downstream hydrocracking catalyst. As mentioned above, the dominant chemistry taking place in this reactor is the removal of nitrogen, sulfur and oxygen heteroatoms contained in the feeds, along with the removal of inorganic metal contaminants over beds of specialized catalysts designed for this purpose. Typical contaminants that are removed in this section are iron, vanadium, nickel, silicon, alkali metals and, more and more frequently, arsenic. Effective strategies for removal of these contaminants are critical for protection of the hydrotreating catalyst in the lower beds of this reactor and have become very important in hydrocracker design due to the increasingly heavy feed being processed in hydrocracker and due to the irreversibility of their impact in depressing catalyst activity downstream. The main HDN and HDS reactions take in lower beds with production of over catalysts optimized especially for HDN due to the strong impact inhibition effect of nitrogen compounds discussed above. Another main reaction taking place in this section is the hydrogenation of aromatics present in the feed to naphthenes and naphtheno-aromatics. Heat release from the aromatics hydrogenation is managed in these reactors and in the hydrocracking reactors by limiting bed sizes and by introducing quench and redistribution zones between beds to cool the mixture before entry into the next bed.

In the single-stage and once-through designs the combined liquid and gas effluents of the first reactor enter the cracking reactor where the majority of hydrocracking and boiling point conversion reactions predominate. The presence of (C-C bond scission) ammonia in the gas phase environment of these reactors imposes a major requirement for a strong enough acid function to enable the hydrocracking catalyst to overcome the inhibition and catalyze the cracking reactions. In terms of catalyst requirements and process design, this is a main distinguishing feature between the single-stage and once-through designs on the one hand and the second-stage and separate-hydrotreat on the other. It should be noted here that that the first-stage reactor of the two-stage process configuration is essentially a once-through type reactor and catalyst considerations for this service are very similar. To deal with the presence of relatively high concentrations of ammonia in this reactor which can range as high as 2000 ppmv, catalysts employed in this zone typically have much stronger acidity than their counterparts in the second-stage reactors. Difference in feed composition entering the cracking reactor must also be considered in this context because they are main distinguishing features between the once-through and the single-stage configurations. The liquid feed entering the cracking reactor in once-through (and the first stage of two-stage) units is the hydrotreating reactor product. As such, the main changes in its molecular composition from the feed entering the unit are the absence of contaminant metals and heterocyclic compounds along with some decrease in the level of aromatics. In contrast, feeds entering the cracking reactor of single-stage recycle units are the combined effluent of the hydrotreating reactor and the unconverted oil, or fractionator bottoms, boiling above the cut-point of the heaviest product and being returned to the cracking zones for further boiling-point conversion. This latter stream is very different in composition from the hydrotreating unit effluent in that it is highly hydrogenated and contains a larger proportion of branched and shorter-chain paraffins. The different reactivities of these two streams may require the

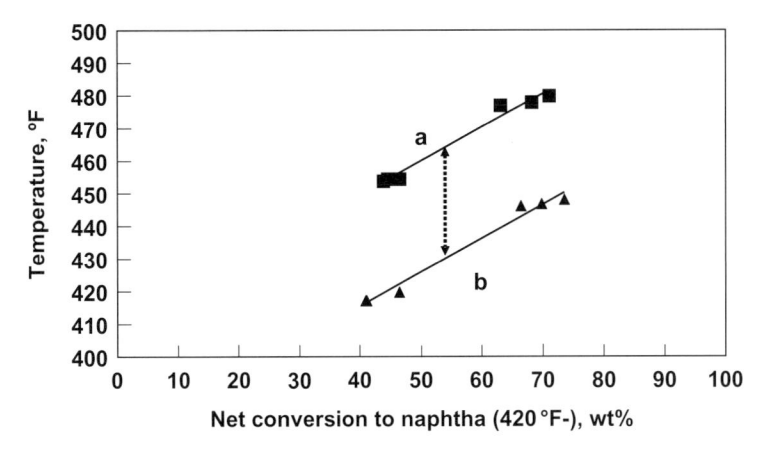

Figure 16.10 Second-stage activity versus metal content of Pd/stabilized Y zeolite catalysts: (a) 0.8% Pd, (b) 2.4% Pd.

selection of different catalyst compositions to more effectively produce an optimized yield structure.

The main distinguishing feature between the first/single-stage designs and the two-stage designs is the presence of a gas–liquid separator in the latter set where the effluents of the first treating or cracking reactors are sent and where the separator overhead gases are scrubbed to remove NH_3 and H_2S contaminants before returning the clean hydrogen back to the inlet of the first- or second-stage reactors. Thus, the presence of a cracking reactor to convert the feed after separation is a common main feature between the two-stage and separate-hydrotreating designs. Here again, there are many subtle differences between the two designs in terms of feed composition entering the second stage reactor which result from various options for directing the separator liquid to fractionation or to the second-stage reactor and where the recycle oil from the fractionator bottoms is sent. In all cases a major factor impacting catalytic requirements and catalyst design for the second stage is that it provides a clean environment for the catalyst requiring much lower acid strength to operate in the absence of ammonia. Since acid strength requirements are typically low here, catalysts with only moderate acidity operating in this environment can be thought of as hydrogenation-limited and the activity of such catalysts has been observed to be directly related to the metal level, Figure 16.10. Another approach for taking advantage of the cleaner environment of the second stage has been to use a smaller reactor in this service due to higher catalyst activity, thereby saving on hardware and catalyst costs.

An implication arising from the above discussion on the impact of ammonia in the first- and second-stage reactor environments is the different requirements for the balance of acidity to hydrogenation functions for catalyst intended for service in different locations in the hydrocracker. Namely, a catalyst intended for use in first- or single-stage service would require a different acid/metal ratio than one in

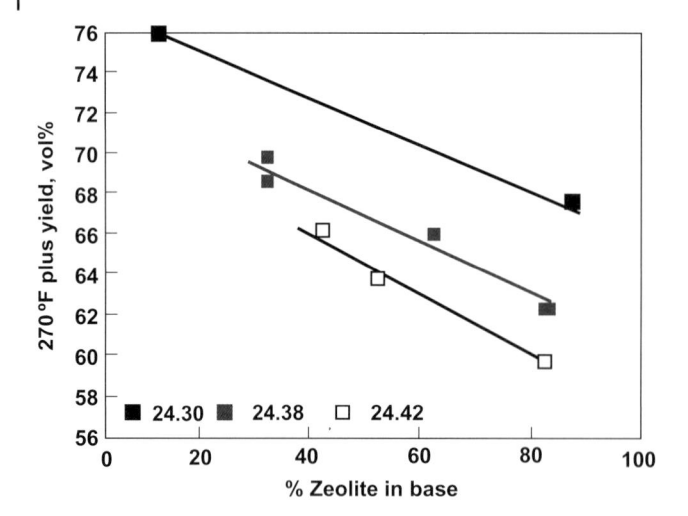

Figure 16.11 Hydrocracking catalyst performance in single stage recycle as a function of zeolite content and unit cell size.

second-stage service. Many methods to tailor these ratios to the intended service are employed by practitioners in this field, including the adjustment of metal types and metal levels along with the adjustment to acid function type and level discussed above.

Figure 16.11 gives performance examples illustrating the impact of the strength of the acid function on catalyst activity and yield structure in a simulated first-stage environment. It shows the results of pilot plant testing of a series of catalysts containing varying amounts of a Y zeolite and intermediate unit cell size 24.35 Å. Test conditions were adjusted to simulate the first-stage environment of distillate hydrocrackers with a heavy gasoil feed doped with NH_3 and H_2S precursors to levels typically present in a commercial reactor. The hydrogenation metals were held constant through the whole series of preparations. A strong dependence of yield of distillate-range products can be observed in these data.

As discussed above, operating conditions prevailing in the second stage are much less severe due to the nearly complete absence of ammonia. However, low levels of ammonia are nearly impossible to avoid due to the difficulty in completely removing it from the gas phase or its generation from the conversion of some unconverted organic nitrogen from the first-stage liquid product. Ammonia management in the second stage is critical to achieving optimum catalyst performance because it has a very large impact on catalyst activity and selectivity at the low levels prevalent in this environment. Figure 16.12 gives an example of its impact on activity, where it can be seen that it exerts a very strong impact in the beginning, which then levels off after a level of several hundred parts per million is reached. The exact level at which the impact tapers off depends, of course, on catalyst composition, feed composition and other variables.

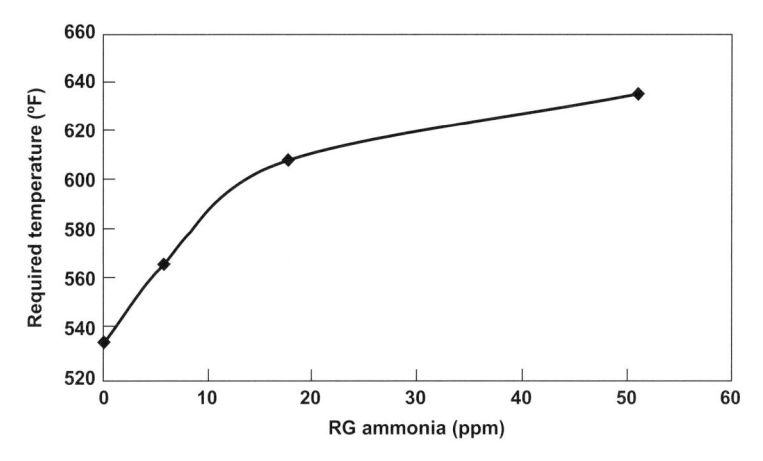

Figure 16.12 Impact of ammonia on hydrocracking catalyst activity.

Finally, we would like to point out that it is critical to consider the impact of the feed and gas composition on the catalyst performance discussed above when conducting scientific studies aimed at elucidating the mechanisms of hydrocracking reactions. Clearly reaction pathways predominant in the hydrocracker can be significantly altered by seemingly subtle changes in feed or gas composition, thereby leading to an unexpected and undesirable outcome. For fundamental studies, neglecting to take these factors into account can greatly diminish the value of the work and our collective ability to build on the knowledge established.

16.5.3
The Changing Role of the Hydrocracker in a Reformulated Fuels Environment

While hydrocracking is historically the main process for production of high quality distillates, this role has become even more prominent in recent years and will into the future due to the increasingly strict environmental regulations on fuel quality. With the adoption of the 10 ppm diesel sulfur standard in the industrialized world and with the rest of the developing world soon to follow, the hydrocracker is well positioned as the premier conversion process to help refiners meet these standards. The combination of its ability to produce ultralow sulfur diesel (ULSD) products with less than 10 ppm sulfur with low aromatic content and high burning qualities (cetane) gives the hydrocracker a distinct advantage over simple hydrotreating. This advantage is derived primarily from its ability to affect the boiling point conversion (C-C bond scission) of heavy gas oils into the distillate boiling range, which significantly enhances profitability over hydrotreating alone since it does not involve significant upgrade in product value but is mainly practiced for regulatory compliance. This bright future for the hydrocracker has spurred intensive research and development activities to enhance the process and optimize its ability to provide high-quality distillates at minimum processing costs.

Both catalyst and process innovations have been brought to market by major technology suppliers, including UOP, Chevron, Axens, Topsoe and Criterion. All of these suppliers have focused much of their recent development effort on the distillate selective catalysts and processes built essentially on manipulation of the principles discussed above. Clearly catalysts optimized for improved distillate selectivity must be built on an acid function of sufficient strength to provide adequate activity to achieve the needed cycle length, but without being so strong as to promote secondary cracking to lighter gasoline range products. These distillate-optimized catalysts must, of course, also maintain the proper metal function to help maintain low sulfur and low aromatics levels in the product. It is expected that much of the development work will maintain this focus in addition to the traditional quest for minimum production of light ends and other less valuable byproducts.

16.6
Conclusions

This chapter tried to illustrate the critically important role for solid acids in general and zeolites in particular in major refining bond breaking and rearrangement processes. A particular emphasis was placed on identifying and linking the key properties of key zeolites to the process chemistry and technology needs. To achieve this goal, the discussion was restricted only to zeolites currently playing a major role in refining today and the examples provided were selected because they represent commercial experience.

References

1 Rabo J.A. (ed.) (1979) *Zeolite Chemistry and Catalysis*, American Chemical Society, Washington, DC
2 Ward J.W. and Qader, S.A. (eds) (1975) *Hydrocracking and Hydrotreating*, ACS Symposium Series, 20, American Chemical Society, Washington, DC, p. 28.
3 Scherzer, J. and Gruia, A.J. (1996) *Hydrocracking Science & Technology*, Marcel Dekker.
4 Venuto, P.B. and Habib, E.T. (1979) *Fluidized Catalytic Cracking with Zeolite Catalysts*, Marcel Dekker, New York.
5 Magee, J.S. and Mitchell, M.M. (1993) Fluid catalytic cracking: science and technology, in Studies in Surface Science and Technology, vol. 76, Elsevier.
6 Coonradt, H.L. and Garwood, W.E. (1964) Mechanism of hydrocracking. reactions of paraffins and olefins. *Ind. Eng. Chem. Proc. Des. Dev.*, **3** (1), 38.
7 Decroocq, D. (1984) *Catalytic Cracking of Heavy Petroleum Fractions*, Editions Technip.
8 Baerlocher, C., McCusker, L.B., and Olson, D.H. (2007) *Atlas of Zeolite Framework Types*, 6th edn, Elsevier.
9 Flanigan, E.M., Lok, B.M., Patton, R.L., and Wilson, S.T. (1988) Molecular sieve compositions and their use in cracking/hydrocracking. US Patent 4,781,814.
10 Breck, D.W. and Flanigen, E.M., (1968) *Molecular Sieves*, Society of Chemical Industry, London, p. 47.
11 Breck, D.W. (1964) US Patent 3,130,007.

12. Vaughan, D.E.W., Edward, G.C., and Barrett, M.G. (1982) US Patent 4,340,573.
13. Wagner, P., Nakagawa, Y., Lee, G.S., Davis, M.E., Elomari, S., Medrud, R.C., and Zones, S.I. (2000) *J. Am. Chem. Soc.*, **122**, 263.
14. Kresge, C.T., Leonowiez, M.E., Roth, W.C., Vartuli, J.C., and Beck, J.S. (1992) *Nature*, **359**, 710
15. Rajagopalan, K. and Peters, A.W. (1985) Preprint of the ACS division of petroleum chemistry. *J. Am. Chem. Soc.*, **30** (3), 538.
16. McDaniel, C.V. and Maher, P.K. (1976) *Zeolite Chemistry and Catalysis* (ed. J.A. Rabo), American Chemistry Society, Washington, p. 285.
17. Ritter, R.E., Creighton, J.E., Roberie, T.G., Chin, D.S., Wear, C.C., (1986) Paper AM-86-45, NPRA Annual Meeting, at Los Angeles, CA.
18. Humphries, A., Yanik, S.J., Gerritsen, L.A., O'Conner, P., and Desai, P.H. (1991) *Hydrocar. Process.*, **70**, 69.
19. Scherzer, J. (1984) *The Preparation and Characterization of Aluminum Deficient Zeolites*, ACS Symposium Series 248, American Chemistry Society, Washington, DC, p. 157.
20. Magee, J.S. and Blazek, J.J. (1976) *Zeolite Chemistry and Catalysis* (ed. J.A. Rabo), American Chemistry Society, Washington, DC, p. 639.
21. Breck, D.W. (1974) *Zeolite Molecular Sieves*, John Wiley & Sons, Inc., New York.
22. Milton, R.M. (1968) *Molecular Sieves*, vol. 68, Soc. Chem. Ind., London.
23. Rabo, J.A., Pickert, P.E., Stamires, D.N., and Boyle, J.E. (1960) Molecular sieve catalysts in hydrocarbon reactions, in *Proceedings of the Second International Congress on Catalysis*, Ed. Tech, Paris, p. 2055.
24. Tanabe, K. (1970) *Solid Acids and Bases*, Academic Press, New York.
25. Plank, C.J. and Rosinski, E.J. (1966) US Patent 3,271,418.
26. Maher, P.K. and McDaniel, C.V. (1966) U.S. Pat. 3,293,192.
27. Maher, P.K. and McDaniel, C.V. (1969) U.S. Pat. 3,449,070.
28. Jacobs, P.A. and Uytterhoeven, J.B., (1972) *J. Chem. Soc. Faraday Trans.*, **69**, 359.
29. Ward, J.W. (1970) *J. Catal.*, **18**, 348.
30. Pine, L., Maher, P.J., and Wachter, W.A. (1984) *J. Catal.*, **85**, 466.
31. Redondo, A. and Hay, J. (1993) *J. Phys. Chem.*, **97**, 11754.
32. Argauer, R.J., Landolt, G.R., and Zeolite, C. (1972) ZSM-5 and method of preparing the same. US Patent 3,702,886.
33. Flanigen, E.M. and Patton, R.L. (1978) Silica polymorph and process for preparing same. US Patent 4,073,865.
34. Degnan, T. (2000) Applications of zeolites in petroleum refining. *Top. Catal.*, 349.
35. Maxwell, I.E. (1987) *Catal. Today*, **1**, 385.
36. Biswas, J. and Maxwell, I.E. (1990) Recent process and catalyst-related development in FCC. *Appl. Catal.*, **63**, 197–258.
37. Butt, J.B., and Petersen, E.E. (1988) *Activation, Deactivation, and Poisoning of Catalysts*, Academic Press.
38. Kerr, G.T. (1964) Synthetic zeolite and method for preparing the same. US Patent 3,247,195.
39. Perez-Pariente, J., Martens, J.A., Jacobs, P.A. (1987) Crystallization of zeolite beta. *Appl. Catal.*, **31**, 35–64.
40. Kotrel, S., Lusford, J.H., and Knozinger, H. (2001) Characterizing zeolite acidity by spectroscopic and catalytic means: a comparison. *J. Phys. Chem. B*, **105** (18), 3917.
41. Wilson, S.T., Lok, B.M., and Flanigen, E. (1982) US Patent 4,310,440.
42. Lok, B.M. Messina, C.A., Patten, R.L., Gajek, R.T., Canaan, T.R., and Flanigen, E.M. (1984) US Patent 4,440,871.
43. Miller, S.J. (1989) US Patent 4,859,312.
44. Kerr, G.T. (1968) *J. Phys. Chem.*, **72**, 2594.
45. Beyer, H.K. and Belenkaja, I. (1980) Stud. Surf. Sci. Catal., vol. 5, Elsevier, Amsterdam, p. 203.
46. Breck, D.W. and Skeels, G.W. (1985) US Patent 4,503,023, LZ210.
47. Skeels, G.W. and Breck, D.W. (1983) *Proc. 6th Int. Zeol. Conf.*, Reno, Nevada.
48. Pellet, R.J., Blackwell, C.S., and Rabo, J.A. (1988) *J. Catal.*, **114**, 71.

49 Bezman, R. and Rabo, J.A. (1983) Midbarrel hydrocracking, US Patent 4,401,556.
50 Jacobs, P.A. (1977) *Carboniogenic Activity of Zeolites*, Elsevier, Amsterdam.
51 Schulz, C.H., and Weitkamp, J. (1971) *Preprints, Div. Petroleum Chem.*, A.C.S., Los Angeles, A-102.
52 Weitkamp, J. (1975) *Hydrocracking and Hydrotreating* (eds J.W. Ward, and S.A. Qader), A.C.S. Symposium Series, 20, ACS, Washington, D.C., p. 1.
53 Pichler, C.J., Schulz, H., Reitemeyer, H.O., and Weitkamp, J. (1972) *Erdol, Kohle-Ergas-Petrochem.*, **25**, 494.
54 Upson, L.L., Lawson, R.J., Cormier, W.E., and Baars, F.J. (1990) Matrix, sieve, binder developments improve FCC catalysts. *Oil Gas J.*, **88**, 64.
55 Frilette, V.J., Haag, W.O., and Lago, R.M. (1981) Introduction of constraint index as a diagnostic test for shape selectivity using cracking rate constants for n-hexane and 3-methylpentane. *J. Catal.*, **67**, 218.
56 Haag, W.O., Lago, R.M., and Weisz, P.B. (1982) Transport and reactivity of hydrocarbon molecules in shape selective zeolites. *J. Chem. Soc., Farad. Disc.*, **72**, 317.
57 Mauleon, J.L. and Courcelle, J.C. (1985) *Oil Gas J.*, **64**, Oct. 21, 1985.
58 Ward, J.W. (1983) *Prepar. Catal.*, III, Studies in Surface Science and Catalysis, **16**, 587.
59 Abdo, S.F., Thakkar, V., and Dziabala B. (2006) Hydrocracking for Clean Fuels Production. AICHE Spring Meeting.
60 Abdo, S.F. (2007) Hydrocracking and Clean Fuels. ACS Spring Meeting, Chicago, IL, USA.
61 Ward, J.W. (1993) *Fuel Process. Technol.*, **35**, 55; Sullivan, R.F. and Scott, J.W. (1993) *Heterogeneous Catalysis: Selected American Histories*, ACS Symposium Series Vol. 222, ACS, Washington, D.C., p. 19.
62 Stork, W.H.J. (1997) *Molecules, Catalysts and Reactors in Hydroprocessing of Oil Fractions*, Studies in Surface Science and catalysis, **16**, 41–67, Elsevier, Amsterdam.
63 Ho, T. (2004) *Catal. Today*, **98**. 3.
64 Houalla, M. and Gates, B.C. (1978) Hydrodesulfurization of dibenzothiophene catalyzed by sulfided CoO-MoO3/γ-Al2O3: the reaction network. *AICHE J.*, **24**, 1015.
65 Topsoe, H., Clausen, B.S., and Massoth, F.E (1996) *Hydrotreating Catalysts – Science and Technology*, Springer-Verlag, Berlin.
66 Krenzke, L.D. and Baron, K. (1995) FCC pretreating to meet new environmental regulations on gasoline. Paper No. AM-95-67 presented at *1995 NPRA Annual Meeting*, San Francisco, Calif., March 1995.
67 Sullivan, R.F. and Meyer, J.A. (1975) *Hydrocracking and Hydrotreating* (eds J.W. Ward, and S.A. Qader), A.C.S. Symposium Series, 20, ACS, Washington, D.C., p. 28.
68 Coonradt, H.L. and Garwood, W.E. (1967) *Preprints Div. Petr. Chemistry*, ACS, **12**, 4.
69 Haag, W.C. and Dessau, R.M. (1984) Duality of mechanism in acid catalyzed paraffin cracking, in *Proceedings of the Eighth International Congress on Catalysis*, vol. 2, Verlag Chemie, Weinheim, p. 305.
70 Sie, S.T. (1881) Acid-catalyzed cracking of paraffinic hydrocarbons: 1. Discussion of existing mechanisms and proposal of a new mechanism. *Ind. Eng. Chem. Res.*, **31**, 1991.
71 Sie, S.T. (1993) Acid-catalyzed cracking of paraffinic hydrocarbons 2. Evidence for protonated cyclopropane mechanism from catalytic cracking experiments. *Ind. Eng. Chem. Res.*, **32**, 397.
72 Sie, S.T. (1993) Acid-catalyzed cracking of paraffinic hydrocarbons 2. Evidence for protonated cyclopropane mechanism from catalytic cracking experiments. *Ind. Eng. Chem. Res.*, **32**, 403.
73 Nace, D.M. (1969) *Ind. Eng. Chem. Prod. Res. Dev.*, **8** (1), 31.
74 Gates, B.C., Katzer, J.R., and Schuit, G.C.A. (1977) *Chemistry of Catalytic Processes*, Academic Press, New York.
75 Munoz Arroyo, J.A., Martens, G.G., Froment, G.F., Marin, G.B., Jacobs, P.A., and Martens, J.A. (2000) *Appl. Catal. A*, **192**, 9.
76 Thakkar, V.P., Abdo, S.F., Gembicki, V.A., and McGehee, J.F. (2005) LOC

Upgrading: A Novel Approach for Greater Added Value and Improved Returns. NPRA Annual Meeting, Paper AM-05-53.

77 McVicker, G.B., Daage, M., Touvelle, M.S., Hudson, C.W., Klein, D.P., Baird, W.C., Cook, B.R., Chen, J.G., Hantzer, S., Vaughan, D.E.W., Ellis, E.S., and Feeley, O.C. (2002) *J. Catal.*, **210**, 137.

78 Egan, C.J., Langlois, G.E., and White, R.J. (1961) *J. Am. Chem. Soc.*, **84**, 1204.

79 Weikamp, J., Ernst, S., and Karge, H. (1984) *Erdol und Kohle*, **37**, 457.

80 Cruz, J.M., Corma, A., and Fornés, V., Framework (1989) Framework and extra-framework aluminium distribution in $(NH_4)_2SiF_6$-dealuminated Y zeolites: relevance to cracking catalysts. *Appl. Catal.*, **50**, 287.

Index

a
ABC-6 family 45
abridged isosteres, water 291
absorbance, IR 127ff.
acetate, cellulose 330
acid-leached mordenites 116ff.
acid sites 157
– density 423ff., 461ff., 527
– quantitation 125ff.
acid strength 421ff., 544
acidic zeolites 72
acidity 420ff.
– Brønsted 124, 417ff.
– external 134ff., 424ff.
– IR spectroscopy 123ff.
– Lewis 124, 423ff.
– low-temperature probes 131ff.
– zeolitic adsorbents 207
acids
– adsorption 269ff.
– citric 206, 270
– fatty 185ff.
– phosphoric 66
– silica-supported phosphoric 443
– solid phosphoric 506
active sites, physicochemical characterization 507, 511ff., 524ff.
activity enhancing 537
additives, electrostabilizing 339ff.
adsorbate loading 209
adsorbates 49
adsorbents 274
– allocation 256ff.
– normal paraffin separation 250ff.
– zeolitic 61ff.
adsorbers, packed bed 275
adsorption 404ff.
– ammonia adsorption/desorption 130
– BET 153
– carbon monoxide 131ff.
– cation exchange 137
– competitive 332, 338
– energetics 411ff.
– equilibrium 276ff.
– equilibrium-selective 211ff.
– heat of 412ff.
– hexane 410
– hydrocarbons 408ff.
– isosteric heat of 405
– isotherms 152ff., 156ff., 209, 414
– isotherms (water) 290
– liquid phase 206ff.
– mixtures 413ff.
– nitrogen 406ff.
– non-regenerable 303
– pressure swing 296ff.
– process design 288
– pure component 408ff.
– pyridine 127ff.
– rate-selective 221ff.
– reactive 224ff.
– selectivity 414
– shape-selective 222ff.
– thermal swing 288ff.
– type I and IV isotherms 406
– vacuum swing 298
– wave shape and length 283
– wave speed 282ff.
adsorption acids, separation 269ff.
adsorption capacity 156ff.
adsorption constant 277
– Langmuir 316
adsorptive separation
– cation exchange 137
– industrial 173ff.
– industrial (gas phase) 229ff., 273ff.

– industrial (liquid non-aromatics)   249ff.
– processes   203ff.
– R&D   176
– zeolites   173ff.
Ag-substituted molecular sieves   225
agents, coupling   340
air separation, PSA   297ff.
Al-free silicalite-1   307
^{27}Al NMR   147ff.
alcohol, aromatic–alcohol alkylation   453ff.
alkanes
– cracking   421, 455
n-alkanes, cracking   459
alkanes, diffusivity   419
n-alkanes, heat of adsorption   412ff.
alkanes, hydroisomerization   436ff.
– light   377
alkenes
– aromatic–alkene alkylation   453ff.
– oligomerization   448ff.
– skeletal isomerization   447ff.
alkoxides   426ff.
alkylate, 2-phenyl   263
alkylation   364ff., 372ff.
– aromatic–alcohol   453ff.
– aromatic–alkene   453ff.
– benzene   454, 512ff.
– biphenyl with propylene   372
– catalysts   452, 509ff.
– ethylene   512ff.
– isobutane   450ff.
– motor fuel   508
– naphthalene   443
– olefins   507ff.
– paraffins   507ff.
– process chemistry   508, 512
– reaction mechanisms   450ff.
– side-chain   516
– trans-, see transalkylation
alkylbenzene
– disproportionation   517ff.
– linear   188ff., 261ff., 512ff.
– process chemistry   517
– transalkylation   517ff.
alpha cavity   29, 90
aluminophosphate (AlPO) molecular sieves   8ff., 334
– important synthesis parameters   67
– oxide ratio   9
– synthesis   62ff.
aluminophosphates, chemical shift   141
aluminosilicates   6ff.
– crystalline   1

– crystalline, see also zeolites
aluminum
– extra-framework   559
– removal   542, 558ff.
ammonia
– acid site quantitation   130ff.
– adsorption/desorption   130
– NH_4^+-FAU zeolite   121
ammonia-free environment   551
ammonium, quaternary cation   15
ammonium exchanged Y-zeolite   153
ammonium fluorosilicate   541
ammonium hexafluorosilicate   542
amorphous carbon   111
anionic surfactants   261ff.
annular dark-field detector   109
aperture, ring   33
applications
– asymmetric mixed-matrix membranes   344
– catalytic   355ff.
– industrial   355ff.
– molecular sieves   16ff.
– NMR spectroscopy   145ff.
– STEM   109ff.
– zeolite/polymer mixed-matrix membranes   346ff.
– zeolites   16ff.
approximation, linear driving force   281
– linear driving force   284ff.
A_8-producing technology   234
Arabian gasoil feed, light   554
aromatic–alcohol alkylation   453ff.
aromatic–alkene alkylation   453ff.
aromatic hydrocarbons, base strengths   214
aromatic petrochemicals, high-volume   230
aromatic ring opening   554ff.
aromatic ring saturation   490
aromatic transformation   462ff.
aromatics
– adsorptive separation   174, 177ff., 190
– C_8-   214ff., 489ff.
– dehydroaromatization to   377, 383
– footprints   431
– isomerization   375ff.
– methane conversion   384
– methanol conversion to   522ff.
– paraffin/olefin conversion to   518ff.
– reactions   369ff.
aromatics complex   232ff., 242
aromatization   382
arrangement, crosswise   507
Arrhenius relationship   496

asymmetric mixed-matrix membranes   342ff.
– flat sheet   343ff.
– hollow fiber   345ff.
asymmetric stretches   117
Atlas of Zeolite Structure Types   3
atomic coordinates, UZM-5   88
attenuated total internal reflection (ATR)   112

b

Ba-X zeolites   95
band shift, hydroxyl   133ff.
base strength, aromatic hydrocarbons   214
basic building unit (BBU)   28
basic probe molecules   126
BEA   453ff.
– effect on cracking selectivity   461
– H-BEA zeolite   132ff., 428
– see also beta polymorphs
beads, zeolitic   69
bed, moving   256ff.
Beer-Lambert law   112
Bellamy-Hallam-Williams (BHW) theory   124
benzene
– alkylation   364, 369, 454, 512ff.
– p-diethyl-   238ff.
– diisopropyl-   223
– ethyl-, see ethylbenzene
– linear alkyl   188ff., 261ff., 512ff.
– poly-alkylated   514
– tetramethyl-   428
– transalkylation of toluene   369
benzothiophenes   547
BET adsorption   153
beta polymorphs (*BEA/BEC)   6ff., 11ff., 537ff.
– framework structure   49ff.
beta scission pathways   455
bifunctional hydrocracking   553
bifunctional isomerization mechanism   480ff.
bifunctional zeolitic catalysts   424
bimolecular reaction mechanism   448
– cracking   455ff.
binary separation   322
binder   238, 253
binder problem   191
biphenyl   372
blending, gasoline   560
boggsite   12
boiling point   480, 551
boiling point distribution   204

bond breaking   535ff.
– FCC   556ff.
bond formation, C—C   505ff.
bond scission processes, chemistry   546ff.
bound zeolites   70
– phase identification   92
branched hexane   410, 414
breakthrough procedure   210ff.
Brønsted acidity   124, 417ff.
Brunauer type I/II/III isotherms   284ff.
buffer zone   256
building units   28ff.
– chain composite   30
– secondary   114
bulk removal   301ff.
i-butane, alkylation   450ff.
butane, aromatization   382
n-butane, cracking   457ff.
butane, isomerization   356ff.
butene, skeletal isomerization   486ff.
butylcyclohexane, cracking   434
2-butylene   450ff.
butylene   403ff.
butylene, isomerization   449
– oligomerization   449
i-butylene, protonation   426ff.
byproduct reactions   483

c

C_8 aromatics   214ff.
– isomerization   375ff., 488
– isomerization catalysts   494ff.
C_5/C_6 alkanes, diffusivity   419
C_5/C_6 boiling point distribution   204
C–C bond formation, industrial   505ff.
$C_2/C_3/C_4$ olefin oligomerization   505ff.
C_3/C_4 olefin   518ff.
C_6/C_7 olefin   520
C_3/C_4 paraffin   518ff.
C_6/C_7 paraffin   520
C_4 Olex process   266
C10+ paraffin, isomerization   358
C_4 separation   265ff.
cages   32ff.
– LTA   37
– sodalite   29, 36, 38, 121
– super-, see supercages
calcined FAU   122ff., 129
calcined Y zeolite   146ff.
capacity
– adsorption   156ff.
– ion exchange   213, 224

– normal paraffin separation adsorbents   251
capillary condensation   279ff.
carbenium ions   426ff.
– propyl   423
– theory   551ff.
carbocations   425ff.
carbohydrates   185ff., 269
carbon
– amorphous   111
– holey   104
carbon dioxide removal   295
carbon monoxide   132ff.
– adsorption   131ff.
carbonium ions   429, 551
carbonyl, nickel   139
carbonylation   438
catalysis
– metal-loading   138ff.
– noble metal-loading   139
– zeolite, see zeolite catalysis
catalysts
– activity (hydrocracking)   565
– alkylation   452
– aromatization   382
– benzene alkylation   512ff.
– bifunctional   424
– biphenyl alkylation   372
– butane isomerization   357
– C_8-aromatic isomerization   375ff., 494ff.
– circulating   511
– commercial   496ff.
– cumene formation   371
– disproportionation   517
– epoxidation   389ff.
– ethylbenzene formation   370
– ethylbenzene production   516
– FCC   93, 152, 378ff., 557ff.
– heavy paraffins isomerization   361
– hexane isomerization   359ff.
– higher-activity   499
– hydrocarbon cracking   4
– hydrocracking   556, 561ff.
– hydroisomerization   561ff.
– isobutane alkylation   368
– isomar ethylbenzene dealkylation   234
– light olefins isomerization   362ff.
– metal-zeolite catalyzed isomerization   479ff.
– methane to aromatics   384
– methanol conversion to olefins   523ff.
– methylation   515
– motor fuel alkylation   509ff.
– new materials   18
– olefin cracking   380ff.
– olefin oligomerization   365ff., 506ff.
– oxidation   391ff.
– paraffin/olefin conversion   519ff.
– Pd/stabilized Y   563
– pentane isomerization   359ff.
– protonation   452
– regeneration   495
– selectivity   359ff.
– SPA   506
– styrene production   516
– toluene transalkylation   373ff.
– transalkylation   517
– UOP Cyclar   519ff.
– zeolitic   61ff.
– zeolitic, see also zeolite catalysis
catalytic cracking, fluid, see fluid catalytic cracking
catalytic membrane reactor (CMR)   323ff.
cation exchange   137, 212ff.
cation sites
– CHA   44
– FAU   40
– LTA   37
– LTL   49
– MFI   47
– MWW   54
– primary   43
cations
– carbo-, see carbocations
– cationic complexes   74
– charge balancing   127
– di-valence   217
– extra-framework   1
– mono-valence   217
– quaternary ammonium   15
caustic solutions   63
cavities   32ff.
– alpha   29, 90
– CHA   43ff.
CBUs, see composite building units
CCR platforming unit   233
CDM, see charge density mismatch
cellulose acetate   330
cement   17
ceramic engineering   69
chabazite (CHA)   8ff.
– framework structure   42ff.
chain-breaking reactions   369
chain composite building units   30
channel system   213
– *BEA/BEC   51
– CHA   43

– FAU   39
– LTA   37
– MFI   46ff.
– MWW   53
– 12-ring   48, 50
– ten-ring   46ff.
– zig-zag   46ff.
channels   32ff.
– intracrystalline   2
– secondary eight-ring   41ff.
characterization
– chemical   152
– compositional   100ff., 108ff.
– electron microscopy   97ff.
– IR spectroscopy   111ff.
– metal-loaded zeolites   136ff.
– morphological   98ff., 106ff.
– NMR spectroscopy   140ff.
– physical   152
– physicochemical   507, 511ff., 524ff.
– structural   105ff.
– X-ray powder diffraction   91ff.
– zeolite membranes   313ff.
– zeolites   85ff.
charge balancing cations   127
charge density mismatch (CDM) concept   14
chemical engineering   68ff.
chemical formulae, zeolites   27
chemical modification, zeolites   72ff.
chemical shift   141
chemicals
– fine   188
– organic   185ff.
chemistry
– bond scission processes   546ff.
– C8 aromatics isomerization   489ff.
– gel   65
chromatographic separation   222, 239
chromium coating   99
CI (constraint index)   432ff.
CI* (modified constraint index)   434
circulating catalyst reactors   511
circumferential mixing   236
cis–trans isomerization   485ff.
citric acid   206
– separation   270
classic cracking mechanism   455ff.
cloverite, gallophosphate   11
clusters
– oriented zeolite crystals   108
– Pt   110
CMR (catalytic membrane reactor)   323ff.
co-boilers   517

co-extrusion method   345
coating, chromium   99
coke analysis   109ff.
coke deposition   527
Collidine   135, 442
colloidal silica   66
colloids   70ff.
columnar growth, Van-der-Drifts   311
columns
– effluent composition   210
– raffinate   237, 243
– zero length   419
combinatorial search   54
commercial catalysts   496ff.
commercially significant zeolites   35
compatibility
– normal paraffin separation desorbents   254
– polymer and zeolite materials   339
competitive adsorption   332, 338
complexes, cationic   74
composite building units (CBUs)   28
– chain   30
composition
– critical   536ff.
– property–function relationship   545ff.
compositional characterization
– SEM   100ff.
– TEM   108ff.
compositional variation   103
condensation, capillary   279ff.
conferences, international   17ff.
configurational bias Monte Carlo (CBMC) simulations   408ff.
configurational diffusion   314ff.
confocal optical microscopy, fluorescent   313
connectivity
– channel systems   46ff.
– T-atom   48
constraint index (CI)   432ff.
continuous synthesis   77
conversion, incomplete   65
corn syrup   269
counter-diffusion CVD   317
countercurrent contact   256
coupling agents   340
covalent organic frameworks (COFs)   393
cracking
– alkanes   421, 455ff.
– n-alkanes   459
– n-butane   457ff.
– butylcyclohexane   434
– catalysts   380ff.

- classic reaction mechanism   455ff.
- cumene   158
- effect of pore size and acid site density   461ff.
- endocyclic protolytic   458
- FCC, *see* fluid catalytic cracking
- hydro-, *see* hydrocracking
- hydrocarbon   4
- ideal   553
- kinetics   458ff.
- naphtha steam   229
- olefin   377, 380ff.
- paraffin   551ff.
- propane   422
- Resid   557
- selectivity   461

p-cresol   245
critical zeolite properties   536ff.
cross polarization   142ff.
crosswise arrangement, *trans*-but-2-ene   507
crystal size effects   446ff.
crystalline aluminosilicates   1
- *see also* zeolites
crystalline materials, liquid separations   177ff.
crystalline phosphates   141
crystallite size   95ff.
crystallization
- hydrothermal   16
- *in situ*   308ff.
- post-forming   71ff.
cube   29
cumene   512ff.
- cracking   158
- formation   371ff.
Cyclar catalyst, UOP   519ff.
cycle length   499, 545
- Sorbex technology   257
cyclic compounds, light cycle oil   555ff.
cyclo-olefins, unsaturated   552
cyclohexane, methyl-   436

d

D-R model   278
dark-field detector, annular   109
Database of Hypothetical Frameworks   55
DDR framework   461
DDR-type zeolite membranes   316ff.
dealkylation   234
dealumination   542, 558ff.
- processes   72ff.
- steaming   93
- Y zeolites   107

n-decane, hydroisomerization   437
defects
- formation   313
- intercrystalline   320
- macroporous   314
degree of exchange   137
deheptanizer   498
dehydration
- industrial   301ff.
- organic   321
- PSA   300ff.
- TSA   289ff.
dehydrators, rotary   301
dehydroaromatization   377ff.
dehydrogenation, Pacol™ process   268
dehydroxylation, zeolite surfaces   424
denitrogenation   546ff.
- hydro-   561ff.
dense films, mixed-matrix   341ff.
density
- acid sites   423ff., 461ff., 527
- density function theory (DFT) methodologies   159
deoxygenation   546ff.
deposition, coke   527
desalination of water   347
desiccant wheels   301ff.
design
- adsorption processes   288
- alkylation reactors   510
- hydrocracker   550
desorbents   207, 219ff.
- "heavy"   255, 261
- normal paraffin separation   253ff.
- zeolite/desorbent combination   237ff., 242ff.
desorption
- ammonia   130
- temperature programmed, *see* temperature programmed desorption
desulfurization, *see* sulfur removal
detergent Molex™ process   249ff., 261ff.
detergent Olex process   268
detergent range Olex   266ff.
detergents, linear alkylbenzene   512ff.
dewatering   67ff.
DFT (density function theory) methodologies   159
di-valence cations   217
diameter, kinetic   33, 309, 431ff.
dibenzothiophenes   547
p-diethylbenzene (PDEB)   238ff.
diffraction, precession electron   106

diffusion   416ff.
– configurational   314ff.
– fluid film   286
– hydrocarbons in zeolites   151ff.
– intra-crystalline   287ff.
– Knudsen   287, 314ff.
– macro-pore   286ff.
– molecular   221
– PFG NMR   144ff.
diffusion-based models   283
diffusion-controlled permeation/separation   318
diffusion mechanisms   286ff., 330
diffusivity
– C5–C6 alkanes   419
– effective   416
– self-   417ff.
diisopropylbenzene isomers   223
dimethylether   438
2,2-dimethylhexane   425
2,6-dimethylnaphthalene   244
dinitrosyl species   139
dipolar recoupling   143ff.
dipolar repulsion   216
dispersed phase   337ff.
disproportionation   462ff.
– alkylbenzene   517ff.
– catalysts   517
– *meta*-xylene   435ff.
– xylene   492
dissolution   64
disulfides   547
double bond
– exocyclic   469
– isomerization   485ff.
double four-rings   28, 36
double six-rings   38
driving force, linear   281
– linear   284ff.
drunken sailor coming out of a bar   417
drying   75
durene/isodurene selectivity   215, 218

e
ease of separation, normal paraffin separation desorbents   254
EDS, *see* energydispersive X-ray spectroscopy
EELS (electron energy loss spectroscopy)   110ff.
EFAL (extra-framework aluminum) species   559
effective diffusivity   416
effective pore diameters   32, 309
effluent column   210

eight-ring channels, secondary   41ff.
eight-ring pore   32
Einstein's expression, self-diffusivity   417
electron diffraction, precession   106
electron energy loss (EELS) spectroscopy   110ff.
electron microscopy
– scanning, *see* scanning electron microscopy
– transmission, *see* transmission electron microscopy
electron microscopy characterization   97ff.
electronegativity, Sanderson   122
– Sanderson   207ff.
electrostabilizing additives   339ff.
endocyclic protolytic cracking   458
energetics, hydrocarbon adsorption   411ff.
energy, Gibbs free   439
energy loss spectroscopy, electron   110ff.
energydispersive X-ray spectroscopy (EDS)   97, 100ff.
engineering, ceramic/chemical   68ff.
enthalpy   415ff.
– balance   281
– control wheels   302
entropy   415ff.
environmental issues (HS&E)   61
EOG (ethylene off gas)   300
epoxidation   387
– catalysts   389ff.
equilibrium-selective adsorption   211ff., 276ff.
– metal cation exchange   212ff.
– operating temperature   220ff.
ether
– dimethyl-   438
– methyl *tert*-butyl   479, 486ff.
ethylbenzene   244, 512ff.
– conversion to xylenes   463ff.
– formation   370
– isomar ethylbenzene dealkylation catalyst   234
– isomerization   369, 492ff.
ethylene alkylation   512ff.
ethylene off gas (EOG)   300
ETS-10   301ff.
EUO framework   497
exocyclic double bond   469
external acidity   134ff., 424ff.
external surface sites   513
extinction coefficients, IR   126
extra-framework aluminum (EFAL) species   559

extra-framework cation   1
extract composition   266ff.
extrusion   345

f

fatty acids   185ff., 270ff.
FAU   3, 12
– calcined   122ff., 129
– cation sites   40
– framework spectrum   115
– framework structure   38ff.
– hydrated sodium   94
– steamed   122ff., 129
– zeolite catalysis   355
FCC, see fluid catalytic cracking
FCOM (fluorescent confocal optical microscopy)   313
feed composition   266ff.
– C8 aromatics isomerization   489ff.
feed production, petrochemical   560
feeds
– alkylation process chemistry   508, 512
– alkylbenzene process chemistry   517
– Arabian gasoil   554
– hydrotreated   550
– methylation process chemistry   514ff.
– oligomerization process chemistry   505ff.
– paraffin/olefin process chemistry   518ff.
FER framework   408, 411ff.
fiber   70ff.
– hollow   345ff.
field gradient, pulsed, see pulsed field gradient
film diffusion   286
films   70ff.
– dense   341ff.
– thin   346
filtration   67ff.
fine chemicals, adsorptive separation   188
finishing   71ff.
firing   75
Fischer–Tropsch reactions   188ff.
flat sheet asymmetric mixed-matrix membranes   343ff.
flow schemes   498ff.
– hydrocracker   550
fluid catalytic cracking (FCC)   16, 93, 152, 355
– bond breaking and rearrangement   556ff.
– catalysts   378ff., 557ff.
– FCC-type feeds   369, 377

– industrial isomerization   486ff.
– product control   544ff.
fluid film diffusion   286
fluorescent confocal optical microscopy (FCOM)   313
fluorosilicate, ammonium   541
footprints   431ff.
formaldehyde   516
formed zeolite particles, mass transfer   280ff.
forming, zeolite powders   68ff.
FOS-5   50
four-rings, double, see double four-rings
Fourier transform IR (FTIR) spectroscopy   98 106, 123
fractional site occupancy   418
fractionation   270ff.
framework IR   114ff.
– zeolites synthesis   118ff.
framework Si/Al   116ff.
frameworks
– ABC-6 family   45
– building units   28ff.
– covalent organic   393
– critical types   536ff.
– DDR   461
– effect on adsorption properties   212
– EUO   497
– FER   408, 411ff.
– flexibility   319ff.
– hypothetical   54ff.
– metal organic   15, 393
– MFI, see MFI
– octahedral–tetrahedral   13
– property–function relationship   545ff.
– SAPO-34   524
– TON   424ff.
– type code   34
– types   27ff.
– ZSM-5   524
free energy, Gibbs   439
free fatty acid separation   270ff.
Freundlich isotherm   279
fructose, high-fructose corn syrup   269
fructose/glucose separation   218ff.
FTC (framework type code)   34
FTIR, see Fourier transform IR spectroscopy
fuels
– alkylation   507ff.
– alkylation catalysts   509ff.
– reformulated   565ff.
– transportation   560

fumed silica 66
function, property–function relationship 542ff.

g
gallophosphate cloverite 11
gas chromatogram–mass spectrophotometry (GCMS) 265
gas-oil conversion processes 539
gas phase adsorptive separation 173
– industrial 229ff., 273ff.
gas phase separation, zeolite/polymer mixed-matrix membranes 347
gases
– EOG 300
– remote natural 521
– ROG 300
gasoil feed 554
gasoline
– blending 560
– Molex™ process 249ff., 258ff.
– range 258
– sulfur removal 176, 188ff.
– upgrading 204
gas/vapor separation 322
GCMS (gas chromatogram–mass spectrophotometry) 265
gel chemistry 65
geometry
– pores 545ff.
– zeolite/polymer mixed-matrix membranes 341ff.
Gibbs free energy 439
glassy polymers 336ff.
glucose, fructose/glucose separation 218ff.
gradient, field, see pulsed field gradient
Grignard reagent 339ff.
growth
– columnar 311
– hydrothermal 310
– polycrystalline membrane 310
– secondary (seeded) 311ff.
gypsum 206

h
H-BEA zeolite 132ff.
– IR spectroscopy 428
H-FAU, IR absorbance 127ff.
H-form zeolites 121
– chemical shift 142
^{1}H NMR 150ff.
Hallam-Williams theory, Bellamy- 124
Hammet titration 157

HDN (hydrodenitrogenation) 561ff.
health, safety and environmental (HS&E) issues 61
heat, latent 292
heat of adsorption 412ff.
– isosteric 405
heavier olefins, oligomerization 364
heavy desorbent 255, 261
heavy paraffins, normal separation 261ff.
Henry coefficient, CBMC simulations 408ff.
Henry's Law 277
heteroatom removal 546ff.
heteroatom substituted zeolites 387
heterocyclic nitrogen compounds 548
hexadecane, n-, see n-hexadecane
hexafluorosilicate, ammonium 542
hexagonal prism 29
hexamethylbenzenium ion 468ff.
n-hexane 410
hexane
– butylcyclo- 434
– 2,2-dimethyl- 425
– equilibrium composition of isomers 480ff.
– isomerization 356ff.
hexenes 486ff.
high-fructose corn syrup 269
high-performance membranes 333
high Si/Al zeolites 6
high-temperature sealing 324
high-volume aromatic petrochemicals 230
higher-activity catalysts 499
higher olefins 383
highly hydrogenated ring-products 561
Hinshelwood, Langmuir–Hinshelwood rate expressions 403ff.
– Langmuir–Hinshelwood rate expressions 458
history
– catalytic applications 355ff.
– international conferences and organizations 17ff.
– molecular sieve materials 5ff.
holey carbon 104
hollow fiber asymmetric mixed-matrix membranes 345ff.
hourly space velocity
– liquid 432, 495ff.
– weight 495ff.
H_2S, removal 294
HS&E issues 61
hydrated samples 147
hydrated sodium faujasites 94

hydride shift 430
hydrocarbons
– adsorption 408ff.
– adsorptive separation 174ff.
– aromatic 214
– conversion over zeolites 429ff.
– cracking catalyst 4
– diffusion in zeolites 151ff.
– "hydrocarbon pool" mechanism 465
– methanol conversion to 527
– packing 239
hydrocracking 383, 387ff.
– bifunctional 553
– bond breaking and rearrangement 560ff.
– catalyst activity 565
– catalysts 556, 561ff.
– n-hexadecane 443ff.
– supports 554
hydrodenitrogenation (HDN) 561ff.
hydrogen, PSA purification 299ff.
hydrogen transfer reactions 544
hydrogenated ring-products 561
hydroisomerization 436ff.
– bond breaking and rearrangement 560ff.
– catalysts 561ff.
– n-decane 437
hydrolysis 64
hydrophilic zeolite membranes 321
hydroprocessing, octadecane 440
hydrothermal crystallization 16
hydrothermal growth 310
hydrothermal synthesis 62ff.
– LTA 118ff.
hydrotreated feed 550
hydrotreating 383, 387ff.
hydroxyl band shift 133ff.
hydroxyl IR 120ff., 543
hydroxyl nests 539
hypothetical frameworks 54ff.
i-butane, alkylation 451ff.
i-butylene, protonation 426ff.

i

ideal cracking 553
impregnation 74
impurities removal 176, 188ff.
in situ crystallization 308ff.
in situ/in operando studies 136
incomplete conversion 65
industrial adsorptive separation 173ff.
– gas phase 229ff., 273ff.
– liquid non-aromatics 249ff.

industrial applications of zeolite catalysis 355ff.
industrial C–C bond formation, processes 505ff.
industrial dehydration 301ff.
industrial isomerization 479ff.
– fluid catalytic cracking 486ff.
– zeolite catalysis processes 483ff.
industrial processes 231
Industrial PSA separations 296ff.
industrial purification 288ff.
industrial TSA separations 288ff.
infrared ... , *see* IR
inorganic polymers 1
inorganic zeolite membranes 76
inter-particle compositional variation 103
interaction, van der Waals 404
interconnected voids 2
interconversion, naphthene 493
intercrystalline defects 320
interface, zeolite/polymer 340
intergrowth 313
intermediate electronegativity, Sanderson 122
– Sanderson 207ff.
intermediate paraffins, normal separation 260ff.
intermediate Si/Al zeolites 6
intermolecular hydride transfer 430
internal reflection, total 112
international conferences 17ff.
International Zeolite Association (IZA) 19, 27, 356
– Structure Commission 3, 96
interzeolitic porosity 311
intracrystalline channels 2
intracrystalline diffusion 287ff.
iodide removal 190
ion exchange 18, 74
– adsorptive separation 223ff.
– isotherm 74
– kinetics 224
– theoretical capacity 213
ionic radius 208
IR absorbance, H-FAU 127ff.
IR extinction coefficients 126
IR spectroscopy
– acidity 123ff.
– Fourier transform, *see* Fourier transform IR spectroscopy
– hydroxyl 543
– hydroxyl IR 120ff.
– OH-stretching vibrations 411
– sample pretreatment 119ff.

– selection of probe molecules   125
– tetramethylbenzene   428
– zeolite characterization   111ff.
isobutane, alkylation   364, 368, 450ff.
isodurene, durene/isodurene selectivity   215, 218
isomar ethylbenzene dealkylation catalyst   234
isomerization   356ff.
– bifunctional mechanism   480ff.
– butylene   449
– byproduct reactions   483
– C_8 aromatics   375ff., 489ff.
– catalyst regeneration   495
– catalysts (C_8 aromatics)   494ff.
– *cis–trans*   485ff.
– double bond   485ff.
– ethylbenzene   492ff.
– feed and product molecules   480
– hydro-, *see* hydroisomerization
– industrial   479ff.
– light paraffins   479ff.
– *meta*-xylene   445ff.
– metal-zeolite catalyzed   479ff.
– olefins   484ff.
– *para*-xylene   135
– SafeCat   482
– side reactions   494ff.
– skeletal   447ff., 486ff.
– xylene   491ff.
isomers, diisopropylbenzene   223
isooctane   262
"ISOSIV" process   4
isosteres   276
– water   291
isosteric heat of adsorption   405
isotherms
– adsorption   152ff., 156ff., 209, 414
– Brunauer type I/II/III   284ff.
– Freundlich   279
– ion exchange   74
– Langmuir   284ff., 404ff.
– Langmuir-Freundlich   279
– linear   277
– Type I and IV   406
– universal   278ff.
– water adsorption   290
IUPAC   27
IZA, *see* International Zeolite Association

k

K-LTA   137
Kelvin equation   279ff., 407
kerosene   261
– *n*-paraffin-depleted   264

kinematic wave equation   282
kinetic diameter   33, 309, 431ff.
kinetics
– cracking   458ff.
– ion exchange   224
– *n*-alkane cracking   459
– propane cracking   422
– reaction   404ff.
Knudsen diffusion   287, 314ff.
Kubelka-Munk relationship   113

l

LAB (linear alkylbenzene)   188ff., 261ff., 512ff.
Lambert–Beer law   112
Langmuir adsorption constant   316
Langmuir equation   277
Langmuir-Freundlich isotherm   279
Langmuir–Hinshelwood rate expressions   403, 405, 458
Langmuir isotherm   284ff.
– reaction kinetics   404ff.
Langmuir plot, linearized   406
Laplace transform   285
large-pore zeolites   393
latent heat   292
laws and equations
– Arrhenius relationship   496
– Beer-Lambert   112
– Henry's Law   277
– Kelvin equation   279ff., 407
– kinematic wave equation   282
– Kubelka-Munk relationship   113
– Langmuir equation   277
– Langmuir–Hinshelwood rate expressions   403, 405, 458
– Maxwell model equation   334ff.
– partial differential equations   281ff.
lead reactor   549
lead/trim bed installation   293
Lewis acidity   124, 423ff.
light alkanes   377
– conversion to aromatics   377
light Arabian gasoil feed   554
light cycle oil, cyclic compounds   555ff.
light olefins   358, 364
light paraffins   258ff., 479ff.
Linde Type A, *see* LTA
Linde Type L, *see* LTL
linear alkylbenzene (LAB)   188ff., 261ff., 512ff.
linear driving force approximation   281, 284ff.
linear hexane   410, 414
linear isotherm   277

linearized Langmuir plot 406
liquid adsorption acid separation 269ff.
liquid adsorptive separation, industrial 249ff.
liquid hourly space velocity (LHSV) 432
liquid-hourly space velocity (LHSV) 495ff.
liquid-liquid separation 320ff.
liquid phase adsorption 206ff.
liquid separation, zeolite/polymer mixed-matrix membranes 347
liquid–solid countercurrent contact 256
liquids, adsorptive separation 174ff.
loading 274
– metal- 138ff.
Loeb-Sourirajan process 342
longer-chain olefins 488
low-density silica 14
low Si/Al zeolites 6
low-temperature acidity probes 131ff.
LRC model 278ff.
LTA 3, 8ff.
– cation sites 37
– characterization 90
– framework structure 36ff.
– K-LTA 137
– Na-LTA 137
LTL 13
– framework structure 47ff.
Lutidene 135
m-xylene, *see meta*-xylene

m

macro-pore diffusion 286ff.
macroporous defects 314
magic angle spinning (MAS) NMR 140ff.
– calcined Y zeolite 146ff.
– methanol conversion 466
magnetic resonance, nuclear, *see* NMR
manufacture, post-forming 71ff.
markets for synthetic zeolites and molecular sieves 17
mass spectrophotometry, gas chromatogram– 265
mass transfer
– in formed zeolite particles 280ff.
– zone length 285ff.
mass transfer coefficients 284
mass transfer rate, normal paraffin separation adsorbents 252
mass transfer resistance 69
– micropore 416

materials
– selection 336ff.
– zeolite 34ff.
materials explosion 7ff.
matrix, mixed-matrix membranes, *see* mixed-matrix membranes
MaxEne™ process 250, 260ff.
Maxwell model 334ff.
McCabe–Thiele diagram 284
MCM-22 12
– framework structure 51ff.
– *see also* MWW
MeAPO, *see* metal aluminophosphate
mechanical analysis 154ff.
membranes
– catalytic membrane reactor 323ff.
– high-performance 333
– inorganic zeolite 76
– mixed matrix 76
– mixed-matrix, *see* mixed-matrix membranes
– polymer 329ff.
– reverse osmosis 346
– separations 307ff.
– UOP Separex™ 330
– zeolite, *see* zeolite membranes
mercaptans 294, 547
mercury removal 296
mesopore network 546
mesoporous molecular sieves 13ff.
meta-xylene
– diamine 241
– disproportionation 435ff.
– industrial processes 241
– isomerization 445ff.
– production 241
metal aluminophosphate (MeAPO) molecular sieves 10
metal cation exchange 212ff.
metal-loaded zeolites 136ff.
metal-loading for catalysis 138ff.
metal organic frameworks (MOFs) 15, 393
metal salt 73
– solutions 73
metal sulfides, microporous 11
metal traps 558
metal-zeolite catalyzed light paraffin isomerization 479ff.
metallosilicate molecular sieves 10ff.
metals, adsorptive separation 190ff.
– in zeolites 109ff.
metastable phases 62ff.

methane, conversion to aromatics 384
– dehydroaromatization 383
methanol
– conversion catalysts 523ff.
– conversion to aromatics 522ff.
– conversion to hydrocarbons 527
– conversion to olefins 383, 385ff., 446, 464ff., 521ff.
– MTG 523
methyl, mono-, see mono-methyl ...
methyl tert-butyl ether (MTBE) 479, 486ff.
methylation
– catalysts 515
– process chemistry 514ff.
– toluene 514ff.
methylcyclohexane 436
methylethylcyclopentene intermediates 465
3-methylpentane 410
MFI 3, 307ff., 332
– bond breaking and rearrangement 537ff.
– CBMC simulations 408ff.
– effect on cracking selectivity 461
– framework structure 45, 45ff.
– methanol conversion 522
– n-butane cracking 457ff.
– see also ZSM-5
micropore mass transfer resistance 416
microporous metal sulfides 11
microscopy, electron 97ff.
– fluorescent confocal optical 313
microwave synthesis 77
mismatch, charge density 14
mixed-matrix dense films 341ff.
mixed-matrix membrane (MMM) 75ff., 329ff.
– asymmetric 342ff.
– composition 332ff.
– thin-film composite 346
mixing, UOP Parex™ process 236
mixtures, adsorption 413ff.
MMP Sorbex process 263
modeling and simulations
– C_8 aromatics isomerization 497
– CBMC 408ff.
– DFT 159
– Gibbs free energy 439
– molecular diffusion in silicalite 221
– Monte Carlo 408ff.
– moving bed operation 256ff.
– resistance modeling 284ff.
models, see theories and models

modes of operation, IR spectroscopy 112ff.
– zeolite separation 208ff.
modified constraint index (CI*) 434
MOFs, see metal organic frameworks
moisture content, zeolites 218ff.
molecular diffusion in silicalite 221
molecular dimensions, heavy organic compounds 253
molecular probing 310, 313
molecular sieves
– Ag-substituted 225
– AlPO, see aluminophosphate molecular sieves
– and zeolites 1ff.
– applications 16ff.
– materials history 5ff.
– mesoporous 13ff.
– metallosilicate 10ff.
– nanoparticles 75ff.
– nomenclature 2ff.
– regeneration 275ff.
– SAPO 9, 334
– silica 6ff.
– synthesis 15ff.
molecular sieving effect 287
Molex™ process 249ff.
– adsorbent allocation 256ff.
monetization, remote natural gas 521
mono-methyl paraffin 222
– separation 263ff.
mono-valence cations 217
monomolecular reaction mechanism 447ff.
– cracking 456ff.
monounsaturated cyclo-olefins 552
Monte Carlo simulations, configurational bias 408ff.
mordenite (MOR) 3
– acid-leached 116ff.
– framework structure 40ff.
morphological characterization, SEM 98ff.
– TEM 106ff.
motor fuel alkylation 508
– catalysts 509ff.
motor octane number (MON), alkene oligomerization 450
– C5–C7 alkane isomers 358
moving bed operation 256ff.
MQMAS 148ff.
MTBE (methyl tert-butyl ether) 479, 486ff.

MTG (methanol to gasoline), production distribution   523
MTO (methanol to olefin), see methanol
MTW, zeolite powders   149
multi-technique methodology   86ff.
Munk relationship, Kubelka-   113
MWW   12
– external surface sites   513
– framework structure   51ff.
– see also MCM-22
MX Sorbex process   241ff.
n-alkanes, cracking   459
– heat of adsorption   412ff.
n-butane, cracking   457ff.
n-decane, hydroisomerization   437
n-hexadecane   443ff.
– hydrocracking   444
n-hexane   410
n-paraffin-depleted kerosene   264
n-pentane   262

n
Na-LTA   137
Na-X, zeolites   95
nanocrystalline zeolite   99
nanoparticles   75ff.
nanosized zeolite particles   339
naphtha steam cracking   229
naphthalene, alkylation   443
naphthene, interconversion   493
– ring opening   554ff.
National Institute of Standards and Technology (NIST)   35
natrium, see sodium
natural gas, remote   521
natural zeolites   4ff.
new catalytic materials   18
nickel carbonyl species   139
NIST (National Institute of Standards and Technology)   35
nitrogen
– adsorption   406ff.
– low-temperature acidity probe   134
– physisorption   152ff.
nitrogen compounds, heterocyclic   548
– relative reactivity   547
nitrogen contaminants   548
nitrogenate removal   190, 546ff.
NMR spectroscopy
– ^{27}Al   147ff.
– ^{1}H   150ff.
– methanol conversion   466
– ^{17}O   151
– ^{31}P   149ff.

– REAPDOR   144
– ^{29}Si   145ff.
– TRAPDOR   144
– zeolite characterization   140ff.
noble metal-loading for catalysis   139
nomenclature, molecular sieve materials   2ff.
non-adiabatic packed bed adsorbers   275
non-aromatic hydrocarbons, adsorptive separation   174ff., 183ff., 249ff.
non-petrochemicals, adsorptive separation   175
non-regenerable adsorption   303
non-zeolite/polymer mixed-matrix membranes   333
normal paraffin separation   249ff.
nuclear magnetic resonance, see NMR
nuclei, quadrupolar   143
– spin-half   142
Nusselt number   291
^{17}O NMR   151

o
octadecane, hydroprocessing   440
octahedral–tetrahedral frameworks   13
octahedron, truncated   29
octane number, research, see RON
octane number-clear, research   480ff.
octane pool, Refiner's   258
OH-stretching vibrations   411
oil, gas-oil conversion processes   539
– light cycle   555ff.
olefins
– alkylation   507ff.
– conversion catalysts   519ff.
– conversion to aromatics   518ff.
– cracking   377, 380ff.
– cyclo-   552
– heavier   364
– isomerization   362ff., 484ff.
– light   358
– liquid separation   265ff.
– longer-chain   488
– methanol conversion to   383, 385ff., 446, 464ff., 521ff.
– olefin-paraffin separation   174, 204ff.
– oligomerization   365ff., 505ff.
– process chemistry   518ff.
Olex, detergent range   266ff.
oligomerization   358ff.
– alkenes   448ff.
– butylene   449
– catalysts   506ff.
– olefins   505ff.

– process chemistry 505ff.
– propylene 441ff.
one-dimensional 12-ring channels 48
one-step co-extrusion method 345
optical microscopy, fluorescent confocal 313
optimization, C_8 aromatics isomerization 497
organic chemicals, adsorptive separation 185ff.
organic compounds, volatile, *see* volatile organic compounds
organic dehydration 321
organic reaction center (ORC) 467ff.
organizations 17ff.
organo-nitrogen species 190
organo-sulfur species removal 176, 188ff.
organosilanes 339ff.
oriented zeolite crystals 108
osmosis, reverse 346
over-digestion 65
oxidation, catalysts 391ff.
– heteroatom substituted zeolites 387
oxide ratio, aluminophosphate molecular sieves 9
oxygen, recovery 298
oxygenate removal 190, 546ff.
^{31}P NMR 149ff.

p

packed bed adsorbers 275
Pacol™ process dehydrogenation 268
para-xylene 174, 514ff.
– industrial processes 231ff.
– isomerization 135
– production 232ff., 220
– separation 220
para-xylene : *ortho*-xylene selectivity 445
n-paraffin-depleted kerosene 264
paraffins
– alkylation 507ff.
– boiling point reduction 551ff.
– C10+ 358
– conversion catalysts 519ff.
– conversion to aromatics 518ff.
– cracking 551ff.
– heavy normal separation 261ff.
– intermediate normal separation 260ff.
– light, *see* light paraffins
– mono-methyl 222, 263ff.
– normal separation 249ff.
– olefin-paraffin separation 174, 204ff.
– process chemistry 518ff.
– recovery 258
Parex™ process, UOP 231
– UOP 235ff.
paring mechanism 468ff., 555
partial differential equations (pde) 281ff.
PDMS (polydimethylsiloxane) 347ff.
Pd/stabilized Y zeolite catalysts 563
PED (precession electron diffraction) 106
pellets 69
– self-supporting pellet technique 119ff.
pentacoordinated carbonium ion 429, 551
pentane 205
– *n*-pentane 262
pentane, isomerization 356ff.
– 3-methyl- 410
pentenes, skeletal isomerization 486ff.
performance parameters 403
permeance 314, 322
permeation 318
– zeolite membranes 314ff.
permporosimetry 313
petrochemicals
– feed production 560
– high-volume aromatic 230
– zeolite catalysis 403ff.
PFG, pulsed field gradient, *see* pulsed field gradient
pharmaceutical chemicals, adsorptive separation 188
phase
– dispersed 337ff.
– identification and quantification 92ff.
– inversion technique 344
– metastable 62ff.
2-phenyl alkylate 263
phosphates, crystalline 141
phosphine, trimethyl 150
phosphoric acid 66
– silica-supported 443
– solid 506
physical characterization 152
physicochemical characterization, active sites 507, 511ff., 524ff.
physisorption, nitrogen 152ff.
– pyridine 127ff.
platforming unit, CCR 233
pockets, 12-ring 53
polarization, cross 142ff.
polishing 104
polyalkylated benzene 514
polycrystalline membrane growth 310
polydimethylsiloxane (PDMS) 347ff.
polyetherimide, Ultem® 339
polyimide 337

polymer membranes   329ff.
polymer upper bound correlation   331
polymers
– compatibility with zeolites   339
– glassy   336ff.
– inorganic   1
– material modification   339ff.
– rubbery   347
– zeolite/polymer interface   340
polysulfone supports, porous   346
polyunsaturated cyclo-olefins   552
poly(vinyl chloride) (PVC)   348
pore diameters, effective   309
pore geometry   545ff.
pore openings   213
pore size, effect on cracking   461ff.
pore volume, selective   239
pores   32ff.
– effective width   32
– eight-ring   32
– ten-ring   52
porosity, interzeolitic   311
porous polysulfone supports   346
post-forming crystallization   71ff.
post-forming manufacturing   71ff.
post-Sorbex process fractionation   270ff.
potential theory   278
powders, zeolite, see zeolite powders
precession electron diffraction (PED)   106
precipitated silica   66
preparative conditions, SAPO-34   526
pressure swing adsorption (PSA)   296ff.
pretreating severity   550
primary cation sites, MOR   42
prism, hexagonal   29
probe molecules, basic   126
probe selection, IR spectroscopy   125
probes, low-temperature   131ff.
– see also samples
probing, molecular   313
processes
– adsorptive separation   203ff.
– alkylbenzene process chemistry   517
– benzene alkylation process chemistry   512
– bond scission   546ff.
– C_8 aromatics isomerization   495
– C_4 Olex   266
– dealumination   72ff.
– design   288
– detergent Molex™   249ff., 261ff.
– detergent Olex   268
– FCC   557ff.
– flow schemes   498ff., 550
– fuel alkylation process chemistry   508

– gas-oil conversion   539
– gasoline Molex™   249ff., 258ff.
– hydrocracking and hydroisomerization   561
– industrial   231, 483ff., 505ff.
– "ISOSIV"   4
– Loeb-Sourirajan   342
– MaxEne™   250, 260ff.
– methylation process chemistry   514ff.
– MMP Sorbex   263
– Molex™   249ff.
– MX Sorbex   241ff.
– olefin oligomerization process chemistry   505ff.
– Pacol™   268
– paraffin/olefin process chemistry   518ff.
– reactive   320
– Sorbex technology   208, 235ff., 249, 253ff.
– Sorbutene   267
– TransPlus   518
– UOP Parex™   231, 235ff.
– zeolite membranes   320
– zeolite separation   203ff.
product control   544
– FCC   544ff.
product shape selectivity   438ff.
production
– 2,6-dimethylnaphthalene   244
– ethylbenzene   244
– m-xylene   241
– p-xylene   232ff.
– petrochemical feed   560
– transportation fuel   560
products
– alkylation process chemistry   508, 512
– alkylbenzene process chemistry   517
– methylation process chemistry   514ff.
– oligomerization process chemistry   505ff.
– paraffin/olefin process chemistry   518ff.
profiles, pseudo-voigt   89
propane   382, 422
propene   442
property–function relationship   542ff.
propyl carbenium ion   423
propylene   372
– oligomerization   441ff.
protolytic cracking, endocyclic   458
protonated pyridine   127ff.
protonation   426ff., 452
PSA (pressure swing adsorption)   296ff.

pseudo-voigt profiles 89
pseudoboehmite 67
Pt clusters 110
pulse test procedure 209ff.
pulsed field gradient (PFG) NMR 144ff., 152
pure component adsorption 408ff.
purification, hydrogen 299ff.
purification zone 240, 256
PVC (poly(vinyl chloride)) 348
PX/MX and PX/OX selectivity 216
pygas 230
pyridine adsorption 127ff.

q
quadrupolar nuclei 143
quantum dots 16
quaternary ammonium cation 15

r
R&D adsorptive separation 176
R&D uses of zeolite catalysis 355ff.
radial mixing 236
radioactive waste storage 18
radius, ionic 208
raffinate column 237, 243
rare-earth exchange 537
rate-selective adsorption 221ff.
reactant shape selectivity 435ff.
reaction center, organic 467ff.
reaction mechanisms 447
– bifunctional isomerization 480ff.
– cracking 455ff.
– methanol conversion to hydrocarbons 527
– paring 468ff., 555
reaction product composition, C_8 aromatics isomerization 490ff.
reactions
– alkylation, *see* alkylation
– aromatics 369ff., 462ff., 490
– aromatization 382
– bond breaking 535ff.
– byproduct 483
– carbonylation 438
– chain-breaking 369
– cracking, *see* cracking
– dealkylation 234
– dealumination 72ff.
– dehydroaromatization 377ff.
– dehydroxylation 424
– denitrogenation 546ff.
– deoxygenation 546ff.
– desulfurization, *see* sulfur removal
– disproportionation 435ff., 462ff., 492, 517ff.
– dissolution 64
– epoxidation 387, 389ff.
– Fischer–Tropsch 188ff.
– heteroatom substituted zeolites 387
– hydrocracking, *see* hydrocracking
– hydrogen transfer 544
– hydroisomerization, *see* hydroisomerization
– hydrolysis 64
– hydrotreating 383, 387ff.
– interconversion 493
– isomerization, *see* isomerization
– kinetics 404ff.
– methylation 514ff.
– oligomerization, *see* oligomerization
– oxidation 387, 391ff.
– Pacol™ process dehydrogenation 268
– process chemistry 505ff., 508, 512, 514ff.
– protonation 426ff.
– rearrangement 535ff.
– ring contraction 436
– ring opening 554ff.
– skeletal isomerization 447ff.
– transalkylation 369, 462ff., 517ff.
– zeolitic reforming 520
reactive adsorption 224ff.
reactive processes, zeolite membranes 320
reactivity, normal paraffin separation desorbents 255
– relative 547
reactors
– catalytic membrane 323ff.
– circulating catalyst 511
– design 510
– lead 549
reagents, Grignard 339ff.
REAPDOR NMR 144
rearrangement 535ff.
– FCC 556ff.
recoupling, dipolar 143ff.
recovery, oxygen 298
– paraffin 258
recycling, single stage 564
reference materials 35
refinement, Rietveld 96ff.
Refiner's octane pool 258
refinery off gas (ROG) 300
refining, zeolite catalysis 403ff.
reforming, zeolitic 520
reformulated fuels 565ff.
regeneration, isomerization catalysts 495
– zeolite molecular sieves 275ff.

relative reactivity 547
remote natural gas, monetization 521
removal
– aluminum 542, 558ff.
– bulk 301ff.
– carbon dioxide 295
– heteroatoms 546ff.
– mercury 296
– sulfur 294ff.
– trace components/impurities 175, 188ff.
– VOC 296
– water, see dehydration
reproducibility, synthesis 324
research and development, see R&D ...
research octane number (RON) 259
– alkene oligomerization 450
– C5–C7 alkane isomers 358
research octane number-clear (RONC) 480ff.
Resid Cracking 557
resistance, mass transfer 69
– mass transfer 416
resistance modeling 284ff.
retention volume 205
reverse osmosis (RO) membranes 346
Reynolds number 291ff.
Rietveld refinement 96ff.
12-ring channels 48, 50
ring contraction 436
ring opening 554ff.
12-ring pockets 53
ring-products, highly hydrogenated 561
RO (reverse osmosis) membranes 346
ROG (refinery off gas) 300
RON (research octane number) 259
– alkene oligomerization 450
– C5–C7 alkane isomers 358
RONC (research octane number-clear) 480ff.
rotary dehydrators 301
rubbery polymers 347

s
SafeCat isomerization 482
safety, hazards 509
– HS&E issues 61
salts, metal 73
sample preparation, TEM 103ff.
sample pretreatment, IR spectroscopy 119ff.
samples, hydrated 147
– see also probes

Sanderson intermediate electronegativity 122, 207ff.
SAPO, see silicoaluminophosphate ...
saturation, aromatic rings 490
scaling-up 324
scanning electron microscopy (SEM) 98ff.
– zeolite membranes 310ff.
scanning transmission electron microscopy (STEM) 104ff.
scission, bond 546ff.
– pathways 455
sealing, high-temperature 324
secondary building units 114
secondary eight-ring channels 41ff.
secondary/seeded growth 311ff.
selective pore volume 239
selectivity
– adsorption 414
– catalysts 359ff.
– cracking 461
– durene/isodurene 215, 218
– influence of spatial distribution 544
– ion exchange 224
– normal paraffin separation adsorbents 250ff.
– normal paraffin separation desorbents 254
– para-xylene : ortho-xylene 445
– product shape 438ff.
– PX/MX and PX/OX 216
– reactant shape 435ff.
– shape 430ff.
– transition state shape 435ff.
self-diffusivity, Einstein's expression 417ff.
self-supporting pellet technique 119ff.
SEM, see scanning electron microscopy
semi-continuous synthesis 77
separation
– adsorption acids 269ff.
– adsorptive, see adsorptive separation
– binary 322
– C_4 265ff.
– carbohydrates 269
– cation exchange 137
– chromatographic 222, 239
– citric acid 270
– diffusion-controlled 318
– fatty acids 270ff.
– fructose/glucose 218ff.
– gas/vapor 322
– heavy normal paraffin 261ff.
– intermediate normal paraffin 260ff.

– light normal paraffin 258ff.
– liquid-liquid 320ff.
– mechanisms 316ff.
– modes of operation 208ff.
– mono-methyl paraffins 263ff.
– olefin 265ff.
– *para*-xylene 220
– paraffin 249ff.
– strong-strong 322ff.
– zeolite membranes 307ff.
– zeolites 203ff.
Separex™ membrane, UOP 330
"shadow" effect 98
shape selectivity 430ff.
– adsorption 222ff.
– product 438ff.
– reactant 435ff.
– transition state 435ff.
sheets 70ff.
Sherwood number 291
shift
– chemical 141
– hydride 430
– hydroxyl band 133ff.
^{29}Si NMR 145ff.
SI (spaciousness index) 434ff.
side reactions, olefin isomerization 493ff.
sieve materials, history 5ff.
sieves, molecular, *see* molecular sieves
silica 66
– low-density 14
silica molecular sieves 6ff.
silica-supported phosphoric acid 443
silicalite 221, 307
silicoaluminophosphate (SAPO) crystal 101
silicoaluminophosphate (SAPO) molecular sieves 9, 334
– SAPO-34 464ff., 524ff.
simulations, *see* modeling and simulations
single permeation 319
single-rings, substituted 556
single stage recycle 564
SiO_2/Al_2O_3 molar ratio 213, 216ff.
site occupancy, fractional 418
site quantitation 125ff.
sites, active 507, 511ff., 524ff.
– external surface 513
six-rings, double 38
size, crystallites 95ff.
– unit cells 94ff.
size effects, crystal 446ff.
skeletal isomerization 486ff.
– alkenes 447ff.

skin layer 345
slurry, zeolite 342
sodalite cage 29
– FAU 38
– LTA 36
– NH_4^+-FAU zeolite 121
sodium faujasites, hydrated 94
solid phosphoric acid (SPA) 506
solutions
– caustic 63
– metal salt 73
– solution-diffusion mechanism 330
solvents, Sulfolane 233
Sorbex technology 208, 235ff.
– non-aromatic liquid separations 249
Sorbex zone parameters 257ff.
Sorbutene process 267
space velocity
– liquid hourly 432, 495ff.
– weight hourly 495ff.
spaciousness index (SI) 434ff.
spatial distribution, influence on selectivity 544
specificity, hydrocarbon adsorption 408ff.
spectroscopy
– EELS 110ff.
– energydispersive X-ray 97, 100ff.
– GCMS 265
– IR, *see* IR spectroscopy
– NMR, *see* NMR spectroscopy
spin-half nuclei 142
spinning, magic angle, *see* magic angle spinning NMR
stability
– normal paraffin separation adsorbents 252
– thermal 154, 308
stabilization 539ff.
– zeolites 72ff.
stacking sequence 43
standards, NIST 35
steam cracking, naphtha 229
steam-stabilized Y zeolites 148, 543
– TEM 546
steamed FAU 122ff., 129
steaming dealumination 93
STEM, *see* scanning transmission electron microscopy
– applications 109ff.
stilbite 3
stretches, T–O–T 118
– X and Y zeolites 117
stretching vibrations, OH- 411

strong-strong separation   322ff.
structural characterization, TEM   105ff.
structural groups, UZM-5   116
Structure Commission, IZA   3
– IZA   96
structure identification   86ff.
structures, zeolites   27ff.
studies, *in situ/in operando*   136
styrene   515ff.
substituted single-rings   556
sulfides   547
– microporous metal   11
Sulfolane solvent   233
sulfur removal   176, 188ff., 294ff., 546ff.
supercages   29
– FAU   39
– MWW   513
– NH_4^+-FAU zeolite   121
supports, hydrocracking   554
– porous polysulfone   346
surface sites, external   513
surfactants, anionic   261ff.
symmetric stretches   117
symmetry-constrained combinatorial search   54
synthesis
– adsorbents   61ff.
– AlPO molecular sieves   62ff.
– aluminophosphate molecular sieves   66ff.
– catalysts   61ff.
– continuous   77
– hydrothermal   62ff., 118ff.
– important parameters   65ff.
– microwave   77
– molecular sieves   15ff.
– reproducibility   324
– semi-continuous   77
– zeolite membranes   308ff.
synthetic zeolites   4
syrup, corn   269

t

T-atoms   90
T–O–T stretch   118
T-atoms, connectivity   48
Tatoray™ unit   242
TEM, *see* transmission electron microscopy
temperature, equilibrium-selective adsorption   220ff.
temperature programmed desorption (TPD)   128, 131, 157ff.
ten-ring channel system, MFI   46ff.
ten-ring pores   52
tert-butyl ether, methyl   479
– methyl   486ff.
tests, pulse test procedure   209ff.
tetraethylorthosilicate   66
tetrahedra, TO_4 28
tetrahedral frameworks, octahedral–   13
tetramethylbenzene, IR spectroscopy   428
theoretical ion exchange capacity   213
theories and models
– Bellamy-Hallam-Williams (BHW) theory   124
– carbenium ion theory   551ff.
– D-R model   278
– diffusion-based models   283
– linear driving force approximation   281, 284ff.
– LRC model   278ff.
– Maxwell model   334ff., 339
– potential theory   278
– Sanderson intermediate electronegativity   122, 207ff.
– zeolite membranes transport theory   314ff.
thermal analysis   154ff.
thermal stability   154
– zeolite membranes   308
thermal swing adsorption (TSA)   288ff.
thermogravimetric (TGA) analysis   155ff.
Thiele diagram, McCabe–   284
thin-film composite mixed-matrix membranes   346
thin-sectioned SAPO crystal   101
thiophenes   547
three-dimensional 12-ring channels   50
titration, Hammet   157
toluene
– methylation   514ff.
– side-chain alkylation   516
– transalkylation   373ff.
TON framework   424ff.
topology, zeolites   408ff.
total internal reflection, attenuated   112
TO_4 tetrahedra   28
TPD, *see* temperature programmed desorption
trace component/impurity removal   175ff., 188ff.
trans-but-2-ene   507
transalkylation   462ff.
– alkylbenzene   517ff.

– catalysts 517
– toluene 369
transformation, aromatic 462ff.
transition state shape selectivity 435ff.
transmission electron microscopy (TEM) 103ff.
– steam-stabilized Y zeolites 546
TransPlus process 518
transport theory, zeolite membranes 314ff.
transportation fuel production 560
TRAPDOR NMR 144
trimethyl phosphine 150
Tropsch reactions, Fischer– 188ff.
truncated octahedron 29
TSA (thermal swing adsorption) 288ff.
tschortnerite 13
type code, framework 34
types and structures of zeolites 27ff.

u
UFI 86ff.
Ultem® polyetherimide 339
ultrahydrophobic Y 540
ultrastable Y (USY) 541
unimolecular reaction mechanism 447ff.
– cracking 456ff.
unit cell size determination 94ff.
units, building, *see* building units
universal isotherm 278ff.
UOP Cyclar catalyst 519ff.
UOP MolexTM process, *see* MolexTM process
UOP Parex™ process 231, 235ff.
UOP SafeCat isomerization 482
UOP Separex™ membrane 330
upgrading, gasoline 204
upper bound correlation, polymer 331
UZM-5 86ff.
– framework spectrum 116

v
vacuum swing adsorption (VSA) 298
Van-der-Drifts columnar growth 311
van der Waals interaction 404
vapor, gas/vapor separation 322
vibrational band assignments 114
vibrational modes 117
vibrations, OH-stretching 411
voids, interconnected 2
volatile organic compounds (VOC) removal 296
VSA (vacuum swing adsorption) 298

w
Waals, van der, *see* Van der Waals interaction
washing 67ff.
waste storage, radioactive 18
water
– abridged isosteres 291
– desalination 347
– removal, *see* dehydration
wave equation 282
wave shape and length 283
wave speed 282ff.
weight-hourly space velocity (WHSV) 495ff.
wheels, desiccant 301ff.
Williams theory, Bellamy-Hallam- 124
Wyckoff notation 88

x
X-ray diffraction (XRD) 310ff.
– powder 91ff.
X-ray spectroscopy, energydispersive 97
– energydispersive 100ff.
X zeolites
– vibrational modes 117
– zeolite/desorbent combination 238
xylene
– disproportionation 492
– ethylbenzene conversion 463ff.
– isomerization 369, 491ff.
– *m/meta-*, *see* meta-xylene
– *p/para-*, *see* para-xylene
– transalkylation of toluene 369

y
Y zeolites
– ammonium exchanged 153
– bond breaking and rearrangement 535ff.
– calcined 146ff.
– dealuminated 107
– Pd/stabilized 563
– steam-stabilized 148, 543, 546
– ultrahydrophobic 540
– vibrational modes 117
– zeolite/desorbent combination 238

z
zeolite catalysis
– applications 355ff.
– bifunctional catalysts 424
– future trends 393
– literature review 356ff.

- metal-zeolite catalyzed isomerization 479ff.
- paraffin isomerization 482ff.
- R&D uses 355ff.
- reaction mechanisms 447
- refining and petrochemicals 403ff.

zeolite membranes
- characterization 313ff.
- DDR-type 316ff.
- hydrophilic 321
- mixed-matrix 331ff.
- permeation 314ff.
- separation capability 314ff.
- separation mechanisms 316ff.
- separations 307ff.
- synthesis 308ff.
- transport theory 314ff.

zeolite powders 68ff.
- MTW 149
- X-ray diffraction characterization 91ff.

zeolite slurry 342
zeolite/polymer interface 340
zeolite/polymer mixed-matrix membranes 333ff.
- applications 346ff.
- geometry 341ff.

zeolites
- acidic 72
- adsorptive separation, see adsorptive separation
- aluminosilicate 6ff.
- and molecular sieves 1ff.
- applications 16ff.
- Atlas of Zeolite Structure Types 3
- Ba-X 95
- beads 69
- beta, see beta ...
- bound forms 70
- CHA, see chabazite
- characterization, see characterization
- chemical formulae 27
- chemical modification 72ff.
- commercially significant 35
- compatibility with polymers 339
- critical properties 536ff.
- dealuminated Y 107
- diffusion mechanisms 286ff.
- early evolution 6
- electron microscopy characterization 97ff.
- FAU, see FAU
- FCC-type feeds 369, 377
- framework building units 28ff.
- framework IR 115ff.
- framework Si/Al 116ff.
- framework types 31
- H-BEA 132ff., 428
- H-form 121, 142
- heteroatom substituted 387
- high Si/Al 6
- hydrocarbon conversion 429ff.
- hydrocarbon diffusion 151ff.
- hydrotreating/hydrocracking 388
- hypothetical frameworks 54ff.
- important synthesis parameters 65ff.
- inorganic membranes 76
- intermediate Si/Al 6
- large-pore 393
- low Si/Al 6
- LTA, see LTA
- LTL, see LTL
- mass transfer in formed particles 280ff.
- material modification 339ff.
- materials 34ff.
- metal-loaded 136ff.
- metals in 109ff.
- moisture content 218ff.
- MOR, see mordenite
- Na-X 95
- nano-sized particles 339
- nanocrystalline 99
- nanoparticles 75ff.
- natural 4ff.
- NH_4^+-FAU 121
- normal paraffin separation adsorbents 252ff.
- oriented crystals 108
- regeneration 275ff.
- separation processes 203ff.
- stabilization 72ff.
- surface dehydroxylation 424
- synthesis framework IR 118ff.
- synthetic 4
- topology 408ff.
- types and structures 27ff.
- typical properties 213
- X, see X zeolites
- X-ray powder diffraction characterization 91ff.
- Y, see Y zeolites
- zeolite/desorbent combination 237ff., 242ff.

zeolitic adsorbents, post-forming manufacturing 71ff.
- synthesis 61ff.

zeolitic catalysts, post-forming manufacturing 71ff.
– synthesis 61ff.
zeolitic reforming 520
zero length column (ZLC) 419
ZIFs 393
zig-zag channel system 46ff.

ZSM-5 3ff., 15ff., 307
– bond breaking and rearrangement 537ff.
– framework structure 45ff., 524
– *see also* MFI
ZSM-57 3ff., 15ff., 307